INTEGRATION OF MODERN TAXONOMIC METHODS FOR *PENICILLIUM* AND *ASPERGILLUS* CLASSIFICATION

D1457882

INTEGRATION OF MODERN TAXONOMIC METHODS FOR *PENICILLIUM* AND *ASPERGILLUS* CLASSIFICATION

Edited by

Robert A. Samson
Centraalbureau voor Schimmelcultures
Baarn, The Netherlands

and

John I. Pitt
CSIRO Division of Food Science
North Ryde, Australia

ho
ap
harwood academic publishers
Australia • Canada • France • Germany • India • Japan • Luxembourg
Malaysia • The Netherlands • Russia • Singapore • Switzerland

Amsteldijk 166
1st Floor
1079 LH Amsterdam
The Netherlands

British Library Cataloguing in Publication Data

A catalogue record for this book is available from the British Library.

ISBN: 90-5823-159-3

CONTENTS

PREFACE

The International Workshop on *Penicillium* and *Aspergillu*s, held in Baarn, The Netherlands from 26–29 May 1997, was the third meeting of scientists working on the systematics of these important fungi. The first workshop was held in May 1985 and the second took place in the same month in 1989, both at the same venue. These international workshops have contributed significantly to a better dialogue between taxonomists, fundamental and applied mycologists. Since many species are so important in biotechnology, food, medicine, biodeterioration and other applied fields, a practical and stable taxonomy is of vital importance. The taxonomy of *Penicillium* and *Aspergillus* has developed since the 1970s from the classical morphological concepts, and the integration of molecular, physiological and biochemical methods is therefore a logical approach to understanding the classification of these fungi. The IUMS International Commission on *Penicillium* and *Aspergillus* (ICPA) has organized well-attended symposia at various international congresses and has shown that international cooperation is essential for good progress.

These Proceedings comprise 36 papers divided into nine chapters. Chapter 1 includes a list of accepted species and their synonyms in the family Trichocomaceae. In addition it also contains the list of names in current use with their type names and a list of published taxa between 1992 and 1999. We hope that these lists will soon become available on the internet and that the databases can be used. The other chapters are devoted to methods for identification, taxonomic and digital information on *Penicillium* and *Aspergillus,* phylogeny, molecular taxonomy, classification and identification. Chapter 7 contains papers on the taxonomy of the Aspergilli of section *Nigri* and section *Flavi.* The taxa of these sections are still attracting interest because of their correct identification and nomenclature while they are important in biotechnology and as toxic food contaminants. The proceedings conclude with two papers on pathogenic Aspergilli and Penicillia and the potential of *Penicillium* and *Aspergillus* in drug lead discovery. The editors sincerely thank all the authors for their contributions.

Local arrangements of the Third Workshop were very efficient and went smoothly. The assistance and support of the Centraalbureau voor Schimmelcultures is greatly appreciated. The preparation of these Proceedings was a difficult job, partially caused by time constraints and busy working schedules. Most chapters have been updated recently. John Pitt planned to start editing manuscripts in Baarn, but he could not foresee that most would be done in the local hospital in Baarn when he was recovering from surgery. Nevertheless we hope that this book will prove another milestone in *Penicillium* and *Aspergillus* systematics. During the preparation of the manuscript Ans Spaapen-De Veer helped with the index and Karin van den Tweel-Vermeulen with collecting the recently published names and proofreading. Their help is greatly appreciated!

CORRESPONDING CONTRIBUTORS

I. Ahmad
Department of Botany
University of Toronto
Ontario
Canada M5S 3B2

B.W. Bainbridge
Life Sciences Division
King's College
London, W8 7AH
UK

S. Banke
Department of Mycology
University of Copenhagen
DK-1353, Copenhagen
Denmark

D.A. Carter
Microbiology Department
University of Sydney
Sydney, NSW 2006
Australia

M. Christensen
Department of Botany
University of Wyoming
WY 82071-3165
USA

J.C. Frisvad
Department of Biotechnology
Technical University of Denmark
2800 Lyngby
Denmark

W. Gams
Centraalbureau voor Schimmelcultures
Postbus 273
3740 AG Baarn
The Netherlands

D.M. Geiser
Department of Plant Pathology
The Pennsylvania State University
PA 16802
USA

M.A. Klich
USDA Southern Regional Research Center
PO Box 19687
New Orleans, LA 70179-0687
USA

A. Kubátová
CCF Department of Botany
Charles University
128 01 Prague 2
Czech Republic

J.P. Latgé
Laboratoire des Aspergillus
Institut Pasteur
75724 Paris Cedex 15
France

H. Ogawa
Center for Information Biology
National Institute of Genetics
Mishima 411-8540
Japan

T. Okuda
Tamagawa University
Institute for Applied Life Science
Tokyo 194-8610
Japan

L. Parenicova
Molecular Genetics of Industrial
Microorganisms
Wageningen Agricultural University
6703 HA Wageningen
The Netherlands

S.W. Peterson
USDA NCAUR
1815 North University Street
Peoria, IL 61604
USA

J.I. Pitt
CSIRO Division of Food Science
PO Box 52
North Ryde, NSW 2113
Australia

B. Rømer-Rassing
Novo Nordisk A/S Novo Allé
DK-2880 Bagsvaerd
Denmark

R.A. Samson
Centraalbureau voor Schimmelcultures
Postbus 273 3740
AG Baarn
The Netherlands

M. Sasa
Screening Biotechnology Enzyme Research
Novo Nordisk A/S Novo Allé
DK-2880 Bagsvaerd
Denmark

J. Scott
Department of Botany
University of Toronto
Ontario
Canada M5S 3B2

K.A. Seifert
Eastern Cereal and Oilseed Research Centre
Agriculture and Agri-Food Canada
Research Branch
Ontario K1A 0C6
Canada

P. Skouboe
Biotechnological Institute
Koglevej 2
DK-2970 Hørsholm
Denmark

J. Sugiyama
Department of Botany
The University of Tokyo
Tokyo 113-0033
Japan

M. Tamura
Research Center for Pathogenic and
Microbial Toxicoses
Chilba University
Chiba 260-8673
Japan

J. Varga
Department of Microbiology
Attile Jozesef University
H-6701 Szeged
Hungary

W.T. Yoder
Novo Nordisk Biotech Inc
1445 Drew Avenue
Davis, CA 95616
USA

Chapter 1

NOMENCLATURE OF
Penicillium and *Aspergillus*
and their teleomorphs

DATABANKING OF NAMES

W. Gams and G. J. Stegehuis
Centraalbureau voor Schimmelcultures, Baarn. The Netherlands

Numerous strategies have been deployed to stabilize fungal nomenclature. Stable names are highly desirable; there is no doubt about this. But changes of names for nomenclatural reasons make up only a fraction of the name changes proposed each year. Most are due to an improved taxonomy and should therefore be accepted (Korf, 1991). One of the nomenclatural reasons for name changes is the reactivation of old, forgotten names, often of doubtful application and not properly typified. By neotypification, it is possible to reintroduce them, and these kinds of name changes are obviously undesirable. The question then is how best to discourage this practice. Unfortunately, it has not been possible to introduce a direct ruling into the Code that forbids reactivation of names through neotypification in situations where this would only serve to destabilize nomenclature. It is impossible to impose a starting point date that felicitously separates names that may be neotypified from uncertain older names that should not.

There are two possible mechanisms to stabilize names: one is the more or less bureaucratic assemblage of lists for various taxonomic groups of Names in Current Use (NCU), which then would get a protected status. The other is the conservation of names over older competing names (Art. 14) or the explicit rejection of undesirable older names (Art. 56).

At the 15th International Botanical Congress in Tokyo, a protected status for lists of Names in Current Use (NCU principle) did not receive sufficient recognition: 54.98% Yes votes were some 5% short of the minimum required (Greuter *et al.*, 1994b: 125). At the 16th Congress in St. Louis in 1999, the NCU principle was soundly defeated. Arguments against granting a protected status to names in current use include the fear that personal views of some influential taxonomists would be imposed and a recognition of the inefficiency of stabilizing the nomenclature of groups of organisms with insufficiently explored taxonomy, where it is impossible to draw up meaningful lists of recognized names. The Pitt-Samson (1993) list of recognized species names in the *Trichocomaceae* is the only NCU list that has received a limited recognition in the Preface to the 1994 ICBN (Tokyo Code, Greuter *et al.*, 1994a). The three prerequisites for this success were: inclusion of all the contemporary experts on the committee that prepared the list, a thorough preparation, and the unanimity of the committee about the names to be recognized.

As an alternative, introduction of a Code of BioNomenclature (Greuter, 1996; Greuter and Nicolson, 1996; Greuter *et al.*, 1996; Hawksworth, 1996) has been explored. This strategy also includes the NCU principle. Introduction of such a Code is also vehemently opposed by many botanists (e.g., Brummitt, 1997). It is felt that it would not make matters easier for nomenclaturists who would have to master one code for the period before 2000 or another date, and another one for the subsequent years. In St. Louis, the time was not deemed mature to prepare for the introduction of this Code. Another approach to standardize the recognition of newly introduced names by registration also did not pass at the St. Louis Congress.

The conservation of names over older competing names (Art. 14) stabilizes a particular, well typified name, rejecting its competing older synonyms or homonyms, while the explicit rejection of undesirable older names (Art. 56) is an easier mechanism that does not require the determination of types. The possibilities for application of both these possibilities were widened significantly at the Tokyo Congress. Under these provisions, every proposal is handled efficiently by the appropriate committee, in our case the "Committee for Fungi" (CF, installed under the auspices of IAPT). In most instances the procedure is short and can be closed within 1–2 years after publication of the proposal in a first published report. From that point on, the situation is normally settled, though the decision requires ratification by the next Botanical Congress. Since the Tokyo decisions, the CF has not been overburdened with proposals involving species names.

Of relevance to *Penicillium/Aspergillus* nomenclature are also the ongoing discussions concerning Article 59, which deals with names of anamorphic and pleomorphic fungi. This article provides an exception to the system of botanical nomenclature, allowing a different naming of anamorphs even when a teleomorph connection may be known; the teleomorph name, however, has preference. This mechanism complies with practical needs, because many fungi normally occur only as anamorphs and must be identified as such. For many of these cases the teleomorph connection does not universally apply to all isolates identified as the corresponding anamorph, unless the criteria for recognition of anamorph species are so precisely defined that this can infallibly be done, an ideal which is often not attained. It has been proposed to abandon Art. 59 because molecular methods nowadays allow an unequivocal identification of every fungus and because Art. 59 is purported to have a destabilizing effect and to be contrary to the spirit of nomenclature (E. Swann and D. R. Reynolds, pers. comm.; Seifert and Samuels, 2000). In opposing such attempts, we must consider how much destabilization would ensue if all anamorph names could compete at equal rank with teleomorph names, e.g., if *Penicillium*, *Talaromyces* and *Eupenicillium* were just nomenclaturally equal genera. *Penicillium* and *Aspergillus* contain many examples of unequivocal anamorph/teleomorph connections, where the teleomorph name really should be used preferentially, because in these fungi the strongest differentiation is observed in ascospore morphology. Separate anamorph names for the same species can easily be done without. The most outstanding example is *Aspergillus nidulans* and *Emericella nidulans*. But who will convince all the users of the anamorph name *A. nidulans* of the necessary change to the teleomorph genus? Moreover, all the recently published species of *Talaromyces*, differentiated mainly by ascospore morphology, should really not overburden nomenclature with the concomitant formal introduction of perfunctory anamorph names. At the St. Louis Congress, a Recommendation was added to Art. 59 that introduction of additional binomials for anamorphs should be avoided if the teleomorph is widely available and the connection is unequivocal. Consequently, introducing separate anamorph binomials in the *Trichocomaceae* is discouraged (but not forbidden); this intention was endorsed by several participants of the Third International

Workshop on *Penicillium* and *Aspergillus*. In certain cases, when the teleomorph is irregularly or tardily formed, the availability of anamorph names continues to be important.

The Pitt-Samson list teaches us several lessons. It has been recognized in the Preface to the Tokyo Code (Greuter *et al.*, 1994a: page X) and people work with it. It has certain merits and some deficiencies. We must consider here the deficiencies.

(1) The list contains a few factual errors. (2) It is not 100% efficient in stabilizing nomenclature: If a certain species was at that time considered as a synonym of another species and therefore not included in the list, but shortly afterwards was found to be specifically distinct, it would have to be reintroduced as a nomenclatural change. According to the philosophy of NCU, it would even be possible to coin an entirely new name for that fungus, which would then automatically become an NCU after a few years, as the former name would have been excluded from nomenclatural protection not being vetted by the NCU list. Nevertheless, it is good if this does not happen and mycologists take the effort to screen the old literature first for available names. (3) The NCU list is very incomplete and so far excludes many well-known names, the application of which is not 100% certain. It is this body of information, i.e., the correct application of these names, that is most urgently needed by all users of nomenclature. Databases of names are in production in many fields of mycology and they are particularly useful if they are accessible on the Internet. Examples are the Aphyllophorales and *Fusarium* databases now accessible through the CBS webpage. The *Index Nominum Genericorum* database also is accessible on the Internet (http://nmnh.si.edu/ing)

The CABI International Mycological Institute (IMI, now CABI Bioscience) at Egham has always been particularly active in maintaining fungal databases. This institute is the seat of the indispensable reference periodicals *Index of Fungi*, *Bibliography of Systematic Mycology*, *Review of Plant Pathology*, and of *Descriptions of Fungi and Bacteria*, *Distribution Maps of Plant Diseases*, and *Mycological Papers*. The institute has been active in promoting the accessibility of databases through the Internet. It coordinates the fungal component of the Species2000 project.

A major sign of progress is the on-line availability of FUNINDEX through CABI Bioscience in January 2000 at http://194.131.255.3/cabipages/. This is a database of over 340,000 names of fungi at the species level and below, derived from authoritative published lists including Saccardo's Sylloge Fungorum, Petrak's Lists, Saccardo's Omissions, Lamb's Index, Zahlbruckner's Catalogue of Lichens, and CABI's Index of Fungi. Search is possible either by genus, binomial or specific epithet. A record usually has a reference to an entry in one of these catalogues and, in addition, more recent entries have the full citation from the source publication. Through another entry, the hierarchy of classification of a genus can be traced as it is treated in the *Dictionary of the Fungi*.

The Pitt-Samson list can be regarded as a welcome first step that should be succeeded by a complete database. This will have a much greater value to the mycological community. Fortunately, the organizers of the Third International Workshop on *Penicillium* and *Aspergillus* were already convinced of this necessity. After the workshop Pitt, Samson and Frisvad worked on compilations of synonyms, accepted species, types and taxa published between 1992—1999, which are published in the subsequent chapters. These will form the basis of a database which will be much more powerful in stabilizing nomenclature than an NCU list.

To work efficiently and contribute to stabilizing the nomenclature we need the following agreements:

- about the data format
- about the procedures of construction (conversion from text files, etc.)
- about granting access
- not to reactivate by neotypification names that have been declared uncertain (a self-constraint by responsible mycologists), if this would lead to a displacement of other, recognized names.

For each name the nomenclatural status [(in)valid, (il)legitimate, (in)correct] is to be entered, and also its taxonomic status [recognized, synonymous, uncertain, or doubtful in various degrees]. Typification or its absence is important information to be included. Much further work is, therefore, clearly required. Synonymy is partly a matter of opinion; it must be possible to trace the responsible author in the database and see the available arguments; conclusions about synonymy need not be imposed dictatorially. Changes of synonym links can be enacted in discussion with the community and are not likely to greatly upset the system of recognized names.

In a first phase, it is possible to build the database only as a tool for systematists. Subsequently it will be easy to relate it to all kinds of information:

- morphological descriptions
- metabolites and mycotoxins
- coenzyme Q
- isoenzymes
- molecular data on:
  - GC content of DNA
  - fingerprints (RAPD etc.)
  - RFLP of different kinds
  - sequences of different DNA fragments.

Each feature is to be listed with methods applied and responsible authority.

A few years ago, a database was constructed for all *Fusarium* names. Gordon Neish (Ottawa) made a first step; then the special committee for the taxonomy of *Fusarium* installed by the International Committee for the Taxonomy of Fungi (ICTF) worked it out and the system is now installed at CBS and is made accessible through the Internet. This system has been structured with the software BASIS, like the CBS strain databank which involves numerous fields, subfields, sub-sub-fields, etc., and is not fully relational. These systems are now being translated into relational ones using INTERBASE software. The *Fusarium* database is complete in respect of the published names, but it is still very incomplete in respect of typification, literature data and much other information. The obvious bottleneck is the lack of time available to competent mycologists to supply the missing information. The recently established access to the database through the Internet has strongly increased its usage and, hopefully, it will also help to direct the flow of information towards CBS, so that the database can be completed.

6

Proposed contents of *Penicillium-Aspergillus* database

- name
  - ▸ genus - authors
    - ▸ subgenus - authors
      - ▸ section - authors
        - ▸ species - authors
          - ▸ infraspecific ranks – authors

- protologue
- year
- recombinations
- basionym
- type - designated by …
- status, taxonomic
  - o   close to … *fide* …
  - o   synonym of … *fide* …
  - o   alternate name of …
  - o   doubtful application
- status, nomenclatural
  - o   (in)valid
  - o   (il)legitimate
  - o   (in)correct
  - o   recognized *fide* …
  - o   conserved / rejected
  - o   doubtful name
- representative strains
- descriptive / taxonomic literature
- remarks
- responsible authority

If the community of experts agrees not to resurrect names marked as uncertain or doubtful in the database by neotypification so that they would displace younger names of unequivocal identity, this will provide the kind of stabilization that is needed and can be implemented now. It need not even be formalized in a code of nomenclature.

During the workshop some discussion arose about the question, whether a databank or a published list of synonyms was more urgent. A printed version seems very desirable, though its distribution will be more efficient if it is published separately from the proceedings of this workshop. An extract of names from the CBS strain database can serve as a first basis for both the printed list and the databank. A databank to be updated regularly and accessible through the Internet was applauded by many participants.

In conclusion, we propose to build a relational database for names in the *Trichocomaceae* that should be based on the structure shown and, as a minimum, contain the same information fields. The following papers by Pitt, Samson and Frisvad are a start for this database. A list of synonyms, the types of names and recently described taxa are presented. Plans are in progress to make these lists of names available on the Internet.

ACKNOWLEDGEMENTS

We are indebted to Dr P. M. Kirk, CABI Bioscience Egham, for additions and improvements on a draft of the paper. Dr R. C. Summerbell contributed to improving the text in a late stage.

REFERENCES

Brummitt, R. K., (1997). Report of the Committee for Spermatophyta: 45. Taxon 46: 323–328.

Greuter, W., (1996). On the new BioCode, harmony, and expediency. Taxon 45: 291–194.

Greuter, W. & Nicolson, D. H., (1966). Introductory comments on the Draft BioCode, from a botanical point of view. Taxon 45: 343–348.

Greuter, W. *et al.* (eds), (1994a). International Code of Botanical Nomenclature (Tokyo Code). Reg. Veget. 131. Königstein.

Greuter, W., McNeill, J., & Barrie, F. R., (1994b). Report on botanical nomenclature – Yokohama 1993. Englera 14: 1–265.

Greuter, W. *et al ,* (1996). Draft BioCode, Taxon 45: 349–372.

Hawksworth, D. L. (ed.), (1996). Draft BioCode. The prospective international rules for the scientific names of organisms. Int. Union Biol. Sci., Paris.

Korf, R. P., (1991). Shall we abandon the principle of priority (and other nomenclatural caveats) for the sake of expediency? Mycotaxon 40: 459–468.

Pitt, J. I. & Samson, R. A., (1993). Species names in current use in the *Trichocomaceae* (Fungi, Eurotiales). Reg. Veget. 128: 13-57.

Seifert, K. A. & Samuels, G. J., (2000). How should we look at anamorphs? Stud. Mycol. 45 (in press).

LIST OF ACCEPTED SPECIES AND THEIR SYNONYMS IN THE FAMILY TRICHOCOMACEAE

J.I. Pitt[1], R.A. Samson[2] and J.C. Frisvad[3]
[1]CSIRO Division of Food Science and Technology, North Ryde, NSW 2113, Australia, [2]Centraalbureau voor Schimmelcultures, Baarn, the Netherlands and [3]Department of Biotechnology, Technical University of Denmark, 2800 Lyngby, Denmark.

Following the publication of the List of "Names in Current Use" (NCU) for the family Trichocomaceae (Pitt and Samson, 1993), it became apparent that a list incorporating synonyms as well as accepted names would be of great value to everyone working with these fungi. The accompanying list has been generated from the literature, with only a few additions or corrections where these were considered to have expert backing. Every attempt has been made to make the list both comprehensive and authoritative. It is to be expected that a small number of names included as accepted may in due course be shown to be synonyms and some accepted names may be split by future taxonomic judgments, but the great majority of names can be expected to remain stable into the foreseeable future.

The list has been generated primarily to give ready access to information concerning the disposition of synonyms, but provides much additional information. Holomorph and anamorph genera are indicated, as well as teleomorph anamorph connections in both directions where these are known to exist. Where names have resulted from recombination, basionyms are included, and in addition details of the original species from which basionyms were derived.

In the list which follows, genera and species which are accepted in the family Trichocomaceae are shown in bold face italics, while synonyms are shown in normal italics. Basionym genera either not accepted or lying outside the family Trichocomaceae are also shown in normal italics.

In all 1083 names are treated here, broken down into 28 accepted genera and 617 accepted species within the family Trichocomaceae as shown in the accompanying table (Table 1).

As is to be expected, *Penicillium* and *Aspergillus* make up the majority of the included species. The 225 species accepted in *Penicillium* represent an increase of 75 (exactly 50%) over the number accepted by Pitt (1980). The increase reflects both continued discovery of species and improved understanding of the concept of species in this large and complex genus. Synonyms of *Penicillium* species make up 75% of all listed synonyms, reflecting both the interest in this genus and the difficulty experienced by taxonomists. It is worth noting that Pitt (1980) listed another 450 species names in *Penicillium* which he could not recognise, and that most of these remain of incertain disposition. Thanks to general acceptance of the NCU list, those names can be disregarded for all time

In *Aspergillus*, the acceptance of 182 species also marks a considerable increase over those given by Raper and Fennell (1965), the last generally accepted complete taxonomy of *Aspergillus*. Only 24 *Aspergillus* synonyms are listed, reflecting both stability in names, and equally importantly, the impact of the NCU list permitting maintenance of many accepted names known or suspected to be of doubtful validity under the provisions of the Botanical Code.

The greatest change from the Raper and Fennell (1965) taxonomy of *Aspergillus* lies in the more recent acceptance of teleomorph names associated with *Aspergillus*. As shown in Table 1, eight holomorphic genera are accepted here as being associated with *Aspergillus* anamorphs: *Chaetosartorya, Emericella, Eurotium, Fennellia, Hemicarpenteles, Neosartorya, Petromyces* and *Sclerocleista*. Only three of these, *Emericella, Eurotium* and *Neosartorya,* include more than three species. In all seventy species are accepted, but only seven synonyms.

The other genus of substantial size (in terms of accepted species) included here is *Paecilomyces*, with 41 accepted species. It is worth noting that not all of these *Paecilomyces* species are regarded as belonging to the family Trichocomaceae, indicating some name changes in the future. In a similar vein, *Geosmithia* has recently been shown to be a genus of mixed origins, and some species may have changed generic dispositions in the future.

Considerable effort has been expended to ensure that this list is as complete as possible, but inevitably it will be less than perfect. Comments on the names included, errors or omissions will be gratefully received by the authors.

Table 1. Genera and species comprising the family *Trichocomaceae*, or providing basionyms to such species.

Genus	Type	Associated with	Accepted	Synonyms
Arachniotis	Holomorph		0	5
Aspergillus	Anamorph	*Eight holomorph genera*	184	24
Byssochlamys	Holomorph	*Paecilomyces*	4	1
Carpenteles	Holomorph	*Penicillium*	0	5
Chaetosartorya	Holomorph	*Aspergillus*	3	0
Citromyces	Anamorph	*Penicillium*	0	11
Coremium	Anamorph	*Penicillium*	0	5
Corethropsis			0	1
Cristaspora	Holomorph	*No anamorph*	1	0
Cylindrodendrum			0	1
Dactylomyces	Holomorph	*Thermoascus*	0	2
Dendrosphaera	Holomorph		1	0
Dichlaena	Holomorph		1	0
Dichotomomyces	Holomorph		1	0
Edyuillia	Holomorph	*Eurotium*	0	1
Eladia	Anamorph	*Penicillium*	0	1
Emericella	Holomorph	*Aspergillus*	27	1
Eupenicillium	Holomorph	*Penicillium*	43	5
Eurotium	Holomorph	*Aspergillus*	19	6
Fennellia	Holomorph	*Aspergillus*	3	0
Geosmithia	Anamorph	*Talaromyces*	8	0
Gymnoascus	Holomorph		0	2
Hamigera	Holomorph	*Merimbla*	1	2
Hemicarpenteles	Holomorph	*Aspergillus*	2	0
Isaria	Anamorph		0	4
Merimbla	Anamorph	*Hamigera*	1	0
Monilia	Anamorph		0	1
Mucor	Holomorph		0	1
Neosartorya	Holomorph	*Aspergillus*	12	0
Oospora	Anamorph		0	1
Paecilomyces	Anamorph	*Talaromyces*	41	1
Penicilliopsis	Holomorph		2	1
Penicillium	Anamorph	*Eupenicillium Talaromyces*	225	347
Petromyces	Holomorph	*Aspergillus*	2	0
Pritzeliella	Anamorph		0	1
Raperia	Anamorph	*Warcupiella*	1	1
Sarophorum	Anamorph	*Penicilliopsis*	1	0
Sclerocleista	Holomorph	*Aspergillus*	2	0
Spicaria	Anamorph		0	1
Sterigmatocystis	Anamorph	*Penicillium*	0	11
Stilbodendron	Anamorph		1	0
Stilbothamnium	Anamorph	*Aspergillus*	0	2
Stilbum	Anamorph		0	1
Talaromyces	Holomorph	*Penicilium, Paecilomyces, Geosmithia*	24	17
Thermoascus	Holomorph	*Paecilomyces*	4	0
Torulomyces	Anamorph	*Eupenicillium*	1	0
Trichocoma	Holomorph		1	0
Ustilago	Holomorph		0	1
Verticillium	Anamorph		0	2
Warcupiella	Holomorph	*Raperia*	1	0
Total			**617**	**466**

ARACHNIOTIS J. SCHRÖT. (GYMNOASCACEAE) (HOLOMORPHS)

Arachniotus indicus Chattop. & C. Das Gupta, Trans. Br. Mycol. Soc. 42: 72, 1959 = *Talaromyces flavus* (Klöcker) Stolk & Samson, Stud. Mycol. 2: 10, 1972 (Stolk and Samson, 1972).

Arachniotus indicus var. *major* Chattop. & C. Das Gupta, Trans. Br. Mycol. Soc. 42: 73, 1959 = *Talaromyces flavus* (Klöcker) Stolk & Samson, Stud. Mycol. 2: 10, 1972 (Stolk and Samson, 1972).

Arachniotus intermedius Apinis, Mycol. Papers 96: 45, 1964 ≡ *Talaromyces intermedius* (Apinis) Stolk & Samson, Stud. Mycol. 2: 21, 1972.

Arachniotus purpureus E. Müll. & Pacha-Aue, Nova Hedwigia 15· 552, 1968 ≡ *Talaromyces purpureus* (E. Müll. & Pacha-Aue) Stolk & Samson, Stud. Mycol. 2: 57, 1972.

Arachniotus trachyspermus Shear, Science, Ser. 2, 16: 138, 1902 ≡ *Talaromyces trachyspermus* (Shear) Stolk & Samson, Stud. Mycol. 2: 32, 1972.

ASPERGILLUS FR.: FR. (ANAMORPHS)

Aspergillus acanthosporus Udagawa & Takada, Bull. Natl Sci. Mus. Tokyo, ser. 2, 14: 503, 1971. [Teleomorph *Hemicarpenteles acanthosporus* Udagawa & Takada].

Aspergillus aeneus Sappa, Allionia 2: 84, 1954.

Aspergillus albertensis J. P. Tewari, Mycologia 77: 114, 1985. [Teleomorph *Petromyces albertensis* J. P. Tewari].

Aspergillus allahabadii B. S. Mehrotra & Agnihotri, Mycologia 54: 400, 1963.

Aspergillus alliaceus Thom & Church, Aspergilli: 163, 1926. [Teleomorph *Petromyces alliaceus* Malloch & Cain].

Aspergillus amazonensis (Henn.) Samson & Seifert, in Samson & Pitt (eds), Adv. *Penicillium Aspergillus* Syst.: 418, 1985. — Basionym: *Stilbothamnium amazonense* Henn., Hedwigia 43: 396, 1904.

Aspergillus ambiguus Sappa, Allionia 2: 254, 1955.

Aspergillus amylovorus Panas. ex Samson, Stud. Mycol. 18: 28, 1979.

Aspergillus anthodesmis Bartoli & Maggi, Trans. Br. Mycol. Soc. 71: 386, 1979.

Aspergillus appendiculatus Blaser, Sydowia 28: 38, 1975. [Teleomorph *Eurotium appendiculatum* Blaser].

Aspergillus arenarius Raper & Fennell, Gen. *Aspergillus*: 475, 1965.

Aspergillus asperescens Stolk, Antonie van Leeuwenhoek 20: 303, 1954.

Aspergillus atheciellus Samson & W. Gams Samson & Pitt (eds), Adv. *Penicillium Aspergillus* Syst.: 34, 1985.

Aspergillus athecius Raper & Fennell, Gen. *Aspergillus*: 183, 1965 (nom. holomorph.) ≡ *Edyuillia athecia* Subram., Curr. Sci. 41: 756, 1972 ≡ *Eurotium athecium* (Raper & Fennell) Arx, Gen. Fungi Sporul. Pure Cult., ed. 2: 91, 1974. [Teleomorph *Eurotium athecium* (Raper & Fennell) Arx].

Aspergillus aureolatus Munt. Cvetk. & Bata, Bull. Inst. Jard. Bot. Univ. Beograd, ser. 2, 1: 196, 1964.

Aspergillus aureoluteus Samson & W. Gams in Samson & Pitt (eds), Adv. *Penicillium Aspergillus* Syst.: 34, 1985. [Teleomorph *Neosartorya aureola* (Fennell & Raper) Malloch & Cain].

Aspergillus auricomus (Guég.) Saito, J. Ferment. Technol. 17: 3, 1939. — Basionym: *Sterigmatocystis auricoma* Guég., Bull. Soc. Mycol. France 15: 171, 1899.

Aspergillus avenaceus G. Sm., Trans. Br. Mycol. Soc. 26: 24, 1943.

Aspergillus awamorii Nakaz., Rep. Gov. Res. Inst. Formosa 1907: 1, 1907.

Aspergillus bicolor M. Chr. & States, Mycologia 70: 337, 1978. [Teleomorph *Emericella bicolor* M. Chr. & States].

Aspergillus biplanus Raper & Fennell, Gen. *Aspergillus*: 434, 1965.

Aspergillus bisporus Kwon-Chung & Fennell, Mycologia 63: 479, 1971.

Aspergillus brevipes G. Sm., Trans. Br. Mycol. Soc. 35: 241, 1952.

Aspergillus bridgeri M. Chr., Mycologia 74: 210, 1982.

Aspergillus brunneouniseriatus Suj. Singh & B. K. Bakshi, Trans. Br. Mycol. Soc. 44: 160, 1961.

Aspergillus brunneus Delacr., Bull. Soc. Mycol. France 9: 185, 1893. [Teleomorph *Eurotium echinulatum* Delacr.].

Aspergillus caesiellus Saito, J. Coll. Sci. Imp. Univ. Tokyo 18(5): 49, 1904.

Aspergillus caespitosus Raper & Thom, Mycologia 36: 563, 1944.

Aspergillus campestris M. Chr., Mycologia 74: 212, 1982.

Aspergillus candidus Link, Ges. Naturf. Freunde Berlin Mag. Neuesten Entdeck. Gesammten Naturk. 3: 16, 1809: Fr.

Aspergillus carbonarius (Bainier) Thom, J. Agric. Res. 7: 12, 1916. — Basionym: *Sterigmatocystis carbonaria* Bainier, Bull. Soc. Bot. France 27: 27, 1880.

Aspergillus carneus Blochwitz, Ann. Mycol. 31: 81, 1933.

Aspergillus cervinus Massee, Bull. Misc. Inform. Kew 1914: 158, 1914.

Aspergillus chevalieri L. Mangin, Ann. Sci. Nat., Bot., ser. 9, 10: 362, 1909. [Teleomorph *Eurotium chevalieri* L. Mangin].

Aspergillus chevalieri var. *intermedius* (Thom & Raper) Malloch & Cain, Can. J. Bot. 50: 64, 1972 (invalid basionym) = *Aspergillus chevalieri* var. *intermedius* Thom & Raper, Misc. Publ. U.S. Dept. Agric. 426: 21, 1941 (invalid name, nom. holomorph.) ≡ *Eurotium intermedium* Blaser, Sydowia 28: 41, 1975.

Aspergillus chevalieri var. *intermedius* Thom & Raper, Misc. Publ. U.S. Dept. Agric. 426: 21, 1941 (invalid name, nom. holomorph.) ≡ *Eurotium intermedium* Blaser, Sydowia 28: 41, 1975.

Aspergillus chryseides Samson & W. Gams in Samson & Pitt (eds), Adv. *Penicillium Aspergillus* Syst.: 36, 1985. [Teleomorph *Chaetosartorya chrysella* (Kwon-Chung & Fennell) Subram.].

Aspergillus citrisporus Höhn., Sitzungsber. Kaiserl. Akad. Wiss., Math.-Naturwiss. Cl., Abt.1, 111: 1036, 1902. [Teleomorph *Sclerocleista thaxteri* Subram.].

Aspergillus clavatoflavus Raper & Fennell, Gen. *Aspergillus*: 378, 1965.

Aspergillus clavatonanicus Bat. *et al.*, Anais Fac. Med. Univ. Recife 15: 197, 1955.

Aspergillus clavatus Desm., Ann. Sci. Nat., Bot., ser. 2, 2: 71, 1834.

Aspergillus compatibilis Samson & W. Gams in Samson & Pitt (eds), Adv. *Penicillium Aspergillus* Syst.: 42, 1985. [Teleomorph *Emericella heterothallica* (Kwon-Chung *et al.*) Malloch & Cain].

Aspergillus conicus Blochwitz, Ann. Mycol. 12: 38, 1914.

Aspergillus conjunctus Kwon-Chung & Fennell in Raper & Fennell, Gen. *Aspergillus*: 552, 1965.

Aspergillus coremiiformis Bartoli & Maggi, Trans. Br. Mycol. Soc. 71: 386, 1979.

Aspergillus corrugatus Udagawa & Y. Horie, Mycotaxon 4: 535, 1976. [Teleomorph *Emericella corrugata* Udagawa & Y. Horie].

Aspergillus cremeoflavus Samson & W. Gams in Samson & Pitt (eds), Adv. *Penicillium Aspergillus* Syst.: 37, 1985. [Teleomorph *Chaetosartorya cremea* (Kwon-Chung & Fennell) Subram.].

Aspergillus cristatellus Kozak., Mycol. Pap.161: 81, 1989. [Teleomorph *Eurotium cristatum* (Raper & Fennell) Malloch & Cain].

Aspergillus cristatus Raper & Fennell, Gen. *Aspergillus*: 169, 1965 (nom. holomorph.) ≡ *Eurotium cristatum* (Raper & Fennell) Malloch & Cain, Can. J. Bot. 50: 64, 1972.

Aspergillus crustosus Raper & Fennell, Gen. *Aspergillus*: 532, 1965.

Aspergillus curviformis H. J. Chowdhery & J. N. Rai, Nova Hedwigia 32: 231, 1980.

Aspergillus crystallinus Kwon-Chung & Fennell in Raper & Fennell, Gen. *Aspergillus*: 471, 1965.

Aspergillus deflectus Fennell & Raper, Mycologia 47: 83, 1955.

Aspergillus dimorphicus B. S. Mehrotra & Prasad, Trans. Br. Mycol. Soc. 52: 331, 1969.

Aspergillus diversus Raper & Fennell, Gen. *Aspergillus*: 437, 1965.

Aspergillus duricaulis Raper & Fennell, Gen. *Aspergillus*: 249, 1965.

Aspergillus dybowskii (Pat.) Samson & Seifert in Samson & Pitt (eds), Adv. *Penicillium Aspergillus* Syst.: 422, 1985. — Basionym: *Penicilliopsis dybowskii* Pat., Bull. Soc. Mycol. France 7: 54, 1891.

Aspergillus eburneocremeus Sappa, Allionia 2: 87, 1954.

Aspergillus echinulatus (Delacr.) Thom & Church, Aspergilli: 107, 1926 (includes teleomorph) ≡ *Eurotium echinulatum* Delacr., Bull. Soc. Mycol. Fr. 9: 266, 1893.

Aspergillus egyptiacus Moub. & Mustafa, Egypt. J. Bot. 15: 153, 1972.

Aspergillus elegans Gasperini, Atti Soc. Tosc. Sci. Nat. Pisa, Processi Verbali 8: 328, 1887.

Aspergillus ellipticus Raper & Fennell, Gen. *Aspergillus*: 319, 1965.

Aspergillus elongatus J. N. Rai & S. C. Agarwal, Can. J. Bot. 48: 791, 1970.

Aspergillus erythrocephalis Berk. & M. A. Curtis, J. Linn. Soc. Bot. 10: 362, 1868.

Aspergillus falconensis Y. Horie, Trans. Mycol. Soc. Japan 30: 257, 1989. [Teleomorph *Emericella falconensis* Y. Horie *et al.*].

Aspergillus fennelliae Kwon-Chung & S. J. Kim, Mycologia 66: 629, 1974. [Teleomorph *Neosartorya fennelliae* Kwon-Chung & S. J. Kim].

Aspergillus flaschentraegeri Stolk, Trans. Br. Mycol. Soc. 47: 123, 1964.

Aspergillus fischeri var. *thermomutatus* Paden, Mycopathol. Mycol. Appl. 36: 161, 1968 ≡ *Aspergillus thermomutatus* (Paden) S. W. Peterson, Mycol. Res. 96: 549, 1992.

Aspergillus fischerianus Samson & W. Gams in Samson & Pitt (eds), Adv. *Penicillium Aspergillus* Syst.: 39, 1985.

Aspergillus flavipes (Bainier & Sartory) Thom & Church, Aspergilli: 155, 1926. — Basionym: *Sterigmatocystis flavipes* Bainier & Sartory, Bull. Soc. Mycol. France 27: 90, 1911. [Teleomorph *Fennellia flavipes* B. J. Wiley & E. G. Simmons].

Aspergillus flavofurcatus Bat. & H. Maia, Anais Soc. Biol. Pernambuco 13: 94-96, 1955.

13

Aspergillus flavus Link, Ges. Naturf. Freunde Berlin Mag. Neuesten Entdeck. Gesammten Naturk. 3: 16, 1809: Fr.

Aspergillus floriformis Samson & Mouch., Antonie van Leeuwenhoek 40: 343, 1975.

Aspergillus foeniculicola Udagawa, Trans. Mycol. Soc. Japan 20: 13, 1979. [Teleomorph *Emericella foeniculicola* Udagawa].

Aspergillus foetidus Thom & Raper, Man. Aspergilli: 219, 1945.

Aspergillus foveolatus Y. Horie, Trans. Mycol. Soc. Japan 19: 313, 1978. [Teleomorph *Emericella foveolata* Y. Horie].

Aspergillus fructiculosus Raper & Fennell, Gen. *Aspergillus*: 506, 1965 (nom. holomorph.) ≡ *Emericella fruticulosa* (Raper & Fennell) Malloch & Cain, Can. J. Bot. 50: 61, 1972.

Aspergillus fruticans Samson & W. Gams in Samson & Pitt (eds), Adv. *Penicillium Aspergillus* Syst.: 40, 1985. [Teleomorph *Emericella fruticulosa* (Raper & Fennell) Malloch & Cain].

Aspergillus fumigatus Fresen., Beitr. Mykol.: 81, 1863.

Aspergillus funiculosus G. Sm., Trans. Br. Mycol. Soc. 39: 111, 1956.

Aspergillus giganteus Wehmer, Mem. Soc. Phys. Genève 33(2): 85, 1901.

Aspergillus glaber Blaser, Sydowia 28: 35, 1975. [Teleomorph *Eurotium glabrum* Blaser].

Aspergillus glaucoaffinis Samson & W. Gams in Samson & Pitt (eds), Adv. *Penicillium Aspergillus* Syst.: 47, 1985. [Teleomorph *Eurotium pseudoglaucum* (Blochwitz) Malloch & Cain].

Aspergillus glauconiveus Samson & W. Gams in Samson & Pitt (eds), Adv. *Penicillium Aspergillus* Syst.: 45, 1985. [Teleomorph *Eurotium niveoglaucum* (Thom & Raper) Malloch & Cain].

Aspergillus glaucus Link, Ges. Naturf. Freunde Berlin Mag. Neuesten Entdeck. Gesammten Naturk. 3: 16, 1809: Fr. [Teleomorph *Eurotium herbariorum* Link].

Aspergillus glaucus var. *repens* Corda, Icon. Fung. 5: 53, 1839 (anamorphic name) ≡ *Eurotium repens* de Bary, Abh. Senckenberg. Naturf. Ges. 7: 379, 1870.

Aspergillus glaucus var. *repens* Corda, Icon. Fung. 5: 53, 1842 [non *Aspergillus repens* (de Bary) E. Fisch., 1897] ≡ *Aspergillus reptans* Samson & W. Gams in Samson & Pitt (eds), Adv. *Penicillium Aspergillus* Syst.: 48, 1985.

Aspergillus globosus H. J. Chowdhery & J. N. Rai, Nova Hedwigia 32: 233, 1980.

Aspergillus gorakhpurensis Kamal & Bhargava, Trans. Br. Mycol. Soc. 52: 338, 1969.

Aspergillus gracilis Bainier, Bull. Soc. Mycol. France 23: 90, 1907.

Aspergillus granulosus Raper & Thom, Mycologia 36: 565, 1944.

Aspergillus halophilicus C. M. Chr. *et al.*, Mycologia 51: 636, 1961. [Teleomorph *Eurotium halophilicum* C. M. Chr. *et al.*].

Aspergillus helicothrix Al-Musallam, Antonie van Leeuwenhoek 46: 407, 1980.

Aspergillus herbariorum ser. *minor* L. Mangin, Ann. Sci. Nat., Bot., ser. 9, 10: 365, 1909 ≡ *Aspergillus manginii* Thom & Raper, Man. Aspergilli: 127, 1945 (includes teleomorph) ≡ *Aspergillus minor* (L. Mangin) Thom & Raper, U.S. Dep. Agric. Misc. Publ. 426: 27, 1941 (includes teleomorph) = *Eurotium herbariorum* Link, Ges. Naturf. Freunde Berlin Mag. Neuesten Entdeck. Gesammten Naturk. 3: 31, 1809 (Malloch and Cain, 1972).

Aspergillus heterocaryoticus C. M. Chr. *et al.*, Mycologia 57: 535, 1965.

Aspergillus heteromorphus Bat. & H. Maia, Anais Soc. Biol. Pernambuco 15: 200, 1957.

Aspergillus heterothallicus Kwon-Chung *et al.* in Raper & Fennell, Gen. *Aspergillus*: 502, 1965 (nom. holomorph.) ≡ *Emericella heterothallica* (Kwon-Chung *et al.*) Malloch & Cain, Can. J. Bot. 50: 62, 1972.

Aspergillus hiratsukae Udagawa *et al.*, Trans. Mycol. Soc. Japan 32: 23, 1991. [Teleomorph *Neosartorya hiratsukae* Udagawa *et al.*].

Aspergillus igneus Kozak., Mycol. Pap. 161: 52, 1989. [Teleomorph *Neosartorya aurata* (Warcup) Malloch & Cain].

Aspergillus insulicola Montem. & A. R. Santiago, Mycopathologia 55: 130, 1975.

Aspergillus intermedius Blaser, Sydowia 28: 41, 1976. [Teleomorph *Eurotium intermedium* Blaser].

Aspergillus itaconicus Kinosh., Bot. Mag. (Tokyo) 45: 60, 1931.

Aspergillus ivoriensis Bartoli & Maggi, Trans. Br. Mycol. Soc. 71: 383, 1979.

Aspergillus janus Raper & Thom, Mycologia 36: 556, 1944.

Aspergillus japonicus Saito, Bot. Mag. (Tokyo) 20: 61, 1906.

Aspergillus kanagawaensis Nehira, J. Jap. Bot. 26: 109, 1951.

Aspergillus lanosus Kamal & Bhargava, Trans. Br. Mycol. Soc. 52: 336, 1969.

Aspergillus leporis States & M. Chr., Mycologia 58: 738, 1966.

Aspergillus leucocarpus Hadlok & Stolk, Antonie van Leeuwenhoek 35: 9, 1969. [Teleomorph *Eurotium leucocarpum* Hadlok & Stolk]

Aspergillus longivesica L. H. Huang & Raper, Mycologia 63: 53, 1971.

Aspergillus lucknowensis J. N. Rai *et al.*, Can. J. Bot. 46: 1483, 1968.

Aspergillus malodoratus Kwon-Chung & Fennell in Raper & Fennell, Gen. *Aspergillus*: 468, 1965.

Aspergillus manginii Thom & Raper, Man. Aspergilli: 127, 1945 (invalid name) = **Eurotium herbariorum** Link, Ges. Naturf. Freunde Berlin Mag. Neuesten Entdeck. Gesammten Naturk. 3: 31, 1809 (Malloch and Cain, 1972).

Aspergillus maritimus Samson & W. Gams in Samson & Pitt (eds), Adv. *Penicillium Aspergillus* Syst.: 43, 1985.

Aspergillus medius R. Meissn., Bot. Zeitung, 2. Abt., 55: 356, 1897. [Teleomorph *Eurotium medium* R. Meissn].

Aspergillus melleus Yukawa, J. Coll. Agric. Imp. Univ. Tokyo 1: 358, 1911.

Aspergillus microcysticus Sappa, Allionia 2: 251, 1955.

Aspergillus microthecius Samson & W. Gams in Samson & Pitt (eds), Adv. *Penicillium Aspergillus* Syst.: 46, 1985. [Teleomorph *Emericella parvathecia* (Raper & Fennell) Malloch & Cain].

Aspergillus minor (L. Mangin) Thom & Raper, U.S. Dep. Agric. Misc. Publ. 426: 27, 1941 (includes teleomorph) = **Eurotium herbariorum** Link, Ges. Naturf. Freunde Berlin Mag. Neuesten Entdeck. Gesammten Naturk. 3: 31, 1809 (Malloch and Cain, 1972).

Aspergillus montevidense Talice & Mackinnon, Compt. Rend. Soc. Biol. Fr. 108: 1007, 1931 (includes teleomorph) = **Eurotium amstelodami** L. Mangin, Ann. Sci. Nat., Bot., ser. 9, 10: 360, 1909 (Pitt, 1985).

Aspergillus multicolor Sappa, Allionia 2: 87, 1954.

Aspergillus navahoensis M. Chr. & States, Mycologia 74: 226, 1982. [Teleomorph *Emericella navahoensis* M. Chr. & States].

Aspergillus neocarnoyi Kozak., Mycol. Pap. 161: 63, 1989. [Teleomorph *Eurotium carnoyi* Malloch & Cain].

Aspergillus neoglaber Kozak., Mycol. Pap. 161: 56, 1989. [Teleomorph *Neosartorya glabra* (Fennell & Raper) Kozak.].

Aspergillus nidulans (Eidam) G. Winter, Rabenh. Krypt.-Fl., ed. 2, 1(2): 62, 1884. — Basionym: *Sterigmatocystis nidulans* Eidam, Beitr. Biol. Pflanzen 3: 392, 1883. [Teleomorph *Emericella nidulans* (Eidam) Vuill.].

Aspergillus niger Tiegh., Ann. Sci. Nat., Bot., ser. 5, 8: 240, 1867, *nom. cons.* (Kozakiewicz *et al.,* 1992).

Aspergillus niveoglaucus Thom & Raper, Misc. Publ. U.S. Dept. Agric. 426: 35, 1941 (nom. holomorph.) = **Eurotium niveoglaucum** (Thom & Raper) Malloch & Cain, Can. J. Bot. 50: 64, 1972.

Aspergillus niveus Blochwitz, Ann. Mycol. 27: 205, 1929. [Teleomorph *Fennellia nivea* (B. J. Wiley & E. G. Simmons) Samson].

Aspergillus nomius Kurtzman *et al.*, Antonie van Leeuwenhoek 53: 151, 1987.

Aspergillus nutans McLennan & Ducker, Aust. J. Bot. 2: 355, 1954.

Aspergillus ochraceoroseus Bartoli & Maggi, Trans. Br. Mycol. Soc. 71: 393, 1979.

Aspergillus ochraceus K. Wilh., Beitr. Kenntn. *Aspergillus*: 66, 1877.

Aspergillus ornatulus Samson & W. Gams in Samson & Pitt (eds), Adv. *Penicillium Aspergillus* Syst.: 45, 1985. [Teleomorph *Sclerocleista ornata* (Raper *et al.*) Subram.].

Aspergillus oryzae (Ahlb.) Cohn, Jahresber. Schles. Ges. Vaterl. Cult. 61: 226, 1884. — Basionym: *Eurotium oryzae* Ahlb., Dingler's Polytechn. J. 230: 330, 1878.

Aspergillus ostianus Wehmer, Bot. Centralbl. 80: 461, 1899.

Aspergillus paleaceus Samson & W. Gams in Samson & Pitt (eds), Adv. *Penicillium Aspergillus* Syst.: 50, 1985. [Teleomorph *Neosartorya stramenia* (R. Novak & Raper) Malloch & Cain].

Aspergillus pallidus Kamyschko, Bot. Mater. Otd. Sporov. Rast.16: 93, 1963.

Aspergillus panamensis Raper & Thom, Mycologia 36: 568, 1944.

Aspergillus paradoxus Fennell & Raper, Mycologia 47: 69, 1955. [Teleomorph *Hemicarpenteles paradoxus* A. K. Sarbhoy & Elphick].

Aspergillus parasiticus Speare, Bull. Div. Pathol. Physiol., Hawaiian Sugar Planters' Assoc. Exp. Sta. 12: 38, 1912.

Aspergillus parvathecius Raper & Fennell, Gen. *Aspergillus*: 509, 1965 (nom. holomorph.) ≡ **Emericella parvathecia** (Raper & Fennell) Malloch & Cain, Can. J. Bot. 50: 62, 1972.

Aspergillus parvulus G. Sm., Trans. Br. Mycol. Soc. 44: 45, 1961.

Aspergillus penicillioides Speg., Revista Fac. Agron. Univ. Nac. La Plata 2: 246, 1896.

Aspergillus petrakii Vörös, Sydowia, ser. 2, Beih., 1: 62, 1957.

Aspergillus peyronelii Sappa, Allionia 2: 248, 1955.

Aspergillus phoenicis (Corda) Thom, J. Agric. Res. 7: 14, 1916. — Basionym: *Ustilago phoenicis* Corda, Icon. Fung. 4: 9, 1840.

Aspergillus protuberus Munt. Cvetk., Mikrobiologija 5: 119, 1968.

Aspergillus pseudodeflectus Samson & Mouch., Antonie van Leeuwenhoek 40: 345, 1975.

Aspergillus pseudoglaucus Blochwitz, Ann. Mycol. 27: 207, 1929 (nom. holomorph.) ≡ **Eurotium pseudoglaucum** (Blochwitz) Malloch & Cain, Can. J. Bot. 50: 64, 1972.

15

Aspergillus pulverulentus (McAlpine) Wehmer, Centralbl. Bakteriol., 2. Abth., 18: 394, 1907. — Basionym: *Sterigmatocystis pulverulenta* McAlpine, Agric. Gaz. N.S.W. 7: 302, 1897.

Aspergillus pulvinus Kwon-Chung & Fennell in Raper & Fennell, Gen. *Aspergillus*: 45, 1965.

Aspergillus puniceus Kwon-Chung & Fennell in Raper & Fennell, Gen. *Aspergillus*: 547, 1965.

Aspergillus purpureus Samson & Mouch., Antonie van Leeuwenhoek 41: 350, 1975. [Teleomorph *Emericella purpurea* Samson & Mouch.].

Aspergillus quadricingens Kozak., Mycol. Pap. 161: 54, 1989. [Teleomorph *Neosartorya quadricincta* (E. Yuill) Malloch & Cain].

Aspergillus quadrilineatus Thom & Raper, Mycologia 31: 660, 1939 (nom. holomorph.) ≡ **Emericella quadrilineata** (Thom & Raper) C. R. Benj., Mycologia 47: 680, 1955.

Aspergillus raperi Stolk, Trans. Br. Mycol. Soc. 40: 190, 1957.

Aspergillus recurvatus Raper & Fennell, Gen. *Aspergillus*: 529, 1965.

Aspergillus reptans Samson & W. Gams in Samson & Pitt (eds), Adv. *Penicillium Aspergillus* Syst.: 48, 1985 ≡ *Aspergillus glaucus* var. *repens* Corda, Icon. Fung. 5: 53, 1842 [non *Aspergillus repens* (de Bary) E. Fisch., Engler & Prantl, Nat. Pflanzenfam. 1(1): 302, 1897]). [Teleomorph *Eurotium repens* de Bary].

Aspergillus restrictus G. Sm., J. Textile Inst. 22: 115, 1931.

Aspergillus robustus M. Chr. & Raper, Mycologia 70: 200, 1978.

Aspergillus rubrobrunneus Samson & W. Gams in Samson & Pitt (eds), Adv. *Penicillium Aspergillus* Syst.: 49, 1985. [Teleomorph *Eurotium rubrum* Jos. König et al.].

Aspergillus rubrum (Jos. König et al.) Thom & Church, *Aspergillus*: 112, 1926 (includes teleomorph) ≡ *Eurotium rubrum* Jos. König et al., Z. Untersuch. Nahrungs- Gen.smittel 4: 726, 1901.

Aspergillus rugulosus Thom & Raper, Mycologia 31: 660, 1939 (nom. holomorph.) ≡ **Emericella rugulosa** (Thom & Raper) C. R. Benj., Mycologia 47: 680, 1955.

Aspergillus rugulovalvus Samson & W. Gams in Samson & Pitt (eds), Adv. *Penicillium Aspergillus* Syst.: 49, 1985. [Teleomorph *Emericella rugulosa* (Thom & Raper) C. R. Benj.].

Aspergillus sclerotiorum G. A. Huber, Phytopathology 23: 306, 1933.

Aspergillus sejunctus Bainier & Sartory, Bull. Trimest. Soc. Mycol. Fr. 27: 346, 1911 (includes teleomorph) = *Eurotium rubrum* Jos. König et al., Z. Untersuch. Nahrungs- Gen.smittel 4: 726, 1901 (Thom & Raper, 1945).

Aspergillus sepultus Tuthill & M. Chr., Mycologia 78: 475, 1986.

Aspergillus silvaticus Fennell & Raper, Mycologia 47: 83, 1955.

Aspergillus sojae Sakag. & K. Yamada ex Murak., Rep. Res. Inst. Brewing 143: 8, 1971.

Aspergillus sparsus Raper & Thom, Mycologia 36: 572, 1944.

Aspergillus spathulatus Takada & Udagawa, Mycotaxon 24: 396, 1985. [Teleomorph *Neosartorya spathulata* Takada & Udagawa].

Aspergillus spectabilis M. Chr. & Raper, Mycologia 70: 333, 1978. [Teleomorph *Emericella spectabilis* M. Chr. & Raper].

Aspergillus speluneus Raper & Fennell, Gen. *Aspergillus*: 457, 1965.

Aspergillus spinosus Kozak., Mycol. Pap.161: 58, 1989. [Teleomorph *Neosartorya spinosa* (Raper & Fennell) Kozak.].

Aspergillus stellifer Samson & W. Gams in Samson & Pitt (eds), Adv. *Penicillium Aspergillus* Syst.: 52, 1985. [Teleomorph *Emericella variecolor* Berk. & Broome].

Aspergillus striatulus Samson & W. Gams in Samson & Pitt (eds), Adv. *Penicillium Aspergillus* Syst.: 50, 1985. [Teleomorph *Emericella striata* (J. N. Rai et al.) Malloch & Cain].

Aspergillus striatus J. N. Rai et al., Can. J. Bot. 42: 1521, 1964 (nom. holomorph.) ≡ **Emericella striata** (J. N. Rai et al.) Malloch & Cain, Can. J. Bot. 50: 62, 1972.

Aspergillus stromatoides Raper & Fennell, Gen. *Aspergillus*: 421, 1965. [Teleomorph *Chaetosartorya stromatoides* B. J. Wiley & E. G. Simmons].

Aspergillus sublatus Y. Horie, Trans. Mycol. Soc. Japan 20: 481, 1979. [Teleomorph *Emericella sublata* Y. Horie].

Aspergillus subolivaceus Raper & Fennell, Gen. *Aspergillus*: 385, 1965.

Aspergillus subsessilis Raper & Fennell, Gen. *Aspergillus*: 530, 1965.

Aspergillus sulphureus (Fresen.) Wehmer, Mem. Soc. Phys. Genève 33(2): 113, 1901. — Basionym: *Sterigmatocystis sulphurea* Fresen., Beitr. Mykol.: 83, 1863.

Aspergillus sydowii (Bainier & Sartory) Thom & Church, Aspergilli: 147, 1926. — Basionym: *Sterigmatocystis sydowii* Bainier & Sartory, Ann. Mycol. 11: 25, 1913.

Aspergillus tamarii Kita, Centralbl. Bakteriol.. 2. Abth.. 37: 433, 1913.

Aspergillus tardus Bissett & Widden, Can. J. Bot. 62: 2521, 1984.

Aspergillus tatenoi Y. Horie et al., Trans. Mycol. Soc. Japan 33: 395, 1992. [Teleomorph *Neosartorya tatenoi* Y. Horie et al.].

Aspergillus terreus Thom, Am. J. Bot.5: 85, 1918.

Aspergillus terricola E. J. Marchal, Rev. Mycol. (Toulouse) 15: 101, 1893.

Aspergillus tetrazonus Samson & W. Gams in Samson & Pitt (eds), Adv. *Penicillium Aspergillus* Syst.: 48, 1985. [Teleomorph *Emericella quadrilineata* (Thom & Raper) C. R. Benj.].

Aspergillus thermomutatus (Paden) S. W. Peterson, Mycol. Res. 96: 549, 1992. — Basionym: *Aspergillus fischeri* var. *thermomutatus* Paden, Mycopathol. Mycol. Appl. 36: 161, 1968. [Teleomorph *Neosartorya pseudofischeri* S. W. Peterson].

Aspergillus togoensis (Henn.) Samson & Seifert in Samson & Pitt (eds), Adv. *Penicillium Aspergillus* Syst.: 419, 1985. — Basionym: *Stilbothamnium togoense* Henn., Bot. Jahrb. Syst. 23: 542, 1897.

Aspergillus tonophilus Ohtsuki, Bot. Mag. (Tokyo) 75: 438, 1962. [Teleomorph *Eurotium tonophilum* Ohtsuki].

Aspergillus umbrosus Bainier & Sartory, Bull. Trimest. Soc. Mycol. Fr. 28: 267, 1912 = *Aspergillus glaucus* Link, Ges. Naturf. Freunde Berlin Mag. Neuesten Entdeck. Gesammten Naturk. 3: 16, 1809 (Blaser, 1975).

Aspergillus undulatus H. Z. Kong & Z. T. Qi, Acta Mycol. Sin. 4: 211, 1985. [Teleomorph *Emericella undulata* H. Z. Kong & Z. T. Qi].

Aspergillus unguis (Emile-Weil & L. Gaudin) Thom & Raper, Mycologia 31: 667, 1939. — Basionym: *Sterigmatocystis unguis* Emile-Weil & L. Gaudin, Arch. Med. Exp. Anat. Pathol. 28: 463, 1918. [Teleomorph *Emericella unguis* Malloch & Cain].

Aspergillus unilateralis Thrower, Aust. J. Bot. 2: 355, 1954.

Aspergillus ustus (Bainier) Thom & Church, Aspergilli: 152, 1926. — Basionym: *Sterigmatocystis usta* Bainier, Bull. Soc. Bot. France 28: 78, 1881.

Aspergillus varians Wehmer, Bot. Centralbl. 80: 460, 1899.

Aspergillus versicolor (Vuill.) Tirab., Ann. Bot. (Roma) 7: 9, 1908. — Basionym: *Sterigmatocystis versicolor* Vuill., Mirsky , Causes Err. Dét. Aspergill.: 15, 1903.

Aspergillus violaceobrunneus Samson & W. Gams in Samson & Pitt (eds), Adv. *Penicillium Aspergillus* Syst.: 53, 1985. [Teleomorph *Emericella violacea* (Fennell & Raper) Malloch & Cain].

Aspergillus violaceofuscus Gasperini, Atti Soc. Tosc. Sci. Nat. Pisa, Processi Verbali 8: 326, 1887.

Aspergillus violaceus Fennell & Raper, Mycologia 47: 75, 1955 (nom. holomorph.) ≡ *Emericella violacea* (Fennell & Raper) Malloch & Cain, Can. J. Bot. 50: 62, 1972.

Aspergillus viridinutans Ducker & Thrower, Aust. J. Bot. 2: 355, 1954.

Aspergillus vitellinus (Massee) Samson & Seifert in Samson & Pitt (eds), Adv. *Penicillium Aspergillus* Syst.: 417, 1985. — Basionym: *Sterigmatocystis vitellina* Ridl. ex Massee, J. Bot. 34: 152, 1896.

Aspergillus vitis Novobr., Novosti Sist. Nizs. Rast. 9: 175, 1972. [Teleomorph *Eurotium amstelodami* L. Mangin].

Aspergillus wentii Wehmer, Centralbl. Bakteriol., 2. Abth., 2: 149, 1896.

Aspergillus xerophilus Samson & Mouch., Antonie van Leeuwenhoek 41: 348, 1975. [Teleomorph *Eurotium xerophilum* Samson & Mouch.].

Aspergillus zonatus Kwon-Chung & Fennell, Raper & Fennell, Gen. *Aspergillus*: 377, 1965.

BYSSOCHLAMYS WESTLING (HOLOMORPHS)

Byssochlamys fulva Olliver & G. Sm., J. Bot. 71: 196, 1933. [Anamorph: *Paecilomyces fulvus* Stolk & Samson].

Byssochlamys nivea Westling, Svensk Bot. Tidskr. 3: 134, 1909. [Anamorph: *Paecilomyces niveus* Stolk & Samson].

Byssochlamys striata (Raper & Fennell) Arx, Mycotaxon 26: 120, 1986 ≡ *Talaromyces striatus* (Raper & Fennell) C. R. Benj., Mycologia 47: 682, 1955.

Byssochlamys verrucosa Samson & Tansey, Trans. Br. Mycol. Soc. 65: 512, 1975. [Anamorph: *Paecilomyces verrucosus* Samson & Tansey].

Byssochlamys zollerniae C. Ram, Nova Hedwigia 16: 312, 1968. [Anamorph: *Paecilomyces zollerniae* Stolk & Samson].

CARPENTELES SHEAR (HOLOMORPHS)

Carpenteles asperum Shear, Mycologia 26: 107, 1934 ≡ *Penicillium asperum* (Shear) Raper & Thom, Man. Penicillia: 263, 1949 (nom. holomorph.) = *Eupenicillium crustaceum* F. Ludw., Lehrb. Nied. Krypt.: 263, 1892 (Stolk and Scott, 1967).

Carpenteles brefeldianum (B. O. Dodge) Shear, Mycologia 26: 107, 1934 ≡ *Eupenicillium brefeldianum* (B. O. Dodge) Stolk & D. B. Scott, Persoonia 4: 400, 1967.

Carpenteles javanicum (J. F. H. Beyma) Shear, Mycologia 26: 107, 1934 ≡ *Eupenicillium javanicum* (J. F. H. Beyma) Stolk & D. B. Scott, Persoonia 4: 398, 1967.

Carpenteles levitum (Raper & Fennell) C. R. Benj., Mycologia 47: 685, 1953 ≡ *Eupenicillium levitum* (Raper & Fennell) Stolk & D. B. Scott, Persoonia 4: 402, 1967.

Carpenteles parvum (Raper & Fennell) Udagawa, Trans. Mycol. Soc. Japan 6: 79, 1965 ≡ *Eupeni-*

cillium parvum (Raper & Fennell) Stolk & D. B. Scott, Persoonia 4: 402, 1967.

<u>CHAETOSARTORYA</u> SUBRAM. (HOLOMORPHS)
Chaetosartorya chrysella (Kwon-Chung & Fennell) Subram., Curr. Sci. 41: 761, 1972. — Basionym: *Aspergillus chrysellus* Kwon-Chung & Fennell in Raper & Fennell, Gen. *Aspergillus*: 424, 1965 (nom. holomorph.). [Anamorph: *Aspergillus chryseides* Samson & W. Gams].
Chaetosartorya cremea (Kwon-Chung & Fennell) Subram., Curr. Sci. 41: 761, 1972. — Basionym: *Aspergillus cremeus* Kwon-Chung & Fennell in Raper & Fennell, Gen. *Aspergillus*: 418, 1965 (nom. holomorph.). [Anamorph: *Aspergillus cremeoflavus* Samson & W. Gams].
Chaetosartorya stromatoides B. J. Wiley & E. G. Simmons, Mycologia 65: 935, 1973. [Anamorph: *Aspergillus stromatoides* Raper & Fennell].

<u>CITROMYCES</u> (ANAMORPHS)
Citromyces cesiae Bainier & Sartory, Bull. Trimest. Soc. Mycol. France 29: 148, 1913 ≡ *Penicillium cesiae* (Bainier & Sartory) Biourge, Cellule 33: 101, 1923 = ***Penicillium roseopurpureum*** Dierckx, Ann. Soc. Sci. Bruxelles 25: 86, 1901 (Biourge, 1923).
Citromyces cyaneus Bainier & Sartory, Bull. Soc. Mycol. France 29: 157, 1913 ≡ ***Penicillium cyaneum*** (Bainier & Sartory) Biourge, Cellule 33: 102, 1923.
Citromyces glaber Wehmer, Beitr. Einh. Pilze 1: 24, 1893 ≡ ***Penicillium glabrum*** (Wehmer) Westling, Ark. Bot. 11(1): 131, 1911.
Citromyces griseus Sopp, Skr. Vidensk.-Selsk. Christiana, Math.-Naturvidensk. Kl. 11: 119, 1912 ≡ *Penicillium griseum* (Sopp) Biourge, Cellule 33: 103, 1923 (non Bonorden, 1951) = ***Penicillium restrictum*** J. C. Gilman & E. V. Abbott, Iowa State Coll. J. Sci. 1: 297, 1927 (Pitt, 1980).
Citromyces pfefferianus Wehmer, Beitr. Einh. Pilze 1: 24, 1893 ≡ *Penicillium pfefferianum* (Wehmer) Westling, Ark. Bot. 11(1): 132, 1911 ≡ *Penicillium pfefferianum* (Wehmer) Pollaci, Atti Ist. Bot. Univ. Pavia, Ser. 2, 16: 135, 1916 ≡ *Penicillium pfefferianum* (Wehmer) Biourge, Cellule 23: 105, 1923 = ***Penicillium glabrum*** (Wehmer) Westling, Ark. Bot. 11(1): 131, 1911 (Pitt, 1980).
Citromyces purpurascens Sopp, Skr. Vidensk.-Selsk. Christiana, Math.-Naturvidensk. Kl. 11: 117, 1912 ≡ ***Penicillium purpurascens*** (Sopp) Biourge, Cellule 33: 105, 1923.
Citromyces sanguifluus Sopp, Skr. Vidensk.-Selsk. Christiana, Math.-Naturvidensk. Kl. 11: 115, 1912 ≡ *Penicillium sanguifluum* (Sopp) Biourge, Cellule 33: 105, 1923 = ***Penicillium roseopur-***

pureum Dierckx, Ann. Soc. Sci. Bruxelles 25: 86, 1901 (Biourge, 1923).
Citromyces sphagnicola Malchevsk., Trudy Pushkin Nauchno-Issled. Lab. Rasv. sel'.-khoz. Zhivot. 13: 23, 1939 = ***Talaromyces thermophilus*** Stolk, Antonie van Leeuwenhoek 31: 268, 1965 (Stolk, 1965).
Citromyces subtilis Bainier & Sartory, Bull. Trimest. Soc. Mycol. Fr. 28: 46, 1912 ≡ *Penicillium subtile* (Bainier & Sartory) Biourge, Cellule 33: 106, 1923 (non Berk. 1841) = ***Penicillium citrinum*** Thom, Bull. Bur. Anim. Ind. U.S. Dep. Agric. 118: 61, 1910 (Pitt, 1980).
Citromyces thomii (Maire) Sacc., Syll. Fung. 25: 683, 1931 ≡ ***Penicillium thomii*** Maire, Bull. Soc. Hist. Nat. Afrique N. 8: 189, 1917.
Citromyces virido-albus Sopp, Skr. Vidensk.-Selsk. Christiana, Math.-Naturvidensk. Kl. 11: 131, 1912 ≡ *Penicillium virido-albus* (Sopp) Biourge, Cellule 33: 106, 1923 = ***Penicillium purpurascens*** (Sopp) Biourge, Cellule 33: 105, 1923 (Raper & Thom, 1949).

<u>COREMIUM</u> (ANAMORPHS)
Coremium coprophilum Berk. & M. A. Curtis, J. Linn. Soc. Bot., 10: 363, 1868 ≡ ***Penicillium coprophilum*** (Berk. & M. A. Curtis) Seifert & Samson in Samson & Pitt (eds), Adv. *Penicillium Aspergillus* Syst.: 145, 1985.
Coremium glandicola Oudem., Ned. Kruidk. Arch., Ser. 3, 2: 918, 1903 ≡ ***Penicillium glandicola*** (Oudem.) Seifert & Samson in Samson & Pitt (eds), Adv. *Penicillium Aspergillus* Syst.: 147, 1985.
Coremium leucopus Pers., Mycol. Eur. 1: 42, 1822 ≡ *Penicillium leucopus* (Pers.) Biourge, C. R. Séanc. Soc. Biol. 82: 877, 1919 = ***Penicillium expansum*** Link, Ges. Naturf. Freunde Berlin Mag. Neuesten Entdeck. Gesammten Naturk. 3: 16, 1809 (Thom, 1930).
Coremium silvaticum Wehmer, Ber. Dt. Bot. Ges. 31: 373, 1914 ≡ ***Penicillium vulpinum*** (Cooke & Massee) Seifert & Samson in Samson & Pitt (eds), Adv. *Penicillium Aspergillus* Syst.: 144, 1985 (Seifert and Samson, 1985).
Coremium vulpinum Cooke & Massee, Grevillea 16: 81, 1888 ≡ ***Penicillium vulpinum*** (Cooke & Massee) Seifert & Samson in Samson & Pitt (eds), Adv. *Penicillium Aspergillus* Syst.: 144, 1985.

<u>CORETHROPSIS</u> CORDA (ANAMORPHS)
Corethropsis puntonii Vuill., Compt. Rend. Hebd. Séances Acad. Sci. 190: 1334, 1930 ≡ ***Paecilomyces puntonii*** (Vuill.) Nann., Repert. Mic. Uomo: 245, 1934.

18

CRISTASPORA FORT & GUARRO (HOLOMORPHS)
Cristaspora arxii Fort & Guarro, Mycologia 76: 1115, 1984.

CYLINDRODENDRUM BONORD. (ANAMORPHS)
Cylindrodendrum suffultum Petch, Trans. Brit. Mycol. Soc. 27: 91, 1944 ≡ **Paecilomyces suffultus** (Petch) Samson, Stud. Mycol. 6: 55, 1974.

DACTYLOMYCES SOPP HOLOMORPHS
Dactylomyces crustaceus Apinis & Chesters, Trans. Br. Mycol. Soc. 47: 428, 1964 ≡ **Thermoascus crustaceus** (Apinis & Chesters) Stolk, Antonie van Leeuwenhoek 31: 272, 1965.
Dactylomyces thermophilus Sopp, Skr. Vidensk.-Selsk. Christiana, Math.-Naturvidensk. Kl. 11: 35, 1912 ≡ **Thermoascus thermophilus** (Sopp) Arx, Gen. Fungi Sporul. Pure Cult., ed. 2: 94, 1974.

DENDROSPHAERA PAT. (HOLOMORPHS)
Dendrosphaera eberhardtii Pat., Bull. Soc. Mycol. France 23: 69, 1907.

DICHLAENA DURIEU & MONT. (HOLOMORPHS)
Dichlaena lentisci Durieu & Mont. in Durieu, Expl. Sci. Algérie, Bot., 1: 405, 1849.

DICHOTOMOMYCES D. B. SCOTT (HOLOMORPHS)
Dichotomomyces cejpii (Milko) D. B. Scott, Trans. Br. Mycol. Soc. 55: 313, 1970. — Basionym: *Talaromyces cejpii* Milko, Novosti Sist. Nizs. Rast. 1964: 208, 1964.

EDYUILLIA SUBRAM. (HOLOMORPHS)
Edyuillia athecia Subram., Curr. Sci. 41: 756, 1972 ≡ **Eurotium athecium** (Raper & Fennell) Arx, Gen. Fungi Sporul. Pure Cult., ed. 2: 91, 1974.

ELADIA G. SM. (ANAMORPHS)
Eladia saccula (E. Dale) G. Sm., Trans. Br. Mycol. Soc. 44: 47, 1961 ≡ **Penicillium sacculum** E. Dale, Ann. Mycol. 24: 137, 1926.

EMERICELLA BERK. (HOLOMORPHS)
Emericella acristata (Fennell & Raper) Y. Horie, Trans. Mycol. Soc. Japan 21: 491, 1980. — Basionym: *Aspergillus nidulans* var. *acristatus* Fennell & Raper, Mycologia 47: 79, 1955 (nom. holomorph.).
Emericella astellata (Fennell & Raper) Y. Horie, Trans. Mycol. Soc. Japan 21: 491, 1980. — Basionym: *Aspergillus variecolor* var. *astellatus* Fennell & Raper, Mycologia 47: 81, 1955 (nom. holomorph.).
Emericella aurantiobrunnea (G. A. Atkins *et al.*) Malloch & Cain, Can. J. Bot. 50: 61, 1972. —

Basionym: *Emericella nidulans* var. *aurantiobrunnea* G. A. Atkins *et al.*, Trans. Br. Mycol. Soc. 41: 504, 1958.
Emericella bicolor M. Chr. & States, Mycologia 70: 337, 1978. [Anamorph: *Aspergillus bicolor* M. Chr. & States].
Emericella corrugata Udagawa & Y. Horie, Mycotaxon 4: 535, 1976. [Anamorph: *Aspergillus corrugatus* Udagawa & Y. Horie].
Emericella dentata (D. K. Sandhu & R. S. Sandhu) Y. Horie, Trans. Mycol. Soc. Japan 21: 491, 1980. — Basionym: *Aspergillus nidulans* var. *dentatus* D. K. Sandhu & R. S. Sandhu, Mycologia 55: 297, 1963 (nom. holomorph.).
Emericella desertorum Samson & Mouch., Antonie van Leeuwenhoek 40: 121, 1974.
Emericella echinulata (Fennell & Raper) Y. Horie, Trans. Mycol. Soc. Japan 21: 492, 1980. — Basionym: *Aspergillus nidulans* var. *echinulatus* Fennell & Raper, Mycologia 47: 79, 1955 (nom. holomorph.).
Emericella falconensis Y. Horie *et al.*, Trans. Mycol. Soc. Japan 30: 257, 1989. [Anamorph: *Aspergillus falconensis* Y. Horie].
Emericella foeniculicola Udagawa, Trans. Mycol. Soc. Japan 20: 13, 1979. [Anamorph: *Aspergillus foeniculicola* Udagawa].
Emericella foveolata Y. Horie, Trans. Mycol. Soc. Japan 19: 313, 1978. [Anamorph: *Aspergillus foveolatus* Y. Horie].
Emericella fruticulosa (Raper & Fennell) Malloch & Cain, Can. J. Bot. 50: 61, 1972. — Basionym: *Aspergillus fructiculosus* Raper & Fennell, Gen. *Aspergillus*: 506, 1965 (nom. holomorph.). [Anamorph: *Aspergillus fruticans* Samson & W. Gams].
Emericella heterothallica (Kwon-Chung *et al.*) Malloch & Cain, Can. J. Bot. 50: 62, 1972. — Basionym: *Aspergillus heterothallicus* Kwon-Chung *et al.* in Raper & Fennell, Gen. *Aspergillus*: 502, 1965 (nom. holomorph.). [Anamorph: *Aspergillus compatibilis* Samson & W. Gams].
Emericella navahoensis M. Chr. & States, Mycologia 74: 226, 1982. [Anamorph: *Aspergillus navahoensis* M. Chr. & States].
Emericella nidulans (Eidam) Vuill., Compt. Rend. Hebd. Séances Acad. Sci. 184: 137, 1927. — Basionym: *Sterigmatocystis nidulans* Eidam, Beitr. Biol. Pflanzen 3: 393, 1883 (nom. holomorph.). [Anamorph: *Aspergillus nidulans* (Eidam) G. Winter].
Emericella nivea B. J. Wiley & E. G. Simmons, Mycologia 65: 934, 1973 ≡ **Fennellia nivea** (B. J. Wiley & E. G. Simmons) Samson, Stud. Mycol. 18: 5, 1979.
Emericella parvathecia (Raper & Fennell) Malloch & Cain, Can. J. Bot. 50: 62, 1972. — Basionym:

Aspergillus parvathecius Raper & Fennell, Gen. *Aspergillus*: 509, 1965 (nom. holomorph.). [Anamorph: *Aspergillus microthecius* Samson & W. Gams].

Emericella purpurea Samson & Mouch., Antonie van Leeuwenhoek 41: 350, 1975. [Anamorph: *Aspergillus purpureus* Samson & Mouch.].

Emericella quadrilineata (Thom & Raper) C. R. Benj., Mycologia 47: 680, 1955. — Basionym: *Aspergillus quadrilineatus* Thom & Raper, Mycologia 31: 660, 1939 (nom. holomorph.). [Anamorph: *Aspergillus tetrazonus* Samson & W. Gams].

Emericella rugulosa (Thom & Raper) C. R. Benj., Mycologia 47: 680, 1955. — Basionym: *Aspergillus rugulosus* Thom & Raper, Mycologia 31: 660, 1939 (nom. holomorph.). [Anamorph: *Aspergillus rugulovalvus* Samson & W. Gams].

Emericella similis Y. Horie *et al.*, Trans. Mycol. Soc. Japan 31: 425, 1990.

Emericella spectabilis M. Chr. & Raper, Mycologia 70: 333, 1978. [Anamorph: *Aspergillus spectabilis* M. Chr. & Raper].

Emericella striata (J. N. Rai *et al.*) Malloch & Cain, Can. J. Bot. 50: 62, 1972. — Basionym: *Aspergillus striatus* J. N. Rai *et al.*, Can. J. Bot. 42: 1521, 1964 (nom. holomorph.). [Anamorph: *Aspergillus striatulus* Samson & W. Gams].

Emericella sublata Y. Horie, Trans. Mycol. Soc. Japan 20: 481, 1979. [Anamorph: *Aspergillus sublatus* Y. Horie].

Emericella undulata H. Z. Kong & Z. T. Qi, Acta Mycol. Sin. 4: 211, 1985. [Anamorph: *Aspergillus undulatus* H. Z. Kong & Z. T. Qi].

Emericella unguis Malloch & Cain, Can. J. Bot. 50: 62, 1972. [Anamorph: *Aspergillus unguis* (Emile-Weil & L. Gaudin) Thom & Raper].

Emericella variecolor Berk. & Broome in Berkeley, Intr. Crypt. Bot.: 340, 1857. [Anamorph: *Aspergillus stellifer* Samson & W. Gams].

Emericella violacea (Fennell & Raper) Malloch & Cain, Can. J. Bot. 50: 62, 1972. — Basionym: *Aspergillus violaceus* Fennell & Raper, Mycologia 47: 75, 1955 (nom. holomorph.). [Anamorph: *Aspergillus violaceobrunneus* Samson & W. Gams].

EUPENICILLIUM F. LUDW. (HOLOMORPHS)

Eupenicillium abidjanum Stolk, Antonie van Leeuwenhoek 34: 49, 1968. [Anamorph: *Penicillium abidjanum* Stolk].

Eupenicillium alutaceum D. B. Scott, Mycopathol. Mycol. Appl. 36: 17, 1968. [Anamorph: *Penicillium alutaceum* D. B. Scott].

Eupenicillium anatolicum Stolk, Antonie van Leeuwenhoek 34: 46, 1968. [Anamorph: *Penicillium anatolicum* Stolk].

Eupenicillium angustiporcatum Takada & Udagawa, Trans. Mycol. Soc. Japan 24: 143, 1983. [Anamorph: *Penicillium angustiporcatum* Takada & Udagawa].

Eupenicillium arvense Udagawa & T. Yokoy. in Udagawa and Horie, Trans Mycol. Soc. Japan 15: 107, 1974 = *Eupenicillium reticulisporum* Udagawa, Trans. Mycol. Soc. Japan 9: 52, 1968 (Pitt, 1980).

Eupenicillium baarnense (J. F. H. Beyma) Stolk & D. B. Scott, Persoonia 4: 401, 1967. — Basionym: *Penicillium baarnense* J. F. H. Beyma, Antonie van Leeuwenhoek 6: 271, 1940 (nom. holomorph.). [Anamorph: *Penicillium vanbeymae* Pitt].

Eupenicillium brefeldianum (B. O. Dodge) Stolk & D. B. Scott, Persoonia 4: 400, 1967. — Basionym: *Penicillium brefeldianum* B. O. Dodge, Mycologia 25: 92, 1933 (nom. holomorph.). [Anamorph: *Penicillium dodgei* Pitt].

Eupenicillium caperatum Udagawa & Y. Horie, Trans. Mycol. Soc. Japan 14: 371, 1973 = *Eupenicillium ornatum* Udagawa in Trans. Mycol. Soc. Japan 9: 49, 1968 (Pitt, 1980).

Eupenicillium catenatum D. B. Scott, Mycopathol. Mycol. Appl. 36: 24, 1968. [Anamorph: *Penicillium catenatum* D. B. Scott].

Eupenicillium cinnamopurpureum D. B. Scott & Stolk, Antonie van Leeuwenhoek 33: 308, 1967. [Anamorph: *Penicillium phoeniceum* J. F. H. Beyma].

Eupenicillium crustaceum F. Ludw., Lehrb. Nied. Krypt.: 263, 1892. [Anamorph: *Penicillium gladioli* L. McCulloch & Thom].

Eupenicillium cryptum Goch., Mycotaxon 26: 349, 1986. [Anamorph: *Penicillium cryptum* Goch.].

Eupenicillium egyptiacum (J. F. H. Beyma) Stolk & D. B. Scott, Persoonia 4: 401, 1967. — Basionym: *Penicillium egyptiacum* J. F. H. Beyma, Zentralbl. Bakteriol., 2. Abt., 88: 137, 1933 (nom. holomorph.). [Anamorph: *Penicillium nilense* Pitt].

Eupenicillium ehrlichii (Kleb.) Stolk & D. B. Scott, Persoonia 4: 400, 1967. — Basionym: *Penicillium ehrlichii* Kleb., Ber. Deutsch. Bot. Ges. 48: 374, 1930 (nom. holomorph.). [Anamorph: *Penicillium klebahnii* Pitt].

Eupenicillium erubescens D. B. Scott, Mycopathol. Mycol. Appl. 36: 14, 1968. [Anamorph: *Penicillium erubescens* D. B. Scott].

Eupenicillium fractum Udagawa, Trans. Mycol. Soc. Japan 9: 51, 1968. [Anamorph: *Penicillium fractum* Udagawa].

Eupenicillium gracilentum Udagawa & Y. Horie, Trans. Mycol. Soc. Japan 14: 373, 1973. [Anamorph: *Penicillium gracilentum* Udagawa & Y. Horie].

20

Eupenicillium hirayamae D. B. Scott & Stolk, Antonie van Leeuwenhoek 33: 305, 1967. [Anamorph: *Penicillium hirayamae* Udagawa].

Eupenicillium inusitatum D. B. Scott, Mycopathol. Mycol. Appl. 36: 20, 1968. [Anamorph: *Penicillium inusitatum* D. B. Scott].

Eupenicillium javanicum (J. F. H. Beyma) Stolk & D. B. Scott, Persoonia 4: 398, 1967. — Basionym: *Penicillium javanicum* J. F. H. Beyma, Verh. Kon. Ned. Akad. Wetensch., Afd. Natuurk., Tweede Sect., 26(4): 17, 1929 (nom. holomorph.). [Anamorph: *Penicillium indonesiae* Pitt].

Eupenicillium katangense Stolk, Antonie van Leenwenhoek 34: 42, 1968. [Anamorph: *Penicillium katangense* Stolk].

Eupenicillium lapidosum D. B. Scott & Stolk, Antonie van Leeuwenhoek 33: 298, 1967. [Anamorph: *Penicillium lapidosum* Raper & Fennell].

Eupenicillium lassenii Paden, Mycopathol. Mycol. Appl. 43: 266, 1971. [Anamorph: *Penicillium lassenii* Paden].

Eupenicillium levitum (Raper & Fennell) Stolk & D. B. Scott, Persoonia 4: 402, 1967. — Basionym: *Penicillium levitum* Raper & Fennell, Mycologia 40: 511, 1948 (nom. holomorph.). [Anamorph: *Penicillium rasile* Pitt].

Eupenicillium limoneum Goch. & Zlattner, Stud. Mycol. 23: 100, 1983. [Anamorph: *Torulomyces lagena* Delitsch].

Eupenicillium lineolatum Udagawa & Y. Horie, Mycotaxon 5: 493, 1977. [Anamorph: *Penicillium lineolatum* Udagawa & Y. Horie].

Eupenicillium ludwigii Udagawa, Trans. Mycol. Soc. Japan 10: 2, 1969. [Anamorph: *Penicillium ludwigii* Udagawa].

Eupenicillium luzoniacum Udagawa & Y. Horie, J. Jap. Bot. 47: 338, 1972 = *Eupenicillium terrenum* D. B. Scott in Mycopathol. Mycol. Appl. 36: 1, 1968 (Pitt, 1980).

Eupenicillium meloforme Udagawa & Y. Horie, Trans. Mycol. Soc. Japan 14: 376, 1973. [Anamorph: *Penicillium meloforme* Udagawa & Y. Horie].

Eupenicillium meridianum D. B. Scott, Mycopathol. Mycol. Appl. 36: 12, 1968. [Anamorph: *Penicillium meridianum* D. B. Scott].

Eupenicillium molle Malloch & Cain, Can. J. Bot. 50: 62, 1972. [Anamorph: *Penicillium molle* Pitt].

Eupenicillium nepalense Takada & Udagawa, Trans. Mycol. Soc. Japan 24: 146, 1983. [Anamorph: *Penicillium nepalense* Takada & Udagawa].

Eupenicillium ochrosalmoneum D. B. Scott & Stolk, Antonie van Leeuwenhoek 33: 302, 1967. [Anamorph: *Penicillium ochrosalmoneum* Udagawa].

Eupenicillium ornatum Udagawa, Trans. Mycol. Soc. Japan 9: 49, 1968. [Anamorph: *Penicillium ornatum* Udagawa].

Eupenicillium osmophilum Stolk & Veenb.-Rijks, Antonie van Leeuwenhoek 40: 1, 1974. [Anamorph: *Penicillium osmophilum* Stolk & Veenb.-Rijks].

Eupenicillium papuanum Udagawa & Y. Horie, Trans. Mycol. Soc. Japan 14: 378, 1973 = *Eupenicillium parvum* (Raper & Fennell) Stolk & D.B. Scott, Persoonia 4: 402, 1967 (Pitt, 1980).

Eupenicillium parvum (Raper & Fennell) Stolk & D. B. Scott, Persoonia 4: 402, 1967. — Basionym: *Penicillium parvum* Raper & Fennell, Mycologia 40: 508, 1948 (nom. holomorph.). [Anamorph: *Penicillium papuanum* Udagawa & Y. Horie].

Eupenicillium pinetorum Stolk, Antonie van Leeuwenhoek 34: 37, 1968. [Anamorph: *Penicillium pinetorum* M. Chr. & Backus].

Eupenicillium reticulisporum Udagawa, Trans. Mycol. Soc. Japan 9: 52, 1968. [Anamorph: *Penicillium reticulisporum* Udagawa].

Eupenicillium rubidurum Udagawa & Y. Horie, Trans. Mycol. Soc. Japan 14: 381, 1973. [Anamorph: *Penicillium rubidurum* Udagawa & Y. Horie].

Eupenicillium senticosum D. B. Scott, Mycopathol. Mycol. Appl. 36: 5, 1968. [Anamorph: *Penicillium senticosum* D. B. Scott].

Eupenicillium shearii Stolk & D. B. Scott, Persoonia 4: 396, 1967. [Anamorph: *Penicillium shearii* Stolk & D. B. Scott].

Eupenicillium sinaicum Udagawa & S. Ueda, Mycotaxon 14: 266, 1982. [Anamorph: *Penicillium sinaicum* Udagawa & S. Ueda].

Eupenicillium stolkiae D. B. Scott, Mycopathol. Mycol. Appl. 36: 8, 1968. [Anamorph: *Penicillium stolkiae* D. B. Scott].

Eupenicillium terrenum D. B. Scott, Mycopathol. Mycol. Appl. 36: 1, 1968. [Anamorph: *Penicillium terrenum* D. B. Scott].

Eupenicillium tularense Paden, Mycopathol. Mycol. Appl. 43: 262, 1971. [Anamorph: *Penicillium tularense* Paden].

Eupenicillium vermiculatum (P. A. Dang.) C. Ram & A. Ram, Broteria, Ser. 3, 41: 106, 1972 = *Talaromyces flavus* (Klöcker) Stolk & Samson, Stud. Mycol. 2: 10, 1972 (Stolk and Samson, 1972).

Eupenicillium zonatum Hodges & J. J. Perry, Mycologia 65: 697, 1973. [Anamorph: *Penicillium zonatum* Hodges & J. J. Perry].

EUROTIUM LINK: FR. (HOLOMORPHS)

Eurotium amstelodami L. Mangin, Ann. Sci. Nat., Bot., ser. 9, 10: 360, 1909. [Anamorph: *Aspergillus vitis* Novobr.]. It was pointed out by Pitt (1985) that the concept of *E. amstelodami sensu* Raper and Thom (1941) and Raper and Fennell (1965) was in error; the neotypification of Gams *et al.* (1985) was also in error therefore. Under the Botanical Code, a new name was needed (Pitt, 1985). However it has been possible to maintain the name commonly used for this species within the NCU framework by neotypification (Pitt and Samson, 1993).

Eurotium appendiculatum Blaser, Sydowia 28: 38, 1976. [Anamorph: *Aspergillus appendiculatus* Blaser].

Eurotium athecium (Raper & Fennell) Arx, Gen. Fungi Sporul. Pure Cult., ed. 2: 91, 1974. — Basionym: *Aspergillus athecius* Raper & Fennell, Gen. *Aspergillus*: 183, 1965 (nom. holomorph.). [Anamorph: *Aspergillus atheciellus* Samson & W. Gams].

Eurotium carnoyi Malloch & Cain, Can. J. Bot. 50: 63, 1972. [Anamorph: *Aspergillus neocarnoyi* Kozak.].

Eurotium chevalieri L. Mangin, Ann. Sci. Nat., Bot., ser. 9, 10: 361, 1909. [Anamorph: *Aspergillus chevalieri* L. Mangin].

Eurotium cristatum (Raper & Fennell) Malloch & Cain, Can. J. Bot. 50: 64, 1972. — Basionym: *Aspergillus cristatus* Raper & Fennell, Gen. *Aspergillus*: 169, 1965 (nom. holomorph.). [Anamorph: *Aspergillus cristatellus* Kozak.].

Eurotium echinulatum Delacr., Bull. Soc. Mycol. France 9: 266, 1893. [Anamorph: *Aspergillus brunneus* Delacr.].

Eurotium glabrum Blaser, Sydowia 28: 35, 1976. [Anamorph: *Aspergillus glaber* Blaser].

Eurotium halophilicum C. M. Chr. *et al.*, Mycologia 51: 636, 1961. [Anamorph: *Aspergillus halophilicus* C. M. Chr. *et al.*].

Eurotium herbariorum Link, Ges. Naturf. Freunde Berlin Mag. Neuesten Entdeck. Gesammten Naturk. 3: 31, 1809: Fr. [Anamorph: *Aspergillus glaucus* Link].

Eurotium herbariorum var. *minor* L. Mangin, Annls Sci. Nat. Bot., Ser. 9, 10: 365, 1909 = *Eurotium herbariorum* Link, Ges. Naturf. Freunde Berlin Mag. Neuesten Entdeck. Gesammten Naturk. 3: 31, 1809 (Malloch and Cain, 1972).

Eurotium heterocaryoticum C. M. Chr. *et al.*, Mycologia 57: 536, 1965 = *Eurotium amstelodami* L. Mangin, Ann. Sci. Nat., Bot., ser. 9, 10: 360, 1909 (Caten, 1979).

Eurotium intermedium Blaser, Sydowia 28: 41, 1976. — Basionym: *Aspergillus chevalieri* var. *intermedius* Thom & Raper, Misc. Publ. U.S. Dept. Agric. 426: 21, 1941 (nom. holomorph.). [Anamorph: *Aspergillus intermedius* Blaser].

Eurotium leucocarpum Hadlok & Stolk, Antonie van Leeuwenhoek 35: 9, 1969. [Anamorph: *Aspergillus leucocarpus* Hadlok & Stolk].

Eurotium medium R. Meissn., Bot. Zeitung, 2. Abt., 55: 356, 1897. [Anamorph: *Aspergillus medius* R. Meissn.].

Eurotium montevidense (Talice & Mackinnon) Malloch & Cain, Can. J. Bot. 50: 64, 1972 ≡ *Aspergillus montevidense* Talice & Mackinnon, Compt. Rend. Soc. Biol. Fr. 108: 1007, 1931 (includes teleomorph) = *Eurotium amstelodami* L. Mangin, Ann. Sci. Nat., Bot., ser. 9, 10: 360, 1909 (Pitt, 1985).

Eurotium niveoglaucum (Thom & Raper) Malloch & Cain, Can. J. Bot. 50: 64, 1972. — Basionym: *Aspergillus niveoglaucus* Thom & Raper, Misc. Publ. U.S. Dept. Agric. 426: 35, 1941 (nom. holomorph.). [Anamorph: *Aspergillus glauconiveus* Samson & W. Gams].

Eurotium oryzae Ahlb., Dingler's Polytechn. J. 230: 330, 1878 ≡ *Aspergillus oryzae* (Ahlb.) Cohn, Jahresber. Schles. Ges. Vaterl. Cult. 61: 226, 1884.

Eurotium pseudoglaucum (Blochwitz) Malloch & Cain, Can. J. Bot. 50: 64, 1972. — Basionym: *Aspergillus pseudoglaucus* Blochwitz, Ann. Mycol. 27: 207, 1929 (nom. holomorph.). [Anamorph: *Aspergillus glaucoaffinis* Samson & W. Gams].

Eurotium repens de Bary, Abh. Senckenberg. Naturf. Ges. 7: 379, 1870. [Anamorph: *Aspergillus reptans* Samson & W. Gams].

Eurotium rubrum Jos. König *et al.*, Z. Untersuch. Nahrungs- Gen.smittel 4: 726, 1901. [Anamorph: *Aspergillus rubrobrunneus* Samson & W. Gams].

Eurotium spiculosum Blaser, Sydowia 28: 42, 1975 = *Eurotium cristatum* (Raper & Fennell) Malloch & Cain, Can. J. Bot. 50: 64, 1972 (Pitt, 1985).

Eurotium tonophilum Ohtsuki, Bot. Mag. (Tokyo) 75: 438, 1962. [Anamorph: *Aspergillus tonophilus* Ohtsuki].

Eurotium umbrosum (Bainier & Sartory) Malloch & Cain, Can. J. Bot. 50: 64, 1972 = *Eurotium herbariorum* Link, Ges. Naturf. Freunde Berlin Mag. Neuesten Entdeck. Gesammten Naturk. 3: 31, 1809 (Blaser, 1975).

Eurotium xerophilum Samson & Mouch., Antonie van Leeuwenhoek 41: 348, 1975. [Anamorph: *Aspergillus xerophilus* Samson & Mouch.].

FENNELLIA B. J. WILEY & E. G. SIMMONS (HOLOMORPHS)

Fennellia flavipes B. J. Wiley & E. G. Simmons, Mycologia 65: 937, 1973. [Anamorph: *Aspergillus flavipes* (Bainier & Sartory) Thom & Church].

Fennellia monodii Locq.-Lin., Mycotaxon 39: 10, 1990.

Fennellia nivea (B. J. Wiley & E. G. Simmons) Samson, Stud. Mycol. 18: 5, 1979. -Basionym: *Emericella nivea* B. J. Wiley & E. G. Simmons, Mycologia 65: 934, 1973. [Anamorph: *Aspergillus niveus* Blochwitz].

GEOSMITHIA PITT (ANAMORPHS)

Geosmithia argillacea (Stolk *et al.*) Pitt, Can. J. Bot. 57: 2026, 1979. — Basionym: *Penicillium argillaceum* Stolk *et al.*, Trans. Br. Mycol. Soc. 53: 307, 1969.

Geosmithia cylindrospora (G. Sm.) Pitt, Can. J. Bot. 57: 2024, 1979. — Basionym: *Penicillium cylindrosporum* G. Sm., Trans. Br. Mycol. Soc. 40: 483, 1957.

Geosmithia emersonii (Stolk) Pitt, Can. J. Bot. 57: 2027, 1979. — Basionym: *Penicillium emersonii* Stolk, Antonie van Leeuwenhoek 31: 262, 1965. [Teleomorph *Talaromyces emersonii* Stolk].

Geosmithia lavendula (Raper & Fennell) Pitt, Can. J. Bot. 57: 2022, 1979. — Basionym: *Penicillium lavendulum* Raper & Fennell, Mycologia 40: 530, 1948.

Geosmithia namyslowskii (K. M. Zalessky) Pitt, Can. J. Bot. 57: 2024, 1979. — Basionym: *Penicillium namyslowskii* K. M. Zalessky, Bull. Int. Acad. Polon. Sci., Cl. Sci. Math., Ser. B, Sci. Nat., 1927: 479, 1927.

Geosmithia putterillii (Thom) Pitt, Can. J. Bot. 57: 2022, 1979. — Basionym: *Penicillium putterillii* Thom, Penicillia: 368, 1930.

Geosmithia swiftii Pitt, Can. J. Bot. 57: 2028, 1979. [Teleomorph *Talaromyces bacillisporus* (Swift) C. R. Benj.].

Geosmithia viridis Pitt & A. D. Hocking, Mycologia 77: 822, 1985.

GYMNOASCUS (HOLOMORPHS)

Gymnoascus flavus Klöcker, Hedwigia 41: 80, 1902 ≡ *Talaromyces flavus* (Klöcker) Stolk & Samson, Stud. Mycol. 2: 10, 1972.

Gymnoascus luteus (Zukal) Sacc., Syll. Fung. 11: 437, 1895 ≡ *Talaromyces luteus* (Zukal) C. R. Benj., Mycologia 47: 681, 1955.

HAMIGERA (TELEOMORPHS)

Hamigera avellanea (Thom & Turesson) Stolk & Samson, Persoonia 6: 345, 1971 ≡ *Talaromyces avellaneus* (Thom & Turesson) C. R. Benj., Mycologia 47: 682, 1955. [Anamorph: *Merimbla ingelheimense* (J. F. H. Beyma) Pitt].

Hamigera spinulosa (Warcup) Arx, Mycotaxon 26: 121, 1986 ≡ *Warcupiella spinulosa* (Warcup) Subram., Curr. Sci. 41: 757, 1972.

Hamigera striata (Raper & Fennell) Stolk & Samson, Persoonia 6: 347, 1971 ≡ *Talaromyces striatus* (Raper & Fennell) C. R. Benj., Mycologia 47: 682, 1955.

HEMICARPENTELES A. K. SARBHOY & ELPHICK (HOLOMORPHS)

Hemicarpenteles acanthosporus Udagawa & Takada, Bull. Natl. Sci. Mus. Tokyo, ser. 2, 14: 503, 1971. [Anamorph: *Aspergillus acanthosporus* Udagawa & Takada].

Hemicarpenteles paradoxus A. K. Sarbhoy & Elphick, Trans. Br. Mycol. Soc. 51: 156, 1968. [Anamorph: *Aspergillus paradoxus* Fennell & Raper].

ISARIA FR. (ANAMORPHS)

Isaria amoenerosea Henn., Hedwigia 41: 66, 1902 ≡ *Paecilomyces amoenroseus* (Henn.) Samson, Stud. Mycol. 6: 37, 1974.

Isaria cicadae Miq., Bull. Sci. Phys. Nat. Néerl. 1838: 86, 1838 ≡ *Paecilomyces cicadae* (Miq.) Samson, Stud. Mycol. 6: 52, 1974.

Isaria fumosorosea Wize, Bull. Int. Acad. Sci. Cracovie, Cl. Sci. Math., 1904: 72, 1904 ≡ *Paecilomyces fumosoroseus* (Wize) A. H. S. Br. & G. Sm., Trans. Br. Mycol. Soc. 40: 67, 1957.

Isaria xylariiformis Lloyd, Mycol. Not. 7: 1200, 1923 ≡ *Paecilomyces xylariiformis* (Lloyd) Samson, Stud. Mycol. 6: 54, 1974.

MERIMBLA PITT (ANAMORPHS)

Merimbla ingelheimensis (J. F. H. Beyma) Pitt, Can. J. Bot. 57: 2395, 1979. — Basionym: *Penicillium ingelheimense* J. F. H. Beyma, Antonie van Leeuwenhoek 8: 109, 1942. [Teleomorph *Hamigera avellanea* Stolk & Samson].

MONILIA BONORD. (ANAMORPHS)

Monilia digitata Pers., Syn. Meth. Fung.: 693, 1801 ≡ *Monilia digitata* Pers.: Fr., Syst. Mycol. 3(2): 411, 1832 ≡ *Penicillium digitatum* (Pers.: Fr.) Sacc., Fung. Ital.: tab. 894, 1881.

MUCOR FRESEN. (MUCORACEAE) (HOLOMORPHS)

Mucor herbariorum Wiggers, Primitiae Florae Holsatiae: 111, 1780 ≡ *Eurotium herbariorum* Link, Ges. Naturf. Freunde Berlin Mag. Neuesten Entdeck. Gesammten Naturk. 3: 31, 1809.

NEOSARTORYA MALLOCH & CAIN (HOLOMORPHS)

Neosartorya aurata (Warcup) Malloch & Cain, Can. J. Bot. 50: 2620, 1973. — Basionym: *Aspergillus auratus* Warcup, Raper & Fennell in Gen. *Aspergillus*: 263, 1965 (nom. holomorph.). [Anamorph: *Aspergillus igneus* Kozak.].

Neosartorya aureola (Fennell & Raper) Malloch & Cain, Can. J. Bot. 50: 2620, 1973. — Basionym: *Aspergillus aureolus* Fennell & Raper, Mycologia 47: 71, 1955 (nom. holomorph.). [Anamorph: *Aspergillus aureoluteus* Samson & W. Gams].

Neosartorya fennelliae Kwon-Chung & S. J. Kim, Mycologia 66: 629, 1974. [Anamorph: *Aspergillus fennelliae* Kwon-Chung & S. J. Kim].

Neosartorya fischeri (Wehmer) Malloch & Cain, Can. J. Bot. 50: 2621, 1973. — Basionym: *Aspergillus fischeri* Wehmer, Centralbl. Bakteriol., 2. Abth., 18: 390, 1907 (nom. holomorph.). [Anamorph: *Aspergillus fischerianus* Samson & W. Gams].

Neosartorya glabra (Fennell & Raper) Kozak., Mycol. Pap. 161: 56, 1989. — Basionym: *Aspergillus fischeri* var. *glaber* Fennell & Raper, Mycologia 47: 74, 1955 (nom. holomorph.). [Anamorph: *Aspergillus neoglaber* (Fennell & Raper) Kozak.]

Neosartorya hiratsukae Udagawa *et al.*, Trans. Mycol. Soc. Japan 32: 23, 1991. [Anamorph: *Aspergillus hiratsukoe* Udagawa *et al.*].

Neosartorya pseudofischeri S. W. Peterson, Mycol. Res. 96: 549, 1992. [Anamorph: *Aspergillus thermomutatus* (Paden) S. W. Peterson].

Neosartorya quadricincta (E. Yuill) Malloch & Cain, Can. J. Bot. 50: 2621, 1973. — Basionym: *Aspergillus quadricinctus* E. Yuill, Trans. Br. Mycol. Soc. 36: 58, 1953 (nom. holomorph.). [Anamorph: *Aspergillus quadricingens* Kozak.].

Neosartorya spathulata Takada & Udagawa, Mycotaxon 24: 396, 1985. [Anamorph: *Aspergillus spathulatus* Takada & Udagawa].

Neosartorya spinosa (Raper & Fennell) Kozak., Mycol. Pap. 161: 58, 1989. — Basionym: *Aspergillus fischeri* var. *spinosus* Raper & Fennell, Gen. *Aspergillus*: 256, 1965 (nom. holomorph.). [Anamorph: *Aspergillus spinosus* Kozak.].

Neosartorya stramenia (R. Novak & Raper) Malloch & Cain, Can. J. Bot. 50: 2622, 1973. — Basionym: *Aspergillus stramenius* R. Novak & Raper in Raper & Fennell, Gen. *Aspergillus*: 260, 1965 (nom. holomorph.). [Anamorph: *Aspergillus paleaceus* Samson & W. Gams].

Neosartorya tatenoi Y. Horie *et al.*, Trans. Mycol. Soc. Japan 33: 395, 1992. [Anamorph: *Aspergillus tatenoi* Horie *et al.*].

OOSPORA WALLR. (ANAMORPHS)

Oospora fasciculata (Grev.) Sacc. & Voglino in Sacchardo, Syll. Fung. 4: 11, 1886 (*non* Sommerf. 1826) = **Penicillium italicum** Wehmer, Hedwigia 33: 211, 1894 (Pitt, 1980).

PAECILOMYCES BAINIER (ANAMORPHS)

Paecilomyces aegyptiacus S. Ueda & Udagawa, Trans. Mycol. Soc. Japan 24: 135, 1983. [Teleomorph *Thermoascus aegyptiacus* S. Ueda & Udagawa].

Paecilomyces aerugineus Samson, Stud. Mycol. 6: 20, 1974.

Paecilomyces amoeneroseus (Henn.) Samson, Stud. Mycol. 6: 37, 1974. — Basionym: *Isaria amoenerosea* Henn., Hedwigia 41: 66, 1902.

Paecilomyces breviramosus Bissett in S. Hughes, Fungi Can.: no. 159, 1979.

Paecilomyces byssochlamydoides Stolk & Samson, Stud. Mycol. 2: 45, 1972. [Teleomorph *Talaromyces byssochlamydoides* Stolk & Samson].

Paecilomyces carneus (Duché & R. Heim) A. H. S. Br. & G. Sm., Trans. Br. Mycol. Soc. 40: 70, 1957. — Basionym: *Spicaria carnea* Duché & R. Heim, Trav. Cryptog. Louis L. Mangin: 454, 1931.

Paecilomyces cateniannulatus Z. Q. Liang, Acta Phytopathol. Sin. 11: 10, 1981.

Paecilomyces cateniobliquus Z. Q. Liang, Acta Phytopathol. Sin. 11: 9, 1981.

Paecilomyces cicadae (Miq.) Samson, Stud. Mycol. 6: 52, 1974. — Basionym: *Isaria cicadae* Miq., Bull. Sci. Phys. Nat. Néerl. 1838: 86, 1838.

Paecilomyces cinnamomeus (Petch) Samson & W. Gams, Stud. Mycol. 6: 62, 1974. — Basionym: *Verticillium cinnamomeum* Petch, Trans. Br. Mycol. Soc. 16: 233, 1932.

Paecilomyces clavisporus Hammill, Mycologia 62: 109, 1970.

Paecilomyces coleopterorum Samson & H. C. Evans, Stud. Mycol. 6: 47, 1974.

Paecilomyces crustaceus Apinis & Chesters, Trans. Br. Mycol. Soc. 47: 429, 1964. [Teleomorph *Thermoascus crustaceus* (Apinis & Chesters) Stolk]

Paecilomyces farinosus (Holmsk.) A. H. S. Br. & G. Sm., Trans. Br. Mycol. Soc. 40: 50, 1957. — Basionym: *Ramaria farinosa* Holmsk., Nye Saml. Kongel. Danske Vidensk. Selsk. Skr. 1: 279, 1781.

Paecilomyces fulvus Stolk & Samson, Persoonia 6: 354, 1971. [Teleomorph *Byssochlamys fulva* Olliver & G. Sm., J. Bot. 70: 196, 1933].

Paecilomyces fumosoroseus (Wize) A. H. S. Br. & G. Sm., Trans. Br. Mycol. Soc. 40: 67, 1957. - Basionym: *Isaria fumosorosea* Wize, Bull. Int. Acad. Sci. Cracovie, Cl. Sci. Math., 1904: 72, 1904.

Paecilomyces ghanensis Samson & H. C. Evans, Stud. Mycol. 6: 46, 1974.

Paecilomyces gunnii Z. Q. Liang, Acta Mycol. Sin. 4: 163, 1985.

24

Paecilomyces hawkesii Z. M. Xiao *et al.*, Acta Mycol. Sin. 3: 109, 1984.

Paecilomyces inflatus (Burnside) J. W. Carmich., Can. J. Bot. 40: 1148, 1962. — Basionym: *Myceliophthora inflata* Burnside, Pap. Michigan Acad. Sci. 8: 82-84, 1927.

Paecilomyces javanicus (Frieder. & W. Bally) A. H. S. Br. & G. Sm., Trans. Br. Mycol. Soc. 40: 65, 1957. — Basionym: *Spicaria javanica* Frieder. & W. Bally, Koffiebessenboeboek-Fonds Meded. 6: 146, 1923.

Paecilomyces leycettanus (Evans & Stolk) Stolk *et al.*, Persoonia 6: 342, 1971. — Basionym: *Penicillium leycettanum* H. C. Evans & Stolk, Trans. Br. Mycol. Soc. 56: 45, 1971. [Teleomorph *Talaromyces leycettanus* H. C. Evans & Stolk].

Paecilomyces lilacinus (Thom) Samson, Stud. Mycol. 6: 58, 1974. — Basionym: *Penicillium lilacinum* Thom, Bull. Bur. Anim. Ind. U.S. Dep. Agric. 118: 73, 1910.

Paecilomyces lineatus (Pitt) Arx, Mycotaxon 26: 120, 1986 ≡ *Penicillium lineatum* Pitt, Genus *Penicillium*: 485, 1980 ["1979"].

Paecilomyces marquandii (Massee) S. Hughes, Mycol. Pap. 45: 30, 1951. — Basionym: *Verticillium marquandii* Massee, Trans. Br. Mycol. Soc. 1: 24, 1898.

Paecilomyces niphetodes Samson, Stud. Mycol. 6: 65, 1974.

Paecilomyces niveus Stolk & Samson, Persoonia 6: 351, 1971. [Teleomorph *Byssochlamys nivea* Westling, Svensk Bot. Tidskr. 3: 134, 1909].

Paecilomyces nostocoides M. T. Dunn, Mycologia 75: 179, 1983.

Paecilomyces pascuus Pitt & A. D. Hocking, Mycologia 77: 822, 1985.

Paecilomyces penicillatus (Höhn.) Samson, Stud. Mycol. 6: 72, 1974. — Basionym: *Spicaria penicillata* Höhn., Ann. Mycol. 2: 56, 1904.

Paecilomyces puntonii (Vuill.) Nann., Repert. Mic. Uomo: 245, 1934. — Basionym: *Corethropsis puntonii* Vuill., Compt. Rend. Hebd. Séances Acad. Sci. 190: 1334, 1930.

Paecilomyces ramosus Samson & H. C. Evans, Stud. Mycol. 6: 44, 1974.

Paecilomyces reniformis Samson & H. C. Evans, Stud. Mycol. 6: 43, 1974.

Paecilomyces sinensis Q. T. Chen *et al.*, Acta Mycol. Sin. 3: 25, 1984.

Paecilomyces suffultus (Petch) Samson, Stud. Mycol. 6: 55, 1974. — Basionym: *ylindrodendrum suffultum* Petch, Trans. Brit. Mycol. Soc. 27: 91, 1944.

Paecilomyces sulphurellus (Sacc.) Samson & W. Gams, Stud. Mycol. 6: 67, 1974. — Basionym: *Verticillium sulphurellum* Sacc., Fung. Ital.: tab. 641, 1881.

Paecilomyces tenuipes (Peck) Samson, Stud. Mycol. 6: 49, 1974. — Basionym: *Isaria tenuipes* Peck, Ann. Rep. New York State Mus. 33: 49, 1883.

Paecilomyces variotii Bainier, Bull. Soc. Mycol. France 23: 26, 1907.

Paecilomyces verrucosus Samson & Tansey, Trans. Brit. Mycol. Soc. 65: 512, 1975. [Teleomorph *Byssochlamys verrucosa* Samson & Tansey, Trans. Br. Mycol. Soc. 65: 512, 1975].

Paecilomyces viridis Segretain ex Samson, Stud. Mycol. 6: 64, 1974.

Paecilomyces xylariiformis (Lloyd) Samson, Stud. Mycol. 6: 54, 1974. — Basionym: *Isaria xylariiformis* Lloyd, Mycol. Not. 7: 1200, 1923.

Paecilomyces zollerniae Stolk & Samson, Persoonia 6: 356, 1971. [Teleomorph *Byssochlamys zollerniae* C. Ram, Nova Hedwigia 16: 312, 1968.]

PENICILLIOPSIS SOLMS (HOLOMORPHS)

Penicilliopsis africana Samson & Seifert in Samson & Pitt (eds), Adv. *Penicillium Aspergillus* Syst.: 408, 1985. [Anamorph: *Stilbodendron cervinum* (Cooke & Massee) Samson & Seifert].

Penicilliopsis clavariiformis Solms, Ann. Jard. Bot. Buitenzorg 6: 53, 1887. [Anamorph: *Sarophorum palmicola* (Henn.) Seifert & Samson].

Penicilliopsis dybowskii Pat., Bull. Soc. Mycol. Fr. 7: 54, 1891 ≡ *Aspergillus dybowskii* (Pat.) Samson & Seifert in Samson & Pitt (eds), Adv. *Penicillium Aspergillus* Syst.: 422, 1985.

PENICILLIUM LINK: FR. (ANAMORPHS)

Penicillium abeanum G. Sm., Trans. Br. Mycol. Soc. 46: 333, 1963 = *Penicillium spinulosum* Thom, Bull. Bur. Anim. Ind. U.S. Dep. Agric. 118: 76, 1910 (Pitt, 1980).

Penicillium abidjanum Stolk, Antonie van Leeuwenhoek 34: 49, 1968. [Teleomorph *Eupenicillium abidjanum* Stolk].

Penicillium aculeatum Raper & Fennell, Mycologia 40: 535, 1948.

Penicillium aculeatum var. *apiculatum* S. Abe, J. Gen. Appl. Microbiol., Tokyo 2: 1956 (nom. inval., Art. 36) = *Penicillium verruculosum* Peyronel, Germi Atmosf. Fung. Micel.: 22, 1913 (Pitt, 1980).

Penicillium adametzii K. M. Zalessky, Bull. Int. Acad. Polon. Sci., Cl. Sci. Math., Sér. B, Sci. Nat., 1927: 507, 1927.

Penicillium adametzioides S. Abe ex G. Sm., Trans. Br. Mycol. Soc. 46: 335, 1963 ≡ *Penicillium adametzioides* S. Abe, J. Gen. Appl. Microbiol., Tokyo 2: 68, 1956 (nom. inval., Art. 36).

Penicillium aeneum G. Sm., Trans. Br. Mycol. Soc. 46: 334, 1963 ≡ *Penicillium citreoviride* var. *aeneum* S. Abe, J. Gen. Appl. Microbiol., Tokyo 2: 58, 1956 (nom. inval., Art. 36) = *Penicillium ci-*

treonigrum Dierckx, Ann. Soc. Sci. Bruxelles 25: 86, 1901 (Pitt, 1980).

Penicillium aeruginosum Demelius, Verh. Zool.-Bot. Ges. Wein 72: 76, 1923 (1922)(*non* Dierckx 1901) = **Penicillium expansum** Link, Ges. Naturf. Freunde Berlin Mag. Neuesten Entdeck. Gesammten Naturk. 3: 16, 1809 (Pitt, 1980).

Penicillium aeruginosum Dierckx, Annls Soc. Scient. Brux. 25: 87, 1901 = **Penicillium italicum** Wehmer, Hedwigia 33: 211, 1894 (Biourge, 1923).

Penicillium aethiopicum Frisvad, Mycologia 81: 848, 1990.

Penicillium albocinerascens Chalab., Bot. Mater. Otd. Sporov. Rast. 6: 166, 1950 [*non* (Maublanc) Biourge 1923] = **Penicillium adametzii** K. M. Zalessky, Bull. Int. Acad. Polon. Sci., Cl. Sci. Math., Sér. B, Sci. Nat., 1927: 507, 1927 (Pitt, 1980).

Penicillium album Epstein, Ark. Hyg. Bakt. 45: 360, 1902 (*non* Preuss, 1851) = **Penicillium camemberti** Thom, Bull. Bur. Anim. Ind. U.S. Dep. Agric. 82: 33, 1906 (Pitt, 1980).

Penicillium alicantinum C. Ramírez & A. T. Martínez, Mycopathologia 72: 185, 1980 = **Penicillium citreonigrum** Dierckx, Ann. Soc. Sci. Bruxelles 25: 86, 1901 (Frisvad *et al.*, 1990b).

Penicillium allahabadense B. S. Mehrotra & D. Kumar, Can. J. Bot. 40: 1399, 1962.

Penicillium allii Vincent & Pitt, Mycologia 81: 300, 1989.

Penicillium alutaceum D. B. Scott, Mycopathol. Mycol. Appl. 36: 17, 1968. [Teleomorph *Eupenicillium alutaceum* D. B. Scott].

Penicillium anatolicum Stolk, Antonie van Leenwenhoek 34: 46, 1968. [Teleomorph *Eupenicillium anatolicum* Stolk].

Penicillium angustiporcatum Takada & Udagawa, Trans. Mycol. Soc. Japan 24: 143, 1983. [Teleomorph *Eupenicillium angustiporcatum* Takada & Udagawa].

Penicillium arabicum Baghdadi, Nov. Sist. Niz. Rast. 5: 105, 1968 = **Penicillium decumbens** Thom, Bull. Bur. Anim. Ind. U.S. Dep. Agric. 118: 71, 1910 (Pitt, 1980).

Penicillium aragonense C. Ramírez & A. T. Martínez, Mycopathologia 74: 41, 1981 = **Penicillium oxalicum** Currie & Thom, J. Biol. Chem. 22: 289, 1915 (Stolk *et al.*, 1990).

Penicillium ardesiacum Novobr., Novosti Sist. Nizs. Rast. 11: 228, 1974.

Penicillium arenicola Chalab., Bot. Mater. Otd. Sporov. Rast. 6: 162, 1950.

Penicillium argillaceum Stolk *et al.*, Trans. Br. Mycol. Soc. 53: 307, 1969 = **Geosmithia argillacea** (Stolk *et al.*) Pitt, Can. J. Bot. 57: 2026, 1979.

Penicillium aromaticum casei Sopp, Zentbl. Bakt. ParasitKde, Abt II, 4: 164, 1898 (nom. inval., Art.

32) = **Penicillium roqueforti** Thom, Bull. Bur. Anim. Ind. U.S. Dep. Agric. 82: 35, 1906 (Pitt, 1980).

Penicillium aromaticum f. *microsporum* Romankova, Uchen. Zap. Leningr. Gos. Univ. (ser. Biol. Nauk. 40:) 191: 102, 1955 (nom. inval., Art. 36) = **Penicillium chrysogenum** Thom, Bull. Bur. Anim. Ind. U.S. Dep. Agric. 118: 58, 1910 (Pitt, 1980).

Penicillium aromaticum Sopp, Skr. Vidensk.-Selsk. Christiana, Math.-Naturvidensk. Kl. 11: 159, 1912 = **Penicillium roqueforti** Thom, Bull. Bur. Anim. Ind. U.S. Dep. Agric. 82: 35, 1906 (Thom, 1930).

Penicillium aromaticum-casei Sopp ex Sacc., Syll. Fung. 22: 1278, 1913 = **Penicillium roqueforti** Thom, Bull. Bur. Anim. Ind. U.S. Dep. Agric. 82: 35, 1906 (Pitt, 1980).

Penicillium arvense Udagawa & T. Yokoy. in Udagawa and Horie, Trans Mycol. Soc. Japan 15: 107, 1974 = **Penicillium reticulisporum** Udagawa, Trans. Mycol. Soc. Japan 9: 52, 1968 (Pitt, 1980).

Penicillium asperosporum G. Sm., Trans. Br. Mycol. Soc. 48: 275, 1965 ≡ *Penicillium echinosporum* G. Sm., Trans. Br. Mycol. Soc. 45: 387, 1962 [*non* Nehira, J. Ferment. Technol. 11: 849, 1933].

Penicillium asperum (Shear) Raper & Thom, Man. Penicillia: 263, 1949 (nom. holomorph.) = *Carpenteles asperum* Shear, Mycologia 26: 107, 1934 = **Eupenicillium crustaceum** F. Ludw., Lehrb. Nied. Krypt.: 263, 1892 (Stolk and Scott, 1967).

Penicillium assiutense Samson & Abdel-Fattah, Persoonia 9: 501, 1978. [Teleomorph *Talaromyces assiutensis* Samson & Abdel-Fattah].

Penicillium asturianum C. Ramírez & A. T. Martínez, Mycopathologia 74: 42, 1981 = **Penicillium oxalicum** Currie & Thom, J. Biol. Chem. 22: 289, 1915 (Stolk *et al.*, 1990; Frisvad *et al.*, 1990b).

Penicillium atramentosum Thom, Bull. Bur. Anim. Ind. U.S. Dep. Agric. 118: 65, 1910.

Penicillium atrosanguineum B.X. Dong, Ceská Mycol. 27: 174, 1973 = **Penicillium miczynskii** K. M. Zalessky, Bull. Int. Acad. Polon. Sci., Cl. Sci. Math., Sér. B, Sci. Nat., 1927: 482, 1927 (Pitt, 1980).

Penicillium atrovenetum G. Sm., Trans. Br. Mycol. Soc. 39: 112, 1956.

Penicillium atrovirens G. Sm., Trans. Br. Mycol. Soc. 46: 334, 1963 = **Penicillium fellutanum** Biourge, Cellule 33: 262, 1923 (Pitt, 1980).

Penicillium atroviride Sopp, Skr. Vidensk.-Selsk. Christiana, Math.-Naturvidensk. Kl. 11: 149, 1912 = **Penicillium roqueforti** Thom, Bull. Bur. Anim. Ind. U.S. Dep. Agric. 82: 35, 1906 (Pitt, 1980).

Penicillium aurantiacum J.H. Mill. *et al.*, Mycologia 49: 797, 1957 = **Penicillium funiculosum** Thom,

Bull. Bur. Anim. Ind. U.S. Dep. Agric. 118: 69, 1910 (Pitt, 1980).

Penicillium aurantiogriseum Dierckx, Ann. Soc. Sci. Bruxelles 25: 88, 1901.

Penicillium aurantio-albidum Biourge, Cellule 33: 197, 1923 = *Penicillium aurantiogriseum* Dierckx, Ann. Soc. Sci. Bruxelles 25: 88, 1901 (Pitt, 1980).

Penicillium aurantiobrunnium Dierckx, Annls Soc. Scient. Brux. 25: 86, 1901 = *Penicillium glabrum* (Wehmer) Westling, Ark. Bot. 11(1): 131, 1911 (Pitt, 1980).

Penicillium aurantiocandidum Dierckx, Ann. Soc. Sci. Bruxelles 25: 88, 1901 = *Penicillium aurantiogriseum* Dierckx, Ann. Soc. Sci. Bruxelles 25: 88, 1901 (Pitt, 1980).

Penicillium aurantioflammiferum C. Ramírez et al., Mycopathologia 72: 28, 1980 = *Penicillium islandicum* Sopp, Skr. Vidensk.-Selsk. Christiana, Math.-Naturvidensk. Kl. 11: 161, 1912 (Frisvad et al., 1990b).

Penicillium aurantiogriseum var. *poznaniense* K. M. Zalessky, Bull. Int. Acad. Polon. Sci., Cl. Sci. Math., Sér. B, Sci. Nat., 1927: 444, 1927 = *Penicillium crustosum* Thom, Penicillia: 399, 1930 (Pitt, 1980).

Penicillium aurantioviolaceum Biourge, Cellule 33: 282, 1923 = *Penicillium thomii* Maire, Bull. Soc. Hist. Nat. Afrique N. 8: 189, 1917 (Pitt, 1980).

Penicillium aurantiovirens Biourge, Cellule 33: 119, 1923 ≡ *Penicillium cyclopium* var. *aurantiovirens* (Biourge) Fassatiová, Acta Univ. Carol., Biol. 12: 326, 1977 = *Penicillium aurantiogriseum* Dierckx, Ann. Soc. Sci. Bruxelles 25: 88, 1901 (Pitt, 1980).

Penicillium aurifluum (Bainier & Sartory) Biourge, Cellule 33: 250, 1923 = *Penicillium citrinum* Thom, Bull. Bur. Anim. Ind. U.S. Dep. Agric. 118: 61, 1910 (Pitt, 1980).

Penicillium australicum Sopp ex J. F. H. Beyma, Antonie van Leeuwenhoek 10: 53, 1944 = *Penicillium commune* Thom, Bull. Bur. Anim. Ind. U.S. Dep. Agric. 118: 56, 1910 (Stolk et al., 1990).

Penicillium avellaneum Thom & Turesson, Mycologia 7: 284, 1915 (nom. holomorph.) = *Talaromyces avellaneus* (Thom & Turesson) C. R. Benj., Mycologia 47: 682, 1955.

Penicillium baarnense J. F. H. Beyma, Antonie van Leeuwenhoek 6: 271, 1940 (nom. holomorph.) ≡ *Eupenicillium baarnense* (J. F. H. Beyma) Stolk & D. B. Scott, Persoonia 4: 401, 1967.

Penicillium bacillisporum Swift, Bull. Torrey Bot. Club 59: 221, 1932 (nom. holomorph.) = *Talaromyces bacillisporus* (Swift) C. R. Benj., Mycologia 47: 684, 1955.

Penicillium baiiolum Biourge, Cellule 33: 305, 1923 = *Penicillium spinulosum* Thom, Bull. Bur.

Anim. Ind. U.S. Dep. Agric. 118: 76, 1910 (Pitt, 1980).

Penicillium bailowiezense K. M. Zalessky, Bull. Int. Acad. Polon. Sci., Cl. Sci. Math., Sér. B, Sci. Nat., 1927: 450, 1927 = *Penicillium brevicompactum* Dierckx, Ann. Soc. Sci. Bruxelles 25: 88, 1901 (Pitt, 1980).

Penicillium baradicum Baghdadi, Nov. Sist. Niz. Rast. 5: 107, 1968 = *Penicillium citrinum* Thom, Bull. Bur. Anim. Ind. U.S. Dep. Agric. 118: 61, 1910 (Pitt, 1980).

Penicillium bertai Talice & Mackinnon, Annls Parasit. Hum. Comp. 7: 97, 1929 = *Penicillium citreonigrum* Dierckx, Ann. Soc. Sci. Bruxelles 25: 86, 1901 (Pitt, 1980).

Penicillium biforme Thom, Bull. Bur. Anim. Ind. U.S. Dep. Agric. 118: 54, 1910 = *Penicillium camemberti* Thom, Bull. Bur. Anim. Ind. U.S. Dep. Agric. 82: 33, 1906 (Pitt, 1980).

Penicillium biforme var. *vitriolum* Tatuzo Sato, J. Agric. Chem. Soc., Japan 15: 77, 1939 = *Penicillium ochrochloron* Biourge, Cellule 33: 269, 1923 (Raper and Thom, 1949).

Penicillium bilaiae Chalab., Bot. Mater. Otd. Sporov. Rast. 6: 165, 1950.

Penicillium biougeianeum K. M. Zalessky, Bull. Int. Acad. Polon. Sci., Cl. Sci. Math., Sér. B, Sci. Nat., 1927: 462, 1927 = *Penicillium brevicompactum* Dierckx, Ann. Soc. Sci. Bruxelles 25: 88, 1901 (Pitt, 1980).

Penicillium biourgei Arnaudi, Boll. Ist. Sieroter. Milan. 6: 27, 1928 (1927)(non Dierckx, 1901) = *Penicillium roqueforti* Thom, Bull. Bur. Anim. Ind. U.S. Dep. Agric. 82: 35, 1906 (Raper and Thom, 1949).

Penicillium blakesleei K. M. Zalessky, Bull. Int. Acad. Polon. Sci., Cl. Sci. Math., Sér. B, Sci. Nat., 1927: 441, 1927 = *Penicillium viridicatum* Westling, Ark. Bot. 11(1): 88, 1911 (Raper and Thom, 1949).

Penicillium botryosum Bat. & H. Maia, Anais Soc. Biol. Pernamb. 15: 159, 1957 = *Penicillium citrinum* Thom, Bull. Bur. Anim. Ind. U.S. Dep. Agric. 118: 61, 1910 (Pitt, 1980).

Penicillium brasilianum Bat., Anais Soc. Biol. Pernambuco 15: 162, 1957.

Penicillium brefeldianum B. O. Dodge, Mycologia 25: 92, 1933 (nom. holomorph.) ≡ *Carpenteles brefeldianum* (B. O. Dodge) Shear, Mycologia 26:107, 1934 ≡ *Eupenicillium brefeldianum* (B. O. Dodge) Stolk & D. B. Scott, Persoonia 4: 400, 1967.

Penicillium brevicompactum Dierckx, Ann. Soc. Sci. Bruxelles 25: 88, 1901.

Penicillium brevicompactum var. *magnum* C. Ramírez, Manual Atlas Penicillia: 398, 1982 =

27

Penicillium olsonii Bainier & Sartory, Ann. Mycol. 10: 398, 1912 (Frisvad *et al.*, 1990b).

Penicillium brevissimum J. N. Rai & Wadhwani, Curr. Sci., India 45: 192, 1976 = *Penicillium capsulatum* Raper & Fennell, Mycologia 40: 528, 1948 (Frisvad *et al.*, 1990b).

Penicillium brunneorubrum Dierckx, Annls Soc. Scient. Brux. 25: 88, 1901 = *Penicillium chrysogenum* Thom *nom. cons.*, Bull. Bur. Anim. Ind. U.S. Dep. Agric. 118: 58, 1910 (Kozakiewicz *et al.*, 1992).

Penicillium brunneostoloniferum S. Abe ex C. Ramírez, Manual Atlas Penicillia: 412, 1982 ≡ *Penicillium brunneostoloniferum* S. Abe, J. Gen. Appl. Microbiol., Tokyo 2: 104, 1956 (nom. inval., Art. 36) = *Penicillium brevicompactum* Dierckx, Ann. Soc. Sci. Bruxelles 25: 88, 1901 (Frisvad *et al.*, 1990b).

Penicillium brunneoviolaceum Biourge, Cellule 33: 145, 1923 = *Penicillium aurantiogriseum* Dierckx, Ann. Soc. Sci. Bruxelles 25: 88, 1901 (Pitt, 1980).

Penicillium brunneoviride Szilvinyi, Zentbl. Bakt. ParasitKde, Abt. II, 103: 144, 1941 = *Penicillium spinulosum* Thom, Bull. Bur. Anim. Ind. U.S. Dep. Agric. 118: 76, 1910 (Pitt, 1980).

Penicillium brunneum Udagawa, J. Agric. Sci. Tokyo Nogyo Daigaku 5: 16, 1959.

Penicillium burgense Quintan., Avances Nutr. Mejora Anim. Aliment. 30: 176, 1990 = *Eupenicillium lapidosum* D. B. Scott & Stolk, Antonie van Leeuwenhoek 33: 298, 1967 (Frisvad *et al.*, 1990b).

Penicillium caerulescens Quintan., Mycopathologia 82: 101, 1983 = *Penicillium raciborskii* K. M. Zalessky, Bull. Int. Acad. Polon. Sci., Cl. Sci. Math., Sér. B, Sci. Nat., 1927: 454, 1927 (Frisvad *et al.*, 1990b).

Penicillium camemberti Sopp, Skr. Vidensk.-Selsk. Christiana, Math.-Naturvidensk. Kl. 11: 179, 1912 (*non* Thom, 1906) = *Penicillium camemberti* Thom, Bull. Bur. Anim. Ind. U.S. Dep. Agric. 82: 33, 1906 (Pitt, 1980).

Penicillium camemberti Thom, Bull. Bur. Anim. Ind. U.S. Dep. Agric. 82: 33, 1906.

Penicillium camemberti var. *rogeri* Thom, Bull. Bur. Anim. Ind. U.S. Dep. Agric. 118: 52, 1910 = *Penicillium camemberti* Thom, Bull. Bur. Anim. Ind. U.S. Dep. Agric. 82: 33, 1906 (Pitt, 1980).

Penicillium camerunense R. Heim in Heim *et al.*, Bull. Acad. R. Belg. Cl. Sci. 35: 42, 1949 (nom. inval., Art. 36) = *Penicillium chrysogenum* Thom *nom. cons.*, Bull. Bur. Anim. Ind. U.S. Dep. Agric. 118: 58, 1910 (Pitt, 1980).

Penicillium canadense G. Sm., Trans. Br. Mycol. Soc. 39: 113, 1956 = *Penicillium arenicola* Chalab., Bot. Mater. Otd. Sporov. Rast. 6: 162, 1950 (Pitt, 1980).

Penicillium candidofulvum Dierckx, Annls Soc. Scient. Brux. 25: 86, 1901 = *Penicillium glabrum* (Wehmer) Westling, Ark. Bot. 11(1): 131, 1911 (Pitt, 1980).

Penicillium candidum Roger in Biourge, Cellule 33: 193, 1923 = *Penicillium camemberti* Thom, Bull. Bur. Anim. Ind. U.S. Dep. Agric. 82: 33, 1906 (Pitt, 1980).

Penicillium canescens Sopp, Skr. Vidensk.-Selsk. Christiana, Math.-Naturvidensk. Kl. 11: 181, 1912.

Penicillium caperatum Udagawa & Y. Horie, Trans. Mycol. Soc. Japan 14: 371, 1973 = *Penicillium ornatum* Udagawa, Trans. Mycol. Soc. Japan 9: 49, 1968 (Pitt, 1980).

Penicillium capsulatum Raper & Fennell, Mycologia 40: 528, 1948.

Penicillium carminoviolaceum Dierckx, Ann. Soc. Sci. Bruxelles 25: 86, 1901 = *Penicillium roseopurpureum* Dierckx, Ann. Soc. Sci. Bruxelles 25: 86, 1901 (Pitt, 1980).

Penicillium carneolutescens G. Sm., Trans. Br. Mycol. Soc. 22: 252, 1939 = *Penicillium aurantiogriseum* Dierckx, Ann. Soc. Sci. Bruxelles 25: 88, 1901 (Stolk *et al.*, 1990).

Penicillium casei Staub, Zentbl. Bakt. ParasitKde, Abt II, 31: 454, 1911 = *Penicillium verrucosum* Dierckx, Ann. Soc. Sci. Bruxelles 25: 88, 1901 (Stolk *et al.*, 1990).

Penicillium casei var. *compactum* S. Abe, J. Gen. Appl. Microbiol., Tokyo 2: 101, 1956 (nom. inval., Art. 36) = *Penicillium solitum* Westling, Ark. Bot. 11(1): 65, 1911 (Stolk *et al.*, 1990).

Penicillium caseicola Bainier, Bull. Trimest. Soc. Mycol. Fr. 23: 94, 1907 = *Penicillium camemberti* Thom, Bull. Bur. Anim. Ind. U.S. Dep. Agric. 82: 33, 1906 (Pitt, 1980).

Penicillium castellae Quintan., Avances Nutr. Mejora Anim. Aliment. 23: 336, 1982 = *Penicillium raistrickii* G. Sm., Trans. Br. Mycol. Soc. 18: 90, 1933 (Frisvad *et al.*, 1990b).

Penicillium castellonense C. Ramírez & A. T. Martínez, Mycopathologia 74: 46, 1981 = *Penicillium madriti* G. Sm., Trans. Br. Mycol. Soc. 44: 44, 1961 (Frisvad *et al.*, 1990b).

Penicillium catenatum D. B. Scott, Mycopathol. Mycol. Appl. 36: 24, 1968. [Teleomorph *Eupenicillium catenatum* D. B. Scott].

Penicillium cesiae (Bainier & Sartory) Biourge, Cellule 33: 101, 1923 = *Penicillium roseopurpureum* Dierckx, Ann. Soc. Sci. Bruxelles 25: 86, 1901 (Pitt, 1980).

Penicillium chalybeum Pitt & A. D. Hocking, Mycotaxon 22: 204, 1985.

Penicillium charlesii G. Sm., Trans. Br. Mycol. Soc. 18: 90, 1933 = ***Penicillium fellutanum*** Biourge, Cellule 33: 262, 1923 (Pitt, 1980).

Penicillium charlesii var. *rapidum* S. Abe, J. Gen. Appl. Microbiol., Tokyo 2: 73, 1956 (nom. inval., Art. 36) = ***Penicillium waksmanii*** K. M. Zalessky, Bull. Int. Acad. Polon. Sci., Cl. Sci. Math., Sér. B, Sci. Nat., 1927: 468, 1927 (Pitt, 1980).

Penicillium chermesinum Biourge, Cellule 33: 284, 1923.

Penicillium chloroleucon Biourge, Cellule 33: 270, 1923 = ***Penicillium corylophilum*** Dierckx, Ann. Soc. Sci. Bruxelles 25: 86, 1901 (Pitt, 1980).

Penicillium chlorophaeum Biourge, Cellule 33: 271, 1923 = ***Penicillium chrysogenum*** Thom, Bull. Bur. Anim. Ind. U.S. Dep. Agric. 118: 58, 1910 (Pitt, 1980).

Penicillium chrysitis Biourge, Cellule 33: 252, 1923 = ***Penicillium rugulosum*** Thom, Bull. Bur. Anim. Ind. U.S. Dep. Agric. 118: 60, 1910 (Pitt, 1980).

Penicillium chrysogenum Thom, Bull. Bur. Anim. Ind. U.S. Dep. Agric. 118: 58, 1910.

Penicillium chrzaszczii K. M. Zalessky, Bull. Int. Acad. Polon. Sci., Cl. Sci. Math., Sér. B, Sci. Nat., 1927: 464, 1927 = ***Penicillium miczynskii*** K. M. Zalessky, Bull. Int. Acad. Polon. Sci., Cl. Sci. Math., Sér. B, Sci. Nat., 1927: 482, 1927 (Pitt, 1980).

Penicillium ciegleri Quintan., Avances Nutr. Mejora Anim. Aliment. 23: 338, 1982 = ***Penicillium pulvillorum*** Turfitt, Trans. Br. Mycol. Soc. 23: 186, 1939 (Frisvad *et al.*, 1990b).

Penicillium cinerascens Biourge, Cellule 33: 308, 1923.

Penicillium cinereoatrum Chalab., Bot. Mater. Otd. Sporov. Rast. 6: 167, 1950 = ***Penicillium citreonigrum*** Dierckx, Ann. Soc. Sci. Bruxelles 25: 86, 1901 (Pitt, 1980).

Penicillium cinnamopurpureum Udagawa, J. Agric. Food Sci., Tokyo 5: 1, 1959 = ***Penicillium phoeniceum*** J. F. H. Beyma, Zentralbl. Bakteriol., 2. Abt., 88: 136, 1933 (Pitt, 1980).

Penicillium citreonigrum Dierckx, Ann. Soc. Sci. Bruxelles 25: 86, 1901.

Penicillium citreosulfuratum Biourge, Cellule 33: 285, 1923 = ***Penicillium citreonigrum*** Dierckx, Ann. Soc. Sci. Bruxelles 25: 86, 1901 (Biourge, 1923).

Penicillium citreovirens S. Abe, J. Gen. Appl. Microbiol., Tokyo 2: 87, 1956 (nom. inval., Art. 36) = ***Penicillium spinulosum*** Thom, Bull. Bur. Anim. Ind. U.S. Dep. Agric. 118: 76, 1910 (Pitt, 1980).

Penicillium citreoviride Biourge, Cellule 33: 297, 1923 = ***Penicillium citreonigrum*** Dierckx, Ann. Soc. Sci. Bruxelles 25: 86, 1901 (Pitt, 1980).

Penicillium citreoviride var. *aeneum* S. Abe, J. Gen. Appl. Microbiol., Tokyo 2: 58, 1956 (nom. inval., Art. 36) = ***Penicillium citreonigrum*** Dierckx, Ann. Soc. Sci. Bruxelles 25: 86, 1901 (Pitt, 1980).

Penicillium citrinum Thom, Bull. Bur. Anim. Ind. U.S. Dep. Agric. 118: 61, 1910.

Penicillium claviforme Bainier, Bull. Trimest. Soc. Mycol. Fr. 21: 127, 1905 = ***Penicillium vulpinum*** (Cooke & Massee) Seifert & Samson in Samson & Pitt (eds), Adv. *Penicillium Aspergillus* Syst.: 144, 1985 (Seifert and Samson, 1985).

Penicillium clavigerum Demelius, Verh. Zool.-Bot. Ges. Wien 72: 74, 1923.

Penicillium cluniae Quintan., Avances Nutr. Mejora Anim. Aliment. 30: 174, 1990 = ***Penicillium cremeogriseum*** Chalab., Bot. Mater. Otd. Sporov. Rast. 6: 168, 1950 (Frisvad *et al.*, 1990b).

Penicillium coalescens Quintan., Mycopathologia 84: 115, 1983.

Penicillium coeruleoviride G. Sm., Trans. Br. Mycol. Soc. 48: 275, 1957 = ***Penicillium atrovenetum*** G. Sm., Trans. Br. Mycol. Soc. 39: 112, 1956 (Frisvad and Filtenborg, 1989).

Penicillium commune Thom, Bull. Bur. Anim. Ind. U.S. Dep. Agric. 118: 56, 1910.

Penicillium concavorugulosum S. Abe, J. Gen. Appl. Microbiol., Tokyo 2: 127, 1956 (nom. inval., Art. 36) = ***Penicillium rugulosum*** Thom, Bull. Bur. Anim. Ind. U.S. Dep. Agric. 118: 60, 1910 (Pitt, 1980).

Penicillium concentricum Samson *et al.*, Stud. Mycol. 11: 17, 1976.

Penicillium conditaneum Westling, Ark. Bot. 11(1): 63, 1911 = ***Penicillium aurantiogriseum*** Dierckx, Ann. Soc. Sci. Bruxelles 25: 88, 1901 (Raper and Thom, 1949).

Penicillium confertum (Frisvad *et al.*) Frisvad, Mycologia 81: 852, 1990. — Basionym: *Penicillium glandicola* var. *confertum* Frisvad *et al.*, Can. J. Bot. 65: 769, 1987.

Penicillium conservandi Novobr., Nov. Sist. Niz. Rast. 11: 233, 1974 = ***Penicillium roqueforti*** Thom, Bull. Bur. Anim. Ind. U.S. Dep. Agric. 82: 35, 1906 (Pitt, 1980).

Penicillium coprobium Frisvad, Mycologia 81: 853, 1990.

Penicillium coprophilum (Berk. & M. A. Curtis) Seifert & Samson in Samson & Pitt (eds), Adv. *Penicillium Aspergillus* Syst.: 145, 1985. — Basionym: *Coremium coprophilum* Berk. & M. A. Curtis, J. Linn. Soc., Bot., 10: 363, 1868.

Penicillium coralligerum Nicot & Pionnat, Bull. Soc. Mycol. France 78: 245, 1963 ["1962"].

Penicillium cordubense C. Ramírez & A. T. Martínez, Mycopathologia 74: 164, 1981 = ***Penicillium aurantiogriseum*** Dierckx, Ann. Soc. Sci. Bruxelles 25: 88, 1901 (Frisvad *et al.*, 1990b).

Penicillium corylophiloides S. Abe, J. Gen. Appl. Microbiol., Tokyo 2: 89, 1956 (nom. inval., Art. 36) = **Penicillium jensenii** K. M. Zalessky, Bull. Int. Acad. Polon. Sci., Cl. Sci. Math., Sér. B, Sci. Nat., 1927: 494, 1927 (Pitt, 1980).

Penicillium corylophilum Dierckx, Ann. Soc. Sci. Bruxelles 25: 86, 1901.

Penicillium corymbiferum Westling, Ark. Bot. 11(1): 92, 1911 = **Penicillium hirsutum** Dierckx, Ann. Soc. Sci. Bruxelles 25: 89, 1901 (Pitt, 1980).

Penicillium corynephorum Pitt & A. D. Hocking, Mycotaxon 22: 202, 1985 = **Penicillium smithii** Quintan., Avances Nutr. Mejora Anim. Aliment. 23: 340, 1982 (Frisvad *et al.*, 1990b).

Penicillium crateriforme J. C. Gilman & L. V. Abbott, Iowa State Coll. J. Sci. 1: 293, 1927.

Penicillium cremeogriseum Chalab., Bot. Mater. Otd. Sporov. Rast. 6: 168, 1950.

Penicillium crocicola W. Yamam. in Yamamoto *et al.*, Scient. Rep. Hyogo Univ. Agric., Agric. Biol. Ser. 2, 2: 28, 1956 = **Penicillium thomii** Maire, Bull. Soc. Hist. Nat. Afrique N. 8: 189, 1917 (Pitt, 1980).

Penicillium crustosum Thom, Penicillia: 399, 1930.

Penicillium crustosum var. *spinulosporum* Yugi Sasaki, J. Fac. Agric. Hokkaido Univ. 49: 158, 1950 (nom. inval., Art. 36) = **Penicillium echinulatum** Raper & Thom ex Fassat., Acta Univ. Carol., Biol. 12: 326, 1977 (Stolk *et al.*, 1990).

Penicillium cryptum Goch., Mycotaxon 26: 349, 1986. [Teleomorph *Eupenicillium cryptum* Goch.].

Penicillium cuprophilum Tatuzo Sato, J. Agric. Chem. Soc., Japan 15: 359, 1939 = **Penicillium ochrochloron** Biourge, Cellule 33: 269, 1923 (Raper and Thom, 1949).

Penicillium cyaneum (Bainier & Sartory) Biourge, Cellule 33: 102, 1923. — Basionym: *Citromyces cyaneus* Bainier & Sartory, Bull. Soc. Mycol. France 29: 157, 1913.

Penicillium cyaneofulvum Biourge, Cellule 33: 174, 1923 = **Penicillium chrysogenum** Thom, Bull. Bur. Anim. Ind. U.S. Dep. Agric. 118: 58, 1910 (Stolk *et al.*, 1990).

Penicillium cyclopium Westling, Ark. Bot. 11(1): 90, 1911 ≡ **Penicillium verrucosum** var. *cyclopium* (Westling) Samson *et al.*, Stud. Mycol. 11: 37, 1976 = **Penicillium aurantiogriseum** Dierckx, Ann. Soc. Sci. Bruxelles 25: 88, 1901 (Pitt, 1980).

Penicillium cyclopium var. *album* G. Sm., Trans. Br. Mycol. Soc. 34: 18, 1951 ≡ *Penicillium cyclopium* f. *album* Fassatiová, Acta Univ. Carol., Biol. 12: 326, 1977 = **Penicillium commune** Thom, Bull. Bur. Anim. Ind. U.S. Dep. Agric. 118: 56, 1910 (Stolk *et al.*, 1990).

Penicillium cyclopium var. *aurantiovirens* (Biourge) Fassatiová, Acta Univ. Carol., Biol. 12: 326, 1977

≡ *Penicillium aurantiovirens* Biourge, Cellule 33: 119, 1923 = **Penicillium aurantiogriseum** Dierckx, Ann. Soc. Sci. Bruxelles 25: 88, 1901 (Pitt, 1980).

Penicillium cyclopium var. *echinulatum* Raper & Thom, Man. Penicillia: 497, 1949 ≡ **Penicillium echinulatum** Raper & Thom ex Fassat., Acta Univ. Carol., Biol. 1974: 326, 1977.

Penicillium cylindrosporum G. Sm., Trans. Br. Myeol. Soe. 40: 483, 1957 ≡ *Geosmithia cylindrospora* (G. Sm.) Pitt, Can. J. Bot. 57: 2024, 1979.

Penicillium daleae K. M. Zalessky, Bull. Int. Acad. Polon. Sci., Cl. Sci. Math., Sér. B, Sci. Nat., 1927: 495, 1927.

Penicillium damascenum Baghdadi, Nov. Sist. Niz. Rast. 5: 101, 1968 = **Penicillium melinii** Thom, Penicillia: 273, 1930 (Pitt, 1980).

Penicillium dangeardii Pitt, Genus *Penicillium*: 472, 1980 ["1979"]. [Teleomorph *Talaromyces flavus* (Klöcker) Stolk & Samson].

Penicillium decumbens Thom, Bull. Bur. Anim. Ind. U.S. Dep. Agric. 118: 71, 1910.

Penicillium decumbens var. *atrovirens* S. Abe, J. Gen. Appl. Microbiol., Tokyo 2: 70, 1956 (nom. ir.val., Art. 36) = **Penicillium fellutanum** Biourge, Cellule 33: 262, 1923 (Pitt, 1980).

Penicillium dendriticum Pitt, Genus *Penicillium*: 413, 1980 ["1979"].

Penicillium derxii Takada & Udagawa, Mycotaxon 31: 418, 1988. [Teleomorph *Talaromyces derxii* Takada & Udagawa].

Penicillium dierckxii Biourge, Cellule 33: 313, 1923.

Penicillium digitatoides Peyronel, Germi Atmosf. Fung. Micel.: 22, 1913 = **Penicillium digitatum** (Pers.: Fr.) Sacc., Fung. Ital.: tab. 894, 1881 (Thom, 1930).

Penicillium digitatum (Pers.: Fr.) Sacc., Fung. Ital.: tab. 894, 1881. — Basionym: *Monilia digitata* Pers., Syn. Meth. Fung.: 693, 1801: Fr., Syst. Mycol. 3(2): 411, 1832.

Penicillium digitatum var. *latum* S. Abe, J. Gen. Appl. Microbiol., Tokyo 2: 97, 1956 (nom. inval., Art. 36) ≡ *Penicillium japonicum* G. Sm., Trans. Br. Mycol. Soc. 46: 333, 1963 (new name) = **Penicillium italicum** Wehmer, Hedwigia 33: 211, 1894 (Cruickshank and Pitt, 1987).

Penicillium dimorphosporum H. Swart, Trans. Br. Mycol. Soc. 55: 310, 1970.

Penicillium divergens Bainier & Sartory, Bull. Trimest. Soc. Mycol. Fr. 28: 270, 1912 = **Penicillium glandicola** (Oudem.) Seifert & Samson in Samson & Pitt (eds), Adv. *Penicillium Aspergillus* Syst.: 147, 1985 (Seifert and Samson, 1985).

Penicillium diversum Raper & Fennell, Mycologia 40: 539, 1948.

Penicillium diversum var. *aereum* Raper & Fennell, Mycologia 40: 541, 1948 (nom. inval., Art. 36) ≡ **Penicillium primulinum** Pitt, Genus *Penicillium*: 455, 1980 ["1979"] (new name).

Penicillium dodgei Pitt, Genus *Penicillium*: 117, 1980 ["1979"]. [Teleomorph *Eupenicillium brefeldianum* (B. O. Dodge) Stolk & D. B. Scott].

Penicillium donkii Stolk, Persoonia 7: 333, 1973.

Penicillium duclauxii Delacr., Bull. Soc. Mycol. France 7: 107, 1891.

Penicillium duninii Sidibe, Mikol. Fitopatol. 8: 371, 1974 = **Penicillium griseofulvum** Dierckx, Ann. Soc. Sci. Bruxelles 25: 88, 1901 (Pitt, 1980).

Penicillium dupontii Griffon & Maubl., Bull. Soc. Mycol. France 27: 73, 1911. [Teleomorph *Talaromyces thermophilus* Stolk].

Penicillium eben-bitarianum Baghdadi, Nov. Sist. Niz. Rast. 5: 106, 1968 = **Penicillium fellutanum** Biourge, Cellule 33: 262, 1923 (Pitt, 1980).

Penicillium echinosporum G. Sm., Trans. Br. Mycol. Soc. 45: 387, 1962 (*non* Nehira) ≡ **Penicillium asperosporum** G. Sm., Trans. Br. Mycol. Soc. 48: 275, 1965 (new name).

Penicillium echinosporum Nehira, J. Ferment. Technol., Osaka 11: 861, 1933 = **Penicillium rugulosum** Thom, Bull. Bur. Anim. Ind. U.S. Dep. Agric. 118: 60, 1910 (Pitt, 1980).

Penicillium echinulatum E. Dale in Biouge, Cellule 33: 278, 1923 (*non* Rivolta, 1873) = **Penicillium janczewskii** K. M. Zalessky, Bull. Int. Acad. Polon. Sci., Cl. Sci. Math., Sér. B, Sci. Nat., 1927: 488, 1927 (Pitt, 1980).

Penicillium echinulatum Raper & Thom ex Fassat., Acta Univ. Carol., Biol. 1974: 326, 1977.

Penicillium echinulonalgiovense S. Abe, J. Gen. Appl. Microbiol., Tokyo 2: 80, 1956 (nom. inval., Art. 36) = **Penicillium janthinellum** Biourge, Cellule 33: 258, 1923 (Pitt, 1980).

Penicillium egyptiacum J. F. H. Beyma, Zentralbl. Bakteriol., 2. Abt., 88: 137, 1933 (nom. holomorph.) ≡ **Eupenicillium egyptiacum** (J. F. H. Beyma) Stolk & D. B. Scott, Persoonia 4: 401, 1967.

Penicillium ehrlichii Kleb., Ber. Deutsch. Bot. Ges. 48: 374, 1930 (nom. holomorph.) ≡ **Eupenicillium ehrlichii** (Kleb.) Stolk & D. B. Scott, Persoonia 4: 400, 1967.

Penicillium elegans Sopp, Skr. Vidensk.-Selsk. Christiana, Math.-Naturvidensk. Kl. 11: 161, 1912 (*non* Corda, 1838) = **Penicillium herquei** Bainier & Sartory, Bull. Soc. Mycol. France 28: 121, 1912 (Raper and Thom, 1949).

Penicillium elongatum Bainier, Bull. Trimest. Soc. Mycol. Fr. 23: 17, 1907 (*non* Dierckx 1901) = **Penicillium rugulosum** Thom, Bull. Bur. Anim. Ind. U.S. Dep. Agric. 118: 60, 1910 (Pitt, 1980).

Penicillium elongatum Dierckx, Annls Soc. Scient. Brux. 25: 87, 1901 = **Penicillium expansum** Link, Ges. Naturf. Freunde Berlin Mag. Neuesten Entdeck. Gesammten Naturk. 3: 16, 1809 (Biourge, 1923).

Penicillium emersonii Stolk, Antonie van Leeuwenhoek 31: 262, 1965 = **Geosmithia emersonii** (Stolk) Pitt, Can. J. Bot. 57: 2027, 1979.

Penicillium emmonsii Pitt, Genus *Penicillium*: 479, 1980 ["1979"]. [Teleomorph *Talaromyces stipitatus* (Thom) C. R. Benj.].

Penicillium epsteinii Lindau, Rabenh. Krypt.-Fl. 1, Abt. 8: 166, 1904 (rejected name, Art. 71) = **Penicillium camemberti** Thom, Bull. Bur. Anim. Ind. U.S. Dep. Agric. 82: 33, 1906 (Pitt, 1980).

Penicillium erubescens D. B. Scott, Mycopathol. Mycol. Appl. 36: 14, 1968. [Teleomorph *Eupenicillium erubescens* D. B. Scott].

Penicillium erythromellis A. D. Hocking, Pitt, Genus *Penicillium*: 459, 1980 ["1979"].

Penicillium es-suveidense Baghdadi, Nov. Sist. Niz. Rast. 5: 108, 1968 = **Penicillium simplicissimum** (Oudem.) Thom, Penicillia: 335, 1930 (Pitt, 1980).

Penicillium estinogenum A. Komatsu & S. Abe ex G. Sm., Trans. Br. Mycol. Soc. 46: 335, 1963.

Penicillium estinogenum A. Komatsu & S. Abe, J. Gen. Appl. Microbiol., Tokyo 2: 132, 1956 (nom. inval., Art. 36) ≡ **Penicillium estinogenum** A. Komatsu & S. Abe ex G. Sm., Trans. Br. Mycol. Soc. 46: 335, 1963.

Penicillium expansum Link, Ges. Naturf. Freunde Berlin Mag. Neuesten Entdeck. Gesammten Naturk. 3: 16, 1809.

Penicillium expansum var. *crustosum* Fassat., Acta Univ. Carol. Biol. 12: 329, 1977 ≡ **Penicillium crustosum** Thom, Penicillia: 399, 1930 (Pitt, 1980).

Penicillium fagi C. Ramírez & A. T. Martínez, Mycopathologia 63: 57, 1978 = **Penicillium raciborskii** K. M. Zalessky, Bull. Int. Acad. Polon. Sci., Cl. Sci. Math., Sér. B, Sci. Nat., 1927: 454, 1927 (Frisvad *et al.*, 1990b).

Penicillium farinosum Novobranova, Nov. Sist. Niz. Rast. 11: 232, 1974 [*non* (Holmsk.) Biourge] = **Penicillium crustosum** Thom, Penicillia: 399, 1930 (Pitt, 1980).

Penicillium fellutanum Biourge, Cellule 33: 262, 1923.

Penicillium fellutanum var. *nigrocastaneum* S. Abe, J. Gen. Appl. Microbiol., Tokyo 2: 71, 1956 (nom. inval., Art. 36) = **Penicillium fellutanum** Biourge, Cellule 33: 262, 1923 (Pitt, 1980).

Penicillium fennelliae Stolk, Antonie van Leeuwenhoek 35: 261, 1969.

Penicillium flavidomarginatum Biourge, Cellule 33: 150, 1923 = **Penicillium chrysogenum** Thom,

31

Bull. Bur. Anim. Ind. U.S. Dep. Agric. 118: 58, 1910 (Pitt, 1980).

Penicillium flavidorsum Biourge, Cellule 23: 290, 1923 = ***Penicillium glabrum*** (Wehmer) Westling, Ark. Bot. 11(1): 131, 1911 (Pitt, 1980).

Penicillium flavidostipitatum C. Ramírez & C. C. González, Mycopathologia 88: 3, 1984.

Penicillium flavocinereum Biourge, Cellule 33: 293, 1923 = ***Penicillium spinulosum*** Thom, Bull. Bur. Anim. Ind. U.S. Dep. Agric. 118: 76, 1910 (Pitt, 1980).

Penicillium flavoglaucum Biourge, Cellule 33: 130, 1923 = ***Penicillium verrucosum*** Dierckx, Ann. Soc. Sci. Bruxelles 25: 88, 1901 (Pitt, 1980).

Penicillium flexuosum E. Dale in Biourge, Cellule 33: 264, 1923 (*non* Preuss, 1851) = ***Penicillium griseofulvum*** Dierckx, Ann. Soc. Sci. Bruxelles 25: 88, 1901 (Pitt, 1980).

Penicillium fluitans Tiegs, Ber. Deut. Bot. Ges. 37: 500, 1919 = ***Penicillium glabrum*** (Wehmer) Westling, Ark. Bot. 11(1): 131, 1911 (Raper & Thom, 1949).

Penicillium formosanum H. M. Hsich et al., Trans. Mycol. Soc. Republ. China 2: 159, 1987.

Penicillium fractum Udagawa, Trans. Mycol. Soc. Japan 9: 51, 1968. [Teleomorph *Eupenicillium fractum* Udagawa].

Penicillium frequentans Westling, Ark. Bot. 11(1): 133, 1911 = ***Penicillium glabrum*** (Wehmer) Westling, Ark. Bot. 11(1): 131, 1911 (Pitt, 1980).

Penicillium funiculosum Thom, Bull. Bur. Anim. Ind. U.S. Dep. Agric. 118: 69, 1910.

Penicillium gaditanum C. Ramírez & A. T. Martínez, Mycopathologia 74: 165, 1981 = ***Penicillium minioluteum*** Dierckx, Ann. Soc. Sci. Bruxelles 25: 87, 1901 (Frisvad et al., 1990a; van Reenan-Hoekstra et al., 1990).

Penicillium galapagense Samson & Mahoney, Trans. Br. Mycol. Soc. 69: 158, 1977. [Teleomorph *Talaromyces galapagensis* Samson & Mahoney].

Penicillium galliacum C. Ramírez et al., Mycopathologia 72: 30, 1980 = ***Penicillium citreonigrum*** Dierckx, Ann. Soc. Sci. Bruxelles 25: 86, 1901 (Frisvad et al., 1990b).

Penicillium gerundense C. Ramírez & A. T. Martínez, Mycopathologia 72: 182, 1980 = ***Penicillium dierckxii*** Biourge, Cellule 33: 313, 1923 (Frisvad et al., 1990b).

Penicillium giganteum R. Y. Roy & G. N. Singh, Trans. Br. Mycol. Soc. 51: 805, 1968 = ***Penicillium megasporum*** Orpurt & Fennell, Mycologia 47: 233, 1955 (Pitt, 1980).

Penicillium gilmanii Thom, Penicillia: 345, 1930 = ***Penicillium restrictum*** J. C. Gilman & E. V. Abbott, Iowa State Coll. J. Sci. 1: 297, 1927 (Pitt, 1980).

Penicillium glabrum (Wehmer) Westling, Ark. Bot. 11(1): 131, 1911. — Basionym: *Citromyces glaber* Wehmer, Beitr. Einh. Pilze 1: 24, 1893.

Penicillium gladioli Machacek, Rep. Queb. Soc. Prot. Pl. 19: 77, 1928 (*nom. illegit.*, Art. 64) = ***Penicillium gladioli*** L. McCulloch & Thom, Science, ser. 2, 67: 217, 1928 (Pitt, 1980).

Penicillium gladioli L. McCulloch & Thom, Science, ser. 2, 67: 217, 1928. [Teleomorph *Eupenicillium crustaceum* F. Ludw.].

Penicillium glandicola (Oudem.) Seifert & Samson in Samson & Pitt (eds), Adv. *Penicillium Aspergillus* Syst.: 147, 1985. — Basionym: *Coremiun glandicola* Oudem., Ned. Kruidk. Arch., ser. 3, 2: 918, 1903.

Penicillium glandicola var. *confertum* Frisvad et al., Can. J. Bot. 65: 769, 1987 ≡ ***Penicillium confertum*** (Frisvad et al.) Frisvad in Frisvad and Filtenborg, Mycologia 81: 851, 1989.

Penicillium glandicola var. *mononematosum* Frisvad et al., Can. J. Bot. 65: 767, 1987 ≡ ***Penicillium mononematosum*** (Frisvad et al.) Frisvad in Frisvad and Filtenborg, Mycologia 81: 857, 1989.

Penicillium glaucolanosum Chalab., Bot. Mater. Otd. Sporov. Rast. 6: 168, 1950 = ***Penicillium decumbens*** Thom, Bull. Bur. Anim. Ind. U.S. Dep. Agric. 118: 71, 1910 (Pitt, 1980).

Penicillium glaucoroseum Demelius, Verh. Zool.-Bot. Ges. Wien 72: 72, 1923 (1922) = ***Penicillium janthinellum*** Biourge, Cellule 33: 258, 1923 (Raper and Thom, 1949).

Penicillium godlewskii K. M. Zalessky, Bull. Int. Acad. Polon. Sci., Cl. Sci. Math., Sér. B, Sci. Nat., 1927: 466, 1927 = ***Penicillium jensenii*** K. M. Zalessky, Bull. Int. Acad. Polon. Sci., Cl. Sci. Math., Sér. B, Sci. Nat., 1927: 494, 1927 (Pitt, 1980).

Penicillium gorgonzolae Weidemann in Biourge, Cellule 33: 204, 1923 (new name) ≡ *Penicillium roqueforti* var. *weidemannii* Westling, Ark. Bot. 11(1): 71, 1911 = ***Penicillium roqueforti*** Thom, Bull. Bur. Anim. Ind. U.S. Dep. Agric. 82: 35, 1906 (Raper and Thom, 1949).

Penicillium gorlenkoanum Baghdadi, Nov. Sist. Niz. Rast. 5: 97, 1968 = ***Penicillium citrinum*** Thom, Bull. Bur. Anim. Ind. U.S. Dep. Agric. 118: 61, 1910 (Pitt, 1980).

Penicillium gossypii Pitt, Gen. *Penicillium*: 500, 1980 (1979) = ***Penicillium assiutense*** Samson & Abdel-Fattah, Persoonia 9: 501, 1978 (Frisvad et al., 1990b).

Penicillium gracilentum Udagawa & Y. Horie, Trans. Mycol. Soc. Japan 14: 373, 1973. [Teleomorph *Eupenicillium gracilentum* Udagawa & Y. Horie].

Penicillium grancanariae C. Ramírez et al., Mycopathologia 66: 79, 1978 = ***Penicillium thomii***

Maire, Bull. Soc. Hist. Nat. Afrique N. 8: 189, 1917 (Frisvad *et al.*, 1990b).

Penicillium granulatum Bainier, Bull. Trimest. Soc. Mycol. Fr. 21: 126, 1905 = **Penicillium glandicola** (Oudem.) Seifert & Samson in Samson & Pitt (eds), Adv. *Penicillium Aspergillus* Syst.: 147, 1985 (Seifert and Samson, 1985).

Penicillium granulatum var. *globosum* Bridge *et al.*, J. Gen. Microbiol. 135: 2958, 1989 = **Penicillium glandicola** (Oudem.) Seifert & Samson in Samson & Pitt (eds), Adv. *Penicillium Aspergillus* Syst.: 147, 1985 (Frisvad *et al.*, 1990b).

Penicillium gratanense C. Ramírez *et al.*, Mycopathologia 72: 31, 1980 = **Penicillium janczewskii** K. M. Zalessky, Bull. Int. Acad. Polon. Sci., Cl. Sci. Math., Sér. B, Sci. Nat., 1927: 488, 1927 (Frisvad *et al.*, 1990b).

Penicillium griseo-azureum C. Moreau & M. Moreau, Revue Mycol. 6: 59, 1941 (nom. inval., Art. 36) = **Penicillium waksmanii** K. M. Zalessky, Bull. Int. Acad. Polon. Sci., Cl. Sci. Math., Sér. B, Sci. Nat., 1927: 468, 1927 (Pitt, 1980).

Penicillium griseobrunneum Dierck, Ann. Soc. Sci. Bruxelles 25: 88, 1901 = **Penicillium brevicompactum** Dierckx, Ann. Soc. Sci. Bruxelles 25: 88, 1901 (Pitt, 1980).

Penicillium griseofulvum Dierckx, Ann. Soc. Sci. Bruxelles 25: 88, 1901.

Penicillium griseolum G. Sm., Trans. Br. Mycol. Soc. 40: 485, 1957 = **Penicillium restrictum** J. C. Gilman & E.V. Abbott, Iowa State Coll. J. Sci. 1: 297, 1927 (Pitt, 1980).

Penicillium griseoroseum Dierckx, Annls Soc. Scient. Brux. 25: 86, 1901 = **Penicillium chrysogenum** Thom *nom. cons.*, Bull. Bur. Anim. Ind. U.S. Dep. Agric. 118: 58, 1910 (Kozakiewicz *et al.*, 1992).

Penicillium griseopurpureum G. Sm., Trans. Br. Mycol. Soc. 48: 275, 1965.

Penicillium griseum (Sopp) Biourge, Cellule 33: 103, 1923 (*non* Bonorden, 1951) = **Penicillium restrictum** J. C. Gilman & E. V. Abbott, Iowa State Coll. J. Sci. 1: 297, 1927 (Pitt, 1980).

Penicillium guttulosum J. C. Gilman & E. V. Abbott, Iowa State Coll. J. Sci. 1: 298, 1927 = **Penicillium vinaceum** J. C. Gilman & E. V. Abbott, Iowa State Coll. J. Sci. 1: 299, 1927 (Pitt, 1980).

Penicillium hagemii K. M. Zalessky, Bull. Int. Acad. Polon. Sci., Cl. Sci. Math., Sér. B, Sci. Nat., 1927: 448, 1927 = **Penicillium brevicompactum** Dierckx, Ann. Soc. Sci. Bruxelles 25: 88, 1901 (Pitt, 1980).

Penicillium harmonense Baghdadi, Nov. Sist. Niz. Rast. 5: 102, 1968 = **Penicillium chrysogenum** Thom, Bull. Bur. Anim. Ind. U.S. Dep. Agric. 118: 58, 1910 (Pitt, 1980).

Penicillium helicum Raper & Fennell, Mycologia 40: 515, 1948 (nom. holomorph.) ≡ **Talaromyces helicus** (Raper & Fennell) C. R. Benj., Mycologia 47: 684, 1955.

Penicillium herquei Bainier & Sartory, Bull. Soc. Mycol. France 28: 121, 1912.

Penicillium heteromorphum H. Z. Kong & Z. T. Qi, Mycosystema 1: 107, 1988.

Penicillium hirayamae Udagawa, J. Agric. Sci. Tokyo Nogyo Daigaku 5: 6, 1959. [Teleomorph *Eupenicillium hirayamae* D. B. Scott & Stolk].

Penicillium hirsutum Dierckx, Ann. Soc. Sci. Bruxelles 25: 89, 1901.

Penicillium hirsutum var. *allii* (Vincent & Pitt) Frisvad in Frisvad and Filtenborg, Mycologia 81: 855, 1989 ≡ **Penicillium allii** Vincent & Pitt, Mycologia 81: 300, 1989 (Pitt and Samson, 1993).

Penicillium hirsutum var. *hordei* (Stolk) Frisvad in Frisvad & Filtenborg, Mycologia 81: 855, 1989 ≡ **Penicillium hordei** Stolk, Antonie van Leeuwenhoek 35: 270, 1969 (Pitt and Samson, 1993).

Penicillium hispanicum C. Ramírez *et al.*, Mycopathologia 66: 77, 1978 = **Penicillium implicatum** Biourge, Cellule 33: 278, 1923 (accepted here). Frisvad *et al.* (1990a) reported that the ex type culture of *P. implicatum* in the CBS collection was a *P. citrinum*, and that *P. hispanicum* was the earliest available name for the concept of *P. fellutanum* used by Raper and Thom (1949) and Pitt (1980). However, as that strain does not match the original protocol of *P. fellutanum*, Pitt and Samson (1993) accepted the neotypification of Pitt (1980). *P. hispanicum* becomes a synonym.

Penicillium hordei Stolk, Antonie van Leeuwenhoek 35: 270, 1969.

Penicillium humuli J. F. H. Beyma, Zentbl. Bakt. ParasitKde, Abt II, 99: 393, 1939 = **Penicillium corylophilum** Dierckx, Ann. Soc. Sci. Bruxelles 25: 86, 1901. (R.H. Cruickshank, unpubl.; accepted here).

Penicillium idahoense Paden, Mycopath. Mycol. Appl. 43: 259, 1971 = **Penicillium phoeniceum** J. F. H. Beyma, Zentralbl. Bakteriol., 2. Abt., 88: 136, 1933 (Pitt, 1980).

Penicillium ilerdanum C. Ramírez & A. T. Martínez, Mycopathologia 72: 32, 1980 = **Penicillium piceum** Raper & Fennell, Mycologia 40: 533, 1948 (Frisvad *et al.*, 1990b).

Penicillium implicatum Biourge, Cellule 33: 278, 1923.

Penicillium implicatum var. *aureomarginatum* Thom, Penicillia: 211, 1930 = **Penicillium sclerotiorum** J. F. H. Beyma, Zentralbl. Bakteriol. ParasitKde, 2. Abt., 96: 418, 1937 (Pitt, 1980).

Penicillium indicum D. K. Sandhu & R. S. Sandhu, Can. J. Bot. 41: 1273, 1963 = **Penicillium cher-**

mesinum Biourge, Cellule 33: 284, 1923 (Pitt, 1980).

Penicillium indonesiae Pitt, Genus *Penicillium*: 114, 1980 ["1979"]. [Teleomorph *Eupenicillium javanicum* (J. F. H. Beyma) Stolk & D. B. Scott].

Penicillium inflatum Stolk & Malla, Persoonia 6: 197, 1971.

Penicillium ingelheimense J. F. H. Beyma, Antonie van Leeuwenhoek 6: 109, 1942 ≡ *Merimbla ingelheimensis* (J. F. H. Beyma) Pitt, Can. J. Bot. 57: 2395, 1979.

Penicillium internascens Szilvinyi, Zentbl. Bakt. ParasitKde, Abt. II, 103: 148, 1941 = *Penicillium purpurascens* (Sopp) Biourge, Cellule 33: 105, 1923 (Pitt, 1980).

Penicillium inusitatum D. B. Scott, Mycopathol. Mycol. Appl. 36: 20, 1968. [Teleomorph *Eupenicillium inusitatum* D. B. Scott].

Penicillium iriense Boretti *et al.*, Arch. Mikrobiol. 92: 190, 1973 (nom. inval., Art. 36) = *Penicillium janthinellum* Biourge, Cellule 33: 258, 1923 (Pitt, 1980).

Penicillium isariiforme Stolk & J. Mey., Trans. Br. Mycol. Soc. 40: 187, 1957.

Penicillium islandicum Sopp, Skr. Vidensk.-Selsk. Christiana, Math.-Naturvidensk. Kl. 11: 161, 1912.

Penicillium italicum Wehmer, Hedwigia 33: 211, 1894.

Penicillium italicum var. *album* C. T. Wei, Nanking J. 9: 241, 1940 = *Penicillium italicum* Wehmer, Hedwigia 33: 211, 1894 (Stolk *et al.*, 1990).

Penicillium italicum var. *avellaneum* Samson & Gutter in Samson *et al.*, Stud. Mycol. 11: 30, 1976 = *Penicillium italicum* Wehmer, Hedwigia 33: 211, 1894 (Stolk *et al.*, 1990).

Penicillium janczewskii K. M. Zalessky, Bull. Int. Acad. Polon. Sci., Cl. Sci. Math., Sér. B, Sci. Nat., 1927: 488, 1927.

Penicillium janthinellum Biourge, Cellule 33: 258, 1923.

Penicillium janthocitrinum Biourge, Cellule 33: 311, 1923 = *Penicillium spinulosum* Thom, Bull. Bur. Anim. Ind. U.S. Dep. Agric. 118: 76, 1910 (Pitt, 1980).

Penicillium janthogenum Biourge, Cellule 33: 143, 1923 = *Penicillium expansum* Link, Ges. Naturf. Freunde Berlin Mag. Neuesten Entdeck. Gesammten Naturk. 3: 16, 1809 (Stolk *et al.*, 1990).

Penicillium japonicum G. Sm., Trans. Br. Mycol. Soc. 46: 333, 1963 (new name) ≡ *Penicillium digitatum* var. *latum* S. Abe, J. Gen. Appl. Microbiol., Tokyo 2: 97, 1956 (nom. inval., Art. 36) = *Penicillium italicum* Wehmer, Hedwigia 33: 211, 1894 (Cruickshank and Pitt, 1987).

Penicillium javanicum J. F. H. Beyma, Verh. Kon. Ned. Akad. Wetensch., Afd. Natuurk., Tweede

Sect., 26(4): 17, 1929 (nom. holomorph.) ≡ *Eupenicillium javanicum* (J. F. H. Beyma) Stolk & D. B. Scott, Persoonia 4: 398, 1967.

Penicillium jensenii K. M. Zalessky, Bull. Int. Acad. Polon. Sci., Cl. Sci. Math., Sér. B, Sci. Nat., 1927: 494, 1927.

Penicillium johanniolii K. M. Zalessky, Bull. Int. Acad. Polon. Sci., Cl. Sci. Math., Sér. B, Sci. Nat., 1927: 453, 1927 = *Penicillium aurantiogriseum* Dierckx, Ann. Soc. Sci. Bruxelles 25: 88, 1901 (Pitt, 1980).

Penicillium jugoslavicum C. Ramírez & Munt.-Cvetk., Mycopathologia 88: 65, 1984.

Penicillium kabunicum Baghd., Novosti Sist. Nizs. Rast.1968: 98, 1968.

Penicillium kap laboratorium Sopp in Biourge, Cellule 36: 454, 1925 = *Penicillium expansum* Link, Ges. Naturf. Freunde Berlin Mag. Neuesten Entdeck. Gesammten Naturk. 3: 16, 1809 (Pitt, 1980).

Penicillium kapuscinskii K. M. Zalessky, Bull. Int. Acad. Polon. Sci., Cl. Sci. Math., Sér. B, Sci. Nat., 1927: 454, 1927 = *Penicillium canescens* Sopp, Skr. Vidensk.-Selsk. Christiana, Math.-Naturvidensk. Kl. 11: 181, 1912 (Pitt, 1980).

Penicillium katangense Stolk, Antonie van Leeuwenhoek 34: 42, 1968. [Teleomorph *Eupenicillium katangense* Stolk].

Penicillium kewense G. Sm., Trans. Br. Mycol. Soc. 44: 42, 1961 (nom. holomorph.) = *Eupenicillium crustaceum* F. Ludw., Lehrb. Nied. Krypt.: 263, 1892 (Stolk and Scott, 1967).

Penicillium klebahnii Pitt, Genus *Penicillium*: 122, 1980 ["1979"]. [Teleomorph *Eupenicillium ehrlichii* (Kleb.) Stolk & D. B. Scott].

Penicillium kloeckeri Pitt, Genus *Penicillium*: 491, 1980 ["1979"].

Penicillium kojigenum G. Sm., Trans. Br. Mycol. Soc. 44: 43, 1961 = *Penicillium lanosum* Westling, Ark. Bot. 11(1): 97, 1911 (Stolk *et al.*, 1990).

Penicillium korosum J. N. Rai *et al.*, Antonie van Leeuwenhoek 35: 430, 1969 = *Penicillium pinophilum* Hedgc. in Thom, Bull. Bur. Anim. Ind. U.S. Dep. Agric. 118: 37, 1910 (Pitt, 1980).

Penicillium krzemieniewskii K. M. Zalessky, Bull. Int. Acad. Polon. Sci., Cl. Sci. Math., Sér. B, Sci. Nat., 1927: 495, 1927 = *Penicillium daleae* K. M. Zalessky, Bull. Int. Acad. Polon. Sci., Cl. Sci. Math., Sér. B, Sci. Nat., 1927: 495, 1927 (Pitt, 1980).

Penicillium kurssanovii Chalab., Bot. Mater. Otd. Sporov. Rast. 6: 168, 1950 = *Penicillium restrictum* J. C. Gilman & E. V. Abbott, Iowa State Coll. J. Sci. 1: 297, 1927 (Pitt, 1980).

Penicillium lacus-sarmientei C. Ramírez, Mycopathologia 96: 29, 1986 = *Penicillium roseopur-*

34

pureum Dierckx, Ann. Soc. Sci. Bruxelles 25: 86, 1901 (Frisvad *et al.*, 1990b).

Penicillium lanoso-coeruleum Thom, Penicillia: 322, 1930 = **Penicillium commune** Thom, Bull. Bur. Anim. Ind. U.S. Dep. Agric. 118: 56, 1910 (Cruickshank and Pitt, 1987).

Penicillium lanosogrisellum Biourge, Cellule 33: 196, 1923 = **Penicillium digitatum** (Pers.: Fr.) Sacc., Fung. Ital.: tab. 894, 1881 (Thom, 1930).

Penicillium lanosogriseum Thom, Penicillia: 327, 1930 = **Penicillium commune** Thom, Bull. Bur. Anim. Ind. U.S. Dep. Agric. 118: 56, 1910 (Cruickshank and Pitt, 1987).

Penicillium lanosoviride Thom. Penicillia: 314, 1930 = **Penicillium commune** Thom, Bull. Bur. Anim. Ind. U.S. Dep. Agric. 118: 56, 1910 (Stolk *et al.*, 1990).

Penicillium lanosum Westling, Ark. Bot. 11(1): 97, 1911.

Penicillium lapatayae C. Ramírez, Mycopathologia 91: 96, 1985.

Penicillium lapidosum Raper & Fennell, Mycologia 40: 524, 1948. [Teleomorph *Eupenicillium lapidosum* D. B. Scott & Stolk].

Penicillium lassenii Paden, Mycopathol. Mycol. Appl. 43: 266, 1971. [Teleomorph *Eupenicillium lassenii* Paden].

Penicillium lavendulum Raper & Fennell, Myeologia 40: 530, 1948 = **Geosmithia lavendula** (Raper & Fennell) Pitt, Can. J. Bot. 57: 2022, 1979.

Penicillium lehmanii Pitt, Genus *Penicillium*: 497, 1980 ["1979"]. [Teleomorph *Talaromyces trachyspermus* (Shear) Stolk & Samson].

Penicillium lemonii Sopp, Skr. Vidensk.-Selsk. Christiana, Math.-Naturvidensk. Kl. 11: 194, 1912 = **Penicillium herquei** Bainier & Sartory, Bull. Soc. Mycol. France 28: 121, 1912 (Raper and Thom, 1949).

Penicillium leucopus (Pers.) Biourge, C. R. Séanc. Coc. Biol. 82: 877, 1919 ≡ *Coremium leucopus* Pers., Mycol. Eur. 1: 42, 1822 = **Penicillium expansum** Link, Ges. Naturf. Freunde Berlin Mag. Neuesten Entdeck. Gesammten Naturk. 3: 16, 1809 (Pitt, 1980).

Penicillium levitum Raper & Fennell, Mycologia 40: 511, 1948 (nom. holomorph.) ≡ **Eupenicillium levitum** (Raper & Fennell) Stolk & D. B. Scott, Persoonia 4: 402, 1967.

Penicillium liani Kamyschko, Bot. Mater. Otd. Sporov. Rast. 15: 86, 1962 (nom. holomorph.) = **Talaromyces flavus** (Klöcker) Stolk & Samson, Stud. Mycol. 2: 10, 1972 (Pitt, 1980).

Penicillium lignorum Stolk, Antonie van Leeuwenhoek 35: 264, 1969.

Penicillium lilacinoechinulatum S. Abe ex G. Sm., Trans. Br. Mycol. Soc. 46: 335, 1963 ≡ *Penicillium lilacinoechinulatum* S. Abe, J. Gen. Appl.

Microbiol., Tokyo 2: 54, 1956 (nom. inval., Art. 36) = **Penicillium bilaiae** Chalab., Bot. Mater. Otd. Sporov. Rast. 6: 165, 1950 (Pitt, 1980).

Penicillium lineatum Pitt, Genus *Penicillium*: 485, 1980 ["1979"]. [Teleomorph *Talaromyces striatus* (Raper & Fennell) C. R. Benj.].

Penicillium lineolatum Udagawa & Y. Horie, Mycotaxon 5: 493, 1977. [Teleomorph *Eupenicillium lineolatum* Udagawa & Y. Horie].

Penicillium lividum Westling, Ark. Bot. 11(1): 134, 1911.

Penicillium loliense Pitt, Genus *Penicillium*: 450, 1980 ["1979"].

Penicillium ludwigii Udagawa, Trans. Mycol. Soc. Japan 10: 2, 1969. [Teleomorph *Eupenicillium ludwigii* Udagawa].

Penicillium luteo-aurantium G. Sm., Trans. Br. Mycol. Soc. 46: 331, 1963 = **Penicillium resedanum** McLennan & Ducker in McLennan *et al.*, Aust. J. Bot. 2: 360, 1954 (Pitt, 1980).

Penicillium luteocoeruleum Saito, J. Ferment. Technol., Osaka 27: 2, 1949 (nom. inval., Art. 36) = **Penicillium herquei** Bainier & Sartory, Bull. Soc. Mycol. France 28: 121, 1912 (Pitt, 1980).

Penicillium luteum Zukal, Sitzungsber. Kaiserl. Akad. Wiss., Math.-Naturwiss. Cl., Abt. 1, 98: 561, 1890 (nom. holomorph.) ≡ **Talaromyces luteus** (Zukal) C. R. Benj., Mycologia 47: 681, 1955.

Penicillium luzoniacum Udagawa & Y. Horie, J. Jap. Bot. 47: 338, 1972 = **Penicillium terrenum** D. B. Scott, Mycopathol. Mycol. Appl. 36: 1, 1968 (Pitt, 1980).

Penicillium macedonense Verona & Mickovski, Mycopath. Mycol. Appl. 18: 292, 1962 = **Penicillium pinetorum** M. Chr. & Backus, Mycologia 53: 457, 1962 (Pitt, 1980).

Penicillium maclennaniae H. Y. Yip, Trans. Br. Mycol. Soc. 77: 202, 1981.

Penicillium macrosporum Frisvad *et al.*, Antonie van Leeuwenhoek 57: 186, 1990. [Teleomorph *Talaromyces macrosporus* (Stolk & Samson) Frisvad *et al.*].

Penicillium madriti G. Sm., Trans. Br. Mycol. Soc. 44: 44, 1961.

Penicillium majusculum Westling, Ark. Bot. 11(1): 60, 1911 = **Penicillium commune** Thom, Bull. Bur. Anim. Ind. U.S. Dep. Agric. 118: 56, 1910 (Stolk *et al.*, 1990).

Penicillium malacaense C. Ramírez & A. T. Martínez, Mycopathologia 72: 186, 1980 = **Penicillium restrictum** J. C. Gilman & E. V. Abbott, Iowa State Coll. J. Sci. 1: 297, 1927 (Frisvad *et al.*, 1990b).

Penicillium mali Gorlenko & Novobr., Mikol. Fitopatol. 17: 464, 1983 = *Penicillium mali* Novobr., Nauch. Dokl. Vssh. Shk. Biol. Nauk. 10, 107,

1972 (nom. inval., Art. 36) = *Penicillium solitum* Westling, Ark. Bot. 11(1): 65, 1911 (Cruickshank and Pitt, 1987).

Penicillium mali Novobr., Nauch. Dokl. Vssh. Shk. Biol. Nauk. 10, 107, 1972 (nom. inval., Art. 36) = *Penicillium solitum* Westling, Ark. Bot. 11(1): 65, 1911 (Cruickshank and Pitt, 1987).

Penicillium maltum M. Hori & T. Yamam. in Hori *et al.*, Jap. J. Bacteriol. 9: 1105, 1954 (nom. inval., Art. 36) = *Penicillium griseofulvum* Dierckx, Ann. Soc. Sci. Bruxelles 25: 88, 1901 (Pitt, 1980).

Penicillium manginii Duché & R. Heim, Trav. Cryptog. Louis L. Mangin: 450, 1931.

Penicillium mariaecrucis Quintan., Avances Nutr. Mejora Anim. Aliment. 23: 334, 1982.

Penicillium marneffei Segretain in Segretain *et al.*, Bull. Soc. Mycol. France 75: 416, 1960.

Penicillium martensii Biourge, Cellule 33: 152, 1923 = *Penicillium aurantiogriseum* Dierckx, Ann. Soc. Sci. Bruxelles 25: 88, 1901 (Pitt, 1980).

Penicillium matris-meae K. M. Zalessky, Bull. Int. Acad. Polon. Sci., Cl. Sci. Math., Sér. B, Sci. Nat., 1927: 477, 1927 = *Penicillium miczynskii* K. M. Zalessky, Bull. Int. Acad. Polon. Sci., Cl. Sci. Math., Sér. B, Sci. Nat., 1927: 482, 1927 (Pitt, 1980).

Penicillium mediocre Stapp & Bortels, Zentbl. Bakt. ParasitKde, Abt. II, 93: 50, 1935 = *Penicillium spinulosum* Thom, Bull. Bur. Anim. Ind. U.S. Dep. Agric. 118: 76, 1910 (Pitt, 1980).

Penicillium mediolanense Dragoni & Cantoni, Ind. Aliment. 155: 281, 1979 (nom. inval., Art. 36) = *Penicillium verrucosum* Dierckx, Ann. Soc. Sci. Bruxelles 25: 88, 1901 (Frisvad *et al.*, 1990b).

Penicillium megasporum Orpurt & Fennell, Mycologia 47: 233, 1955.

Penicillium melinii Thom, Penicillia: 273, 1930.

Penicillium melanochlorum (Samson *et al.*) Frisvad in Samson & Pitt (eds), Adv. *Penicillium Aspergillus* Syst.: 330, 1985 = *Penicillium solitum* Westling, Ark. Bot. 11(1): 65, 1911 (Pitt and Cruickshank, 1990; Stolk *et al.*, 1990).

Penicillium meleagrinum Biourge, Cellule 33: 147, 1923 = *Penicillium chrysogenum* Thom, Bull. Bur. Anim. Ind. U.S. Dep. Agric. 118: 58, 1910 (Pitt, 1980).

Penicillium meleagrinum var. *viridiflavum* S. Abe, J. Gen. Appl. Microbiol., Tokyo 2: 92, 1956 (nom. inval., Art. 36) = *Penicillium janthinellum* Biourge, Cellule 33: 258, 1923 (Pitt, 1980).

Penicillium meloforme Udagawa & Y. Horie, Trans. Mycol. Soc. Japan 14: 376, 1973. [Teleomorph *Eupenicillium meloforme* Udagawa & Y. Horie].

Penicillium meridianum D. B. Scott, Mycopathol. Mycol. Appl. 36: 12, 1968. [Teleomorph *Eupenicillium meridianum* D. B. Scott].

Penicillium michaelis Quintan., Mycopathologia 80: 79, 1982 = *Penicillium soppii* K. M. Zalessky, Bull. Int. Acad. Polon. Sci., Cl. Sci. Math., Sér. B, Sci. Nat., 1927: 476, 1927 (Frisvad *et al.*, 1990b).

Penicillium miczynskii K. M. Zalessky, Bull. Int. Acad. Polon. Sci., Cl. Sci. Math., Sér. B, Sci. Nat., 1927: 482, 1927.

Penicillium mimosinum A. D. Hocking in Pitt, Genus *Penicillium*: 507, 1980 ["1979"]. [Teleomorph *Talaromyces mimosinus* A. D. Hocking].

Penicillium minioluteum Dierckx, Ann. Soc. Sci. Bruxelles 25: 87, 1901.

Penicillium mirabile Beliakova & Milko, Mikol. Fitopatol. 6: 145, 1972.

Penicillium moldavicum Milko & Beliakova, Novosti Sist. Nizs. Rast. 1967: 255, 1967.

Penicillium molle Pitt, Genus *Penicillium*: 148, 1980 ["1979"]. [Teleomorph *Eupenicillium molle* Malloch & Cain].

Penicillium mononematosum (Frisvad *et al.*) Frisvad, Mycologia 81: 857, 1990. — Basionym: *Penicillium glandicola* var. *mononematosum* Frisvad *et al.*, Can. J. Bot. 65: 767, 1987.

Penicillium montanense M. Chr. & Backus, Mycologia 54: 574, 1963.

Penicillium monstrosum Sopp, Skr. Vidensk.-Selsk. Christiana, Math.-Naturvidensk. Kl. 11: 150, 1912 = *Penicillium brevicompactum* Dierckx, Ann. Soc. Sci. Bruxelles 25: 88, 1901 (Pitt, 1980).

Penicillium mucosum Stapp & Bortels, Zentbl. Bakt. ParasitKde, Abt. II, 93: 51, 1935 = *Penicillium spinulosum* Thom, Bull. Bur. Anim. Ind. U.S. Dep. Agric. 118: 76, 1910 (Pitt, 1980).

Penicillium murcianum C. Ramírez & A. T. Martínez, Mycopathologia 74: 37, 1981 = *Penicillium canescens* Sopp, Skr. Vidensk.-Selsk. Christiana, Math.-Naturvidensk. Kl. 11: 181, 1912 (Frisvad *et al.*, 1990b).

Penicillium musae Weidemann, Zentbl. Bakt. ParasitKde, Abt II 19: 687, 1907 = *Penicillium expansum* Link, Ges. Naturf. Freunde Berlin Mag. Neuesten Entdeck. Gesammten Naturk. 3: 16, 1809 (Pitt, 1980).

Penicillium nalgiovense Laxa, Zentralbl. Bakteriol., 2. Abt., 86: 160, 1932.

Penicillium namyslowskii K. M. Zalessky, Bull. Int. Aead. Polonc. Sci., Cl. Sci. Math., Sér. B, Sci. Nat., 1927: 479, 1927 = *Geosmithia namyslowskii* (K. M. Zalessky) Pitt, Can. J. Bot. 57: 2024, 1979.

Penicillium nepalense Takada & Udagawa, Trans. Mycol. Soc. Japan 24: 146, 1983. [Teleomorph *Eupenicillium nepalense* Takada & Udagawa].

Penicillium nigricans Bainier in Thom, Penicillia: 351, 1930 = *Penicillium janczewskii* K. M. Zalessky, Bull. Int. Acad. Polon. Sci., Cl. Sci. Math., Sér. B, Sci. Nat., 1927: 488, 1927 (Pitt, 1980).

36

Penicillium nigricans var. *sulphureum* S. Abe, J. Gen. Appl. Microbiol., Tokyo 2: 83, 1956 (nom. inval., Art. 36) = **Penicillium janczewskii** K. M. Zalessky, Bull. Int. Acad. Polon. Sci., Cl. Sci. Math., Sér. B, Sci. Nat., 1927: 488, 1927 (Pitt, 1980).

Penicillium niklewskii K. M. Zalessky, Bull. Int. Acad. Polon. Sci., Cl. Sci. Math., Sér. B, Sci. Nat., 1927: 504, 1927 = **Penicillium adametzii** K. M. Zalessky, Bull. Int. Acad. Polon. Sci., Cl. Sci. Math., Sér. B, Sci. Nat., 1927: 507, 1927 (Pitt, 1980).

Penicillium nilense Pitt, Genus *Penicillium*: 145, 1980 ["1979"]. [Teleomorph *Eupenicillium egyptiacum* (J. F. H. Beyma) Stolk & D. B. Scott].

Penicillium nodositanum Valla, Pl. Soil 114: 146, 1989.

Penicillium nodulum H. Z. Kong & Z. T. Qi, Mycosystema 1: 108, 1988.

Penicillium nordicum Dragoni & Cantoni ex C. Ramírez in Samson & Pitt (eds), Adv. *Penicillium Aspergillus* Syst.: 139, 1986 ≡ *Penicillium nordicum* Dragoni & Cantoni, Ind. Aliment. 155: 283, 1979 (nom. inval., Art. 36) = **Penicillium verrucosum** Dierckx, Ann. Soc. Sci. Bruxelles 25: 88, 1901 (Frisvad *et al.*, 1990b).

Penicillium notatum Westling, Ark. Bot. 11(1): 95, 1911 = **Penicillium chrysogenum** Thom, Bull. Bur. Anim. Ind. U.S. Dep. Agric. 118: 58, 1910 (Pitt, 1980).

Penicillium novae-caledoniae G. Sm., Trans. Br. Mycol. Soc. 48: 273, 1965 = **Penicillium pulvillorum** Turfitt, Trans. Br. Mycol. Soc. 23: 186, 1939 (Frisvad *et al.*, 1990b).

Penicillium novae-caledoniae var. *album* C. Ramírez & A. T. Martínez, Mycopathologia 74: 47, 1981 = **Penicillium pulvillorum** Turfitt, Trans. Br. Mycol. Soc. 23: 186, 1939 (Frisvad *et al.*, 1990b).

Penicillium novae-zeelandiae J. F. H. Beyma, Antonie van Leeuwenhoek 6: 275, 1940.

Penicillium oblatum Pitt & A. D. Hocking, Mycologia 77: 819, 1985.

Penicillium obscurum Biourge, Cellule 33: 267, 1923 = **Penicillium corylophilum** Dierckx, Ann. Soc. Sci. Bruxelles 25: 86, 1901 (Pitt, 1980).

Penicillium ochraceum Bainier in Thom, Penicillia: 309, 1930 [*non* (Corda) Biourge 1840 *nec* (Boudier) Biourge 1903 *nec* Raillo 1929] ≡ *Penicillium olivicolor* Pitt, Gen. *Penicillium*: 369, 1980 ["1979"] = **Penicillium viridicatum** Westling, Ark. Bot. 11(1): 88, 1911 (Cruickshank and Pitt, 1987).

Penicillium ochraceum var. *macrosporum* Thom, Penicillia: 310, 1930 = **Penicillium commune** Thom, Bull. Bur. Anim. Ind. U.S. Dep. Agric. 118: 56, 1910 (Cruickshank and Pitt, 1987).

Penicillium ochrochloron Biourge, Cellule 33: 269, 1923.

Penicillium ochrosalmoneum Udagawa, J. Agric. Sci. Tokyo Nogyo Daigaku 5: 10, 1959. [Teleomorph *Eupenicillium ochrosalmoneum* D. B. Scott & Stolk].

Penicillium odoratum M. Chr. & Backus, Mycologia 53: 459, 1962 = **Penicillium lividum** Westling, Ark. Bot. 11(1): 134, 1911 (Pitt, 1980).

Penicillium ohiense L. H. Huang & J. A. Schmitt, Ohio J. Sci. 75: 78, 1975. [Teleomorph *Talaromyces ohiensis* Pitt].

Penicillium oledskii K. M. Zalessky, Bull. Int. Acad. Polon. Sci., Cl. Sci. Math., Sér. B, Sci. Nat., 1927: 499, 1927 = **Penicillium glabrum** (Wehmer) Westling, Ark. Bot. 11(1): 131, 1911 (Pitt, 1980).

Penicillium oligosporum Saito & Minoura, J. Ferment. Technol., Osaka 26: 5, 1948 (nom. holomorph.; nom. inval., Art 36) = **Eupenicillium javanicum** (J. F. H. Beyma) Stolk & D. B. Scott, Persoonia 4: 398, 1967 (Pitt, 1980).

Penicillium olivaceum Sopp, Skr. Vidensk.-Selsk. Christiana, Math.-Naturvidensk. Kl. 11: 176, 1912 (*non* Corda, 1939 *nec* Wehmer, 1895) = **Penicillium digitatum** (Pers.: Fr.) Sacc., Fung. Ital.: tab. 894, 1881 (Pitt, 1980).

Penicillium olivaceum Wehmer, Beitr. Kennt. einh. Pilze 2: 73, 1895 (*non* Corda, 1939) = **Penicillium digitatum** (Pers.: Fr.) Sacc., Fung. Ital.: tab. 894, 1881 (Pitt, 1980).

Penicillium olivaceum var. *italicum* Sopp, Skr. Vidensk.-Selsk. Christiana, Math.-Naturvidensk. Kl. 11: 179, 1912 = **Penicillium digitatum** (Pers.: Fr.) Sacc., Fung. Ital.: tab. 894, 1881 (Pitt, 1980).

Penicillium olivaceum var. *norvegicum* Sopp, Skr. Vidensk.-Selsk. Christiana, Math.-Naturvidensk. Kl. 11: 177, 1912 = **Penicillium digitatum** (Pers.: Fr.) Sacc., Fung. Ital.: tab. 894, 1881 (Pitt, 1980).

Penicillium olivicolor Pitt, Gen. *Penicillium*: 369, 1980 (1979) = **Penicillium viridicatum** Westling, Ark. Bot. 11(1): 88, 1911 (Cruickshank & Pitt, 1987).

Penicillium olivinoviride Biourge, Cellule 33: 132, 1923 = **Penicillium viridicatum** Westling, Ark. Bot. 11(1): 88, 1911 (Pitt, 1980).

Penicillium olsonii Bainier & Sartory, Ann. Mycol. 10: 398, 1912.

Penicillium onobense C. Ramírez & A. T. Martínez, Mycopathologia 74: 44, 1981.

Penicillium ornatum Udagawa, Trans. Mycol. Soc. Japan 9: 49, 1968. [Teleomorph *Eupenicillium ornatum* Udagawa].

Penicillium osmophilum Stolk & Veenb.-Rijks, Antonie van Leeuwenhoek 40: 1, 1974. [Teleomorph *Eupenicillium osmophilum* Stolk & Veenb.-Rijks].

Penicillium ovetense C. Ramírez & A. T. Martínez, Mycopathologia 74: 39, 1981 = ***Penicillium phoeniceum*** J. F. H. Beyma, Zentralbl. Bakteriol., 2. Abt., 88: 136, 1933 (Frisvad *et al.*, 1990b).

Penicillium oxalicum Currie & Thom, J. Biol. Chem. 22: 289, 1915.

Penicillium paczoskii K. M. Zalessky, Bull. Int. Acad. Polon. Sci., Cl. Sci. Math., Sér. B, Sci. Nat., 1927: 505, 1927 = ***Penicillium spinulosum*** Thom, Bull. Bur. Anim. Ind. U.S. Dep. Agric. 118: 76, 1910 (Pitt, 1980).

Penicillium paecilomyceforme Szilvinyi, Zentbl. Bakt. ParasitKde, Abt. II, 103: 156, 1941 = ***Penicillium camemberti*** Thom, Bull. Bur. Anim. Ind. U.S. Dep. Agric. 82: 33, 1906 (Raper and Thom, 1949).

Penicillium palitans Westling, Ark Bot. 11(1): 83, 1911.

Penicillium palitans var. *echinoconidium* S. Abe, J. Gen. Appl. Microbiol., Tokyo 2: 111, 1956 (nom. inval., Art. 36) = ***Penicillium echinulatum*** Raper & Thom ex Fassat., Acta Univ. Carol., Biol. 1974: 326, 1977 (Pitt, 1980).

Penicillium pallidum G. Sm., Trans. Br. Mycol. Soc. 18: 88, 1933 = ***Geosmithia putterillii*** (Thom) Pitt, Can. J. Bot. 57: 2022, 1979 (Pitt, 1979).

Penicillium palmae Samson *et al.*, Stud. Mycol. 31: 135, 1989.

Penicillium palmense C. Ramírez *et al.*, Mycopathologia 66: 80, 1978.

Penicillium panamense Samson *et al.*, Stud. Mycol. 31: 136, 1989.

Penicillium panasenkoi Pitt, Genus *Penicillium*: 482, 1980 ["1979"]. [Teleomorph *Talaromyces panasenkoi* Pitt].

Penicillium papuanum Udagawa & Y. Horie, Trans. Mycol. Soc. Japan 14: 378, 1973. [Teleomorph *Eupenicillium parvum* (Raper & Fennell) Stolk & D. B. Scott].

Penicillium paraherquei S. Abe, J. Gen. Appl. Microbiol., Tokyo 2: 131, 1956 (nom. inval., Art. 36) ≡ ***Penicillium paraherquei*** S. Abe ex G. Sm., Trans. Br. Mycol. Soc. 46: 335, 1963.

Penicillium paraherquei S. Abe ex G. Sm., Trans. Br. Mycol. Soc. 46: 335, 1963.

Penicillium parallelosporum Y. Sasaki, J. Fac. Agric. Hokkaido Univ. 49: 147, 1950 (nom. inval., Art. 36) = ***Penicillium thomii*** Maire, Bull. Soc. Hist. Nat. Afrique N. 8: 189, 1917 (Pitt, 1980).

Penicillium parvum Raper & Fennell, Mycologia 40: 508, 1948 (nom. holomorph.) ≡ *Eupenicillium parvum* (Raper & Fennell) Stolk & D. B. Scott, Persoonia 4: 402, 1967.

Penicillium pascuum (Pitt & A. D. Hocking) Frisvad *et al.*, Persoonia 14: 229, 1990 = ***Paecilomyces pascuus*** Pitt & A.D. Hocking, Mycologia 77: 822, 1985 (accepted here). The protologue for this spe-

cies clearly indicates that phialide necks are often bent away from the phialide axis, the characteristic which separates *Paecilomyces* from *Penicillium* (Pitt and Hocking, 1985). To include this species in *Penicillium* would require an emended genus description.

Penicillium patens Pitt & A. D. Hocking, Mycotaxon 22: 205, 1985.

Penicillium patris-mei K. M. Zalessky, Bull. Int. Acad. Polon. Sci., Cl. Sci. Math., Sér. B, Sci. Nat., 1927: 496, 1927 = ***Penicillium brevicompactum*** Dierckx, Ann. Soc. Sci. Bruxelles 25: 88, 1901 (Pitt, 1980).

Penicillium patulum Bainier, Bull. Trimest. Soc. Mycol. Fr. 22: 208, 1906 = ***Penicillium griseofulvum*** Dierckx, Ann. Soc. Sci. Bruxelles 25: 88, 1901 (Pitt, 1980).

Penicillium paxilli Bainier, Bull. Soc. Mycol. France 23: 95, 1907.

Penicillium pedemontanum Mosca & A. Fontana, Allionia 9: 40, 1963.

Penicillium pfefferianum (Wehmer) Westling, Ark. Bot. 11(1): 132, 1911 ≡ *Penicillium pfefferianum* (Wehmer) Pollaci, Atti Ist. Bot. Univ. Pavia, Ser. 2, 16: 135, 1916 ≡ *Penicillium pfefferianum* (Wehmer) Biourge, Cellule 23: 105, 1923 = ***Penicillium glabrum*** (Wehmer) Westling, Ark. Bot. 11(1): 131, 1911 (Pitt, 1980).

Penicillium phaeo-janthinellum Biourge, Cellule 33: 289, 1923 = ***Penicillium fellutanum*** Biourge, Cellule 33: 262, 1923 (Pitt, 1980).

Penicillium phialosporum Udagawa, J. Agric. Sci., Tokyo 5: 11, 1959 = ***Penicillium rugulosum*** Thom, Bull. Bur. Anim. Ind. U.S. Dep. Agric. 118: 60, 1910 (Pitt, 1980).

Penicillium philippinense Udagawa & Y. Horie, J. Jap. Bot. 47: 341, 1972 = ***Penicillium alutaceum*** D. B. Scott, Mycopathol. Mycol. Appl. 36: 17, 1968 (Pitt, 1980).

Penicillium phoeniceum J. F. H. Beyma, Zentralbl. Bakteriol., 2. Abt., 88: 136, 1933. [Teleomorph *Eupenicillium cinnamopurpureum* D. B. Scott & Stolk].

Penicillium piceum Raper & Fennell, Mycologia 40: 533, 1948.

Penicillium pinetorum M. Chr. & Backus, Mycologia 53: 457, 1962. [Teleomorph *Eupenicillium pinetorum* Stolk].

Penicillium pinophilum Hedgc. In Thom, Bull. Bur. Anim. Ind. U.S. Dep. Agric. 118: 37, 1910.

Penicillium piscarium Westling, Ark. Bot. 11(1): 86, 1911.

Penicillium pittii Quintan., Mycopathologia 91: 75, 1985.

Penicillium plumiferum Demelius, Verh. Zool.-Bot. Ges. Wein 72: 76, 1923 ["1922"] = ***Penicillium expansum*** Link, Ges. Naturf. Freunde Berlin

Mag. Neuesten Entdeck. Gesammten Naturk. 3: 16, 1809 (Thom, 1930).

Penicillium polonicum K. M. Zalessky, Bull. Int. Acad. Polon. Sci., Cl. Sci. Math., Sér. B, Sci. Nat., 1927: 445, 1927 = **Penicillium aurantiogriseum** Dierckx, Ann. Soc. Sci. Bruxelles 25: 88, 1901 (Raper and Thom, 1949).

Penicillium populi J. F. H. Beyma, Zentralbl. Bakteriol. ParasitKde, Abt. II, 96: 421, 1937 = **Penicillium simplicissimum** (Oudem.) Thom, Penicillia: 335, 1930 (Pitt, 1980).

Penicillium porraceum Biourge, Cellule 33: 188, 1923 = **Penicillium aurantiogriseum** Dierckx, Ann. Soc. Sci. Bruxelles 25: 88, 1901 (Cruickshank and Pitt, 1987).

Penicillium primulinum Pitt, Genus *Penicillium*: 455, 1980 ["1979"].

Penicillium proteolyticum Kamyschko, Bot. Mater. Otd. Sporov. Rast. 14: 228, 1961.

Penicillium pseudocasei S. Abe, J. Gen. Appl. Microbiol., Tokyo 2: 102, 1956 (nom. inval., Art. 36) ≡ *Penicillium pseudocasei* S. Abe ex G. Sm., Trans. Br. Mycol. Soc. 46: 335, 1963 = **Penicillium crustosum** Thom, Penicillia: 399, 1930 (Pitt, 1980).

Penicillium pseudostromaticum Hodges *et al.*, Mycologia 62: 1106, 1971.

Penicillium psittacinum Thom, Penicillia: 369, 1930 = **Penicillium viridicatum** Westling, Ark. Bot. 11(1): 88, 1911 (Pitt, 1980).

Penicillium puberulum Bainier, Bull. Trimest. Soc. Mycol. Fr. 23: 16, 1907 = **Penicillium aurantiogriseum** Dierckx, Ann. Soc. Sci. Bruxelles 25: 88, 1901 (Pitt *et al.*, 1986; Cruickshank and Pitt, 1987).

Penicillium pulvillorum Turfitt, Trans. Br. Mycol. Soc. 23: 186, 1939.

Penicillium purpurascens (Sopp) Biourge, Cellule 33: 105, 1923. — Basionym: *Citromyces purpurascens* Sopp, Skr. Vidensk.-Selsk. Christiana, Math.-Naturvidensk. Kl. 11: 117, 1912.

Penicillium purpureum Stolk & Samson, Stud. Mycol. 2: 57, 1972. [Teleomorph *Talaromyces purpureus* (E. Müll. & Pacha-Aue) Stolk & Samson].

Penicillium purpurogenum Stoll, Beitr. Charakt. Penicill.: 32, 1904.

Penicillium purpurogenum var. *rubrisclerotium* Thom, Mycologia 7: 142, 1915 = **Penicillium pinophilum** Hedgc. in Thom, Bull. Bur. Anim. Ind. U.S. Dep. Agric. 118: 37, 1910 (Pitt, 1980).

Penicillium pusillum G. Sm., Trans. Br. Mycol. Soc. 22: 254, 1939 = **Penicillium phoeniceum** J. F. H. Beyma, Zentralbl. Bakteriol., 2. Abt., 88: 136, 1933 (Pitt, 1980).

Penicillium putterillii Thom, Penicillia: 368, 1930 ≡ **Geosmithia putterillii** (Thom) Pitt, Can. J. Bot. 57: 2022, 1979.

Penicillium quercetorum Baghdadi, Nov. Sist. Niz. Rast. 5:, 110, 1968 = **Penicillium thomii** Maire, Bull. Soc. Hist. Nat. Afrique N. 8: 189, 1917 (Frisvad *et al.*, 1990b).

Penicillium raciborskii K. M. Zalessky, Bull. Int. Acad. Polon. Sci., Cl. Sci. Math., Sér. B, Sci. Nat., 1927: 454, 1927.

Penicillium rademiricii Quintan., Mycopathologia 91: 72, 1985.

Penicillium radiatolobatum Lörinczi, Publ. Soc. Nat. Rom. Pent. Stiinta Sol. 10B: 435, 1972 = **Penicillium canescens** Sopp, Skr. Vidensk.-Selsk. Christiana, Math.-Naturvidensk. Kl. 11: 181, 1912 (Frisvad *et al.*, 1990b).

Penicillium radulatum G. Sm., Trans. Br. Mycol. Soc. 40: 484, 1957 = **Penicillium melinii** Thom, Penicillia: 273, 1930 (Pitt, 1980).

Penicillium ramusculum Bat. & H. Maia, Anais Soc. Biol. Pernamb. 13: 27, 1955 = **Penicillium sublateritium** Biourge, Cellule 33: 315, 1923 (Pitt, 1980).

Penicillium raperi G. Sm., Trans. Br. Mycol. Soc. 40: 486, 1957.

Penicillium raistrickii G. Sm., Trans. Br. Mycol. Soc.18: 90, 1933.

Penicillium rasile Pitt, Genus *Penicillium*: 120, 1980 ["1979"]. [Teleomorph *Eupenicillium levitum* (Raper & Fennell) Stolk & D. B. Scott].

Penicillium resedanum McLennan & Ducker, Aust. J. Bot. 2: 360, 1954.

Penicillium resinae Z. T. Qi & H. Z. Kong, Acta Mycol. Sin. 1: 103, 1982 = **Penicillium asperosporum** G. Sm., Trans. Br. Mycol. Soc. 48: 275, 1965 (Frisvad *et al.*, 1990b).

Penicillium resticulosum Birkinshaw *et al.*, Biochem. J. 36: 830, 1942 = **Penicillium expansum** Link, Ges. Naturf. Freunde Berlin Mag. Neuesten Entdeck. Gesammten Naturk. 3: 16, 1809 (Cruickshank and Pitt, 1987).

Penicillium restrictum J. C. Gilman & E. V. Abbott, Iowa State Coll. J. Sci. 1: 297, 1927.

Penicillium reticulisporum Udagawa, Trans. Mycol. Soc. Japan 9: 52, 1968. [Teleomorph *Eupenicillium reticulisporum* Udagawa].

Penicillium rivolii K. M. Zalessky, Bull. Int. Acad. Polon. Sci., Cl. Sci. Math., Sér. B, Sci. Nat., 1927: 471, 1927 = **Penicillium jensenii** K. M. Zalessky, Bull. Int. Acad. Polon. Sci., Cl. Sci. Math., Sér. B, Sci. Nat., 1927: 494, 1927 (Pitt, 1980).

Penicillium rogeri Wehmer in Lafar, Hendb. Tech. Mykol. 4: 226, 1906 ≡ *Penicillium camemberti* var. *rogeri* (Wehmer) Thom, Bull. Bur. Anim. Ind. U.S. Dep. Agric. 118: 52, 1910 = **Penicillium** *camemberti* Thom, Bull. Bur. Anim. Ind. U.S. Dep. Agric. 82: 33, 1906 (Pitt, 1980).

Penicillium rolfsii Thom, Penicillia: 489, 1930.

Penicillium rolfsii var. *sclerotiale* Novobranova, Nov. Sist. Niz. Rast. 11: 230, 1974 = ***Penicillium gladioli*** L. McCulloch & Thom, Science, ser. 2, 67: 217, 1928 (Pitt, 1980).

Penicillium roqueforti Sopp, Skr. Vidensk.-Selsk. Christiana, Math.-Naturvidensk. Kl. 11: 156, 1912 (*non* Thom 1906) = ***Penicillium roqueforti*** Thom, Bull. Bur. Anim. Ind. U.S. Dep. Agric. 82: 35, 1906 (Pitt, 1980).

Penicillium roqueforti Thom, Bull. Bur. Anim. Ind. U.S. Dep. Agric. 82: 35, 1906.

Penicillium roqueforti var. *punctatum* S. Abe, J. Gen. Appl. Microbiol., Tokyo 2: 99, 1956 (nom. inval., Art. 36) = ***Penicillium verrucosum*** Dierckx, Ann. Soc. Sci. Bruxelles 25: 88, 1901 (Pitt, 1980).

Penicillium roqueforti var. *viride* Dattilo-Rubbo, Trans. Br. Mycol. Soc. 22: 178, 1938 = ***Penicillium roqueforti*** Thom, Bull. Bur. Anim. Ind. U.S. Dep. Agric. 82: 35, 1906 (Pitt, 1980).

Penicillium roqueforti var. *weidemannii* Westling, Ark. Bot. 11(1): 71, 1911 = ***Penicillium roqueforti*** Thom, Bull. Bur. Anim. Ind. U.S. Dep. Agric. 82: 35, 1906 (Pitt, 1980).

Penicillium roseocitreum Biourge, Cellule 33: 184, 1923 = ***Penicillium chrysogenum*** Thom, Bull. Bur. Anim. Ind. U.S. Dep. Agric. 118: 58, 1910 (Stolk *et al.*, 1990).

Penicillium roseololilacinum Novobranova, Novosti Sist. Nizs. Rast. 11: 226, 1974 = ***Penicillium phoeniceum*** J. F. H. Beyma, Zentralbl. Bakteriol., 2. Abt., 88: 136, 1933 (Pitt, 1980).

Penicillium roseomaculatum Biourge, Cellule 33: 301, 1923 = ***Penicillium spinulosum*** Thom, Bull. Bur. Anim. Ind. U.S. Dep. Agric. 118: 76, 1910 (Pitt, 1980).

Penicillium roseopurpureum Dierckx, Ann. Soc. Sci. Bruxelles 25: 86, 1901.

Penicillium roseoviride Stapp & Bortels, Zentbl. Bakt. ParasitKde, Abt. II, 93: 51, 1935 = ***Penicillium thomii*** Maire, Bull. Soc. Hist. Nat. Afrique N. 8: 189, 1917 (Pitt, 1980).

Penicillium rotundum Raper & Fennell, Mycologia 40: 518, 1948 (nom. holomorph.) ≡ ***Talaromyces rotundus*** (Raper & Fennell) C. R. Benj., Mycologia 47: 683, 1955.

Penicillium rubefaciens Quintan., Mycopathologia 80: 73, 1982.

Penicillium rubens Biourge, Cellule 33: 265, 1923 = ***Penicillium chrysogenum*** Thom, Bull. Bur. Anim. Ind. U.S. Dep. Agric. 118: 58, 1910 (Pitt, 1980).

Penicillium rubicundum J. H. Mill. *et al.*, Mycologia 49: 797, 1957 = ***Penicillium funiculosum*** Thom, Bull. Bur. Anim. Ind. U.S. Dep. Agric. 118: 69, 1910 (Pitt, 1980).

Penicillium rubidurum Udagawa & Y. Horie, Trans. Mycol. Soc. Japan 14: 381, 1973. [Teleomorph *Eupenicillium rubidurum* Udagawa & Y. Horie].

Penicillium rubrum Stoll, Beitr. Charakt. Penicill.: 35, 1904 = ***Penicillium purpurogenum*** Stoll, Beitr. Charakt. Penicill.: 32, 1904 (Pitt, 1980).

Penicillium rugulosum Thom, Bull. Bur. Anim. Ind. U.S. Dep. Agric. 118: 60, 1910.

Penicillium rugulosum var. *atricolum* Thom, Penicillia: 474, 1930 = ***Penicillium rugulosum*** Thom, Bull. Bur. Anim. Ind. U.S. Dep. Agric. 118: 60, 1910 (Pitt, 1980).

Penicillium sabulosum Pitt & A. D. Hocking, Mycologia 77: 818, 1985.

Penicillium sacculum E. Dale, Ann. Mycol. 24: 137, 1926.

Penicillium sajarovii Quintan., Avances Nutr. Mejora Anim. Aliment. 22: 539, 1981.

Penicillium samsonii Quintanilla, Mycopathologia 91: 69, 1985 = ***Penicillium minioluteum*** Dierckx, Ann. Soc. Sci. Bruxelles 25: 87, 1901 (van Reenan-Hoekstra *et al.*, 1990).

Penicillium sanguifluum (Sopp) Biourge, Cellule 33: 105, 1923 = ***Penicillium roseopurpureum*** Dierckx, Ann. Soc. Sci. Bruxelles 25: 86, 1901 (Pitt, 1980).

Penicillium sanguineum Sopp, Skr. Vidensk.-Selsk. Christiana, Math.-Naturvidensk. Kl. 11: 175, 1912 = ***Penicillium purpurogenum*** Stoll, Beitr. Charakt. Penicill.: 32, 1904 (Biourge, 1923).

Penicillium sartoryi Thom, Penicillia: 233, 1930 = ***Penicillium citrinum*** Thom, Bull. Bur. Anim. Ind. U.S. Dep. Agric. 118: 61, 1910 (Pitt, 1980).

Penicillium scabrosum Frisvad *et al.*, Persoonia 14: 177, 1990.

Penicillium schmidtii Szilvinyi, Zentbl. Bakt. ParasitKde, Abt. II, 103: 148, 1941 = ***Penicillium crustosum*** Thom, Penicillia: 399, 1930 (Pitt, 1980).

Penicillium schneggii Boas, Mykol. Zentbl. 5: 73, 1914 = ***Penicillium glandicola*** (Oudem.) Seifert & Samson in Samson & Pitt (eds), Adv. *Penicillium Aspergillus* Syst.: 147, 1985 (Seifert and Samson, 1985).

Penicillium sclerotigenum W. Yamam., Sci. Rep. Hyogo Univ. Agric., Ser. Agric. Biol., ser. 2, 1: 69, 1955.

Penicillium sclerotiorum J. F. H. Beyma, Zentralbl. Bakteriol., 2. Abt., 96: 418, 1937.

Penicillium scorteum Takedo *et al.*, J. Agric. Chem. Soc., Japan 10: 103, 1934 = ***Penicillium rugulosum*** Thom, Bull. Bur. Anim. Ind. U.S. Dep. Agric. 118: 60, 1910 (Pitt, 1980).

Penicillium senticosum D. B. Scott, Mycopathol. Mycol. Appl. 36: 5, 1968. [Teleomorph *Eupenicillium senticosum* D. B. Scott].

Penicillium severskii Schekh., Microbiologia 43: 122, 1981 = **Penicillium soppii** K. M. Zalessky, Bull. Int. Acad. Polon. Sci., Cl. Sci. Math., Sér. B, Sci. Nat.,1927: 476, 1927 (Frisvad *et al.*, 1990b).

Penicillium shearii Stolk & D. B. Scott, Persoonia 4: 396, 1967. [Teleomorph *Eupenicillium shearii* Stolk & D. B. Scott].

Penicillium shennangjianum H. Z. Kong & Z. T. Qi, Mycosystema 1: 110, 1988.

Penicillium siamense Manoch & C. Ramírez, Mycopathologia 101: 32, 1988.

Penicillium siemaszkii K. M. Zalessky, Bull. Int. Acad. Polon. Sci., Cl. Sci. Math., Sér. B, Sci. Nat., 1927: 487, 1927 = **Penicillium jensenii** K. M. Zalessky, Bull. Int. Acad. Polon. Sci., Cl. Sci. Math., Sér. B, Sci. Nat., 1927: 494, 1927 (Thom, 1930).

Penicillium silvaticum Suprun, Byull. Mosk. Obshch. Ispyrt. Prir. 61: 90, 1956 (*non* Oudem. 1902) = **Penicillium pinetorum** M. Chr. & Backus, Mycologia 53: 457, 1962 (Pitt, 1980).

Penicillium silvaticum (Wehmer) Gäumann, Vergl. Morph. Pilze: 177, 1926 (*non* Oudem. 1902) ≡ *Penicillium silvaticum* (Wehmer) Biourge, Cellule 33: 105, 1923 (*non* Oudem. 1902) ≡ *Coremium silvaticum* Wehmer, Ber. Dt. Bot. Ges. 31: 373, 1914 = **Penicillium vulpinum** (Cooke & Massee) Seifert & Samson in Samson & Pitt (eds), Adv. *Penicillium Aspergillus* Syst.: 144, 1985 (Seifert and Samson, 1985).

Penicillium simplicissimum (Oudem.) Thom, Penicillia: 335., 1930. — Basionym: *Spicaria simplicissima* Oudem., Ned. Kruidk. Arch., ser. 3, 2: 763, 1902.

Penicillium sinaicum Udagawa & S. Ueda, Mycotaxon 14: 266, 1982. [Teleomorph *Eupenicillium sinaicum* Udagawa & S. Ueda].

Penicillium sizovae Baghd., Novosti Sist. Nizs. Rast. 1968: 103, 1968.

Penicillium skrjabinii Schmotina & Golovleva, Mikol. Fitopatol. 8: 530, 1974.

Penicillium smithii Quintan., Avances Nutr. Mejora Anim. Aliment. 23: 340, 1982.

Penicillium solitum Westling, Ark. Bot. 11(1): 65, 1911.

Penicillium solitum var. *crustosum* (Thom) Bridge *et al.*, J. Gen. Microbiol. 135: 2957, 1989 ≡ **Penicillium crustosum** Thom, Penicillia: 399, 1930 (Frisvad *et al.*, 1990b).

Penicillium soppii K. M. Zalessky, Bull. Int. Acad. Polon. Sci., Cl. Sci. Math., Sér. B, Sci. Nat., 1927: 476, 1927.

Penicillium sphaerum Pitt, Genus *Penicillium*: 494, 1980 ["1979"]. [Teleomorph *Talaromyces rotundus* (Raper & Fennell) C. R. Benj.].

Penicillium spiculisporum Lehman, Mycologia 12: 271, 1920 (nom. holomorph.) = **Talaromyces trachyspermus** (Shear) Stolk & Samson, Stud. Mycol. 2: 32, 1972 (Stolk and Samson, 1972).

Penicillium spinuloramigenum Y. Sasaki, J. Fac. Agric. Hokkaido Univ. 49: 153, 1950 (nom. inval., Art. 36) = **Penicillium spinulosum** Thom, Bull. Bur. Anim. Ind. U.S. Dep. Agric. 118: 76, 1910 (Pitt, 1980).

Penillium spinulosum Thom, Bull. Bur. Anim. Ind. U.S. Dep. Agric. 118: 76, 1910.

Penicillium spirillum Pitt, Genus *Penicillium*: 476, 1980 ["1979"]. [Teleomorph *Talaromyces helicus* (Raper & Fennell) C. R. Benj.].

Penicillium steckii K. M. Zalessky, Bull. Int. Acad. Polon. Sci., Cl. Sci. Math., Sér. B, Sci. Nat., 1927: 469, 1927.

Penicillium stephaniae K. M. Zalessky, Bull. Int. Acad. Polon. Sci., Cl. Sci. Math., Sér. B, Sci. Nat., 1927: 451, 1927 = **Penicillium viridicatum** Westling, Ark. Bot. 11(1): 88, 1911 (Raper and Thom, 1949).

Penicillium stilton Biourge, Cellule 33: 204, 1923 = **Penicillium roqueforti** Thom, Bull. Bur. Anim. Ind. U.S. Dep. Agric. 82: 35, 1906 (Raper and Thom, 1949).

Penicillium stipitatum Thom, Mycologia 27: 138, 1935 (nom. holomorph.) ≡ **Talaromyces stipitatus** (Thom) C. R. Benj., Mycologia 47: 684, 1955.

Penicillium stolkiae D. B. Scott, Mycopathol. Mycol. Appl. 36: 8, 1968. [Teleomorph *Eupenicillium stolkiae* D. B. Scott].

Penicillium stoloniferum Thom, Bull. Bur. Anim. Ind. U.S. Dep. Agric. 118: 68, 1910 = **Penicillium brevicompactum** Dierckx, Ann. Soc. Sci. Bruxelles 25: 88, 1901 (Pitt, 1980).

Penicillium striatisporum Stolk, Antonie van Leeuwenhoek 35: 268, 1969.

Penicillium striatum Raper & Fennell, Mycologia 40: 521, 1948 (nom. holomorph.) ≡ **Talaromyces striatus** (Raper & Fennell) C. R. Benj., Mycologia 47: 682, 1955.

Penicillium suaveolens Biourge, Cellule 33: 200, 1923 = **Penicillium roqueforti** Thom, Bull. Bur. Anim. Ind. U.S. Dep. Agric. 82: 35, 1906 (Raper and Thom, 1949).

Penicillium subcinearum Westling, Ark. Bot. 11(1): 137, 1911 = **Penicillium citreonigrum** Dierckx, Ann. Soc. Sci. Bruxelles 25: 86, 1901 (Biourge, 1923).

Penicillium sublateritium Biourge, Cellule 33: 315, 1923.

Penicillium subtile (Bainier & Sartory) Biourge, Cellule 33: 106, 1923 (*non* Berk. 1841) = **Penicillium citrinum** Thom, Bull. Bur. Anim. Ind. U.S. Dep. Agric. 118: 61, 1910 (Pitt, 1980).

Penicillium sumatrense Szilvinyi, Archiv. Hydrobiol. 14, Suppl. 6: 535, 1936 = ***Penicillium corylophilum*** Dierckx, Ann. Soc. Sci. Bruxelles 25: 86, 1901 (Pitt, 1980).

Penicillium swiecickii K. M. Zalessky, Bull. Int. Acad. Polon. Sci., Cl. Sci. Math., Sér. B, Sci. Nat., 1927: 474, 1927 = ***Penicillium janczewskii*** K. M. Zalessky, Bull. Int. Acad. Polon. Sci., Cl. Sci. Math., Sér. B, Sci. Nat., 1927: 488, 1927 (Pitt, 1980).

Penicillium syriacum Baghd., Novosti Sist. Nizs. Rast. 1968: 111, 1968.

Penicillium szaferi K. M. Zalessky, Bull. Int. Acad. Polon. Sci., Cl. Sci. Math., Sér. B, Sci. Nat., 1927: 447, 1927 = ***Penicillium brevicompactum*** Dierckx, Ann. Soc. Sci. Bruxelles 25: 88, 1901 (Thom, 1930).

Penicillium tabescens Westling, Ark. Bot. 11(1): 100, 1911 = ***Penicillium brevicompactum*** Dierckx, Ann. Soc. Sci. Bruxelles 25: 88, 1901 (Thom, 1930).

Penicillium tannophagum Stapp & Bortels, Zentbl. Bakt. ParasitKde, Abt. II, 93: 52, 1935 = ***Penicillium spinulosum*** Thom, Bull. Bur. Anim. Ind. U.S. Dep. Agric. 118: 76, 1910 (Pitt, 1980).

Penicillium tannophilum Stapp & Bortels, Zentbl. Bakt. ParasitKde, Abt. II, 93: 52, 1935 = ***Penicillium spinulosum*** Thom, Bull. Bur. Anim. Ind. U.S. Dep. Agric. 118: 76, 1910 (Pitt, 1980).

Penicillium tardum Thom, Penicillia: 485, 1930.

Penicillium terlikowskii K. M. Zalessky, Bull. Int. Acad. Polon. Sci., Cl. Sci. Math., Sér. B, Sci. Nat., 1927: 501, 1927.

Penicillium terraconense C. Ramírez & A. T. Martínez, Mycopathologia 72: 188, 1980 = ***Penicillium digitatum*** (Pers.: Fr.) Sacc., Fung. Ital.: tab. 894, 1881 (Stolk *et al.*, 1990).

Penicillium terrenum D. B. Scott, Mycopathol. Mycol. Appl. 36: 1, 1968. [Teleomorph *Eupenicillium terrenum* D. B. Scott].

Penicillium thomii Maire, Bull. Soc. Hist. Nat. Afrique N. 8: 189, 1917.

Penicillium thomii var. *flavescens* S. Abe, J. Gen. Appl. Microbiol., Tokyo 2: 50, 1956 (nom. inval., Art. 36) = ***Penicillium thomii*** Maire, Bull. Soc. Hist. Nat. Afrique N. 8: 189, 1917 (Pitt, 1980).

Penicillium toxicarium L. Miyake in Miyake *et al.*, Rep. Res. Inst. Rice Improvement 1: 1, 1940 (nom. inval., Art. 36) = ***Penicillium citreonigrum*** Dierckx, Ann. Soc. Sci. Bruxelles 25: 86, 1901 (Pitt, 1980).

Penicillium trzebinskianum S. Abe, J. Gen. Appl. Microbiol., Tokyo 2: 63, 1956 (nom. inval., Art. 36) = ***Penicillium lividum*** Westling, Ark. Bot. 11(1): 134, 1911 (Pitt, 1980).

Penicillium trzebinskii K. M. Zalessky, Bull. Int. Acad. Polon. Sci., Cl. Sci. Math., Sér. B, Sci. Nat., 1927: 498, 1927 = ***Penicillium spinulosum*** Thom, Bull. Bur. Anim. Ind. U.S. Dep. Agric. 118: 76, 1910 (Pitt, 1980).

Penicillium trzebinskii var. *magnum* Sakag. & S. Abe in Abe, J. Gen. Appl. Microbiol., Tokyo 2: 62, 1956 (nom. inval., Art. 36) = ***Penicillium spinulosum*** Thom, Bull. Bur. Anim. Ind. U.S. Dep. Agric. 118: 76, 1910 (Pitt, 1980).

Penicillium tularense Paden, Mycopathol. Mycol. Appl. 43: 264, 1971. [Teleomorph *Eupenicillium tularense* Paden].

Penicillium turbatum Westling, Ark. Bot. 11(1): 128, 1911.

Penicillium turolense C. Ramírez & A. T. Martínez, Mycopathologia 74: 36, 1981 = ***Penicillium westlingii*** K. M. Zalessky, Bull. Int. Acad. Polon. Sci., Cl. Sci. Math., Sér. B, Sci. Nat., 1927: 473, 1927 (Frisvad *et al.*, 1990b).

Penicillium turris-painense C. Ramírez, Mycopathologia 91: 93, 1985 = ***Geosmithia namyslowski*** (K. M. Zalessky) Pitt, Can. J. Bot. 57: 2024, 1979 (Frisvad *et al.*, 1990b).

Penicillium ucrainicum Panas., Mycologia 56: 59, 1964 (nom. holomorph.) ≡ ***Talaromyces panasenkoi*** Pitt, Genus *Penicillium*: 482, 1980 ["1979"](new name).

Penicillium udagawae Stolk & Samson, Stud. Mycol. 2: 36, 1972. [Teleomorph *Talaromyces udagawae* Stolk & Samson].

Penicillium ulaiense H. M. Hsieh *et al.*, Trans. Mycol. Soc. Republ. China 2: 161, 1987.

Penicillium unicum Tzean *et al.*, Mycologia 84: 739, 1992. [Teleomorph *Talaromyces unicus* Tzean *et al.*].

Penicillium urticae Bainier, Bull. Trimest. Soc. Mycol. Fr. 23: 15, 1907 = ***Penicillium griseofulvum*** Dierckx, Ann. Soc. Sci. Bruxelles 25: 88, 1901 (Pitt, 1980).

Penicillium vaccaeorum Quintan., Mycopathologia 80: 77, 1982 = ***Penicillium roseopurpureum*** Dierckx, Ann. Soc. Sci. Bruxelles 25: 86, 1901 (Frisvad *et al.*, 1990b).

Penicillium valentinum C. Ramírez & A. T. Martínez, Mycopathologia 72: 183, 1980 = ***Penicillium thomii*** Maire, Bull. Soc. Hist. Nat. Afrique N. 8: 189, 1917 (Frisvad *et al.*, 1990b).

Penicillium vanbeymae Pitt, Genus *Penicillium*: 142, 1980 ["1979"]. [Teleomorph *Eupenicillium baarnense* (J. F. H. Beyma) Stolk & D. B. Scott].

Penicillium vanilliae Bouriquet, Bull. Acad. Malgache 24: 68, 1941 (nom. inval., Art. 36) = ***Penicillium purpurogenum*** Stoll, Beitr. Charakt. Penicill.: 32, 1904 (Pitt, 1980).

Penicillium variabile Sopp, Skr. Vidensk.-Selsk. Christiana, Math.-Naturvidensk. Kl. 11: 169, 1912.

Penicillium variabile Wehmer, Mykol. Zentbl. 2: 195, 1913 (*non* Sopp 1912) = *Penicillium expansum* Link, Ges. Naturf. Freunde Berlin Mag. Neuesten Entdeck. Gesammten Naturk. 3: 16, 1809 (Biourge, 1923).

Penicillium varians G. Sm., Trans. Br. Mycol. Soc.18: 89, 1933.

Penicillium vasconiae C. Ramírez & A. T. Martínez, Mycopathologia 72: 189, 1980.

Penicillium velutinum J. F. H. Beyma, Zentralbl. Bakteriol., 2. Abt., 91: 353, 1935.

Penicillium ventruosum Westling, Ark. Bot. 11(1): 112, 1911 = *Penicillium italicum* Wehmer, Hedwigia 33: 211, 1894 (Thom, 1930).

Penicillium vermiculatum P. A. Dang., Botaniste 10: 123, 1907 (nom. holomorph.) = *Talaromyces flavus* (Klöcker) Stolk & Samson, Stud. Mycol. 2: 10, 1972 (Stolk and Samson, 1972).

Penicillium verrucosum Dierckx, Ann. Soc. Sci. Bruxelles 25: 88, 1901.

Penicillium verrucosum var. *corymbiferum* (Westling) Samson *et al.*, Stud. Mycol. 11: 36, 1976 ≡ *Penicillium corymbiferum* Westling, Ark. Bot. 11(1): 92, 1911 = *Penicillium hirsutum* Dierckx, Ann. Soc. Sci. Bruxelles 25: 89, 1901 (Pitt, 1980).

Penicillium verrucosum var. *cyclopium* (Westling) Samson *et al.*, Stud. Mycol. 11: 37, 1976 ≡ *Penicillium cyclopium* Westling, Ark. Bot. 11(1): 90, 1911 = *Penicillium aurantiogriseum* Dierckx, Ann. Soc. Sci. Bruxelles 25: 88, 1901 (Pitt, 1980).

Penicillium verrucosum var. *melanochlorum* Samson *et al.*, Stud. Mycol. 11: 41, 1976 ≡ *Penicillium melanochlorum* (Samson *et al.*) Frisvad in Samson & Pitt (eds), Adv. *Penicillium Aspergillus* Syst.: 330, 1985 = *Penicillium solitum* Westling, Ark. Bot. 11(1): 65, 1911 (Pitt and Cruickshank, 1990; Stolk *et al.*, 1990).

Penicillium verrucosum var. *ochraceum* (Bainier) Samson *et al.*, Stud. Mycol. 11: 42, 1976 ≡ *Penicillium ochraceum* Bainier in Thom, Penicillia: 309, 1930 [*non* (Corda) Biourge 1840 *nec* (Boudier) Biourge 1903 *nec* Raillo 1929] = *Penicillium aurantiogriseum* Dierckx, Ann. Soc. Sci. Bruxelles 25: 88, 1901 (Stolk *et al.*, 1990).

Penicillium verruculosum Peyronel, Germi Atmosf. Fung. Micel.: 22, 1913.

Penicillium vesiculosum Bainier, Bull. Trimest. Soc. Mycol. Fr. 23: 10, 1907 = *Penicillium roqueforti* Thom, Bull. Bur. Anim. Ind. U.S. Dep. Agric. 82: 35, 1906 (Thom, 1930).

Penicillium victoriae Szilvinyi, Archiv. Hydrobiol. 14, Suppl. 6: 535, 1936 ≡ *Paecilomyces victoriae* (Szilvinyi) A.H.S. Brown & G. Sm., Trans. Br. Mycol. Soc. 40: 60, 1957 = *Penicillium janthinellum* Biourge, Cellule 33: 258, 1923 (Pitt, 1980).

Penicillium vinaceum J. C. Gilman & E. V. Abbott, Iowa State Coll. J. Sci. 1: 299, 1927.

Penicillium virescens Sopp, Skr. Vidensk.-Selsk. Christiana, Math.-Naturvidensk. Kl. 11: 157, 1912 (*non* Bainier, 1907) = *Penicillium roqueforti* Thom, Bull. Bur. Anim. Ind. U.S. Dep. Agric. 82: 35, 1906 (Pitt, 1980).

Penicillium viride (Pitt & A. D. Hocking) Frisvad *et al.*, Persoonia 14: 229, 1990 (*non* Fres. 1851 *nec* Rivera 1873 *nec* Sopp 1912 *nec* (Matr.) Biourge 1923) ≡ *Geosmithia viridis* Pitt & A. D. Hocking, Mycologia 77: 822, 1985.

Penicillium viridicatum Westling, Ark. Bot. 11(1): 88, 1911.

Penicillium viridicyclopium S. Abe, J. Gen. Appl. Microbiol., Tokyo 2: 107, 1956 (nom. inval., Art. 36) = *Penicillium aurantiogriseum* Dierckx, Ann. Soc. Sci. Bruxelles 25: 88, 1901 (Pitt, 1980).

Penicillium virididorsum Biourge, Cellule 33: 306, 1923 = *Penicillium spinulosum* Thom, Bull. Bur. Anim. Ind. U.S. Dep. Agric. 118: 76, 1910 (Pitt, 1980).

Penicillium virido-albus (Sopp) Biourge, Cellule 33: 106, 1923 = *Penicillium purpurascens* (Sopp) Biourge, Cellule 33: 105, 1923 (Raper & Thom, 1949).

Penicillium vitale Pidoplichko & Bilai in Bilai, Antibiotic-prod. fung.: 44, 1961 (nom. inval., Art. 36) = *Penicillium janthinellum* Biourge, Cellule 33: 258, 1923 (Pitt, 1980).

Penicillium volgaense Beliakova & Milko, Mikol. Fitopatol. 6: 147, 1972 = *Penicillium olsonii* Bainier & Sartory, Ann. Mycol. 10: 398, 1912 (Stolk *et al.*, 1990).

Penicillium vulpinum (Cooke & Massee) Seifert & Samson in Samson & Pitt (eds), Adv. *Penicillium Aspergillus* Syst.: 144, 1985. — Basionym: *Coremium vulpinum* Cooke & Massee, Grevillea 16: 81, 1888.

Penicillium waksmanii K. M. Zalessky, Bull. Int. Acad. Polon. Sci., Cl. Sci. Math., Sér. B, Sci. Nat., 1927: 468, 1927.

Penicillium weidemannii (Westling) Biourge, Cellule 33: 204, 1923 = *Penicillium roqueforti* Thom, Bull. Bur. Anim. Ind. U.S. Dep. Agric. 82: 35, 1906 (Pitt, 1980).

Penicillium weidemannii var. *fuscum* Arnaudi, Boll. Ist. Sieroter. Milan. 6: 27, 1928 ["1927"] = *Penicillium roqueforti* Thom, Bull. Bur. Anim. Ind. U.S. Dep. Agric. 82: 35, 1906 (Thom, 1930).

Penicillium westlingii K. M. Zalessky, Bull. Int. Acad. Polon. Sci., Cl. Sci. Math., Sér. B, Sci. Nat., 1927: 473, 1927.

Penicillium wortmannii Klöcker, Compt. Rend. Trav. Carlsberg Lab. 6: 100, 1903 (nom. holomorph.) ≡ *Talaromyces wortmannii* (Klöcker) C. R. Benj., Mycologia 47: 683, 1955.

Penicillium yarmokense Baghdadi, Nov. Sist. Niz. Rast. 5: 99, 1968 = *Penicillium canescens* Sopp,

Skr. Vidensk.-Selsk. Christiana, Math.-Naturvidensk. Kl. 11: 181, 1912 (Pitt, 1980).

Penicillium yezoense Hanzawa in Sasaki and Nakane, J. Agric. Chem. Soc., Japan 19: 774, 1943 (nom. inval., Art. 36) = **Penicillium thomii** Maire, Bull. Soc. Hist. Nat. Afrique N. 8: 189, 1917 (Pitt, 1980).

Penicillium zacinthae C. Ramírez & A. T. Martínez, Mycopathologia 74: 167, 1981 = **Penicillium allahabadense** B. S. Mehrotra & D. Kumar, Can. J. Bot. 40: 1399, 1962 (Frisvad *et al.*, 1990b).

Penicillium zonatum Hodges & J. J. Perry, Mycologia 65: 697, 1973. [Teleomorph *Eupenicillium zonatum* Hodges & J. J. Perry].

PETROMYCES MALLOCH & CAIN (HOLOMORPHS)

Petromyces albertensis J. P. Tewari, Mycologia 77: 114, 1985. [Anamorph: *Aspergillus albertensis* J. P. Tewari].

Petromyces alliaceus Malloch & Cain, Can. J. Bot. 50: 2623, 1973. [Anamorph: *Aspergillus alliaceus* Thom & Church].

PRITZELIELLA HENN. (ANAMORPHS)

Pritzeliella caerulea Henn., Hedwigia Baiblatt 42: 88, 1903 = **Penicillium coprophilum** (Berk. & M. A. Curtis) Seifert & Samson in Samson & Pitt (eds), Adv. *Penicillium Aspergillus* Syst.: 145, 1985 (Seifert and Samson, 1985).

RAPERIA SUBRAM. & RAJENDRAN (ANAMORPHS)

Raperia ingelheimensis (J.F.H. Beyma) Arx, Mycotaxon 26: 121, 1986 ≡ **Merimbla ingelheimensis** (J. F. H. Beyma) Pitt, Can. J. Bot. 57: 2395, 1979.

Raperia spinulosa Subram. & Rajendran, Kavaka 3: 129, 1975 ≡ *Aspergillus warcupii* Samson & W. Gams in Samson & Pitt (eds), Adv. *Penicillium Aspergillus* Syst.: 50, 1985. [Teleomorph *Warcupiella spinulosa* (Warcup) Subram.].

SAROPHORUM SYD. & P. SYD. (ANAMORPHS)

Sarophorum palmicola (Henn.) Seifert & Samson in Samson & Pitt (eds), Adv. *Penicillium Aspergillus* Syst.: 403, 1985. — Basionym: *Penicilliopsis palmicola* Henn., Hedwigia 43: 352, 1904. [Teleomorph *Penicilliopsis clavariiformis* Solms].

SCLEROCLEISTA SUBRAM. (HOLOMORPHS)

Sclerocleista ornata (Raper *et al.*) Subram., Curr. Sci. 41: 757, 1972. — Basionym: *Aspergillus ornatus* Raper *et al.*, Mycologia 45: 678, 1953 (nom. holomorph.). [Anamorph: *Aspergillus ornatulus* Samson & W. Gams].

Sclerocleista thaxteri Subram., Curr. Sci. 41: 757, 1972. [Anamorph: *Aspergillus citrisporus* Höhn.].

SPICARIA HARTING (ANAMORPHS)

Spicaria simplicissima Oudem., Ned. Kruidk. Arch., Ser. 3, 2: 763, 1903 ≡ **Penicillium simplicissimum** (Oudem.) Thom, Penicillia: 335, 1930.

STERIGMATOCYSTIS CRAMER *(ANAMORPHS)*

Sterigmatocystis auricoma Guég., Bull. Soc. Mycol. Fr. 15: 171, 1899 ≡ **Aspergillus auricomus** (Guég.) Saito, J. Ferment. Technol. 17: 3, 1939.

Sterigmatocystis carbonaria Bainier, Bull. Soc. Bot. Fr, 27: 27, 1880 ≡ *Aspergillus carbonarius* (Bainier) Thom, J. Agric. Res. 7: 12, 1916.

Sterigmatocystis flavipes Bainier & Sartory, Bull. Soc. Mycol. Fr. 27: 90, 1911 ≡ *Aspergillus flavipes* (Bainier & Sartory) Thom & Church, Aspergilli: 155, 1926.

Sterigmatocystis nidulans Eidam, Beitr. Biol. Pflanzen 3: 393, 1883 (nom. holomorph.) ≡ **Emericella nidulans** (Eidam) Vuill., Compt. Rend. Hebd. Séances Acad. Sci. 184: 137, 1927.

Sterigmatocystis pulverulenta McAlpine, Agric. Gaz. New South Wales 7: 302, 1897 ≡ *Aspergillus pulverulentus* (McAlpine) Wehmer, Centralbl. Bakteriol., 2. Abth., 18: 394, 1907.

Sterigmatocystis sulphurea Fresen., Beitr. Mykol.: 83, 1863 ≡ *Aspergillus sulphureus* (Fresen.) Wehmer, Mem. Soc. Phys. Genève 33(2): 113, 1901.

Sterigmatocystis sydowii Bainier & Sartory, Ann. Mycol. 11: 25, 1913 ≡ *Aspergillus sydowii* (Bainier & Sartory) Thom & Church, Aspergilli: 147, 1926.

Sterigmatocystis unguis Emile-Weil & L. Gaudin, Arch. Med. Exp. Anat. Pathol. 28: 463, 1918 ≡ **Aspergillus unguis** (Emile-Weil & L. Gaudin) Thom & Raper, Mycologia 31: 667, 1939.

Sterigmatocystis usta Bainier, Bull. Soc. Bot. Fr. 28: 78, 1881 ≡ *Aspergillus ustus* (Bainier) Thom & Church, Aspergilli: 152, 1926.

Sterigmatocystis versicolor Vuill., Mirsky, Causes Err. Det. Aspergill.: 15, 1903 ≡ *Aspergillus versicolor* (Vuill.) Tirab., Ann. Bot. (Roma) 7: 9, 1908.

Sterigmatocystis vitellina Ridl. ex Massee, J. Bot. 34: 152, 1896 ≡ *Aspergillus vitellinus* (Massee) Samson & Seifert in Samson & Pitt (eds), Adv. *Penicillium Aspergillus* Syst.: 417, 1985.

STILBODENDRON SYD. & P. SYD. (ANAMORPHS)

Stilbodendron cervinum (Cooke & Massee) Samson & Seifert in Samson & Pitt (eds), Adv. *Penicillium Aspergillus* Syst.: 408, 1985. — Basionym: *Corallodendron cervinum* Cooke & Massee, Grevillea 16: 71, 1888. [Teleomorph *Penicilliopsis africana* Samson & Seifert].

STILBOTHAMNIUM HENN. *(ANAMORPHS)*

Stilbothamnium amazonense Henn., Hedwigia 43: 396, 1904 ≡ *Aspergillus amazonensis* (Henn.) Samson & Seifert in Samson & Pitt (eds), Adv. *Penicillium Aspergillus* Syst.: 418, 1985.

Stilbothamnium togoense Henn., Bot. Jahrb. Syst. 23: 542, 1897 ≡ *Aspergillus togoensis* (Henn.) Samson & Seifert in Samson & Pitt (eds), Adv. *Penicillium Aspergillus* Syst.: 419, 1985.

STILBUM TODE (ANAMORPHS)

Stilbum humanum P. Karst, Rev. Mycol. 1888: 75, 1888 = *Penicillium coprophilum* (Berk. & M. A. Curtis) Seifert & Samson in Samson & Pitt (eds), Adv. *Penicillium Aspergillus* Syst.: 145, 1985 (Seifert and Samson, 1985).

TALAROMYCES C. R. BENJ. (HOLOMORPHS)

Talaromyces assiutensis Samson & Abdel-Fattah, Persoonia 9: 501, 1978. [Anamorph: *Penicillium assiutense* Samson & Abdel-Fattah].

Talaromyces avellaneus (Thom & Turesson) C. R. Benj., Mycologia 47: 682, 1955 ≡ *Hamigera avellanea* (Thom & Turesson) Stolk & Samson, Persoonia 6: 345, 1971. — Basionym: *Penicillium avellaneum* Thom & Turesson, Mycologia 7: 284, 1915 (nom. holomorph.). [Anamorph: *Merimbla ingelheimense* (J. F. H. Beyma) Pitt].

Talaromyces bacillisporus (Swift) C. R. Benj., Mycologia 47: 684, 1955. — Basionym: *Penicillium bacillisporum* Swift, Bull. Torrey Bot. Club 59: 221, 1932 (nom. holomorph.). [Anamorph: *Geosmithia swiftii* Pitt].

Talaromyces byssochlamydoides Stolk & Samson, Stud. Mycol. 2: 45, 1972. [Anamorph: *Paecilomyces byssochlamydoides* Stolk & Samson].

Talaromyces derxii Takada & Udagawa, Mycotaxon 31: 418, 1988. [Anamorph: *Penicillium derxii* Takada & Udagawa].

Talaromyces dupontii (Griffin & Maubl.) Apinis, Nova Hedwigia 5: 72, 1963 (nom. inval., Art. 36) = *Talaromyces dupontii* (Griffin & Maubl.) R. Emers. in Fergus, Mycologia 56: 277, 1964 (nom. inval., Art. 36) = *Talaromyces dupontii* (Griffin & Maubl.) Cooney & R. Emers., Thermophilic Fungi: 38, 1964 (nom. inval., Art. 36) ≡ *Talaromyces thermophilus* Stolk, Antonie van Leeuwenhoek 31: 268, 1965 (new name).

Talaromyces dupontii (Griffin & Maubl.) Cooney & R. Emers., Thermophilic Fungi: 38, 1964 (nom. inval., Art. 36) ≡ *Talaromyces thermophilus* Stolk, Antonie van Leeuwenhoek 31: 268, 1965 (new name).

Talaromyces dupontii (Griffin & Maubl.) R. Emers. in Fergus, Mycologia 56: 277, 1964 (nom. inval., Art. 36) = *Talaromyces dupontii* (Griffin & Maubl.) Cooney & R. Emers., Thermophilic

Fungi: 38, 1964 (nom. inval., Art. 36) ≡ *Talaromyces thermophilus* Stolk, Antonie van Leeuwenhoek 31: 268, 1965 (new name).

Talaromyces emersonii Stolk, Antonie van Leeuwenhoek 31: 262, 1965. [Anamorph: *Geosmithia emersonii* (Stolk) Pitt].

Talaromyces flavus (Klöcker) Stolk & Samson, Stud. Mycol. 2: 10, 1972. — Basionym: *Gymnoascus flavus* Klöcker, Hedwigia 41: 80, 1902. [Anamorph: *Penicillium dangeardii* Pitt].

Talaromyces flavus var. *macrosporus* Stolk & Samson, Stud. Mycol. 2: 15, 1972 ≡ *Talaromyces macrosporus* (Stolk & Samson) Frisvad *et al.*, Antonie van Leeuwenhoek 57: 186, 1990.

Talaromyces galapagensis Samson & Mahoney, Trans. Br. Mycol. Soc. 69: 158, 1977. [Anamorph: *Penicillium galapagense* Samson & Mahoney].

Talaromyces gossypii Pitt, Gen. *Penicillium*: 500, 1980 ["1979'"] = *Talaromyces assiutensis* Samson & Abdel-Fattah, Persoonia 9: 501, 1978 (Frisvad *et al.*, 1990).

Talaromyces helicus (Raper & Fennell) C. R. Benj., Mycologia 47: 684, 1955. — Basionym: *Penicillium helicum* Raper & Fennell, Mycologia 40: 515, 1948 (nom. holomorph.). [Anamorph: *Penicillium spirillum* Pitt].

Talaromyces helicus Stolk & Samson, Stud. Mycol. 2: 16, 1972 ≡ *Talaromyces helicus* (Raper & Fennell) C. R. Benj., Mycologia 47: 684, 1955.

Talaromyces helicus var. *major* Stolk & Samson, Stud. Mycol. 2: 19, 1972 = *Talaromyces helicus* (Raper & Fennell) C. R. Benj., Mycologia 47: 684, 1955 (Pitt, 1980).

Talaromyces intermedius (Apinis) Stolk & Samson, Stud. Mycol. 2: 21, 1972 (accepted here).

Talaromyces leycettanus H. C. Evans & Stolk, Trans. Br. Mycol. Soc. 56: 45, 1971. [Anamorph: *Paecilomyces leycettanus* (H. C. Evans & Stolk) Stolk *et al.*].

Talaromyces luteus (Zukal) C. R. Benj., Mycologia 47: 681, 1955. — Basionym: *Penicillium luteum* Zukal, Sitzungsber. Kaiserl. Akad. Wiss., Math.-Naturwiss. Cl., Abt. 1, 98: 561, 1890 (nom. holomorph.). [Anamorph: no valid name].

Talaromyces luteus (Zukal) Stolk & Samson, Stud. Mycol. 2: 23, 1972 ≡ *Talaromyces luteus* (Zukal) C. R. Benj., Mycologia 47: 681, 1955.

Talaromyces macrosporus (Stolk & Samson) Frisvad *et al.*, Antonie van Leeuwenhoek 57: 186, 1990. — Basionym: *Talaromyces flavus* var. *macrosporus* Stolk & Samson, Stud. Mycol. 2: 15, 1972. [Anamorph: *Penicillium macrosporum* Frisvad *et al.*].

Talaromyces mimosinus A. D. Hocking in Pitt, Genus *Penicillium*: 507, 1980 ["1979"]. [Anamorph: *Penicillium mimosinum* A. D. Hocking].

Talaromyces ohiensis Pitt, Genus *Penicillium*: 502, 1980 ["1979"]. [Anamorph: *Penicillium ohiense* L. H. Huang & J. A. Schmitt].

Talaromyces panasenkoi Pitt, Genus *Penicillium*: 482, 1980 ["1979"]. [Anamorph: *Penicillium panasenkoi* Pitt].

Talaromyces purpureus (E. Müll. & Pacha-Aue) Stolk & Samson, Stud. Mycol. 2: 57, 1972. — Basionym: *Arachniotus purpureus* E. Müll. & Pacha-Aue, Nova Hedwigia 15: 552, 1968. [Anamorph: *Penicillium purpureum* Stolk & Samson].

Talaromyces rotundus (Raper & Fennell) C. R. Benj., Mycologia 47: 683, 1955. — Basionym: *Penicillium rotundum* Raper & Fennell, Mycologia 40: 518, 1948 (nom. holomorph.). [Anamorph: *Penicillium sphaerum* Pitt].

Talaromyces rotundus Stolk & Samson, Stud. Mycol. 2: 27, 1972 ≡ *Talaromyces rotundus* (Raper & Fennell) C. R. Benj., Mycologia 47: 683, 1955.

Talaromyces spiculisporus (Lehman) C. R. Benj., Mycologia 47: 683, 1955 = *Talaromyces trachyspermus* (Shear) Stolk & Samson, Stud. Mycol. 2: 32, 1972 (Stolk and Samson, 1972).

Talaromyces spiculisporus var. *macrocarpus* J. E. Wright & Loewenb. *in* Bertoni *et al.*, Boln Soc. Argent. Bot. 15: 100, 1973 = *Talaromyces trachyspermus* (Shear) Stolk & Samson, Stud. Mycol. 2: 32, 1972 (Pitt, 1980).

Talaromyces stipitatus (Thom) C. R. Benj., Mycologia 47: 684, 1955. — Basionym: *Penicillium stipitatum* Thom, Mycologia 27: 138, 1935 (nom. holomorph.). [Anamorph: *Penicillium emmonsii* Pitt].

Talaromyces stipitatus Stolk & Samson, Stud. Mycol. 2: 29, 1972 = *Talaromyces stipitatus* Thom) C. R. Benj., Mycologia 47: 684, 1955 (Pitt, 1980).

Talaromyces striatus (Raper & Fennell) C. R. Benj., Mycologia 47: 682, 1955. — Basionym: *Penicillium striatum* Raper & Fennell, Mycologia 40: 521, 1948 (nom. holomorph.). [Anamorph: *Penicillium lineatum* Pitt].

Talaromyces thermophilus Stolk, Antonie van Leeuwenhoek 31: 268, 1965. [Anamorph: *Penicillium dupontii* Griffon & Maubl.].

Talaromyces trachyspermus (Shear) Stolk & Samson, Stud. Mycol. 2: 32, 1972. — Basionym: *Arachniotus trachyspermus* Shear, Science, ser. 2, 16: 138, 1902. [Anamorph: *Penicillium lehmanii* Pitt].

Talaromyces ucrainicus (Panas.) Udagawa, Trans. Mycol. Soc. Japan 7: 94, 1966 (misapplied name) = *Talaromyces panasenkoi* Pitt, Genus *Penicillium*: 482, 1980 ["1979"].

Talaromyces ucrainicus Stolk & Samson, Stud. Mycol. 2: 34, 1972 = *Talaromyces ohiensis* Pitt, Genus *Penicillium*: 502, 1980 ["1979"].

Talaromyces udagawae Stolk & Samson, Stud. Mycol. 2: 36, 1972. [Anamorph: *Penicillium udagawae* Stolk & Samson].

Talaromyces unicus Tzean *et al.*, Mycologia 84: 739, 1992. [Anamorph: *Penicillium unicum* Tzean *et al.*].

Talaromyces vermiculatus (P. A. Dang.) C. R. Benj., Mycologia 47: 684, 1955 = *Talaromyces flavus* (Klöcker) Stolk & Samson, Stud. Mycol. 2: 10, 1972 (Stolk and Samson, 1972).

Talaromyces wortmannii (Klöcker) C. R. Benj., Mycologia 47: 683, 1955. — Basionym: *Penicillium wortmannii* Klöcker, Compt.-Rend. Trav. Carlsberg Lab. 6: 100, 1903 (nom. holomorph.). [Anamorph: *Penicillium kloeckeri* Pitt].

Talaromyces wortmannii Stolk & Samson, Stud. Mycol. 2: 39, 1972 ≡ *Talaromyces wortmannii* (Klöcker) C. R. Benj., Mycologia 47: 683, 1955.

THERMOASCUS MIEHE (HOLOMORPHS)

Thermoascus aurantiacus Miehe, Selbsterhitzung Heus: 70, 1907.

Thermoascus aegyptiacus S. Ueda & Udagawa, Trans. Mycol. Soc. Japan 24: 135, 1983. [Anamorph: *Paecilomyces aegyptiacus* S. Ueda & Udagawa]

Thermoascus crustaceus (Apinis & Chesters) Stolk, Antonie van Leeuwenhoek 31: 272, 1965. — Basionym: *Dactylomyces crustaceus* Apinis & Chesters, Trans. Br. Mycol. Soc. 47: 428, 1964. [Anamorph: *Paecilomyces crustaceus* Apinis & Chesters].

Thermoascus thermophilus (Sopp) Arx, Gen. Fungi Sporul. Pure Cult., ed. 2: 94, 1974. — Basionym: *Dactylomyces thermophilus* Sopp, Skr. Vidensk.-Selsk. Christiana, Math.-Naturvidensk. Kl. 11: 35, 1912.

Torulomyces Delitsch (anamorphs)

Torulomyces lagena Delitsch, Syst. Schimmelpilze: 91, 1943. [Teleomorph *Eupenicillium limoneum* Goch. & Zlattner].

TRICHOCOMA JUNGH. (HOLOMORPHS)

Trichocoma paradoxa Jungh., Praem. Fl. Crypt. Java 1: 9, 1838.

USTILAGO (PERS.) ROUSSEL (USTILAGINACEAE) (HOLOMORPHS)

Ustilago phoenicis Corda, Icon. Fung. 4: 9, 1840 ≡ *Aspergillus phoenicis* (Corda) Thom., J. Agric. Res. 7: 14, 1916.

VERTICILLIUM NEES (ANAMORPHS)

Verticillium cinnamomeum Petch, Trans. Br. Mycol. Soc. 16: 233, 1932 ≡ *Paecilomyces cinnamomeus* (Petch) Samson & W. Gams, Stud. Mycol. 6: 62, 1974.

Verticillium sulphurellum Sacc., Fung. Ital.: tab. 641, 1881 ≡ **Paecilomyces sulphurellus** (Sacc.) Samson & W. Gams, Stud. Mycol. 6: 67, 1974.

WARCUPIELLA SUBRAM. (HOLOMORPHS)

Warcupiella spinulosa (Warcup) Subram., Curr. Sci. 41: 757, 1972. — Basionym: *Aspergillus spinulosus* Warcup *et al.*, Genus *Aspergillus*: 204, 1965 (nom. holomorph.). [Anamorph: *Raperia spinulosa* Subram. & Rajendran].

REFERENCES

Biourge, P. 1923. Les moissisures du groupe *Penicillium* Link. Cellule 33: 7-331.

Blaser, P. 1975. Taxonomische und physiologische Untersuchungen über die Gattung *Eurotium* Link ex Fries. Sydowia 28: 1-49.

Caten, C.E. 1979. Genetic determination of conidial colour in *Aspergillus heterocaryoticus* and relationship of this species to *Aspergillus amstelodami*. Trans. Br. Mycol. Soc. 73: 65-74.

Cruickshank, R.H. and Pitt, J.I. 1987. The zymogram technigue: isoenzyme patterns as an aid in *Penicillium* classification. Microbiol. Sci. 4: 14-17

Frisvad, J.C. and Filtenborg, O. 1989. Terverticillate Penicillia: chemotaxonomy and mycotoxin production. Mycologia 81: 837-861.

Frisvad, J.C., Samson, R.A. and Stolk, 1990a. Notes on the typification of some species of *Penicillium*. Persoonia 14: 193-202.

Frisvad, J.C., Samson, R.A. and Stolk, 1990b. Disposition of some recently described species of *Penicillium*. Persoonia 14: 209-232.

Gams, W., Christensen, M., Onions, A.H., Pitt, J.I., and Samson, R.A. 1985. Infrageneric taxa of *Aspergillus*. *In* Advances in *Penicillium* and *Aspergillus* Systematics, R.A. Samson and J.I. Pitt, eds, pp. 55-62. New York: Plenum Press.

Kozakiewicz, Z., Frisvad, J.C., Hawksworth, D.L., Pitt, J.I., Samson, R.A. and Stolk, A.C. 1992. Proposals for nomina specifica conservanda and rejicienda in *Aspergillus* and *Penicillium* (Fungi). Taxon 41: 109-113.

Malloch, D. and Cain, R.F. 1972. The Trichocomataceae: Ascomycetes with *Aspergillus*, *Paecilomyces* and *Penicillium* imperfect states. Can. J. Bot. 50: 2613-2628.

Pitt, J.I. 1980. The Genus *Penicillium* and Its Teleomorphic States *Eupenicillium* and *Talaromyces*. London: Academic Press. ["1979"].

Pitt, J.I. 1985. Nomenclatorial and taxonomic problems in the genus *Eurotium*. *In* "Advances in *Penicillium* and *Aspergillus* Systematics", R.A. Samson and J.I. Pitt, eds., pp. 383-396. New York: Plenum Press.

Pitt, J.I. and Cruickshank, R.H. 1990. Speciation and synonymy in *Penicillium* subgenus *Penicillium*. *In* "Modern Concepts in *Penicillium* and *Aspergillus* Classification", R. A. Samson and J.I. Pitt, eds., pp. 103-119. New York: Plenum Press.

Pitt, J.I. and Samson, R.A. 1993. Species names in current use in the *Trichocomaceae* (Fungi, Eurotiales). *In* "Names in current use in the families *Trichocomaceae*, *Cladoniaceae*, *Pinaceae*, and *Lemnaceae*", W. Greuter, ed. Regnum Vegetabile 128: 13-57. Königstein, Germany: Koeltz Scientific Books.

Pitt, J.I., Cruickshank, R.H. and Leistner, L. 1986. *Penicillium commune*, *P. camembertii*, the origin of white cheese moulds, and the production of cyclopiazonic acid. Food Microbiol. 3: 363-371.

Raper, K.B. and Fennell, D.I. 1965. The Genus *Aspergillus*. Baltimore, Maryland: Williams and Wilkins.

Raper, K.B. and Thom, C. 1949. A Manual of the Penicillia. Baltimore, Maryland: Williams and Wilkins

Samson, R.A. and Gams, W. 1985. Typification of the species of *Aspergillus* and associated teleomorphs. *In* Advances in *Penicillium* and *Aspergillus* Systematics, eds R.A. Samson and J.I. Pitt. New York: Plenum Press. pp. 31-54.

Samson, R.A. and Seifert, K.A.. 1985. The Ascomycete genus *Penicilliopsis* and its anamorphs. *In* Advances in *Penicillium* and *Aspergillus* Systematics, eds R.A. Samson and J.I. Pitt. New York: Plenum Press. pp. 397-428.

Seifert, K.A. and Samson, R.A. 1985. The genus *Coremium* and the synnematous Penicillia. *In* Advances in *Penicillium* and *Aspergillus* Systematics, eds R.A. Samson and J.I. Pitt. New York: Plenum Press. pp. 143-154.

Stolk, A.C. 1965. Thermophilic species of *Talaromyces* Benjamin and *Thermoascus* Miehe. Antonie van Leeuwenhoek 31: 262-276.

Stolk, A.C. and Samson, R.A. 1972. The genus *Talaromyces*. Studies on *Talaromyces* and related genera. II. Stud. Mycol., Baarn 2: 1-65.

Stolk, A.C. and Scott, D.B. 1967. Studies on the genus *Eupenicillium* Ludwig. I. Taxonomy and nomenclature of Penicillia in relation to their sclerotioid ascocarpic states. Persoonia 4: 391-405.

Stolk, A.C., Samson, R.A., Frisvad, J.C. and Filtenborg, O. 1990. The systematics of the terverticillate Penicillia. *In* "Modern Concepts in *Penicillium* and *Aspergillus* Classification (Eds Samson, R. A. and Pitt, J.I.), pp. 121-137. New York: Plenum Press

Thom, C. 1930. The Penicillia. Baltimore, Maryland: Williams and Wilkins.

Thom, C. and Raper, K.B. 1941. The *Aspergillus glaucus* group. U.S. Dept Agric. Misc. Pub. 426: 1-26.

Thom, C. and Raper, K.B. 1945. A Manual of the Aspergilli. Baltimore, Maryland: Williams and Wilkins.

Van Reenen-Hoekstra, E.S., Frisvad, J.C., Samson, R.A., and Stolk, A.C. 1990. The *Penicillium funiculosum* complex — well defined species and problematic taxa. *In* "Modern Concepts in *Penicillium* and *Aspergillus* Classification", R. A. Samson and J.I. Pitt, eds., pp. 173-192. New York: Plenum Press

Types of *Aspergillus* and *Penicillium* and their teleomorphs in current use.

The following list is a slightly modified compilation of the NCU list published by Pitt and Samson (1993). Types of *Paecilomyces* and its teleomorphs are here excluded.

ASPERGILLUS FR.: FR. (ANAMORPHS)

Aspergillus acanthosporus Udagawa & Takada in Bull. Natl. Sci. Mus. Tokyo, ser. 2, 14: 503. 30 Sep 1971. — Holotype: No. 22462 p.p. (NHL)

Aspergillus aeneus Sappa in Allionia 2: 84. 1954. — Neotype (Samson & Gams, 1985): No.128.58 (CBS).

Aspergillus albertensis J. P. Tewari in Mycologia 77: 114. 15 Feb 1985. — Holotype: No. 2976 p.p. (UAMH)

Aspergillus allahabadii B. S. Mehrotra & Agnihotri in Mycologia 54: 400. 7 Jan 1963. — Neotype (Samson & Gams, 1985): No. 164.63 (CBS).

Aspergillus alliaceus Thom & Church, Aspergilli: 163. Mar 1926. — Neotype (Malloch & Cain, 1972): No. 46232 p.p. (TRTC).

Aspergillus amazonensis (Henn.) Samson & Seifert in Samson & Pitt, Adv. Penicillium Aspergillus Syst.: 418. 1985. — Basionym: *Stilbothamnium amazonens* Henn. in Hedwigia 43: 396. 3 Sep 1904. — Holotype: Brazil, Jurna, Jul 1907, *Ule* in herb. Hennings (S)

Aspergillus ambiguus Sappa in Allionia 2: 254. 1955. — Neotype (Samson & Gams, 1985): No. 117.58 (CBS).

Aspergillus amylovorus Panas. ex Samson in Stud. Mycol. 18: 28. 3 Jan 1979. — Holotype: No. 600.67 (CBS).

Aspergillus anthodesmis Bartoli & Maggi in Trans. Brit. Mycol. Soc. 71: 386. 11 Jan 1979. — Holotype: No. 103 S (RO).

Aspergillus appendiculatus Blaser in Sydowia 28: 38. Dec 1975. — Holotype: No. 8286 (ZT).

Aspergillus arenarius Raper & Fennell, Genus Aspergillus: 475. 1965. — Neotype (Samson & Gams, 1985): No.55632 (IMI).

Aspergillus asperescens Stolk in Antonie van Leeuwenhoek J. Microbiol. Serol. 20: 303. 1954. — Neotype (designated here): No. 46813 (IMI).

Aspergillus atheciellus Samson & W. Gams in Samson & Pitt, Adv. Penicillium Aspergillus Syst.: 34. 1985. — Holotype: No. 32048 p.p. (IMI).

Aspergillus aureolatus Munt. Cvetk. & Bata in Bull. Inst. Jard. Bot. Univ. Beograd, ser. 2, 1: 196.

1964. — Neotype (Samson & Gams, 1985): No. 190.65 (CBS).

Aspergillus aureoluteus Samson & W. Gams in Samson & Pitt, Adv. Penicillium Aspergillus Syst.: 34. 1985. — Holotype: No.105.55 p.p. (CBS).

Aspergillus auricomus (Guég.) Saito in J. Ferment. Technol. 17: 3. 1939. — Basionym: *Sterigmatocystis auricoma* Guég. in Bull. Soc. Mycol. France 15: 171. 31 Jul 1899. — Neotype (Samson & Gams, 1985): No. 467.65 (CBS).

Aspergillus avenaceus G. Sm. in Trans. Brit. Mycol. Soc. 26: 24. 8 Apr 1943. — Neotype (Samson & Gams, 1985): No. 109.46 (CBS).

Aspergillus awamorii Nakaz. in Rep. Gov. Res. Inst. Formosa 1907: 1. 1907. — Neotype (Al-Musallam, 1980): No. 557.65 (CBS).

Aspergillus bicolor M. Chr. & States in Mycologia 70: 337. 25 Mai 1978. — Holotype: No. RMF 2058 p.p. (NY).

Aspergillus biplanus Raper & Fennell, Genus Aspergillus: 434. 1965. — Neotype (Samson & Gams, 1985): No. 235602 (IMI).

Aspergillus bisporus Kwon-Chung & Fennell in Mycologia 63: 479.4 Jun 1971. — Holotype: No. NRRL 3693 p.p. (BPI).

Aspergillus brevipes G. Sm. in Trans. Brit. Mycol. Soc. 35: 241. 19 Dec 1952. — Holotype: [dried culture from soil from] Australia, Mt. Kosciusko, *s.coll.* (K).

Aspergillus bridgeri M. Chr. in Mycologia 74: 210. 2 Apr 1982. — Holotype: No. JB 26-1-2 (NY).

Aspergillus brunneouniseriatus Suj. Singh & B. K. Bakshi in Trans. Brit. Mycol. Soc 44: 160 20 Jun 1961. — Neotype (Samson & Gams, 1985): No. 227677 (IMI).

Aspergillus brunneus Delacr. in Bull. Soc. Mycol. France 9: 185. 1893. — Neotype (Samson & Gams, 1985): No. 211378 p.p. (IMI).

Aspergillus caesiellus Saito in J. Coll. Sci. Imp. Univ. Tokyo 18(5): 49.28 Jan 1904. — Neotype (Samson & Gams, 1985): No. 172278 (IMI).

Aspergillus caespitosus Raper & Thom in Mycologia 36: 563. Nov Dec 1944. — Neotype (Samson & Gams, 1985): No. 16034ii (IMI).

51

Aspergillus campestris M. Chr. in Mycologia 74: 212. 2 Apr 1982. — Holotype: No. ST 2-3-1 (NY).

Aspergillus candidus Link in Ges. Naturf. Freunde Berlin Mag. Neuesten Entdeck. Gesammten Naturk. 3: 16. 1809 : Fr. — Neotype (Samson & Gams, 1985): No. 567.65 (CBS).

Aspergillus carbonarius (Bainier) Thom in J. Agric. Res.7: 12.2 Oct 1916. — Basionym: *Sterigmatocystis carbonaria* Bainier in Bull. Soc. Bot. France 27: 27. 1880. — Neotype (Al-Musallam, 1980): No. 556.65 (CBS).

Aspergillus carneus Blochwitz in Ann. Mycol. 31: 81.25 Jan 1933. — Neotype (Samson & Gams, 1985): No. 1358818 (IMI).

Aspergillus cervinus Massee in Bull. Misc. Inform Kew 1914: 158. 10 Jun 1914. — Neotype (Christensen & Fennell, 1964): No. WISC WT 540 (WIS).

Aspergillus chevalieri L. Mangin in Ann. Sci. Nat., Bot., ser. 9, 10: 362. 1909. — Neotype (Kozakiewicz, 1989): No. 211382 p.p. (IMI).

Aspergillus chryseides Samson & W. Gams in Samson & Pitt, Adv. Penicillium Aspergillus Syst.: 36. 1985. — Holotype: No. 238612 (IMI).

Aspergillus citrisporus Höhn. in Sitzungsber. Kaiserl. Akad. Wiss., Math.-Naturwiss. Cl., Abt. l,111: 1036. 1902. — Holotype (Subramanian, 1972): ex caterpillar dung, Kittery Point, *R. Thaxter* (FH).

Aspergillus clavatoflavus Raper & Fennell, Genus Aspergillus: 378. 1965. — Neotype (Samson & Gams, 1985): No. 124937 (IMI).

Aspergillus clavatonanicus Bat. & al. in Anais Fac. Med. Univ. Recife 15: 197. 1955. — Neotype (Samson & Gams, 1985): No. 235352 (IMI).

Aspergillus clavatus Desm. in Ann. Sci. Nat., Bot., ser. 2, 2: 71. 1834. — Neotype (Samson & Gams, 1985): No. 15949 (IMI).

Aspergillus compatibilis Samson & W. Gams in Samson & Pitt, Adv. Penicillium Aspergillus Syst.: 42. 1985. — Holotype: No. 488.65 (CBS)

Aspergillus conicus Blochwitz in Ann. Mycol. 12: 38. 10 Feb 1914. — Neotype (Samson & Gams, 1985): No. 172281 (IMI).

Aspergillus conjunctus Kwon-Chung & Fennell in Raper & Fennell, Genus Aspergillus: 552. 1965. — Neotype (Samson & Gams, 1985): No. 135421 (IMI).

Aspergillus coremiiformis Bartoli & Maggi in Trans. Brit. Mycol. Soc. 71: 386. 11 Jan 1979. — Holotype: No. 102 S (RO).

Aspergillus corrugatus Udagawa & Y. Horie in Mycotaxon 4: 535. 18 Dec 1976. — Holotype: No. 2763 p.p. (NHL).

Aspergillus cremeoflavus Samson & W. Gams in Samson & Pitt, Adv. Penicillium Aspergillus Syst.: 37. 1985. — Holotype: No. 123749ii (IMI).

Aspergillus cristatellus Kozak. in Mycol Pap.161:81. 30 Jun 1989.- Holotype: No. 172280 p.p. (IMI).

Aspergillus crustosus Raper & Fennell, Genus Aspergillus: 532. 1965. — Neotype (Samson & Gams, 1985): No. 135819 (IMI).

Aspergillus curviformis H. J. Chowdhery & J. N. Rai in Nova Hedwigia 32: 231.11 Feb 1980. — Holotype: [dried culture from soil from] India, Kagh Islands, *s.coll.* (LWG).

Aspergillus crystallinus Kwon-Chung & Fennell in Raper & Fennell, Genus Aspergillus: 471. 1965. — Neotype (Samson & Gams, 1985): No. 139270 (IMI).

Aspergillus deflectus Fennell & Raper in Mycologia 47: 83. 8 Mar 1955. — Neotype (Samson & Gams, 1985): No. 61448 (IMI).

Aspergillus dimorphicus B. S. Mehrotra & Prasad in Trans. Brit. Mycol. Soc.52: 331 25 Apr 1969. — Neotype (Kozakiewicz, 1989): No. 131553 (IMI).

Aspergillus diversus Raper & Fennell, Genus Aspergillus: 437. 1965. — Neotype (Samson & Gams, 1985): No. 232882 (IMI).

Aspergillus duricaulis Raper & Fennell, Genus Aspergillus: 249. 1965. — Neotype (Samson & Gams, 1985): No. 172282 (IMI).

Aspergillus dybowskii (Pat.) Samson & Seifert in Samson & Pitt, Adv. Penicillium Aspergillus Syst.: 422. 1985. — Basionym: *Penicilliopsis dybowskii* Pat. in Bull. Soc. Mycol. France 7: 54. 1891. — Lectotype (designated here; see also Samson & Seifert, 1985): Congo, Jan 1894, *Dybowski* in herb. Bresadola (S).

Aspergillus eburneocremeus Sappa in Allionia 2: 87. 1954. — Neotype (Samson & Gams, 1985): No. 69856 (TMI).

Aspergillus egyptiacus Moub. & Mustafa in Egypt. J. Bot. 15: 153. 1972. — Neotype (Samson & Gams, 1985): No. 141415 (IMI).

Aspergillus elegans Gasperini in Atti Soc. Tosc. Sci. Nat. Pisa, Processi Verbali 8: 328.1887. — Neotype (Samson & Gams, 1985): No. 102.14 (CBS).

Aspergillus ellipticus Raper & Fennell, Genus Aspergillus: 319. 1965. — Neotype (Al-Musallam, 1980): No.707.79 (CBS).

Aspergillus elongatus J. N. Rai & S. C. Agarwal in Canad. J. Bot. 48: 791. 29 Mai 1970. — Neotype (Samson & Gams, 1985): No. 387.75 (CBS).

Aspergillus erythrocephalus Berk. — & M. A. Curtis in J. Linn. Soc., Bot., 10: 362. 16 Jun 1868. — Holotype: Cuba, *Wright 764* (K).

Aspergillus falconensis Y. Horie in Trans. Mycol. Soc. Japan 30: 257. Oct 1989. — Holotype: No. 10001 P.p. (CBM).

Aspergillus fennelliae Kwon-Chung & S. J. Kim in Mycologia 66: 629. 28 Aug 1974. — Neotype (Kozakiewicz, 1989): No. 278382 (IMI).

Aspergillus flaschentraegeri Stolk in Trans. Brit. Mycol. Soc. 47: 123. 10 Apr 1964. — Neotype (Samson & Gams, 1985): No. CBS 108.63 (K).

Aspergillus fischerianus Samson & W. Gams in Samson & Pitt, Adv. Penicillium Aspergillus Syst.: 39. 1985. — Holotype: No. 21139ii p.p. (IMI).

Aspergillus flavipes (Bainier & Sartory) Thom & Church, Aspergilli: 155. Mar 1926. — Basionym: *Sterigmatocystis flavipes* Bainier & Sartory in Bull. Soc. Mycol. France 27: 90. 1911. — Neotype (Samson & Gams, 1985): No. 171885 (IMI).

Aspergillus flavofurcatus Bat. & H. Maia in Anais Soc. Biol. Pernambuco 13: 94-96. 1955. — Neotype (Samson & Gams, 1985): No. 124938 (IMI).

Aspergillus flavus Link in Ges. Naturf. Freunde Berlin Mag. Neuesten Entdeck. Gesammten Naturk. 3: 16. 1809: Fr. — Neotype (Kozakiewicz, 1982): No. 124930 (IMI).

Aspergillus floriformis Samson & Mouch. in Antonie van Leeuwenhoek J. Microbiol. Serol. 40: 343. 1975. — Holotype: No. 937.73 (CBS).

Aspergillus foeniculicola Udagawa in Trans. Mycol. Soc. Japan 20: 13. Jun 1979. — Holotype: No. 2777 p.p. (NHL).

Aspergillus foetidus Thom & Raper, Man. Aspergilli: 219. 1945. — Neotype (Samson & Gams, 1985): No. 121.28 (CBS).

Aspergillus foveolatus Y. Horie in Trans. Mycol. Soc. Japan 19: 313. Oct 1978. — Holotype: No. 4547 p.p. (IFM).

Aspergillus fruticans Samson & W. Gams in Samson & Pitt, Adv. Penicillium Aspergillus Syst.: 40. 1985. — Holotype: No. 139279 p.p. (IMI).

Aspergillus fumigatus Fresen., Beitr. Mykol.: 81. 18 Aug 1863. — Neotype (Samson & Gams, 1985): No. 16152 (IMI).

Aspergillus funiculosus G. Sm. in Trans. Brit. Mycol. Soc. 39: 111. 20 Mar 1956. — Neotype (Samson & Gams, 1985): No. 44397 (IMI).

Aspergillus giganteus Wehmer in Mém. Soc. Phys. Genève 33(2): 85. Aug 1901. — Neotype (Samson & Gams, 1985): No. 227678 (IMI).

Aspergillus glaber Blaser in Sydowia 28: 35. Dec 1975. — Holotype: No. 8218T p.p. (ZT).

Aspergillus glaucoaffinis Samson & W. Gams in Samson & Pitt, Adv. Penicillium Aspergillus Syst.: 47. 1985. — Holotype: No. 16122ii (IMI).

Aspergillus glauconiveus Samson & W. Gams in Samson & Pitt, Adv. Penicillium Aspergillus Syst.: 45. 1985. — Holotype: No. 32050ii (IMI).

Aspergillus glaucus Link, in Ges. Naturf. Freunde Berlin Mag. Neuesten Entdeck. Gesammten Naturk. 3: 16. 1809: Fr. — Neotype (designated here): No. 211383 (IMI).

Aspergillus globosus H. J. Chowdhery & J. N. Rai in Nova Hedwigia 32: 233. 11 Feb 1980. — Holotype: [dried culture from soil from] India, Kagh Islands, *s.coll.* (LWG).

Aspergillus gorakhpurensis Kamal & Bhargava in Trans. Brit. Mycol. Soc. 52: 338. 25 Apr 1969. — Neotype (Samson & Gams, 1985): No. 130728 (IMI).

Aspergillus gracilis Bainier in Bull. Soc. Mycol. France 23: 90. 1907. — Neotype (Samson & Gams, 1985): No. 211393 (IMI).

Aspergillus granulosus Raper & Thom in Mycologia 36: 565. NovDec 1944. — Neotype (Kozakiewicz, 1989; incorrectly listed as No. 172278 (IMI) by Samson & Gams, 1985): No. 17278ii (IMI).

Aspergillus halophilicus C. M. Chr. & al. in Mycologia 51: 636. 17 Mar 1961. — Neotype (Samson & Gams, 1985): No. NRRL 2739 p.p. (BPI).

Aspergillus helicothrix Al-Musallam in Antonie van Leeuwenhoek J. Microbiol. Serol.46: 407. 1980. — Holotype: No. 677.79 (CBS).

Aspergillus heterocaryoticus C. M. Chr. & al. in Mycologia 57: 535.2 Aug 1965. — Holotype: No. NCF C-100 p.p. (BPI).

Aspergillus heteromorphus Bat. & H. Maia in Anais Soc. Biol. Pernambuco 15: 200. 1957. — Neotype (Samson & Gams, 1985): No. 172288 (IMI).

Aspergillus hiratsukae Udagawa & al. in Trans. Mycol. Soc. Japan 32: 23. Apr 1991. — Holotype: No. 3008 p.p. (NHL).

Aspergillus igneus Kozak. in Mycol. Pap. 161: 52. 30 Jun 1989. — Holotype: No. 75886 (IMI).

Aspergillus insulicola Montem. & A. R. Santiago in Mycopathologia 55: 130. 30 Apr 1975. — Neotype (Samson & Gams, 1985): No. 382.75 (CBS).

Aspergillus intermedius Blaser in Sydowia 28:41. Dec 1976.-Neotype (Kozakiewicz, 1989): No. 89278 p.p. (IMI).

Aspergillus itaconicus Kinosh. in Bot. Mag. (Tokyo) 45: 60.1931. — Neotype (Samson & Gams, 1985): No. 16119 (IMI).

Aspergillus ivoriensis Bartoli & Maggi in Trans. Brit. Mycol. Soc. 71: 383. 11 Jan 1979. — Holotype: No. 101 S (RO).

Aspergillus janus Raper & Thom in Mycologia 36: 556. Nov-Dec 1944. — Neotype (Samson & Gams, 1985): No. 16065 (IMI).

Aspergillus japonicus Saito in Bot. Mag. (Tokyo) 20: 61. Jun 1906. — Neotype (Samson & Gams, 1985): No. 114.51 (CBS).

Aspergillus kanagawaënsis Nehira in J. Jap. Bot. 26: 109. Apr 1951. — Neotype (Samson & Gams, 1985): No. 126690 (IMI).

Aspergillus lanosus Kamal & Bhargava in Trans. Brit. Mycol. Soc. 52: 336. 25 Apr 1969. — Neotype (Samson & Gams, 1985): No. 130727 (IMI).

Aspergillus leporis States & M. Chr. in Mycologia 58: 738. 9 Nov 1966. — Holotype: No. RMF 99 (NY).

Aspergillus leucocarpus Hadlok & Stolk in Antonie van Leeuwenhoek J. Microbiol. Serol. 35: 9. 1969. — Holotype: No. 353.68 p.p. (CBS).

Aspergillus longivesica L. H. Huang & Raper in Mycologia 63: 53. 10 Mar 1971. — Holotype: No. Nl l79 (WIS).

Aspergillus lucknowensis J. N. Rai & al. in Canad. J. Bot. 46: 1483. 20 Dec 1968. — Neotype (designated here): No. 449.75 (CBS).

Aspergillus malodoratus Kwon-Chung & Fennell in Raper & Fennell, Genus Aspergillus: 468. 1965. — Neotype (Samson & Garns, 1985): No. 172289 (IMI).

Aspergillus maritimus Samson & W. Gams in Samson & Pitt, Adv. Penicillium Aspergillus Syst.: 43. 1985. — Holotype: India, Kagh Islands, *s.coll.* (LWG).

Aspergillus medius R. Meissn. in Bot. Zeitung, 2. Abt., 55: 356. 1 Dec 1897. — Neotype (Samson & Gams, 1985): No. 113.27 (CBS).

Aspergillus melleus Yukawa in J. Coll. Agric. Imp. Univ. Tokyo 1: 358. 28 Mar 1911. — Neotype (Samson & Gams, 1985): No. 546.65 (CBS).

Aspergillus microcysticus Sappa in Allionia 2: 251. 1955. — Neotype (Samson & Gams, 1985): No. 139275 (IMI).

Aspergillus microthecius Samson & W. Gams in Samson & Pitt, Adv. Penicillium Aspergillus Syst.: 46. 1985. — Holotype: No. 139280 p.p. (IMI).

Aspergillus multicolor Sappa in Allionia 2: 87. 1954. — Neotype (Samson & Gams, 1985): No. 69875 (IMI).

Aspergillus navahoënsis M. Chr. & States in Mycologia 74: 226. 2 Apr 1982. — Holotype: No. SD-5 p.p. (NY).

Aspergillus neocarnoyi Kozak. in Mycol. Pap.161: 63. 30 Jun 1989. — Holotype: No. 172279 p.p. (IMI).

Aspergillus neoglaber Kozak. in Mycol. Pap.161: 56. 30 Jun 1989. — Holotype: No. 61447 (IMI).

Aspergillus nidulans (Eidam) G. Winter in Rabenh. Krypt.-Fl., ed. 2, 1(2): 62. Mar 1884, *nom. cons. prop.* — Basionym: *Sterigmatocystis nidulans* Eidam in Beitr. Biol. Pflanzen 3: 392. 1883. — Neotype (Kozakiewicz & al., 1992): No. 86806 (IMI).

Aspergillus niger Tiegh. in Ann. Sci. Nat., Bot., ser. 5, 8: 240. Oct 1867, *nom. cons. prop* — Neotype (Al-Musallam, 1980): No. 554.65 (CBS).

Aspergillus niveus Blochwitz in Ann. Mycol. 27: 205. 25 Jun 1929. — Neotype (Samson & Gams, 1985): No. 171878 (IMI).

Aspergillus nomius Kurtzman & al. in Antonie van Leeuwenhoek J. Microbiol. Serol. 53: 151. 1987. — Holotype: No. NRRL 13137 (BPI).

Aspergillus nutans McLennan & Ducker in Austral. J. Bot. 2: 355. Nov 1954. — Neotype (Samson & Gams, 1985): No. 62874ii (IMI).

Aspergillus ochraceoroseus Bartoli & Maggi in Trans. Brit. Mycol. Soc. 71: 393. 11 Jan 1979. — Holotype: No. 104 S (RO).

Aspergillus ochraceus K. Wilh., Beitr. Kenntn. Aspergillus: 66. 28 Apr 1877. — Neotype (designated here): No. 16247iv (IMI).

Aspergillus ornatulus Samson & W. Gams in Samson & Pitt, Adv. Penicillium Aspergillus Syst.: 45. 1985. — Holotype: No. 55295 p.p. (IMI).

Aspergillus oryzae (Ahlb.) Cohn in Jahresber. Schles. Ges. Vaterl. Cult. 61: 226. 1884. — Basionym: *Eurotium oryzae* Ahlb. in Dingler's Polytechn. J.230: 330. 1878. — Neotype (Samson & Gams, 1985): No. 16266 (IMI).

Aspergillus ostianus Wehmer in Bot. Centralbl. 80: 461. 13 Dec 1899. — Neotype (Samson & Gams, 1985): No. 15960 (IMI).

Aspergillus paleaceus Samson & W. Gams in Samson & Pitt, Adv. Penicillium Aspergillus Syst.: 50. 1985. — Neotype (Samson & Gams, 1985): No. 172293 p.p. (IMI).

Aspergillus pallidus Kamyschko in Bot. Mater. Otd. Sporov. Rast.16: 93. 19 Jun 1963. — Neotype (Samson & Gams, 1985): No. 129967 (IMI).

Aspergillus panamensis Raper & Thom in Mycologia 36: 568. Nov Dec 1944. — Neotype (Samson & Gams, 1985): No. 19393iii (IMI).

Aspergillus paradoxus Fennell & Raper in Mycologia 47: 69. 8 Mar 1955. — Neotype (Samson & Gams, 1985): No. 117502 (IMI).

Aspergillus parasiticus Speare in Bull. Div. Pathol. Physiol., Hawaiian Sugar Planters' Assoc. Exp. Sta. 12: 38. 1912. — Neotype (Kozakiewicz, 1982): No. 15957ix (IMI).

Aspergillus parvulus G. Sm. in Trans. Brit. Mycol. Soc. 44: 45. 21 Mar 1961. — Holotype: No. 86558 (IMI).

Aspergillus penicillioides Speg. in Revista Fac. Agron. Univ. Nac. La Plata 2: 246. 1896. — Neotype (Samson & Gams, 1985): No. 211342 (IMI).

Aspergillus petrakii Vörös in Sydowia, ser. 2, Beih., 1: 62. Jan-Aug 1957. — Neotype (Samson & Gams, 1985): No. 172291 (IMI).

Aspergillus peyronelii Sappa in Allionia 2: 248. 1955. — Neotype (Samson & Gams, 1985): No. 139271 (IMI).

Aspergillus phoenicis (Corda) Thom. in J. Agric. Res. 7: 14. 2 Oct 1916. — Basionym: *Ustilago phoenicis* Corda, Icon. Fung. 4: 9. Sep 1840. — Holotype: [on dates from] Turkey, Istanbul, 1837, *Schmidt* (PRM).

Aspergillus protuberus Munt.-Cvetk. in Mikrobiologija 5: 119. 1968. — Neotype (Samson & Gams, 1985): No. 602.74 (CBS).

Aspergillus pseudodeflectus Samson & Mouch. in Antonie van Leeuwenhoek J. Microbiol. Serol. 40: 345. 1975. — Holotype: No. 756.74 (CBS).

Aspergillus pulverulentus (McAlpine) Wehmer in Centralbl. Bakteriol., 2. Abth., 18: 394. 22 Apr 1907.— Basionym: *Sterigmatocystis pulverulenta* McAlpine in Agric. Gaz. New South Wales 7: 302. 1897. — Holotype: [on *Phaseolus vulgaris* from] Australia, Victoria, Burnley Bot. Garden, *McAlpine* (VPRI).

Aspergillus pulvinus Kwon-Chung & Fennell in Raper & Fennell, Genus Aspergillus: 45. 1965. — Neotype (Samson & Gams, 1985): No. 139628 (IMI).

Aspergillus puniceus Kwon-Chung & Fennell in Raper & Fennell, Genus Aspergillus: 547. 1965. — Neotype (Samson & Gams, 1985): No. 126692 (IMI).

Aspergillus purpureus Samson & Mouch. in Antonie van Leeuwenhoek J. Microbiol. Serol. 41: 350. 1975. — Holotype: No. 754.74 p.p. (CBS).

Aspergillus quadricingens Kozak. in Mycol. Pap. 161: 54. 30 Jun 1989. — Holotype: No. 48583ii p.p. (IMI).

Aspergillus raperi Stolk in Trans. Brit. Mycol. Soc. 40: 190. 1 Jul 1957. — Holotype: [dried culture from soil from] Zaire, Yangambi, *Meyer* (K).

Aspergillus recurvatus Raper & Fennell, Genus Aspergillus: 529. 1965. — Neotype (Samson & Gams, 1985): No. 36528 (IMI).

Aspergillus reptans Samson & W. Gams in Samson & Pitt, Adv. Penicillium Aspergillus Syst.: 48. 1985 (= *Aspergillus glaucus* var. *repens* Corda, Icon. Fung. 5: 53. Jun 1842 [non *Aspergillus repens* (de Bary) E. Fisch. in Engler & Prantl, Nat. Pflanzenfam. 1(1); 302.Feb 1897]).— Neotype (Samson & Gams, 1985): No. 529.65 p.p. (CBS).

Aspergillus restrictus G. Sm. in J. Textile Inst.22: 115.1931.— Neotype (Samson & Gams, 1985): No. 16267 (IMI).

Aspergillus robustus M. Chr. & Raper in Mycologia 70: 200. 27 Mar 1978. — Holotype: No. WB 5286 (NY).

Aspergillus rubrobrunneus Samson & W. Gams in Samson & Pitt, Adv. Penicillium Aspergillus Syst.: 49. 1985. — Holotype: No. 530.65 p.p. (CBS).

Aspergillus rugulovalvus Samson & W. Gams in Samson & Pitt, Adv. Penicillium Aspergillus Syst.: 49. 1985. — Holotype: No. 136775 p.p. (IMI).

Aspergillus sclerotiorum G. A. Huber in Phytopathology 23: 306. Mar 1933. — Neotype (Samson & Gams, 1985): No. 56673 (IMI).

Aspergillus sepultus Tuthill & M. Chr. in Mycologia 78: 475. 16 Jun 1986. — Holotype: No. RMF 7602 (NY).

Aspergillus silvaticus Fennell & Raper in Mycologia 47: 83. 8 Mar 1955. — Neotype (Samson & Gams, 1985): No. 61456 (IMI).

Aspergillus sojae Sakag. & K. Yamada ex Murak. in Rep. Res. Inst. Brewing 143: 8. 1971.— Neotype (designated here): No. 191300 (IMI).

Aspergillus sparsus Raper & Thom in Mycologia 36: 572. Nov-Dec 1944. — Neotype (Samson & Gams, 1985): No. 19394 (IMI).

Aspergillus spathulatus Takada & Udagawa in Mycotaxon 24: 396. 11 Dec 1985. — Holotype: No. 2947 p.p. (NHL).

Aspergillus spectabilis M. Chr. & Raper in Mycologia 70: 333. 25 Mai 1978. — Holotype: No. RMFH 429 p.p. (NY).

Aspergillus spelunceus Raper & Fennell, Genus Aspergillus: 457. 1965. — Neotype (Samson Gams, 1985): No. 211389 (IMI).

Aspergillus spinosus Kozak. in Mycol. Pap.161: 58.30 Jun 1989. — Holotype: No. 211390 (IMI).

Aspergillus stellifer Samson & W. Gams in Samson & Pitt, Adv. Penicillium Aspergillus Syst: 52. 1985. — Holotype: Bowenpilly near Secundarabad, *s. coll.,* (K).

Aspergillus striatulus Samson & W. Gams in Samson & Pitt, Adv. Penicillium Aspergillus Syst.: 50. 1985. — Holotype: No. 96679 (IMI).

Aspergillus stromatoides Raper & Fennell, Genus Aspergillus: 421. 1965. — Neotype (Samson & Gams, 1985): No. 123750 (IMI).

Aspergillus sublatus Y. Horie in Trans. Mycol. Soc. Japan 20: 481. Dec 1979. — Holotype: No. 4553 p.p. (IFM).

Aspergillus subolivaceus Raper & Fennell, Genus Aspergillus: 385. 1965. — Neotype (Samson & Gams, 1985): No. 44882 (IMI).

Aspergillus subsessilis Raper & Fennell, Genus Aspergillus: 530. 1965. — Neotype (Samson & Gams, 1985): No. 135820 (IMI).

Aspergillus sulphureus (Fresen.) Wehmer in Mém. Soc. Phys. Geneve 33(2): 113. Aug 1901. — Basionym: *Sterigmatocystis sulphurea* Fresen., Beitr. Mykol.: 83. 18 Aug 1863. — Neotype (Kozakiewicz, 1989): No. 211397 (IMI).

Aspergillus sydowii (Bainier & Sartory) Thom & Church, Aspergilli: 147. Mar 1926. — Basionym: *Sterigmatocystis sydowii* Bainier & Sartory in Ann. Mycol. 11: 25. 15 Mar 1913. — Neotype (Samson & Gams, 1985): No. 211384 (IMI).

Aspergillus tamarii Kita in Centralbl. Bakteriol., 2. Abth., 37: 433. 22 Mai 1913. — Neotype (Samson & Gams, 1985): No. 104.13 (CBS).

Aspergillus tardus Bissett & Widden in Canad. J. Bot. 62: 2521. 19 Dec 1984. — Holotype: No. 183872 (DAOM).

Aspergillus tatenoi Y. Horie & al. in Trans. Mycol. Soc. Japan 33: 395. 1992. — Holotype: No. FA 0022 p.p. (CBM).

Aspergillus terreus Thom in Amer. J. Bot. 5: 85. Feb 1918.— Neotype (Samson & Gams, 1985): No. 17294 (IMI).

Aspergillus terricola E. J. Marchal in Rev. Mycol. (Toulouse) 15: 101. 1893.— Neotype (Samson & Gams, 1985): No. 172294 (IMI).

Aspergillus tetrazonus Samson & W. Gams in Samson & Pitt, Adv. Penicillium Aspergillus Syst.: 48. 1985. — Holotype: No. 89351 p.p. (IMI).

Aspergillus thermomutatus (Paden) S.W. Peterson in Mycol. Res. 96: 549. Jul 1992. — Basionym: *Aspergillus fischeri* var. *thermomutatus* Paden in Mycopathol. Mycol. Appl. 36: 161. 13 Nov 1968. — Holotype: No. 1108305 p.p. (BPI).

Aspergillus togoënsis (Henn.) Samson & Seifert in Samson & Pitt, Adv. Penicillium Aspergillus Syst.: 419. 1985. — Basionym: *Stilbothamnium togoënse* Henn. in Bot. Jahrb. Syst. 23: 542. 25 Mai 1897. — Neotype (Samson & Seifert, 1985): Zaire, *Louis 6190*, No. B 1009 (BR).

Aspergillus tonophilus Ohtsuki in Bot. Mag. (Tokyo) 75: 438. 25 Nov 1962. — Neotype (Samson & Gams, 1985): No. 108299 p.p. (IMI).

Aspergillus undulatus H. Z. Kong & Z. T. Qi in Acta Mycol. Sin. 4: 211 Nov 1985. — Holotype: No. 47644 p.p. (HMAS).

Aspergillus unguis (Emile-Weil & L. Gaudin) Thom & Raper in Mycologia 31: 667. Nov-Dec 1939. — Basionym: *Sterigmatocystis unguis* Emile-Weil & L. Gaudin in Arch. Méd. Exp. Anat. Pathol.28: 463. 1918. — Neotype (Samson & Gams, 1985): No. 136526 (IMI).

Aspergillus unilateralis Thrower in Austral. J. Bot. 2. 355. Nov 1954. — Neotype (Samson & Gams, 1985): No. 62876 (IMI).

Aspergillus ustus (Bainier) Thom & Church, Aspergilli: 152. Mar 1926. — Basionym: *Sterigmatocystis usta* Bainier in Bull. Soc. Bot. France 28: 78. 1881. — Neotype (Samson & Gams, 1985): No. 211805 (IMI).

Aspergillus varians Wehmer in Bot. Centralbl. 80: 460. 13 Dec 1899. — Neotype (Samson & Gams, 1985): No. 172297 (IMI).

Aspergillus versicolor (Vuill.) Tirab. in Ann. Bot. (Roma) 7: 9. 31 Aug 1908.— Basionym: *Sterigmatocystis versicolor* Vuill. in Mirsky, Causes Err. Dét. Aspergill.: 15. 1903. — Neotype (Samson & Gams, 1985): No. 538.65 (CBS).

Aspergillus violaceobrunneus Samson & W. Gams in Samson & Pitt, Adv. Penicillium Aspergillus Syst.: 53. 1985. — Holotype: No. 61449 p.p. (IMI).

Aspergillus violaceofuscus Gasperini in Atti Soc. Tosc. Sci. Nat. Pisa, Processi Verbali 8: 326. 1887. — Neotype (Samson & Gams, 1985): No. 123.27 (CBS).

Aspergillus viridinutans Ducker & Thrower in Austral. J. Bot.2: 355. Nov 1954. — Neotype (Samson & Gams, 1985): No. 62875 (IMI).

Aspergillus vitellinus (Massee) Samson & Seifert in Samson & Pitt, Adv. Penicillium Aspergillus Syst.: 417.1985.—Basionym: *Sterigmatocystis vitellina* Ridl. ex Massee in J. Bot. 34: 152. Apr 1896. — Holotype: Singapore, 1894, *Ridley 2970* (K).

Aspergillus vitis Novobr. In Novosti Sist. Nizš. Rast. 9: 175. 4 Aug 1972. — Neotype (Kozakiewicz, 1989): No. 174724 (IMI).

Aspergillus warcupii Samson & W. Gams in Samson & Pitt, Adv. Penicillium Aspergillus Syst.: 50. 1985. — Holotype: No. 75885 p.p. (IMI).

Aspergillus wentii Wehmer in Centralbl. Bakteriol., 2. Abth., 2: 149. 27 Mar 1896. — Neotype (Samson & Gams, 1985): No. 17295 (IMI).

Aspergillus xerophilus Samson & Mouch. in Antonie van Leeuwenhoek J. Microbiol. Serol. 41: 348. 1975. — Holotype: No. 938.73 p.p. (CBS).

Aspergillus zonatus Kwon-Chung & Fennell in Raper & Fennell, Genus Aspergillus: 377. 1965.— Neotype (Samson & Gams, 1985): No. 506.65 (CBS).

BYSSOCHLAMYS WESTLING (HOLOMORPHS)

Byssochlamys fulva Olliver & G. Sm. in J. Bot. 71: 196. Jul 1933. — Neotype (designated here): No. 40021 (IMI). [Anamorph: *Paecilomyces fulvus* Stolk & Samson].

Byssochlamys nivea Westling in Svensk Bot. Tidskr.3: 134.28 Jun 1909. — Neotype (designated here): No. 100.11 (CBS). [Anamorph: *Paecilomyces niveus* Stolk & Samson].

Byssochlamys verrucosa Samson & Tansey in Trans. Brit. Mycol. Soc. 65: 512. Dec 1975. — Holotype: No. 605.74 (CBS). [Anamorph: *Paecilomyces verrucosus* Samson & Tansey].

Byssochlamys zollerniae C. Ram in Nova Hedwigia 16: 312. 5 Dec 1968. — Lectotype (designated here): icon in Nova Hedwigia 16: tab. 107. 5 Dec 1968. [Anamorph: *Paecilomyces zollerniae* Stolk & Samson].

CHAETOSARTORYA SUBRAM. (HOLOMORPHS)

Chaetosartorya chrysella (Kwon-Chung & Fennell) Subram. in Curr. Sci. 41: 761. 5 Nov 1972. Basionym: *Aspergillus chrysellus* Kwon-Chung & Fennell in Raper & Fennell, Genus Aspergillus: 424. 1965 (nom. holomorph.). — Neotype (Samson & Gams, 1985): No. 238612 (IMI). [Anamorph: *Aspergillus chryseides* Samson & W. Gams].

Chaetosartorya cremea (Kwon-Chung & Fennell) Subram. in Curr. Sci. 41: 761. 5 Nov 1972. — Basionym: *Aspergillus cremeus* Kwon-Chung & Fennell in Raper & Fennell, Genus Aspergillus: 418. 1965 (nom. holomorph.). — Neotype (Samson & Gams, 1985): No. 123749ii (IMI). [Anamorph: *Aspergillus cremeoflavus* Samson & W. Gams].

Chaetosartorya stromatoides B. J. Wiley & E. G. Simmons in Mycologia 65: 935. 25 Sep 1973. — Holotype: No. 8944 (QM). [Anamorph: *Aspergillus stromatoides* Raper & Fennell].

CRISTASPORA FORT & GUARRO (HOLOMORPHS)

Cristaspora arxii Fort & Guarro in Mycologia 76: 1115. 5 Dec 1984. — Holotype: No. 525.83 (CBS).

DENDROSPHAERA PAT. (HOLOMORPHS)

Dendrosphaera eberhardtii Pat. in Bull. Soc. Mycol. France 23: 69. 1907. — Type not known.

DICHLAENA DURIEU & MONT. (HOLOMORPHS)

Dichlaena lentisci Durieu & Mont. in Durieu, Expl. Sci. Algérie, Bot., 1: 405. 1849. — Holotype: *von Höhnel* (FH).

DICHOTOMOMYCES *D. B.* SCOTT (HOLOMORPHS)

Dichotomomyces cejpii (Milko) D. B. Scott in Trans. Brit. Mycol. Soc. 55: 313. 19 Oct 1970. — Basionym: *Talaromyces cejpii* Milko in Novosti Sist. Nizš. Rast. 1964: 208. 12 Aug 1964. — Neotype (designated here): No. 157.66 (CBS).

EMERICELLA BERK. (HOLOMORPHS)

Emericella acristata (Fennell & Raper) Y. Horie in Trans. Mycol. Soc. Japan 21: 491. Dec 1980. — Basionym: *Aspergillus nidulans* var. *acristatus* Fennell & Raper in Mycologia 47: 79. 8 Mar 1955 (nom. holomorph.). — Neotype (designated here): No. 61453 (IMI).

Emericella astellata (Fennell & Raper) Y. Horie in Trans. Mycol. Soc. Japan 21: 491. Dec 1980. — Basionym: *Aspergillus varicolor* var. *astellatus* Fennell & Raper in Mycologia 47: 81. 8 Mar 1955 (nom. holomorph.). — Neotype (designated here): No. 61455 (IMI).

Emericella aurantiobrunnea (G. A. Atkins & al.) Malloch & Cain in Canad. J. Bot. 50: 61. 8 Feb 1972. — Basionym: *Emericella nidulans* var. *aurantiobrunnea* G. A. Atkins & al. in Trans. Brit. Mycol. Soc. 41: 504. 19 Dec 1958. — Holotype: No. 74897 (IMI).

Emericella bicolor M. Chr. & States in Mycologia 70: 337. 25 Mai 1978. — Holotype: No. RMF 2058 (NY). [Anamorph: *Aspergillus bicolor* M. Chr. & States].

Emericella corrugata Udagawa & Y. Horie in Mycotaxon 4: 535. 18 Dec 1976. — Holotype: No. 2763 (NHL). [Anamorph: *Aspergillus corrugatus* Udagawa & Y. Horie].

Emericella dentata (D. K. Sandhu & R. S. Sandhu) Y. Horie in Trans. Mycol. Soc. Japan 21: 491. Dec 1980. — Basionym: *Aspergillus nidulans* var. *dentatus* D. K. Sandhu & R. S. Sandhu in Mycologia 55: 297. 7 Jun 1963 (nom. holomorph.). — Neotype (designated here): No. 126693 (IMI).

Emericella desertorum Samson & Mouch. in Antonie van Leeuwenhoek J. Microbiol. Serol. 40: 121. 1974. — Holotype: No. 653.73 (CBS).

Emericella echinulata (Fennell & Raper) Y. Horie in Trans. Mycol. Soc. Japan 21: 492. Dec 1980. — Basionym: *Aspergillus nidulans* var. *echinulatus* Fennell & Raper in Mycologia 47: 79. 8 Mar 1955 (nom. holomorph.). — Neotype (designated here): No. 61454 (IMI).

Emericella falconensis Y. Horie & al. in Trans. Mycol. Soc. Japan 30: 257. Oct 1989. — Holotype: No. 10001 p.p. (CBM). [Anamorph: *Aspergillus falconensis* Y. Horie].

Emericella foeniculicola Udagawa in Trans. Mycol. Soc. Japan 20: 13. Jun 1979. — Holotype: No. 2777 (NHL). [Anamorph: *Aspergillus foeniculicola* Udagawa].

Emericella foveolata Y. Horie in Trans. Mycol. Soc. Japan 19: 313. Oct 1978. — Holotype: No. 4547 (IFM). [Anamorph: *Aspergillus foveolatus* Y. Horie].

Emericella fruticulosa (Raper & Fennell) Malloch & Cain in Canad. J. Bot.50: 61. 8 Feb 1972. — Basionym: *Aspergillus fructiculosus* Raper & Fennell, Genus Aspergillus: 506. 1965 (nom. holomorph.). — Neotype (Samson & Gams, 1985): No. 139279 (IMI). [Anamorph: *Aspergillus fruticans* Samson & W. Gams].

Emericella heterothallica (Kwon-Chung & al.) Malloch & Cain in Canad. J. Bot.50: 62.8 Feb 1972. — Basionym: *Aspergillus heterothallicus* Kwon-Chung & al. in Raper & Fennell, Genus Aspergillus: 502. 1965 (nom. holomorph.). — Neotype (Samson & Gams, 1985): No. 488.65 x 489.65 (CBS). [Anamorph: *Aspergillus compatibilis* Samson & W. Gams].

Emericella navahoënsis M. Chr. & States in Mycologia 74: 226. 2 Apr 1982. — Holotype: No. SD-5 (NY). [Anamorph: *Aspergillus navahoënsis* M. Chr. & States].

Emericella nidulans (Eidam) Vuill. in Compt. Rend. Hebd. Séances Acad. Sci. 184: 137. 1927. — Basionym: *Sterigmatocystis nidulans* Eidam in Beitr. Biol. Pflanzen 3: 393. 1883 (nom. holomorph.). — Neotype (Samson & Gams, 1985): No. 86806 (IMI). [Anamorph: *Aspergillus nidulans* (Eidam) G. Winter].

Emericella parvathecia (Raper & Fennell) Malloch & Cain in Canad. J. Bot. 50: 62. 8 Feb 1972. — Basionym: *Aspergillus parvathecius* Raper & Fennell, Genus Aspergillus: 509. 1965 (nom. holomorph.). — Neotype (Samson & Gams, 1985): No. 139280 (IMI). [Anamorph: *Aspergillus microthecius* Samson & W. Gams].

Emericella purpurea Samson & Mouch. in Antonie van Leeuwenhoek J. Microbiol. Serol. 41: 350. 1975. — Holotype: No. 754.74 (CBS). [Anamorph: *Aspergillus purpureus* Samson & Mouch.].

Emericella quadrilineata (Thom & Raper) C. R. Benj. in Mycologia 47: 680. 7 Oct 1955. — Basionym: *Aspergillus quadrilineatus* Thom & Raper in Mycologia 31: 660. Nov-Dec 1939 (nom. holomorph.). — Neotype (Samson & Gams, 1985): No. 89351 (IMI). [Anamorph: *Aspergillus tetrazonus* Samson & W. Gams].

Emericella rugulosa (Thom & Raper) C. R. Benj. in Mycologia 47: 680. 7 Oct 1955. — Basionym: *Aspergillus rugulosus* Thom & Raper in Mycologia 31: 660. Nov-Dec 1939 (nom. holomorph.). — Neotype (Samson & Gams, 1985): No. 136775 (IMI). [Anamorph: *Aspergillus rugulovalvus* Samson & W. Gams].

Emericella similis Y. Horie & al. in Trans. Mycol. Soc. Japan 31: 425. Dec 1990. — Holotype: No. 10007 (CBM).

Emericella spectabilis M. Chr. & Raper in Mycologia 70: 333. 25 Mai 1978. — Holotype: No. RMFH 429 (NY). [Anamorph: *Aspergillus spectabilis* M. Chr. & Raper].

Emericella striata (J. N. Rai & al.) Malloch & Cain in Canad. J. Bot. 50: 62. 8 Feb 1972. — Basionym: *Aspergillus striatus* J. N. Rai & al. in Canad. J. Bot. 42: 1521. 1964 (nom. holomorph.). — Neotype (Samson & Gams, 1985): No. 96679 (IMI). [Anamorph: *Aspergillus striatulus* Samson & W. Gams].

Emericella sublata Y. Horie in Trans. Mycol. Soc. Japan 20: 481. Dec 1979. — Holotype: No. 4553 (IFM). [Anamorph: *Aspergillus sublatus* Y. Horie].

Emericella undulata H. Z. Kong & Z. T. Qi in Acta Mycol. Sin. 4: 211. Nov 1985. — Holotype: No. 47644 (HMAS). [Anamorph: *Aspergillus undulatus* H. Z. Kong & Z. T. Qi].

Emericella unguis Malloch & Cain in Canad. J. Bot. 50: 62. 8 Feb 1972. — Holotype: No. 2393 (NRRL). [Anamorph: *Aspergillus unguis* (Emile-Weil & L. Gaudin) Thom & Raper].

Emericella variecolor Berk. & Broome in Berkeley, Intr. Crypt. Bot.: 340. Mar-Jun 1857. — Holotype: India, Sacundrabad, 3 Jul 1855, *s.coll.* (K). [Anamorph: *Aspergillus stellifer* Samson & W. Gams].

Emericella violacea (Fennell & Raper) Malloch & Cain in Canad. J. Bot. 50: 62. 8 Feb 1972. — Basionym: *Aspergillus violaceus* Fennell & Raper in Mycologia 47: 75. 8 Mar 1955 (nom. holomorph.). — Neotype (Samson & Gams,

1985): No. 61449 (IMI). [Anamorph: *Aspergillus violaceobrunneus* Samson & W. Gams].

EUPENICILLIUM F. LUDW. (HOLOMORPHS)

Eupenicillium abidjanum Stolk in Antonie van Leeuwenhoek J. Microbiol. Serol. 34: 49. 1968. — Holotype: No. 247.67 p.p. (CBS). [Anamorph: *Penicillium abidjanum* Stolk].

Eupenicillium alutaceum D. B. Scott in Mycopathol. Mycol. Appl. 36: 17. 31 Oct 1968. — Holotype: No. 317.67 p.p. (CBS). [Anamorph: *Penicillium alutaceum* D. B. Scott].

Eupenicillium anatolicum Stolk in Antonie van Leeuwenhoek J. Microbiol. Serol. 34: 46. 1968. — Holotype: No. 479.66 p.p. (CBS). [Anamorph: *Penicillium anatolicum* Stolk].

Eupenicillium angustiporcatum Takada & Udagawa in Trans. Mycol. Soc. Japan 24: 143. Jul 1983. — Holotype: No. 6481 p.p. (NHL). [Anamorph: *Penicillium angustiporcatum* Takada & Udagawa].

Eupenicillium baarnense (J. F. H. Beyma) Stolk & D. B. Scott in Persoonia 4: 401. 1 Aug 1967. — Basionym: *Penicillium baarnense* J. F. H. Beyma in Antonie van Leeuwenhoek J. Microbiol. Serol. 6: 271.1940 (nom. holomorph.). — Neotype (Pitt, 1980): No. 134.41 (CBS). [Anamorph: *Penicillium vanbeymae* Pitt].

Eupenicillium brefeldianum (B. O. Dodge) Stolk & D. B. Scott in Persoonia 4: 400. 1 Aug 1967. — Basionym: *Penicillium brefeldianum* B. O. Dodge in Mycologia 25: 92. Mar-Apr 1933 (nom. holomorph.). — Neotype (Stolk & D. B. Scott, 1967): No. 216895 (IMI). [Anamorph: *Penicillium dodgei* Pitt].

Eupenicillium catenatum D. B. Scott in Mycopathol. Mycol. Appl. 36: 24. 31 Oct 1968. — Holotype: No. 352.67 p.p. (CBS). [Anamorph: *Penicillium catenatum* D. B. Scott].

Eupenicillium cinnamopurpureum D. B. Scott & Stolk in Antonie van Leeuwenhoek J. Microbiol. Serol. 33: 308. 1967. — Holotype (Pitt, 1980): No. 114483 (IMI). [Anamorph: *Penicillium phoeniceum* J. F. H. Beyma].

Eupenicillium crustaceum F. Ludw., Lehrb. Nied. Krypt.: 263. Jul 1892. — Lectotype (Stolk & D. B. Scott, 1967): icon of '*P. glaucum*' in Brefeld, Bot. Unters. Schimmelpilze 2: fig. 10-54. Feb 1874. [Anamorph: *Penicillium gladioli* L. McCulloch & Thom].

Eupenicillium cryptum Goch. in Mycotaxon 26: 349. 15 Jul 1986. — Holotype: No. 769 p.p. (NY). [Anamorph: *Penicillium cryptum* Goch.].

Eupenicillium egyptiacum (J. F. H. Beyma) Stolk & D. B. Scott in Persoonia 4: 401. 1 Aug 1967. — Basionym: *Penicillium egyptiacum* J. F. H. Beyma in Zentralbl. Bakteriol., 2. Abt., 88: 137. 24 Apr 1933 (nom. holomorph.). — Neotype (Pitt, 1980): No. 40580 p.p. (IMI). [Anamorph: *Penicillium nilense* Pitt].

Eupenicillium ehrlichii (Kleb.) Stolk & D. B. Scott in Persoonia 4: 400. 1 Aug 1967. — Basionym: *Penicillium ehrlichii* Kleb. in Ber. Deutsch. Bot. Ges. 48: 374. 29 Dec 1930 (nom. holomorph.). — Neotype (Pitt, 1980): No. 39737 (IMI). [Anamorph: *Penicillium klebahnii* Pitt].

Eupenicillium erubescens D. B. Scott in Mycopathol. Mycol. Appl. 36: 14. 31 Oct 1968. — Holotype: No. 318.67 p.p. (CBS). [Anamorph: *Penicillium erubescens* D. B. Scott].

Eupenicillium fractum Udagawa in Trans. Mycol. Soc. Japan 9: 51. 1 Nov 1968. — Holotype: No. 6104 p.p. (NHL). [Anamorph: *Penicillium fractum* Udagawa].

Eupenicillium gracilentum Udagawa & Y. Horie in Trans. Mycol. Soc. Japan 14: 373. 20 Dec 1973. — Holotype: No. 6452 p.p. (NHL). [Anamorph: *Penicillium gracilentum* Udagawa & Y. Horie].

Eupenicillium hirayamae D. B. Scott & Stolk in Antonie van Leeuwenhoek J. Microbiol. Serol. 33: 305. 1967. — Holotype: No. 229.60 (CBS). [Anamorph: *Penicillium hirayamae* Udagawa].

Eupenicillium inusitatum D. B. Scott in Mycopathol. Mycol. Appl. 36: 20. 31 Oct 1968. — Holotype: No. 351.67 p.p. (CBS). [Anamorph: *Penicillium inusitatum* D. B. Scott].

Eupenicillium javanicum (J. F. H. Beyma) Stolk & D. B. Scott in Persoonia 4: 398. 1 Aug 1967. — Basionym: *Penicillium javanicum* J. F. H. Beyma in Verh. Kon. Ned. Akad. Wetensch., Afd. Natuurk., Tweede Sect., 26(4): 17. 1929 (nom. holomorph.). — Neotype (Pitt, 1980): No.39733 p.p. (IMI). [Anamorph: *Penicillium indonesiae* Pitt].

Eupenicillium katangense Stolk in Antonie van Leeuwenhoek J. Microbiol. Serol. 34: 42. 1968. — Holotype: No. 247.67 p.p. (CBS). [Anamorph: *Penicillium katangense* Stolk].

Eupenicillium lapidosum D. B. Scott & Stolk in Antonie van Leeuwenhoek J. Microbiol. Serol. 33: 298. 1967. — Holotype: No. 343.48 (CBS). [Anamorph: *Penicillium lapidosum* Raper & Fennell].

Eupenicillium lassenii Paden in Mycopathol. Mycol. Appl. 43: 266. 25 Mar 1971. — Holotype: No. JWP 69-26 p.p. (UVIC). [Anamorph: *Penicillium lassenii* Paden].

Eupenicillium levitum (Raper & Fennell) Stolk & D. B. Scott in Persoonia 4: 402. 1 Aug 1967. — Basionym: *Penicillium levitum* Raper & Fennell in Mycologia 40: 511. Sep-Oct 1948 (nom. holomorph.). — Neotype (Pitt, 1980): No. 39735 p.p. (IMI). [Anamorph: *Penicillium rasile* Pitt].

Eupenicillium limoneum Goch. & Zlattner in Stud. Mycol. 23: 100. 15 Mar 1983. — Holotype: No. 650.82 (CBS). [Anamorph: *Torulomyces lagena* Delitsch].

Eupenicillium lineolatum Udagawa & Y. Horie in Mycotaxon 5: 493. 6 Mai 1977. — Holotype: No. 2776 p.p. (NHL). [Anamorph: *Penicillium lineolatum* Udagawa & Y. Horie].

Eupenicillium ludwigii Udagawa in Trans. Mycol. Soc. Japan 10: 2. 1 Aug 1969.   Holotype: No.6118 p.p. (NHL). [Anamorph: *Penicillium ludwigii* Udagawa].

Eupenicillium meliforme Udagawa & Y. Horie in Trans. Mycol. Soc. Japan 14: 376. 20 Dec 1973. — Holotype: No. 6468 p.p. (NHL). [Anamorph: *Penicillium meliforme* Udagawa & Y. Horie]. [Anamorph: *Penicillium meliforme* Udagawa & Y. Horie].

Eupenicillium meridianum D. B. Scott in Mycopathol. Mycol. Appl. 36: 12. 31 Oct 1968. — Holotype: No. 314.67 p.p. (CBS). [Anamorph: *Penicillium meridianum* D.B. Scott].

Eupenicillium molle Malloch & Cain in Canad. J. Bot. 50: 62. 8 Feb 1972. — Holotype: No. 45714 (TRTC). [Anamorph: *Penicillium molle* Pitt].

Eupenicillium nepalense Takada & Udagawa in Trans. Mycol. Soc. Japan 24: 146. Jul 1983. — Holotype: No. 6482 p.p. (NHL). [Anamorph: *Penicillium nepalense* Takada & Udagawa].

Eupenicillium ochrosalmoneum D B. Scott & Stolk in Antonie van Leeuwenhoek J. Microbiol. Serol. 33: 302. 1967. — Holotype: No. 489.66 (CBS). Anamorph: *Penicillium ochrosalmoneum* Udagawa].

Eupenicillium ornatum Udagawa in Trans. Mycol. Soc. Japan 9: 49. 1 Nov 1968. — Holotype: No.6101 p.p. (NHL). [Anamorph: *Penicillium ornatum* Udagawa].

Eupenicillium osmophilum Stolk & Veenb.-Rijks in Antonie van Leeuwenhoek J. Microbiol. Serol. 40: 1. 1974. — Holotype: No. 462.72 p.p. (CBS). [Anamorph: *Penicillium osmophilum* Stolk & Veenb.-Rijks].

Eupenicillium parvum (Raper & Fennell) Stolk & D. B. Scott in Persoonia 4: 402. 1 Aug 1967. — Basionym: *Penicillium parvum* Raper & Fennell in Mycologia 40: 508. Sep-Oct 1948 (nom. holo-

morph.). — Neotype (Pitt, 1980): No. 359.48 (CBS). [Anamorph: *Penicillium papuanum* Udagawa & Y. Horie].

Eupenicillium pinetorum Stolk in Antonie van Leeuwenhoek J. Microbiol. Serol. 34: 37. 1968.— Holotype: No. 295.62 (CBS). [Anamorph: *Penicillium pinetorum* M. Chr. & Backus].

Eupenicillium reticulisporum Udagawa in Trans. Mycol. Soc. Japan 9: 52. 1 Nov 1968. — Holotype: No. 6105 p.p. (NHL). [Anamorph: *Penicillium reticulisporum* Udagawa].

Eupenicillium rubidurum Udagawa & Y. Horie in Trans. Mycol. Soc. Japan 14: 381. 20 Dec 1973. — Holotype: No.6460 p.p. (NHL). [Anamorph: *Penicillium rubidurum* Udagawa & Y. Horie].

Eupenicillium senticosum D. B. Scott in Mycopathol. Mycol. Appl. 36: 5. 31 Oct 1968. — Holotype: No. 316.67 p.p. (CBS). [Anamorph: *Penicillium senticosum* D. B. Scott].

Eupenicillium shearii Stolk & D. B. Scott in Persoonia 4: 396. 1 Aug 1967. — Holotype: No. 290.48 p.p. (CBS). [Anamorph: *Penicillium shearii* Stolk & D. B. Scott].

Eupenicillium sinaicum Udagawa & S. Ueda in Mycotaxon 14: 266. 22 Jan 1982. — Holotype: No. 2894 p.p. (NHL). [Anamorph: *Penicillium sinaicum* Udagawa & S. Ueda].

Eupenicillium stolkiae D. B. Scott in Mycopathol. Mycol. Appl. 36: 8. 31 Oct 1968. — Holotype: No. 315.67 p.p. (CBS). [Anamorph: *Penicillium stolkiae* D. B. Scott].

Eupenicillium terrenum D. B. Scott in Mycopathol. Mycol. Appl. 36: 1. 31 Oct 1968. — Holotype: No. 313.67 p.p. (CBS). [Anamorph: *Penicillium terrenum* D. B. Scott].

Eupenicillium tularense Paden in Mycopathol. Mycol. Appl. 43: 262. 25 Mar 1971. — Holotype: No. JWP 68-31 p.p. (UVIC). [Anamorph: *Penicillium tularense* Paden].

Eupenicillium zonatum Hodges & J. J. Perry in Mycologia 65: 697. 19 Jul 1973. — Holotype: No. FSL 525 p.p. (BPI). [Anamorph: *Penicillium zonatum* Hodges & J. J. Perry]

EUROTIUM LINK: FR (HOLOMORPHS)

Eurotium amstelodami L. Mangin in Ann. Sci. Nat., Bot., ser. 9, 10: 360. 1909. — Neotype (Samson & Gams, 1985): No. 518.65 (CBS). [Anamorph: *Aspergillus vitis* Novobr.].

Eurotium appendiculatum Blaser in Sydowia 28: 38. Dec 1976. — Holotype: No. 8286 (ZT). [Anamorph: *Aspergillus appendiculatus* Blaser].

Eurotium athecium (Raper & Fennell) Arx, Gen. Fungi Sporul. Pure Cult., ed. 2: 91. 1974. — Basionym: *Aspergillus athecius* Raper & Fennell, Genus Aspergillus: 183. 1965 (nom. holomorph.). — Neotype (Samson & Gams, 1985): No. 32048 (IMI). [Anamorph: *Aspergillus atheciellus* Samson & W. Gams].

Eurotium carnoyi Malloch & Cain in Canad. J. Bot. 50: 63. 8 Feb 1972. — Neotype (Samson & Gams, 1985): No. 172279 (IMI). [Anamorph: *Aspergillus neocarnoyi* Kozak.].

Eurotium chevalieri L. Mangin in Ann. Sci. Nat., Bot., ser. 9, 10: 361. 1909. — Neotype (Blaser, 1975): No. 211382 (IMI). [Anamorph: *Aspergillus chevalieri* L. Mangin].

Eurotium cristatum (Raper & Fennell) Malloch & Cain in Canad. J. Bot. 50: 64. 8 Feb 1972. — Basionym: *Aspergillus cristatus* Raper & Fennell, Genus Aspergillus: 169. 1965 (nom. holomorph.). — Neotype (Samson & Gams, 1985): No. 172280 (IMI). [Anamorph: *Aspergillus cristatellus* Kozak.].

Eurotium echinulatum Delacr. in Bull. Soc. Mycol. France 9: 266. 1893. — Neotype (Blaser, 1975): No. 211378 (IMI). [Anamorph: *Aspergillus brunneus* Delacr.].

Eurotium glabrum Blaser in Sydowia 28: 35. Dec 1976. — Holotype: No. 8218 T (ZT). [Anamorph: *Aspergillus glaber* Blaser].

Eurotium halophilicum C. M. Chr. & al. in Mycologia 51: 636. 17 Mar 1961. — Neotype (Samson & Gams, 1985); No. NRRL 2739 (BPI). [Anamorph: *Aspergillus halophilicus* C. M. Chr. & al.].

Eurotium herbariorum Link in Ges. Naturf. Freunde Berlin Mag. Neuesten Entdeck. Gesammten Naturk. 3: 31. 1809: Fr. — Neotype (Malloch & Cain, 1972): No. 137960 (DAOM). [Anamorph: *Aspergillus glaucus* Link].

Eurotium intermedium Blaser in Sydowia 28: 41. Dec 1976. — Basionym: *Aspergillus chevalieri* var. *intermedius* Thom & Raper in Misc. Publ. U.S. Dept. Agric.426: 21. 1941 (nom. holomorph.). — Neotype (Kozakiewicz, 1989): No. 89278 (IMI). [Anamorph: *Aspergillus intermedius* Blaser].

Eurotium leucocarpum Hadlok & Stolk in Antonie van Leeuwenhoek J. Microbiol. Serol. 35: 9. 1969. — Holotype: No. 353.68 (CBS). [Anamorph: *Aspergillus leucocarpus* Hadlok & Stolk].

Eurotium medium R. Meissn. in Bot. Zeitung, 2. Abt., 55: 356. 1 Dec 1897. — Neotype (Samson & Gams, 1985): No. 113.27 (CBS). [Anamorph: *Aspergillus medius* R. Meissn.].

Eurotium niveoglaucum (Thom & Raper) Malloch & Cain in Canad. J. Bot. 50: 64. 8 Feb 1972. —

Basionym: *Aspergillus niveoglaucus* Thom & Raper in Misc. Publ. U.S. Dept. Agric. 426: 35. 1941 (nom. holomorph.). — Neotype (Samson & Gams, 1985): No. 32050ii (IMI). [Anamorph: *Aspergillus glauconiveus* Samson & W. Gams].

Eurotium pseudoglaucum (Blochwitz) Malloch & Cain in Canad. J. Bot. 50: 64. 8 Feb 1972. — Basionym: *Aspergillus pseudoglaucus* Blochwitz in Ann. Mycol. 27: 207. 25 Jun 1929 (nom. holomorph.). — Neotype (Samson & Gams, 1985): No. 16122ii (IMI). [Anamorph: *Aspergillus glaucoaffinis* Samson & W. Gams].

Eurotium repens de Bary in Abh. Senckenberg. Naturf. Ges. 7: 379. 1870. — Neotype (Samson & Gams, 1985): No. 529.65 (CBS). [Anamorph: *Aspergillus reptans* Samson & W. Gams].

Eurotium rubrum Jos. König & al. in Z. Untersuch. Nahrungs- Genussmittel 4: 726. 1901. — Neotype (Samson & Gams, 1985): No. 530.65 (CBS). [Anamorph: *Aspergillus rubrobrunneus* Samson & W. Gams].

Eurotium tonophilum Ohtsuki in Bot. Mag. (Tokyo) 75: 438. 25 Nov 1962. — Neotype (Samson & Gams, 1985): No. 108299 (IMI). [Anamorph: *Aspergillus tonophilus* Ohtsuki].

Eurotium xerophilum Samson & Mouch. in Antonie van Leeuwenhoek J. Microbiol. Serol. 41: 348. 1975. — Holotype: No. 938.73 (CBS). [Anamorph: *Aspergillus xerophilus* Samson & Mouch.].

FENNELLIA B. J. WILEY & E. G. SIMMONS (HOLOMORPHS)

Fennellia flavipes B. J. Wiley & E. G. Simmons in Mycologia 65: 937 25 Sep 1973. — Holotype: No. 9131 (QM). [Anamorph: *Aspergillus flavipes* (Bainier & Sartory) Thom & Church].

Fennellia monodii Locq.-Lin. in Mycotaxon 39: 10. 26 Nov 1990. — Holotype: No. 89-3570 LCP (PC).

Fennellia nivea (B. J. Wiley & E. G. Simmons) Samson in Stud. Mycol.18: 5.3 Jan 1979. — Basionym: *Emericella nivea* B. J. Wiley & E. G. Simmons in Mycologia 65: 934.25 Sep 1973. — Holotype: No. 8942 (QM). [Anamorph: *Aspergillus niveus* Blochwitz].

GEOSMITHIA PITT (ANAMORPHS)

Geosmithia argillacea (Stolk & al.) Pitt in Canad. J. Bot. 57: 2026. 1 Oct 1979. — Basionym: *Penicillium argillaceum* Stolk & al. in Trans. Brit. Mycol. Soc. 53: 307. 8 Oct 1969. — Holotype: No. 101.69 (CBS).

Geosmithia cylindrospora (G. Sm.) Pitt in Canad. J. Bot. 57: 2024. 1 Oct 1979. — Basionym *Penicil-*

61

lium cylindrosporum G. Sm. in Trans. Brit. Mycol. Soc. 40: 483. 20 Dec 1957. — Neotype (Pitt, 1979): No. 71623 (IMI).

Geosmithia emersonii (Stolk) Pitt in Canad. J. Bot. 57: 2027. 1 Oct 1979. — Basionym: *Penicillium emersonii* Stolk in Antonie van Leeuwenhoek J. Microbiol. Serol. 31: 262. 1965. — Holotype (Stolk, 1965): No.393.64 p.p. (CBS).

Geosmithia lavendula (Raper & Fennell) Pitt in Canad. J. Bot. 57: 2022. 1 Oct 1979. — Basionym: *Penicillium lavendulum* Raper & Fennell in Mycologia 40: 530. Sep-Oct 1948. — Neotype (Pitt, 1979): No. 40570 (IMI).

Geosmithia namyslowskii (K. M. Zalessky) Pitt in Canad. J. Bot.57: 2024. 1 Oct 1979. — Basionym: *Penicillium namyslowskii* K. M. Zalessky in Bull. Int. Acad. Polon. Sci., Cl. Sci. Math., Sér. B, Sci. Nat., 1927: 479. 1927. — Neotype (Pitt, 1979): No. 40033 (IMI).

Geosmithia putterillii (Thom) Pitt in Canad. J.-Bot. 57: 2022. 1 Oct 1979. — Basionym: *Penicillium putterillii* Thom, Penicillia: 368. Jan 1930. — Neotype (Pitt, 1979): No. 40212 (IMI).

Geosmithia swiftii Pitt in Canad. J. Bot. 57: 2028. 1 Oct 1979. — Holotype: No. 40045 (IMI).

Geosmithia viridis Pitt & A. D. Hocking in Mycologia 77: 822.15 Oct 1985. — Holotype: No. 1863 (FRR).

HEMICARPENTELES A. K. SARBHOY & ELPHICK (HOLOMORPHS)

Hemicarpenteles acanthosporus Udagawa & Takada in Bull. Natl. Sci. Mus. Tokyo, ser. 2, 14: 503. 30 Sep 1971. — Holotype: No. 22462 (NHL). [Anamorph: *Aspergillus acanthosporus* Udagawa & Takada].

Hemicarpenteles paradoxus A. K. Sarbhoy & Elphick in Trans. Brit. Mycol. Soc. 51: 156. 30 Mar 1968. — Neotype (Samson & Gams, 1985): No. 61446 (IMI). [Anamorph: *Aspergillus paradoxus* Fennell & Raper].

MERIMBLA PITT (ANAMORPHS)

Merimbla ingelheimensis (J. F. H. Beyma) Pitt in Canad. J. Bot. 57: 2395. 1 Nov 1979. — Basionym: *Penicillium ingelheimense* J. F. H. Beyma in Antonie van Leeuwenhoek J. Microbiol. Serol. 8: 109. 1942. — Neotype (Pitt & Hocking, 1979): No. 234977 (IMI).

NEOSARTORYA MALLOCH & CAIN (HOLOMORPHS)

Neosartorya aurata (Warcup) Malloch & Cain in Canad. J. Bot. 50: 2620. 26 Jan 1973. — Basionym: *Aspergillus auratus* Warcup in Raper & Fennell, Genus Aspergillus: 263. 1965 (nom.

holomorph.). — Neotype (Kozakiewicz, 1989): No. 75886 (IMI). [Anamorph: *Aspergillus igneus* Kozak.].

Neosartorya aureola (Fennell & Raper) Malloch & Cain in Canad. J. Bot. 50: 2620. 26 Jan 1973. Basionym: *Aspergillus aureolus* Fennell & Raper in Mycologia 47: 71. 8 Mar 1955 (nom. holomorph.). — Neotype (Samson & Gams, 1985): No. 105.55 (CBS). [Anamorph: *Aspergillus aureoluteus* Samson & W. Gams].

Neosartorya fennelliae Kwon-Chung & S. J. Kim in Mycologia 66: 629. 28 Aug 1974. — Neotype (Kozakiewicz, 1989): No. 278382 (IMI). [Anamorph: *Aspergillus fennelliae* Kwon-Chung & S. J. Kim].

Neosartorya fischeri (Wehmer) Malloch & Cain in Canad. J. Bot. 50: 2621. 26 Jan 1973. — Basionym: *Aspergillus fischeri* Wehmer in Centralbl. Bakteriol., 2. Abth.,18: 390. 22 Apr 1907 (nom. holomorph.). — Neotype (Samson & Gams, 1985): No. 21139ii (IMI). [Anamorph: *Aspergillus fischerianus* Samson & W. Gams].

Neosartorya glabra (Fennell & Raper) Kozak. in Mycol. Pap. 161: 56. 30 Jun 1989. — Basionym: *Aspergillus fischeri* var. *glaber* Fennell & Raper in Mycologia 47: 74. 8 Mar 1955 (nom. holomorph.). — Neotype (Samson & Gams, 1985): No. 61447 (IMI). [Anamorph: *Aspergillus neoglaber* (Fennell & Raper) Kozak.].

Neosartorya hiratsukae Udagawa & al. in Trans. Mycol. Soc. Japan 32: 23. Apr 1991. — Holotype: No. 3008 (NHL). [Anamorph: *Aspergillus hiratsukae* Udagawa & al.].

Neosartorya pseudofischeri S. W. Peterson in Mycol. Res. 96: 549. Jul 1992. — Holotype: 1108305 p.p. (BPI). [Anamorph: *Aspergillus thermomutatus* (Paden) S. W Peterson].

Neosartorya quadricincta (E. Yuill) Malloch & Cain in Canad. J. Bot. 50: 2621. 26 Jan 1973. — Basionym: *Aspergillus quadricinctus* E. Yuill in Trans. Brit. Mycol. Soc. 36: 58. 28 Mar 1953 (nom. holomorph.). — Holotype: No. 48583ii (IMI). [Anamorph: *Aspergillus quadricingens* Kozak.].

Neosartorya spathulata Takada & Udagawa in Mycotaxon 24: 396. 11 Dec 1985. — Holotype: No. 2947 (NHL). [Anamorph: *Aspergillus spathulatus* Takada & Udagawa].

Neosartorya spinosa (Raper & Fennell) Kozak. in Mycol. Pap. 161: 58. 30 Jun 1989. — Basionym: *Aspergillus fischeri* var. *spinosus* Raper & Fennell, Genus Aspergillus: 256. 1965 (nom. holomorph.). — Neotype (Samson & Gams, 1985): No. 211390 (IMI). [Anamorph: *Aspergillus spinosus* Kozak].

Neosartorya stramenia (R. Novak & Raper) Malloch & Cain in Canad. J. Bot. 50: 2622. 26 Jan 1973. — Basionym: *Aspergillus stramenius* R. Novak & Raper in Raper & Fennell, Genus Aspergillus: 260. 1965 (nom. holomorph.). — Neotype (Samson & Gams, 1985): No. 172293 (IMI). [Anamorph: *Aspergillus paleaceus* Samson & W. Gams].

Neosartorya tatenoi Y. Horie & al. in Trans. Mycol. Soc. Japan 33: 395. 1992. — Holotype: No. FA 0022 p.p. (CBM). [Anamorph: *Aspergillus tatenoi* Horie & al.].

PENICILLIOPSIS SOLMS (HOLOMORPHS)

Penicilliopsis africana Samson & Seifert in Samson & Pitt, Adv. Penicillium Aspergillus Syst.: 408. 1985. — Holotype: Metiquette, *Louis 6275* (BR). [Anamorph: *Stilbodendron cervinum* (Cooke & Massee) Samson & Seifert].

Penicilliopsis clavariiformis Solms in Ann. Jard. Bot. Buitenzorg 6: 53. 1887. — Lectotype (designated here; see also Samson & Seifert, 1985): Bot. Garden Bogor, *Solms-Laubach* in herb. Hauman (BR). [Anamorph: *Saprophorum palmicola* (Henn.) Seifert & Samson].

PENICILLIUM LINK: FR. (ANAMORPHS)

Penicillium abidjanum Stolk in Antonie van Leeuwenhoek J. Microbiol. Serol. 34: 49. 1968.— Holotype: No. 247.67 p.p. (CBS).

Penicillium aculeatum Raper & Fennell in Mycologia 40: 535. Sep-Oct 1948. — Neotype (Pitt, 1980): No. 40588 (IMI).

Penicillium adametzii K. M. Zalessky in Bull. Int. Acad. Polon. Sci., Cl. Sci. Math., Sér. B, Sci. Nat., 1927: 507. 1927. — Neotype (Pitt, 1980): No. 39751 (IMI).

Penicillium aethiopicum Frisvad in Mycologia 81: 848. 11 Jan 1990. — Holotype: No. 285524 (IMI).

Penicillium allahabadense B. S. Mehrotra & D. Kumar in Canad. J. Bot. 40: 1399. 19 Oct 1962. Neotype (designated here): No. 304.63 (CBS).

Penicillium allii Vincent & Pitt in Mycologia 81: 300. 19 Apr 1989. — Holotype: Vincent 114 (MU).

Penicillium alutaceum D. B. Scott in Mycopathol. Mycol. Appl. 36: 17. 31 Oct 1968. — Holotype: No. 317.67 p.p. (CBS).

Penicillium anatolicum Stolk in Antonie van Lecuwenhoek J. Microbiol. Serol. 34: 46. 1968.Holotype: No. 479.66 p.p. (CBS).

Penicillium angustiporcatum Takada & Udagawa in Trans. Mycol. Soc. Japan 24: 143. Jul 1983.- Holotype: No. 6481 p.p. (NHL).

Penicillium ardesiacum Novobr. in Novosti Sist. Nizš. Rast. 11: 228. 11 Nov 1974. — Neotype (designated here): No. 174719 (IMI).

Penicillium arenicola Chalab. in Bot. Mater. Otd. Sporov. Rast. 6: 162. 28 Jun 1950. — Neotype (Pitt, 1980): No. 117658 (IMI).

Penicillium asperosporum G. Sm. in Trans. Brit. Mycol. Soc. 48: 275. 22 Jun 1965. (=*Penicillium echinosporum* G. Sm. in Trans. Brit. Mycol. Soc. 45: 387. 1962 [non Nehira in J. Ferment. Technol. 11: 849. 1933]). — Holotype: No. 80450 (IMI).

Penicillium assiutense Samson & Abdel-Fattah in Persoonia 9: 501. 13 Jul 1978. — Holotype: No. 147.78 p.p. (CBS).

Penicillium atramentosum Thom in U.S.D.A. Bur. Anim. Industr. Bull. 118: 65. 1910. — Neotype (Pitt, 1980): No. 39752 (IMI).

Penicillium atrovenetum G. Sm. in Trans. Brit. Mycol. Soc. 39: 112. 20 Mar 1956. — Neotype (designated here): No. 61837 (IMI).

Penicillium aurantiogriseum Dierckx in Ann. Soc. Sci. Bruxelles 25: 88. 1901. — Neotype (Pitt, 1980): No. 195050 (IMI).

Penicillium bilaiae Chalab. in Bot. Mater. Otd. Sporov. Rast. 6: 165. 28 Jun 1950. — Neotype (Pitt, 1980): No. 113677 (IMI).

Penicillium brasilianum Bat. in Anais Soc. Biol. Pernambuco 15: 162. 1957. — Holotype: No. IMUR 56 (URM).

Penicillium brevicompactum Dierckx in Ann. Soc. Sci. Bruxelles 25: 88. 1901. — Neotype (Pitt, 1980): No. 40225 (IMI).

Penicillium brunneum Udagawa in J. Agric. Sci. Tokyo Nogyo Daigaku 5: 16. 1959. — Holotype: No. 6054 (NHL).

Penicillium camemberti Thom in U.S.D.A. Bur. Anim. Industr. Bull. 82: 33. 1906. — Neotype (Pitt, 1980): No. 27831 (IMI).

Penicillium canescens Sopp in Skr. Vidensk.-Selsk. Christiana, Math.-Naturvidensk. K1. 11: 181. 1912. — Neotype (Pitt, 1980): No. 28260 (IMI).

Penicillium capsulatum Raper & Fennell in Mycologia 40: 528. Sep-Oct 1948. — Neotype (Pitt, 1980): No. 40576 (IMI).

Penicillium catenatum D. B. Scott in Mycopathol. Mycol. Appl. 36: 24. 31 Oct 1968. — Holotype: No. 352.67 p.p. (CBS).

Penicillium chalybeum Pitt & A. D. Hocking in Mycotaxon 22: 204. 8 Feb 1985. — Holotype: No. 2660 (FRR).

Penicillium chermesinum Biourge in Cellule 33: 284. 1923. — Neotype (Pitt, 1980): No. 191730 (IMI).

63

Penicillium chrysogenum Thom in U.S.D.A. Bur. Anim. Industr. Bull. 118: 58. 1910. — Neotype (Pitt, 1980): No. 24314 (IMI).

Penicillium cinerascens Biourge in Cellule 33: 308. 1923. — Neotype (designated here): No. 92234 (IMI).

Penicillium citreonigrum Dierckx in Ann. Soc. Sci. Bruxelles 25: 86. 1901. — Neotype (Pitt, 1980): No. 92209i (IMI).

Penicillium citrinum Thom in U.S.D.A. Bur. Anim. Industr. Bull. 118: 61. 1910. — Neotype (Pitt, 1980): No. 92196ii (IMI).

Penicillium clavigerum Demelius in Verh. Zool.-Bot. Ges. Wien 72: 74. 1923. — Neotype (designated here): No. 39807 (IMI).

Penicillium coalescens Quintan. in Mycopathologia 84: 115. 1983. — Neotype (designated here): No. 103.83 (CBS).

Penicillium commune Thom in U.S.D.A. Bur. Anim. Industr. Bull. 118: 56. 1910. — Neotype (designated here): No. 39812 (IMI).

Penicillium concentricum Samson & al. in Stud. Mycol. 11: 17. 30 Jan 1976. — Holotype: No. 477.75 (CBS).

Penicillium confertum (Frisvad & al.) Frisvad in Mycologia 81: 852. 11 Jan 1990. — Basionym: *Penicillium glandicola* var. *confertum* Frisvad & al. in Canad. J. Bot. 65: 769. 23 Apr 1987. — Holotype (Frisvad & al., 1987): No. 296930 (IMI).

Penicillium coprobium Frisvad in Mycologia 81: 853. 11 Jan 1990. — Holotype: No. 293209 (IMI).

Penicillium coprophilum (Berk. & M. A. Curtis) Seifert & Samson in Samson & Pitt, Adv. Penicillium Aspergillus Syst.: 145. 1985. — Basionym: *Coremium coprophilum* Berk. & M. A. Curtis in J. Linn. Soc., Bot., 10: 363. 16 Jun 1868. — Holotype: Cuba, *Wright 666* (K).

Penicillium coralligerum Nicot & Pionnat in Bull. Soc. Mycol. France 78: 245. 10 Jan 1963 ["1962"]. — Neotype (designated here): No. 99159 (IMI).

Penicillium corylophilum Dierckx in Ann. Soc. Sci. Bruxelles 25: 86. 1901. — Neotype (Pitt, 1980): No. 39754 (IMI).

Penicillium crateriforme J. C. Gilman & E. V. Abbott in Iowa State Coll. J. Sci. 1: 293. 1 Apr 1927. — Neotype (designated here): No. 94165 (IMI).

Penicillium cremeogriseum Chalab. in Bot. Mater. Otd. Sporov. Rast. 6: 168. 28 Jun 1950. — Neotype (designated here): No. 223.66 (CBS).

Penicillium crustosum Thom, Penicillia: 399. Jan 1930. — Neotype (Pitt, 1980): No. 91917 (IMI).

Penicillium cryptum Goch. in Mycotaxon 26: 349. 15 Jul 1986. — Holotype: No. 769 p.p. (NY).

Penicillium cyaneum (Bainier & Sartory) Biourge in Cellule 33: 102. 1923. — Basionym: *Citromyces cyaneus* Bainier & Sartory in Bull. Soc. Mycol. France 29: 157. 1913. — Neotype (Pitt, 1980): No. 39744 (IMI).

Penicillium daleae K. M. Zalessky in Bull. Int. Acad. Polon. Sci., Cl. Sci. Math., Sér. B, Sci. Nat., 1927: 495. 1927. — Neotype (Pitt, 1980): No. 89338 (IMI).

Penicillium dangeardii Pitt, Genus Penicillium: 472. 1980 ["1979"]. — Holotype: No. 197477 (IMI).

Penicillium decumbens Thom in U.S.D.A. Bur. Anim. Industr. Bull. 118: 71. 1910. — Neotype (Pitt, 1980): No. 190875 (IMI).

Penicillium dendriticum Pitt, Genus Penicillium: 413.1980 ["1979"].— Holotype: No.216897 (IMI).

Penicillium derxii Takada & Udagawa in Mycotaxon 31: 418. 6 Mai 1988. — Holotype: No. 2980 p.p. (NHL).

Penicillium dierckxii Biourge in Cellule 33: 313. 1923. — Neotype (designated here): No. 92216 (IMI).

Penicillium digitatum (Pers.: Fr.) Sacc., Fung. Ital.: tab. 894. Jul 1881. — Basionym: *Monilia digitata* Pers., Syn. Meth. Fung.: 693. 31 Dec 1801: Fr., Syst. Mycol. 3(2): 411. 1832. — Lectotype (Pitt, 1980): icon in Saccardo, Fung. Ital.: tab. 894. Jul 1881.

Penicillium dimorphosporum H. Swart in Trans. Brit. Mycol. Soc. 55: 310. 19 Oct 1970. — Holotype: No. 456.70 (CBS).

Penicillium diversum Raper & Fennell in Mycologia 40: 539. Sep-Oct 1948. — Neotype (Pitt, 1980): No. 40579 (IMI).

Penicillium dodgei Pitt, Genus Penicillium: 117. 1980 ["1979"]. — Holotype: No. 216896 (IMI).

Penicillium donkii Stolk in Persoonia 7: 333. 20 Jul 1973. — Holotype: No. 188.72 (CBS).

Penicillium duclauxii Delacr. in Bull. Soc. Mycol. France 7: 107. 1891. — Neotype (Pitt, 1980): No. 24312 (IMI).

Penicillium dupontii Griffon & Maubl. in Bull. Soc. Mycol. France 27: 73 1911. — Neotype (Pitt, 1980): No. 236.58 (CBS).

Penicillium echinulatum Raper & Thom ex Fassat. in Acta Univ. Carol., Biol.1974: 326. Dec 1977. — Holotype: No. 778523 (PRM).

Penicillium emmonsii Pitt, Genus Penicillium 479. 1980 ["1979"]. — Holotype: No.39805 p.p. (IMI).

Penicillium erubescens D. B. Scott in Mycopathol. Mycol. Appl. 36: 14. 31 Oct 1968. — Holotype: No. 318.67 p.p. (CBS).

Penicillium erythromellis A. D. Hocking in Pitt, Genus Penicillium: 459. 1980 ["1979"].-Holotype: No. 216899 (IMI).

Penicillium estinogenum A. Komatsu & S. Abe ex G. Sm. in Trans. Brit. Mycol. Soc. 46: 335. 21 Oct 1963. — Neotype (designated here): No. 68241 (IMI).

Penicillium expansum Link, in Ges. Naturf. Freunde Berlin Mag. Neuesten Entdeck. Gesammten Naturk. 3: 16. 1809. — Neotype (Samson & al., 1976): No. 325.48 (CBS).

Penicillium fellutanum Biourge in Cellule 33: 262. 1923. — Neotype (Pitt, 1980): No. 39734 (IMI).

Penicillium fennelliae Stolk in Antonie van Leeuwenhoek J. Microbiol. Serol.35: 261. 1969.— Holotype: No. 711.68 (CBS).

Penicillium flavidostipitatum C. Ramírez & C. C. González in Mycopathologia 88: 3. 1984. — Neotype (Frisvad & al., 1990b): No. 202.87 (CBS).

Penicillium formosanum H. M. Hsieh & al. in Trans. Mycol. Soc. Republ. China 2: 159.1987. — Holotype: No. 10001 (PPEH).

Penicillium fractum Udagawa in Trans. Mycol. Soc. Japan 9: 51. 1 Nov 1968. — Holotype: No. 6104 p.p. (NHL).

Penicillium funiculosum Thom in U.S.D.A. Bur. Anim. Industr. Bull. 118: 69. 1910. — Neotype (Pitt, 1980): No. 193019 (IMI).

Penicillium galapagense Samson & Mahoney in Trans. Brit. Mycol. Soc. 69: 158. 19 Aug 1977. — Holotype: No.751.74 p.p. (CBS).

Penicillium glabrum (Wehmer) Westling in Ark. Bot. 11(1): 131. 1911. — Basionym: *Citromyces glaber* Wehmer, Beitr. Einh. Pilze 1: 24. Jul 1893. — Neotype (Pitt, 1980): No. 91944 (IMI).

Penicillium gladioli L. McCulloch & Thom in Science, sen 2, 67: 217.1928.— Neotype (Pitt, 1980): No. 34911 (IMI).

Penicillium glandicola (Oudem.) Seifert & Samson in Samson & Pitt, Adv. Penicillium Aspergillus Syst.: 147. 1985. — Basionym: *Coremiun glandicola* Oudem. in Ned. Kruidk. Arch., sen 3,2: 918. Jun 1903. — Holotype: Netherlands, Valkenburg, Jul 1901, *Rick* in herb. Oudemans (L).

Penicillium gracilentum Udagawa & Y. Horie in Trans. Mycol. Soc. Japan 14: 373. 20 Dec 1973. — Holotype: No. 6452 p.p. (NHL).

Penicillium griseofulvum Dierckx in Ann. Soc. Sci. Bruxelles 25: 88. 1901. — Neotype (Pitt, 1980): No. 75832 (IMI).

Penicillium griseopurpureum G. Sm. in Trans. Brit. Mycol. Soc. 48: 275. 22 Jun 1965. — Neotype (designated here): No. 96157 (IMI).

Penicillium herquei Bainier & Sartory in Bull. Soc. Mycol. France 28: 121. 1912. — Neotype (Pitt, 1980): No. 28809 (IMI).

Penicillium heteromorphum H. Z. Kong & Z. T. Qi in Mycosystema 1: 107. 29 Feb 1988. — Neotype (Frisvad & al., 1990b): No. 226.89 (CBS).

Penicillium hirayamae Udagawa in J. Agric. Sci. Tokyo Nogyo Daigaku 5: 6. 1959. — Neotype (Pitt, 1980): No. 78255 (IMI).

Penicillium hirsutum Dierckx in Ann. Soc. Sci. Bruxelles 25: 89. 1901. — Neotype (Pitt, 1980): No. 40213 (IMI).

Penicillium hordei Stolk in Antonie van Lecuwenhoek J. Microbiol. Serol. 35: 270. 1969. — Holotype: No. 701.68 (CBS).

Penicillium implicatum Biourge in Cellule 33: 278. 1923. — Neotype (Pitt, 1980): No. 190235 (IMI).

Penicillium inflatum Stolk & Malla in Persoonia 6: 197. 23 Mar 1971. — Holotype: No. 682.70 (CBS).

Penicillium indonesiae Pitt, Genus Penicillium: 114. 1980 ["1979"]. — Holotype: No. 39733 (IMI).

Penicillium inusitatum D. B. Scott in Mycopathol. Mycol. Appl. 36: 20. 31 Oct 1968. — Holotype: No. 351.67 p.p. (CBS).

Penicillium isariiforme Stolk & J. Mey. in Trans Brit. Mycol. Soc. 40: 187. 1 Jul 1957. — Neotype (Pitt, 1980): No. 60371 (IMI).

Penicillium islandicum Sopp in Skr. Vidensk.-Selsk. Christiana, Math.-Naturvidensk. K1. 11: 161. 1912. — Neotype (Pitt, 1980): No. 40042 (IMI).

Penicillium italicum Wehmer in Hedwigia 33: 211. 1 Aug 1894. — Neotype (Samson & al., 1976): No. 339.48 (CBS).

Penicillium janczewskii K. M. Zalessky in Bull. Int. Acad. Polon. Sci., Cl. Sci. Math., Sér. B, Sci. Nat., 1927: 488. 1927. — Neotype (Pitt, 1980): No. 191499 (IMI).

Penicillium janthinellum Biourge in Cellule 33: 258. 1923. — Neotype (Pitt, 1980): No. 40238 (IMI).

Penicillium jensenii K. M. Zalessky in Bull. Int. Acad. Polon. Sci., Cl. Sci. Math., Sér. B, Sci. Nat., 1927: 494. 1927. — Neotype (Pitt, 1980): No. 39768 (IMI).

Penicillium jugoslavicum C. Ramírez & Munt.-Cvetk. in Mycopathologia 88: 65. 1984. — Neotype (Frisvad & al., 1990b): No. 192.87 (CBS).

65

Penicillium kabunicum Baghd. in Novosti Sist. Nizš. Rast.1968: 98. 16 Oct 1968. — Neotype (designated here): No. 409.69 (CBS).

Penicillium katangense Stolk in Antonie van Leeuwenhoek J. Microbiol. Serol. 34: 42. 1968.— Holotype: No. 247.67 p.p. (CBS).

Penicillium klebahnii Pitt, Genus Penicillium: 122. 1980 ["1979"]. — Holotype: No. 39737 (IMI).

Penicillium kloeckeri Pitt, Genus Penicillium: 491. 1980 ["1979"]. — Holotype (Pitt, 1980): No. 40047 p.p. (IMI).

Penicillium lanosum Westling in Ark. Bot. 11(1): 97. 1911. — Neotype (designated here): No.40224 (IMI).

Penicillium lapatayae C. Ramírez in Mycopathologia 91: 96.1985. — Neotype (Frisvad & al., 1990b): No. 203.87 (CBS).

Penicillium lapidosum Raper & Fennell in Mycologia 40: 524. Sep-Oct 1948. — Neotype (Pitt, 1980): No. 39743 (IMI).

Penicillium lassenii Paden in Mycopathol. Mycol. Appl. 43: 266. 25 Mar 1971. — Holotype: No. JWP 69-26 (UVIC).

Penicillium lehmanii Pitt, Genus Penicillium: 497. 1980 ["1979"]. — Holotype: No. 40043 (IMI).

Penicillium lignorum Stolk in Antonie van Leeuwenhoek J. Microbiol. Serol. 35: 264.1969. — Holotype: No. 709.68 (CBS).

Penicillium lineatum Pitt, Genus Penicillium: 485. 1980 ["1979"]. — Holotype: No.39741 p.p. (IMI).

Penicillium lineolatum Udagawa & Y. Horie in Mycotaxon 5: 493. 6 Mai 1977. — Holotype: No.2776 p.p. (NHL).

Penicillium lividum Westling in Ark. Bot. 11(1): 134. 1911. — Neotype (Pitt, 1980): No. 39736 (IMI).

Penicillium loliense Pitt, Genus Penicillium: 450. 1980 ["1979"]. — Holotype: No. 216901 (IMI).

Penicillium ludwigii Udagawa in Trans. Mycol. Soc. Japan 10: 2. 1 Aug 1969. — Holotype: No. 6118 p.p. NHL).

Penicillium maclennaniae H. Y. Yip in Trans. Brit. Mycol. Soc. 77: 202. 6 Aug 1981. — Holotype: No. 35238 (DAR).

Penicillium macrosporum Frisvad & al. in Antonie van Leeuwenhoek J. Microbiol. Serol. 57: 186. 1990. — Holotype: No.317.63 p.p. (CBS).

Penicillium madriti G. Sm. in Trans. Brit. Mycol. Soc. 44: 44. 21 Mar 1961. — Holotype (Pitt, 1980): No. 86563 (IMI).

Penicillium manginii Duché & R. Heim in Trav. Cryptog. Louis L. Mangin: 450. Sep 1931. — Neotype (designated here): No. 253.31 (CBS).

Penicillium mariaecrucis Quintan. in Avances Nutr. Mejora Anim. Aliment. 23: 334. 1982. — Neotype (Frisvad & al., 1990b): No. 270.83 (CBS).

Penicillium marneffei Segretain & al. in Bull. Soc. Mycol. France 75: 416.28 Feb 1960. — Neotype (Pitt, 1980): No. 68794iii (IMI).

Penicillium megasporum Orpurt & Fennell in Mycologia 47: 233. 29 Apr 1955. — Neotype (Pitt, 1980): No. 216904 (IMI).

Penicillium melinii Thom, Penicillia: 273. Jan 1930.— Neotype (Pitt, 1980): No. 40216 (IMI).

Penicillium meliforme Udagawa & Y. Horie in Trans. Mycol. Soc. Japan 14: 376. 20 Dec 1973. — Holotype: No. 6468 p.p. (NHL).

Penicillium meridianum D. B. Scott in Mycopathol. Mycol. Appl. 36: 12. 31 Oct 1968. — Holotype: No. 314.67 p.p. (CBS).

Penicillium miczynskii K. M. Zalessky in Bull. Int. Acad. Polon. Sci., Cl. Sci. Math., Sér. B, Sci. Nat., 1927: 482. 1927. — Neotype (Pitt, 1980): No. 40030 (IMI).

Penicillium mimosinum A. D. Hocking in Pitt, Genus Penicillium: 507.1980 ["1979"]. — Holotype: No. 223991 p.p. (IMI).

Penicillium minioluteum Dierckx in Ann. Soc. Sci. Bruxelles 25: 87. 1901. — Neotype (designated here): No. 642.68 (CBS).

Penicillium mirabile Beliakova & Milko in Mikol. & Fitopatol. 6: 145. 1972. — Holotype: No. F1328 (BKM).

Penicillium moldavicum Milko & Beliakova in Novosti Sist. Nizš. Rast. 1967: 255. 28 Feb 1967. — Neotype (designated here): No. 129966 (IMI).

Penicillium molle Pitt, Genus Penicillium: 148. 1980 ["1979"]. — Holotype: No. 84589 (IMI).

Penlcillium mononematosum (Frisvad & al.) Frisvad in Mycologia 81: 857.11 Jan 1990. — Basionym: *Penicillium glandicola* var. *mononematosum* Frisvad & al. in Canad. J. Bot. 65: 767. 23 Apr 1987. — Holotype (Frisvad & al., 1987): No. 296925 (IMI).

Penicillium montanense M. Chr. & Backus in Mycologia 54: 574. 25 Jan 1963. — Holotype: Cryptogamic Herb. No. GW1-6 (WIS).

Penicillium nalgiovense Laxa in Zentralbl. Bakteriol., 2. Abt., 86: 160.22 Jun 1932. — Neotype (designated here): No. 352.48 (CBS).

Penicillium nepalense Takada & Udagawa in Trans. Mycol Soc. Japan 24: 146. Jul 1983. — Holotype: No. 6482 p.p. (NHL).

Penicillium nilense Pitt, Genus Penicillium: 145. 1980 ["1979"]. — Holotype: No. 40580 (IMI).

Penicillium nodositanum Valla in Pl. & Soil 114: 146. 1989. — Neotype (designated here): No. 330.90 (CBS).

Penicillium nodulum H. Z. Kong & Z. T. Qi in Mycosystema 1: 108. 29 Feb 1988. — Neotype (Frisvad & al., 1990b): No. 227.89 (CBS).

Penicillium novae-zeelandiae J. F. H. Beyma in Antonie van Leeuwenhoek J. Microbiol. Serol. 6: 275. 1940. — Neotype (Pitt, 1980): No. 40584ii (IMI).

Penicillium ochrochloron Biourge in Cellule 33: 269. 1923. — Neotype (Pitt, 1980): No. 39806 (IMI).

Penicillium ochrosalmoneum Udagawa in J. Agric. Sci. Tokyo Nogyo Daigaku 5: 10. 1959.— Holotype: No. 6048 (NHL).

Penicillium oblatum Pitt & A. D. Hocking in Mycologia 77: 819.15 Oct 1985. — Holotype: No. 2234 (FRR).

Penicillium ohiense L. H. Huang & J. A. Schmitt in Ohio J. Sci. 75: 78. 1975. — Holotype: No. 6086 p.p.(NHL).

Penicillium olsonii Bainier & Sartory in Ann. Mycol. 10: 398. 10 Aug 1912. — Neotype (Pitt, 1980): No. 192502 (IMI).

Penicillium onobense C. Ramírez & A. T. Martínez in Mycopathologia 74: 44. 1981. — Neotype (Frisvad & al., 1990b): No. 174.81 (CBS).

Penicillium ornatum Udagawa in Trans. Mycol. Soc. Japan 9: 49. 1 Nov 1968. — Holotype: No. 6101 p.p.(NHL).

Penicillium osmophilum Stolk & Veenb.-Rijks in Antonie van Leeuwenhoek J. Microbiol. Serol. 40: 1. 1974. — Holotype: No. 462.72 p.p. (CBS).

Penicillium oxalicum Currie & Thom in J. Biol. Chem. 22: 289. 1915. — Neotype (Pitt, 1980): No. 192332 (IMI).

Penicillium palmae Samson & al. in Stud. Mycol. 31: 135. 1 Jul 1989. — Holotype: No. 442.88 (CBS).

Penicillium palmense C. Ramírez & al. in Mycopathologia 66: 80. 29 Dec 1978. — Neotype (Frisvad & al., 1990b): No. 336.79 (CBS).

Penicillium panamense Samson & al. in Stud. Mycol. 31: 136. 1 Jul 1989. — Holotype: No. 128.89 (CBS).

Penicillium panasenkoi Pitt, Genus Penicillium: 482. 1980 ["1979"]. — Holotype: No. 129962 (IMI).

Penicillium papuanum Udagawa & Y. Horie in Trans. Mycol. Soc. Japan 14: 378. 20 Dec 1973. — Holotype: No. 6463 (NHL).

Penicillium paraherquei S. Abe ex G. Sm. in Trans. Brit. Mycol. Soc. 46: 335. 21 Oct 1963. — Neotype (designated here): No. 68220 (IMI).

Penicillium patens Pitt & A. D. Hocking in Mycotaxon 22: 205. 8 Feb 1985. — Holotype: No. 2661 (FRR).

Penicillium paxilli Bainier in Bull. Soc. Mycol. France 23: 95. 1907. — Neotype (Pitt, 1980): No.40226 (IMI).

Penicillium pedemontanum Mosca & A. Fontana in Allionia 9: 40. 1963. — Neotype (designated here): No. 265.65 (CBS).

Penicillium phoeniceum J F. H. Beyma in Zentralbl. Baktenol., 2. Abt., 88: 136. 24 Apr 1933. — Neotype (Pitt, 1980): No. 40585 (IMI).

Penicillium piceum Raper & Fennell in Mycologia 40: 533. Sep-Oct 1948. — Neotype (Pitt, 1980): No. 40038 (IMI).

Penicillium pinetorum M. Chr. & Backus in Mycologia 53: 457. 21 Jun 1962. — Holotype: No. WSF 15-C (WIS).

Penicillium pinophilum Hedgc. in U.S.D.A. Bur. Anim. Industr. Bull. 118: 37. 1910. — Neotype (Pitt, 1980): No. 114933 (IMI).

Penicillium piscarium Westling in Ark. Bot; 11(1): 86. 1911. — Neotype (designated here): No.40032 (IMI).

Penicillium pittii Quintan. in Mycopathologia 91: 75. 1985. — Neotype (Frisvad & al., 1990b): No. 139.84 (CBS).

Penicillium primulinum Pitt, Genus Penicillium: 455. 1980 ["1979"]. — Holotype: No. 40031 (IMI).

Penicillium proteolyticum Kamyschko in Bot. Mater. Otd. Sporov. Rast. 14: 228. 29 Mar 1961. — Neotype (designated here): No. 303.67 (CBS).

Penicillium pseudostromaticum Hodges & al. in Mycologia 62: 1106. 10 Feb 1971. — Holotype: Warner 18 (NY).

Penicillium pulvillorum Turfitt in Trans. Brit. Mycol. Soc. 23: 186. 31 Jul 1939. — Neotype (designated here): No. 280.39 (CBS).

Penicillium purpurascens (Sopp) Biourge in Cellule 33: 105. 1923. — Basionym: *Citromyces purpurascens* Sopp in Skr. Vidensk Selsk. Christiana, Math.-Naturvidensk. K1. 11: 117. 1912. — Neotype (Pitt, 1980): No. 39745 (IMI).

Penicillium purpureum Stolk & Samson in Stud. Mycol. 2: 57. 1 Nov 1972. — Lectotype (Pitt, 1980): icon in Stud. Mycol. 2: 59. 1 Nov 1972.

Penicillium purpurogenum Stoll, Beitr. Charakt. Penicill.: 32. 1904. — Neotype(Pitt, 1980):No.91926 (IMI).

Penicillium raciborskii K. M. Zalessky in Bull. Int. Acad. Polon. Sci., C1. Sci. Math., Sér. B, Sci.

Nat., 1927: 454. 1927. — Neotype (designated here): No. 40568 (IMI).

Penicillium rademiricii Quintan. in Mycopathologia 91: 72. 1985. — Neotype (Frisvad & al., 1990b): No. 140.84 (CBS).

Penicillium raperi G. Sm. in Trans. Brit. Mycol. Soc. 40: 486. 20 Dec 1957. — Neotype (designated here): No. 71625 (IMI).

Penicillium raistrickii G. Sm. in Trans. Brit. Mycol. Soc.18: 90.16 Aug 1933. — Neotype (Pitt, 1980): No. 40221 (IMI).

Penicillium rasile Pitt, Genus Penicillium: 120. 1980 ["1979"]. — Holotype: No. 39735 (IMI).

Penicillium resedanum McLennan & Ducker in Austral. J. Bot. 2: 360. Nov 1954. — Neotype (Pitt, 1980): No. 62877 (IMI)

Penicillium restrictum J. C. Gilman & E. V. Abbott in Iowa State Coll. J. Sci. 1: 297. 1 Apr 1927. — Neotype (Pitt,- 1980): No. 40228 (IMI).

Penicillium reticulisporum Udagawa in Trans. Mycol. Soc. Japan 9: 52. 1 Nov 1968. — Holotype: No. 6105 p.p. (NHL).

Penicillium rolfsii Thom, Penicillia: 489. Jan 1930. — Neotype (Pitt, 1980): No. 40029 (IMI).

Penicillium roqueforti Thom in U.S.D.A. Bur. Anim. Industr. Bull. 82: 35. 1906. — Neotype (Pitt, 1980): No. 24313 (IMI).

Penicillium roseopurpureum Dierckx in Ann. Soc. Sci. Bruxelles 25: 86. 1901. — Neotype (Pitt, 1980): No. 40573 (IMI).

Penicillium rubefaciens Quintan. in Mycopathologia 80: 73.1982. — Neotype (Frisvad & al., l990b): No. 145.83 (CBS).

Penicillium rubidurum Udagawa & Y. Horie in Trans. Mycol. Soc. Japan 14: 381. 20 Dec 1973. — Holotype: No. 6460 p.p. (NHL).

Penicillium rugulosum Thom in U.S.D.A. Bur. Anim. Industr. Bull. 118: 60. 1910. — Neotype (Pitt, 1980): No. 40041 (IMI).

Penicillium sabulosum Pitt & A. D. Hocking in Mycologia 77: 818.15 Oct 1985. — Holotype: No. 2743 (FRR).

Penicillium sacculum E. Dale in Ann. Mycol. 24: 137. 20 Jun 1926. — Neotype (designated here): No. 231.61 (CBS).

Penicillium sajarovii Quintan. in Avances Nutr. Mejora Anim. Aliment. 22: 539. 1981. — Neotype (Frisvad & al., 1990b): No. 277.83 (CBS).

Penicillium scabrosum Frisvad & al. in Persoonia 14: 177. 27 Jun 1990. — Holotype: No. 285533 (IMI).

Penicillium sclerotigenum W. Yamam. in Sci. Rep. Hyogo Univ. Agnc., Ser. Agnc. Biol., ser. 2, 1:

69. 1955. — Neotype (Pitt, 1980): No. 68616 (IMI).

Penicillium sclerotiorum J. F. H. Beyma in Zentralbl. Bakteriol., 2. Abt., 96: 418. 10 Aug 1937. — Neotype (Pitt, 1980): No. 40569 (IMI).

Penicillium senticosum D. B. Scott in Mycopathol. Mycol. Appl. 36: 5. 31 Oct 1968. — Holotype: No. 316.67 p.p. (CBS).

Penicillium shearii Stolk & D. B. Scott in Persoonia 4: 396. 1 Aug 1967. — Holotype: No. 290.48 p.p. (CBS).

Penicillium shennangjianum H. Z. Kong & Z. T. Qi in Mycosystema 1: 110. 29 Feb l988.— Neotype (Frisvad & al., 1990b): No. 228.89 (CBS).

Penicillium siamense Manoch & C. Ramírez in Mycopathologia 101: 32. Jan 1988. — Neotype (designated here): No. 475.88 (CBS).

Penicillium simplicissimum (Oudem.) Thom, Penicillia: 335. Jan 1930. — Basionym: *Spicaria simplicissima* Oudem. in Ned. Kruidk. Arch., sen 3, 2: 763. Jun 1902. — Neotype (Jensen, 1912): No. 5921 (CUP).

Penicillium sinaicum Udagawa & S. Ueda in Mycotaxon 14: 266. 22 Jan 1982. — Holotype: No. 2894 p.p. (NHL).

Penicillium sizovae Baghd. in Novosti Sist. Nizš. Rast. 1968: 103. 16 Oct 1968. — Neotype (designated here): No. 413.69 (CBS).

Penicillium skrjabinii Schmotina & Golovleva in Mikol. & Fitopatol. 8: 530. 1974. — Neotype (designated here): No. 196528 (IMI).

Penicillium smithii Quintan. in Avances Nutr. Mejora Anim. Aliment. 23: 340. 1982. — Neotype (designated here): No. 276.83 (CBS).

Penicillium solitum Westling in Ark. Bot. 11(1): 65. 1911. — Neotype (Frisvad & al., 1990a): No. 424.89 (CBS).

Penicillium soppii K. M. Zalessky in Bull. Int. Acad. Polon. Sci., Cl. Sci. Math., Sér. B, Sci. Nat., 1927: 476. 1927. — Neotype (designated here): No. 40217 (IMI).

Penicillium sphaerum Pitt, Genus Penicillium: 494. 1980 ["1979"]. — Holotype: No.40589 p.p. (IMI).

Penicillium spinulosum Thom in U.S.D.A. Bur. Anim. Industr. Bull. 118: 76. 1910. — Neotype (Pitt, 1980): No. 24316i (IMI).

Penicillium spirillum Pitt, Genus Penicillium: 476. 1980 ["1979"]. — Holotype: No.40593 p.p. (IMI).

Penicillium steckii K. M. Zalessky in Bull. Int. Acad. Polon. Sci., Cl. Sci. Math., Sér. B, Sci. Nat., 1927: 469. 1927. — Neotype (designated here): No. 40583 (IMI).

Penicillium stolkiae D. B. Scott in Mycopathol. Mycol. Appl. 36: 8. 31 Oct 1968. — Holotype: No. 315.67 p.p. (CBS).

Penicillium striatisporum Stolk in Antonie van Leeuwenhoek J. Microbiol. Serol. 35: 268. 1969. — Holotype: No. 705.68 (CBS).

Penicillium sublateritium Biourge in Cellule 33: 315. 1923. — Neotype (Pitt, 1980): No. 40594 (IMI).

Penicillium syriacum Baghd. in Novosti Sist. Nizš. Rast. 1968: 111. 16 Oct 1968. — Neotype (designated here): No. 418.69 (CBS).

Penicillium tardum Thom, Penicillia: 485. Jan 1930. — Neotype (designated here): No. 40034 (IMI).

Penicillium terlikowskii K. M. Zalessky in Bull. Int. Acad. Polon. Sci., Cl. Sci. Math., Sér. B, Sci. Nat., 1927: 501. 1927. — Neotype (designated here): No. 228.28 (CBS).

Penicillium terrenum D. B. Scott in Mycopathol. Mycol. Appl. 36: 1. 31 Oct 1968. — Holotype: No. 313.67 p.p. (CBS).

Penicillium thomii Maire in Bull. Soc. Hist. Nat. Afrique N. 8: 189. 1917. — Neotype (Pitt, 1980): No. 189694 (IMI).

Penicillium tularense Paden in Mycopathol. Mycol. Appl. 43: 264. 25 Mar 1971. — Holotype: No. JWP 68-31 p.p. (UVIC).

Penicillium turbatum Westling in Ark. Bot. 11(1): 128. 1911. — Neotype (Pitt, 1980): No. 39738 (IMI).

Penicillium udagawae Stolk & Samson in Stud. Mycol. 2: 36. 1 Nov 1972. — Holotype: No. 579.72 p.p. (CBS).

Penicillium ulaiense H. M. Hsieh & al. in Trans. Mycol. Soc. Republ. China 2: 161. 1987. — Holotype: No. 29001.87 (PPEH).

Penicillium unicum Tzean & al. in Mycologia 84: 739. 21 Oct 1992. — Holotype: No. PPH16 p.p. (PPEH).

Penicillium vanbeymae Pitt, Genus Penicillium: 142. 1980 ["1979"]. — Holotype: No. 40590 (IMI).

Penicillium variabile Sopp in Skr. Vidensk.-Selsk. Christiana, Math.-Naturvidensk. Kl. 11: 169. 1912. — Neotype (Pitt, 1980): No. 40040 (IMI).

Penicillium varians G. Sm. in Trans. Brit. Mycol. Soc.18: 89.16 Aug 1933. — Neotype (designated here): No. 40586 (IMI).

Penicillium vasconiae C. Ramírez & A. T. Martínez in Mycopathologia 72: 189. 28 Nov 1980. — Neotype (Frisvad & al., 1990b): No. 339.79 (CBS).

Penicillium velutinum J. F. H. Beyma in Zentralbl. Bakteriol., 2. Abt., 91: 353. 28 Feb 1935. — Neotype (Pitt, 1980): No. 40571 (IMI).

Penicillium verrucosum Dierckx in Ann. Soc. Sci. Bruxelles 25: 88. 1901. — Neotype (Pitt, 1980): No. 200310 (IMI).

Penicillium verruculosum Peyronel, Germi Atmosf. Fung. Micel.: 22. 1913. — Neotype (Pitt, 1980): No. 40039 (IMI).

Penicillium vinaceum J. C. Gilman & E. V. Abbott in Iowa State Coll. J. Sci. 1: 299. 1 Apr 1927. — Neotype (Pitt, 1980): No. 29189 (IMI).

Penicillium viridicatum Westling in Ark. Bot. 11(1): 88. 1911. — Neotype (Pitt, 1980): No. 39758ii (IMI).

Penicillium vulpinum (Cooke & Massee) Seifert & Samson in Samson & Pitt, Adv. Penicillium Aspergillus Syst.: 144. 1985. — Basionym: *Coremium vulpinum* Cooke & Massee in Grevillea 16: 81. Mar 1888. — Holotype (see Seifert & Samson, 1985): "on dung", *s. coll.,* in herb. Cooke (K).

Penicillium waksmanii K. M. Zalessky in Bull. Int. Acad. Polon. Sci., Cl. Sci. Math., Sér. B, Sci. Nat., 1927: 468. 1927. — Neotype (Pitt, 1980): No. 39746i (IMI).

Penicillium westlingii K. M. Zalessky in Bull. Int. Acad. Polon. Sci., Cl. Sci. Math., Sér. B, Sci. Nat., 1927: 473. 1927. — Neotype (designated here): No. 92272 (IMI).

Penicillium zonatum Hodges & J. J. Perry in Mycologia 65: 697. 19 Jul 1973. — Holotype: No. FSL 525 p.p. (BPI).

PETROMYCES MALLOCH & CAIN (HOLOMORPHS)

Petromyces albertensis J. P. Tewari in Mycologia 77: 114. 15 Feb 1985. — Holotype: No. 2976 (UAMH). [Anamorph: *Aspergillus albertensis* J. P. Tewari].

Petromyces alliaceus Malloch & Cain in Canad. J. Bot. 50: 2623. 26 Jan 1973. — Holotype: No. 46232 (TRTC). [Anamorph: *Aspergillus alliaceus* Thom & Church].

SAROPHORUM SYD. & P. SYD. (ANAMORPHS)

Sarophorum palmicola (Henn.) Seifert & Samson in Samson & Pitt, Adv. Penicillium Aspergillus Syst.: 403. 1985. — Basionym: *Penicilliopsis palmicola* Henn. in Hedwigia 43: 352. 15 Jul 1904. — Lectotype (designated here; see also Seifert & Samson, 1985): Brazil, Jurna, Jul 1901, *Ule 2834* (L).

SCLEROCLEISTA SUBRAM. (HOLOMORPHS)

Sclerocleista ornata (Raper & al.) Subram. in Curr. Sci. 41: 757. 5 Nov 1972. — Basionym: *Aspergillus ornatus* Raper & al. in Mycologia 45: 678. 9 Oct 1953 (nom. holomorph.). —

Neotype (Samson & Gams, 1985): No. 55295 (IMI). [Anamorph: *Aspergillus ornatulus* Samson & W. Gams].

Sclerocleista thaxteri Subram. in Curr. Sci. 41: 757. 5 Nov 1972.— Holotype (Subramanian, 1972): ex caterpillar dung, Kittery Point, *R. Thaxter* (FH). [Anamorph: *Aspergillus citrisporus* Höhn.].

STILBODENDRON SYD. & P. SYD. (ANAMORPHS)

Stilbodendron cervinum (Cooke & Massee) Samson & Seifert in Samson & Pitt, Adv. Penicillium Aspergillus Syst.: 408. 1985. — Basionym: *Corallodendron cervinum* Cooke & Massee in Grevillea 16: 71. Mar 1888. — Holotype (see Samson & Seifert, 1985): Africa, *Holmes* (K).

TALAROMYCES C. R. BENJ. (HOLOMORPHS)

Talaromyces assiutensis Samson & Abdel-Fattah in Persoonia 9: 501. 13 Jul 1978. — Holotype: No. 147.78 p.p. (CBS). [Anamorph: *Penicillium assiutense* Samson & Abdel-Fattah].

Talaromyces avellaneus (Thom & Turesson) C. R. Benj. in Mycologia 47: 682. 7 Oct 1955. — Basionym: *Penicillium avellaneum* Thom & Turesson in Mycologia 7: 284. Sep 1915 (nom. holomorph.). — Neotype (designated here): No. 40230 (IMI). [Anamorph: *Merimbla ingelheimense* (J. F. H. Beyma) Pitt].

Talaromyces bacillisporus (Swift) C. R. Benj. in Mycologia 47: 684. 7 Oct 1955. — Basionym: *Penicillium bacillisporum* Swift in Bull. Torrey Bot. Club 59: 221. 4 Mai 1932 (nom. holomorph.). — Holotype (Stolk & Samson, 1972): No. 296.48 (CBS). [Anamorph: *Geosmithia swiftii* Pitt].

Talaromyces byssochlamydoides Stolk & Samson in Stud. Mycol. 2: 45. 1 Nov 1972. — Holotype: No. 413.74 (CBS). [Anamorph: *Paecilomyces byssochlamydoides* Stolk & Samson].

Talaromyces derxii Takada & Udagawa in Mycotaxon 31: 418. 6 Mai 1988. — Holotype: No. 2980 p.p. (NHL). [Anamorph: *Penicillium derxii* Takada & Udagawa].

Talaromyces emersonii Stolk in Antonie van Leeuwenhoek J. Microbiol. Serol. 31: 262. 1965. — Holotype: No. 393.64 p.p. (CBS). [Anamorph: *Geosmithia emersonii* (Stolk) Pitt].

Talaromyces flavus (Klöcker) Stolk & Samson in Stud. Mycol. 2: 10. 1 Nov 1972. — Basionym: *Gymnoascus flavus* Klöcker in Hedwigia 41: 80. 24 Apr 1902. — Neotype (Stolk & Samson, 1972): No. 310.38 p.p. (CBS). [Anamorph: *Penicillium dangeardii* Pitt].

Talaromyces galapagensis Samson & Mahoney in Trans. Brit. Mycol. Soc. 69: 158. 19 Aug 1977. — Holotype: No. 751.74 p.p. (CBS). [Anamorph: *Penicillium galapagense* Samson & Mahoney].

Talaromyces helicus (Raper & Fennell) C. R. Benj. in Mycologia 47: 684. 7 Oct 1955. — Basionym: *Penicillium helicum* Raper & Fennell in Mycologia 40: 515. Sep-Oct 1948 (nom. holomorph.). — Neotype (Pitt, 1980): No. 40593 p.p. (IMI). [Anamorph: *Penicillium spirillum* Pitt].

Talaromyces leycettanus H. C. Evans & Stolk in Trans. Brit. Mycol. Soc. 56: 45. 29 Feb 1971. — Holotype: No. 398.68 (CBS). [Anamorph: *Paecilomyces leycettanus* (H. C. Evans & Stolk) Stolk & al.].

Talaromyces luteus (Zukal) C. R. Benj. in Mycologia 47: 681.7 Oct 1955. — Basionym: *Penicillium luteum* Zukal in Sitzungsber. Kaiserl. Akad. Wiss., Math.-Naturwiss. Cl., Abt. 1, 98: 561. 1890 ["1890"] (nom. holomorph.). — Neotype (Pitt, 1980): No. 89305 (IMI).

Talaromyces macrosporus (Stolk & Samson) Frisvad & al. in Antonie van Leeuwenhoek J. Microbiol. Serol. 57: 186. 1990. Basionym: *Talaromyces flavus* var. *macrosporus* Stolk & Samson in Stud. Mycol. 2: 15. 1 Nov 1972. — Holotype (Stolk & Samson, 1972): No. 317.63 p.p. (CBS). [Anamorph: *Penicillium macrosporum* Frisvad & al.].

Talaromyces mimosinus A. D. Hocking in Pitt, Genus Penicillium: 507.1980 ["1979"]. — Holotype: No. 223991 p.p. (IMI). [Anamorph: *Penicillium mimosinum* A. D. Hocking].

Talaromyces ohiensis Pitt, Genus Penicillium: 502. 1980 ["1979"]. — Holotype: No. 6086 p.p. (NHL). [Anamorph: *Penicillium ohiense* L. H. Huang & J. A. Schmitt].

Talaromyces panasenkoi Pitt, Genus Penicillium: 482. 1980 ["1979"]. — Holotype: icon of '*Penicillium ucrainicum*' in Mycologia 56: 59. 20 Feb 1964. [Anamorph: *Penicillium panasenkoi* Pitt].

Talaromyces purpureus (E. Müll. & Pacha-Aue) Stolk & Samson in Stud. Mycol. 2: 57. 1 Nov 1972. — Basionym: *Arachniotus purpureus* E. Müll. & Pacha-Aue in Nova Hedwigia 15: 552. 5 Dec 1968. — Lectotype (Pitt, 1980): icon in Stud. Mycol. 2: 59. 1 Nov 1972. [Anamorph: *Penicillium purpureum* Stolk & Samson].

Talaromyces rotundus (Raper & Fennell) C. R. Benj. in Mycologia 47: 683. 7 Oct 1955. — Basionym: *Penicillium rotundum* Raper & Fennell in Mycologia 40: 518. Sep-Oct 1948 (nom.

holomorph.). — Neotype (Pitt, 1980): No. 40589 p.p. (IMI). [Anamorph: *Penicillium sphaerum* Pitt].

Talaromyces stipitatus (Thom) C. R. Benj. in Mycologia 47: 684. 7 Oct 1955. — Basionym: *Penicillium stipitatum* Thom in Mycologia 27: 138. Mar-Apr 1935 (nom. holomorph.). — Neotype (Pitt, 1980): No. 39805 p.p. (IMI). [Anamorph: *Penicillium emmonsii* Pitt].

Talaromyces striatus (Raper & Fennell) C. R. Benj. in Mycologia 47: 682. 7 Oct 1955. — Basionym: *Penicillium striatum* Raper & Fennell in Mycologia 40: 521. Sep-Oct 1948 (nom. holomorph.). — Neotype (Pitt, 1980): No.39741 p.p. (IMI). [Anamorph: *Penicillium lineatum* Pitt].

Talaromyces thermophilus Stolk in Antonie van Leeuwenhoek J. Microbiol. Serol.31: 268. 1965. — Holotype: No. 236.58 (CBS). [Anamorph: *Penicillium dupontii* Griffon & Maubl.].

Talaromyces trachyspermus (Shear) Stolk & Samson in Stud. Mycol. 2: 32. 1 Nov 1972. — Basionym: *Arachniotus trachyspermus* Shear in Science, ser. 2,16: 138. 1902. — Holotype (Shear, 1902): No. 5798 (BPI). [Anamorph: *Penicillium lehmanii* Pitt].

Talaromyces udagawae Stolk & Samson in Stud. Mycol. 2: 36. 1 Nov 1972. — Holotype: No. 579.72 p.p. (CBS). [Anamorph: *Penicillium udagawae* Stolk & Samson].

Talaromyces unicus Tzean & al. in Mycologia 84: 739. 21 Oct 1992. — Holotype: No. PPH16 p.p. (PPEH). [Anamorph: *Penicillium unicum* Tzean & al.].

Talaromyces wortmannii (Klöcker) C. R. Benj. in Mycologia 47: 683. 7 Oct 1955. — Basionym: *Penicillium wortmannii* Klöcker in Compt.-Rend. Trav. Carlsberg Lab. 6: 100. 1903 (nom. holomorph.). — Neotype (Pitt, 1980): No. 40047 p.p. (IMI). [Anamorph: *Penicillium kloeckeri* Pitt].

THERMOASCUS MIEHE (HOLOMORPHS)

Thermoascus aurantiacus Miehe, Selbsterhitzung Heus: 70. 1907. — Neotype (Cooney & Emerson, 1964): No. M206516 (UC).

Thermoascus aegyptiacus S. Ueda & Udagawa in Trans. Mycol. Soc. Japan 24: 135. Jul 1983. — Holotype: No. 2914 (NHL). [Anamorph: *Paecilomyces aegyptiacus* S. Ueda & Udagawa].

Thermoascus crustaceus (Apinis & Chesters) Stolk in Antonie van Leeuwenhoek J. Microbiol. Serol. 31: 272. 1965. — Basionym: *Dactylomyces crustaceus* Apinis & Chesters in Trans. Brit. Mycol. Soc.47: 428. 23 Sep 1964. — Neotype (designated here): Herb. 102470 (IMI). [Anamorph: *Paecilomyces crustaceus* Apinis & Chesters].

Thermoascus thermophilus (Sopp) Arx, Gen. Fungi Sporul. Pure Cult., ed. 2: 94. 1974. — Basionym: *Dactylomyces thermophilus* Sopp in Skr. Vidensk.-Selsk. Christiana, Math.-Naturvidensk. K1. 11: 35. 1912. — Neotype (designated here): No. 528.71 (CBS).

TORULOMYCES DELITSCH (ANAMORPHS)

Torulomyces lagena Delitsch, Syst. Schimmelpilze: 91. 1943. — Neotype (Stolk & Samson, 1983): No. 185.65 (CBS).

TRICHOCOMA JUNGH. (HOLOMORPHS)

Trichocoma paradoxa Jungh., Praem. Fl. Crypt. Java 1: 9. 1838. — Holotype: *"Trichocoma paradoxa", Junghuhn* (BO).

WARCUPIELLA SUBRAM. (HOLOMORPHS)

Warcupiella spinulosa (Warcup) Subram. in Curr. Sci. 41: 757. 5 Nov 1972. — Basionym: *Aspergillus spinulosus* Warcup in Raper & Fennell, Genus Aspergillus: 204. 1965 (nom. holomorph.) — Neotype (Samson & Gams, 1985): No. 75885 (IMI). [Anamorph: *Aspergillus warcupii* Samson & W. Gams].

REFERENCES

Al-Musallam, A. (1980). Revision of the black *Aspergillus* species. Utrecht.

Blaser, P. (1975). Taxonomische und physiologische Untersuchungen über die Gattung *Eurotium* Link ex Fries. Sydowia 28: 1-49.

Christensen, M. & Fennell, D. I. (1964). The rediscovery of *Aspergillus cervinus*. Mycologia 56: 350-361.

Cooney, D. G. & Emerson, R. (1964). Thermophilic fungi. San Francisco.

Eriksson, O. E. & Hawksworth, D. L. (1991). Outline of the Ascomycetes-1990. Syst. Ascomycetum 9: 39-271.

Frisvad, J. C., Filtenborg, O. & Wicklow, D. T. (1987). Terverticillate penicillia isolated from underground seed caches and cheek pouches of banner-tailed kangaroo rats (*Dipodomys spectabilis*). Canad. J. Bot. 65: 765-773.

Frisvad, J. C., Hawksworth, D. L., Pitt, J. I., Samson, R. A. & Stolk, A. C. (1992). Proposals for nomina specifica conservanda and rejicienda in *Aspergillus* and *Penicillium* (fungi). Taxon 41: 109-113.

Frisvad, J. C., Samson, R. A. & Stolk, A. C. (1990a). Notes on the typification of some species of *Penicillium*. Persoonia 14: 193-202.

Frisvad, J. C, Samson, R. A. & Stolk, A. C. (1990b). Disposition of recently described species of *Penicillium* . Persoonia 14: 209-232.

Holmgren, P. K., Holmgren, N. H. & Barnett, L. C. (1990). Index herbariorum. Part. I: the herbaria of the world. Eighth edition. Regnum Veg. 120.

Hughes, S. J. (1951). Studies of microfungi XI. Some Hyphomycetes which produce phialides. Mycol. Pap. 45: 1-36.

Jensen, C. N. (1912). Fungous flora of the soil. Cornell lJniv. Agric. Exp. Sta. Bull. 315: 415-501.

Kozakiewicz, Z. (1982). The identity and typification of *Aspergillus parasiticus*. Mycotaxon 15: 293-305.

Kozakiewicz, Z. (1989). Aspergillus species on stored products. Mycol. Pap. 161: 1-188 ,

Malloch, D. & Cain, R. F. (1972). The Trichocomataceae: ascomycetes with *Aspergillus, Paecilomyces* and *Penicillium* imperfect states. Canad. J. Bot. 50: 2613-2628. Pitt, J. I. 1979. *Geosmithia*, gen. nov. for *Penicillium lavendulum* and related species. Canad. J. Bot. 57: 2021-2030.

Pitt, J. I. (1980). The genus Penicillium and its teleomorphic states *Eupenicillium* and *Talaromyces*. London.

Pitt. J.I.& Hocking, A. D. (1979). *Merimbla* gen. nov. for the anamorphic state of *Talaromyces avellaneus*. Canad. J. Bot. 57: 2394-2398.

Pitt, J.I. and Samson, R.A. (1993). Species names in current use in the *Trichocomaceae* (Fungi, Eurotiales). *In* "Names in current use in the families *Trichocomaceae, Cladoniaceae, Pinaceae,* and *Lemnaceae*", W. Greuter, ed. Regnum Vegetabile 128: 13-57. Königstein, Germany: Koeltz Scientific Books.

Samson, R. A. (1974). Paecilomyces and some allied Hyphomycetes. Stud. Mycol. 6: 1-119.

Samson, R. A. & Gams, W. (1985). Typification of the species of *Aspergillus* and associated teleomorphs. Pp. 31-54 in: Samson, R. A. & Pitt, J. I. (ed.) Advances in *Penicillium* and *Aspergillus* Systematics. New York .

Samson, R. A. & Seifert, K. A. (1985). The ascomycete genus Penicilliopsis and its anamorphs. Pp. 397-428 in: Samson, R. A. & Pitt, J. I. (ed.) Advances in *Penicillium* and *Aspergillus* Sys~ematics. New York.

Stolk, A. C. & Hadlok, R. (1976). Revision of the Subsection Fasciculata of *Penicillium* and some allied species. Stud. Mycol. 11: 1-47.

Seifert, K. A. & Samson, R. A. (1985). The genus *Coremium* and the synnematous penicillia. Pp. 143-154 in: Samson, R. A. & Pitt, J. I. (ed.) Advances in *Penicillium* and *Aspergillus* Systematics. New York .

Shear, C. L. (1902). *Arachniotus trachyspermus*, a new species of the Gymnoascaceae. Science, ser. 2, 16: 138.

Stolk, A. C. (1965). Thermophilic species of *Talaromyces* Benjamin and *Thermoascus* Miehe. Antonie van Leeuwenhoek J. Microbiol. Serol. 31: 262-276.

Stolk, A. C. & Samson, R. A. (1972). The genus Talaromyces. Studies on *Talaromyce*s and related genera II. Stud. Mycol. 2: 1-65.

Stolk, A. C. & Samson, R. A. (1983). The Ascomycete genus *Eupenicillium* and related *Penicillium* anamorphs. Stud. Mycol. 23: 1-147.

Stolk, A.C. & Scott, D. B. (1967). Studies on the genus *Eupenicillium* Ludwig. I. Taxonomy and nomenclature of Penicillia in relation to their sclerotioid ascocarpic states. Persoonia 4: 391-405.

Subramanian, C. V. (1972). The perfect states of *Aspergillus*. Curr. Sci. 41: 75-761.

Takishima, Y., Shimura, J., Udagawa, Y. & Sugawara, A. (1989). Guide to worlddata center on microorganisms, with a list of culture collections in the world. Saitama (Japan).

List of Names of Trichocomaceae published between 1992 and 1999

R.A. Samson

Centraalbureau voor Schimmelcultures, Baarn, The Netherlands

The following list are names of *Penicillium* and *Aspergillus* with their teleomorphs published between 1992 and 1999. These names are not sanctioned by the International Commission of *Penicillium* and *Aspergillus* and is merely meant for a complete and updated compilation of names. Of all published taxa attempts were made to obtain the holotype or ex type culture and this is indicated by the ex type CBS accession number. However, for many new taxa type material could not be obtained although the type designation has been correctly made with the species description.

ASPERGILLUS

Aspergillus acristatulus M.A. Ismail, Abdel-Sater & Zohri, Mycotaxon 53: 396 (1995). — Holotype not indicated. (Teleomorph *Emericella acristata* (Fennell & Raper) Y. Horie). [Nom.inval., Arts 36.1, 37.1. Published as 'st. nov.'.]

Aspergillus appendiculatus Y. Horie & D.M. Li, in Horie, Li, Fukihara, Li, Abliz, Nishimura & Wang, Mycoscience 39(2): 161 (1998) from damp grassland soil, Xinjiang, China. — Holotype FA-865 (CBM). (Teleomorph *Emericella appendiculata* Y. Horie & D.M. Li). Homonym of *Aspergillus appendiculatus* Blaser, Sydowia 28: 38, 1975. (Teleomorph *Eurotium appendiculatum* Blaser).

Aspergillus arvii R. Aho, Y. Horie, Nishim. & Miyaji, Mycoses 37(11/12): 390 (1994) from liver abscesses of cow, Finland. — Holotype Chiba University, Japan. [Nom.inval., Art. 37.3.]

Aspergillus aurantiobrunnellus M.A. Ismail, Abdel-Sater & Zohri [as '*aurantiobrunneullus*'], Mycotaxon 53: 397 (1995). — Holotype not indicated. (Teleomorph *Emericella aurantiobrunnea* (Atkins, Hindson & Russell) Malloch & Cain). [Nom.inval., Arts 36.1, 37.1. Published as 'st. nov.'.]

Aspergillus beijingensis D.M. Li, Y. Horie, Y.X. Wang & R. Li, Mycoscience 39: 299 (1998) from maxillary sinusitis of 38 year old man, Beijing, China. — Holotype FD-285 (CBM).

Aspergillus botucatensis Y. Horie, Miyaji & Nishim., in Horie, Miyaji, Nishimura, Franco & Coelho, Mycoscience 36(2): 159 (1995) from soil, São Paulo, Brazil. — Holotype FA-0672 (CBM). (Teleomorph *Neosartorya botucatensis* Y. Horie, Miyaji & Nishim.).

Aspergillus caelatus B.W. Horn, Mycotaxon 61: 186 (1997) from soil in field of *Arachis hypogaea*, Georgia, USA. — Holotype BPI 737601, ex type CBS 763.97.

Aspergillus costiformis H.Z. Kong & Z.T. Qi, Acta Mycologica Sinica 14(1): 10 (1995) from rotten paper, Hebei, China. — Holotype HMAS 62766. (Teleomorph *Eurotium costiforme* H.Z. Kong & Z.T. Qi).

Aspergillus delicatus H.Z. Kong, Mycotaxon 62: 429 (1997) from fruit, Yunnan, China. — Holotype HMAS 71159. (Teleomorph *Neosartorya delicata* H.Z. Kong).

Aspergillus dentatulus M.A. Ismail, Abdel-Sater & Zohri, Mycotaxon 53: 397 (1995), ? — Holotype not indicated. (Teleomorph *Emericella dentata* (D.K. Sandhu & R.S. Sandhu) Y. Horie). [Nom.inval., Arts 36.1, 37.1.]

Aspergillus dorothicus Varshney & A.K. Sarbhoy, Microbiol. Research 151(3): 231 (1996) from soil, Tamil Nadu, India. Typification: 2600 and IMI 357698, Indian Type Culture Collection, New Dehli. [Nom.inval., Arts 37.3, 37.4, 37.5.]

Aspergillus flavus var. *parvisclerotigenus* Mich. Saito & Tsuruta, Proc. Jap. Assoc. Mycotoxicol. 37: 32 (1993) from soil of maize field, Thailand. — Holotype NFRI 1538, National Food Research Institute, Tsukuba.

Aspergillus fumisynnematus Y. Horie, Miyaji, Nishim., Taguchi & Udagawa, Trans. Mycol. Soc. Japan 34(1): 3 (1993) from soil, Venezuela. — Holotype FD-0001 (CBM).

Aspergillus heyangensis Z.T. Qi, Z.M. Sun & Wang [*sic*], in Sun & Qi, Acta Mycol. Sinica 13(2): 81 (1994) from placenta of *Gossypium,* Shaanxi, China. — Holotype HMAS 58982.

Aspergillus homomorphus Steiman, Guiraud, Sage & Seigle-Mur., Systematic and Applied Microbiology 17(4): 621 (1995) in soil, Israel. — Holotype Collection Mycologie Pharmacie Grenoble, Aug. 1992, ex type CBS 101889. [Nom.inval., Art. 37.4.]

Aspergillus implicatus A.M. Persiani & O. Maggi, in Maggi & Persiani, Mycol. Research 98(8): 871 (1994) from forest soil, Ivory Coast. — Holotype ROHB 110 S, ex type CBS 484.95.

Aspergillus ingratus Yaguchi, Someya & Udagawa, in Yaguchi, Someya, Miyadoh & Udagawa, Trans. Mycol. Soc. Japan 34(2): 305 (1993) from soil, California, USA. — Holotype PF-1116 (CBM), ex type CBS 643.95.

Aspergillus miyajii Y. Horie, in Horie, Fukihara, Nishimura, Taguchi, Wang & Li, Mycoscience 37(3): 323, 1996 (1997) from soil of vine field, Ningxia, China. — Holotype FA-716 (CBM). (Teleomorph *Emericella miyajii* Y. Horie).

Aspergillus montenegroi Y. Horie, Miyaji & Nishim., in Horie, Miyaji, Nishimura, Franco & Coelho, Mycoscience 37(2): 137 (1996) from soil, São Paulo, Brazil. — Holotype FA-0669 (CBM). (Teleomorph *Emericella montenegroi* Y. Horie, Miyaji & Nishim.).

Aspergillus multiplicatus Yaguchi, Someya & Udagawa, Mycoscience 35(4): 310 (1994) on soil, Taiwan. — Holotype PF-1154 (CBM). (Teleomorph *Neosartorya multiplicata* Yaguchi, Someya & Udagawa).

Aspergillus muricatus Udagawa, Uchiy. & Kamiya, Mycotaxon 52(1): 210 (1994) from grassland soil, Philippines. — Holotype BF-42515 (CBM). (Teleomorph *Petromyces muricatus* Udagawa, Uchiy & Kamiya).

Aspergillus omanensis Y. Horie & Udagawa, Mycoscience 36(4): 391 (1995) from forest soil, Oman. — Holotype FA-700 (CBM). (Teleomorph *Emericella omanensis* Y. Horie & Udagawa).

Aspergillus parviverruculosus H.Z. Kong & Z.T. Qi, Acta Mycol. Sinica 14(1): 12 (1995) from soil, Hebei, China. — Holotype HMAS 62767. (Teleomorph *Eurotium parviverruculosum* H.Z. Kong & Z.T. Qi).

Aspergillus paulistensis Y. Horie, Miyaji & Nishim., in Horie, Miyaji, Nishimura, Franco & Coelho, Mycoscience 36(2): 164 (1995) from soil, São Paulo, Brazil. — Holotype FA-0690 (CBM).

(Teleomorph *Neosartorya paulistensis* Y. Horie, Miyaji & Nishim.).

Aspergillus primulinus Udagawa, Toyaz. & Tsub., Mycotaxon 47: 360 (1993) from canned oolong tea beverage, Japan. — Holotype SUM-3014 (CBM). (Teleomorph *Neosartorya primulina* Udagawa, Toyaz. & Tsub.).

Aspergillus pseudo-heteromorphus Steiman, Guiraud, Sage & Seigle-Mur., Systematic and Applied Microbiology 17(4): 622 (1995) in soil, Israel. — Holotype Collection Mycologie Pharmacie Grenoble, Aug. 1992, ex type CBS 101888. [Nom.inval., Art. 37.4.]

Aspergillus qizutongii D.M. Li, Y. Horie, Y.X. Wang & R. Li, Mycoscience 39: 301 (1998) from maxillary sinusitis of 48 year old woman, Beijing, China. — Holotype FD-284 (CBM).

Aspergillus salviicola Udagawa, Kamiya & Tsub., Mycoscience 35(3): 245 (1994) on dried leaves of *Salvia officinalis,* Turkey. — Holotype NCI - 2090 (CBM).

Aspergillus sublevisporus Someya, Yaguchi & Udgawa, Mycoscience 40: 405 (1999) from soil, Yokohama City, Kanagawa pref., Japan. — Holotype PF-1207 (CBM). (Teleomorph *Neosartorya sublevispora* Someya, Yaguchi, & Udagawa).

Aspergillus sunderbanii Varshney & A.K. Sarbhoy, Microbiological Research 151(3): 232 (1996) from soil, West Bengal, India. Typification: 2630 and IMI 357699, Indian Type Culture Collection, New Dehli. [Nom.inval., Arts 37.3, 37.4, 37.5.]

Aspergillus taichungensis Yaguchi, Someya & Udagawa, Mycoscience 36(4): 421 (1995) from soil, Taiwan. — Holotype PF-1167 (CBM).

Aspergillus tuberculatus Z.T. Qi & Z.M. Sun, in Sun & Qi, Acta Mycologica Sinica 13(2): 86 (1994) from soil, Shaanxi, China. — Holotype HMAS 65948. (Teleomorph *Eurotium tuberculatum* Z.T. Qi & Z.M. Sun).

Aspergillus udagawae Y. Horie, Miyaji & Nishim., in Horie, Miyaji, Nishimura, Franco & Coelho, Mycoscience 36(2): 199 (1995) from soil, São Paulo, Brazil. — Holotype FA-0702 (CBM). (Teleomorph *Neosartorya udagawae* Y. Horie, Miyaji & Nishim.).

Aspergillus vinosobubalinus Udagawa, Kamiya & Kaori Osada, trans. Mycol. Soc. Japan 34(2): 255 (1993) from cultivated soil, Japan. — Holotype BF-33501 (CBM).

Aspergillus wangduanlii D.M. Li, Y.Horie, Y.X. Wang & R. Li, Mycoscience 39: 299 (1998) from maxillary sinusitis of 35 year old woman, Beijing, China. — Holotype FD-283 (CBM).

Aspergillus wentii var. *fumeus* Z.T. Qi & Z.M. Sun, in Sun & Qi, Acta Mycologica Sinica 13(2): 84 (1994) from *Oryza*, Guangxi, China. — Holotype HMAS 58983.

CHROMOCLEISTA

Chromocleista cinnabrina Yaguchi & Udagawa in Yaguchi, Miyadoh & Udagawa, Transactions of the Mycological Society of Japan 34(1): 105 (1993) from pepperfield soil, Japan. — Holotype NHL-2673 (CBM). (Anamorph *Paecilomyces cinnabarinus* S.C. Jong & E.E. Davis).

Chromocleista malachitea Yaguchi & Udagawa in Yaguchi, Miyadoh & Udagawa, Transactions of the Mycological Society of Japan 34(1): 102 (1993) from soil, Japan. — Holotype PF-1073 (CBM), ex type CBS 647.95. (Anamorph *Geosmithia malachitea* Yaguchi & Udagawa).

CORDYCEPS

Cordyceps loushanensis Z.Q. Liang & A.Y. Liu, in Liang, Liu, Huang & Jiao, Mycosystema 16(1): 61 (1997) on larva of Coleoptera, Guizhou, China. — Holotype GACP [as 'CGAC'] 89-7071. (Anamorph *Paecilomyces loushanensis* Z.Q. Liang & A.Y. Liu).

EMERICELLA

Emericella appendiculata Y. Horie & D.M. Li, in Horie, Li, Fukihara, Li, Abliz, Nishimura & Wang, Mycoscience 39(2): 161 (1998) from damp grassland soil, Xinjiang, China. — Holotype FA-865 (CBM). (Anamorph *Aspergillus appendiculatus* Y. Horie & D.M. Li).

Emericella miyajii Y. Horie, in Horie, Fukihara, Nishimura, Taguchi, Wang & Li, Mycoscience 37(3): 323, 1996 (1997) from soil, Ningxia, China. — Holotype FA-716 (CBM). (Anamorph *Aspergillus miyajii* Y. Horie).

Emericella montenegroi Y. Horie, Miyaji & Nishim., in Horie, Miyaji, Nishimura, Franco & Coelho, Mycoscience 37(2): 137 (1996) from soil, São Paulo, Brazil. — Holotype FA-0669(CBM). (Anamorph *Aspergillus montenegroi* Y. Horie, Miyaji & Nishim.).

Emericella omanensis Y. Horie & Udagawa, Mycoscience 36(4): 391 (1995) from forest soil, Oman. — Holotype FA-700 (CBM). (Anamorph *Aspergillus omanensis* Y. Horie & Udagawa).

Emericella pluriseminata Stchigel & Guarro, Mycologia 89(6): 937 (1997) from soil, Rajasthan, India. — Holotype FMR 5588; isotype IMI 370867, Universitat Rovira i Virgili, Reus.

Emericella rugulosa var. *lazulina* Y. Horie Miyaji & Nishim., in Horie, Miyaji, Nishimura, Franco & Coelho, Mycoscience 37(2): 140 (1996) from soil, São Paulo, Brazil. — Holotype FA-0710 (CBM). (Anamorph *Aspergillus rugulovalvus* Samson & W. Gams).

EUPENICILLIUM

Eupenicillium limosum S. Ueda, Mycoscience 36(4): 451 (1995) from marine sediment, Japan. — Holotype NEI-5220 (CBM), ex type CBS 339.97. (Anamorph *Penicillium limosum* S.Ueda).

EUROTIUM

Eurotium aridicola H.Z. Kong & Z.T. Qi, Acta Mycol. Sinica 14(2): 87 (1995) from sheep dung, Xizang, China. — Holotype HMAS 62768, ex type CBS 101746.

Eurotium costiforme H.Z. Kong & Z.T. Qi, Acta Mycol. Sinica 14(1): 10 (1995) from rotten paper, Hebei, China. — Holotype HMAS 62766, ex type CBS 101749. (Anamorph *Aspergillus costiformis* H.Z. Kong & Z.T. Qi).

Eurotium fimicola H.Z. Kong & Z.T. Qi, Acta Mycologica Sinica 14(2): 86 (1995) from animal dung, Xizang, China. — Holotype HMAS 62769, ex type CBS 101747.

Eurotium parviverruculosum H.Z. Kong & Z.T. Qi, Acta Mycologica Sinica 14(2): 12 (1995) from soil, Hebei, China. — Holotype HMAS 62767, ex type CBS 101750. (Anamorph *Aspergillus parviverruculosus* H.Z. Kong & Z.T. Qi).

Eurotium tuberculatum Z.T. Qi & Z.M. Sun, in Sun & Qi, Acta Mycol. Sinica 13(2): 85 (1994) from soil, Shaanxi, China. — Holotype HMAS 65948, ex type CBS 101748. (Anamorph *Aspergillus tuberculatus* Z.T. Qi & Z.M. Sun).

GEOSMITHIA

Geosmithia eburnea Yaguchi, Someya & Udagawa, Mycoscience 35(3): 249 (1994) from soil, Taiwan. — Holotype PF-1151 (CBM), ex type CBS 100538. (Teleomorph *Talaromyces eburneus* Yaguchi, Someya & Udagawa).

Geosmithia malachitea Yaguchi & Udagawa in Yaguchi, Miyadoh & Udagawa, Transactions of the Mycological Society of Japan 34(1): 102 (1993) from soil, Japan. — Holotype PF-1073 (CBM). (Teleomorph *Chromocleista malachitea* Yaguchi & Udagawa).

MERIMBLA

Merimbla brevicompacta H.Z. Kong, Mycosystema 18: 9 (1999) from moulded vegetable, Wolong,

Sichuan Province, China. — Holotype HMAS 62770, Institute of Microbiology Academia Sinica, Beijing, China. (Teleomorph *Talaromyces brevicompactus* H.Z. Kong).

NEOSARTORYA

Neosartorya botucatensis Y. Horie, Miyaji & Nishim., in Horie, Miyaji, Nishimura, Franco & Coelho, Mycoscience 36(2): 159 (1995) from soil, São Paulo, Brazil. — Holotype FA-0672 (CBM). (Anamorph *Aspergillus botucatensis* Y. Horie, Miyaji & Nishim.).

Neosartorya delicata H.Z. Kong, Mycotaxon 62: 429 (1997) from fruit, Yunnan, China. — Holotype HMAS 71159, ex type CBS 101754. (Anamorph *Aspergillus delicatus* H.Z. Kong).

Neosartorya multiplicata Yaguchi, Someya & Udagawa, Mycoscience 35(4): 309 (1994) from soil, Taiwan. — Holotype PF-1154 (CBM), ex type CBS 646.95. (Anamorph *Aspergillus multiplicatus* Yaguchi, Someya & Udagawa).

Neosartorya paulistensis Y. Horie, Miyaji & Nishim., in Horie, Miyaji, Nishimura, Franco & Coelho, Mycoscience 36(2): 163 (1995) from soil, São Paulo, Brazil. — Holotype FA-0690 (CBM). (Anamorph *Aspergillus paulistensis* Y. Horie, Miyaji & Nishim.).

Neosartorya primulina Udagawa, Toyaz. & Tsub., Mycotaxon 47: 360 (1993) from canned oolong tea beverage, Japan. — Holotype SUM-3014 (CBM), ex type CBS 253.94. (Anamorph *Aspergillus primulinus* Udagawa, Toyaz. & Tsub.).

Neosartorya sublevispora Someya, Yaguchi, & Udagawa, Mycoscience 40: 405 (1999) from soil, Yokohama City, Kanagawa pref., Japan. — Holotype PF-1207 (CBM). (Anamorph *Aspergillus sublevisporus* Someya, Yaguchi & Udagawa).

Neosartorya udagawae Y. Horie, Miyaji & Nishim., in Horie, Miyaji, Nishimura, Franco & Coelho, Mycoscience 36(2): 199 (1995) from soil, São Paulo, Brazil. — Holotype FA-0771 (CBM). (Anamorph *Aspergillus udagawae* Y. Horie, Miyaji & Nishim.).

PAECILOMYCES

Paecilomyces atrovirens Z.Q. Liang & A.Y. Liu in Liang, Liu & Feng, Acta Mycologica Sinica 12(2): 110 (1993) on *Telligonia*, Guizhou, China. — Holotype CGAC86-36D, Guizhou Agricultural College, Guiyang.

Paecilomyces borysthenicus Borisov& Tarasov, Mikologiya i Fitopatologiya 31(5): 16 (1997) in larva of *Tipula*, Ukraina. — Holotype CMPPB-99, Lomonosov Moscow State University, Moscow.

Paecilomyces crustaceus (Apinis & Chesters) Yaguchi, Someya & Udagawa, Mycoscience 36(2): 151 (1995). [Syn. *Dactylomyces crustaceus*]. (Teleomorph *Thermoascus crustaceus* (Apinis & Chesters) Stolk). [Nom.inval., Art. 33.2, basionym not indicated and reference omitted. First published here.] CBS 374.62 AUT of *Thermoascus crustaceus*, CBS 181.67 T of *Thermoascus crustaceus*.

Paecilomyces loushanensis Z.Q. Liang & A.Y. Liu, in Liang, Liu, Huang & Jiao, Mycosystema 16(1): 62 (1997) from larva of Coleoptera, Guizhou, China. — Holotype GACP [as 'CGAC'] 89-7071. (Telcomorph *Cordyceps loushanensis* Z.Q. Liang & A.Y. Liu).

Paecilomyces odonatae Z.Y. Liu, Z.Q. Liang & A.Y. Liu, Mycosystema 8-9: 84, 1995-1996 (1996) from stroma of *Cordyceps odonatae,* Guizhou, China. — Holotype 9489, Guizhou Agricultural College, Guiyang.

Paecilomyces spectabilis Udagawa & Shoji Suzuki, Mycotaxon 50: 82 (1994) from heat processed fruit beverage, Japan. — Holotype SUM- 3030 (CBM). (Teleomorph *Talaromyces spectabilis* Udagawa & Shoji Suzuki).

Paecilomyces taitungiacus K.Y. Chen & Z.C. Chen, Mycotaxon 60: 226 (1996) from field soil, Taiwan. — Holotype 8709-2, TAI (Mycology), Chen. (Teleomorph *Thermoascus taitungiacus* K.Y. Chen & Z.C. Chen).

PENICILLIUM

Penicillium austrocalifornicum Yaguchi & Udagawa, in Yaguchi, Miyadoh & Udagawa, Trans. Mycol. Soc. Japan 34(2): 245 (1993) from soil, California, USA. — Holotype PF-1117 (CBM). (Teleomorph *Talaromyces austrocalifornicus* Yaguchi & Udagawa).

Penicillium barcinense Yaguchi & Udagawa in Yaguchi, Miyadoh & Udagawa, Trans. Mycol. Soc. Japan 34(1): 15 (1993) from soil, Spain. — Holotype PF-1081 (CBM). (Teleomorph *Talaromyces barcinensis* Yaguchi & Udagawa).

Penicillium carneum (Frisvad) Frisvad, in Boysen, Skouboe, Frisvad & Rossen, Microbiology Reading 142(3): 546 (1996). (Syn. *Penicillium roqueforti* var. *carneum*).

Penicillium convolutum Udagawa, Mycotaxon 48: 142 (1993) from soil, Nepal. — Holotype SUM-3018 (CBM), ex type CBS 100537. (Teleomorph *Talaromyces convolutus* Udagawa).

Penicillium dipodomyis (Frisvad, Filt. & Wicklow) Banke, Frisvad & S. Rosend., Mycol. Research

101(5): 622 (1997). (Syn. *Penicillium chrysogenum* var. *dipodomyis*). [Nom.inval., Art.33.2, see Note 1, direct reference to basionym omitted (page spread given). Published as 'Frisvad, Filt. & Wicklow comb. nov.']

Penicillium discolor Frisvad & Samson, in Frisvad, Samson, Rassing, Horst, Rijn & Stark, Antonie van Leeuwenhoek 72(2): 120 (1997) from *Raphanus sativus*, Israel. — Holotype IMI 285513.

Penicillium emodense Udagawa, Mycotaxon 48: 146 (1993) from paddy soil, Nepal. — Holotype SUM- 3025. (CBM). (Teleomorph *Talaromyces emodensis* Udagawa).

Penicillium euchlorocarpium Yaguchi, Someya & Udagawa, Mycoscience 40: 133 (1999) from soil, Yokohama-city, Kanagawa Pref., Japan. — Holotype PF-1203 (CBM). (Teleomorph *Talaromyces euchlorocarpius* Yaguchi, Someya & Udagawa).

Penicillium flavigenum Frisvad & Samson, in Banke, Frisvad & Rosendahl, Mycological Research 101(5): 622 (1997), — Holotype CBS 419.89. [Published as 'Frisvad & Samson'.]

Penicillium freii Frisvad & Samson, in Lund & Frisvad, Mycol. Research 98(5): 488 (1994), [Nom.inval., Arts 36.1, 37.1. Published as 'in press'.]

Penicillium incoloratum L.Q. Huang & Z.T. Qi, Acta Mycologica Sinica 13(4): 264 (1994) from seed of *Phaseolus angularis*, Beijing, China. — Holotype HMAS 65949.

Penicillium indigoticum Takada & Udagawa, Mycotaxon 46: 129 (1993) from soil, Nepal. — Holotype SUM-3010 (CBM), ex type CBS 100534. (Teleomorph *Talaromyces indigoticus* Takada & Udagawa).

Penicillium kananaskense Seifert, Frisvad & McLean, Canadian Journal of Botany 72(1): 20 (1994) from soil under *Pinus contorta* var. *latifolia*, Alberta, Canada. — Holotype DAOM 216105, ex type CBS 530.93.

Penicillium lagunense Udagawa, Uchiy. & Kamiya, Mycoscience 35(4): 403 (1994) from soil, Philippines. — Holotype BF-49341 (CBM). (Teleomorph *Talaromyces lagunensis* Udagawa, Uchiy. & Kamiya).

Penicillium limosum S. Ueda, Mycoscience 36(4): 451 (1995) from marine sediment, Japan. — Holotype NEI-5220 (CBM). (Teleomorph *Eupenicillium limosum* S. Ueda).

Penicillium melanoconidium (Frisvad) Frisvad & Samson, in Lund & Frisvad, Mycol. Research 98(5): 489 (1994). [Syn. Not indicated].

[Nom.inval., Art. 33.2, basionym not indicated and reference omitted. Published as 'in press'.]

Penicillium neoechinulatum (Frisvad, Filt. & Wicklow) Frisvad & Samson, in Lund & Frisvad, Mycological Research 98(5): 489 (1994), = ex type CBS 169.87 (*Penicillium aurantiogriseum* var. *neoechinulatum*). [Syn. Not indicated]. [Nom.inval., Art. 33.2, basionym not indicated and reference omitted. Published as 'in press'.]

Penicillium pimiteouiense S.W. Peterson, Mycologia 91(2): 271 (1999) from kidney epithelical cell culture flask, Peoria, Illinois, USA. — Holotype 806262 (BPI), ex type NRRL 25542 = CBS 102479.

Penicillium radicum A.D. Hocking & Whitelaw, in Hocking, Whitelaw & Harden, Mycol. Research 102(7): 802 (1998) from root of seedling *Triticum aestivum*, New South Wales, Australia. — Holotype DAR 72374, ex type CBS 100489.

Penicillium retardatum Udagawa, Kamiya & Kaori Osada, Trans. Mycol. Soc. Japan 34(1): 9 (1993) from rotten bark, Japan. — Holotype BF-24811 (CBM). (Teleomorph *Talaromyces retardatus* Udagawa, Kamiya & Kaori Osada).

Penicillium subinflatum Yaguchi & Udagawa, in Yaguchi, Miyadoh & Udagawa, Trans. Mycol Soc. Japan 34(2): 249 (1993) from soil, Japan. — Holotype PF-1113 (CBM). (Teleomorph *Talaromyces subinflatus* Yaguchi & Udagawa).

Penicillium tardifaciens Udagawa, Mycotaxon 48: 151 (1993) from paddy soil, Nepal. — Holotype SUM-3017 (CBM). (Teleomorph *Talaromyces tardifaciens* Udagawa).

Penicillium tricolor Frisvad, Seifert, Samson & John T. Mills, Can. J. Bot. 72(7): 937 (1994) on stored grain of red spring wheat, Saskatchewan, Canada. — Holotype DAOM 216240, ex type CBS 635.93.

Penicillium unicum Tzean, J.L. Chen & Shiu, Mycologia 84(5): 739 (1992) on soil, Taiwan. — Holotype PPH16E, National Taiwan University, Taipei, ex type CBS 100535. (Teleomorph *Talaromyces unicus* Tzean, J.L. Chen & Shiu).

PETROMYCES

Petromyces muricatus Udagawa, Uchiy. & Kamiya, Mycotaxon 52(1): 208 (1994) from grassland soil, Philippines. — Holotype BF-42515 (CBM). (Anamorph *Aspergillus muricatus* Udagawa, Uchiy. & Kamiya).

TALAROMYCES

Talaromyces austrocalifornicus Yaguchi & Udagawa, in Yaguchi, Miyadoh & Udagawa,

Trans. Mycol. Soc. of Japan 34(2): 245 (1993) from soil, California, USA. — Holotype PF-1117 (CBM), ex type CBS 644.95. (Anamorph *Penicillium austrocalifornicum* Yaguchi & Udagawa).

Talaromyces barcinensis Yaguchi & Udagawa in Yaguchi, Miyadoh & Udagawa, Trans. Mycol. Soc. of Japan 34(1): 15 (1993) from soil, Spain. — Holotype PF-1081 (CBM), ex type CBS 649.95. (Anamorph *Penicillium barcinense* Yaguchi & Udagawa).

Talaromyces brevicompactus H.Z. Kong, Mycosystema 18: 9 (1999) from moulded vegetable, Wolong, Sichuan Province, China. — Holotype HMAS 62770, Institute of Microbiology Academia Sinica, Beijing. (Anamorph *Merimbla brevicompacta* H.Z. Kong).

Talaromyces convolutus Udagawa, Mycotaxon 48: 141 (1993) from soil, Nepal. — Holotype SUM-3018 (CBM), ex type CBS 100537. (Anamorph *Penicillium convolutum* Udagawa).

Talaromyces eburneus Yaguchi, Someya & Udagawa, Mycoscience 35(3): 249 (1994) from soil, Taiwan. — Holotype, PF-1151 (CBM), ex type CBS 100538. (Anamorph *Geosmithia eburnea* Yaguchi, Someya & Udagawa).

Talaromyces emodensis Udagawa, Mycotaxon 48: 146 (1993) from paddy soil, Nepal. — Holotype SUM-3025 (CBM), ex type CBS 100536. (Anamorph *Penicillium emodense* Udagawa).

Talaromyces euchlorocarpius Yaguchi, Someya & Udagawa, Mycoscience 40: 133 (1999) from soil, Yokohama-city, Kanagawa Pref., Japan. — Holotype PF-1203 (CBM). (Anamorph *Penicillium euchlorocarpium* Yaguchi, Someya & Udagawa).

Talaromyces hachijoensis Yaguchi, Someya & Udagawa, Mycoscience 37(1): 157 (1996) from cultivated soil, Japan. — Holotype PF-1174 (CBM).

Talaromyces helicus var. *boninensis* Yaguchi & Udagawa, in Yaguchi, Imai & Udagawa, Trans. Mycol. Soc. Japan 33(4): 511 (1992) from soil, Japan. — Holotype PF-1103 (CBM). (Anamorph *Penicillium spirillum* Pitt, pro parte).

Talaromyces indigoticus Takada & Udagawa, Mycotaxon 46: 129 (1993) from soil, Nepal. — Holotype SUM-3010 (CBM), ex type CBS 100534. (Anamorph *Penicillium indigoticum* Takada & Udagawa).

Talaromyces lagunensis Udagawa, Uchiy. & Kamiya, Mycoscience 35(4): 403 (1994) from soil, Philippines. — Holotype BF-49341 (CBM).

(Anamorph *Penicillium lagunense* Udagawa, Uchiy. & Kamiya).

Talaromyces muroii Yaguchi, Someya & Udagawa, Mycoscience 35(3): 252 (1994) from soil, Taiwan. — Holotype PF-1153 (CBM), ex type CBS 756.96.

Talaromyces retardatus Udagawa, Kamiya & Kaori Osada, Trans. Mycol. Soc. Japan 34(1): 9 (1993) from rotten bark, Japan. — Holotype BF-24811 (CBM). (Anamorph *Penicillium retardatum* Udagawa, Kamiya & Kaori Osada).

Talaromyces sect. TRACHYSPERMUS Yaguchi & Udagawa, in Yaguchi, Someya & Udagawa, Mycoscience 37(1): 57 (1996). Sp. typ. *T. trachyspermus*.

Talaromyces spectabilis Udagawa & Shoji Suzuki, Mycotaxon 50: 82 (1994) from heat processed fruit beverage, Japan. — Holotype SUM-3030 (CBM), ex type CBS 101075. (Anamorph *Paecilomyces spectabilis* Udagawa & Shoji Suzuki).

Talaromyces subinflatus Yaguchi & Udagawa, in Yaguchi, Miyadoh & Udagawa, Trans. Mycol. Soc. Japan 34(2): 249 (1993) from soil, Japan. — Holotype PF-1113 (CBM), ex type CBS 652.95. (Anamorph *Penicillium subinflatum* Yaguchi & Udagawa).

Talaromyces tardifaciens Udagawa, Mycotaxon 48: 150 (1993) from paddy soil, Nepal. — Holotype SUM-3017 (CBM). (Anamorph *Penicillium tardifaciens* Udagawa).

Talaromyces trachyspermus var. *assiutensis* (Samson & Abdel-Fattah) Yaguchi & Udagawa, in Yaguchi, Someya, Miyadoh & Udagawa, Mycoscience 35(1): 65 (1994). (Syn. *Talaromyces assiutensis*). (Anamorph *Penicillium lehmanii* Pitt, pro parte).

Talaromyces unicus Tzean, J.L. Chen & Shiu, Mycologia 84(5): 739 (1992) from soil, Taiwan. — Holotype PPH16 National Taiwan University, Taipei, ex type CBS 100535. (Anamorph *Penicillium unicum* Tzean, J.L. Chen & Shiu).

Talaromyces wortmannii var. *sublevisporus* Yaguchi & Udagawa, in Yaguchi, Someya, Miyadoh & Udagawa, Mycoscience 35(1): 63 (1994) from soil, Japan. — Holotype PF-1130 (CBM). (Anamorph *Penicillium kloeckeri* Pitt, pro parte).

THERMOASCUS

Thermoascus crustaceus var. *verrucosus* Yaguchi, Someya & Udagawa, Mycoscience 36(2): 151 (1995) from soil, Guangdong, China. — Holotype PF-1160 (CBM). (Anamorph *Paecilomyces crustaceus* (Apinis & Chesters) Yaguchi, Someya & Udagawa).

Thermoascus taitungiacus K.Y. Chen & Z.C. Chen, Mycotaxon 60: 226 (1996) from field soil, Taiwan. — Holotype 8709-2 TAI (Mycology), Chen. (Anamorph *Paecilomyces taitungiacus* K.Y. Chen & Z.C. Chen).

Chapter 2

METHODS FOR IDENTIFICATION OF *PENICILLIUM* AND *ASPERGILLUS*

MEDIA AND INCUBATION EFFECTS ON MORPHOLOGICAL CHARACTERISTICS OF *PENICILLIUM* AND *ASPERGILLUS*

Toru Okuda[1], Maren A. Klich[2], Keith A. Seifert[3] and Katsuhiko Ando[4]
[1]Tamagawa University, Institute for Applied Life Science, Machida, Tokyo 194-8610, Japan, [2]USDA Southern Regional Research Center, New Orleans, LA 70179-0687 USA, [3]Eastern Cereal and Oilseed Research Centre, Agriculture and Agri-Food Canada, Research Branch, Ottawa, Ontario K1A 0C6, Canada and [4]Kyowa Hakko Kogyo Co., Ltd.. Tokyo Research Laboratories, 3-6-6 Asahimachi, Machida, 194 Japan

Every scientist involved in *Penicillium/Aspergillus* identification is aware that a variety of known and unknown factors affect colony and microscopic characteristics. Since Thom and Raper, reproducibility in taxonomic work has been sought by standardizing cultivation parameters. Introduction of new media and methods has resulted in a diversity of protocols employed in different laboratories. Few intensive studies on these factors or variables have been carried out. To summarize the variability in routine methods for *Penicillium* and *Aspergillus* taxonomy, we distributed a questionnaire to the members of the International Commission on *Penicillium* and *Aspergillus*. In this study we considered the following variables: medium ingredients, types of Petri dishes, volume of media, method of inoculation, and incubation conditions including temperature, air exchange, duration, and sporadic exposure to light. In our pilot studies, the most important factors strikingly affecting colony appearance and growth were volume of media and air exchange. Inoculum size and occasional light influenced the colonies to a lesser extent. Differences in water or brands of yeast extract or agar did not greatly affect cultural characteristics.

INTRODUCTION

In taxonomic studies on *Penicillium* and *Aspergillus*, strains are cultured and examined under standardized laboratory conditions. This has been a well recognized procedure since Raper and Thom (1949), and Raper and Fennell (1965) compiled their monographs using Czapek's solution agar (CZ). Pitt (1973, 1979) developed Czapek yeast extract agar (CYA) to improve growth over that on CZ, on which some strains do not grow well. He also introduced 25% glycerol nitrate agar for diagnostic purposes. Several other taxonomists have subsequently described new useful media and methods in their monographs or studies (Abe, 1956; Frisvad and Filtenborg 1989; Klich and Pitt, 1988; Ramirez, 1982; Tzean *et al.,* 1990; 1994). A number of "standard" media and methods have thus been described during the past half century.

Most taxonomists agree that there are deviations in cultural and micro-morphological characters within a species of these genera, and that standardization will permit more accurate identification. Several attempts have been made to describe character variability and standard methods (Pitt, 1985; Constantinescu, 1990; Okuda, 1994). However, a consensus on standardized media or methods has not yet been reached. Attempts to date have been based on small data sets and a new standard will require a great deal of tedious testing. As well, people have their own personal preferences for media or methods.

The International Commission on *Penicillium* and *Aspergillus* (ICPA) assembled a small project team to optimise standardization. We first listed possible factors affecting colony characteristics then distributed a questionnaire in order to survey media compositions and different inoculation or cultivation methods that ICPA members employ. Our pilot studies on media and incubation effects are discussed here, along with an analysis of a variety of methods employed by the members.

MATERIALS AND METHODS

Questionnaire: A comprehensive questionnaire to survey media compositions and different inoculation or incubation methods routinely used in the taxonomy of *Penicillium* and *Aspergillus* was prepared and circulated it to 13 members of ICPA. The questions covered agar media routinely used, their composition, manufacturers' name for all ingredients, pH, autoclaving parameters, type of water, type and size of Petri dishes used, volume of media, source of inoculum, inoculation method and tools, incubation conditions such as temperature, type of wrapping, lighting conditions, and placement of dishes, colour determination and standards, source of microscopic slides, wet mount preparation methods, and objective lenses used.

Strains used: Two strains were used for comparisons of yeast extracts, agars, and water between laboratories in Toda, Japan; New Orleans, United States; and Ottawa, Canada: *Aspergillus flavus* SRRC 2375 (=IMI 124930) and *Penicillium aurantiogriseum* SRRC 1379 (= CBS 324.89). In addition to those cultures, the following strains were used in further experiments in Japan: *A. terreus* KY 16507 and LG 0027; *A. versicolor* KY 16511 and LG 0038; *P. citrinum* IFO 6352 and LG 0033; and *P. sclerotiorum* KY 12388 and LG 0037. Strains with SRRC numbers were obtained from USDA Southern Regional Research Center, New Orleans, USA; IFO numbers from Institute for Fermentation, Osaka, Japan; KY from Kyowa Hakko, Co., Tokyo, Japan; and LG numbers from Tanabe Seiyaku Co., Saitama, Japan.

General procedure: General procedures used were primarily based on those described by Pitt (1979, 1988). The following standard conditions were used for most work done in Japan: Prior to the preparation of a conidial suspension, the preserved culture was removed from preservation at $-80°C$, inoculated on modified malt extract agar (MMA) consisting of 1% malt extract (Difco), 0.1% yeast extract (Difco), 0.1% soytone (Difco), 1% glucose, and 2% agar, and incubated at 25°C for 7 days. Potato dextrose agar (PDA) or Blakeslee's malt extract agar (MEA) were also used sometimes for growing slant cultures. Czapek yeast extract agar (CYA) was prepared according to the recipe of Pitt (1988) using Oxoid yeast extract and Oxoid agar with addition of the trace metal ions Cu^{2+} and Zn^{2+}. Agar media (20 mL) were poured into standard plastic Petri dishes (86 mm inner diameter, 12.9 mm deep, Sterile Auto Schale, Eiken Kizai Co, Tokyo, Japan). A conidial suspension was prepared by adding a small portion of conidia from a 7-day old slant culture into 500 FL detergent agar consisting of 0.2% agar and 0.05% Tween 80

(Pitt, 1988). This suspension was used to inoculate the agar media, with 2 FL at each three points with a micropipettor. The inoculated agar media were incubated at 25°C for 7 days"6 hours. Colony diameters were measured with slide callipers from the reverse side, recording six colonies from two plates. Sporulation rate, colour of the conidial area, mycelial colour, exudate production, pigment production, and colour of the reverse were also recorded.

Manufacturers of medium ingredients: Yeast extracts from different manufacturers were compared in otherwise uniform formulations of CYA, in two separate sets of experiments. In the first experiment conducted in Toda, New Orleans and Ottawa, Oxoid Yeast Extract (Unipath Ltd., Hampshire, England) was compared with two independently purchased lots of Bacto Yeast Extract (Difco Laboratories, Detroit, USA). Further experiments in Japan compared Dried Yeast Extract-S (Nippon Seiyaku Co., Tokyo, Japan), Yeast Extract (Oriental Yeast Industry, Tokyo, Japan), Powdered Yeast Extract (Kyokuto Seiyaku Co., Tokyo, Japan), and Yeast Extract (Asahi Brewery, Tokyo, Japan).

Agars from different suppliers were also compared in otherwise uniform formulations of CYA in two separate sets of experiments. The first experiment was conducted in Toda, New Orleans and Ottawa, and compared Oxoid Bacteriological Agar (Agar No.1) (Unipath Ltd., Hampshire, England), USP grade bulk gellidium, and BDH "Mikrobiologie" grade agar (British Drug House, VWR, Montreal, Canada). Further experiments in Japan included Bacto Agar (Difco Laboratories, Detroit, USA), Agar Powder (Wako Pure Chemicals, Osaka, Japan), Agarose-1 (Dojindo, Kumamoto, Japan), and Agar (Moorhead & Co., Van Nuys, USA).

Comparison of water: To determine the effect of water, various commercial brands of non-carbonated mineral waters available in Japan as well as distilled or tap water were compared. The mineral waters were Azumino (Nagano, Japan), Mitsutoge (Yamanashi, Japan), Dewa-sanzan (Yamagata, Japan), South Alps (Yamanashi, Japan), Rokko (Hyogo, Japan), Tanagashima (Kagoshima, Japan), ValVert (Ardennes, Belgium), and Volvic (Volvic, France), California (California, USA). Experiments comparing water between countries employed tap water and nanopure water by MilliQ (Millipore) from Toda, Saitama, Japan, house deionized water in New Orleans, USA, and water obtained through reverse osmosis in Ottawa, Canada.

Effect of Petri dishes and media volume: Various polystyrene Petri dishes were compared: Sterilin (85.9 mm in inner diam., 13.0 mm in inner depth; Bibby Sterilin Ltd., UK), Tissue culture dish (85.3 mm x 17.9; Iwaki Glass Co., Japan), Sterile dish (86.4 mm X 17.5 mm; Iwaki Glass Co, Japan), Sterile Auto Schale (86.0 mm x 12.9 mm; Eiken Kizai Co., Japan), Sterile H Schale (88.7 mm x 19.4 mm; Eiken Kizai Co., Japan), and glass Petri dishes (85.5 mm x 18.5 mm). Effects of medium volume were compared from 10 mL to 40 mL per Petri dish (using Eiken Kizai Co. Petri dishes). Freshly prepared plates were compared with 4 week old and 8 week old ones. Plates were stored at 5°C in the original plastic sleeves, 18 plates per bag, sealed at the top with a rubber band. The pH and water activity (a_w) of the media were measured after storage in New Orleans. Water activity was measured at 25°C with a CX-1 system (Decagon Devices Inc, Pullman Washington) according to the manufacturer's instructions. For each medium, on each date, 9 readings were taken using three 1 cm diameter plugs taken from each of 3 plates.

The pH was measured from each of three plates of each medium using a Corning 360i pH system (Corning Inc. Corning NY USA).

Effect of inoculum size: A spore mass taken from a well-grown slant was suspended in detergent agar (Pitt 1988) and diluted from x1 to x10,000, or from 10^8 to 10^2 conidia/mL. Spore concentration was determined microscopically using a haemocytometer, and appropriate dilutions were made with sterile detergent agar. Pin point inoculation with a sterile bamboo stick was compared to inoculation (2 FL) with a micropipettor.

Incubation conditions: The effects of air exchange was tested by comparing dishes tightly wrapped with two polyethylene bags with those unwrapped during incubation. To determine the effect of incidental light, dishes were loosely wrapped with a black polyethylene bag and growth was compared with plates covered in transparent wrapping. The incubator door was opened routinely, once to several times a day, so that sporadic exposure to light occurred.

RESULTS AND DISCUSSION

Media composition and ingredients

Questionnaire results: Eleven of the 13 questionnaires circulated were filled out and collected. For *Penicillium*, most ICPA members used Czapek yeast extract agar (CYA; Pitt 1988; 10 members) and Blakeslee's malt extract agar (MEA; Raper and Thom, 1949; 7 members) as basic media (Table 1). Czapek's solution agar (CZ; Raper and Thom, 1949) was used as an alternative to CYA or as an additional medium by 5 members. CYA without addition of Zn^{2+} and Cu^{2+} (Pitt, 1979) was the choice of one member. Some members used 2% malt extract agar (MA2) or Oxoid malt extract agar as an alternative to MEA or as an additional medium. Ten members preferred to use two to three media at the same time. Additional media used were creatine sucrose agar (CREA; Frisvad, 1985), CSN (Pitt, 1993), Czapek yeast extract agar with 20% sucrose (CY20S; Klich and Pitt, 1988), 25% glycerol nitrate agar (G25N; Pitt 1988), oatmeal agar, yeast extract agar, and yeast extract sucrose agar (YES; Frisvad and Filtenborg, 1989).

For *Aspergillus*, the most frequently used media were CYA (7 members), MEA (7 members), and CY20S (6 members). CZ was an additional or alternative medium used by 5 members. MA or Oxoid malt extract agar was also selected as an alternative to MEA. Additional media sometimes used were Czapek solution agar with 20% sucrose (CZ20; Raper and Thom, 1949), G25N, M40Y (Raper and Fennell, 1965), MY20 (Raper and Fennell, 1965), oatmeal agar, and YES.

The yeast extract, malt extract, and peptone products used came from only a few manufacturers (Table 2) and were interestingly quite uniform. Bacto Yeast Extract (Difco Laboratories) was most frequently used (6 members), whereas Oxoid Yeast Extract (Unipath Ltd.) was the second most popular (3 members). Other sources of yeast extracts were Wako Pure Chemicals, Japan, and British Drug House, UK (BDH). In contrast, agar sources were rather diverse (Table 2); Difco (3 members), Oxoid (2 members), BDH (1), So-Bi-Gel (1), Ferdiwo (1), Moorhead (1), and commercial pharmaceutical grade (1).

Table 1. Media used by members of the International Commission on *Penicillium* and *Aspergillus*

Media	1	2	3	4	5	6	7	8	9	10	11	Total
For *Penicillium*												
Czapek yeast extract agar	x	x	x	x	x	x	x	x	x		x	10
Blakeslee's malt extract agar	x		x	x	x	x	x			x		7
Czapek's solution agar				x	x			x	x		x	5
25% Glycerol nitrate agar	x				x		x			x		4
Oatmeal agar				x		x		x	x			4
2% Malt extract agar		x						x	x			3
Yeast extract-sucrose agar				x		x						2
CREA				x		x						2
Malt extract agar (Oxoid)								x	x			1
CSN						x						1
CY20S or Yeast extract agar									x			1
For *Aspergilus*												
Czapek yeast extract agar	x	x	x	x		x	x			x		7
Blakeslee's malt extract agar	x		x	x	x	x	x			x		7
Czapek yeast extract agar with 20% sucrose	x	x		x		x	x			x		6
Czapek's solution agar				x	x			x	x		x	5
2% Malt Extract Agar		x						x	x			3
CZ20					x			x	x			3
M40Y					x						x	2
Malt extract agar (Oxoid)								x	x			2
25% Glycerol nitrate agar							x					1
MY20										x		1
Oatmeal agar						x						1
Yeast extract sucrose agar				x								1

x, selected by the members.

For malt extract, 5 members used Difco, and 4 used Oxoid. BDH and Brewers grade were each used by one member. Five members used Difco's peptone, 3 used Oxoid and one member used peptone from Kyokuto. With regard to the other components such as phosphate salt, metal ions, nitrate, and sugar, no uniformity was observed; each member purchased products from local manufacturers. However, the use of analytical grade products should ensure a relatively constant quality of chemical ingredients in all laboratories.

Water was prepared by distillation (5 members) or deionization (5 members). In every case, the pH of CYA and MEA was not adjusted. One member however adjusted the pH of CZ to 6-7 before autoclaving. Although most members used stock solutions for preparing CYA or CZ, the composition of stocks was not always identical. Six members followed the procedure of Pitt (1988), while four members used Difco Czapek broth concentrate with addition of trace elements separately.

For CZ, one member divided the formula into three parts; solution A containing nitrate and solution B containing phosphate were finally combined with the sugar solution. One member autoclaved glucose separately when MEA was prepared.

Table 2. Manufacturers of medium ingredients used by ICPA members.

Members Ingredients	1	2	3	4	5	6	7	8	9	10	11
Yeast extract	O	D	W	D	D	BDH	O	D	D	D	O
Malt extract	O	D	D	D	D	BDH	Bg	O	O	D	O
Peptone	O	D	Ky	D	D	D	O			D	O
Sucrose	N	Ka	Kn	D	DS	BDH	Fg	D	D	B	BDH
Glucose	Kn	M	N	BDH	D	BDH	Fg	-	-	B	BDH
K2HPO4	N	M	Kn	D	BAR	JT B	A	D	D	B	BDH
NaNO3	Ko	Ka	Kn	D	BAR	BDH	BDH	D	D	B	BDH
Agar	O	D	D	SBG	D ·	BDH	Pg	F	F	Mo	O
Water	MQ	-	Dist	Dd	Dei	Dis	MQ	Dei	Dei	Dei	Dis

A = Ajax; B = Baker; BAR = Baker AR; BDH; Bg= Brewers grade; Dei = Deionized; D= Difco; Dis= Distilled; DS= Domino Sugar; Dd= Double distilled; F= Ferdiwo; Fg= Food grade; JTB= JT Baker; Kn= Kanto; Ka=Katayama; Ko= Kokusan; Ky= Kyokuto; M = Merck ; MQ= Milli-Q; Mo= Moorhead; N = Nacalai; O= Oxoid; Pg= Pharmaceutical grade; SBG= So-Bi-Gel; W = Wako;

All members autoclaved CYA or CZ at 121°C for 15 to 20 min. Variations on the standard media preparation and storage procedure were reported as follows:

- Ingredients were thoroughly dissolved before adding the next one in the order of ingredient lists.
- Agar was dissolved by heating at 100-105 for 15 min before autoclaving.
- Plates were air dried for 20-30 min, 120-150 min, or never.
- Plates were stored at 0-5°C, but used up within 60-90 days, or 40 days.
- Plates were stored at room temperature, but used up within 2 days.

Experimental results: Some minor variability in growth was observed on CYA with different yeast extracts, agars, or water. However, the difference in growth were inconspicuous except for the case of agarose, as shown below.

Effect of yeast extract: Our comparisons of yeast extracts purchased from different manufacturers showed that although differences observed in colony diameters were insignificant (within 3.2 mm in diam.), colony appearance was affected sometimes. Aerial mycelium of *P. aurantiogriseum* over-grew the conidial area when yeast extracts other than Difco were used. Nippon Seiyaku's yeast extract reduced sporulation of *P. citrinum*. *Aspergillus flavus* produced a more greenish conidial area with Nippon Seiyaku, and sporulated poorly on Asahi Brewery's yeast extract.

Effect of agar: As expected, colonies on agarose were quite distinctive compared with colonies on normal agar. All species grew more slowly on CYA with agarose (Fig. 1). *Penicillium aurantiogriseum* grew slightly more slowly on Oxoid agar than on Moorhead agar or BDH agar. It showed better sporulation on Difco agar, but sporulated poorly on Wako or Moorhead agar. *Penicillium citrinum* and *A. versicolor* also sporulated poorly on Wako agar. The reverse of *A. versicolor* was more yellowish on Difco agar. *Aspergillus flavus* formed abundant mycelia, fewer sclerotia, and a more reddish reverse on agarose.

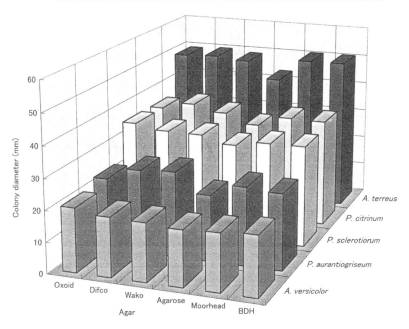

Fig. 1. Effect of different brands of agar on colony diameters of five *Penicillium* and *Aspergillus* species after 7 days on CYA at 25°C.

Effect of water: The analysis of the commercial non-carbonated mineral waters used compared with tap water and MilliQ distilled water are given in Table 3. Dewa-sanzan water characteristically contains a high concentration of potassium ion (16 mg/L). Sodium content was the highest in Rokko water (18 mg/L). ValVert contains an extremely high concentration of calcium ions (68 mg/L). However, these differences in mineral water composition has no significant effect on growth rate (less than 3.4 mm in colony diam.).

Penicillium aurantiogriseum sporulated poorly on a medium with Tanegashima water while *P. citrinum* had reduced sporulation on media with Mitsutoge or Dewa-sanzan water. *Aspergillus versicolor* produced more abundant exudate on CYA with ValVert water, while ValVert induced poor sporulation in *A. terreus*. Water derived by reverse osmosis (in Canada), deionization (in the United States), and MilliQ (distillation and deionization, in Japan) did not affect colony diameters of *P. aurantiogriseum* or *A. flavus*.

Effect of Petri dishes and medium volume

Questionnaire results: Nine members used plastic Petri dishes (Table 4), but manufacturers varied (Bacto, Eiken, Falcon, Fisher, Greiner, Sterilin, and VWR). The inner diameter of the dishes varied from 85 to 100 mm (average, 89.4 mm), and the depth from 12 to 17 mm (average 13.9 mm). Two members used glass Petri dishes 90 x 17 mm. Volumes of media poured varied from laboratory to laboratory, from 15 to 25 mL (average 20.5 mL). Depth of media therefore varied from 2.4 to 4.4 mm depending on the size of dishes and volume of media.

89

Table 3. Analyses data of water used in preparation of media.

Name	Collection site etc.	Ca	Mg	K	Na	pH
Azumino	gano, Japan	5,3	2,4	0,9	4,6	7,18
Mitsutoge	Yamanashi, Japan	16	6,7	1,1	7,3	6,98
Dewa-sanzan	Yamagata, Japan	12	5,2	16	0,6	8,23
South Alps	Yamanashi, Japan	10	1,5	2,6	4,7	6,96
Rokko	Kobe, Japan	24	5,7	0,3	18	7,58
Tanegashima Island	Kagoshima, Japan	NT	NT	NT	NT	7,87
Valvert	Ardennes, Belgium	67,5	2	0,7	1,9	7,52
Volvic	Volvic, France	9,9	6,1	5,7	9,4	7,52
California	Olancha Pk., USA	16,5	1,6	1,7	NT	7,89
Tap water	Toda, Saitama, Japan	NT	NT	NT	NT	7,26
Nano pure water	MilliQ	NT	NT	NT	NT	8,36

Metal ion contents (mg/liter) are based on the data from suppliers; NT, not tested.

Experimental results: Various plastic Petri dishes were compared for their effect on culture characteristics. The inner diameter of dishes varied from 85.3 to 88.7 mm, and depth from 12.9 to 19.4 mm. No significant colony or growth differences with the *Penicillium* and *Aspergillus* strains tested were attributed to differences in the actual Petri dishes used.

Table 4. Dishes and medium volume used by the members

Members	1	2	3	4	5	6	7	8	9	10	11	Average
Dishes	Eiken	Eiken	Greiner	Falcon	Fisher	Bacto				VWR	Sterilin	
Materials	P	P	P/G	P	P	P	P	G	G	P	P	
Inner diameter (mm)	85	87	85	88	90	100	90	90	90	88	90	89,4
Depth (mm)	12	14	12	14	10	15	14	17	17	13	15	13,9
Media volume (ml)	20	25	25	17	15	25	18	15	15	25	25	20,5
Medium depth (mm)	3,5	4,2	4,4	2,8	2,4	3,2	2,8	2,4	2,4	4,1	3,9	3,3

P, plastic dishes; G, glass dishes; Participant 3 answered 20-30 ml; ; Participant.6 answered not meassured;. ; Participant.7 answered 16-20 ml; Medium depth, by calculation

As noted by Raper and Fennell (1965), the volume of media had a remarkable effect on colony diameter as well as various other cultural characteristics. In our experiments with CYA, there was a strong positive correlation (correlation coefficient >0.95) between growth rate and volume of media, except for *P. sclerotiorum* (Table 5). As medium volume increased, *P. aurantiogriseum* sporulated better and formed a darker reverse. Similarly, *P. citrinum* produced more abundant exudate and a darker reverse (Fig. 2), while *P. sclerotiorum* formed more bluish colonies, more orange mycelia, more abundant exudate, more yellow pigment, and darker reverse (Fig. 3) with increased medium volume. *Aspergillus flavus* produced more abundant sclerotia and a slightly darker reverse (Fig. 4), and *A. versicolor* also produced a slightly darker reverse. In contrast, *A. terreus* sporulated less, and has a darker reverse side (Fig. 5) with larger volumes of media. Similar depth effects were also confirmed on MEA even when a single

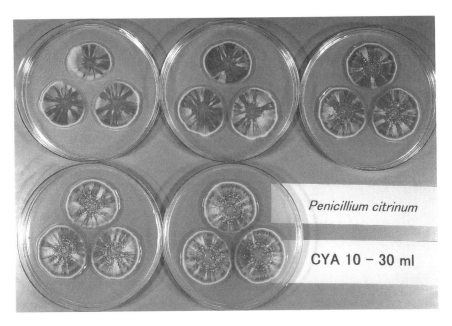

Fig. 2.. Colonies of *Penicillium citrinum* after 7 days on CYA at 25°C with different media volumes, showing effect on soluble pigment and sporulation. From top left to bottom right, 10, 15, 20, 25 and 30 mL

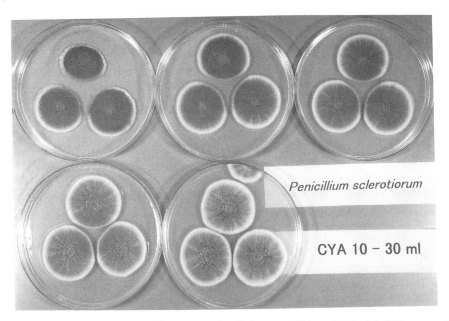

Fig. 3. Colonies of *Penicillium sclerotiorum* after 7 days on CYA at 25°C with different media volumes, showing effect on production of sclerotia and exudate. From top left to bottom right, 10, 15, 20, 25 and 30 mL.

91

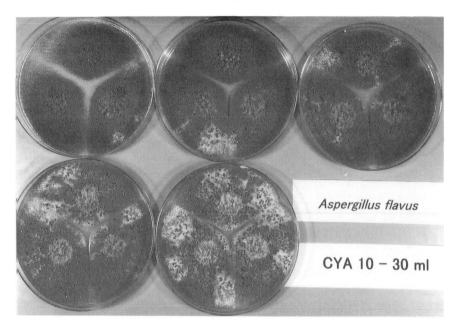

Fig. 4. Colonies of *Aspergillus flavus* after 7 days on CYA at 25°C with different media volumes, showing effect on sporulation and sclerotium production. From top left to bottom right, 10, 15, 20, 25 and 30 mL.

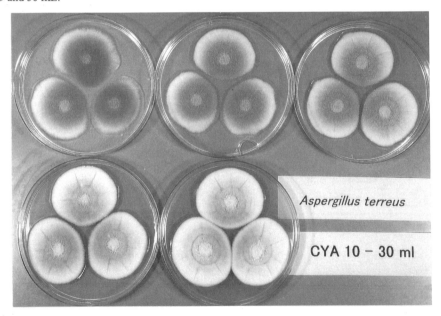

Fig. 5. Colonies of *Aspergillus terreus* after 7 days on CYA at 25°C with different media volumes, showing effect on sporulation. From top left to bottom right, 10, 15, 20, 25 and 30 mL.

conidium was used for inoculation. Deeper MEA enhanced the growth of *P. aurantiogriseum, P. citrinum, P. sclerotiorum, A. terreus* and *A. versicolor* (Fig. 6). The effect was pronounced particularly for slow growers such as *P. aurantiogriseum, P. sclerotiorum* and *A. versicolor* when incubation was prolonged.

Table 5. Effect of medium volume in CYA plates on colony diameters (mm) for five *Penicillium* and *Aspergillus* species.

Volume of media (ml)	10	15	20	25	30	Pearson**
Medium depth (mm)	1,7	2,6	3,5	4,4	5,3	
*P. aurantiogriseum**	14,5	18,2	19,6	20,7	21,4	0,95
P. citrinum	30,7	32,6	33,6	34,3	34,8	0,97
P. sclerotiorum	32,4	35,6	35,2	35,6	36,7	0,84
A. versicolor	15,6	17,5	18,7	19,2	19,7	0,96
A. terreus	42,8	42,7	43,5	45,0	46,0	0,95

Medium depth, by calculation; * Colony diameter in mm; Pearson**, Pearson's correlation coefficient between volume and colony diameter

Several possible reasons exist for the effect of variations in medium volume described above. One possibility is availability of nutrients for growth. Sporulation and production of pigments or exudate require a certain amount of vegetative growth during trophophase. Thicker media may provide the fungus with increased nutrients or may conserve the pH of the media for a longer period. Another possibility is changes in head space above the agar, which could serve as a reservoir of oxygen or gaseous wastes such as carbon dioxide and volatile compounds given off by the fungus. This may repress luxurious vegetative growth so that sporulation and metabolite production can be initiated. The third possible reason would be water potential. Finally, thicker media may limit the changes in water activity noted below.

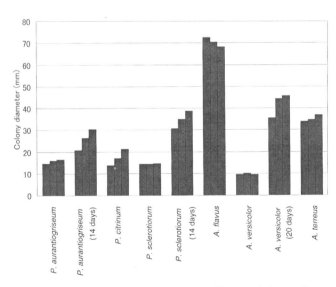

Fig. 6. Effect of volume on colony diameters of *Penicillium* and *Aspergillus* species grown on MEA for 7 days or longer at 25°C.

Effect of Medium Storage

In our experiments, there was no change in the pH of CYA (6.8-6.9) during 8 weeks of storage at 5°C. There were, however, changes in water potential, and the changes were larger during the second four week period than the first (Fig. 7). *Penicillium aurantiogriseum* produced smaller colony diameters with older media. The age of the media somehow affected colony diameters increasing with time probably for the more xerophilic species. Although Constantinescu (1990) suggested that the sealing effect (see below) might be a consequence of preventing the reduction of water potential by evaporation, this seems unlikely. Although storage of plates for 8 weeks caused a slight change of water potential, there were no drastic differences in colonies between new and old plates. The effect of wrapping during cultivation for 7 days was much greater (see below).

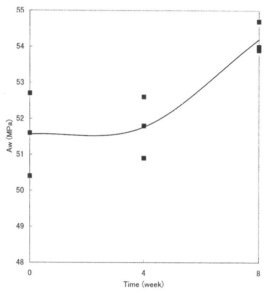

Fig. 7. Changes in water potential of CYA plates wrapped in polyethylene bags and stored at 5°C.

Effect of inoculation method

Questionnaire results: For the culture used as the inoculum source, five ICPA members used MEA, two used CYA, two CZ, one PDA and one MMA. Slant cultures were incubated for 7 to 21 days by most members before being used as inoculum. Conidial suspensions were used for inoculum by nine members, agar and Tween 80 by five, agar only by two, 0.5% Kodak Photoflo by one and water only by one. Two members used dry spores for inoculation. As an inoculation tool, five members used a pin point needle (0.5-1 mm or 25 gauge), five members used a loop with 2-5 mm in diam., and two members inoculated 2 FL with a micropipettor. Ten members inoculated each plate with three points equidistant from each other, and one point inoculation was also used by two members. The following variations on these inoculation procedure were noted:

- Material for inoculation was taken from cultures dried on a silica gel crystals, poured onto a CYA plate, then used for inoculum 5-7 days later.
- Plates were kept upside down during inoculation near the flame.
- Plates were left upright for 1 hour before they are put upside down into incubator.

The effect of spore numbers in the inoculum on resulting growth, described originally by Okuda (1994), was confirmed. As the degree of dilution increased, all species apparently formed smaller colonies except *A. terreus* and *A. flavus* (Fig. 8). *Aspergillus flavus* showed an opposite effect, with somewhat larger colony diameters as spore concentration decreased. In most strains, a pin-point inoculation gave smaller colonies than inoculation of 2 FL with a micropipettor. The micropipettor, as well as wire or plastic loops, give a 2-5 mm diameter of inoculation.

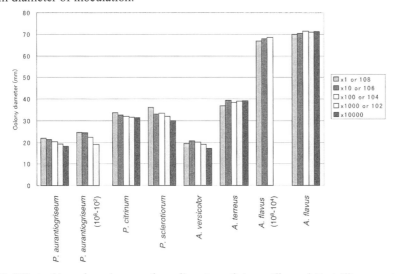

Fig. 8. Effect of inoculum size on colony diameters of *Aspergillus* and *Penicillium* species after 7 days on CYA at 25°C. Dilution of conidial suspension from left to right, x1, x10, x100, x1000, and x10000.

Incubation conditions

Questionnaire results: Seven members wrapped their Petri plates in bags or put them in a box during incubation, while two members placed unwrapped dishes directly on the incubator shelf. One member used perforated plastic bags with 0.5 cm holes 1 cm apart. Another member half-sealed the plates with tape to prevent mite contamination. Five members placed the plates upside down in the incubator, but three members did not. Nine members incubated the plates in the dark, but the incubator door was sometimes opened during incubation. Two members incubated the dishes in complete darkness.

Experimental results: Light and air exchange were expected to influence the growth of *Penicillium* and *Aspergillus* species. No significant differences were observed in colony diameters (less than 2.4 mm) between cultures incubated in complete darkness and with sporadic exposure to light. In complete darkness, however, *P. sclerotiorum* produced a more bluish conidial area, paler mycelia, and more abundant yellow exudate. *Aspergillus versicolor* sporulated slightly less in the dark. The effect of wrapping, causing reduced air exchange, was more obvious. Growth rates were positively or negatively affected by wrapping, depending on the species (difference: 1.5-14.4 mm in diameter). *Penicillium citrinum, P. sclerotiorum* and *A. versicolor* grew better without wrapping, whereas *P. aurantiogriseum* and *A. terreus* grew better when plates were sealed, as shown in Fig. 9. *Penicillium aurantiogriseum* also sporulated better when wrapped. Fig. 10 shows *P. citrinum* had more bluish colonies, less sporulation, more abundant pigments, and more yellow reverse when wrapped. *Penicillium sclerotiorum* showed paler conidial areas, no orange mycelia, more exudate, less pigment, and paler colouration in reverse when wrapped. *Aspergillus flavus* produced no sclerotia when wrapped while *A. versicolor* had paler conidial areas, poor sporulation, less pigment, and a paler reverse. The conidial colouration of *A. terreus* was paler, and colonies sporulated poorly, and had a darker reverse when wrapped. These distinctive changes seem reasonable from the viewpoint of secondary metabolite production. Degree of aeration may affect production of secondary metabolites, positively or negatively; some enzymatic reactions requires oxygen molecules while other enzymes react more effectively in a low oxygen environment.

Colour characterization and morphological examination

Questionnaire results: Eight members determined colour under daylight, while four observed cultures under fluorescent lamps. The *Methuen Handbook of Colour* (Kornerup and Wanscher, 1978) was used by five members, and Ridgeway's *Color Standards and Color Nomenclature* (1912) was still used by five members.

Interestingly, nine members used MEA cultures for microscopic observation, whereas only four used CYA for this purpose and six used CZ. This is probably because CYA sometimes shows atypical microscopic structures (Constantinescu, 1990). When slide mounts were prepared, ten members washed a small amount of culture with ethanol. One member used PhotoFlo (Kodak), although another member noted that this mountant sometimes causes collapse of cells. Slides were prepared with lactic acid by seven members, Shear's solution by two members, and distilled water by one member. Seven members stained preparations with fuchsin, phloxine, aniline blue or cotton blue. All members used the x100 objective lens for characterization of conidiophore and conidium surface texture; eight used differential interference contrast (=Nomarski), five used bright field and three used phase contrast. One member emphasized that microscopic features were more stable taxonomic criteria than colony features. Another member suggested that the appearance of stipe and conidium roughening were highly dependent on the quality of the optics employed.

CONCLUSIONS

Precise standardized manuals or practical guides such as Pitt (1988), Klich and Pitt (1988), and Yaguchi (1997) are helpful for industrial microbiologists and food taxonomists identifying *Penicillium* or *Aspergillus* cultures. The results of our questionnaire demonstrate that despite this, actual methods employed in different laboratories are still variable; even ICPA members apply quite diverse methodologies.

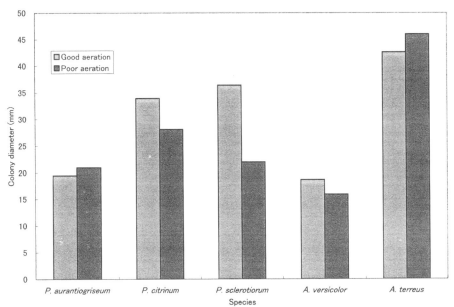

Fig. 9. Effect of air exchange on colony diameters of *Aspergillus* and *Penicillium* species after 7 days on CYA at 25°C. Left bars, good air exchange without wrapping; right bars, poor air exchange with tight wrapping in two polyethylene bags.

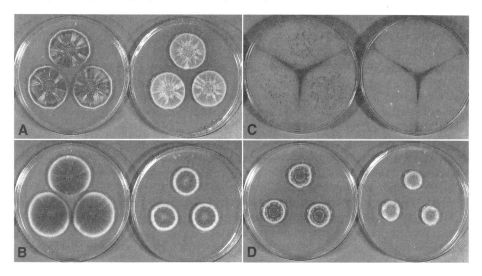

Fig. 10 Effect of air exchange on colony appearance of *Aspergillus* and *Penicillium* species after 7 days on CYA at 25°C. Left, colonies without wrapping; right, colonies wrapped in two polyethylene bags; A, *P. citrinum,* showing changes in sporulation and exudate; B, *P. sclerotiorum,* showing reduced growth and sclerotium production; C, *A. flavus* reverse showing effect on sclerotium production; D, *A. versicolor* showing reduction in sporulation and soluble pigment.

97

Because of the diversity of suppliers of medium ingredients, and the restrictions imposed by the existing practises and resources of functioning laboratories, it would be difficult to establish a rigidly standardized system on which everyone agrees. Until methods can be standardized, we strongly recommend that *Aspergillus* researchers compare results obtained on their media to those from the standard strains available in the "*Aspergillus* Reference Cultures" (Klich *et al.*, 1996). This will ensure that the media are yielding the necessary colony and microscopic characters, and allow calibration of growth rates according to a standard. The book and the accompanying cultures are available from major collections around the world.

In our experiments, we found that the most important factors strikingly affecting colony appearance and growth were volume of media and air exchange. Inoculum size and sporadic light also influenced the colonies to a lesser extent. Difference in yeast extract, agar or water did not greatly affect cultural characteristics. Although we examined the effect of various parameters on cultural characteristics, the inherent phenotypic plasticity in many species may mask minor medium effects.

The experiments reported here are preliminary and it is not possible to recommend a standardized protocol at this time. However, it is clearly important that researchers working with cultures of *Penicillium* and *Aspergillus* employ a uniform, reproducible agar volume in their work, and that this volume be recorded in their publications. Present indications are that agar volumes of 20 or 25 mL may be suitable, but it is essential to collect more comprehensive data on additional strains and species before settling on a standard. We also recommend that growing cultures not be wrapped or incubated in boxes, so that adequate air exchange can occur.

ACKNOWLEDGMENTS

We thank ICPA members for their helpful cooperation in answering our comprehensive questionnaire. We also thank Ms. Sherry Humphrey, Ms. Kyoko Kato, Mr. Noboru Kishi and Mr. Gerry Louis-Seize for their technical assistance. The manuscript was critically reviewed by J. Bissett and C. Babcock.

REFERENCES

Abe, S. (1956). Studies on the classification of the Penicillia. J. Gen. Appl. Microbiol. 2: 1-344.

Costantinescu, O. (1990). Standardization in *Penicillium* identification. In Modern concepts in *Penicillium* and *Aspergillus*. (Eds. R. A. Samson and J. I. Pitt). New York, Plenum Press. pp. 17-25.

Frisvad, J. C. (1985). Creatine sucrose agar, a differential medium for mycotoxin producing terverticillate *Penicillium* species. Let. Appl. Microbiol. 1: 109-113.

Frisvad, J. C. & Filtenborg, O. (1989). Terverticillate penicillia: chemotaxonomy and mycotoxin production. Mycologia 81: 837-861.

Klich, M. A. & Pitt, J. I. (1988). A Laboratory Guide to Common *Aspergillus* Species and Their Teleomorphs. North Ryde, N.S.W., CSIRO, Division of Food Processing.

Klich, M. A., Samson, R. A., and members of the International Commission on *Penicillium* and *Aspergillus*. (1996). *Aspergillus* Reference Cultures. International Union of Microbiological Societies, available from Centraalbureau voor Schimmelcultures, Baarn.

Kornerup, A. & Wanscher, J. H. (1978). Methuen Handbook of Colour. London, Eyre Methuen Ltd.

Okuda, T. (1994). Variation in colony characteristics of *Penicillium* strains resulting from minor variations in culture conditions. Mycologia 86: 259-262.

Pitt, J. I. (1973). An appraisal of identification methods for *Penicillium* species: a novel taxonomic criteria based on temperature and water relations. Mycologia 65: 1135-1157.

Pitt, J. I. (1979). The Genus *Penicillium* and its Teleomorphic States *Eupenicillium* and *Talaromyces*. London, Academic Press.

Pitt, J. I. (1985). Media and incubation conditions for *Penicillium* and *Aspergillus* taxonomy. In Advances in *Penicillium* and *Aspergillus* Systematics (Eds. Samson, R. A. & Pitt, J. I.). New York, London, Plenum Press. pp. 93-103.

Pitt, J. I. (1988). A Laboratory Guide to Common *Penicillium* Species, 2nd Edition. North Ryde, N.S.W., CSIRO, Division of Food Processing.

Pitt, J. I. (1993). A modified creatine sucrose medium for differentiation of species in *Penicillium* subgenus *Penicillium*. J. Appl. Bacteriol. 75: 559-563.

Ramirez, C. (1982). Manual and Atlas of the Penicillia. Amsterdam, Elsevier Biomedical Press.

Raper, K. B. & Fennell, D. I. (1965). The Genus *Aspergillus*. Baltimore, Williams and Wilkins.

Raper, K. B. & Thom, C. (1949). A Manual of the Penicillia. Baltimore, Williams and Wilkins.

Ridgeway, R. (1912). Color Standards and Nomenclature. Washington, D. C., published by the author.

Tzean, S. S., Chen, J. L., Liou, G. Y., Chen. C. C., & Hsu, W. H. (1990). *Aspergillus* and Related Teleomorphs from Taiwan. Hsinchu, Taiwan, Food Industry Research and Development Institute.

Tzean, S. S., Chiu, S. C., Chen, J. L., Hseu, S. H., Lin, G. H., Liou, G. Y., Chen. C. C., & Hsu, W. H. (1994). *Penicillium* and Related Teleomorphs from Taiwan. Hsinchu, Taiwan, Food Industry Research and Development Institute.

Yaguchi, T. (1997). Observation Method for Fungal Colonies (in Japanese). Nippon Kingakukai Kaiho 38: 41-46.

THE USE OF PRIMARY METABOLISM FOR IDENTIFICATION OF *PENICILLIUM* SPECIES

Iftikhar Ahmad and David Malloch
Department of Botany, University of Toronto, Toronto, Ontario, Canada M5S 3B2

INTRODUCTION

Solid media have been used for taxonomic comparison and identification of *Penicillium* species for nearly a century. In the early part of this century a great variety of media was used by workers on this genus, often without much regard for colony characteristics. Charles Thom (Thom, 1910), in his classic study of cultural variability in *Penicillium,* stressed the physiological individuality of all *Penicillium* species and how this individuality is expressed uniquely on each medium employed. Thom believed this individuality could be used as a taxonomic or diagnostic tool and advocated the use of standardized cultural media. He was critical of media having ingredients inconsistent from batch to batch and pointed out that even small differences in contents could lead to significant differences in colonial appearance.

Thom (1910) advocated the use of five media for growing *Penicillium* species. Four of these, Potato Agar, Bean Agar, Peptone-milk sugar-gelatin and 15% gelatin in distilled water would be called "natural media" today and were certainly not absolutely repeatable from lab to lab in 1910, or even with today's purer ingredients. However, Thom believed that these were consistent enough to allow reliable comparisons. Later, Thom (1930) and Raper and Thom (1949) abandoned these media and settled upon the use of Malt Extract Agar (Blakeslee, 1915) and Steep Agar (Raper and Thom, 1949). Malt Extract Agar continues to be used today, but Steep Agar, which includes corn steep liquor, a by-product of cornstarch manufacture, is seldom employed.

Thom's fifth medium, developed by A.W. Dox (Dox, 1910), a chemist working with Thom on the *Penicillium* project, is completely synthetic. This medium, based loosely on one published by F. Czapek (Czapek, 1902), contains sucrose and sodium nitrate as sources of carbon and nitrogen. Although it lacks trace elements considered necessary for fungal growth, most *Penicillium* species are able to grow and reproduce on it. The Czapek/Dox formula was used by Thom (1930), his associates, and those who followed in his tradition, for work on *Penicillium* well into the 1970s.

Biourge (1923) was also concerned with reproducibility in cultural media and used Dierckx' apparently unpublished modification of Raulin's Solution (Raulin, 1869). This medium, further modified by Mosseray (1934), is a synthetic medium lacking some essential elements but more complete than the Czapek/Dox formula. In spite of its apparent superiority, Raulin's Solution never gained the popularity of Czapek/Dox Agar, probably because Thom and Raper did not use it in their monumental works on *Penicillium*.

The agar media most commonly used for the cultivation and identification of *Penicillium* species have attracted much criticism. These media have unnecessarily high C/N ratios and lack a number of essential elements. Recently users of these media have noticed differences in colony colours from those published in the classic works, notably a loss of green. Frisvad (in discussion following Pitt, 1985) stated that the loss of green is due to the lack of copper in Czapek/Dox. Pitt (1973) proposed the addition of yeast extract to Czapek/Dox yielding a formulation he called Czapek Yeast Autolysate Agar (CYA). Pitt (1979) and Ramirez (1982) went on to use CYA as a standard in their monographs of *Penicillium*.

The general acceptance of Pitt's monograph and his modification of the Czapek/Dox formula has resulted in the abandonment of synthetic media for routine work. This has caused some concern, but is not easily remedied without a re-examination of representative isolates of many species on new media.

Our dissatisfaction with present cultural media arose from the need to identify a large number of Penicillia from indoor environments. Czapek/Dox and CYA yield a wide variety of colonial responses but are in some ways too good. Multiple isolates of the same species vary on these media greatly and can be assumed to be different on this basis. Microscopic structures are often distorted. Growth on Czapek/Dox and CYA usually takes the form of a wrinkled and folded basal mat covered with abundant vegetative growth, making the preparation of microscopic mounts difficult. Finally, the original concept of Czapek/Dox from Thom (1910) as useful in studying assimilation of a variety of nutrients must itself be questioned. With so many minerals lacking in this formulation, its use in physiological studies seems inadvisable.

Malt Extract Agar (MEA) is a second medium surviving up to the present. This medium is excellent for the production of microscopic structures but often yields nondescript colonies. This was recognized by Raper and Thom (1949) who developed Steep Agar because it allowed ". fairly luxuriant growth without losing characteristic colony patterns as on malt agar."

More recently Frisvad (1981, 1985) introduced Creatine Sucrose Agar (CREA), a medium finding wide acceptance as an aid in identification of *Penicillium* species, especially those in Subgenus *Penicillium*. CREA is a synthetic medium containing creatine as a source of nitrogen. Frisvad's formulation is strongly alkaline at first but often becomes acid around and under colonies of certain *Penicillium* species. Recognizing these changes, Frisvad incorporated bromcresol purple into the medium, allowing acidification to be viewed as a yellow colour change. Pitt (1993) presented a neutral formulation of CREA (Neutral Creatine Sucrose Agar; CSN) where both alkaline and acid changes could be observed. In discussing Frisvad's original formulation, Pitt (1993) stated that "The requirement for a medium such as CREA is urgent..." We agree. The usefulness of CREA, underlined by its rapid acceptance, demonstrates the real need for chemically defined diagnostic media.

In this work we present alternative synthetic media for taxonomic and physiological work with *Penicillium* species. Our goal was to develop formulations supporting stable and taxonomically diagnostic colony morphologies coupled with the ability to assess assimilation of a variety of nutrients. We agree with Thom (1910) that physiology offers a powerful taxonomic tool for *Penicillium* taxonomists and have carried out this work in is long shadow.

MATERIALS AND METHODS

Media: The media used in this investigation consist of a complete medium, called the *Penicillium* Reference Medium (PRM), containing ingredients judged to be suitable for the cultivation of all *Penicillium* species, and variants on this medium containing substitute sources of carbon and nitrogen or, in one, a metabolic inhibitor. Variants are labelled according to the ingredient changed and whether this is a carbon or nitrogen source. Labels follow the form: variant ingredient-nutrient type, e.g. Mannitol-C indicates the PRM with the glucose carbon source replaced by mannitol. An additional medium, called *Penicillium* Descriptive Medium (PDM), was used to grow colonies for macro- and microscopic descriptions.

In our study, colonies on PRM and its variants were incubated at room temperature, in fluctuating diurnal light cycles in plastic Petri dishes sealed with paraffin film. Our building has year round temperature control set at approximately 20-23° C and ceiling-mounted fluorescent lights, turned on during the day and off at night. On weekends the lights are not turned on but there is still considerable light from the large windows. Although plastic dishes are probably not preferable to glass (Constantinescu, 1990), they are commonly used in routine work by most laboratories. Sealing with paraffin is a safety requirement in some institutions and we believe should be practised in all microbiological laboratories. We believe the above conditions of incubation to be representative of those in most modern mycological laboratories.

Penicillium Reference Medium: The *Penicillium* Reference Medium (see page 00) played a central role in our studies. All *Penicillium* species tested were able to grow on this formula, most very well. Glucose and nitrate serve as the carbon and nitrogen sources; both are assimilated into metabolic pathways with little pre-processing. Sucrose, a disaccharide used in CYA, requires transformation via invertase before it can be assimilated. Although invertase may be present in most *Penicillium* species, Gochenaur (discussion following Pitt, 1985) reported that a "significant number of isolates" from soil are unable to utilize sucrose. Nitrate is not directly assimilated by fungi and must be reduced to ammonia by nitrate reductase. Although it is not unlikely that nitrate reductase could be lacking in some species as well; we have not yet encountered one. The compelling reason for using nitrate in the Primary Reference Medium is that pH can be controlled more easily than it can be with ammonium salts.

Vitamins were included in PRM: thiamin, biotin and vitamin B12. These take the role played by yeast extract in CYA. Yeast extract, although an excellent source of vitamins may introduce unknown components to a medium, leading to difficulties in interpreting results.

The initial pH of PRM is 7.6 and is maintained during early growth by the Tris buffer. Many species of *Penicillium* soon produce large amounts of acid as secondary or excess primary metabolites, causing the pH to fall. Changes in pH, constant characteristics of many isolates of *Penicillium*, are visualized by the addition of bromcresol purple to the medium, causing a change from dark purple from yellow when the pH falls below about 5.5 (Pitt 1993). Production of acid on PRM and many of its variants yields useful taxonomic information. As with CREA and CSN, acidity may be confined to the reverse of the colony or may extend well beyond its margin as a yellow halo. The colour of the acid zone may range from a barely detectable yellow cast to a pure bright yellow. Ultimately we found acidity to be quite variable and placed most emphasis on the presence or absence of acidity and whether the acid was confined to the colony reverse or spread out beyond the colony. Some colonies produce coloured pigments on PRM. When these are yellow, care must be taken not to interpret these as pH changes in the medium.

Formulation Basal Medium

The basal medium was prepared to contain in a final volume of 1 litre 15 g bacterial grade agar, 100 ml each of major salts stock, micronutrient stock and 10 ml of buffer stock:

1. Major salts stock: 4.85 g KCl + 4.93 g MgSO4.7H2O + 4.41 mg CaCl2.2H2O + 0.88 g NaCl in 1 litre.
2. Buffer stock: 60.6 g/l Tris adjusted to pH 7.8 with HCl.
3. Micronutrients + vitamins stock: One litre stock prepared by mixing 5 ml each of 54.4 g/l NaH2PO4.H2O, 12.6 g/l FeCl3.6H2O chelated with 17 g/l of disodium salt of EDTA, 12.2 g/l H3BO3, 732 mg/l MnCl2.6H2O, 922 mg/l ZnSO4.7H2O, 29 mg/l Na2MoO4.2H2O, 47.6 mg/l CoCl2.6H2O, 40 mg CuSO4.5H2O, 40 mg/l thiamine chloride, 2 mg/l biotin, 2 mg/l vitamin B12.

The basal medium was supplemented with different nitrogen and carbon sources, and where indicated bromcresol purple (BCP) or phenol red (PR) was also added to monitor either acidification or alkalinization of the agar. Both pH indicators were solubilized by heating 250 mg in about 50 ml. For parahydroxybenzoic acid additions, 100 ml of 69.05 g/l parahydroxybenzoic acid solubilized with NaOH was added to the basal medium and the pH of the medium was adjusted to 6.8 with HCl. For fatty acid additions, 1 g palmitic acid solubilized by heating in 5 ml ethanol was slowly added to melted agar media at 80°C. All media were sterilised by autoclaving except for the additions of glufosinate-ammonium (Ignite), urea and bovine serum albumin (BSA) where filter-sterilised solutions were added to autoclaved media held at 50°C.

The following media formulations give various additions to the basal medium in a final volume of 1 litre.

A. Nitrate-N and glucose-C reference medium (NITR-N) (= PRM): 425 mg NaNO3 + 9.1 g glucose + 50 mg BCP
B. High calcium medium (+ CALC): 425 mg NaNO3 + 9.1 g glucose + 1 g Ca2Cl + 50 mg BCP
C: N-Inhibitor (N-INHIB): 425 mg NaNO3 + 9.1 g glucose + 1 g glufosinate-ammonium
D. Ammonium-N (AMM-N): 286 mg NH4Cl + 9.1 g glucose + 50 mg BCP (without BCP in the PDM variant)
E. Urea-N medium (UREA-N): 300 mg urea + 9.1 g glucose + 50 mg BCP
F. Phenylalanine-N medium (PHEN-N): 826 mg L-phenylalanine + 9.1 g glucose
G. Betaine-N medium (BETA-N): 586 mg glycinebetaine + 9.1 g glucose + 50 mg BCP
H. Choline-N medium (CHOL-N): 698 mg Choline chloride + 9.1 g glucose + 50 mg BCP
I. Sorbitol-C medium (SORB-C): 9.1 g sorbitol + 425 mg NaNO3 + 50 mg BCP
J. Proline-N & Sorbitol-C, calcium free medium (PROSOCF): 9.1 g
 sorbitol + 575 mg proline in a calcium free basal medium prepared by adding 373 mg KCl + 1.232 g MgSO4.7H2 and 10ml of buffer and 100 ml of micronutrient + vitamin stocks.
K. Mannitol-C medium (MANN-C): 9.1 g mannitol + 425 mg NaNO3 + 50 mg BCP
L. Cellobiose-C medium (CELB-C): 8.1 g cellobiose + 425 mg NaNO3 + 50 mg BCP
M. Parahydroxybenzoate-C medium (PARA-C): 6.9 g parahydroxybenzoate + 425 mg NaNO3 + 50 mg phenol red.
N. Cellulose-C medium (CELL-C): 8.1 g alphacel + 425 mg NaNO3
O. BSA-N and palmitic acid-C medium (BSA-N): 1 g BSA + 1 g Palmitic acid + 50 mg BCP

Growth on PRM and its variants is quite different from that on CYA. Colonies grow more slowly, are thinner, produce less vegetative mycelium, and sporulate more heavily. First impressions with PRM are that growth is considerably poorer than it is on CYA and MEA. Closer study, however, reveals colonies on PRM to be healthy and easy to manipulate. As with CYA, results can be recorded in 7 days. Many Penicillia produce their best growth on PRM. Although the necessity of converting of nitrate to ammonium should reduce the growth rate of colonies on this medium relative to the Ammonium-N variant this was often not the case. Rapidly increasing acidity on the Ammonium-N variant often slowed growth to less than that of PRM.

Ammonium-N. As ammonium is assimilated by colonies growing on Ammonium-N, chloride ion from the ammonium chloride becomes associated with hydrogen ions of the water, resulting in a marked lowering of pH, initially set at 7.6. As the decrease in pH is the result of ammonium assimilation and not metabolite excretion, growth of all species of *Penicillium* will cause this decrease. Isolates excreting additional acid will accelerate the drop in pH. The increasing acidity of Ammonium-N sets it apart from the other variants. Not all species tolerate the acidity equally; some may respond by producing pigments or unusual and characteristic colonies. Although Ammonium-N is a medium supporting the growth of all Penicillia, it may not be the one promoting the fastest or heaviest growth. For reasons discussed below, Ammonium-N was chosen as the basis for our *Penicillium* Descriptive Medium (PDM).

High Calcium. Many Penicillia are known to excrete large amounts of organic acids such as citric acid and oxalic acid. These acids vary greatly in solubility; oxalic acid readily precipitated in the presence of a calcium salt, producing highly recognizable large crystals of calcium oxalate. This medium was formulated to examine changes in the size of the acid zone around the colony resulting from the precipitation of organic acids and for the detection of calcium oxalate crystals

N-Inhibitor. In our previous studies (Ahmad and Malloch, 1995), we showed that *Penicillium* species varied greatly in their resistance to the presence of glufosinate, a commercially available herbicide disrupting nitrogen assimilation both in plants and microorganisms by inhibiting glutamine synthetase.

UREA-N. An increase in pH is considered a positive test for urea utilization by bacteria having urease activity (Atlas, 1993). In our preliminary studies Penicillia growing on urea showed a wide range of pH changes. This is a good trait for species differentiation.

Phenylalanine-N. The studies of Hunter and Segel (1971) Wolfinbarger (1980) and Roos (1989) indicated a considerable diversity in processes associated with the uptake of organic nitrogen by species of *Penicillium*. Phenylalanine, a precursor in the tryptophan and shikimic acid pathways, is an important nitrogenous metabolite in the secondary metabolism of fungi. In many *Penicillium* species, the diffusion of distinct pigments around the margins of growing colonies is indicative of a release of certain secondary metabolites by the fungus. This variant was formulated to investigate correlations between phenylalanine availability and diffusion of pigments by *Penicillium* species.

Glycine betaine-N. Glycine betaine is a major osmotic solute synthesized by plants living in salty and dry conditions (Stewart and Ahmad, 1983) but not by fungi (Brown, 1976). A number of bacterial species have been reported to use glycine betaine as a nitrogen source. Our preliminary studies showed glycine betaine to be an adequate source of nitrogen for certain species of *Penicillium*.

Choline-N. Choline is a precursor of glycine betaine and is abundant in nature as an integral part of cell membranes. The known pathway for the breakdown of glycine beatine to choline suggests a close linkage between the catabolic degradation of these two quaternary ammonium compounds in microorganisms.

Sorbitol-C. The polyol sorbitol belongs to another category of compatible osmotic solutes synthesized by both plants and some microorganisms under water stress conditions. According to Jennings (1995), only some animal cells and a limited number of fungal species are able to utilize sorbitol as a carbon source. Thom (1910) demonstrated the utilization of sorbitol by certain *Penicillium* species.

Mannitol-C. Mannitol is another polyol used by plants and microorganisms as a compatible solute that can be degraded only by certain animal cell and fungal species in nature (Jennings, 1995). Two different metabolic pathways are thought to be involved in the conversion of sorbitol and mannitol to hexose sugars.

Cellobiose-C. Cellobiose is a key product of enzymatic degradation of cellulose, and as such is available to microorganisms where cellulolytic activity is high. A comparison between growth on cellobiose and cellulose may distinguish between Penicillia with varying abilities to utilize breakdown products of cellulose.
PARA-HYDROXYBENZOATE-C. Parahydroxybenzoate is permitted in some countries as a preservative against microbial contamination in food products. The microbial inhibition caused by this compound is only effective under acidic conditions. It is therefore interesting to compare the response of *Penicillium* species to the presence of parahydroxybenzoate in media initially adjusted to a slightly acidic pH values of 6.8.

Bovine serum albumin-C. Penicillia with the ability to grow on highly enriched organic substrates are likely to utilize proteins and fatty acids as good nitrogen and carbon sources. Bovine serum albumin (BSA) is a highly soluble protein adequate for agar media. Highly purified forms of both BSA and the fatty acid palmitic acid are readily available.

Penicillium Descriptive medium. *Penicillium* Descriptive Medium (PDM) has the same formulation as PRM: Ammonia-N except the bromcresol purple indicator is omitted (see Appendix). It has been used in our studies as the standard medium for the preparation of macro- and microscopic descriptions.

Colonies on PDM rarely produce the thick basal felts, dense aerial hyphae and abundant exudate characteristic growth on of CYA. In general they are velvety and heavily sporulating. Marginal zones are usually not well developed on PDM but concentric zonation often becomes pronounce. Synnemata and fascicle formation is usually well developed. Members *Penicillium* Section *Penicillium* are readily recognized on PDM by their characteristic appearance; eroded and often almost dendritic at the margin and more granular toward the centre. We have found most species of *Penicillium* to have a characteristic colonial appearance on PDM. These characteristics appear to be constant among different strains. In direct comparisons we have found colony characteristics on PDM to be less variable than those on CYA.

Because colonies on PDM commonly produce little mycelial overgrowth, colours are easy to judge. On CYA, in contrast, strain to strain variation in exudate production and sterile overgrowth lead to inconsistent and diffuse colours that are very hard to compare.

106

In our earliest trials of PDM the vivid and consistent colours produced by colonies on this medium struck us. We still believe this to be one of its greatest assets.

Colonies on PDM produce a variety of diffusing pigments and reverse colours, as do those on CYA. The two media sometimes differ in this regard, with each eliciting some colours the other cannot. Other substances are expressed equally on both. On PDM *Penicillium polonicum* and related species produce a characteristic yellow diffusing pigment bounded by an outer red halo. Several species, not necessarily related, produce these two rings on PDM but not on CYA or any of the variants of PRM.

Microscopic structures produced on PDM appear to be free of the distortion commonly produced by CYA. We have done all our microscopic observations on this medium, with results comparable to those published by others.

Fungal strains: Thirty strains derived from type, obtained from American Type Culture Collection (ATCC), Centraalbureau voor Schimmelcultures (CBS), International Mycological Institute (IMI) and U.S. Department of Agriculture (NRRL) were included in this study: *Penicillium aethiopicum* Frisvad, CBS 484.84, ATCC 18330 *P. allii* Vincent & Pitt, ATCC 64868 *P. arenicola* Chalabuda, NRRL 795 *P. atramentosum* Thom, NRRL 971 *P. aurantiogriseum* Dierckx, NRRL 863 P. *brevicompactum* Dierckx, IBT 6603 (not a type), *P. camemberti* Thom, NRRL 807 *P. chrysogenum* Thom, NRRL 13485 *P. dipodomys* Frisvad et al., (Frisvad), NRRL 1003 *P. clavigerum* Demelius, NRRL 890 *P. commune* Thom, CBS 477.75 *P. concentricum* Samson et al., NRRL 13448 *P. confertum* (Frisvad et al.,) Frisvad, ATCC 58615 *P. coprobium* Frisvad, CBS 189.89 *P. coprophilum* (Berk. & Curt.) Seifert & Samson, IMI 91917 *P. crustosum* Thom, NRRL 1888 *P. cyclopium* Westling, NRRL 786 *P. digitatum* (Pers.:Fr.) Sacc., NRRL 1151 *P. echinulatum* ex Fassat., NRRL 976 *P. expansum* Link, ATCC 22050 *P. fennelliae* Stolk, IMI 321509 *P. flavigenum* Frisvad et al., NRRL 951 *P. freii* Frisvad and Samson, NRRL 2036 *P. glandicola* (Oudem.) Seifert & Samson, NRRL 13487 *P. dipodomyicola* (Frisvad et al.,) Frisvad, NRRL 2152 *P. griseofulvum* Dierckx *griseofulvum*, CBS 187.88 *P. hirsutum* var. *albocoremium* Frisvad, NRRL 2032 *P. hirsutum* Dierckx var. *hirsutum*, ATCC 16025 *P. hirsutum* var. *venetum* Frisvad, CBS 701.68 *P. hordei* Stolk.

Inoculation and incubation: In all experiments, triplicate plates were inoculated in the centre with conidia from 7-10 day old cultures dispersed in a 1 cm agar section. Plates were sealed with parafilm and incubated at 20-22° C. All cultures except those grown on cellulose as the carbon source (Cellobiose-C) were examined after 7 d incubation. Cellulose cultures were examined after three weeks.

Culture examinations: The diameter (d) of the fungal colony was measured and used to calculate colony area (CA). Because the hyphal density per unit colony area for a given species varied on different media with that on Ammonia-N medium being consistently dense, the hyphal density of the colony was recorded as either good, fair or poor, where good corresponded to at least 70%, fair corresponded to 30-70% and poor to less than 30% of the visual estimate of hyphal density on Ammonia-N medium. To allow a quantitative estimate of colony growth, the hyphal densities in the three categories were rounded up in order to hyphal density (HD) values of 1.0 for good, 0.7 for fair and 0.3 for poor. Growth values (G) of each colony was calculated as a product of CA x HD. Relative growth rates of the species on a given medium (G_x) was calculated as G_x / G_{NITR-N} where G_{NITR-N} is the value of CA x HD on Nitrate-N medium.

In media containing bromcresol purple as a pH indicator, the diameter (d) of the yellow acid zone was obtained from the reverse of the fungal colony, and the area of yellow zone (YZA) calculated as. Where the acidification of the agar across the plate was less than complete, the area covered by remaining purple alkaline zone (PZA) was calculated as 56.7 cm^2 - YZA where 56.7 cm^2 represented total plate area. The colour intensity of both the yellow acid zone (YCR) and purple alkaline zone (PCR) were recorded on a scale of 10, where a rating of yellow 10 indicated maximum acidification of the agar plate, an interchangeable rating of either purple 0 or yellow 1 indicated the beginning of acid appearance, and a rating of purple 10 indicated no acidification. The acidification level of both yellow and purple zones was obtained by expressing yellow colour rating (YCE) as YCR/10 and purple colour rating (PCE) as (10-PCR)/100. The percentage acidification of the agar plate was calculated as (YZA x YCE + PZA x PCE)/56.7 x 100. For measuring alkalinization of phenol red containing PARAB media, the diameter (d) of red alkaline zone was obtained from the reverse of fungal colony, and the area alkaline zone (RZA) was calculated as πr^2. The intensity of red alkaline zone (RCR) was rated on a scale of 0-10 and expressed (RCE) as RCR/10. The percent alkalinization of agar plate was calculated as 100 x RZA x RCE/56.7.

RESULTS

The results of the physiological experiments are summarised in Table 1. Production of acid or base was found to be easily interpreted and was used for the initial separation of the strains into 31 groups. When more than one isolate occurred in a group they could be separated on the basis of relative growth rate.

DISCUSSION

Although the type strains examined in our experiments could be differentiated using their physiological profiles the question of character constancy still remains. If another isolate of one of the taxa were grown on these media, would it exhibit the same characteristic or would it vary? An answer to this question requires the same approach as a question relating to morphological constancy. Over the last three years we have grown nearly 1500 isolates of indoor and outdoor *Penicillium* species on our diagnostic media. Many easily recognised species, such as *P. atramentosum* or *P. expansum* exhibit physiological profiles similar to those of the type strains. Often, however, one of the tests will yield variable results among the strains tested while the other tests remain constant. It is clear that the tests cannot be accepted as universal truths; however the exercise of good taxonomic judgement should separate those tests found to be variable from those that are

Table 1a. Physiological profiles of selected type strains of *Penicillium* species

Type strains	NITR-N Growth	AMMO-N Growth	Relative Growth Rate								Acid Index INORG-N	Base Index PARA-C
			N-INHIB	BETA-N	CHOL-N	SORB-C	MANN-C	CELB-C	PARA-C	BSA-N		
No acid on BCP containing media, alkalinity on PARA-C												
P. atramentosum	7,5	2,6	good	fair	poor	poor	fair	fair	good	poor	high	very high
Acid on BSA-N & PALM-C, alkalinity on PARA-C												
P. italicum	1,8	7,1	fair	good	poor	poor	poor	good	good	fair	high	high
Acid on CHOL-N, alkalinity on PARA-C												
P. lanosum	3,4	1,3	poor	good	excellent	fair	poor	fair	poor	poor	low	low
Acid on UREA-N, CHOL-N, alkalinity on PARA-C												
P. paxilli	2,5	4,2	good	poor	good	fair	poor	good	fair	fair	low	high
Acid on CHOL-N, BSA-N & PALM-C, alkalinity on PARA-C												
P. coprophilum	3,9	2,0	fair	good	good	good	poor	fair	good	poor	high	very high
Acid on UREA-N, BSA-N & PALM-C, no alkalinity on PARA-C												
P. digitatum	1,4	11,1	fair	poor	excellent	poor	poor	poor	poor	poor	very high	very low
Acid on UREA-N, BETA-N, alkalinity on PARA-C												
P. oxalicum	6,7	11,5	fair	good	fair	poor	poor	fair	good	fair	very high	high
Acid on UREA-N, BETA-N, CHOL-N, no alkalinity on PARA-C												
P. clavigerum	2,6	2,0	poor	good	good	poor	poor	poor	poor	poor	high	very low
Acid on BETA-N, CHOL-N, alkalinity on PARA-C												
P. arenicola	3,7	1,6	poor	good	good	good	fair	fair	good	poor	high	very high
Acid on UREA-N, BETA-N, CHOL-N, alkalinity on PARA-C												
P. glandicola	1,8	1,2	poor	excellent	excellent	fair	poor	excellent	excellent	fair	high	high
P. roque var roqueforti	9,1	12,2	fair	good	fair	good	fair	fair	good	fair	very high	very high
Acid on UREA-N, BETA-N, CHOL-N, CELB-C, alkalinity on PARA-C												
P. roque var carneum	8,9	8,7	good	fair	fair	fair	poor	poor	good	fair	high	very high
Acid on UREA-N, BETA-N, CHOL-N, SORB-C, MANN-C, CELB-C, alkalinity on PARA-C												
P. griseof var dipodomyicola	1,7	1,5	poor	good	good	good	fair	fair	fair	fair	high	very low
Acid on NITR-N, UREA-N, CELB-C, alkalinity on PARA-C												
P. claviforme	7,5	3,2	poor	poor	poor	fair	poor	fair	fair	poor	high	low
Acid on NITR-N, UREA-N, CHOL-N, alkalinity on PARA-C												
P. griseof var griseof	1,3	2,5	poor	excellent	excellent	excellent	excellent	excellent	excellent	fair	low	very high
Acid on NITR-N, UREA-N, BETA-N, alkalinity on PARA-C												
P. commune	6,8	4,7	poor	good	fair	good	poor	fair	good	poor	very high	very high

Table 1b. Physiological profiles of selected type strains of *A72* species (continued)

Type strains	NITR-N Growth	AMMO-N Growth	N-INHIB	BETA-N	CHOL-N	SORB-C	MANN-C	CELB-C	PARA-C	BSA-N	Acid Index INORG-N	Base Index PARA-C
						Relative Growth Rate						
Acid on NITR-N, UREA-N, BETA-N, CHOL-N, CELB-C, alkalinity on PARA-C												
P. olsoni	3,6	4,9	fair	fair	fair	good	fair	good	excellent	poor	high	very high
Acid on NITR-N, UREA-N, BETA-N, CHOL-N, alkalinity on PARA-C												
P. cameberti	2,9	3,2	good	fair	good	fair	poor	fair	fair	poor	high	high
P. chrys var diodomys	1,5	1,5	fair	good	fair	excellent	good	good	excellent	poor	high	very high
P. verrucosum	2,8	1,8	fair	good	good	good	fair	good	good	poor	high	very high
P. vulpinum	7,5	3,1	poor	poor	poor	poor	poor	fair	fair	poor	high	very high
P. fennelliae	1,3	2,3	poor	good	good	poor	poor	fair	fair	poor	low	high
Acid on NITR-N, UREA-N, BETA-N, CELB-C, BSA-N & PALM-C, alkalinity on PARA-C												
P. coprobium	3,3	2,5	fair	fair	fair	poor	fair	fair	fair	poor	high	low
P. flavigenum	2,9	3,8	fair	good	good	good	fair	good	fair	fair	very high	very high
Acid on NITR-N, UREA-N, BETA-N, CHOL-N, no alkalinity on PARA-C												
P. nalgiovense	2,9	1,3	fair	good	good	good	fair	good	poor	poor	low	very low
Acid on NITR-N, UREA-N, BETA-N, CHOL-N, BSA-N & PALM-C, alkalinity on PARA-C												
P. crustosum	3,2	2,7	good	good	excellent	good	fair	good	excellent	poor	high	very high
P. polonicum	8,8	6,8	poor	good	fair	fair	fair	fair	fair	poor	very high	very high
Acid on NITR-N, UREA-N, BETA-N, CHOL-N, BSA-N & PALM-C, no alkalinity on PARA-C												
P. concentricum	1,9	2,0	fair	good	good	poor	poor	fair	poor	fair	high	very low
Acid on NITR-N, UREA-N, BETA-N, CHOL-N, CELB-C, BSA-N & PALM-C, alkalinity on PARA-C												
P. allii	7,2	8,2	poor	fair	poor	poor	poor	good	good	poor	very high	very high
P. cyclopium	5,2	3,9	poor	poor	fair	fair	poor	good	fair	poor	very high	very high
Acid on NITR-N, UREA-N, BETA-N, CHOL-N, SORB-C, alkalinity on PARA-C												
P. chrys var chrys	8,0	4,3	fair	good	fair	good	poor	good	good	poor	very high	very high
P. viridicatum	6,8	4,9	poor	fair	fair	fair	fair	fair	fair	poor	high	high

110

Table 1c. Physiological profiles of selected type strains of *Penicillium* species (continued)

Type strains	NITR-N Growth	AMMO-N Growth	Relative Growth Rate N-INHIB	BETA-N	CHOL-N	SORB-C	MANN-C	CELB-C	PARA-C	BSA-N	Acid Index INORG-N	Base Index PARA-C
Acid on NITR-N, UREA-N, BETA-N, CHOL-N, SORB-C, CELB-C, alkalinity on PARA-C												
P. aethiopicum	6,5	3,1	poor	fair	poor	good	fair	good	good	poor	very high	very high
P. neoechinulatum	4,5	2,3	poor	good	good	fair	fair	fair	fair	poor	very high	high
P. tricolor	4,3	2,4	poor	fair	fair	good	fair	good	good	poor	very high	low
P. solitum	6,0	3,5	poor	fair	fair	good	fair	good	good	poor	high	high
P. brevicompactum	1,8	1,3	poor	good	fair	excellent	good	good	fair	fair	high	very low
Acid on NITR-N, UREA-N, BETA-N, CHOL-N, SORB-C, CELB-C, BSA-N & PALM-C, no alkalinity on PARA-C												
P. melanoconidium	2,2	1,7	poor	excellent	fair	good	good	good	good	poor	very high	very low
Acid on NITR-N, UREA-N, BETA-N, CHOL-N, SORB-C, CELB-C, BSA-N & PALM-C, alkalinity on PARA-C												
P. confertum	3,4	1,5	poor	poor	fair	excellent	fair	good	good	poor	very high	very high
P. expansum	8,0	9,4	poor	good	fair	good	poor	good	fair	poor	very high	high
P. hirsutum var. albocoremium	8,6	7,4	poor	fair	fair	good	fair	fair	good	poor	very high	very high
P. hirsutum var. hirsutum	9,6	6,0	poor	good	good	fair	poor	good	good	poor	very high	low
P. hirsutum var. venetum	9,6	6,4	poor	good	good	fair	poor	poor	fair	poor	very high	low
P. hordei	5,4	2,7	poor	good	good	fair	fair	good	good	poor	very high	very high
Acid on NITR-N, UREA-N, BETA-N, CHOL-N, SORB-C, MANN-C, CELB-C, alkalinity on PARA-C												
P. freii	6,3	4,7	poor	fair	fair	good	fair	fair	fair	poor	very high	high
P. mononematosum	2,8	2,5	poor	fair	good	good	good	fair	good	fair	high	high
P. echinulatum	3,6	2,8	poor	good	good	excellent	good	good	excellent	fair	very high	very high
Acid on NITR-N, UREA-N, BETA-N, CHOL-N, SORB-C, MANN-C, CELB-C, BSA-N & PALM-C, no alkalinity on PARA-C												
P. aurantiogriseum	3,1	2,5	poor	good	good	good	good	good	good	poor	very high	very low

constant for any taxon. Attempts to produce a purely physiological taxonomy would be greatly at odds with traditional schemes and, we should add, most likely not the truth. We are most comfortable with an approach rooted in tradition, whereby new approaches are added to those already in use; a process of accumulated wisdom. By combining our data on primary metabolism with that on secondary metabolism, macro- and micromorphology and genetic homology we should arrive at a consensus satisfactory to all.

REFERENCES

Ahmad, I. & Malloch, D. (1995). Interaction of soil microflora with the herbicide phosphinothricin. Agric., Ecosyst., Environ. 54: 165-174.

Atlas, R. M. (1993). Handbook of Microbiological Media. CRC Press, Boca Raton, Florida.

Biourge, P. (1923). Les moisissures du groupe *Penicillium* Link. Étude monographique. Cellule 33: 1-331.

Blakeslee, A. F. (1915). Lindner's roll tube method of separation cultures. Phytopathology 5: 68-69.

Brown, A.D. (1976). Microbial water stress. Bacteriol. Rev. 40: 803-846.

Constantinescu, O. (1990). Standardization in *Penicillium* identification *Ia* Modern concepts in *Penicillium* and *Aspergillus* classification (Eds Samson, R.A. & Pitt, J.I.) Plenum Press, New York, pp. 17-25.

Czapek, F. (1902). Untersuchungen über die Stickstoffgewinnung und Eiweissbildung der Pflanzen. Beitr. Chem. Phys. Pathol, 1(10-12): 540-560.

Dox, A.W. (1910). The intracellular enzymes of *Penicillium* and *Aspergillus*. Bull. Bur. Anim. Ind. US Dept. Agric. 120: 1-70.

Frisvad, J. C. (1981). Physiological criteria and mycotoxin production as aids in identification of common asymmetric Penicillia. Appl. Environ Microbiol. 41: 568-579.

Frisvad, J. C. (1985). Creatine sucrose agar, a differential medium for mycotoxin producing terverticillate *Penicillium* species. Lett. Appl. Microbiol. 1: 109-113.

Hunter, D. R. & Segel, I. H. (1971). Acidic and basic amino acid transport system of *Penicillium chrysogenum*. Arch. Biochem. Biophys. 144: 168-183.

Jennings, D. H. (1995). The Physiology of Fungal Nutrition. Cambridge Univ. Press, Cambridge.

Mosseray, R. (1934). Les *Aspergillus* de la section "Niger" Thom et Church. Cellule 43: 201-286.

Pitt, J. I. (1973). An appraisal of identification methods for *Penicillium* species: novel taxonomic criteria based on temperature and water relations. Mycologia 65: 1137-1157.

Pitt, J. I. (1979). The Genus *Penicillium* and Its Teleomorphic States *Eupenicillium* and *Talaromyces*. Academic Press, London.

Pitt, J. I. (1985). Media and incubation conditions for *Penicillium* and *Aspergillus* taxonomy. Advances in *Penicillium* and *Aspergillus* systematics. (Eds. Samson, R.A. & Pitt, J.I.) Plenum Press, New York. pp.. 93-103.

Pitt, J. I. (1993). A modified creatine sucrose medium for differentiation of speciesinf *Penicillium* subgenus *Penicillium*. J. Appl. Bacteriol. 75: 559-563.

Ramirez, C. (1982). Manual and Atlas of the Penicillia. Elsevier Biomedical, Amsterdam.

Raper, K. B. & Thom , C. (1949). A Manual of the Penicillia. Williams and Wilkins, Baltimore.

Raulin, (1869). Etudes chimiques sur la végétation des Mucinidées. Ann. Sci. Nat. Bot., Sér. 5, 11:93.

Roos, W. (1989). Kinetic properties, nutrient-dependent regulation and energy coupling of amino acid transport in *Penicillium cyclopium*. Biochem. Biophys. Acta 978: 119-133.

Thom, C. (1910). Cultural studies of species of *Penicillium*. Bull. Bur. Anim. Ind. US Dept. Agric. 118: 1-109.

Stewart, G. R. & Ahmad, I. (1983). Adaptation to salinity in angiosperm halophytes. Metals and Micronutrients: Uptake and Utilization by Plants. (Eds. Robb, D.A. & Pierpoint, W.S. Academic Press, London. pp.

Thom, C. (1930). The Penicillia. Williams and Wilkins, Baltimore.

Wolfinbarger, L. Jr. (1980). Transport and utilisation of amino acids by fungi. In Microorganisms and Nitrogen Sources (Ed. Payne, J.W.). John Wiley, Chichester.

COLLABORATIVE STUDY ON STIPE ROUGHNESS AND CONIDIUM FORM IN SOME TERVERTICILLATE PENICILLIA

Jens C. Frisvad, Ole Filtenborg, Ulf Thrane, Robert A. Samson[1] and Keith Seifert[2]
Department of Biotechnology, Technical University of Denmark, 2800 Lyngby, Denmark, [1]Centraalbureau voor Schimmelcultures, Baarn, the Netherlands and [2]Eastern Cereal and Oilseed Research Centre,Agriculture and Agri-Food Canada, Research Branch,Ottawa, Ontario, Canada K1A 0C6

Four species of *Penicillium*, with two isolates of each, were examined in an international collaborative study with 15 participants from 8 countries,. The important morphological attributes studied were stipe ornamentation and conidium form. All 8 selected isolates were inoculated on Czapek yeast autolysate (CYA) agar, Blakeslee malt extract agar (MEA), the favourite medium for morphological examinations in the individual laboratory and on MEA at airtight conditions. The identity of the cultures was not revealed, to rule out prejudice. The cultures were examined by slightly different techniques, but all used 1000 x oil immersion microscopy. There was a good agreement on stipe ornamentation between experts even if some experts decided stipes to be rough when at least 5% were rough, whereas other preferred a majority rule decision. There was a significant influence of growth medium on micromorphological features, stipes on MEA were typically more consistently and distinctly rough than those on CYA. Conidium form was also perceived differently by experts and less experienced microscopists. Some of the latter had problems seeing differences between elliptical and globose conidia. Most participants agreed that *Penicillium paneum* has globose conidia, albeit two participants described them as subglobose. Most experienced taxonomist described the conidia of *P. griseofulvum* to be elliptical (with some subglobose conidia), whereas less experienced microscopists often regarded these conidia as globose to subglobose. The influence of medium on conidium form was insignificant. It is concluded that standard cultures and an accurate and detailed description protocol for microscopic examinations is generally needed. A training set of three Penicillia with different morphological features for the micromorphological features evaluated here is suggested: *P. griseofulvum*, *P. paneum* and *P. echinulatum*. It appears that substrates which result in poor mycelium formation and maximal stipe and conidium ornamentation are best for micromorphological examinations and a majority of participants prefer malt extract based media for the purpose.

INTRODUCTION

Stipe roughness and conidium roughness and form are important diagnostic characters for the classification and identification of the difficult group of the terverticillate Penicillia. These criteria have been included in nearly every taxonomic treatment of these fungi, and were regarded as the major characters for taxon delimitation by Samson *et al.* (1976) in their revision of the fasciculate terverticillate species together with conidiophore fascicu-

113

lation and phialide shape and length. Even though these criteria are regarded as important key characters, they have proved to be difficult to assess for non-experts who encounter Penicillia. Furthermore monographers do not always agree on some of these criteria. For example the species typifying the genus *Penicillium*, *P. expansum* was claimed to be "with walls smooth or finely roughened" by Raper and Thom (1949) and illustrated on their fig. 129, A1, A3 and A4 as distinctly rough. The stipes of *P. expansum* were characterized as "usually smooth-walled, sometimes very finely roughened, this latter feature more pronounced on malt agar" by Samson *et al.* (1976), while Pitt (1979) regarded *P. expansum* as having smooth-walled stipes and based the series *Expansa* on this character state. Berny and Hennebert (1990) obtained one mutant type among 9 types of *P. expansum* with rough conidiophore walls and globose conidia, while the other types and the original culture had smooth stipes and elliptical conidia.

Experts have also characterized conidium form differently. For example the conidia of the closely related species *P. crustosum* have been described as "subglobose to slightly elliptical" by Raper and Thom (1949), but as "ellipsoidal, less commonly subsphaeroidal" by Pitt (1979). Many characters, such as colony texture, conidium colour and odour, have been claimed to be subjective, evanescent or emphemeral and depending on the medium used (Thom, 1930; Ciegler *et al.*, 1973, 1981; Pitt, 1979). If even the micromorphological criteria are difficult to record objectively, there are few traditional consistent criteria left for identification purposes. The aim of this collaborative study was to compare how different microbiologists perceive stipe ornamentation and conidium form and suggest ways of improving and standardizing the microscopical techniques used by scientists encountering *Penicillium* isolates.

MATERIALS AND METHODS

The following isolates were used for the collaborative study *Penicillium aethiopicum* IBT 16873 and IBT 15747; *P.coprophilum* IBT 19279 and IBT 18704; *P. paneum* IBT 19862 and IBT 19298; *P. griseofulvum* IBT 19442 and IBT 18340. These isolates were sent to 20 participants. 15 participants, from 8 countries sent a reply. The first seven participants are experts, who have worked with *Penicillium* identification for a decade or more, the last 8 participants have occasionally identified *Penicillium* strains. Most participants tabulated their results as required, evaluating stipe roughness and conidium form at four different culture conditions: After growth for one week at 25°C on the media Czapek yeast autolysate agar (CYA)(Pitt, 1979), Blakeslee Malt extract agar (MEA)(Raper and Thom, 1949; Pitt, 1979), and their own favourite medium for microscopic examinations. There have been some indications in our laboratories that lack of oxygen may influence morphological characters and so cultures grown on MEA at "airtight" conditions (wrapped in two plastic bags) were also compared to the other media. Some participants also tabulated conidium ornamentation of the isolates. The media were added trace metals (Copper and Zinc salts) (Smith, 1949; Constatinescu, 1990), if the water used for the media was of high purity. The stipes and conidia were examined at 1000 times magnification (100 x oil immersion objective and a 10-15 x ocular using different optics: Normarski interference contrast, phase contrast or bright field).

RESULTS

Conidiophore stipe roughness (Fig. 1-4)

The experts all agreed that the conidiophore stipes of the two *Penicillium griseofulvum* isolates were smooth on MEA, while two non-experts reported that the stipes of *P. griseofulvum* were rough in one or the other of the two isolates. It is possible that some non-experts may perceive stipes as rough if they contain granules in the stipes. Published descriptions have never indicated that the stipes of *P. griseofulvum* could be rough (Raper and Thom, 1949; Samson *et al.*, 1976; Pitt, 1979). Similarly the stipes of *P. paneum* (morphologically very close to *P. roqueforti*, Boysen *et al.*, 1996; Pitt and Hocking, 1997) were reported as having rough-walled stipes by the experts and by most non-experts. Even though this fungus (under the name *P. roqueforti*) has always been regarded as a good example of a *Penicillium* species with very rough (tuberculate) stipes, the stipes were perceived as finely roughened by one expert (one of the isolates) and both isolates were regarded as having smooth stipe walls on MEA by two non-experts (both isolates). The degree of roughness was perceived differently by experts from finely roughened to very rough. In summary 12 out of 30 *P. paneum* preparations were reported to contain very rough stipes, while 23 out of 30 contained rough or very rough stipes on MEA.

The results for *P. aethiopicum* are less clear in agreement with published data. The original description stated it had rough stipes on MEA, but smooth stipes on CYA (Frisvad and Filtenborg, 1989). It had roughened stipes according to Bridge et al. (1989) and was identified as *P. expansum*, whereas Pitt and Hocking (1997) in a more full description of the species reported the stipe walls to be "smooth or at most very finely roughened".

In this investigation the stipes are regarded as smooth on MEA in 9 of 30 answers, and as rough in 21 answers (very rough in 11 answers). Of 7 experts, one regarded the stipes in both isolates as smooth, but noted few rough stipes in one strain, while another regarded one of the strains as having smooth stipes.

P. coprophilum has been regarded as having smooth stipes in all published descriptions (Samson et al., 1976 (as *P. concentricum*); Seifert and Samson, 1985; Pitt, 1979 (as *P. italicum*)). In the present study all experts, except one, agreed that the two isolates of *P. coprophilum* had smooth stipes on MEA. The outlying expert reported that it had smooth to finely roughened and finely roughened to rough stipes respectively. In contrast most non-experts claimed that *P. coprophilum* has rough stipes.

On CYA the results are generally less clear. The stipe ornamentation results on MEA and CYA which includes 102 comparisons showed that in 42 cases the stipes were perceived as more rough on MEA than CYA, in 11 cases the opposite was the case and in the remaining 49 cases the results were similar on MEA and CYA.

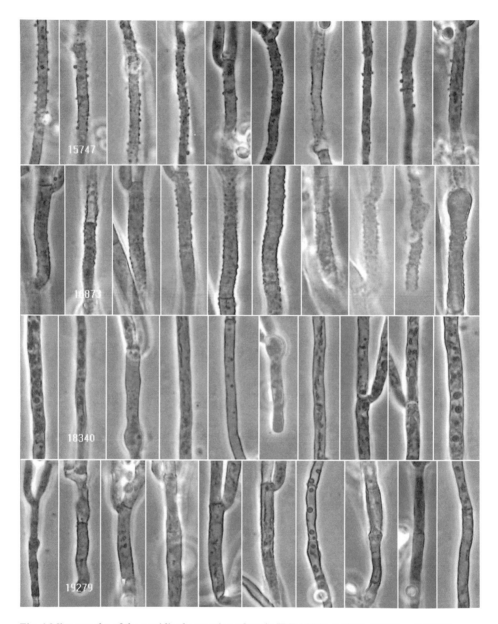

Fig. 1 Micrographs of the conidiophores stipe of strain IBT 15747, 16873, 18340 and 19279

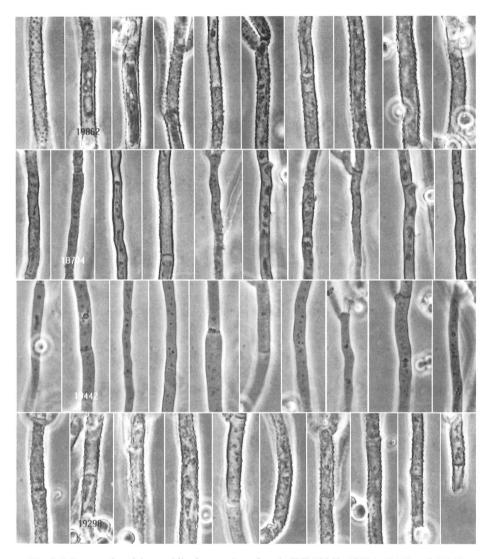

Fig. 2. Micrographs of the conidiophores stipe of strain IBT 19862, 18704, 19442 and 19298

117

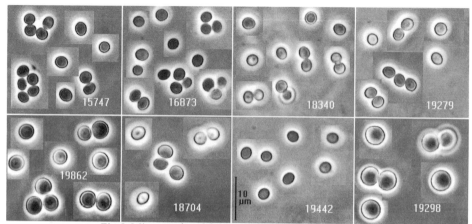

Fig 5. Micrographs of the conidia of the strains examined.

In general the results show that CYA is less suitable for the study of stipe roughness, as the results are often inconsistent within the isolates of the same species and isolates regarding as having typically rough stipes often appear to have smooth stipe walls on CYA.

Nine participants reported on the difference in stipe ornamentation between MEA and the same medium kept at "airtight" conditions. Of the 72 answers 15 observations showed that stipes were rougher on MEA than MEA kept airtight, 9 observations showed the opposite and 48 observations equated the stipe roughness under the tow conditions. In general however it is known that carbon dioxide accumulation or lack of oxygen will inhibit fungi (Pitt and Hocking, 1997), albeit the influence on stipe roughness is not significant in this study.

The other media used for micromorphological examinations such as wort-beer agar, 2% malt agar, Czapek agar, oatmeal agar or tryptic soy agar, were only used by one or two laboratories and so could not be compared in a proper way, although it was commented that the malt based media gave the same results as MEA in the respective individual studies.

Conidium form (Fig. 5-7)

There were only few insignificant differences between the responses regarding conidium form between media (MEA, CYA, and other preferred media for microscopical examinations). The experts were more consistent in assigning the same conidium form to conidia as observed on MEA and CYA than the non-experts. Some participants stated the whole range from globose to ellipsoidal for some conidial preparations while other either decided which type was dominating or even stated approximate percentages of each form in a preparation. Some participants stated that if more than 20 % of the conidia were ellipsoidal, it should be regarded as ellipsoidal.

There was a good agreement among all participants that the conidia of *P. paneum* were globose. Two participants considered those conidia to be subglobose and they stated so for conidia from both MEA and CYA. It has been stated in published descriptions that *P. roqueforti* has perfectly globose conidia (Pitt, 1979) or globose to subglobose conidia (Raper and Thom, 1949; Samson *et al.*, 1977) and this species is close micromorphologically to *P. paneum*.

118

The conidia of *P. aethiopicum* were regarded as subglobose to broadly ellipsoidal to ellipsoidal by Raper and Thom (1949), as broadly ellipsoidal by Pitt and Hocking (1997) and as subglobose to ellipsoidal by Frisvad and Filtenborg (1989). The experts mostly agreed that the conidium form of *P. aethiopicum* is (in average) broadly ellipsoidal, albeit somewhat variable in shape, but in two cases globose conidia were mentioned. The non-experts in several cases regarded them as solely globose.

The conidia of *P. coprophilum* have been regarded as typically ellipsoidal in all *Penicillium* monographs (Samson *et al.*, 1976; Pitt, 1979). The experts most often stated they were indeed ellipsoidal, occasionally with some variation towards broadly ellipsoidal, subglobose or even globose. Non-experts often regarded them as globose to subglobose, with few answers stating they were elliptical.

Stipe roughness on MEA

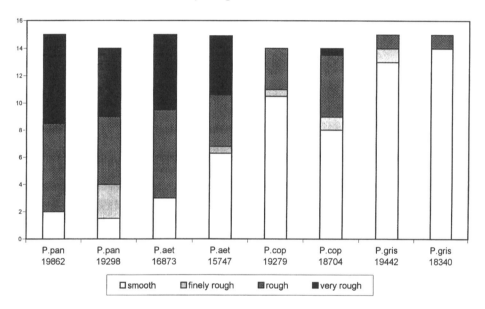

Fig. 3. Sums of participants reporting on each of the degrees of conidiophore stipe roughness on MEA for each isolate of *Penicillium* examined. Fuzzy answers were subdivided into equal parts.

119

Stipe roughness on CYA

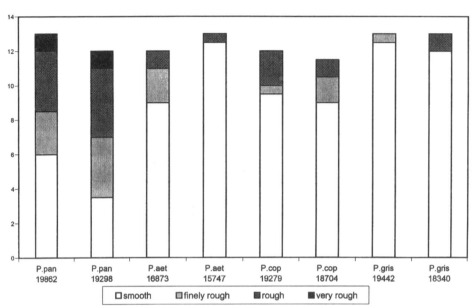

Fig. 4. Sums of participants reporting on each of the degrees of conidiophore stipe roughness on CYA for each isolate of *Penicillium* examined. Fuzzy answers were subdivided into equal parts.

Conidium form on MEA

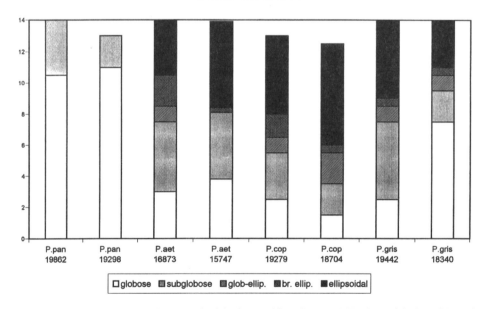

Fig 6. Sums of participants reporting on each of the four conidium forms on MEA for each isolate of *Penicillium* examined. Fuzzy answers were subdivided into equal parts.

120

Conidium form on CYA

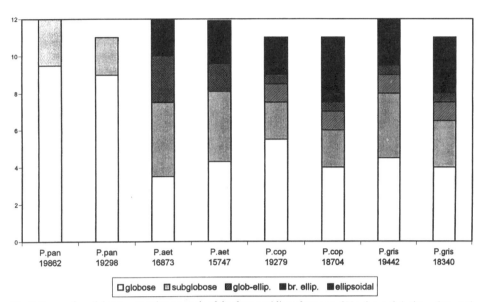

Fig 7. Sums of participants reporting on each of the four conidium forms on CYA for each isolate of *Penicillium* examined. Fuzzy answers were subdivided into equal parts.

The conidia of *P. griseofulvum* have been regarded as "elliptical to tardily subglobose" by Raper and Thom (1949), as ellipsoidal by Pitt (1979) and as "ellipsoidal, sometimes subglobose" by Samson *et al.* (1976). As was the case with *P. coprophilum* most experts regarded the conidia of *P. griseofulvum* as elliptical, but with some subglobose and few globose conidia present (Tab. 3). Non-experts again regarded them as predominantly globose or subglobose with few answers mentioning ellipsoidal at all.

Conidium ornamentation

Even though the participants were not asked to comment on conidium ornamentation, five participants also reported on that attribute. In published descriptions all the four taxa examined have been claimed to have smooth conidia (Raper and Thom, 1949; Samson *et al.*, 1976; Pitt, 1979, Pitt and Hocking, 1997). All five participants considered the conidia smooth or smooth to finely roughened, except one participant that regarded *Penicillium paneum* as having finely roughened conidia. The five participants all regarded the conidia of *P. coprophilum* as smooth, but in three out of ten responses *P. griseofulvum* was regarded as having conidia smooth to finely roughened.

DISCUSSION

Figure 3 shows that it may be wise to use fuzzy classifiers in keys to indicate that the stipes may under certain circumstances be perceived as different from generally accepted expert descriptions. It is seen in Fig. 3 that the stipes of *P. paneum* and *P. aethiopicum* are perceived as rough walled by most participants, and that the stipes of *P. coprophilum*

121

and *P. griseofulvum* are mostly perceived as smooth-walled on MEA. This is also how the stipes are described in monographs, except for *P. aethiopicum*, where agreement on stipe ornamentation is less clear. To optimize the uniformity of the microscopic observations it is recommendable to use standard cultures as examples and "controls" and standardize the culture conditions so that ordinary microbiologists using optimal materials and methods will always be able to use these criteria. Furthermore we suggest using the rule of thumb that if one definitely rough stipe (the most characteristic or differentiated form) is present among 20 investigated seen in at least two fields, the isolate has the ability to produce them and should be keyed as such. The latter alternative seems to be necessary if Czapek agar or CYA is the preferred as the primary medium for microscopical examinations (Fig. 4) (as in Raper and Thom, 1949 and Pitt, 1979). The results could also be reported in a very detailed way, i.e. in giving percentages of each state present, for example 10% stipes smooth, 50% finely roughened, 30% rough and 10% very rough, or the roughness as such can be detailed according to the description in the Dictionary of the fungi (Hawksworth *et al.*, 1995, p. 325). For normal identification routines this is too time consuming and ultimately require a scanning electron microscope. A built in flexibility in the identification keys and a more gross description of roughness may be the best way to proceed in future identification routines. Good classifications and species descriptions must, in contrast, require the best possible means of detailed characterization.

In general the agreement on conidium form is less clear than that on stipe roughness (Fig. 6 and 7). One problem could be that a cylinder or an ellipsoid seen from the end would occasionally appear globose in the two-dimensional microscope view. Furthermore, in spite of the figure in Hawksworth *et al.* (1995, p. 420), conidium forms are difficult to record objectively. Raper and Thom claimed, however, that conidium form and ornamentation are very constant features if only mature conidia are observed. On the other hand Stolk and Samson (1983) reported on variable and smooth walled ellipsoidal younger conidia but some globose to subglobose rough-walled conidia in older cultures in for example *P. turbatum* and *P. sclerotigenum*.

The most clear result of this collaborative study is that experts often give answers close to those in published descriptions, whereas non-experts often had difficulties in seeing differences between smooth, finely roughened, and rough or between globose, subglobose, broadly ellipsoidal and ellipsoidal and so deviated quite seriously from published descriptions. This may have severe consequences for using analytical identification keys or comparison with published descriptions and may result in several misidentifications by the users of monographs. Even within the 7 experts there are a few remarkably different evaluations of conidiophore stipe roughness or conidium form. This may be caused by several factors that can influence the results. It was already noted by Raper and Thom (1949) that conidiophore roughness could be influenced by the medium used and that more typical micromorphological structures were obtained on MEA and compared to Czapek based agars and this has later been confirmed by Pitt (1985) and Constantinescu (1990). Most participants used MEA for microscopic examinations, but some participants used wort-beer agar, tryptic soy agar, CYA, or oatmeal agar in addition to MEA. Two participants used either Czapek agar or Oxoid malt agar exclusively in routine identifications for microscopic data. Most participants stated that MEA showed ornamentations or roughenings more clearly than CYA, but that differences in ornamentation between cultures grown on MEA and other cereal based media (OA) were small. It is clear that MEA is an excellent medium for microscopic observations, and this medium

122

should be the reference medium all newly developed media should be tested against. There are, however, problems with MEA because 3 of its 5 ingredients (malt extract, peptone, agar) may vary with batch and manufacturer (Odds *et al.*, 1978; Filtenborg *et al.*, 1990, Frisvad and Filtenborg, unpubl., Samson, unpubl.). For example one expert participant reported the stipes of *P. aethiopicum* as smooth on MEA in contrast to nearly all other experts, and reported that the malt extract used was a particularly rich malt directly purchased at a brewery, whereas other participants used either Difco, Oxoid or Sigma malt extract. Most participants used lactic acid for microscopical examinations, especially if the used Normarski interference contrast optics. For other optics different dyes were used to enhance contrast: cotton blue, aniline blue or acid fuchsin. One participant used ethanol, one used water, and one used Shear's mountant (potassium acetate/glycerol/ethanol; 5:2:3). The participant that used water as mounting medium saw nearly all stipes as smooth and nearly all conidia as globose, indicating that this mountant is unsuitable for *Penicillium* micromorphology.

No participant used statistics directly in recording their results, but five participants stated a minimum number of structures to be examined. One examined more than 30 stipes and more than 100 conidia for each culture examined and stated that the conidia should be called elliptical if more than 20% of the conidia were of that shape and the stipes should be called rough if more than 5% of the stipes were rough. Another participant examined at least 10 stipes and 50 conidia and used a majority rule decision. A third participant examined more than 10 stipes and "many" conidia and emphasized the high value of drawings. A fourth participant examined more than 10 stipes and 15 conidia. A fifth participant examined all stipes and conidia in the slide prepared and used a majority rule for deciding on ornamentation and shape. While illustrations for micromorphological ornamentation and conidium form are given by Hawksworth *et al.* (1995), there are no general recommendations on how many individual structures should be examined and how accurately they should be recorded. For identification purposes, which is the primary aim in this paper, a majority rule could be used for conidium form, an average (or median) decision could be used for conidium size and a "if roughness is found among 20 mature structures then the isolate is considered to have a rough ornamentation" decision could be used when entering an identification key. Because of the big variability seen in this study a series of suggestions for standardization are suggested (Tab. 4): Apart from obvious advices on aligning the microscope carefully and using standard nomenclature (Hawksworth et al., 1995) and recording schemes, we suggest that only mature micromorphological structures should be used, that at least 20 of these structures should be examined, that a standard set of three Penicillia should be used, at least by non-experts, for comparison, that the isolates should be inoculated on MEA and incubated aerobically for one week at approximately 25 °C and the structures mounted in lactic acid (with a dye if so wished) and examined in a high quality microscope at least 1000 x magnification. Phase contrast optics appears to be useful, but Normarski interference contrast optics may prove to be even better.

In the future we recommend that standard cultures are used, that chemically defined media are developed (see Ahmad and Malloch, this book), that conditions favouring more consistent expression of micromorphological features can be found, and that image analysis methods will eventually be used routinely in mycological laboratories. Keys should be constructed that allow several kinds of entry and with flexibility on this input, for example synoptic computer keys or probabilistic computer keys.

A new collaborative study must be performed that is based on the recommendations in Table 1.

Table 1. Suggestions for standardizing and improving the consistency in recording form and ornamentation micromorphological key character states (quantitative measurements of dimensions not directly considered).

1. Use standard isolates that consistently exhibit the most important character states under "normal" conditions as controls (especially for non-experts)
   Suggestion: *Penicillium paneum*: globose smooth conidia, conspicuously roughened (tuberculate) stipes (CBS 167.91). *Penicillium griseofulvum*: predominantly elliptical smooth conidia, smooth stipes (CBS 485.84). *Penicillium echinulatum*: globose to subglobose rough conidia and rough stipes (CBS 317.48)
2. Examine mature micromorphological structures
   Suggestion: The conidium is usually characteristic by the time 6 newer conidia have been cut off between it and the phialide: Do not examine conidia just borne on a young phialide (Raper and Thom, 1949). Do not examine structures at the outer margin of the colony.
3. Examine many structures
   Suggestion: For identification, examine more than 20 conidia/stipes etc. and use a majority rule decision for conidium form, and a 5% rule for roughness ("if a rough structure is seen among 20 then key the isolate as rough")
4. Incubate for 5 to 7 days, at 25°C, in the dark, before examining the isolate
5. Incubate and examine structures on Blakeslee Malt extract agar (MEA) (Raper and Thom, 1949, p. 41, 334; Pitt, 1979) and use approx. 17 ml of medium in a 9 cm Petri dish with ribs.
6. Transfer part of the colony to the microscope slide with a drop of ethanol, mount in 85% lactic acid and examine at 1000 x magnification using Normarski Interference or phase contrast microscopy or use equivalent materials and methods.
7. Align the microscope carefully before examination. For details use a small aperture.
8. Use a standard scheme for recording data
9. Use standard nomenclature for ornamentation (Hawksworth *et al.*, 1995, p. 325) and conidium form (Hawksworth et al., 1995, p. 420-421)
10. Use flexible keys, for example computer synoptic or probabilistic keys

ACKNOWLEDGEMENTS

We warmly thank the following participants for all the work they put into the study: Birgitte Andersen, Flemming Boisen, Martha Christensen, Jytte Hansen, Anni Jensen, Maren A. Klich, Zofia Kozakiewicz, Alena Kubatová, Jette Schakinger Larsen, Jean-Paul Latgé, Flemming Lund, Ellen Kirstine Lyhne, Toru Okuda and John I. Pitt.

REFERENCES

Berny, J.F. & Hennebert, G.L. (1990). Variants of *Penicillium expansum*: an analysis of cultural and microscopic characters as taxonomic criteria. In: Samson, R.A. & Pitt, J.I. (eds.): Modern concepts in *Penicillium* and *Aspergillus* classification. Plenum Press, New York, pp. 49-63.

Boysen, M., Skouboe, P., Frisvad, J.C. & Rossen, L. (1996). Reclassification of the *Penicillium roqueforti* group into three species on the basis of molecular genetic and biochemical profiles. Microbioloy (UK) 142: 541-549.

Ciegler, A., Fennell, D.I., Sansing, G.A., Detroy, R.W. & Bennett, G.A. (1973). Mycotoxin producing strains of *Penicillum viridicatum*: classification into subgroups. Applied Microbiology 26: 271-278.

Constatinescu, O. (1990). Standardization in *Penicillium* identification. In: Samson, R.A. & Pitt, J.I. (eds.): Modern concepts in *Penicillium* and *Aspergillus* classification. Plenum Press, New York, pp. 17-24.

Filtenborg, O., Frisvad, J.C. & Thrane, U. (1990). The significance of yeast extract composition on metabolite production in *Penicillium*. In: Samson, R.A. & Pitt, J.I. (eds.): Modern concepts in *Penicillium* and *Aspergillus* classification. Plenum Press, New York, pp. 433-440.

Hawksworth, D.L., Kirk, P.M., Sutton, B.C. & Pegler, D.N. (1995). Ainsworth and Bisby's dictionary of the fungi. CAB International, Wallingford.

Odds, F.C., Hall, C.A. & Abbott, A.B. (1978). Peptones and mycological reproducibility. Sabouraudia 16: 237-246.

Pitt, J.I.(1979). The genus *Penicillium* and its teleomorphic states *Eupenicillium* and *Talaromyces*. Academic Press, London.

Pitt, J.I. (1985). Media and incubation conditions for *Penicillium* and *Aspergillus* taxonomy. In: Samson, R.A. & Pitt, J.I., eds.: Advances in *Penicillium* and *Aspergillus* systematics. Plenum Press, New York, pp. 93-100.

Pitt, J.I. & Hocking, A.D. (1997). Fungi and food spoilage. Second ed. Blackie Academic & professional, London.

Raper, K.B. & Thom, C. (1949). A manual of the Penicillia. Williams and Wilkins, Baltimore.

Samson, R.A., Stolk, A.C. & Hadlok, R. (1976). Revision of the subsection *Fasciculata* of *Penicillium* and some allied species. Stud Mycol. (Baarn) 11: 1-47.

Samson, R.A., Eckardt, C. & Orth, R. (1977). The taxonomy of *Penicillium* species from fermented cheeses. Antonie van Leeuwenhoek 43: 341-350.

Seifert, K.A. & Samson, R.A. (1985). The genus *Coremium* and the synnematous Penicillia. In: Samson, R.A. & Pitt, J.I., eds.: Advances in *Penicillium* and *Aspergillus* systematics. Plenum Press, New York, pp. 143-154.

Stolk, A.C. & Samson, R.A. (1983). The ascomycete genus *Eupenicillium* and related *Penicillium* anamorphs. Stud. Mycol. (Baarn) 23: 1-149.

Thom, C. (1930). The Penicillia. Williams and Wilkins, Baltimore

Chapter 3

TAXONOMIC AND DIGITAL INFORMATION ON *PENICILLIUM* AND *ASPERGILLUS*

TAXONOMIC INITIATIVES, TAXONOMIC TOOLS AND THE INFORMATION AGE

Keith A. Seifert and Larry Speers,
Eastern Cereal and Oilseed Research Centre, Agriculture and Agri-Food Canada, Research Branch, Ottawa, Ontario, Canada K1A 0C6

Developments in computer networking, combined with international initiatives on biodiversity, provide new opportunities and obligations for taxonomists, including those working on *Penicillium* and *Aspergillus*. Several international and national agencies are initiating projects designed to document the planet's biota, which will result in the development of huge taxonomic and nomenclatural databases. Computerized identification keys are becoming viewed as the most effective means of identifying organisms. They are presently available in the form of shells (such as DELTA, Linnaeus II, or CABI-KEY), that can be used to develop keys that can be distributed as executable programs on diskettes or downloaded over the Internet, or as interactive keys that are used over the Internet. Some practical aspects of electronic imaging are briefly discussed.

INTRODUCTION

Since the second workshop on *Penicillium* and *Aspergillus* in 1989 (Samson and Pitt, 1990), the world of computers has continued to develop in surprising ways. Klich (1990) provided the overview for the 1989 workshop that is the counterpart of this review. She discussed the development of databases for organizing taxonomic information, the initiation of culture collection databases, advances in numerical taxonomy and other kinds of statistical analyses for identification and classification, and the various types of keys that had been attempted at that time. The remaining papers in that section of the proceedings dealt with the practicalities and philosophy of computerized identification keys (Bridge, 1990; Pitt, 1990; Williams, 1990).

At that time, no one could foresee the development of the Internet or the influence that the World Wide Web (WWW) would have on the way information is packaged and delivered, or the impact of the Rio Convention on Biodiversity. These changes are reflected in this article, which has more URLs than conventional references. Although all of the URLs were active at the time of preparation of this paper, it is likely that many of them will change. However, most of the sites can still be located using Internet search engines.

In this paper, we review initiatives concerning biodiversity, and attempt to set the stage for the contribution that *Penicillium* and *Aspergillus* taxonomists could make. These initiatives are intimately associated with developments in bioinformatics and computerized identification keys.

BIODIVERSITY AND THE DEVELOPMENT OF TAXONOMIC DATABASES

Following the signing of the Rio Convention, many international agencies and national governments initiated projects to meet their obligations under the biodiversity convention. Systematics Agenda 2000 (Anonymous 1994) is a consortium of primarily American taxonomic societies conceived as a lobbying effort to raise funds for systematic research. Three missions were identified: (1) discover, describe and inventory global species diversity, (2) analyse and synthesize the information from this global discovery effort into phylogenetic classification systems, and (3) organize the information in an efficiently retrievable form.

The National Science Foundation Partnerships for Enhancing Expertise in Taxonomy (PEET) (http://www.nhm.ukans.edu/~peet/) is a competitive awards program designed to support research projects targeting groups of poorly known organisms. The program aims to support training of new generations of taxonomists, and to ensure that current taxonomic expertise is transferred into electronic databases. The first PEET competition resulted in 21 awards, some of which were in excess of a half million dollars, of which four pertain to fungi; information on these projects is just beginning to appear on the WWW. The awards must be administered from academic institutions in the United States, but collaboration with non-American research institutions is possible via subcontracting.

The anticipated increase in systematic research will also increase the demand for biological informatics support, particularly in the development and standardization of databases. At the moment, most of these initiatives are in the planning stages, but it is noteworthy how much emphasis is being placed on the computerization and accessibility of the data. In general, two categories of databases are envisioned: 1) nomenclatural databases that are information based; and 2) taxonomic databases that are ultimately specimen based.

Many taxonomic and nomenclatural databases already exist (Table 1). A list of these can be accessed at the National Biological Information Infrastructure (NBII) WWW site (http://www.nbs.gov/nbii/index.html). Most of these have been custom programmed, but in addition to the key programs outlined below, there are other applications available designed specifically for managing nomenclatural information. For example, Platypus is a Windows program produced by CSIRO in Australia for managing taxonomic, geographic, ecological and bibliographic information. A demo is available at the following address http://www.ento.csiro.au/platypus/platypus.html.

The GenBank at the US National Institute of Health has developed a hierarchical, taxonomic database as a component of the genetic and protein sequence data software deposited there (http://www3.ncbi.nlm.nih.gov/Taxonomy/tax.html). In this classification, *Penicillium* and *Aspergillus* are classified as "mitosporic Trichocomaceae". The less detailed Tree of Life project (http://phylogeny.arizona.edu/tree/phylogeny.html) is a collaborative academic effort aimed primarily at students and the general public. The end points of the linked phylogenetic trees are information pages and illustrations demonstrating the diversity of each included group. At present, there is a page on the Ascomycota by Taylor *et al.* (http://phylogeny.arizona.edu/tree/eukaryotes/fungi/ascomycota /ascomycota.html) that includes some information on *Penicillium* and *Aspergillus*.

130

Two ambitious projects are now underway to develop databases with the names of all known organisms, Species 2000 and the Interagency Taxonomic Information System. Species 2000 (http://www.sp2000.org/) is an initiative of the International Union of Biological Sciences (IUBS) begun in 1994, in collaboration with the Committee on Data for Science and Technology (CODATA) and the International Union of Microbiological Sciences (IUMS, of which ICPA is a commission). The objective is to enumerate all known species of plants, animals, fungi and microbes on Earth to use as baseline data for studies of global biodiversity. The intention is to produce a federation of separate databases rather than a single, unified database, but to produce an annually updated species checklist for distribution on the Internet and by CD-ROM.

Table 1. Fungal nomenclatural databases on the Internet.	
Aphyllophorales	http://www.cbs.knaw.nl/www/aphyllo/database.html
Ascomycota- ordinal classification	http://www.ekbot.umu.se/pmg/outline.html
Dictionary of Fungi - list of fungal genera	http://www.cabi.org/cgi-bin/cabi/dictax.pl
Fusarium	http://www.cbs.knaw.nl/www/fusarium/database.html
Index of Fungi	http://nt.ars-grin.gov/indxfun/frmIndF.htm
Index of Fungi vols 1-4	telnet://fungi.ars-grin.gov/ (type login user, password: USER)
Index Nominum Genericorum (Plantarum)	http://www.nmnh.si.edu/ing/
Saccardo Index	telnet://fungi.ars-grin.gov/ (type login user, password: USER)

The web site now contains several model databases, but as of June 1997 no information on fungi. The Integrated Taxonomic Information System (ITIS; http://www.itis.usda.gov/itis/index. html) is intended to be an easily accessible database including species names and hierarchical classification for all taxonomic groups of organisms in the world, focusing on the United States, and probably also Canada and Mexico. It is a partnership between the US Departments of Commerce, Interior, Agriculture, National Oceanic and Atmospheric Administration, Geological Survey, Environmental Protection Agency, Agriculture Research Service, Natural Resources Conservation Service, the Smithsonian Institution and National Museum of Natural History. A Database Work Group has established the database model structure, and a separate Taxonomy Work Group will be responsible for ensuring the quality and integrity of the taxonomic data. Data will be entered using a Windows based program called ITIS Taxonomic Workbench, developed by the USDA using Visual Basic and Microsoft Access. The data entered into the databases is intended to be peer reviewed. ITIS participants are now testing the latest version of Taxonomic Workbench (1.12).

ICPA has been a leader in databasing names germane to its area of expertise, through the creation of the list of Names in Current Use in the Trichocomaceae (Pitt and Samson, 1993). Unfortunately, the database is now only available as the published paper, probably the least useful form possible. Copyright and ownership of this database should be resolved quickly so that it can become incorporated or cross-referenced in the larger nomenclatural databases now planned. Ideally, ICPA we should offer this database as a model to be included among the first entries to the proposed world databases.

131

COMPUTERIZING OF COLLECTIONS

As part of the global biodiversity initiative, several all taxon biological inventories (AT-BIs) have been proposed. Probably the best known of these is the Guanacaste project proposed by the Instituto Nacional de Biodiversidad (INBio) in Costa Rica, which unfortunately has been cancelled (http://www.inbio.ac.cr/en/default.html). Nevertheless, lobbying efforts for ATBI's are continuing, and mycologists are actively seeking to be included (Rossman, http://NT.ars-grin.gov/sbmlweb/documents/Ross-Biod/RSBiod Frame.htm). ATBI's will require the development of large collection or specimen based databases far larger than those of existing herbaria and culture collections. ATBI's will have large requirements for identification of cultures of prolifically sporulating mould genera. Given the increased diversity of *Aspergillus* already noticed in tropical areas, we can anticipate that the true richness of this genus will be revealed by ATBI's in tropical areas.

Herbaria and culture collections have traditionally been as database intensive as other kinds of collections, such as libraries. Consequently, they have been early targets for computerization. Culture collections have been particularly active, and Internet based searchable catalogues are becoming the norm (see Table 2). Most of these collections have developed their own database formats and interfaces, but the development of larger biodiversity databases will probably require standardization of specimen or culture based collection databases. At issue is the perceived need to permit users to assemble data from separate collection databases without having to cope with heterogeneous software and data structures.

Table 2. Culture collection databases on the Internet with significant holdings of *Penicillium* or *Aspergillus* cultures

All-Russian Collection of Microorganisms (VKM)	http://www.bdt.org.br/bdt/msdn/vkm/strain/
American Type Culture Collection (ATCC)	http://www.atcc.org/
Belgium Co-Ordinated Collections of Microorganisms (BCCM)	http://www.belspo.be:80/bccm/
Cabi BioScience (formerly IMI)	http://www.cabi.org/institut/imi/grc.htm
Canadian Collection of Fungus Cultures (CCFC or DAOM)	http://res.agr.ca/brd/ccc/
Centraalbureau voor Schimmelcultures (CBS)	http://www.cbs.knaw.nl/
Czech Collection of Fungi (CCF)	http://www.bdt.org.br/bdt/msdn/ccf/
Deutsche Sammlung von Mikroorganismen (DSM)	http://www.gbf-braunschweig.de/DSMZ/dsmzhome.html
Fungal Genetics Stock Centre[1] (FGSC)	http://www.kumc.edu/research/fgsc/main.html
Japan Collection of Microorganisms (JCM)	http://www.jcm.riken.go.jp/
Microbial Germplasm Database[2]	http://mgd.cordley.orst.edu/
Uppsala University Culture Collection (UPSC)	http://ups.fyto.uu.se/mykotek/index.html
World Data Center for Microorganisms (WFCC)	http://www.wdcm.riken.go.jp/

[1]Contains cultures and information primarily on *Aspergillus nidulans* only
[2]This database contains information about numerous smaller collections, mostly in the United States

An International Working Group on Taxonomic Databases (TDWG) has been established to co-ordinate these activities. Their web site (http://bgbm3.bgbm.fu-berlin.de/TDWG/acc/default.htm) includes detailed discussions of the complexities of this issue and a bibliography. One initiative, begun in 1992 by the Computerization and Networking Committee of the Association of Systematic Collections (http://www.ascoll.org/about.html) is the "Modelling Project for Biological Collections" (http://www.bishop.hawaii.org/asc-cnc/). Their model encompasses taxa of all taxonomic groups and suggests the development of a database standard that will enable specimen data to be combined across institutions. A consequence of this is that collections that want to be computerized can take advantage of the experience and data models already developed by other collections (http://gizmo.lbl.gov/DM_TOOLS/DMTools.html).

Databases of taxonomic scientists are being developed independently by ITIS (see above) and by ETI (see below). Data gathering for both of these initiatives is passive and specialists are expected to enter their own data voluntarily through the WWW.

THE CURRENT STATE OF COMPUTERIZED KEYS

It is hard to imagine that it has only been about 10 years since the production of one of the first computerized keys in mycology, PENKEY (Klich and Pitt, 1988), a DOS-based, line entry type key to common *Penicillium* species. This was followed by Pitt's (1990) PENNAME, a data-sheet style key to the species included in his laboratory guide (Pitt, 1988). This key was derived from a database containing information on approximately 2500 isolates, and was programmed in Clipper to provide a series of data entry screens. The presentation and user-friendliness of PENNAME was a great improvement over PENKEY, and despite its DOS interface, the program operates well under Windows 3.1 and remains a valuable tool for those who purchased it. This was followed by PENIMAT (Bridge et al., 1992), a menu-based DOS program incorporating more sophisticated, probabilistic identification algorithms, and data including physiological and biochemical characters.

The chronology of these programs shows a progressive increase in sophistication and user-friendliness. The relatively short life span of some of these keys is remarkable, compared to the taxonomic publications from which they were derived. For example, although DOS based programs still operate under Windows, most users now prefer the graphical user interfaces that Windows and the Mac operating systems provide. Another problem is that all of these programs are custom programmed and each has its own inherent learning curve.

A final factor in the apparent early obsolescence of some of the existing keys is the development of the WWW, and keys that can be accessed directly without the need to distribute software. Table 3 lists examples of Internet-based key currently available for fungi. Most of these are traditional dichotomous keys of varying degrees of sophistication, programmed using Hypertext Markup Language (HTML). A variant of these kinds of keys is the picture-key, in which the user clicks through a set of comparative pictures. An attractive example is the key to the trees of the Pacific Northwest (http://www.orst.edu/instruct/for241/). There are far fewer of the more effective synoptic keys (e.g. keys to *Fusarium* and to Lichens in Table 1) for which the database search components are custom programmed.

Scientists with an interest in developing computerized keys will probably want to investigate existing key-shell programs, which allow them to enter their own taxonomic data so that interactive keys can then be distributed to interested users. Anyone considering developing computerized keys should consult M. Dallwitz's article "Desirable Attributes for Interactive Identification Programs" available on the DELTA web site (see below).

The best known key-shell program is probably DELTA (Descriptive Language for Taxonomy) (Dalwitz, 1974, 1980), widely used by botanists and entomologists and accepted as the standard for the exchange of taxonomic data by the International Taxonomic Databases Working Group (vide infra). DELTA, originally a DOS package and now available for Windows, is an integrated set of programmes that can produce natural-language descriptions, keys, cladistic and phenetic classifications, interactive identification systems and information retrieval systems. Products developed with DELTA vary from commercially released CD-ROMs to datasets freely available on the Internet. A recent development from the University of Toronto is a program that provides an interface between web browsers and DELTA databases called Polyclave (http://prod.library.utoronto.ca/polyclave). An email discussion list and a newsletter for DELTA users are also available; for information see the web site (http://kaw. keil.ukans.edu/delta/). Users developing DELTA databases are now expected to pay for a licence to use the program, and those who wish to use INTKEY, the component of DELTA that runs completed keys, are also expected to pay a licence fee.

Table 3. Examples of interactive computerized keys for fungi available for use over the World Wide Web.

Agaric Genera of the British Isles	http://www.personal.u-net.com/~ivyhouse/normkey.htm
Agaricales of Costa Rica	http://www.nybg.org/bsci/res/hall/costaric.html
Agaricales of the Hawaiian Islands	http://www.mycena.sfsu.edu/hawaiian/Keys.html
Armillaria	http://www.wisc.edu/botany/fungi/arm.html
Botryobasidium	http://www.uni-tuebingen.de/uni/bbm/mycology/botrkey.htm
Fusarium	http://res.agr.ca/brd/fusarium
Hyphodontia, Hyphodontiella,	
Schizopora, Echinoporia	http://www.uni-tuebingen.de/uni/bbm/mycology/hydokey. htm
Lichens Synoptic Key	http://mgd.orst.edu/hyperSQL/lichenland/index.html
Myxomycetes	http://www.wvonline.com/myxo/id.htm
Tofispora	http://www.uni-tuebingen.de/uni/bbm/mycology/tofilist.htm
Tree Fruit Diseases	http://www.caf.wvu.edu/kearneysville/wvufarm6.html

The Expert Center for Taxonomic Identification (ETI) at the Research School of Biodiversity, University of Amsterdam, the Netherlands, is a non-profit organization funded by UNESCO, the Dutch government, and other agencies. Their mandate is to publish interactive, multimedia CD-ROMs that operate on Windows and Macintosh platforms. The identification software employed was developed in-house and is now available in its second version, Linnaeus II version 2. There are three principal modules including a multimedia database application, three types of identification software, a geographical information system, and several support modules such as a literature database and interactive glossary. The package is provided at no cost to scientists in exchange for the right to publish and sell the completed identification systems on CD-ROMs. The taxonomic information is also intended to be included in an online World Biodiversity Database, accessible from the ETI web site (http://wwweti.eti.bio.uva.nl/default.shtml).

CABI-KEY is DOS-based software developed by the Commonwealth Agricultural Bureaux as a presentation vehicle for taxonomic information developed at its own institutions. Identification programs developed to date are for insects or other arthropods and are distributed commercially on floppy disks or CD-ROMs. More information is available at http://www.cabi.org/ catalog/cabikey/cabikey.htm.

Other key-shell programs are listed at http://www.geocities.com/RainForest/Vines/ 8695/software.html), included are established programs such as MEKA, now available in a Windows version (http://www.mip.berkeley.edu/meka/), and a number of shareware and commercial programs. A promising newcomer in the key-shell game is LucID. The package comprises two modules, LucID Builder and LucID Player, which are used to assemble and run error-tolerant synoptic keys. The user friendly, simple interface reflects the target market for the software, which seems to be aimed at users who are less concerned about taxonomic minutae, such as educators and ecologists. A demonstration version can be down loaded from the web site (http://www.ctpm.uq.au/software/lucid/ LucIDProfessionalMain.htm).

By changing the way that software is used or distributed, the Internet has interfered with economic expectations for marketing of keys and key-shells. This has been compounded by the increased needs in scientific institutions for cost recovery, which has pushed some formerly free products, such as DELTA, into the commercial realm. The availability of free keys and free key-shells on the Internet may discourage people from purchasing commercial key products and we anticipate a period of rather aggressive competition, similar to the competition between other brands in the software market. The relatively small size of the available market is likely to favour products with institutional support, because the products are unlikely to be commercially viable on their own. A challenge for the promoters of shell programs will be the development of products that can be used interactively over the WWW, something that has so far been accomplished only for DELTA.

ELECTRONIC IMAGING

The development of computerized keys implies the need for electronic images. Line drawings or photographs can be composed directly on computers or scanned from pre-existing images using flatbed or hand held scanners. Two bitmapped image file formats are commonly used on the WWW. The Graphics Interlaced File format (.gif) was developed by CompuServe, and is now usually recommended for line art, although many photographs are also transmitted in that way. The format is restricted to 256 colours, but has several features that make it attractive for some uses, such as the ability to have a transparent background. The JPEG format offer full colour (24-bit) and variable compression ratios and is now the format of choice for photographs. Uncompressed bitmapped image file formats, such as the Windows .bmp format, and the .tif format, are not supported by Web browsers, but are used by many common image-handling programs. There are many shareware programs that are capable of converting image files from one format to another.

We have had some experience with direct capture of images from a microscope using video cameras and so-called grabber boards. In our experience, most general computer stores do not have sufficiently knowledgeable staff to assist in assembling a working system. The companies that specialize in imaging often sell high-end equipment, but because of the small market size and the perceived technical audience, the hardware often requires very specific computer configurations, and/or the software is often not user-friendly. Incompatibilities between the video cards that run the PC video monitors and video grabber boards seem to be common. A certain amount of trial and error is required to assemble a functional system, and it therefore helps to buy from a company that understands this and will accept the return of incompatible equipment.

Image resolution is a complex subject because each component (the video camera, the grabber board, the computer monitor, and the image output device) has its own theoretical resolution, all affecting the final image quality. For true digital photography, equivalent to film images, the resolution of all components should be around 2000 x 2000 pixels. Achieving this is presently very expensive, and results in high storage costs because of the file sizes of the resulting images. For most applications, however, this level of resolution probably exceeds real needs. The more critical need is for a fast computer with a large hard disk, high quality video adapter board, as much RAM as possible, and operating in 24-bit (16 million) colour rather than the normal 8-bit (256) colour used as a default by Windows.

There are many kinds of digital video cameras and corresponding software available, but we have had good results even from old 256 line analog video cameras. CCD based video cameras are now the norm, and we are using a single CCD colour camera manufactured by COHU. The video grabber board is a consumer product (rather than a product made specifically for scientific applications) called the ATI Video Basic board, with a theoretical resolution of 640 x 480 pixels, but which we usually operate at 320 x 240 pixels. Several other similar systems have been assembled in our institute using different components, with varying degrees of frustration, some of which have higher effective resolutions. The images produced by our system are perfectly suitable for the WWW, and in some cases adequate for paper publication, although they are not as clear as film based images. Individual images are captured instantly, saved as files, and composed into electronic plates using the Windows Paintbrush program. The resulting images (e.g. Fig. 1) can be used on the WWW, printed onto slide or black and white film, or printed directly using laser or bubble-jet printers. In our experience, this is a very efficient means of image-based note keeping, producing plates for a variety of purposes, and has the further advantage of resulting in full colour images with a high memnonic impact.

FINAL THOUGHTS

It is easy to engage in hyperbole when talking about computers or the Internet, but we do not need a crystal ball to know that this interconnectivity is going to bring about profound changes. One point seems clear. Workers in *Penicillium* and *Aspergillus* taxonomy have been near the forefront of the application of computers to taxonomy, at least in the mycological world. Momentum is now gathering for large multi-taxon projects of national and international scope. Powerful political forces will propel this wave.

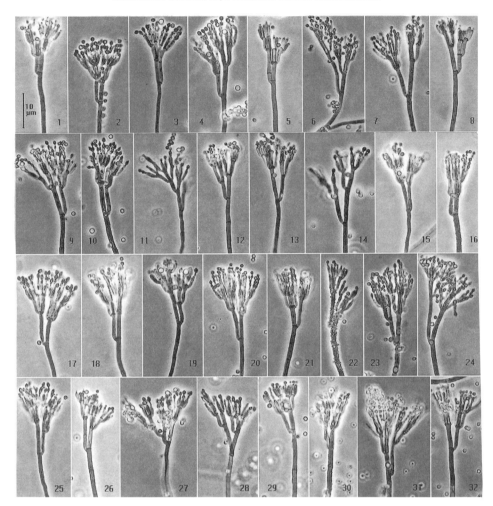

Fig. 1. A comparison of conidiophores of 32 strains of the *Penicillium aurantiogriseum* complex from the ICPA round robin study. The image was assembled in Windows Paintshop from individual images cropped from 320 x 240 pixel originals obtained using an ATI Video Basic grabber board and a COHU colour CCD camera, converted to grey scale using the computer program L-View, and printed from that program using a 300 x 300 dpi bubble jet printer on high quality glossy paper. No film was used in the processing of the image.

Although we can be proud of what we have accomplished, our contribution will become a small subset of a much larger whole. In order not to drown in this wave, we will have to be prepared.

The success experienced by molecular biologists in developing co-operative, community-developed databases should encourage those wanting to develop other kinds of large biological databases. The developments in taxonomic and nomenclatural software brought about by biodiversity needs will give us the tools to create databases incorporating all kinds of information. Taylor (1995) suggested the development of a PhenBank, including phenetic taxonomic information, as a parallel to GenBank. Will the taxonomic world ever demand the deposit of phenetic information in databases as a prerequisite to publication, the way that the molecular world demands the deposit of sequence data before publication? *Penicillium* and *Aspergillus* taxonomists should be active in the planning and development of such a system.

REFERENCES

Anonymous. (1994). Systematics Agenda 2000: Charting the biosphere. Systematics Agenda 2000, New York. (2 reports are available, one general and one subtitled "Technical Report).

Bridge, P. D. (1990). Identification of terverticillate Penicillia from a matrix of positive test results, In *Modern Concepts in Penicillium and Aspergillus Classification*. (Eds R. A. Samson & Pitt, J. I.)., Plenum Press, New York, . pp. 283- 287.

Bridge, P. D., Z. Kozakiewicz & R. R. M. Paterson. (1992). PENIMAT: A computer assisted identification scheme for terverticillate *Penicillium* isolates. Mycol. Pap. 165: 1-64 (+floppy disk).

Dallwitz, M. J. (1974). A flexible computer program for generating identification keys. Syst. Zool. 23, 50-7.

Dallwitz, M. J. (1980). A general system for coding taxonomic descriptions. Taxon 29, 41-6.

Klich, M. A. (1990). Computer applications in *Penicillium & Aspergillus* systematics. pp. 269-278 in *Modern Concepts in Penicillium and Aspergillus Classification*. (Eds by R. A. Samson & Pitt, J. I.) Plenum Press, New York.

Klich, M. I. & J. I. Pitt. (1988). A computer-assisted synoptic key to common *Penicillium* species and their teleomorphs. New Orleans, Louisiana: privately published.

Pitt, J. I. (1988). A laboratory guide to common *Penicillium* species. CSIRO Division of Food Research, North Ryde, N.S.W.

Pitt, J. I. (1990). PENNAME, a new computer key to common *Penicillium* species. pp. 279-281 in in *Modern Concepts in Penicillium and Aspergillus Classification*. (Eds R. A. Samson & Pitt, J. I.), Plenum Press, New York.

Pitt, J. I. & R. A. Samson. (1993). Species names in current us in the *Trichocomaceae* (Fungi: Eurotiales). Regnum Vegetabile 128: 13-57.

Samson, R.A. & J.I. Pitt (1990). Modern Methods on Penicillium and Aspergillus Classification. New York, Plenum Press.

Taylor, J. W. (1995). Making the Deuteromycota redundant: a practical integration of mitosporic and meiosporic fungi. Can. J. Bot. 73 (Suppl. 1): S754-S759.

Williams, A. P. (1990). Identification of *Penicillium* and *Aspergillus*: Computer-assisted keying. pp. 289-297 in in *Modern Concepts in Penicillium and Aspergillus Classification*. (Eds R. A. Samson & Pitt, J. I). Plenum Press, New York.

ASP45, A SYNOPTIC KEY TO COMMON SPECIES OF *ASPERGILLUS*.

Keith A. Seifert,
Eastern Cereal and Oilseed Research Centre, Agriculture and Agri-Food Canada, Research Branch, Ottawa, Ontario K1A 0C6 Canada.

A synoptic, computerized key to 45 common species of *Aspergillus* was developed for use with "Laboratory Guide to Common *Aspergillus* species and their teleomorphs", based on Klich and Pitt (1988).The program was written using the synoptic key shell program SYNOPKEY developed by David Malloch at the University of Toronto.A data set of 35 characters and 184 character states of both anamorphs and teleomorphs was developed from published descriptions. Because the characters used to separate common *Aspergillus* species are often discrete and easily interpreted, it is often possible to use only three or four characters to arrive at a tentative identification. Several years of laboratory experience have demonstrated the efficacy of the keys as an identification aid.

INTRODUCTION

The synoptic key ASP 45 includes the 45 common species of *Aspergillus* and related teleomorphs described and illustrated by Klich and Pitt (1988), which should be consulted for details of medium composition, incubation regimes and character interpretation for identification. Most of the data used in the key were extracted from this book. In some cases, the data has been supplemented with information from Raper and Fennell (1965) or original protologues to complete the data matrix for each taxon. Although I have used data from these sources, the key was compiled without the collaboration of these authors, and the responsibility for errors or misinterpretations is mine.

Synoptic keys were introduced to mycology by Korf (1972) and the reader is referred to that publication for a detailed explanation. There are many advantages to this type of key, primarily the possibility for the user to exploit the diagnostic characters that are most readily interpreted. Many characters used in *Aspergillus* taxonomy are discrete and easily interpreted, therefore the synoptic key is an ideal tool for identifying species of this genus. Furthermore, synoptic keys are readily computerized. The key printed below was created using the DOS program SYNOPKEY written by David Malloch of the University of Toronto. The computerized version of this key can be downloaded from the World Wide Web at the following URL: http://res.agr.ca/brd/ecorc/staff/seif-k.html.

Selection of Characters

Several characters, such as colony diameters on standard media, stipe width, conidial and vesicle dimensions, are readily measured. Many of the qualitative characters are unambiguous, including the colours of exudates, soluble pigments, stipes and conidia, the presence or absence of metulae, the ornamentation of conidia and stipes, and arrangement of the conidial masses. For quantitative characters, average values should be used rather than extreme values.

Some species of *Aspergillus* are differentiated primarily on differences in conidial colour. In compiling this key, I deliberately used rather broad colour categories. More precise colour determinations require the use of a colour guide, and this may be necessary for final decisions. Precise colour data is included in Klich and Pitt (1988) and Raper and Fennell (1965).

Some characters (e.g. white mycelium, absence of sclerotia, absence of exudate, absence of soluble pigment) have not been included in the character list because they have no diagnostic value in this kind of key. Some characters should also be treated with caution. In particular, data for the "stipe with thick walls" character may be incomplete.

In several years of experience using this key, I have found that accurate identifications can often be obtained using 3 to 5 reliably determined characters. The exceptions to this tend to be species that are difficult to separate no matter what the means of identification, in particular *Aspergillus flavus* and related taxa.

List of Taxa (Note: KP refers to page numbers in Klich and Pitt (1988) and RF to page numbers in Raper and Fennell, 1965).

1. *Aspergillus alliaceus* (KP p. 16, RF p. 273)
2. *Eurotium amstelodami* (KP p. 18, RF p. 166)
3. *Aspergillus auricomus* (KP p. 20, RF p. 277)
4. *Aspergillus caespitosus* (KP p 22, RF p. 462)
5. *Aspergillus candidus* (KP p. 24, RF p. 347)
6. *Aspergillus carbonarius* (KP p. 26, RF p. 301)
7. *Aspergillus carneus* (KP p. 28, RF p. 564)
8. *Aspergillus cervinus* (KP p. 30, RF p. 214)
9. *Aspergillus (Eurotium) chevalieri* (KP p. 32, RF p. 163)
10. *Aspergillus clavatus* (KP p. 34, RF p. 140)
11. *Chaetosartorya cremea* (KP p. 36, RF p. 418)
12. *Neosartorya fischeri* (KP p. 38, RF p. 252)
13. *Aspergillus flavipes* (KP p. 40, RF p. 559)
14. *Aspergillus flavus* (KP p. 42, RF p. 361)
15. *Aspergillus foetidus* (KP p. 44, RF p. 323)
16. *Aspergillus fumigatus* (KP p. 46, RF p. 242)
17. *Aspergillus japonicus* var. *aculeatus* (KP p. 48, RF p. 328)
18. *Aspergillus japonicus* var. *japonicus* (KP p. 50, RF p. 327)
19. *Aspergillus kanagawaensis* (KP p. 52, RF p. 217)
20. *Emericella nidulans* (KP p. 54, RF p. 495)
21. *Aspergillus niger* var. *awamori* (KP p. 56, RF p. 315)
22. *Aspergillus niger* var. *niger* (KP p. 58, RF p. 309)
23. *Aspergillus niveus* (KP p. 60, RF p. 562)
24. *Aspergillus ochraceus* (KP p. 62, RF p. 281)
25. *Sclerocleista ornata* (KP p. 64, RF p. 199)
26. *Aspergillus oryzae* (KP p. 66, RF p. 370)
27. *Aspergillus ostianus* (KP p. 68, RF p. 284)
28. *Aspergillus paradoxus* (KP p. 70, RF p. 206)
29. *Aspergillus parasiticus* (KP p. 72, RF p. 369)
30. *Aspergillus penicillioides* (KP p. 74, RF p. 232)
31. *Aspergillus puniceus* (KP p. 76, RF p. 547)
32. *Emericella quadrilineata* (KP p. 78, RF p. 521)
33. *Aspergillus restrictus* (KP p. 80, RF p. 226)
34. *Eurotium rubrum* (KP p. 82, RF p.160)
35. *Emericella rugulosa* (KP p. 84, RF p. 514)
36. *Aspergillus sclerotiorum* (KP p. 86, RF p. 272)
37. *Aspergillus sojae* (KP p. 88)
38. *Aspergillus sparsus* (KP p. 90, RF p. 433)
39. *Aspergillus sydowii* (KP p. 92, RF p. 450)
40. *Aspergillus tamarii* (KP p. 94, RF p. 381)
41. *Aspergillus terreus* (KP p. 96, RF p. 568)
42. *Aspergillus unguis* (KP p. 98, RF p. 525)
43. *Aspergillus ustus* (KP p. 100, RF p. 545)
44. *Aspergillus versicolor* (KP p. 102, RF p. 445)
45. *Aspergillus wentii* (KP p. 104, RF p. 407)

140

SYNOPTIC KEY

Characters on CYA after 7 days growth

Diameter
    0-10mm   8, 9, 11, 13, 19, 23, 25, 30, 33, 35
    10-20 mm   2, 5, 7, 8, 19, 20, 25, 32, 33, 34, 39, 42, 44
    20-30 mm   5, 7, 9, 11, 13, 23, 28, 31, 34, 38, 39, 41, 42, 43, 44, 45
    30-40 mm   3, 4, 10, 16, 23, 24, 27, 28, 31, 32, 36, 41, 42, 43, 45
    40-50 mm   1, 3, 10, 12, 14, 15, 16, 18, 20, 24, 27, 28, 32, 36, 40, 41
    50-60 mm   1, 6, 12, 14, 15, 16, 18, 22, 24, 26, 29, 32, 36, 37, 40
    60-70 mm   1, 6, 12, 14, 15, 16, 17, 18, 21, 22, 26, 29, 37, 40
    more than 70 mm 1, 6, 14, 17, 26

Colour of mycelium
    yellow   2, 4, 7, 9, 11, 12, 13, 15, 20, 21, 22, 23, 26, 28, 31, 34, 35, 38, 45
    grey   2, 20, 31, 32, 43
    tan or brown 31, 44
    pink or orange44, 45
    brown   2, 7, 12, 14, 20, 21, 23, 34, 35, 39, 43, 44

Colour of exudate
    yellow   3, 5, 6, 23, 24, 27, 28, 36, 43
    red   20, 21, 23, 24, 31, 35, 39, 43, 44
    uncoloured   3, 5, 6, 7, 10, 12, 13, 14, 16, 21, 23, 24, 27, 28, 29, 32, 35, 36, 43,44, 45

Colour of soluble pigment
    yellow   7, 15, 23, 24, 28, 31, 32, 35, 41, 43
    brown   7, 16, 20, 24, 32, 35, 39, 44
    red or pink   4, 16, 20, 24, 27, 32, 39, 44

Colour of reverse
    yellow   1, 2, 3, 4, 5, 6, 7, 8, 9, 10, 11, 12, 13, 15, 16, 17, 18, 19, 21, 22, 23, 24, 25, 26,
                28, 29,34, 35, 36, 40, 41, 43, 45
    orange   5, 7, 8, 11, 14, 18, 20, 27, 32, 34, 35, 39, 43
    green   2, 15, 16, 18, 23, 33
    tan or brown 1, 2, 4, 7, 9, 10, 12, 13, 14, 15, 17, 18, 19, 20, 21, 23, 24, 27, 29, 30, 31, 32, 33,
                34, 35, 37, 38, 39, 41, 42, 43, 44, 45
    red   3, 4, 16, 20, 24, 29, 31, 32, 34, 35, 39, 42, 44
    uncoloured   2, 5, 7, 8, 10, 12, 13, 14, 16, 19, 22, 25, 26, 27, 29, 30, 33, 35, 37, 40, 42, 44, 45
Colony sulcate   4, 5, 7, 9, 10, 11, 12, 13, 15, 16, 18, 20, 22, 23, 24, 27, 28, 29, 31, 32, 35, 39,
                40, 41, 42, 43, 44, 45

Diameter at 37°C
    no growth   3, 5, 7, 8, 9, 11, 19, 25, 28, 30, 33, 34, 39, 43, 45
    0-10 mm   2, 4, 5, 7, 8, 9, 13, 17, 18, 19, 23, 24, 27, 31, 38, 39, 42, 43, 44
    10-20 mm   2, 3, 4, 5, 6, 7, 10, 13, 17, 18, 19, 23, 24, 27, 36, 42, 43
    20-30 mm   1, 3, 4, 5, 6, 7, 10, 17, 18, 23, 24, 27, 36, 43
    30-40 mm   1, 5, 7, 10, 18, 23, 24, 27, 36, 41, 43
    40-50 mm   1, 7, 15, 18, 23, 26, 29, 37, 40, 41, 43
    50-60 mm   1, 14, 15, 18, 20, 22, 23, 26, 29, 32, 35, 37, 40, 41
    60-70 mm   12, 14, 16, 20, 21, 22, 26, 29, 32, 37, 40, 41
    more than 70 mm   12, 16

Colony characters on other media after 7 days growth

Diameter on MEA

0-10 mm	5, 9, 30, 33, 34
10-20 mm	2, 5, 7, 9, 11, 13, 23, 33, 34, 35, 38, 39, 44
20-30 mm	2, 5, 7, 9, 11, 13, 20, 23, 28, 31, 32, 34, 39, 42, 44, 45
30-40 mm	3, 4, 7, 10, 13, 16, 19, 20, 22, 23, 28, 31, 32, 41, 42, 43, 45
40-50 mm	3, 4, 10, 14, 16, 17, 19, 20, 22, 24, 25, 27, 28, 29, 32, 36, 37, 41, 43, 45
50-60 mm	3, 4, 6, 8, 14, 15, 16, 17, 19, 20, 21, 22, 24, 25, 27, 29, 32, 36, 37, 40, 41
60-70 mm	1, 4, 6, 8, 12, 14, 15, 16, 17, 20, 21, 22, 25, 26, 29, 37, 40
more than 70 mm	1, 12, 14, 18, 21, 22, 29, 40

Diameter on CY20S

0-10 mm	19, 25, 30, 43
10-20 mm	5, 8, 23, 25, 28, 30, 33, 35, 43, 44
20-30 mm	5, 7, 10, 13, 23, 25, 28, 31, 33, 34, 35, 38, 39, 41, 42, 43, 44
30-40 mm	2, 3, 5, 7, 10, 11, 15, 16, 20, 23, 28, 31, 32, 34, 41, 42, 43, 44
40-50 mm	2, 3, 4, 9, 10, 11, 15, 16, 20, 23, 24, 27, 31, 32, 34, 41, 43, 45
50-60 mm	2, 3, 4, 9, 11, 15, 16, 23, 24, 27, 32, 34, 36, 41, 45
60-70 mm	1, 3, 6, 9, 11, 12, 14, 15, 16, 21, 22, 24, 26, 27, 29, 36, 37, 40, 41, 45
more than 70 mm	6, 12, 14, 17, 18, 22, 26, 29

Conidial heads

Colour in mass

black	6, 15, 17, 18, 21, 22
dark brown	15, 17, 18, 21, 22
brown	9, 24, 26, 27, 29, 37, 40, 41, 43, 44
yellow	1, 3, 5, 23, 24, 26, 27, 36, 37, 40, 44
white	1, 5, 8, 19, 23
green	2, 4, 9, 10, 11, 12, 16, 20, 28, 29, 30, 32, 33, 34, 35, 38, 39, 42, 44
yellow-green	14, 25, 26, 29, 37, 40
blue-green	11, 12, 16, 39
grey-green	2, 4, 9, 10, 12, 16, 28, 32, 33, 34, 35, 39, 42, 44
pink or reddish	7, 19, 27, 31
orange	8, 12, 19, 31

Arrangement

radiating	1, 2, 3, 4, 5, 6, 7, 8, 9, 10, 11, 12, 13, 14, 15, 17, 18, 19, 20, 21, 22, 23,24, 25, 26, 27, 28, 29, 30, 31, 32, 34, 35, 36, 37, 38, 39, 40, 42, 43, 44, 45
in compact columns	4, 16, 20, 32, 33, 35, 41, 42
in loose columns	1, 2, 4, 7, 12, 13, 14, 20, 23, 26, 28, 31, 32, 37, 42, 43
in radiating columns	3, 6, 10, 12, 14, 15, 17, 21, 36, 40, 45

Stipes

Length

less than 50 µm	1, 20, 31, 32, 35, 43
50-100 µm	1, 4, 7, 19, 20, 23, 30, 31, 32, 33, 35, 41, 42, 43
100-250 µm	1, 2, 4, 5, 7, 8, 9, 12, 13, 16, 17, 18, 19, 20, 23, 24, 25, 28, 29, 30, 31, 32, 33, 34, 35, 37, 39, 41, 42, 43, 44, 45

142

250-500 µm	1, 2, 3, 4, 5, 7, 8, 9, 10, 12, 13, 14, 15, 16, 17, 18, 19, 21, 22, 23, 24, 25, 27, 28, 29, 30, 31, 33, 34, 36, 37, 38, 39, 40, 41, 42, 43, 44, 45
500-1000 µm	1, 3, 5, 6, 8, 9, 10, 12, 13, 14, 15, 17, 18, 19, 21, 22, 23, 24, 25, 26, 27, 28, 29, 30, 34, 36, 37, 38, 40, 44, 45
1000-2000 µm	1, 3, 6, 10, 11, 14, 15, 17, 18, 21, 22, 24, 26, 27, 36, 38, 40, 45
2000-5000 µm	3, 6, 10, 11, 14, 17, 22, 26, 27, 40, 45
more than 5000 µm	11

Width

3-5 µm	4, 7, 13, 19, 20, 23, 30, 31, 32, 33, 35, 39, 41, 42, 43
5-7 µm	1, 4, 7, 8, 12, 13, 16, 18, 19, 20, 23, 30, 31, 32, 33, 35, 39, 41, 42, 43, 44
7-10 µm	1, 2, 3, 5, 8, 9, 12, 13, 15, 16, 17, 18, 21, 22, 25, 27, 30, 33, 34, 36, 39, 43
10-15 µm	2, 3, 5, 6, 9, 10, 11, 14, 15, 16, 17, 21, 22, 24, 25, 26, 27, 28, 29, 34, 37, 38, 40, 45
more than 15 µm	3, 5, 6, 10, 11, 14, 15, 21, 22, 25, 26, 28, 40

Walls

smooth	1, 2, 3, 4, 5, 6, 7, 8, 9, 10, 11, 12, 13, 15, 16, 17, 18, 19, 20, 21, 22, 23, 25, 26, 28, 30, 31, 32, 33, 34, 35, 37, 38, 39, 42, 43, 44, 45
finely roughened	6, 13, 14, 24, 26, 28, 29, 33, 34, 37, 38, 45
rough	3, 14, 24, 26, 27, 29, 34, 36, 37, 38, 40
thick	4, 6, 8, 14, 22, 23, 24, 27, 38, 39, 42

Colour

pale brown	2, 3, 4, 6, 7, 8, 9, 10, 13, 14, 15, 17, 18, 19, 20, 21, 22, 24, 34, 35, 36, 39, 42, 44
brown	4, 6, 17, 20, 21, 27, 31, 32, 34, 35, 42, 43
yellow	11, 22, 24, 27, 36, 44
red-brown	38

Vesicles

Shape

globose	1, 2, 3, 5, 6, 7, 8, 9, 11, 13, 14, 15, 17, 18, 19, 21, 22, 24, 26, 27, 29, 31, 36, 37, 38, 39, 40, 41, 43, 45
hemispherical	4, 20, 23, 30, 32, 33, 34, 35, 41
elongate	3, 5, 14, 15, 24, 27, 29, 36, 44, 45
pyriform	1, 12, 16, 25, 32, 33, 37, 38, 44
spathulate	2, 4, 9, 16, 20, 23, 25, 28, 30, 35, 39, 42, 44
clavate	7, 10, 26, 28, 37, 39, 42, 44

Width

less than 10 µm	4, 7, 12, 13, 19, 20, 23, 26, 30, 31, 32, 33, 34, 35, 39, 41, 42, 43, 44, 45
10-25 µm	1, 2, 3, 4, 5, 7, 8, 9, 10, 12, 13, 14, 15, 16, 17, 18, 19, 20, 21, 22, 23, 24, 25, 26, 27, 28, 29, 30, 31, 32, 33, 34, 35, 36, 37, 38, 39, 40, 41, 42, 43, 44, 45
25-50 µm	1, 2, 3, 5, 6, 8, 9, 10, 11, 14, 15, 16, 17, 18, 19, 21, 22, 24, 25, 26, 27, 28, 29, 34, 36, 37, 38, 40, 45
50-100 µm	1, 3, 6, 10, 11, 14, 15, 17, 21, 22, 24, 26

Metulae and Phialides

Metulae present 1, 3, 4, 5, 6, 7, 11, 13, 14, 15, 20, 21, 22, 23, 24, 26, 27, 29, 31, 32, 35, 36, 37, 38, 39, 40, 41, 42, 43, 44, 45

Metulae absent 1, 2, 5, 8, 9, 10, 12, 14, 16, 17, 18, 19, 23, 25, 26, 28, 29, 30, 33, 34, 37, 39, 40

A mixture of vesicles with and vesicles lacking metulae 1, 5, 14, 15, 21, 23, 26, 29, 37, 39, 40

Length of metulae

5-7 µm	1, 3, 4, 5, 7, 13, 14, 15, 20, 21, 23, 24, 26, 31, 32, 35, 36, 39, 41, 42, 43, 44
7-10 µm	1, 3, 4, 5, 7, 11, 13, 14, 15, 20, 21, 23, 24, 26, 27, 29, 36, 37, 38, 43, 44
10-15 µm	1, 3, 5, 11, 14, 15, 21, 22, 24, 26, 27, 36, 37, 40, 45
15-20 µm	1, 3, 5, 6, 14, 15, 21, 22, 24, 27, 40, 45
more than 20 µm	5, 6, 21, 22, 24, 27

Phialide length

5-7 µm	1, 2, 3, 4, 5, 7, 8, 9, 12, 13, 16, 17, 18, 19, 20, 21, 23, 30, 31, 32, 35, 36, 38, 39, 41, 42, 43, 44
7-10 µm	1, 2, 3, 4, 5, 6, 9, 10, 11, 12, 13, 14, 15, 16, 17, 18, 20, 21, 22, 23, 24, 25, 26, 27, 28, 29, 30, 32, 33, 34, 35, 36, 37, 38, 39, 40, 41, 42, 43, 44, 45
10-15 µm	1, 5, 6, 9, 11, 14, 24, 25, 26, 27, 28, 29, 30, 33, 34, 37, 40, 45
15-20 µm	11

Coverage of vesicle

less than ½	7, 13, 20, 23, 32, 33, 35
½-¾	1, 2, 3, 4, 7, 12, 13, 15, 16, 18, 21, 23, 28, 30, 35, 37, 41, 42, 43
more than ¾	5, 6, 8, 9, 10, 11, 14, 17, 19, 22, 24, 25, 26, 27, 29, 31, 34, 36, 38, 39, 40, 44, 45

Conidia

Shape

globose	1, 2, 3, 4, 5, 6, 7, 8, 12, 13, 14, 15, 16, 17, 18, 19, 20, 21, 22, 23, 24, 26, 27, 28, 29, 31, 32, 34, 35, 36, 37, 38, 39, 40, 41, 42, 43, 44, 45
subglobose	1, 2, 3, 9, 11, 12, 14, 16, 17, 18, 19, 23, 24, 26, 27, 30, 34, 35, 38, 45
ellipsoidal	3, 9, 10, 11, 12, 14, 17, 18, 24, 26, 27, 28, 30, 33, 34, 38, 45
oval	1, 5, 9, 16, 25, 26, 27, 28
pyriform	9, 10, 25, 27, 33
doliiform	9

Length

2-3 µm	1, 3, 5, 7, 8, 12, 13, 16, 19, 23, 24, 31, 32, 36, 39, 41, 42, 44
3-4 µm	1, 2, 3, 4, 5, 7, 8, 9, 10, 12, 14, 16, 17, 19, 20, 21, 22, 23, 24, 26, 27, 29, 30, 31, 32, 35, 36, 38, 39, 40, 42, 43, 44, 45
4-5 µm	2, 4, 5, 9, 10, 14, 15, 17, 18, 19, 21, 22, 23, 24, 26, 27, 28, 29, 30, 31, 33, 34, 40, 43, 45
5-7 µm	2, 6, 9, 10, 11, 14, 15, 17, 18, 21, 25, 26, 27, 28, 29, 33, 34, 37, 40, 45
7-10 µm	6, 11, 25, 26, 33, 34, 37, 40
more than 10 µm	6, 25, 34, 37, 40

Width

2-3 µm	1, 3, 5, 7, 8, 10, 12, 13, 16, 19, 23, 24, 31, 32, 36, 39, 41, 42, 44
3-4 µm	1, 2, 3, 4, 5, 7, 8, 9, 10, 11, 14, 16, 17, 19, 20, 21, 22, 23, 24, 26, 27, 28, 29, 30, 31, 32, 33, 35, 36, 38, 39, 40, 42, 43, 44, 45
4-5 µm	2, 4, 5, 11, 14, 15, 18, 19, 21, 22, 23, 24, 26, 27, 28, 29, 31, 33, 40, 43, 45
5-7 µm	2, 6, 14, 15, 21, 25, 26, 27, 29, 33, 34, 37, 40, 45
7-10 µm	6, 26, 37, 40
more than 10 µm	6, 37, 40

Walls

smooth	1, 3, 5, 7, 8, 10, 12, 13, 14, 15, 16, 19, 20, 21, 23, 24, 26, 27, 28, 36, 38, 41, 42, 45
slightly rough	2, 4, 9, 12, 14, 16, 20, 21, 22, 23, 24, 26, 27, 28, 31, 32, 35, 36, 38, 42, 44, 45
rough	4, 6, 9, 20, 22, 25, 29, 30, 31, 33, 35, 37, 39, 40, 42, 43, 44, 45

144

spiny   2, 6, 9, 11, 15, 16, 17, 18, 26, 30, 31, 34, 39

Sclerotia and Hülle cells

Sclerotia present 1, 3, 5, 6, 14, 17, 22, 24, 26, 27, 28, 29, 36

Sclerotia absent 2, 4, 5, 6, 7, 8, 9, 10, 11, 12, 13, 14, 15, 16, 17, 18, 19, 20, 21, 22, 23, 24, 25, 26, 27, 28, 29, 30, 31, 32, 33, 34, 35, 37, 38, 39, 40, 41, 42, 43, 44, 45

Sclerotial colour when mature

    white   27, 36

    yellow   3, 6, 22, 27, 28, 36

    orange   3

    pink   5, 6, 17, 22, 24

    grey   1

    dark brown to black 1, 5, 14, 26, 29

Hülle cells

    present   4, 13, 20, 23, 31, 32, 35, 39, 42, 43, 44

    globose   4, 20, 23, 32, 35, 39, 42, 44

    ellipsoidal or elongate 4, 23, 31, 43

Ascomata

Ascomata sclerotial   1, 28

Ascomata cleistothecial 2, 9, 11, 12, 13, 20, 23, 25, 32, 34, 35, 42

Ascoma diameter

    100-250 µm   2, 9, 11, 12, 13, 20, 23, 25, 32, 34, 35, 42

    250-500 µm   1, 11, 12, 20, 25, 28, 32, 35

    more than 500 µm 1, 12, 25

Ascoma colour

    yellow   2, 9, 11, 13, 23, 28, 34, 35

    grey or black   1

    white   12, 25

    cream or buff   11, 12, 28

    red   20, 35, 42

    brownish red   32

    purple   25

Peridium composition

    one layer of cells   2, 9, 11, 20, 23, 34, 35

    angular cells   2, 9, 12, 34, 35

    interwoven hyphae 13, 20, 23, 32

    sclerenchyma-like cells   1, 28

    both angular cells and interwoven hyphae   25, 34, 35

Ascospores

Shape

    ellipsoidal   1

    lenticular   2, 9, 11, 12, 20, 23, 25, 28, 32, 34, 35, 42

    globose   12

    subglobose   13

    walnut-shaped   35

Ornamentation

    single furrow   1, 2, 13, 23, 34

2 or 3 ridges, flanges or crests   2, 9, 11, 12, 20, 23, 25, 28, 34, 35, 42
4 ridges or crests   32

Length
3-5 μm   2, 9, 11, 20, 23, 28, 32, 42
5-7 μm   1, 2, 9, 11, 12, 13, 20, 23, 28, 32, 34, 35, 42
7-10 μm   1, 12, 13, 25
3-4 μm   2, 9, 20, 23, 28, 32, 34, 35, 42

Width
4-5 μm   11, 12, 23, 34, 35
5-7 μm   1, 12, 13, 25

Walls
smooth   1, 9, 12, 13, 20, 25, 28, 32, 34, 42
rough   2, 12, 35
spinose   11, 23

Colour
uncoloured 1, 2, 9, 11, 12, 13, 23, 25, 28, 34
yellowish   13, 23
pale brown 25
red or purple 20, 32, 35, 42

ACKNOWLEDGMENTS

I am grateful to David Malloch, University of Toronto, for allowing me to use his computer program SYNOPKEY, and my colleagues Scott Redhead, Bob Shoemaker and John Bissett for their reviews of the manuscript.

REFERENCES

Klich, M. A. & Pitt, J. I. (1988). A laboratory guide to common *Aspergillus* species and their teleomorphs.CSIRO Division of Food Processing, North Ryde, N.S.W, 116 pp.

Korf, R. P. 1972. Synoptic key to the genera of the Pezizales.Mycologia 64: 937-994.

Raper, K. B. & Fennell, D. I. (1965).The genus *Aspergillus*.Williams & Wilkins, Baltimore. 686 pp.

Chapter 4

PHYLOGENY AND MOLECULAR
TAXONOMY OF *PENICILLIUM*

EVOLUTIONARY RELATIONSHIPS OF THE CLEISTOTHECIAL GENERA WITH *PENICILLIUM, GEOSMITHIA, MERIMBLA* AND *SAROPHORUM* ANAMORPHS AS INFERRED FROM 18S RDNA SEQUENCE DIVERGENCE

Hiroyuki Ogawa and Junta Sugiyama
Institute of Molecular and Cellular Biosciences, The University of Tokyo, 1-1, Yayoi 1-chome, Bunkyo-ku, Tokyo 113-0032, Japan

Nucleotide sequences were obtained from the 18S rRNA gene for eight species of *Trichocoma, Penicilliopsis, Talaromyces, Hamigera, Chromocleista, Nectria,* and *Tritirachium.* The bootstrapped neighbor joining analysis of plectomycetous taxa showed that ten genera of the Trichocomaceae, *Monascus* (Monascaceae), and *Elaphomyces* (Elaphomycetaceae) form a monophyletic group in the plectomycete lineage. The branch separating these trichocomaceous genera and *Monascus* from *Elaphomyces* received 64% bootstrap support. *Trichocoma* is placed at a position relatively close to *Talaromyces,* including the *Penicillium*-producing *Talaromyces luteus.* However, *T. luteus,* showing similar ascospore and anamorph morphology as *Trichocoma paradoxa* is not included within the *Trichocoma* branch. The *Sarophorum*-producing *Penicilliopsis clavariiformis* is rather close to the *Eupenicillium* group and *Aspergillus* teleomorphs

Hamigera* with a *Merimbla* and a *Penicillium* anamorph is closer related to *Eupenicillium* and the *Aspergillus*-teleomorphs rather than *Talaromyces.* The type species *Chromocleista malachitea* is placed within the *Eupenicillium* group, whereas the second species *Chromocleista cinnabarina* with a *Paecilomyces* anamorph and two species of *Talaromyces* cluster together in 100% of bootstrap replications.

Our phylogenetic analysis of *Geosmithia* and associated teleomorphs shows that the species examined are polyphyletic and form three different lineages in the Hypocreales, *Talaromyces,* and *Eupenicillium.* The anamorph-teleomorph connections and ubiquinone systems in the Trichocomaceae are also discussed in the light of the rRNA gene sequence analysis.

INTRODUCTION

The ascomycetous family Trichocomaceae accommodates cleistothecial genera with *Aspergillus, Penicillium, Geosmithia, Merimbla, Paecilomyces,* and related anamorphs. (cf. Fig. 1 in Tamura *et al.,* 2000).

Traditionally, relationships among the Trichocomaceae and with other ascomycetes have been investigated mainly using morphological and ontogenetic characters such as cleistothecial initials, cleistothecial centre, cleistothecial peridium, stromatic structures, ascus structure, and anamorph associations (Benny and Kimbrough, 1980; Fennell, 1973; Malloch, 1985 a, b; Malloch and Cain, 1972; Subramanian, 1979). Malloch and Cain

149

(1972) recognized 16 teleomorphic genera with *Aspergillus, Paecilomyces,* and *Penicillium* anamorphic states in the family Trichocomataceae (sic) E. Fischer (1897), with a synonym Eurotiaceae Clements & Shear. They paid attention to the similarities between the ascospores of *Trichocoma paradoxa* (Trichocomaceae) and *Talaromyces luteus* (traditionally placed in the Eurotiaceae or Aspergillaceae) suggested by Kominami *et al.* (1952), resulting in unification of these families. Malloch and Cain (l.c.) also suggested that the trichocomaceous genera evolved from a stromatic ancestor similar to present-day Hypocreaceae, with several important evolutionary simplifications and modifications. Subsequently Malloch and Cain (1973) recognized that three genera proposed by themselves were predated a couple of months earlier by those of Subramanian (1972), and published the correct family name Trichocomaceae instead of Trichocomataceae in addition to the nomenclatural corrections for the three genera.

Benny and Kimbrough (1980) particularly emphasized the importance of the centrum development as a valid characteristic and agreed in general with the concept of the Trichocomaceae (= Eurotiaceae) defined by Malloch and Cain (1972, 1973) and accepted 22 genera. Malloch (1981) divided 20 genera of the Trichocomaceae into five groups on the basis of progressive stages of reduction in ascoma complexity. Furthermore Malloch (1985 a, b; cf. Berbee *et al.,* 1995, and Currah, 1994) proposed two subfamilies Trichocomoideae and Dichlaenoideae based on ascospore, ascoma, and substrate preference.

Phylogenetic analysis among distantly related fungal taxa, using 18S rRNA gene sequences, has contributed to well-resolved and statistically supported conclusions (e.g., Bruns *et al.,* 1992; Berbee and Taylor, 1992; Nishida and Sugiyama, 1993, 1994; Nishida *et al.,* 1995; Reynolds and Taylor, 1993). Initially, analyses on the basis of the small or large rRNA sequence comparisons improved our understanding of phylogenetic relationships among selected *Aspergillus, Penicillium* and related teleomorphs in the Trichocomaceae (e.g., Chang *et al.,* 1991; Dupont *et al.,* 1990; Logrieco *et al.,* 1990; Peterson, 1993). Because the partial sequences (i.e., only ca. 600 nucleotides) from 18S rRNA are available for comparison, however, resolution is limited (Chang *et al.,* 1991; cf. Berbee *et al.,* 1995). In recent years, in case of the 18S rRNA gene, phylogenetic analysis shifted to the full sequence (Sugiyama *et al.,* 1996).

To improve our understanding of the phylogenetic relationships among trichocomaceous taxa, we determined 18S rDNA sequences from species of *Penicilliopsis, Talaromyces, Trichocoma, Hamigera, Chromocleista,* and two hypocrealean fungi *Nectria cinnabarina* and *N. ochroleuca.* Based on 11 new sequences in addition to our previous sequence data (Ogawa *et al.,* 1997), in this paper we analyze the evolutionary affinities of cleistothecial genera producing *Penicillium, Geosmithia, Merimbla* and *Sarophorum* anamorphs.

MATERIALS AND METHODS

Fungal strains. The new sequences determined in this study are listed in Table 1. The following rDNA sequences for other fungal species retrieved from the DNA data bank were used in this analysis:

Table 1. Fungal isolates and GenBank/EMBL/DDBJ accession numbers of 18S rRNA gene sequences

Classification/Taxon [a]	IAM no. [b]	DDBJ no [c]	Major ubiquinone system [d]
Family Trichocomaceae, Order Eurotiales			
Trichocoma paradoxa Junghuhn	IAM 14601	AB003943	Q-(10(H_2)(67%)+10(H_4)(33%)
	IAM 14602	AB003944	Q-(10(H_2)(68%)+10(H_4)(32%)
Penicilliopsis clavariiformis Solms-Laubach	IAM 14604	AB003945	Q-9
	IAM 14605	AB003946	No data
Talaromyces luteus (Zukal) C. R. Benjamin	IAM 14715	AB006716	No data
	IAM 14724(T)	AB003947	Q-10(H_2)
Hamigera striata (Raper & Fennell)	IAM 14606(T)	AB003948	Q-10
Stolk & Samson			
Chromocleista cinnabarina Yaguchi &	IAM 14712(T)	AB006747	No data
Udagawa	IAM 14713	AB003952	Q-10(H_2)
Family Hypocreaceae, Order Hypocreales			
Nectria cinnabarina (Tode: Fries) Fries	IAM 14568	AB003949	No data
Nectria ochroleuca (Schweinitz) Berkeley	IAM 14569	AB003950	No data
Mitosporic fungi			
Tritirachium sp. [e]	IAM 14522	AB003951	No data

[a] Fungal classification in the family and order levels is based on Hawksworth *et al*. (1995); [b] IAM, Institute of Molecular and Cellular Biosciences, The University of Tokyo, Japan; IFO, Institute for Fermentation, Osaka, Japan; GJS, Gary Samuels, USDA, ARS, Beltsville, USA. DNA Data Bank of Japan (DDBJ), Mishima, Japan. [d] Ubiquinone data from Kuraishi et al. (1991 and unpublished data); [e] Isolated as a contaminant from a culture received as *Chromocleista cinnabarina* IFO 9598, and identified by Mr. T. Ito (IAM); (T) Ex type strain.

Cultivation and isolation of chromosomal DNA. The strains listed in Table 1 were grown at 25°C in 5 ml of YM medium (Wickerham, 1951), on a rotary shaker (200 rpm) for 2-3 days. Cultured mycelia were gathered and lyophilized. The template DNA was extracted from the freeze-dried mycelia following the method of Lee and Taylor (1990).

PCR amplification of the 18S rDNA fragments. Each diluted sample was used for PCR (Mullis and Fallona, 1987; Saiki *et al.,* 1988). A fragment of DNA spanning approximately 1.7-2.1 kbp of the 18S rRNA gene, was symmetrically amplified using Takara *Taq* DNA polymerase (Takara Shuzo Co., Ltd., Shiga, Japan) with the primer N1 and N2 (Nishida and Sugiyama, 1993). The PCR reaction (100µl) included 0.1 µg of fungal genomic DNA as template, 1x PCR buffer, 0.8 µM each primer, 2.5U *Taq* polymerase, and 0.2 mM dNTPs. Samples were amplified through 30 cycles on a thermal cycler (Takara PCR Thermal Cycler TP-2000; Takara Shuzo Co., Ltd., Shiga, Japan) using the following parameters: DNA denaturation, 1 min at 94°C; primer annealing, 1 min at 55°C; primer extension, 2 min at 72°C. An initial denaturation step for 2 min at 94°C and a final extension step for 8 min at 72°C were carried out.

Sequencing. PCR fragments were purified using ULTRAFREE-MC (30000 NMWL filter unit; Millipore Corp., Bedford, MA, USA). The PCR product (100 µl) and 300 µl of distilled water were loaded onto the column and centrifuged at 3000 x *g* with a fixed-angle rotor for 5-10 min. The sample yielded approximately 40-60 µl and was then moved to a new 1.5 ml microtube. The purified samples were used for templates of the cycle sequencing reactions. Sequencing reaction (20 µl) contains 8.0 µl of Terminator Ready Reaction Mix (Perkin-Elmer Corp., CA, USA), 3.2 pmol of a primer, and 30-180 ng of

151

the template DNA. Primers used for sequencing were N1, N2, NS2, NS3, NS4, NS5, NS6 and NS7 (Nishida and Sugiyama, 1993; White *et al.,* 1990), and our newly designed primers, EC (5´-GTTCTATTTTGTTGG-3'), ECR (5'-CCAACAAAATAGAAC-3'; a reverse compliment of EC), G (5´-CTCGTTCGTTATCGC-3'), 2 (5'-TAACTA ACG-3') and SP1R (5'-CTGCGAATGGCTCATTA-3'). The cycle sequencing reaction was performed on a Takara PCR Thermal Cycler TP-2000 for 25 cycles using the following parameters: 96°C for 30 sec, 50°C for 15 sec and 60°C for 4 min. Samples were analysed in an Applied Biosystems 373S DNA sequencer (Perkin-Elmer Corp., CA, USA) using 4% Long Ranger acrylamide gels (FMC BioProducts, ME, USA) run in 1x TBE buffer.

Sequences retrieved

Neurospora crassa Shear & B. Dodge, X04971;
Microascus cirrosus Curzı, M89994;
Pseudallescheria boydii (Shear) McGinnis *et al.,* M89782;
Ophiostoma ulmi (Buisman) Nannfeldt, M83261;
Leucostoma persoonii (Nicheke) Höhnel, M83259;
Hypomyces chrysospermus Tulasne, M89993;
Colletotrichum gloeosporioides Penzig, M55640;
Chaetomium elatum Kunze & Schmidt, M83257;
Xylaria carpophila , Z49785;
Hypocrea lutea (Tode) Petch, D14407;
Byssochlamys nivea Westling, M83256;
Thermoascus crustaceus (Apinis & Chester) Stolk, M83263;
Talaromyces macrosporus (Stolk & Samson) Frisvad et al., M83262;
Talaromyces bacillisporus (Swift) C. R. Benjamin, D14409;
Talaromyces emersoni Stolk, D88321;
Talaromyces eburneus Yaguchi *et al.,* D88322;
Penicillium chrysogenum Thom (as *P. notatum* Westling), M55628;
Eupenicillium crustaceum Ludwig, D88324;
Eupenicillium javanicum (van Beyma) Stok & Scott, U21298;
Monascus purpureus Went, M83260;
Chromocleista malachitea Yaguchi & Udagawa, D88323;
Neosartorya fischeri (Wehmer) Malloch & Cain, U21299;
Eurotium rubrum König et al., U00970;
Aspergillus fumigatus Fres., M55626;
Aspergillus flavus Link, D63696;
Aspergillus niger van Tieghem, D63697;
Hamigera avellanea (Thom & Turesson) Stolk & Samson, D14406;
Merimbla ingelheimensis (van Beyma) Pitt, D14408;
Geosmithia lavendula (Raper & Fennell) Pitt, D14405;
Geosmithia putterillii (Thom) Pitt, D88318;
Geosmithia namyslowskii (Zaleski) Pitt, D88319;
Geosmithia cylindrospora (G. Smith) Pitt, D88320;
Elaphomyces maculatus Vittadini, U45440;
Elaphomyces leveillei Tulasne, U45441;
Ascosphaera apis (Maasen ex Claussen) Olive & Spiltoir, M83264;
Eremascus albus Eidam, M83258;
Onygena equina (Wildenow) Persoon, U45442;

Ajellomyces dermatitidis McDonough & Lewis (as *Blastomyces dermatitidis* Gilchrist & Stokes), X59420;
Ajellomyces capsulatus (Kwon-Chung) McGinnis & Katz (as *Histoplasma capsulatum* Darling), X58572;
Coccidioides immitis Rixford & Gilchrist, M55627;
Uncinocarpus reesii Sigler & Orr, L27991;
Trichophyton rubrum (Castellani) Sabouraud, X58570;
Ctenomyces serratus Eidam, U29391;
Gymnoascoideus petalosporus Orr et al., U29392;
Renispora flavissima Sigler *et al.,* U29393;
Malbranchea dendritica Sigler & Carmichael, U29389;
Chrysosporium parvum (Emmons & Ashburn) Carmichael, U29390;
Saccharomyces cerevisiae Meyer ex Hansen, V01335 for an outgroup taxon.

For the 5S rRNA-based tree construction we used the following sequences:
Aspergillus flavus X00690;
Aspergillus nidulans (Eidam) Winter, K03162;
Aspergillus niger X00691;
Penicillium chrysogenum X00692;
Penicillium griseofulvum Dierckx (as *P. patulum* Bainier), X00693;
Thermomyces lanuginosus Tsiklinsky, J01890;
Acremonium chrysogenum (Thirumalachar & Sukapure) W. Gams, X00867;
Acremonium persicinum (Nicot) W. Gams, (1) X00864, (2) X00865, (3) X00866;
Neurospora crassa, alpha K02469, beta K02476;
Candida albicans (Robin) Berkhout, K02366;
Candida utilis (Henneberg) Lodder & Kreger-van Rij (as *Torulopsis utilis* Henneberg), V01417;
Pichia membranaefaciens Hansen, M19950;
Kluyveromyces lactis (Dombrowski) van der Walt, M19949;
Saccharomyces cerevisiae, K01047;
Hamigera avellanea, D88371;
Geosmithia lavendula, type 1 D88366, type 2 D88367;
Geosmithia namyslowskii, D88372;
Talaromyces emersonii, D88369;
Talaromyces eburneus, D88370;
Eupenicilliuim crustaceum, D88373;
Hypocrea lutea, D88368

Phylogenetic analyses. A data set comprising 58 taxa including our new 12 sequence data was used in this analysis. DNA sequences were aligned with Clustal W 1.71, the latest version of Clustal W (cf. Thompson *et al.,* 1994), and the alignment was visually corrected. Phylogenetic relationships were estimated from the alignment using the neighbor-joining method (Saitou and Nei, 1987) by Clustal W 1.71 on a Power Macintosh 8500/120 computer. Evolutionary distances were calculated by using Kimura's two-parameter model (Kimura, 1980; transition/ transversion = 2.0). All positions with gaps within the alignment were excluded for calculation. To measure the relative support and stability of each lineage, bootstrap values (Efron, 1982; Felsenstein, 1985) were calculated using Clustal W 1.71 for 1000 replications.

In addition to the 18S rDNA sequence-based phylogeny, the 5S rRNA sequence-based phylogeny was estimated based on the sequence data set previously reported by Hori and Osawa (1987), and Ogawa *et al.* (1997). In this paper we report on the neighbor-joining tree (Saitou and Nei, 1987) of *Geosmithia* and associated teleomorphs, inferred from 69 aligned sites of 5S rRNA gene sequence from 21 fungal taxa using Clustal W 1.71 (Thompson *et al.,* 1994). Support for the topologies was obtained with bootstrap analysis using 1000 replications (Felsenstein, 1985).

RESULTS AND DISCUSSION

Trichocomaceae, Monascaceae and Elaphomycetaceae
Our 18S rDNA sequence based neighbor joining tree (Fig. 1), inferred from 1710 aligned sites for 59 fungal taxa indicated almost the same phylogeny as that reported by Berbee *et al.* (1995), Landvik *et al.* (1996), and Tamura *et al.* (2000). The Monascaceae, t7ypified by *Monascus*, is included within the Trichocomaceae, whereas the Elaphomycetaceae and Trichocomacae are separate families. The branch separating Trichocomaceae from Elaphomycetaceae received 64% bootstrap support. These three families form a monophyletic group (in 99% bootstrap support) within the Plectomycetes (cf. Tamura *et al.,* 2000). As already noted by Berbee *et al.* (1995) and Ogawa *et al.* (1997), *Monascus* is a member of Trichocomaceae, and therefore, the family Monascaceae is not a phylogenetically appropriate taxon. Molecular data for other genera in the Monascaceae are needed to assess the status of members of this family.

The hypogeous genus *Elaphomyces* has a quite different ecology from the other genera of the Trichocomaceae. Species of *Elaphomyces* also resemble those in the order Onygenales in that spore dispersal is by animals and is like those of the apothecial ascomycete order Tuberales. Two *Elaphomyces* species were basal to the Trichocomaceae (Eurotiales) and also close to it rather than the Onygenales. Our phylogenetic placement for *Elaphomyces* is similar to that of Landvik *et al.* (1996), LoBuglio *et al.* (1996), and Tamura *et al.* (2000).

The hypothesized relationships among genera related to the Trichocomaceae are shown in Fig. 2 based on our molecular phylogeny (Fig. 1) and on Alexopoulos *et al.* (1996), with the progressive stages of reduction in ascoma complexity and the ascospore shape. Tamura et al. (2000; Fig. 2) demonstrated a morphology-based schema for Trichocomaceous genera, but our phylogram (Fig. 2) did not agree with this morphology-based phylogenetic speculation. The molecular phylogeny (Figs. 1 and 2) does reflect the ascospore shape (prolate ascospores vs. oblate, bivalved ascospores) rather than the diversity of ascomatal type in the

153

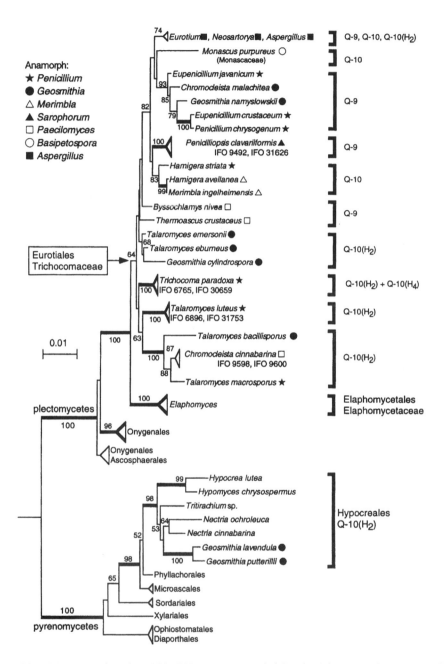

Fig. 1. Neighbor-joining tree based on 18S rDNA sequences of cleistothecial taxa producing *Penicillium*, *Geosmithia*, *Merimbla* and *Sarophorum* anamorphs, and related teleomorphs. The scale bar indicates one base change per 100 nucleotide positions. Bootstrap percentages (> 50%) derived from 1000 replicates are shown at the respective nodes.

154

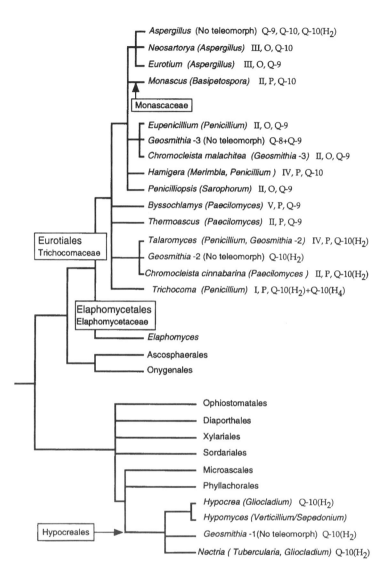

Fig. 2. Hypothesized evolutionary relationships of teleomorphic genera in the Trichocomaceae inferred from 18S rDNA sequence based analysis (cf. Alexopoulos *et al.*, 1996; Ogawa *et al.*, 1997). Roman numerals indicate stages of reduction in ascoma complexity (Malloch, 1981): I, stromata with cleistothecia; II, stromata with cleistothecia lacking; III, stroma lacking, cleistothecial pseudoparenchymatous; IV, stromata lacking, cleistothecia (gymnothecia) hyphal; V, stromata and cleistothecia lacking. O, oblate, bivalved ascospores. P, prolate ascospores (cf. Berbee *et al.*, 1995). Data on the major ubiquinone system from Kuraishi et al. (1985, 1990, 1991), Ogawa et al. (1997), and Yaguchi *et al.*, (1993, 1994).

Trichocomaceae, except for a few genera (e.g., *Hamigera*). The major ubiquinone system reflects their genealogical relationships (cf. Kuraishi *et al.*, 1990, 1991).

Penicillium and its associated teleomorphs

The neighbor-joining phylogeny (Fig. 1) clearly demonstrates that *Penicillium* is not mono-phyletic. It shows that *Eupenicillium* species groups with *Neosartorya* and *Eurotium (Asper-gillus)*, the *Merimbla-* or *Penicillium*-producing *Hamigera*, and the *Sarophorum*-producing *Penicilliopsis* rather than the *Penicillium-* or *Geosmithia*-producing *Talaromyces*, the *Peni-cillium*-producing *Trichocoma*, and the *Paecilomyces*-producing *Byssochlamys*. The branch separating *Eupenicillium* and *Penicilliopsis* from *Byssochlamys* and *Talaromyces* received 82% bootstrap support. Our molecular phylogeny of *Penicillium* was basically identical with Berbee *et al.* (1995) and Ogawa *et al.* (1997).

Trichocoma is the type genus of the family Trichocomaceae (Benny and Kimrough, 1980; Malloch and Cain, 1972, 1973). It is characterized by large brush-like ascomata, stromata with cleistothecia, prolate ascospores, association of a biverticillate *Penicillium* anamorph and 70% of Q-10 (H$_2$) + 30% of Q-10 (H$_4$) as the major ubiquinone system (Kominami *et al.*, 1952; Kuraishi *et al.*, 1991; Malloch, 1981; Malloch and Cain, 1972).

Subsequently Malloch (1985b) placed *Trichocoma* in the group which is characterized by asci borne in hyphal masses or tufts and lack of cleistothecia; i.e., he included *Trichocoma*, *Byssochlamys*, *Dendrosphaera*, *Sagenoma*, and *Talaromyces* in the same group based on the ascomatal type. On the other hand, *Talaromyces luteus* is characterized by lack of stromata, hyphal (gymnothecial) cleistothecia, prolate ascospores, association of *Penicillium luteum* (biverticillate, symmetrical penicilli), and Q-10 (H$_2$) as the major ubiquinone system (Komi-nami *et al.*, 1952; Kuraishi *et al.*, 1991; Malloch, 1981; Pitt, 1979). As stated above, Komi-nami *et al.* (1952) suggested that *Talaromyces luteus* is the closest relative to *Trichocoma paradoxa*.

We determined the 18S rDNA sequence from *Trichocoma paradoxa* IFO 6765 and IFO 30659 and *Talaromyces luteus* IFO 6896 and IFO 37153 (ex type). The sequence difference was seven sites between two strains of the former species, whereas that were five sites in the latter species. As a result, the molecular phylogeny (Fig. 1) indicates that *Trichocoma para-doxa* and *Talaromyces luteus* are closely related, but nor congeneric. Consequently the 18S rDNA sequence data support Malloch's grouping of the ascomatal type for *Trichocoma* (Malloch, 1985b). The major ubiquinone systems of both species do not basically contradict with each other.

Geosmithia and its associated teleomorphs

The neighbor-joining tree (Fig. 1) indicates clear separation between the Pyrenomycetes and Plectomycetes with 100% bootstrap supports. Six species of *Geosmithia* and related teleo-morphs grouped within *Penicillium*, two were related to *Eupenicillium* and four to *Talaro-myces*. The remaining two strictly anamorphs *Geosmithia lavendula* (the type species) and *G. putterillii*, were placed within the Pyrenomycete lineage, in which the two taxa grouped with the Hypocrealean fungi. The Hypocreales (Rehner and Samuels, 1994; Ogawa *et al.*, 1997), appear as a monophyletic group in 98% of bootstrap replications.

In order to analyse the polyphyletic origins of *Geosmithia* species, we evaluated the phy-logenies using the neighbor-joining analysis of the 5S rRNA gene sequences. The tree is shown in Fig. 3. As a result, the bootstrapped 5S rDNA sequence-based neighbor-joining

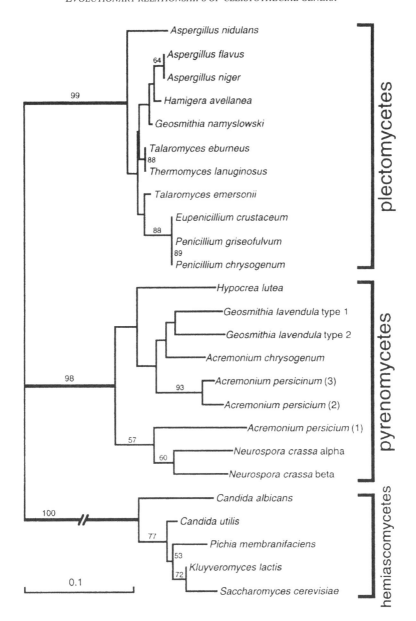

Fig. 3.Neighbor-joining tree based on 5S rDNA sequences of the selected taxa. Numbers below the branches indicate bootstrap percentages (1000 replications). The scale bar indicates one base change per ten nucleotide positions.

tree (Fig. 1) supported polyphyletic origins of *Geosmithia* species previously detected in the 18S rDNA sequence-based trees (cf. Ogawa *et al.,* 1997). *Chromocleista* was segregated from *Eupenicillium* by the production of sclerotioid monolocular stroma and very slow development of asci and ascospores in the ascoma centrum (Yaguchi *et al.,* 1993). The type species *C. malachitea* is characterized by malachite-green sclerotioid ascomata, one-celled lenticular (probably oblate, bivalved) ascospores with two equatorial crests, and Q-9 as the major ubiquinone system. It is further described with a *Geosmithia* anamorph, However, re-examination of the type culture and additional isolates showed that this anamorph is identical with *Penicillium herquei* and *Chromocleista* is probably a further synonym of *Eupenicillium* (J.C. Frisvad and R.A. Samson pers. commun.). This is confirmed in our 18S rDNA sequence-based neighbor-joining tree (Fig. 1) which demonstrates that *Chromocleista* is not monophyletic. The *Penicillium* (*Geosmithia*)-producing *C. malachitea* groups with *Eupenicillium* and the strictly anamorphs *G. namyslowskii* and *P. chrysogenum* in 93% of bootstrap replications.

The second species *C. cinnabarina* was proposed on isolates incorrectly identified as *Aphanoascus cinnabarinus* (Yaguchi *et al.,* 1993). This fungus is characterized by reddish orange sclerotioid ascomata, ellipsoidal (prolate) ascospores with irregular multiple ridges, a *Paecilomyces* anamorph, and the Q-10 (H_2) as the major ubiquinone system (Kuraishi *et al.,* unpublished data; Yaguchi *et al.,* 1993). *C. cinnabarina* forms a monophyletic group with *Talaromyces macrosporus* and *T. bacillisporus* with strong bootstrap support (100 %). The ascomatal type seen in *Chromocleista cinnabarina* may be a specialized form of hyphal cleistothecia.

Merimbla and Hamigera

Hamigera was proposed by Stolk and Samson (1971) but not accepted by Pitt and Samson (1993) in "Names in Current Use in the Trichocomaceae". However, Stolk and Samson's segregation of *Hamigera* from *Talaromyces* is supported by our molecular data as shown in Fig. 1. The type species *H. avellanea* with the *Merimbla ingelheimensis* anamorph groups with *H. striata* with a *Penicillium* anamorph in 83% of the bootstrap replications, indicating that these taxa are congeneric. Their major ubiquinone systems (Kuraishi *et al.,* 1991) support the molecular phylogeny.

Sarophorum and Penicilliopsis

Penicilliopsis clavariiformis is characterized by seed-borne, stipitate stromata containing fertile locules, lack of cleistothecia, a *Sarophorum* anamorph, and the Q-9 ubiquinone system (Kuraishi *et al.,* 1990; Malloch, 1981, 1985b; Malloch and Cain, 1972; Samson and Seifert, 1985). Strains of *P. clavariiformis* IFO 9492 and IFO 31626 (with only three base differences) appear as an independent branch within the group represented by the *Aspergillus* teleomorphs, *Eupenicillium*, and *Hamigera*. Malloch's grouping based on the ascomatal type (Malloch, 1985b; cf. 1981) supports the molecular phylogeny of *Penicilliopsis* shown in Fig. 1. For further phylogenetic speculation, we need more molecular data for its relatives, e.g. the *Stilbodendron*-producing *Penicilliopsis africana* and synnematous *Aspergillus* spp. (cf. Samson and Seifert, 1985).

158

ACKNOWLEDGEMENTS

We thank Dr. M. Takeuchi (IFO, Osaka) and Dr. Gary J. Samuels (USDA, ARS, Beltsville) for providing fungal cultures. We also thank Dr. H. Kuraishi and Mr. T. Ito (IFO, Osaka) for providing unpublished data on the ubiquinone system of *Paecilomyces* and for identifying a contaminant of *Chromocleista cinnabarina* IFO 9598, respectively.

REFERENCES

Alexopoulos, C. J., Mims, C. W. & Blackwell, M. (1996). Introductory mycology, 4th edition. New York, John Wiley & Sons.

Benny, G. L. & Kimbrough, J. W. (1980). A synopsis of the order and families of Plectomycetes with keys to genera. Mycotaxon 12: 1-91.

Berbee, M. L. & Taylor, J. W. (1992). Two ascomycete classes based on fruit-body characters and ribosomal DNA sequences. Mol. Biol. Evol. 9: 278-284.

Berbee, M. L., Yoshimura, A., Sugiyama, J. & Taylor, J. W. (1995). Is *Penicillium* monophyletic? An evaluation in the family Trichocomaceae from 18S, 5.8S and ITS DNA sequence data. Mycologia 87: 210-222.

Bruns, T. D., Vilgalys, R., Barns, S. M., Gonzalez, D., Hibbett, D. S., Lane, D. J., Simon, L., Stickel, S., Szaro, T. M., Weisburg, W. G. & Sogin, M. L. (1992). Evolutionary relationships within the Fungi: Analyses of nuclear small subunit rRNA sequences. Mol. Phylogenet. Evol. 1: 231-241.

Chang, J.-M., Oyaizu, H. & Sugiyama, J. (1991). Phylogenetic relationships among eleven selected species of *Aspergillus* and associated teleomorph genera. J. Gen. Appl. Microbiol. 37: 289-308.

Currah, R. S. (1994). Peridial morphology and evolution in the protunicate ascomycetes. In Ascomycete systematics: problems and perspectives in the nineties (Ed Hawksworth, D. L.). New York, Plenum Press. pp. 281-293.

Dupont, J., Dulertre, M., Lafay, J. F., Roquebert, M.-F. & Bryoo, Y. (1990). A molecular assessment of the position of *Stilbothamnium* in the genus *Aspergillus*. In Modern concepts in *Penicillium* and *Aspergillus* classification (Eds Samson, R. & Pitt, J. I.). New York, Plenum Press. pp. 335-342.

Efron, B. (1982). The jackknife, the bootstrap and other resampling plans. CBMS-NSF Regional Conference Series in Applied Mathematics, Monograph 38, SIAM, Philadelphia. 92 pp.

Felsenstein, J. (1985). Confidence limits on phylogenetics: an approach using the bootstrap. Evolution 39: 783-791.

Fennell, D. I. (1973). Plectomycetes; Eurotiales. In The fungi, an advanced treatise, vol. 4A (Eds Ainsworth, G. C., Sparrow, F. K. & Sussman, A. S.), New York, Academic Press. pp. 45-68.

Fischer, E. (1897). The Trichocomaceae. In Die natürlichen Pflanzenfamilien (Eds Engler, A. & Prantl, K.). Leipzig, Engelmann Verl. pp. 290-320.

Hawksworth, D. L., Kirk, P. M., Sutton, B. C. & Pegler, D. N. (1995). Ainsworth & Bisby's dictionary of the fungi, 8th edition. Wallingford, CAB International.

Hori, H. & Osawa, S. (1987). Origin, and evolution of organisms as deduced from 5S RNA sequences. Mol. Biol. Evol. 4: 445-472.

Kimura, M. (1980). A simple method for estimating evolutionary rate of base substitution through comparative studies of nucleotide sequences. J. Mol. Evol. 16: 11-120.

Kominami, K., Kobayasi, Y. & Tubaki, K. (1952). Is *Trichocoma paradoxa* conspecific with *Penicillium luteum*. Nagaoa 2: 18-23.

Kuraishi, H., Aoki, M., Itoh, M., Katayama, Y., Sugiyama, J. & Pitt, J. I. (1991). Distribution of ubiquinones in *Penicillium*, and related genera. Mycol. Res. 95: 701-711.

Kuraishi, H., Ito, M., Tsuzaki, N., Katayama, Y., Yokoyama, T. & Sugiyama, J. (1990). The ubiquinone system as a taxonomic aid in *Aspergillus* and its teleomorphs. In Modern concepts in *Penicillium* and *Aspergillus* classification (Eds Samson, R. A. & Pitt, J. I.). New York, Plenum Press. pp. 407-421.

Kuraishi, H., Katayama-Fujimura, Y., Sugiyama, J. & Yokoyama, T. (1985). Ubiquinone system in fungi. I. Distribution of ubiquinones in the major families of ascomycetes, basidiomycetes, and deuteromycetes, and their taxonomic implications. Trans. Mycol. Soc. Japan 26: 383-395.

Landvik, S., Shailer, N. F. J. & Eriksson, O. E. (1996). SSU rDNA sequence support for a close relationship between the Elaphomycetales and the Eurotiales and Onygenales. Mycoscience 37: 237-241.

Lee, S. & Taylor, J. W. (1990). Isolation of DNA from fungal mycelia and single spores. In PCR Protocols (Eds Inns, M. A., Gelfand, D. H., Shincky, J. J. & White, T. J.). New York, Academic Press. pp. 282-287.

LoBuglio, K. F., Berbee, M. L. & Taylor, J. W. (1996). Phylogenetic origins of the asexual mycorrhizal symbiont *Cenococcum geophilum* Fr. and other mycorrhizal fungi among the ascomycetes. Mol. Phylogenet. Evol. 6: 287-294.

Logrieco, A., Peterson, S. W. & Wicklow, D. T. (1990). Ribosomal RNA comparisons among taxa of the terverticillate Penicillia. In Modern concepts in *Penicillium* and *Aspergillus* classification (Eds Samson, R. A. & Pitt, J. I.). New York, Plenum Press. pp. 343-355.

Malloch, D. (1981). The plectomycete centrum. In Ascomycete systematics, the Luttrellian concept (Ed Reynolds, D. R.). New York, Springer-Verlag. pp. 73-91.

Malloch, D. (1985a). Taxonomy of the Trichocomaceae. In Filamentous microorganisms, biomedical aspects (Ed Arai, T.). Tokyo, Japan Scientific Societies Press. pp. 37-45.

Malloch, D. (1985b). The Trichocomaceae: relationships with other Ascomycetes. In Advances in *Penicillium* and *Aspergillus* systematics (Eds Samson, R. A. & Pitt, J. I.). New York, Plenum Press. pp. 365-382.

Malloch, D. & Cain, R. F. (1972). The Trichocomactaceae: Ascomycetes with *Aspergillus, Paecilomyces*, and *Penicillium* imperfect states. Can. J. Bot. 50: 2613-2628.

Malloch, D. & Cain, R. F. (1973). The Trichocomaceae (Ascomycetes): synonyms in recent publications. Can. J. Bot. 51: 1647-1648.

Mullis, K. B. & Fallona, F. A. (1987). Specific synthesis of DNA in vitro via a polymerase-catalyzed chain reaction. Methods Enzymol. 155: 335-350.

Nishida, H. & Sugiyama, J. (1993). Phylogenetic relationships among *Taphrina, Saitoella*, and other higher fungi. Mol. Biol. Evol. 10: 431-436.

Nishida, H. & Sugiyama, J. (1994). Archiascomycetes: detection of a major new lineage within the Ascomycota. Mycoscience 35: 361-366.

Nishida, H., Ando, K., Ando, Y., Hirata, A. & Sugiyama, J. (1995). *Mixia osmundae*: transfer from the Ascomycota to the Basidiomycota based on evidence from molecules and morphology. Can. J. Bot. 73 (Suppl. 1): S 660-S 666.

Ogawa, H., Yoshimura, A. & Sugiyama, J. (1997). Phylogenetic origins of the anamorphic genus *Geosmithia* and the relationships of the cleistothecial genera: Evidence from 18S, 5S and 28S sequence analyses. Mycologia 89: 756-771.

Peterson, S.W. (1993). Molecular genetic assessment of relatedness of *Penicillium* subgenus *Penicillium*. In The fungal holomorph: mitotic, meiotic and pleomorphic speciation in fungal systematics (Eds Reynolds, D.R. & Taylor, J.W.). Wallingford, CAB International, pp 121-128.

Pitt, J. I. (1980) ['1979']. The genus *Penicillium* and its teleomorphic states *Eupenicillium* and *Talaromyces*. Academic Press, London, United Kingdom.

Pitt, J. I. & Samson, R. A. (1993). Species names in current use in the Trichocomaceae (Fungi, Eurotiales). In: Names in current use in the families Trichocomaceae, Cladoniaceae, Pinaceae, and Lemnaceae. (Ed W. Greuter). Königstein, Koeltz Scientific Books. pp. 13-57.

Rehner, S. A. & Samuels, G. J. (1994). Taxonomy and phylogeny of *Gliocladium* analysed from nuclear large subunit DNA sequences. Mycol. Res. 98: 625-634.

Reynolds, D. R. & Taylor, J. W. (Eds) (1993). The fungal holomorph: metotic, meiotic and pleomorphic speciation in fungal systematics. Wallingford, CAB International.

Saiki, R. K., Gelfand, D. H., Stoffel, S., Scharf, S. J., Higuchi, R., Horn, G. T., Mullis, K. B. & Erlich, H. A. (1988). Primer-directed enzymatic amplification of DNA with thermostable DNA polymerase. Science 239: 487-491.

Saitou, N. & Nei, M. (1987). The neighbor-joining method: a new method for reconstructing phylogenetic trees. Mol. Biol. Evol. 4: 406-425.

Samson, R. A. & Seifert, K. A. (1985). The ascomycete genus *Penicilliopsis* and its anamorphs. In Advances in *Penicillium* and *Aspergillus* systematics (Eds Samson, R. A. & Pitt, J. I.). New York, Plenum Publishing Corp. pp. 397-428.

Stolk, A. C. & Samson, R. A. (1971). Studies on *Talaromyces* and related genera. I. *Hamigera* gen. nov. and *Byssochlamys*. Persoonia 6: 341-357.

Subramanian, C. V. (1972). The perfect states of *Aspergillus*. Curr. Sci. 41: 755-761.

Subramanian, C. V. (1979). Phialidic hyphomycetes and their teleomorphs - an analysis. In The whole fungus, the sexual-asexual synthesis (Ed Kendrick, B.). Ottawa, National Museum of Canada, pp. 125-151.

Sugiyama, J., Nagahama, T. & Nishida, H. (1996). Fungal diversity and phylogeny with emphasis on 18S ribosomal DNA sequence divergence. In Microbial diversity in time and space (Eds Colwell, R. R., Simidu, U. & Ohwada, K.). New York, Plenum Press. pp. 41-51.

Tamura, M., Kawahara, K., & Sugiyama, J. (2000). Molecular phylogeny of *Aspergillus* and associated teleomorphs in the Trichocomaceae (Eurotiales). In Classification of *Penicillium* and *Aspergillus*: Integration of modern taxonomic methods. (Eds. Samson R. A. & Pitt, J. I.), Reading, Harwood Academic Publishers, pp xxxxxxx

Thompson, J. D., Higgins, D. G. & Gibbson, T. D. (1994). CLUSTAL W: Improving the sensitivity of progressive multiple sequence alignment through sequence weighting, position specific gap penalties, and weight matrix choice. Nucl. Acids Res. 22: 4673-4680.

White, T. J., Bruns, T. D., Lee, S. & Taylor, J. W. (1990). Amplification and direct sequencing of fungal RNA genes for phylogenetics. In PCR protocols (Eds Innis, M. A & Gelfan, D. H.). San Diego, Academic Press, pp. 315-322.

Wickerham, L. J. (1951). Taxonomy of yeasts. United States Department of Agriculture Technical Bulletin 1029: 1-56.

Yaguchi, T., Miyadoh, S. & Udagawa, S. (1993). *Chromocleista*, a new cleistothecial genus with a *Geosmithia* anamorph. Trans. Mycol. Soc. Japan 34: 101-108.

Yaguchi, T., Someya, A. & Udagawa, S. (1994). Two new species of *Talaromyces* from Taiwan and Japan. Mycoscience 35: 249-255.

Present address of the authors:

Dr **H. Ogawa**, Center from Information Biology, National Insititute of Genetics, 1111 Yata, Mishima 411-8540 Japan

Prof. Emeritus **J. Sugiyama,** Department of Botany, The University Museum, The University of Tokyo, 3-1, Hongo 7 chome, Bunkyo-ku Tokyo 113-0033 and NCIMB, Kaminakazato Office, 9-2, Sakae-cho, Kita-ku, Tokyo 114-0005 Japan

PHYLOGENETIC ANALYSIS OF *PENICILLIUM* SPECIES BASED ON ITS AND LSU-RDNA NU-CLEOTIDE SEQUENCES

Stephen W. Peterson
Microbial Properties Research, National Center for Agricultural Utilization Research, Agricultural Research Service, U. S. Department of Agriculture, Peoria, IL 61604

Species of *Eupenicillium* and *Penicillium* from subgenera *Aspergilloides*, *Furcatum* and *Penicillium* were examined morphologically and rDNA sequences were determined and analyzed phylogenetically to determine if taxa in current monographic treatments of *Penicillium* are monophyletic. Parsimony analysis of the rDNA sequences shows that many subgeneric taxa in *Penicillium* are polyphyletic, and that the emphasis placed on penicillus structure in *Penicillium* taxonomy led to polyphyletic classifications. Statistically, the relationships of some of the taxa are not strongly supported by the current DNA sequence data, but with further information from additional genes, monophyletic subgeneric taxa in *Penicillium* will be defined.

INTRODUCTION

Thom (1930) published the first comprehensive monograph of the genus *Penicillium* and wrote a useful key to the accepted species. His criteria for identification of the species included growth rate, colony colour, colony texture, colour in colony reverse, presence of exudate, and the complexity and symmetry of the penicillus. In their *Penicillium* monograph Raper and Thom (1949) largely used the conceptual framework of Thom (1930), increasing the number of accepted species. Pitt (1979) produced a monographic study of *Penicillium* in which nomenclaturally correct subgeneric names were added to the genus, replacing the group and series concepts of prior authors with subgenera and sections, and broadening some species concepts. The monograph by Ramirez (1982) largely followed Raper and Thom (1949), while updating the genus with species described after 1949. Frisvad (1981, 1989) added secondary metabolite profiles to the list of characteristics useful in *Penicillium* identification.

Some *Penicillium* species produce teleomorphs which are either in *Talaromyces* or *Eupenicillium*. On the basis of 18S rDNA sequences Berbee *et al*. (1995) demonstrated, that *Penicillium* is not monophyletic. The genus forms two branches; one includes *Talaromyces* species and *Penicillium* anamorphs of subgenus *Biverticillium* (Lobuglio *et al*., 1993, 1994; Lobuglio and Taylor, 1993), while the other branch contains *Eupenicillium* species and *Penicillium* anamorphs from subgenera *Penicillium*, *Furcatum* and *Aspergilloides*.

In this papers transcribed spacer (ITS) and large subunit ribosomal DNA (lsu-rDNA) sequences for species of *Geosmithia*, *Eupenicillium*, and *Penicillium* subgenera *Aspergilloides*, *Furcatum* and *Penicillium*, have been determined. From the analysis, monophyly of the subgeneric and specific taxa was tested, and the ability of currently used identification criteria to associate strains into monophyletic groups was examined.

MATERIALS AND METHODS

The culture collection accession numbers, GenBank sequence numbers, isolation data, and identification of the fungal isolates examined are presented in Table 1. These isolates are maintained as lyophilized preparations in the NRRL Culture Collection Agricultural Research Service, Peoria, Illiunois, USA.

Table 1. Species and authority names of fungal isolates used in this study, NRRL accession number, equivalent strain numbers for other collections, origin of the strains and GenBank accession numbers for the rDNA sequences.

Aspergillus crystallinus Kwon & Fennell 5082. Ex type, isolated from soil, Costa Rica, 1965. ATCC 16833, CBS 479.65, IMI 139270. GB# AF033346.

Aspergillus malodoratus Kwon & Fennell 5083. Ex type, isolated from soil, Costa Rica, 1965. ATCC 16834, CBS 490.65, IMI 172289. GB # AF033485.

Eupenicillium alutaceum D. B. Scott 5812. Isolated from soil, South Africa, 1967. ATCC 18542, CBS 317.67, IMI 136243. GB # AF033454.

Eupenicillium anatolicum Stolk 5820. Isolated from soil, South Africa, 1967. CBS 467.67. GB # AF033425.

Eupenicillium baarnense (Beyma) Stolk & D. B. Scott 2086. Isolated from soil, The Netherlands, 1941. ATCC 10415, CBS 134.41, IMI 40590. GB # AF033481.

Eupenicillium brefeldianum (B. Dodge) Stolk & D. B. Scott 710 Clinical isolate, 1932. CBS 235.81, IMI 216896. GB # AF033435.

Eupenicillium cinnamopurpureum D. B. Scott & Stolk 3326. From India, 1968. ATCC 18337, CBS 490.66, IMI 114483. GB # AF033414.

Eupenicillium crustaceum F. Ludwig 3332. Contaminant, isolated in U.K., 1968. ATCC 18240, CBS 344.61, IMI 86561. GB # AF033466.

Eupenicillium egyptiacum (Beyma) Stolk & D. B. Scott 2090. Isolated from soil, Egypt, 1946. ATCC 10441, CBS 244.32, IMI 40580. GB # AF033467.

Eupenicillium ehrlichii (Klebahn) Stolk & D. B. Scott 708. Ex type. CBS 324.48, ATCC 10442, IMI 39737. GB # AF033432.

Eupenicillium erubescens D. B. Scott 6223. Isolated from soil, South Africa, 1969. ATCC 18544, CBS 318.67, IMI 136204. GB # AF033464.

Eupenicillium hirayamae D. B. Scott & Stolk 143. Isolated from milled Thai rice, 1960. ATCC 18312, CBS 229.60, IMI 78255. GB # AF033418.

Eupenicillium inusitatum D. B. Scott 5810. Isolated from soil, South Africa, 1969. ATCC 18622, CBS 351.67, IMI 136214. GB # AF033431.

Eupenicillium katangense Stolk 5182. Isolated from soil, Zaire, 1970. ATCC 18388, CBS 247.67, IMI 135206. GB # AF033458.

Eupenicillium lapidosum D. B. Scott & Stolk 718. Isolated from spoiled, canned blueberries, USA, 1929. ATCC 10462, CBS 343.48, IMI 39743. GB # AF033409.

Eupenicillium lassenii Paden 5272. Isolated from soil, California, 1970. ATCC 22054, CBS 277.70, IMI 48395. GB # AF033430.

Eupenicillium levitum (Raper & D. I. Fennell) Stolk & D. B. Scott 705. Isolated from modeling clay, USA, 1935. ATCC 10464, CBS 345.58, IMI 39735. GB # AF033436.

Eupenicillium meridianum D. B. Scott 5814. Isolated from soil, South Africa, 1967. ATCC 18545, CBS 314.67, IMI 136209. GB # AF033451.

Eupenicillium parvum (Raper & D. I. Fennell) Stolk & D. B. Scott 2095. Isolated from soil, Nicaragua, 1945. ATCC 10479, CBS 359.48, IMI 40587. GB # AF033460.

Eupenicillium pinetorum Stolk 3008. Isolated from forest soil, Wisconsin, 1962. ATCC 14770, CBS 295.62, IMI 94209. GB # AF033411.

Eupenicillium reticulisporum Udagawa 3447. Isolated from soil, Japan, 1968. ATCC 18566, CBS 122.68, IMI 136700. GB # AF033437.

Eupenicillium rubidurum Udagawa & Horie 6033. Isolated from soil, Papua, 1973. ATCC 28051, CBS 609.73, IMI 228551. GB # AF033462.

Eupenicillium shearii Stolk & D. B. Scott 715. Ex type. GB # AF033420.

Eupenicillium stolkiae D. B. Scott 5816. Isolated from soil, South Africa, 1969. ATCC 18546, CBS 315.67, IMI 136210. GB # AF033444.

Eupenicillium terrenum D. B. Scott 5824. Isolated from forest soil, Luzon, Philippines, 1973. ATCC 48205, CBS 622.72. GB # AF033446.

Eupenicillium tularense Paden 5273. Isolated from soil, California, 1971. ATCC 22056, CBS 430.69, IMI 148394. GB # AF033487.

Geosmithia argillacea (Stolk, Evans & Nilsson) Pitt 5177. Ex type, isolated from mine tips, U. K., 1967. CBS 101.69, IMI 156096. GB # AF033389.

Geosmithia cylindrospora (G. Smith) Pitt 2673. Ex type, isolated as culture contaminant, U.K., 1958. ATCC 18223, CBS 275.58, IMI 71623. GB # AF033386.

Geosmithia emersonii (Stolk) Pitt 3221. Ex type, isolated from compost, Italy, 1966. ATCC 16479, CBS 393.64, IMI 116815. GB # AF033387.

Geosmithia lavendula (Raper & D. I. Fennell) Pitt 2146. Ex type, isolated as culture contaminant, USA, 1948. ATCC 10463, CBS 344.48, IMI 40570. GB # AF033385.

Geosmithia putterillii (Thom) Pitt 2024. Ex neotype, isolated from discoloured wood, NZ, 1946. ATCC 10487, CBS 233.38, IMI 40212. GB # AF033384.

Geosmithia swiftii Pitt & A. D. Hocking 1025. Ex type, isolated from *Begonia* leaf, New York, 1932. ATCC 10126, CBS 296.48, IMI 40045. GB # AF033388.

Hemicarpenteles paradoxus Sarbhoy & Elphick 2162. Ex type of *Aspergillus paradoxus* Fennell & Raper. Isolated by J. H. Warcup, from opossum dung, N.Z. GB # AF033484.

Monascus purpureus Tiegh. 1596. Ex type, received from Harvard University, 1940. GB # AF033394.

Paecilomyces varioti Bainier 1115. Ex type of *P. divaricatum,* isolated from mucilage glue, 1904. ATCC 10121, CBS 284.48, IMI 40025. GB # AF033395.

Penicilliopsis clavariiformis Solms-Laubach 2482. Received from CBS, 1955. GB # AF033391.

Penicillium aculeatum Raper & D. I. Fennell 2129. Ex type, isolated from square biscuits, Florida,

1946. ATCC 10409, CBS 289.48, IMI 40588. GB # AF033397.

Penicillium adametzii Zaleski 737. Ex type, isolated from forest soil, Poland, 1928. ATCC 10407, CBS 209.28, IMI 39751. GB # AF033401.

Penicillium adametzioides Abe ex G. Smith 3405. Ex type, from Japan, 1968. ATCC 18306, CBS 313.59, IMI 68227. GB # AF033403.

Penicillium asperosporum G. Smith 3411. Ex type, isolated from wood pulp, U.K., 1969. ATCC 18319, CBS 293.63, IMI 80450. GB # AF033412.

Penicillium atramentosum Thom 795. Ex type, isolated from French camembert cheese, 1924. ATCC 10104, CBS 291.48, IMI 39752. GB # AF033483.

Penicillium atrovenetum G. Smith 2571. Ex type, isolated from soil, U.K., 1956.. ATCC 13352, CBS 241.56, IMI 61837. GB # AF033492.

Penicillium aurantiogriseum Dierckx 971. Ex neotype, ATCC 48920, IMI 195050. GB # AF033476.

Penicillium bilaiae Chalabuda 3391. Ex type, isolated from soil, Kiev, Ukraine, 1968. ATCC 48731, CBS 221.66, IMI 113677. GB # AF033402.

Penicillium camemberti Thom 874. Ex type, isolated from camembert cheese, USA, 1904. ATCC 10837, CBS 123.08, IMI 91932. GB # AF034453. 875. Ex type of *P. caseicolum.* Isolated from camembert cheese, 1922. ATCC 10423, CBS 303.48, IMI 28810. GB # AF033474.

Penicillium canescens Sopp 910. Ex neotype, isolated from soil, U.K., 1912. ATCC 10419, CBS 300.48, IMI 28260. GB # AF033493.

Penicillium capsulatum Raper & D. I. Fennell 2056. Ex type, isolated from a camera lens, Panama, 1945. ATCC 10420, CBS 301.48, IMI 40576. GB # AF033429.

Penicillium charlesii Zaleski 778. Ex type, received from G. Smith, 1932. ATCC 8730, CBS 342.51, IMI 40232. GB # AF033400.

Penicillium chermesinum Biourge 735. Received from Biourge, 1924. GB # AF033413.

Penicillium chrysogenum C. Thom 807. Ex type, isolated from cheese, Connecticut, 1904. ATCC 10106, CBS 306.48, IMI 24314. GB # AF033465. 820. Ex type of *P. griseoroseum.* Received from Biourge, 1924. ATCC 48653, IMI 9220. GB # AF033 821. Ex type of *P. notatum.* Received from Westling, 1911. GB # AF034451. 824. Received from A. Fleming, U.K., 1930. ATCC 8537, CBS 205.57, IMI 15378. GB # AF034450. 832. Received from Biourge, 1936. ATCC 7813, CBS 197.46, IMI 17968. GB # AF034449.

165

Penicillium cinerascens Biourge 748. Received from Biourge, 1924. ATCC 48693, IMI 92234. GB # AF033455.

Penicillium citreonigrum Dierckx 761. Ex neotype, received from Biourge, 1924, as type of *P. subcinereum*. ATCC 48736, CBS 258.29, IMI 92209. GB # AF033456.

Penicillium citrinum Thom 1841. Received from G. Smith, U.K., 1929. Type strain of *P. aurifluum*. ATCC 36382, CBS 139.45, IMI 91961. GB # AF033422.

Penicillium coprophilum (Berkeley & M. A. Curtis) Seifert & Samson 13627. Isolated from sorghum, Africa, 1988. GB # AF033469.

Penicillium corylophilum Dierckx 802. Ex neotype, received from Biourge, 1924. ATCC 9784, CBS 312.48, IMI 39754. GB # AF033450.

Penicillium crustosum Thom 968. Isolated from rice, China, 1934. ATCC 10430, CBS 313.48, IMI 39811. GB # AF033472.

Penicillium cyaneum (Bainier & Sartory) Biourge 775. Ex type, ATCC 10432, CBS 315.48, IMI 39744. GB # AF033427.

Penicillium daleae Zaleski 922. Ex type of *P. krzemieniewskii*, isolated from pine forest soil, Poland, 1928. CBS 212.28. GB # AF033442.

Penicillium decumbens Thom 741. Ex type, received from Biourge, 1924. ATCC 48470, CBS 230.81, IMI 190875. GB # AF033453.

Penicillium digitatum (Persoon:Fries) Saccardo 786. Received from L. B. Lockwood, Arlington Farm, Maryland, 1940. GB # AF033471.

Penicillium donkii Stolk 5562. Ex type, isolated from soil, Alaska, 1973. ATCC 48439, CBS 188.72, IMI 197489. GB # AF033445.

Penicillium echinulatum Raper & Thom ex Fassatiova 1151. Ex type, isolated from contaminated Petri-dish by G. A. Ledingham, U.K. GB # AF033473.

Penicillium expansum Link 974. Received from da Fonseca, 1922. GB # AF033479.

Penicillium fellutanum Biourge 746. Ex neotype, isolated in Massachusetts, 1935. ATCC 10443, CBS 326.48, IMI 39734. GB # AF033399.

Penicillium fuscum (Sopp) Raper & Thom 721. Ex type, received from M. B. Morrow. Isolated from soil, Texas, 1930. ATCC 10447, CBS 331.48, IMI 39747. GB # AF033443.

Penicillium glabrum (Wehmer) Westling 766. Ex type, received from Biourge, 1924, as *P. aurantiobrunneum*. ATCC 10103, IMI 92195. GB # AF033407.

Penicillium gladioli McCulloch & Thom 939. Ex type, isolated from *Gladiolus* sp. corms, 1926. ATCC 10448, CBS 332.48, IMI 34911. GB # AF033480.

Penicillium griseofulvum Dierckx 2300. Ex neotype, isolated 1949. ATCC 11885, CBS 185.27, IMI 75832. GB # AF033468.

Penicillium herquei Bainier & Sartory 1040. Ex type, isolated from *Agauria* leaf, France, 1922. ATCC 10118, CBS 336.48, IMI 28809. GB # AF033405.

Penicillium implicatum Biourge 2061. Ex neotype, isolated from soil, India, 1944. ATCC 48445, CBS 184.81, IMI 190235. GB # AF033428.

Penicillium inflatum Stolk & Malla 5179. Ex type, isolated from spruce root, Denmark, 1970. ATCC 48994, CNS 682.70, IMI 191498. GB # AF033393.

Penicillium janthinellum Biourge 2016. Ex neotype, isolated from soil, Nicaragua, 1945. ATCC 10455, CBS 340.48, IMI 40238. GB # AF033434.

Penicillium kojigenum Abe ex Smith 3442. Ex type, isolated from soil, Scotland, 1969. ATCC 18227, CBS 345.61, IMI 86562. GB # AF033489.

Penicillium lividum Westling 754. Ex neotype. ATCC 10102, CBS 347.48, IMI 39736. GB # AF033406.

Penicillium madriti G. Smith 3452. Ex type, isolated from garden soil, Spain, 1961. ATCC 18233, CBS 347.61, IMI 86563. GB # AF033482.

Penicillium megasporum Orput & D. I. Fennell 2232. Ex type, isolated from soil, U.K., 1950. ATCC 12322, CBS 256.55, IMI 216904. GB # AF033494.

Penicillium melinii C. Thom 2041. Ex type, isolated from forest soil, USA, 1946. GB # AF033449.

Penicillium miczynskii Zaleski 1077. Ex type, isolated from conifer forest soil, Poland, 1928. ATCC 10470, CBS 220.28, IMI 40030. GB # AF033416.

Penicillium namyslowskii Zaleski 1070. Ex type, isolated from soil in pine forest, Poland, 1928. ATCC 11127, CBS 353.48, IMI 40033. GB # AF033463.

Penicillium ochrochloron Biourge 926. Ex neotype, isolated from 2% copper sulfate solution, Washington, 1940. ATCC 10540, CBS 357.48, IMI 39806. GB # AF033441.

Penicillium oxalicum Currie & C. Thom 787. Ex type, isolated from soil, Connecticut, 1914. ATCC 1126, CBS 219.30, IMI 192332. GB # AF033438.

Penicillium paxilli Bainier 2008. Ex neotype, isolated from photographic film, Panama, 1945. ATCC 10480, CBS 360.48, IMI 40226. GB # AF033426.

Penicillium polonicum 995. Received from Westerdijk, 1928. CBS 222.28. GB # AF033475.

Penicillium purpurescens (Sopp) Biourge 720. Ex neotype, isolated from soil, Canada, 1932.

ATCC 10485, CBS 366.48, IMI 39745. GB # AF033408.

Penicillium raciborskii Zaleski 2150. Ex type, isolated from soil, Poland, 1946. ATCC 10488, CBS 224.28, IMI 40568. GB # AF033447.

Penicillium raistrickii G. Smith 2039. Ex type, isolated from cotton yarn, U.K., 1945. GB # AF033491.

Penicillium raperi G. Smith 2674. Ex type, isolated from soil, U.K., 1958. ATCC 22355, CBS 281.58, IMI 71625. GB # AF033433.

Penicillium resedanum McLennan & Ducker 578. Ex type, isolated from soil, Australia. ATCC 22356, CBS 181.71, IMI 62877. GB # AF033398.

Penicillium restrictum Gilman & Abbott 1748. Ex neotype, isolated from soil, Honduras, 1940. ATCC 11257, CBS 367.48, IMI 40228. GB # AF033457. 25744. Isolated from corn-field soil near Kilbourne, IL, 1995, by D. T. Wicklow. GB # AF033459.

Penicillium rivolii Zaleski 906. Ex type, isolated from pine forest soil, Poland, 1928. ATCC 48699. GB # AF033419.

Penicillium rolfsii C. Thom 1078. Ex type, isolated from pineapple, Florida, 1905. ATCC 10491, CBS 368.48, IMI 40029. GB # AF033439.

Penicillium roseopurpureum Dierckx 2064. Ex neotype. ATCC 10492, CBS 266.29, IMI 40573. GB # AF033415.

Penicillium sartoryi Thom 783. Ex type. Received from Lewis, University of Texas, 1929. GB # AF033421.

Penicillium sclerotigenum K. Yamamoto 3461. Ex type, isolated from sweet potato, Japan, 1959. ATCC 18488, CBS 343.59, IMI 68616. GB # AF033470.

Penicillium sclerotiorum Beyma 2074. Received from CBS, 1945. Ex type, isolated from air, Indonesia. ATCC 10494, CBS 287.36, IMI 40569. GB # AF033404.

Penicillium simplicissimum (Oudem.) Thom 1075. Ex type of *P. piscarum,* chosen by Pitt as ex neotype of *P. simplicissimum.* Isolated from cod liver oil emulsion, 1911. ATCC 10482, CBS 362.48, IMI 40032. GB # AF033440.

Penicillium soppii Zaleski 2023. Ex type, isolated from woodland soil, Poland, 1946. ATCC 10496, CBS 226.28, IMI 40217. GB # AF033488.

Penicillium spinulosum Thom 1750. Ex type, isolated as culture contaminant, Germany. ATCC 10498, CBS 374.48, IMI 24316. GB # AF033410.

Penicillium sumatrense van Szilvinyi 779. Ex type. Received from Westerdijk, Baarn, The Netherlands, 1936. GB # AF033424.

Penicillium swiecickii Zaleski 918. Ex type, isolated from pine forest soil, Poland, 1928. IMI 191500. GB # AF033490.

Penicillium thomii Maire 2077. Ex neotype, isolated from a pine cone, USA. ATCC 48218, IMI 189694. GB # AF034448.

Penicillium turbatum Westling 757. Ex type, isolated from yew leaf. ATCC 9782, CBS 237.60, IMI 39738. GB # AF034454. 759. Isolated from mouldy soap, Chicago, Illinois, 1929. GB # AF033452.

Penicillium velutinum Beyma 2069. Ex type, isolated from sputum, Netherlands. ATCC 10510, CBS 250.32, IMI 40571. GB # AF033448.

Penicillium vinaceum Gilman & Abbott 739. Ex type, isolated from soil, Utah, 1927. ATCC 10514, CBS 389.48, IMI 29189. GB # AF033461.

Penicillium viridicatum Westling 958. Isolated in Washington, 1927. ATCC 46511. GB # AF033477. 961. Received from C. J. Carrera, Argentina, 1940. ATCC 48916, IMI 192905. GB # AF033478. 5880. Isolated in South Africa. GB # AF033390.

Penicillium waksmanii Zaleski 777. Ex type, isolated from woodland soil, Poland, 1928. ATCC 10516, CBS 230.28, IMI 39746. GB # AF033417.

Penicillium westlingii Zaleski 800. Received from R. Thaxter, Massachusetts, 1917. GB # AF033423.

Sclerocleista ornata (Raper & al.) Subramanian 2256. Ex type, isolated from soil, Wisconsin, 1951. ATCC 16921, CBS 124.53, IMI 55295. GB# AF033392.

Talaromyces helicus (Raper & D. I. Fennell) C. R. Benjamin 2106. Ex type, isolated from soil, Sweden, 1945. ATCC 10451, CBS 335.48, IMI 40593. GB # AF033396.

DNA isolation: Fungal isolates were revived from lyophilized storage and grown on malt extract agar (Raper and Fennell, 1965) slants for seven days. One ml of sterile 0.1% Triton X-100 was added to each slant, and dislodged conidia were pipetted into a 500-ml Erlenmeyer flask containing 100 ml of malt extract broth (Raper and Fennell, 1965) and incubated on a rotary shaker platform (200 rpm, 25°C) for 36-48 h, until 1-2 g of biomass was produced. Biomass was harvested by filtration over cheesecloth. Cells (0.2 g) were suspended in 3 ml of breaking buffer (100 mM Tris, 50 mM EDTA, 1% sarcosyl, pH 8.0) in a 15-ml screw-cap disposable centrifuge tube, and 1.5 g of glass beads (0.5 mm diam) was added. Cell walls were broken by vortexing for 30-45 seconds. Phenol:chloroform (1:1, w:v) was added to the tubes (3 ml) and an emulsion maintained by gentle rocking for 15 minutes. The aqueous and organic phases were separated by low speed centrifugation (ca. 2000 x g) for 5 minutes. The aqueous phase was transferred to a fresh tube, and 0.1 volume of 3 M sodium acetate pH 6.0 and 1.3 volumes 95% ethanol were added. The contents of the tube were mixed, and the nucleic acids pelleted by low speed centrifugation as above. Nucleic acids were dissolved in TE/10 buffer (1 mM Tris, 0.1 mM EDTA, pH 8.0) and further purified by adsorption to a silica matrix (Geneclean, Bio101, LaJolla, CA) according to the manufacturer's instructions. DNA was desorbed from the matrix in 500µl TE/10 and stored frozen (-20°C) until used for PCR amplification.

Table 2. Primers used for amplification and sequencing. The ITS primers are described in White *et al.* (1990), and the lsu primers are described in Peterson (1993).

ITS 1 5'-TCCGTAGGTGAACCTGCGG
ITS 2 5'-GCTGCGTTCTTCATCGATGC
ITS 3 5'-GCATCGATGAAGAACGCAGC
ITS 4 5'-TCCTCCGCTTATTGATATGC
D1 5'-GCATATCAATAAGCGGAGGA
D1R 5'-ACTCTCTTTTCAAAGTGCTTTTC
D2 5'-GAAAAGCACTTTGAAAAGAGAGT
D2R 5'-AACCAGGCACAAAGTTCTGC

PCR amplification and sequencing: A *ca.*1200 nucleotide fragment of DNA that included regions ITS1 and ITS2, the 5.8S rDNA, and about 635 bases from the 5' end of the large subunit (lsu) rDNA was amplified from the genomic DNA using PCR. The PCR amplification tubes contained 5 µl of the genomic DNA preparation, 10 µl of 10X buffer (White *et al.*, 1990), 1 µl of deoxynucleotide triphosphate mix (each dNTP present in a master mix at 1 mM concentration), 1 µl of primer ITS1 (50 µM solution), 1 µl of primer D2R (50 µM solution; Table 2), 0.5 µl of Taq polymerase (5 U per µl) and 81.5 µl of sterile distilled water. The solution was overlaid with *ca.* 50 µl of mineral oil and amplified during 30 thermal cycles of 96°C, 30 sec; 51°C, 30 sec; 72°C, 150 sec, followed by 10 min at 72°C. The amplified DNA fragment was purified by adsorption to a silica matrix (Geneclean), desorbed into TE/10, and stored frozen (-20EC) until used in sequencing reactions. DNA sequences were determined using Applied Biosystems DyeDeoxy sequencing kits (fluorescent labeling, Taq polymerase) and the ABI 373 DNA sequencer. Sequence reaction conditions were those recommended by the manufacturer, using primers ITS1, ITS2, ITS3 and ITS4 and primers D1, D1R, D2 and D2R (Table 2). These primers allow sequencing of both strands of the entire fragment. The DNA sequences of the isolates were aligned using ClustalW (Higgins & Sharp, 1988, 1989) and an ASCII text editor and compared using programs in PAUP 3.1.1 (Swofford, 1993) and from the PHYLIP package (Felsenstein, 1993).

RESULTS

The contiguous aligned segment sequenced contained 1241 nucleotide positions, with 678 positions constant, 99 positions variable but parsimony uninformative, and 464 positions were parsimony informative. Three subsets of the segment were also made into data sets, these were the ITS1, ITS2 regions, and 5.8S plus lsu-rDNA. The ITS1 data set contained 232 aligned nucleotide positions: 72 were constant, 30 were uninformative, and 130 were parsimony informative. The ITS2 data set contained 205 nucleotide positions, of which 41 were constant, 30 were uninformative and 134 were parsimony informative. Each data set was analyzed using the neighbor joining method in 1000 bootstrap samples. The bootstrap trees were compared for congruence, considering only branches with bootstrap values higher than 70% (Hillis and Bull, 1993; Felsenstein and Kishino, 1993). Sixteen strongly supported nodes were common to all of the trees. One of those nodes had a different arrangement of species in the ITS2 tree than it had in the ITS1 and lsu-5.8s rDNA trees. The bootstrap values of the ITS2 tree at the contradictory branch was 71. No other nodes in any of the trees were contradictory. Bootstrap analysis randomly modifies data sets, and at bootstrap values near 70%, the probability that random errors account for the particular branch in the tree is 5%. If 20 common tree branches had been found, it would be expected that one of them would be incorrect based on the probability of error. Because of congruence of the data when analyzed and compared as data subsets, further analysis was based on the complete data set.

The ITS regions have been used to analyze closely related species in many experiments and intraspecific variation has been characterized as low (e.g., Kuhls *et al.*, 1997). Peterson *et al.* (2000) examined nucleotide sequence variation in *Aspergillus tamarii* and in *A. caelatus*. They found a single base substitution, in the ITS 2 region, among 153 isolates of *A. tamarii* collected at different locations around the world. Among 41 strains of *A. caelatus* collected in North America, South America and Asia, 39 strains had identical ITS and lsu-rDNA sequences, while two strains showed two base substitutions in the ITS 2 region. Therefore, for the purpose of interpreting this set of data, strains with identical lsu-rDNA and ITS sequences will be considered synonymous. Species with 1 or 2 substitutions in the ITS regions will be treated as strains of unknown relationship, that is, they might be conspecific or might represent very closely related species. Additional data (e.g., DNA complementarity measurements, isozyme studies, nucleotide sequences of additional genes and patterns of heteromorphism in those genes) is needed in order to establish conspecificity of those strains. When strains differ by 5 or more ITS substitutions, they represent different species.

In a tree generated from a heuristic search using PAUP 3.1.1 that included *Aspergillus* and *Penicillium* species, the majority of *Penicillium* species formed a single monophyletic branch (74% bootstrap value) (Peterson, 2000). This also includes all species of *Eupenicillium*, *Aspergillus crystallinus*, *A. malodoratus* and *Hemicarpenteles paradoxus*. The species of *Penicillium* studied, which are not on this branch are *P. resedanum, P. aculeatum, Talaromyces helicus* and *P. inflatum. Talaromyces helicus* and *P. aculeatum* are representative of the *Talaromyces* clade. *P. resedanum* was placed in subgenus *Aspergilloides* in recent monographs (Pitt, 1979; Ramirez, 1982) although McLennan *et al.* (1954) and Pitt (1979) noted that the yellow hyphae and acerose phialides indicated accommodation in subgenus *Biverticillium*. This is confirmed here. *P. inflatum* makes a very small, *Aspergillus*-like penicillus which is readily recognizable (Pitt, 1979) and was placed in series *Citrina* by Pitt (1979). However, he expressed reservations, while Ramirez (1982) placed it in series *P. nigricans*. The data presented in this study are unable to satisfactorily place the species, but can exclude it from phylogenetic association with either classification.

The species of *Geosmithia* included in this study form two clades that have high bootstrap values. *G. lavendula* and *G. putterillii* are outside the Trichocomaceae (Fig. 1) and are most closely related to species near *Nectria* (Ogawa *et al.*, 1997). *G. namyslowskii* is phylogenetically part of *Penicillium* and was originally described as a *Penicillium* species. It is proposed to maintain the name *Penicillium namyslowskii* Zaleski. The species is most closely related to *P. rolfsii* in subgenus *Furcatum*. The remainder of the *Geosmithia* species form a monophyletic group (80% bootstrap value) along with *T. helicus*, *P. aculeatum* and *P. resedanum*. This second group includes *G. swiftii*, *G. emersonii*, *G. cylindrospora*, and *G. argillacea*.

Other *Penicillium* species formed seven branches that occurred in 86-100% of the bootstrap samples. One group (Fig 1A) including *Eupenicillium* species and taxa from subgenera *Aspergilloides* and *Furcatum*, has 95% bootstrap support. In this clade, *P. citrinum*, *P. sartoryi*, *P. westlingii*, *P. sumatrense*, *P. paxilli* and *P. miczynskii* are classified in subgenus *Furcatum*, section *Furcatum*, series *Citrina*, while *P. waksmanii* and *P. rivolii* are classified in subgenus *Furcatum*, section *Divaricata*, series *Fellutana*. *P. roscopurpureum* is classified in subgenus *Aspergilloides*, section *Exilicaulis*, series *Exilicaulis*. Because the species in this clade are classified in different subgeneric categories, and all of the species of any particular subgeneric group are not present in this clade, the subgenera and sections that these species belong to are not monophyletic.

P. sartoryi was accepted by Thom (1930) but synonymized by later authors under the concept of *P. citrinum*. These two species are closely related but have different sequences in the ITS1 and ITS2 regions (2 and 1 differences respectively), suggesting that they are distinct species, and the name *P. sartoryi* should be revived. *P. westlingi* was accepted by Thom (1930) but was placed in *P. waksmanii* by Raper and Thom (1949) and subsequent authors. Numerous (99 total) DNA sequence differences between the sequences of *P. waksmanii* and *P. westlingi* demonstrate that the two taxa are distinct. *P. rivolii* also differs from *P. waksmanii* at two positions in the ITS2 region. These differences suggest that both *P. westlingii* and *P. rivolii* are distinct species. *P. sumatrense* was regarded as being a synonym of *P. corylophilum* by Raper and Thom (1949), but the DNA sequence of this species differs from *P. corylophilum* at 83 nucleotide positions and *P. sumatrense* should be maintained.

Stolk and Samson (1983) reduced *E. anatolicum* to synonymy with *E. euglaucum* and *P. citreonigrum* as the anamorphic state. *E. anatolicum* is, however, phylogenetically distinct from *E. euglaucum*, and is not closely related to *P. citreonigrum* (Fig. 1A). These authors also assigned *P. soppii* as the anamorph of *E. shearii*. In the phylogenetic tree, these two species are in distinct clades showing that this is not correct.

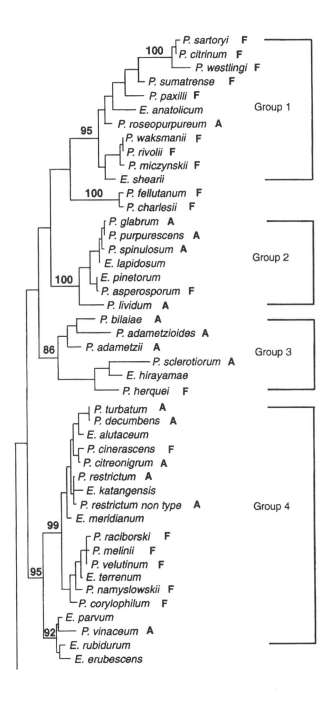

Fig 1 A

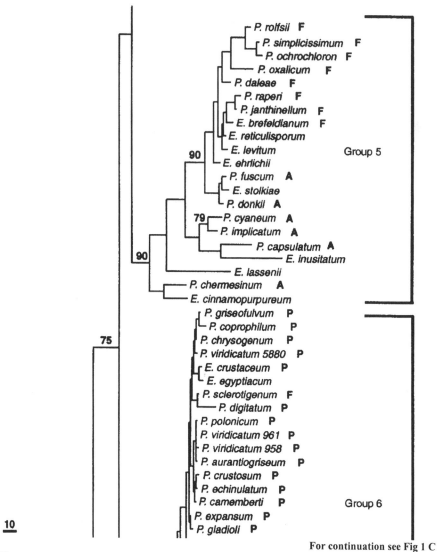

Fig 1B

For continuation see **Fig 1 C**

Figure 1 A-C. Phylogram representing more than 2000 equally parsimonious trees generated in a heuristic search of the complete ITS-lsu data set using PAUP 3.1.1 and random addition order. Each taxon, except *Geosmithia* species, is followed by a bold capital letter indicating what subgenus it is classified in Pitt (1979). (A = *Aspergilloides*, F = *Furcatum*, P = *Penicillium*) Branch lengths in the tree are proportional to the number of steps between nodes. Bootstrap values based on 100 replicates and heuristic search with random addition order (PAUP 3.1.1) are placed on some of the branches. Bootstrap values less than 70% are not included. The groups indicated on the tree refer to discussion in the text and are placed on the tree only to facilitate the discussion. None of the subgenera of *Penicillium* (Pitt, 1979) are monophyletic. All of the species from sub-genus *Penicillium*, along with some species from subgenus *Furcatum* occur on a single branch with 100% bootstrap value. Species from subgenera *Aspergilloides* and *Furcatum* are interspersed on the branches of the tree along with species of *Eupenicillium*.

172

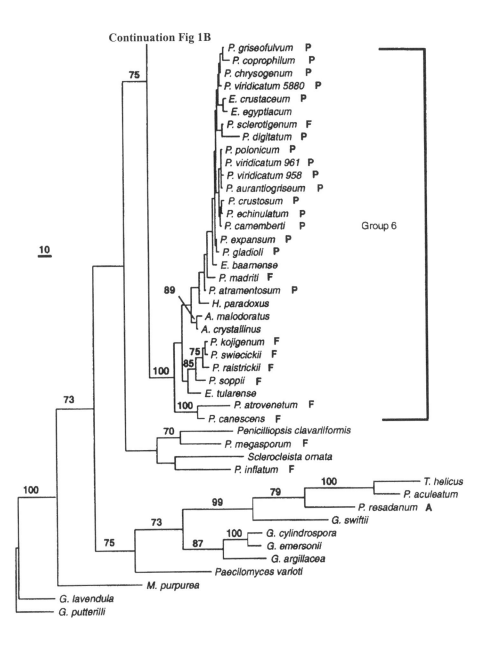

Fig 1C

A second clade (100% bootstrap value) contains primarily taxa that have been assigned to subgenus *Aspergilloides* (Fig 1A). However in addition to the monoverticillate species, *P. asperosporum* from subgenus *Furcatum* is included in the group. Frisvad and Filtenborg (1990) placed *P. asperosporum* in subgenus *Biverticillium* on the basis of secondary metabolite production, but phylogenetically it does not belong in *Biverticillium*. *Eupenicillium lapidosum* forms a terminal group with *P. glabrum*, *P. spinulosum*, and *P. purpurescens*. The sequence of *P. thomii* is identical with that of *E. lapidosum* indicating that is the anamorph of this species.

Stolk and Samson (1983) regarded *E. pinetorum* and *P asperosporum* as synonyms. They are closely related and *E. pinetorum* forms a terminal branch with *P. asperosporum*. Both taxa differ from each other at a single nucleotide position and form a monophyletic group. As discussed earlier, a single base difference may represent intraspecific variation, but additional data are needed to prove the relationship of these taxa.

Taxonomic experts were mostly able to distinguish the four species *P. spinulosum*, *P. glabrum*, *P. montanense*, and *P. purpurescens* (Pitt *et al.*, 1990). The rDNA sequence data suggest that these are closely related but distinct species, displaying 4-8 nucleotide substitutions between each of the pairings. *P. thomii* and *Eupenicillium lapidosum* also fit in this monophyletic group. The first four species mentioned above show DNA sequence differences, but show very few morphological differences, while *P. thomii* and *E. lapidosum* are morphologically distinct, but have evolved no DNA sequence differences.

A third group (86% bootstrap value) contains species from three series of subgenus *Aspergilloides*, and *P. herquei* from subgenus *Furcatum*. *Eupenicillium hirayamae* branches with *P. sclerotiorum*, while *P. adametezioides*, *P. adametzii* and *P. bilaiae* make up the rest of the branch. Stolk and Samson (1983) reduced *E. hirayamae*, *E. lassenii*, *E. anatolicum* and *E. katangense* to synonymy under *E. euglaucum*, with the anamorph *P. citreonigrum*. The four teleomorph species however, are distinct species in the phylogram and are diversely spread among other *Penicillium* and *Eupenicillium* species. Because the *Eupenicillium* species reduced to synonymy displayed variation in their anamorphs, Stolk and Samson (1983) broadened their concept of *P. citreonigrum* to include the variations seen in those anamorphic states. The broadening of that species concept is here rejected. *P. citreonigrum* occurs in a different clade than *E. hirayamae* and shows no close relationship to any particular species of *Eupenicillium*.

A fourth cluster of species (95% bootstrap support) includes a mixture of species from *Eupenicillium*, subgenus *Aspergilloides* and subgenus *Furcatum*. *E. alutaceum* is most closely related to *P. decumbens* and *P. turbatum*, which differ at a single nucleotide position, but are distinguishable morphologically. The species are here maintained until more data are available on their possible synonymy. *E. katangense* and *E. meridianum* are most closely related to strains of *P. restrictum*. *E. rubidurum*, *E. erubescens* and *E. parvum* form a clade (bootstrap value 92%) along with *P. vinaceum*.

Geosmithia namyslowskii occurs in this clade near *P. corylophilum*. Ogawa *et al.* (1997) places *G. namyslowskii* close to *E. crustaceum*, but did not include a large number of species from subgenera *Furcatum* or *Penicillium* in their study. As mentioned above, the name *P. namyslowskii* Zaleski should be taken up again for this species. No other anamorphic species are closely related.

A fifth cluster (100% bootstrap value) includes the terverticillate species placed in subgenus *Penicillium*, some species of *Eupenicillium*, a few species assigned to subgenus *Furcatum*, and *Aspergillus malodoratus*, *A. crystallinus* and *Hemicarpenteles paradoxus*

174

(Fig. 1B). The *Aspergillus* species form aspergilli when cultures are young, but with age they produce structures that Raper and Fennell (1965) equated with terverticillate penicilli. The rDNA sequence data show that the true phylogenetic placement of these species is in subgenus *Penicillium*. *H. paradoxus* is probably indistinguishable from that of *Eupenicillium* (Samson, 1979) but originally described with an *Aspergillus* anamorph, but phylogenetically it belongs in subgenus *Penicillium*.

Also included in the cluster of subgenus *Penicillium* clade are *P. canescens, P. atrovenetum, P. soppii, P. raistrickii, P. swiecickii* and *P. kojigenum*. These species are usually placed in subgenus *Furcatum*, but the association of these species with the terverticillate species has support in 100% of the bootstrap samples. The synonymy of *P. chrysogenum* and *P. griseoroseum* has been generally accepted (Pitt and Cruickshank, 1990), even though those species were originally placed in subgenera *Penicillium* and *Furcatum*, respectively (Pitt, 1979).

Among the many agriculturally important toxin forming species described in subgenus *Penicillium*, there is very little genetic differentiation (Peterson, 1993). Frisvad and Filtenborg (1989) distinguished 38 taxa in subgenus *Penicillium* on the basis of secondary metabolite profiles. Some of their 38 taxa are chemotypes that are indistinguishable by other means (e.g., phenotypic analysis). In the phylogram (Fig. 1B), there are very few substitutions between many of the taxa in the subgenus *Penicillium* clade and some of the species are indistinguishable on the basis of rDNA data (Peterson, 1993). The data analyzed in the current study suggest that the chemotypes defined by Frisvad and Filtenborg (1989) are variant forms rather than distinct species, but additional data from different genes will allow a definitive answer.

Subgenus *Penicillium* has been thoroughly studied because of its impact on human culture. The species from other subgenera are not as important economically and therefore have been less intensely studied. The branches they fall on are also less densely packed with species. Among the strains examined in this study are three isolates identified as *P. fellutanum* by Pitt (1979). Each of these isolates has a distinct rDNA sequence and these strains are placed in two separate clades of the phylogenetic tree as the species *P. fellutanum, P. charlesii* and *P. cinerascens*. This reinforces the evidence presented above showing that the complexity of the penicillus is not an optimal indicator of phylogenetic relatedness, and shows that the latter two species should be revived as species distinct from *P. fellutanum*.

P. kojigenum and *P. rivolii* were considered synonymous with *P. jensenii* by Pitt (1979), but *P. kojigenum* is in the clade with species from subgenus *Penicillium*, while *P. rivolii* is closely related to *P. miczynskii* and P. *waksmanii*. Some species from sections *Furcatum* and *Aspergilloides* that are treated as synonyms on the basis of phenotypic data are genetically distinct species. It will be valuable to our understanding of *Penicillium* to see whether more intensive sequencing reveals that these other subgenera are as rich in species diversity as subgenus *Penicillium*.

The sixth cluster (bootstrap value 90%) includes species from subgenus *Furcatum* but also some from subgenus *Aspergilloides* and *Eupenicillium*. Stolk and Samson (1983) felt that the anamorph of *E. stolkiae* was *P. velutinum*. In the phylogenetic tree, these two species are not closely related, rendering Stolk and Samson's (1983) hypothesis untenable. *E. brefeldianum* and *E. ehrlichii* were reduced to synonymy with *E. javanicum* by Stolk and Samson (1983), and *E. levitum* was reduced to varietal status. However each of these species is genetically distinct from other *Penicillium* and *Eupenicillium* species.

175

Pitt (1979) divided *Penicillium* the genus into four subgenera, *Penicillium*, *Furcatum*, *Aspergilloides* and *Biverticillium* and removed some species to the genus *Geosmithia*. The current data are sufficient to show that classification has polyphyletic taxa, but does not fully define the relationships among the taxa. Berbee *et al.* (1995) and Lobuglio *et al.* (1993, 1994) showed that species from Subgenus *Biverticillium* and *Talaromyces* species form a clade distinct from that of the other three subgenera of *Penicillium* and *Eupenicillium*. Lobuglio *et al.* (1983) further showed classification of subgenus *Biverticillium* as proposed by Pitt (1979) contained polyphyletic taxa.

The phylogenetic tree of *Penicillium*, including *Geosmithia* species shows that penicillus branching is a poor character for defining monophyletic taxa. For example, *P. chrysogenum* and *P. griseoroseum* are sister taxa, but were classified in subgenus *Penicillium* and *Furcatum* respectively by Pitt (1979). Similarly, *P. sclerotiorum* (*Furcatum*) and *P. digitatum* (*Penicillium*), and *P. madriti* (*Furcatum*) and *P. atramentosum* (*Penicillium*) are pairs of sister taxa that are placed in different subgenera. There are also sister taxa that derive from subgenera *Furcatum* and *Aspergilloides*, such as *P. citreonigrum* and *P. fellutanum*. *P. resedanum* (subgenus *Aspergilloides*) is phylogenetically part of the *Talaromyces* / subgenus Biverticillium clade.

When sufficient data are obtained to detail the relationships of all *Penicillium* species, a new subgeneric classification system will be formed that de-emphasizes penicillus complexity, and makes the taxa in the genus monophyletic.

ACKNOWLEDGMENTS

The author appreciates the skillful help of Paul A. Bonneau in accomplishing many technical aspects of this research. Larry Tjarks, Kristin Braun and Eleanor Basehoar-Powers participated in parts of this research project providing skilled technical assistance.

Names are necessary to report factually on available data; however, the USDA neither guarantees nor warrants the standard of the product, and the use of the name by USDA implies no approval of the product to the exclusion of others that may also be suitable.

REFERENCES

Abe, S. (1956). Studies on the classification of Penicillia. J. Gen. Appl. Microbiol. 2: 1-344.

Berbee, M. L., Yoshimura, A., Sugiyama, J. & Taylor, J. W. (1995). Is *Penicillium* monophyletic? An evaluation of phylogeny in the family Trichocomaceae from 18S, 5.8S and ITS ribosomal DNA sequence data. Mycologia 87: 210-222.

Bruns, T. D., White, T. J. & Taylor, J. W. (1991). Fungal molecular systematics. Ann. Rev. Ecol. Syst. 22: 525-564.

Felsenstein, J. (1993). PHYLIP (Phylogeny Inference Package) version 3.5c. Distributed by the author. Department of Genetics, University of Washington, Seattle.

Felsenstein, J. & Kishino, J. (1993). Is there something wrong with the bootstrap on phylogenies? A reply to Hillis and Bull. System. Biol. 42: 193-200.

Frisvad, J. C. (1981). Physiological criteria and mycotoxin production as aids in identification of common asymmetric Penicillia. Appl. Environ. Microbiol. 41: 568-579.

Frisvad, J. C. (1989). The connection between the Penicillia and Aspergilli and mycotoxins with special emphasis on misidentified isolates. Archives Environ. Contamination Toxicology 18: 452-467.

Frisvad, J. C. & Filtenborg, O. (1989). Terverticillate Penicillia:Chemotaxonomy and mycotoxin production. Mycologia 81: 837-861.

Frisvad, J. C. & Filtenborg, O. (1990). Revision of *Penicillium* subgenus *Furcatum* based on secondary metabolites and conventional characters. In: Modern concepts in *Penicillium* and *Aspergillus* classification (Eds Samson, R. A. & Pitt, J. I.). New York, Plenum Press. pp. 159-172 ..

Greuter, W., Barrie, F. R., Burdet, H. M., Chaloner, W. G., Demoulin, V., Hawksworth, D. L., Jorgensen, P. M., Nicolson, D. H., Silva, P. C., Trehane, P. & McNeil, J. (1994). International Code of Botanical Nomenclature. Koeltz Scientific Books, Königstein. 387pp.

Guého, E., Kurtzman, C. P. & Peterson, S. W. (1989). Evolutionary affinities of heterobasidiomycetous yeasts estimated from 18S and 25S ribosomal RNA sequence divergence. System. Appl. Microbiol. 12: 230-236.

Guého, E., Kurtzman, C. P. & Peterson, S. W. (1990). Phylogenetic relationships among species of *Sterigmatomyces* and *Fellomyces* as determined from partial rRNA sequences. Int. J. System. Bacteriol. 40: 60-65.

Higgins, D. G. & Sharp, P. M. (1988). CLUSTAL: a package for performing multiple sequence alignments on a microcomputer. Gene 73: 237

Higgins, D. G. & Sharp, P. M. (1989). Fast and sensitive multiple sequence alignments on a microcomputer. CABIOS 5: 151

Hillis, D. M. & Bull J. J. (1993). An empirical test of bootstrapping as a method for assessing confidence in phylogenetic analysis. System. Biol. 42: 182-192.

Kuhls, K., Lieckfeldt, E., Samuels, G. J., Meyer, W., Kubicek, C. P. & B`rner, T. (1997). Revision of Trichoderma sect. Longibrachiatum including related teleomorphs based on analysis of ribosomal DNA internal transcribed spacer sequences. Mycologia 89: 442-460.

Lobuglio, K. F., Pitt, J. I. & Taylor, J. W. (1993). Phylogenetic analysis of two ribosomal DNA regions indicates multiple independent losses of a sexual *Talaromyces* state among asexual *Penicillium* species in subgenus Biverticillium. Mycologia 85: 592-604.

Lobuglio, K. F., Pitt, J. I. & Taylor, J. W. (1994). Independent origins of the synnematous *Penicillium* species, *P. duclauxii, P. clavigerum*, and *P. vulpinum*, as assessed by two ribosomal DNA regions. Mycological Res. 98: 250-256.

Lobuglio, K. F. & Taylor, J. W. (1993). Molecular phylogeny of *Talaromyces* and *Penicillium* species in subgenus Biverticillium. In: The fungal holomorph: mitotic, meiotic and pleomorphic speciation in fungal systematics, (Eds Reynolds, D. R. & Taylor, J. W.) Oxon, U. K., CAB International, pp. 115-120.

McLennan, E. I., Ducker, S. C. & Thrower, L. B. (1954). New soil fungi from Australian heathland: *Aspergillus, Penicillium, Spegazzinia.* Aust. J. Bot. 2: 360-361.

O'Donnell, K. (1992). Ribosomal DNA internal transcribed spacers are highly divergent in the phytopathogenic ascomycete *Fusarium sambucinum (Gibberella pulicaris)*. Curr. Genet. 22: 213-220.

O'Donnell, K. & Cigelnik, E. (1997). Two divergent intragenomic rDNA ITS2 types within a monophyletic lineage of the fungus *Fusarium* are nonorthologous. Mol. Phylogenetics Evol. 7: 103-116.

Ogawa, H., Yoshimura, A. & Sugiyama, J. (1997). Polyphyletic origins of species of the anamorphic genus *Geosmithia* and the relationships of the cleistothecial genera: Evidence from 18S, 5S and 28S rDNA sequence analysis. Mycologia 89: 756-771.

Peterson, S. W. (1993). Molecular genetic assessment of relatedness of *Penicillium* subgenus *Penicillium*. In: The fungal holomorph: mitotic, meiotic and pleomorphic speciation in fungal systematics (Eds Reynolds, D. R. & Taylor, J. W.). Wallingford, United Kingdom, CAB International. pp. 121-128.

Peterson, S. W. (1995). Phylogenetic analysis of *Aspergillus* sections *Cremei* and *Wentii*, based on ribosomal DNA sequences. Mycological Res. 99: 1349-1355.

Peterson, S. W. (2000). Phylogenetic analysis of *Aspergillus* based on rDNA sequence analysis. In Integration of modern taxonomic methods for *Penicillium* and *Aspergillus* classification (Eds. Samson, R.A. & Pitt, J.I). Harwood Publishers, Amsterdam, pp 323-355.

Peterson, S. W. & Kurtzman, C. P. (1991). Ribosomal RNA sequence divergence among sibling species of yeasts. System. Appl. Microbiol. 14: 124-129.

Peterson, S. W. & Logrieco, A. (1991). Ribosomal RNA sequence variation among interfertile strains of some *Gibberella* species. Mycologia 83: 397-402.

Peterson, S. W., Horn, B. W., Ito, Y. & Goto, T. (2000). Genetic variation and aflatoxin production in *Aspergillus tamarii* and *A. caelatus*. In Integration of modern taxonomic methods for *Penicillium* and *Aspergillus* classification (Eds. Samson, R.A. & Pitt, J.I). Harwood Publishers, Amsterdam, pp 447-458.

Pitt, J. I. (1976). *Geosmithia* gen. nov. for *Penicillium lavendulum* and related species. Canad. J. Bot. 57: 2021-2030.

Pitt, J. I. (1979). The Genus *Penicillium* and its teleomorphic states *Eupenicillium* and *Talaromyces*. Academic Press, New York. 634p.

Pitt, J. I. & Cruickshank, R. H. (1990). Speciation and synonymy in *Penicillium* subgenus *Penicillium*-toward a definitive taxonomy. In: Modern concepts in *Penicillium* and *Aspergillus* classification (Eds Samson, R. A. & Pitt, J. I.). New York, Plenum Press. pp. 103-120

Pitt, J. I., Klich, M. A., Shaffer, G. P., Cruickshank, R. H., Frisvad, J. C., Mullaney, E. J., Onions, A. H. S., Samson, R. A. & Williams, A. P. (1990). Differentiation of *Penicillium glabrum* from *Penicillium spinulosum* and other closely related species: an integrated taxonomic approach. System. Appl. Microbiol. 13: 304-309.

Ramirez, C. (1982). Manual and Atlas of the Penicillia, Elsevier Biomedical Press, New York. 874p.

Raper, K. B. & Fennell, D. I. (1965). The Genus *Aspergillus*. Williams and Wilkins, Baltimore, MD.

Raper, K. B. & Thom, C. (1949). A manual of the Penicillia. The Williams and Wilkins Company, Baltimore, Maryland. 875p.

Samson, R. A. (1979). A compilation of the Aspergilli described since 1965. Stud. Mycol. 18: 1-38.

Scott, D. B. (1968). The genus *Eupenicillium* Ludwig. CSIR Research Report no. 272, pp. 1-150.

Stolk, A. C. & Samson, R. A. (1983). The Ascomycete genus *Eupenicillium* and related *Penicillium* anamorphs. Stud. Mycol. 23: 1-149.

Swofford, D. L. (1993). PAUP: Phylogenetic Analysis Using Parsimony, Version 3.1.1 Computer program distributed by the Illinois Natural History Survey, Urbana, IL.

Thom, C. (1930). The Penicillia. The Williams and Wilkes Company, Baltimore, Maryland. 644p.

White T. J., Bruns, T. D., Lee, S. B. & Taylor, J. W. (1990). Amplification and direct sequencing of fungal ribosomal DNA for phylogenetics. In: PCR protocols: A guide to the methods and applications (Eds Innes, M. A., Gelfand, D. H., Sninsky, J. J. & White, T. J.). New York, Academic Press. pp. 315

178

MOLECULAR METHODS FOR DIFFERENTIATION OF CLOSELY RELATED *PENICILLIUM* SPECIES

Pernille Skouboe, John W. Taylor[1], Jens C. Frisvad[2], Dorte Lauritsen, Lise Larsen, Charlotte Albæk, Marianne Boysen and Lone Rossen
Biotechnological Institute, Hørsholm, Denmark, [1]Department of Plant and Microbial Biology, University of California, Berkeley, CA, USA, [2]Department of Biotechnology, Technical University of Denmark, DK-2800 Lyngby, Denmark

The complete 18S ribosomal RNA gene of *Penicillium freii* (1806 bp) has been sequenced and compared to the 18S rDNA sequence of *Aspergillus fumigatus* (Barns *et al.*, 1991). Comparative sequence analysis showed only 31 (1.7%) nucleotide differences between these two species representing different genera. Nucleotide sequence variation within the ribosomal internal transcribed spacers (ITS 1 and ITS 2) and the 5.8S ribosomal RNA gene was examined in 101 isolates belonging to 31 taxa within *Penicillium* subgenus *Penicillium*. A 600 bp region spanning the ITS spacers and the 5.8S ribosomal RNA gene has been analysed by PCR amplification and direct sequencing. The ITS 1-5.8S-ITS 2 region is extremely conserved between closely related *Penicillium* species.

INTRODUCTION

The small subunit ribosomal RNA gene has been used extensively for phylogenetic analysis and for identification purposes at genus or species level in bacteria (Fox *et al.*, 1980; Weisburg *et al.*, 1991) and eukaryotes (Hendriks *et al.*, 1991). The small subunit ribosomal RNA gene comprises a number of highly conserved regions separated by variable regions. This makes it an ideal target for molecular systematic studies. The conserved regions have been selected for the design of primers that can be used in PCR amplifications followed by analysis of ribosomal DNA sequence variation. Studies of the sequence variation in the internally transcribed spacer (ITS) regions have been possible using primers from conserved regions in the 18S, 5.8S and the 28S ribosomal RNA genes (White *et al.*, 1990; Logieco *et al.*, 1990; Gardes and Bruns, 1993; Peterson, 1993). The small subunit ribosomal RNA gene has primarily been used for phylogenetic analysis at or above the level of families and orders whereas the ITS regions have proven useful for studying phylogenetic relationships between close taxonomic relatives (Bowman *et al.*, 1992; Berbee *et al.*, 1995). This paper evaluates the use of 18S rDNA and ITS sequences for identification of closely related terverticillate *Penicillium* species.

179

METHODS

*Strains:*The strains used in this study are listed in Table 1. The strains are maintained at the IBT Culture Collection, Department of Biotechnology, Technical University of Denmark, Lyngby, Denmark. Whenever possible, at least two isolates of each species were analysed. The interspecific ITS variability were studied by analysing larger numbers (4-12) of isolates from a selection of species.

Isolation of genomic DNA: Genomic DNA was obtained from fungal cultures following a simplified procedure (Yelton *et al.*, 1984). Fungi were grown in 200 ml Czapek Yeast Autolysate (CYA) broth (Pitt, 1979) for 3 days at 25°C, with agitation (200 rpm). The mycelium was separated from the medium by filtration, and the mycelial mat washed in 0.9% NaCl and frozen in dry ice/ethanol while rotating the tube. For extraction of high molecular weight DNA, approximately 500 mg of mycelium was repeatedly pulverized with a glass spatula in an Eppendorf tube while frozen in dry ice/ethanol. The mycelium was resuspended in 800 µl extraction buffer (50 mM EDTA, pH 8.5; 0.2% SDS) and incubated at 68°C for 30 minutes. After cooling to room temperature and centrifugation (10000 x g for 15 min.), the supernatant was transferred to another tube and 80 µl 5M KAc was added by mixing gently. The sample was placed on ice for one hour and recentrifuged at 20000 x g for 15 min. at 4°C. After centrifugation, the supernatant was transferred to a fresh tube and the DNA precipitated with an equal volume of isopropanol. After at least 5 min. at room temperature, the precipitated DNA was pelleted by centrifugation (10000 x g for 15 min.) and washed with 80% ethanol. The pellet was dried and resuspended in 50 µl of TE buffer (10 M Tris.Cl; pH 8.0, 1M EDTA). An aliquot (2 µg) was run on a 1% (w/v) agarose gel stained with ethidium bromide for estimation of DNA concentration.

PCR amplification of ribosomal DNA regions: Primers NS1-NS8, ITS4 and ITS5 described by White *et al.* (1990) were used for PCR amplification of the 18S ribosomal RNA gene and the ITS regions, respectively. DNA amplification was performed using an optimized amplification mix containing 50 mM KCl, 50 mM Tris.Cl; pH 8.3, 0.1 mg/ml bovine serum albumin (BSA), 3 mM $MgCl_2$, 200 µM each of dATP, dCTP, dGTP, and dTTP, 10% dimethylsulfoxide (DMSO), 0.25% Tween 20 and 2.5 units of AmpliTaq DNA polymerase (Perkin Elmer Corp., Norwalk, CT, USA.).Alternativily, single-stranded DNA templates were prepared by the asymmetric primer ratio method (Gyllensten and Erlich, 1988). Templates of both strands from each sample were generated using primer ratios of 20 pmol to 0.4 pmol in separate reactions in a total volume of 50 µl. Finally, 10 ng genomic DNA was added to each sample. Amplification was performed in a DNA thermal cycler 480 (Perkin Elmer Corp., Norwalk, CT, USA.). The temperature cycling parameters for the NS primer set were denaturation at 94°C for 1 min. followed by 40 cycles of 94°C for 5 sec., 53°C for 30 sec. and 72°C for 1 min. plus a final extension step at 72°C for 10 min. The temperature cycling parameters for the ITS primers were denaturation at 93°C for 30 sec. followed by primer annealing at 53°C for 30 sec. and primer extension at 72°C for 1 min. for a total of 40 cycles plus a final extension step at 72°C for 10 min. Following asymmetric amplification, excess primers and deoxynucleotide triphosphates were removed from samples containing single-stranded PCR product by precipitation in 2.5 M ammonium acetate plus 2 volumes of ethanol for 10 min. at room temperature and pelleted in a microcentrifuge (10000 x g, 10 min. room temperature). The PCR products were finally resuspended in TE buffer.

Table 1. List of *Penicillium* strains sequenced from each species. The ITS sequence have been determined for all isolates; isolates indicated with an asterisk have been partially sequenced in the 18S region. Current nomenclature is indicated as well as sources and sequence accession numbers.

Species	IBT no.	Source
P. aurantiogriseum	6215	Barley, Denmark
P. aurantiogriseum	10047	Sunflower, Denmark
P. freii*	3464	Wheat, UK
P. freii	11306	Barley, Denmark
P. tricolor	12471	Wheat, Canada ?
P. tricolor	12493	Wheat, Canada ?
P. tricolor	12494	Wheat, Canada ?
P. sp. 1	12708	Kangaroo rat, New Mexico
P. polonicum	11388	Barley, Canada
P. polonicum	11239	Mixed feed, Norway
P. aurantiovirens	11330	Barley, Denmark
P. sp. 2	12396	Kangaroo rat, New Mexico
P. melanoconidium*	3442	Barley, Denmark
P. melanoconidium*	6794	Salami, Germany
P. viridicatum	5273	Barley, Denmark
P. viridicatum	10057	Sesame seed, USA
P. cyclopium	5311	Amaranthus flower, USA
P. cyclopium	10085	Oats, Denmark
P. neoechinulatum*	3439	Seed cache of kangaroo rat, USA
P. neoechinulatum*	3462	Seed cache of kangaroo rat, USA
P. expansum	15658	Margarine, Australia
P. expansum	16943	Wheat, Germany
P. expansum	10061	Grapefruit, Honduras
P. expansum	10084	Walnuts, France
P. expansum	13554	Canada
P. expansum	13660	Cone, Denmark
P. expansum	13701	Russia
P. expansum	13703	Wheat, India
P. expansum	15622	Apple, Hungary
P. expansum	15626	Apple, Turkey
P. verrucosum*	5010	Barley, Denmark
*	11621	Wheat, Canada
P. camemberti	3505	Brie cheese, France
*	11568	Cheese, Germany
P. commune*	6327	Turnip, Denmark
	10763	Cheese, France
	6359	Cheese, Denmark
P. roqueforti*	12845	Cheese, Denmark
P. roqueforti *	12846	Pickled acid preserved pumpkins, Denmark
P. roqueforti	14429	Starter culture, Denmark
P. roqueforti	14431	Starter culture, Denmark
P. roqueforti	914430	Starter culture, Denmark
P. roqueforti	914433	Starter culture, Denmark
P. roqueforti	914432	Starter culture, Denmark

P. roqueforti	14408	Silage
P. roqueforti	14412	Silage
P. roqueforti	14420	Silage
P. roqueforti	14425	Silage
P. roqueforti	6754	Cheese
P. carneum	6884	Rye bread, Denmark
P. carneum	3477	Rye bread, Denmark
P. carneum	6885	Meat
P. carneum	6753	Cheese
P. carneum	14042	Rye bread, Denmark
P. paneum	12392	Chocolate sauce, Norway
P. paneum	12407	Rye bread, Denmark
P. paneum	11839	Rye bread, Denmark
P. paneum	13321	Soda water
P. paneum	13929	Bakers yeast
P. chrysogenum*	3182	Vulcano dust, Italy
P. chrysogenum*	5848	Sesame seed, Korea
P. nalgiovense*	5746	Desert sand, USA
P. chrysogenum*	12108	Cheese, Denmark
P. chrysogenum var. dipodomyis	5324	Seed cachekangaroo rat, USA
P. aethiopicum	5753	Salami, Germany
P. aethiopicum	5903	Barley, Ethiopia
P. aethiopicum	5748	Soil, Brazil
P. aethiopicum	5902	Sorgum, Zimbabwe
P. aethiopicum	5906	Grape, India
P. aethiopicum	6813	Africa
P. aethiopicum	11494	Carrageenan, Philippines
P. aethiopicum	15747	Apple, South Africa
P. aethiopicum	16873	Soil, Canada
P. aethiopicum	3901	Maize, India
P. aethiopicum	3913	Sorgum, Zimbabwe
P. crustosum	13049	Rye bread, DK
P. crustosum	13769	Cheese, DK
P. solitum	14859	
P. solitum	15170	
P. solitum	3948	
P. echinulatum*	3171	Pomelo, Israel
P. echinulatum*	3238	Mouldy gravy, Denmark
P. echinulatum	12879	
P. echinulatum	7000	
P. discolor	3086	Jerusalem artichoke, DK
P. discolor	3087	Radish
P. discolor	11512	
P. discolor	13522	
P. discolor	16126	
P. discolor	5738	
P. discolor	15185	
P. hirsutum var. hirsutum	10623	Apple, Czech Republic
P. hirsutum var. hirsutum	10628	Aphid, Netherlands
P. hordei	3083	Onion, Denmark
P. hordei	4900	Greenhouse fern, Netherlands

P. hirsutum var. albocoremium*	10682	Salami, Denmark
P. hirsutum var. albocoremium*	10697	Soil, Denmark
P. allii	3056	Food item, UK
P. allii	3058	Garlic, Denmark
P. hirsutum var. venetum	5464	Unknown source
P. hirsutum var. venetum	10594	Water (hot water tank), Denmark

DNA sequencing: Using single-stranded PCR products as templates and the oligonucleotides ITS1, ITS2, ITS3 and ITS4 (primer sequences are listed in White *et al.*, 1990) and ITS6 (5'-GGATCATTACCGAGTGAG) as primers, the ITS 1-5.8S-ITS 2 region was sequenced on both strands using the Sanger dideoxy method (Sanger *et al.*, 1977) and the Sequenase kit Version 2.0 (U.S. Biochemicals, Cleveland, Ohio) plus ^{35}S-labelled dATP (Amersham Int.) according to the manufacturer's instructions, with the following minor modifications: Approximately 1 µg of PCR-amplified DNA and 50 ng primer was combined with 0.7 µl 1 M NaOH and dH$_2$O to a total volume of 10 µl and mixed carefully; annealing template and primer was performed by heating 10 min. at 68°C, adding 2.8 µl TDMN (0.28 M TES (Sigma Chemical Co. No. T-4152), 0.12 M HCl, 0.05 M DTT (dithiothreitol), 0.08 M MgCl$_2$, and 0.2 M NaCl) and incubating at room temperature for 10 min. (Del Sal *et al.*, 1989). After annealing, the labeling and termination reactions were performed as described for the Sequenase kit. The fragments were separated on a denaturating 7 M urea/6% polyacrylamide standard premixed gel cast in a GIBCO BRL sequence gel model S2 (30 x 40 x 0.04 cm) using wedged spacers. The gels were dried and exposed to Kodak X-OMAT AR films for 1-3 days.

Automated DNA sequencing: ITS sequence data was generated using an automated sequencer according to the following procedure: The ITS1, ITS2, ITS3 and ITS4 primers were used for direct cycle sequencing of the amplified PCR products (Murray, 1989). Cycle sequencing reactions were performed using the Thermo Sequenase fluorescent labelled primer cycle sequencing kit (Amersham) according to the manufacturer's instructions, and analysed on an automated DNA sequencer (ALF express, Pharmacia Biotech, Uppsala, S.). DNA sequencing was done in both directions.

Sequence assession numbers: The nucleotide sequence data reported in this paper will appear in the EMBL, Genbank and DDBJ Nucleotide Sequence Databases under the accession numbers shown in Table 1.

Sequence analysis: Computer-aided alignment of the 18S rDNA and ITS sequences were performed using the CLUSTAL option in the IntelliGenetics software package, PC Gene Screen Device, Version 5.03 (IntelliGenetics, Inc., Mountain View, CA). Gaps were introduced manually into the sequences to increase their aligned similarity.

RESULTS AND DISCUSSION

DNA sequence of the 18S ribosomal RNA gene of Penicillium freii
The DNA sequence of the major part of the *P. freii* (IBT 3464) 18S ribosomal RNA gene was obtained using a series of overlapping asymmetrical amplifications to generate sin-

gle-stranded DNA for direct sequencing. The 5′ end of the 18S ribosomal DNA gene was sequenced following an arbitrary double stranded amplification procedure using one primer (NS2).

The entire sequence was compared to the 18S ribosomal RNA gene from *Aspergillus fumigatus* (Barns *et al.*, 1991) and *P. chrysogenum* (Sogin *et al.*, 1990). A total of 31 (1.7%) differences was observed between *A. fumigatus* and *P. freii* and four differences were seen when comparing *P. freii* with *P. chrysogenum*. It was noted that in three of the four positions the *P. chrysogenum* sequence was identical to that of *A. fumigatus*. In Table 2 the relative positions of sequence differences are listed, with numbering given as for the *A. fumigatus* sequence.

Table 2. DNA sequence differences between the 18S rDNA sequences of *Aspergillus fumigatus* (M60300; Barns *et al.*, 1991) , *Penicillium chrysogenum* (M55628, Sogin *et al.*, 1990) and *P. freii*.

Position[a]	Aspergillus fumigatus	Penicillium freii IBT 3464	Penicillium chrysogenum
73	T	C	C
76	A	G	G
80	G	T	T
183	T	C	C
190b	-	A	A
223	T	C	C
451	C	T	T
515	T	C	A
521	C	T	T
642	T	C	T
686	A	G	A
820	C	T	T
1047	G	A	A
1051	T	G	-
1053	T	A	A
1060	G	A	A
1069	C	T	T
1381	G	A	G
1497	G	A	A
1498	G	A	A
1508	C	T	T
1666	G	A	A
1677	C	G	G
1681	C	T	T
1687	C	T	T
1691	G	A	A
1693	A	G	G
1706	T	C	C
1718	G	A	A
1722	G	C	C
1733	C	T	T

[a] *Aspergillus fumigatus* 18S rDNA sequence position number

18S sequence comparisons between Aspergillus and Penicillium species

The comparative sequence analysis of 18S rDNA sequences from *A. fumigatus* and two *Penicillium* species revealed several regions containing ribosomal DNA variation. Two regions of 330 bp and 100 bp, respectively, spanning position 500-830 and position 1650-1750 were selected for further sequence analysis. The 330 bp region spanned three of the four differences between *P. freii* and *P. chrysogenum* and the 100 bp region covered 10 differences between *Aspergillus* and *Penicillium* (see Table 2). Two isolates of each of nine *Penicillium* species were sequenced in the two regions (Table 1). The nine species were chosen to cover as wide a spectrum of *Penicillium* subgenus *Penicillium* as possible. None of the nine species showed any variation in the 100 bp region. In the 330 bp region spanning position 500-830 differences were found in two positions. *P. freii*, (IBT 3464), *P. neoechinulatum*, (IBT 3462) and *P. verrucosum*, (IBT 11621) all had a C in position 642 and a G in position 686, while the rest of the strains had a T in position 642 and an A in position 686.

It was concluded that the 18S ribosomal RNA gene was too conserved for separation of terverticillate Penicillia, and further studies are necessary to ascertain if the region between position 1650 and 1750 can be used for designing a primer specific for *Penicillium* subgenus *Penicillium*. The ITS region was chosen for further studies as this region was previously shown to contain sufficient homology to allow for sequence comparisons (Gardes and Bruns, 1993).

Amplification and sequence analysis of the ITS regions

Using the primers ITS4 and ITS5, the ribosomal ITS 1-5.8S-ITS 2 region was amplified from 101 *Penicillium* strains belonging to 31 taxa (Table 1). Using molar excess of ITS4, the amplification product was in all cases a unique fragment of approximately 610 bp, indicating that the size of the region was conserved (data not shown). However, using molar excess of ITS5, some additional minor bands were amplified.

To investigate the degree of ITS sequence variability between the 31 *Penicillium* taxa, the sequence of an approximately 550 nucleotide region was determined in both directions. The degree of variation within the ITS regions of the *Penicillium* species studied was small, considering the amount of variation found in other fungal genera and in the corresponding region in bacteria (O'Donnell, 1992; Barry *et al.*, 1991). Individual sequence differences between the 31 taxa were found in only 34 positions, and four of these were insertions or deletions. Twenty three of the nucleotide differences were found within the ITS 1 region, and the rest (11) in the ITS 2 region. Some closely related taxa exhibited identical sequences. Interspecific ITS variability was examined and no differences were observed between isolates belonging to the same species, except for the three taxa *P. verrucosum, P. discolor* and *P. nalgiovense*, where a single nucleotide difference was observed within the sequenced region. No explanation could be found for the differences in these three taxa. This is in contrast to the report on *Trichoderma harzianum* (Muthumeenakshi *et al.*, 1994) where up to 22.9% difference in the ITS 1 region was observed between 6 isolates.

To illustrate the extent of ITS sequence variation among well characterised *Penicillium* species a matrix of ITS nucleotide differences were generated and compared by counting the numbers of differences separating each taxon from the others. As can be seen from Table 3 the number of sequence differences varies from 5 to 20 when comparing such species as *P. echinulatum* to *P. chrysogenum* or *P. aurantiogriseum* to *P. roqueforti*.

Table 3. Number of ITS sequence differences between selected species representing series within the *Penicillium* subgenus *Penicillium*.

	P. expansum	P. echinulatum	P. chrysogenum	P. roqueforti
P aurantiogriseum	8	5	8	20
P. expansum	0	8	9	18
P. echinulatum		0	5	17
P. chrysogenum			0	17

Table 4. Number of ITS sequence differences observed between selected taxa related to *Penicillium aurantiogriseum*.

	P. freii	P. tricolor	P. cyclopium	P. melanoconidium
P aurantiogriseum	2	4	5	3
P. freii	0	4	5	3
P. tricolor		0	1	3
P. cyclopium			0	4
P. melanoconidium				0

Table 5. Number of ITS sequence differences observed between selected taxa related to *P. chrysogenum* group.

	P. dipodomyis	P. nalgiovense	P. aethiopicum
P chrysogenum	1	1	3
P. dipodomyis	0	1	2
P. nalgiovense		1	2

Table 6. Number of ITS sequence differences observed between selected taxa related to *P. solitum*.

	P. echinulatum	P. discolor	P. commune
P. solitum	0	0	1
P. echinulatum	0	0	1
P. discolor		0	1

The numbers of sequence differences are even lower when taxa believed to be more closely related are compared as can be seen in Tables 4 to 6 where species believed to be related to *P. aurantiogriseum, P. solitum* and *P. discolor*, and P. chrysogenum are compared individually. As shown in Table 6, the ITS sequences of *P. solitum, P. echinulatum* and *P. discolor* were identical. These three creatine-positive species resembles each other morphologically (e.g. dark green conidia) but were separated on the basis of secondary metabolite profiles (Svendsen and Frisvad, 1994; Frisvad *et al.*, 1997).

The sequence differences were also analyzed phylogenetically using neighbor-joining and bootstrapping. The results showed that the majority of the species belonged to the same clade, but it was possible to separate *P. roqueforti* and related species (*P. roqueforti, P. carneum* and *P. paneum*, Boysen *et al.*, 1996) from other terverticillate Penicillia (data not shown).

186

In conclusion, the ITS region of the terverticillate Penicillia has been shown to be extremely conserved, however it was possible to differentiate between series of taxa when examining the ITS sequence variability.

REFERENCES

Barns, S.M., Lane, D.J., Sogin, M.L., Bibeau, C. & Weisburg, W.G. (1991). Evolutionary relationships among pathogenic *Candida* species and relatives. J. Bacteriol. 173, 2250-2255.

Barry, T., Colleran, G., Glennon, M., Dunican, L. K. & Gannon, F. (1991). The 16S/23S ribosomal spacer region as target for DNA probes to identify Eubacteria. PCR Methods and Applications 1, 51-56.

Berbee, M.L., Yoshimura, A., Sugiyama, J. & Taylor, J.W. (1995). Is *Penicillium* monophyletic? An evaluation of phylogeny in the family Trichocomaceae from 18S, 5.8S and ITS ribosomal DNA sequence data. Mycologia 87, 210-222.

Bowman, B.H., Taylor, J.W., Brownlee, A.G., Lee, J., Lu, S-D.& White, T.J. (1992) Molecular evolution of the fungi: relationship of the Basidiomycetes, Ascomycetes and Chytridiomycetes. Mol. Biol. Evol. 9, 285-296.

Boysen, M., Skouboe, P., Frisvad, J., & Rossen, L. (1996). Reclassification of the *Penicillium roqueforti* group into three species on the basis of molecular genetic and biochemical profiles. Microbiology 142, 541-549.

Del Sal, G., Manfioletti, G. & Schneider, C. (1989). The CTAB-DNA precipitation method: a common miniscale preparation of template DNA from phagemids, phages or plasmids suitable for sequencing. Bio-Techniques 7, 514-519.

Fox, G.E., Stackebrandt, E., Hespell, R.B., Gibson, J., Maniloff, J., Dyer, T.A., Wolfe, R.S., Balch, W.E., Tanner, R.S., Magrum, L.J., Zablen, L.B., Blakemore, R., Gupta, R., Bonen, L., Lewis, B.J., Stahl, D.A., Luehrsen, K.R., Chen, K.N. & Woese, C.R. (1980) .The phylogeny of prokaryotes. Science, 209, 457-463.

Frisvad, J.C., Samson, R.A., Rassing, B.R., van der Horst, M.I., van Rijn, F.T. & Stark, J. (1997). *Penicillium discolor*, a new species from cheese, nuts and vegetables. Antonie Van Leeuwenhoek 72, 119-126.

Gardes, M. & Bruns, T.D. (1993). ITS primers with enhanced specificity for basidiomycetes - application to the identification of mycorrhizae and rusts. Molec. Ecol. 2, 113-118.

Gyllensten, U.B. & Erlich, H.A. (1988). Generation of single stranded DNA by the polymerase chain reaction and its direct sequencing of the HLA-DQA locus. Proc. Natl. Acad. Sci. USA 85, 7652-7656.

Hendriks, L., Goris, A., Van de Peer, Y., Neefs, J-M., Vancanneyt, M., Kersters, K., Hennebert, G.L. & De Wachter, R. (1991). Phylogenetic analysis of five medically important *Candida* species as deduced on the basis of small ribosomal subunit RNA sequences. J. Gen. Microbiol. 137, 1223-1230.

Logrieco, A., Peterson, S.W. & Wicklow, D.T. (1990). Ribosomal RNA comparisons among taxa of the terverticillate penicillia. In Modern Concepts in *Penicillium* and *Aspergillus* Systematics. Eds. Samson, R.A. & Pitt, J.I. New York, Plenum Press, pp. 343-355.

Murray, V. (1989). Improved double-stranded DNA sequencing using the linear polymerase chain reaction. Nucl. Acids Res. 17, 8889.

Muthumeenakshi, S., Mills, P.R., Brown, A.E. & Seaby, D.A. (1994). Intraspecific molecular variation among *Trichoderma harzianum* isolates colonizing mushroom compost in the British Isles. Microbiology 140, 769-777.

O'Donnell, K. (1992). Ribosomal DNA internal transcribed spacers are highly divergent in the phytopathogenic ascomycete *Fusarium sambucinum* (*Gibberella pulicaris*). Cur. Genet. 22, 213-220.

Peterson, S.W. (1993). Molecular genetic assessment of relatedness of *Penicillium* subgenus *Penicillium*. In The Fungal Holomorph: Mitotic, Meiotic and Pleomorphic Speciation in Fungal Systematics. Eds. Reynolds. D.R. & Taylor, J.W. Wallingford, UK, CAB International, pp. 121-128.

Pitt, J.I. (1979). The Genus *Penicillium* and its Teleomorphic States *Eupenicillium* and *Talaromyces*. London, Academic Press.

Sanger F., Nicklen, S. & Coulsen, A.R. (1977). DNA sequencing with chain-terminating inhibitors. Proc. Natl. Acad. Sci. USA 85, 3608-3612.

Sogin, M.L., Bineau, C., Elwood, H., Stickel, S., Weisburg, W., Barnes, S. & Lane, D.L. (1990). Phylogenic relationships between major classes of fungi (unpublished – Accession: M55628).

Svendsen, A. & Frisvad, J.C. (1994). A chemotaxonomic study of the terverticillate Penicillia based on high performance liquid chromatography of secondary metabolites. Mycol. Res. 98, 1317-1328.

Weisburg, W.G, Barns, S.M., Pelletier, D.A. & Lane, D.J. (1991). 16S ribosomal DNA amplification for phylogenetic study. J. Bacteriol. 173, 697-703.

White, T.J., Bruns, T., Lee, S., & Taylor, J. (1990). Amplification and direct sequencing of fungal ribosomal RNA genes for phylogenetics. In PCR Protocols: A Guide to Methods and Applications. Eds. Innis, M.A., Gelfand, D.H., Sninsky, J.J. & White, T.J. London, Academic Press, pp 315-322.

Yelton, M.M., Hamer, J.E. & Timberlake, W.E. (1984). Transformation of Aspergillus nidulans by using a trpC plasmid. Proc. Natl. Acad. Sci. USA 81, 1470-1474.

PHYLOGENY AND SPECIES CONCEPTS IN THE *PENICILLIUM AURANTIOGRISEUM* COMPLEX AS INFERRED FROM PARTIAL β-TUBULIN GENE DNA SEQUENCES [1]

Keith A. Seifert and Gerry Louis-Seize,
Eastern Cereal and Oilseed Research Centre, Agriculture and Agri-Food Canada, Research Branch, Ottawa, Ontario, Canada K1A 0C6

A 360 base pair region of the β-tubulin gene was sequenced for 45 strains of *Penicillium*, mostly members of the *P. aurantiogriseum* complex. Multiple strains were sequenced for critical species to sample infraspecific variation. Phylogenetic analysis of the data revealed five major clades, but the relationship between them was unresolved. The distinction between the species with large conidia (*P. commune, P. crustosum, P. solitum*) and the small-spored species of the *P. aurantiogriseum* complex was confirmed. *Penicillium polonicum* and *P. cyclopium* formed discrete species within one of the main clades. *Penicillium verrucosum* and *P. melanoconidium* formed discrete clades. The relationships within the final clade, between *P. viridicatum, P. freii* and *P. aurantiogriseum* were unresolved using this data, although there was no evidence that any of these species was poly- or paraphyletic. In general, the β-tubulin partial sequences provided support for species concepts based on mycotoxin profiles, physiological and micromorphological characters. In contrast, the nuclear large ribosomal gene subunit (28S) and the mitochondrial small ribosomal gene subunit (mtSSU) provided only limited phylogenetically useful information for resolving species in this complex.

INTRODUCTION

Species concepts in the *Penicillium aurantiogriseum* complex have been a topic of much debate over the past twenty years. The economic and health importance of these fungi relates to their mycotoxin producing capabilities, in particular the ability of some species to produce nephrotoxins such as ochratoxin A. Given the relative frequency with which these species are isolated from stored grain in most temperate countries, resolution of their taxonomy at the species level is a critical issue. Biological species concepts cannot be derived in the absence of known teleomorphs, so other approaches to delimiting the species must be developed.

The complex has been treated in many different ways, with treatments varying from a single species with four varieties (Samson *et al.*, 1976) to as many as fifteen species (Samson *et al.*, 1995). These taxonomies reflect different interpretations and weighting of phenetic characters such as colony growth, pigmentation and texture on standard or relatively specialized agar media, differences in mycotoxin profiles, and similarities in micromorphology. In recent years, the idea that differences in mycotoxin profiles may be useful delimiters of phylogenetic species has become more widely accepted. Thus, many of the taxa formerly designated as chemotypes (e.g. Frisvad and Filtenborg, 1983) or varieties (e,g. Frisvad and Filtenborg, 1989) have tended to be raised to the rank of species.

Species concepts became an important part of our work on Penicillia contaminating stored grain in Canada (Mills *et al.*, 1995) because of our desire to develop molecular based diagnostics. We felt that if the taxa delineated by mycotoxin profiles could also be detected by an analysis of an independent character set, that this would provide increased support for these species concepts. Our initial experiments with the subunits of the ribosomal gene, which have been widely used in fungal taxonomy, including *Penicillium*, provided limited useful information at the species level. Therefore, we decided to explore other genes. Glass and Donaldson (1995) developed a series of primers for amplifying introns from conserved genes, such as histone and β-tubulin, from filamentous ascomycetes. A preliminary screening of these lead us to sequencing part of the β-tubulin gene, the results of which are reported in this paper.

MATERIALS & METHODS

Cultures. The fungal cultures used in this study are listed in Table 1. Wherever possible, ex-type (or ex-neotype) cultures were included, as well as cultures with known mycotoxin profiles. There were two groups of species of prime interest, the so-called *Penicillium aurantiogriseum* group (*P. aurantiogriseum*, *P. freii*, *P. melanoconidium*, *P. polonicum*, *P. verrucosum* and *P. viridicatum*) and the large-conidium counterparts of these species (*P. commune*, *P. crustosum* and *P. solitum*), which were represented by 2-4 strains for each species. Other similar taxa from subgenus *Penicillium* with rough-walled stipes were represented by single strains. *Penicillium citrinum*, from subgenus *Furcatum* was chosen as the outgroup, with *P. expansum* used as a representative of subgenus *Penicillium* with smooth-walled stipes.

DNA isolation, amplification, purification and sequencing. Cultures were grown and DNA was isolated using the methods given in Seifert *et al.* (1997). Some minipreps were prepared using glass beads in a BioSpec Products Beadbeater in place of manual grinding. The specimens were homogenized for 5 minutes in a 2 mL vial containing lysis buffer and 0.5 mm glass beads. The Bt2 fragment was amplified in a 50 μL PCR reaction with the following profile: incubation at 94°C for 3 min, followed by 30 cycles with denaturation at 94°C for 1 minute, annealing at 56°C for 1.5 minutes, extension at 72°C for 2 minutes, and a final incubation at 72°C for 10 minutes after cycling. Some PCR reactions were run using the premixed "PCR Super Mix" from Life Technologies. PCR products were purified and sequenced following the methods detailed by Seifert *et al.* (1997).

Selection of genes for sequencing. To select a novel gene area for sequencing, we assessed primer sets developed by Glass and Donaldson (1995) for amplifying introns from the histone-3 (H3-1a/H3-1b), histone- 4 (H4-1a/H4-ab), and β-tubulin (Bt1a/Bt1b and

190

Bt2a/Bt2b) genes of filamentous ascomycetes. The primer sets were tested for their ability to produce single PCR products from DNA preparations from selected *Penicillium* strains, then screened using restriction fragment length polymorphism analysis to determine whether a sufficient amount of variation existed amongst closely related isolates to be taxonomically useful. The PCR profile followed that given by Glass and Donaldson (1995).

Table 1. Cultures employed for β-tubulin sequencing and the corresponding GenBank accession numbers for their partial β-tubulin sequences. Strains lacking GenBank numbers had identical sequences to other strains of the same species and are marked n/a in the table.

Species	Accession numbers	Origin	GgenBank number
P. allii Vincent & Pitt	ATCC 64868 = CBS 131.89	T, Egypt, *Allium*	AF004156
P. aurantiogriseum Dierckx	ATCC 48920 = CBS 324.89 = IMI 195050 = NRRL 971	NT, Belgium	AF004157
	DAOM 213189	Canada, *Hordeum*	AF004158
	DAOM 216724 = IBT 3471	USA, *Hordeum*	n/a
P. citrinum Thom	DAOM 216702	Japan	AF003539
P. commune Thom	ATCC 1111 = CBS 104.28 = IMI 38812 = NRRL 890	T, USA, cheese	AF003538
	DAOM 214793 = IBT 10501	France, goat cheese	AF003537
	MayoDF A21	Canada, *Pseudotsuga*	n/a
P. crustosum Thom	IMI 91917	T	AF003536
	DAOM 171024	Canada, salami	n/a
	CBS 471.84 = DAOM 214800 = IBT 3425 = IMI 285510	Denmark, *Thymus*	n/a
	DAOM 215343	Canada, *Picea*	n/a
P. cyclopium Westling	ATCC 8731 = CBS 144.45 = IMI 89372 = NRRL 1888	T	AF004159
	DAOM 216703= IBT 3221?		n/a
	DAOM 216704 = IBT 3452	Ethiopia, *Hordeum*	AF003248
P. echinulatum Raper & Thom ex Fassatiová	ATCC 10453 = CBS 317.48 = IMI 40028 = NRRL 1151	T, Canada, air	AF003247
P. expansum Link	ATCC 7861 = CBS 325.48 = IMI 39761 = NRRL 976	NT, USA, Malus	AF003246
	DAOM 215351	Canada, wood	AF003245
P. freii Frisvad & Samson	CBS 348.48 = NRRL 951	T?, Belgium	AF003244
	DAOM 216705 = IBT 3464	England, *Triticum*	n/a
	DAOM 216706 = IBT 5147	Denmark, *Hordeum*	AF003243
P. griseofulvum Dierckx	ATCC 11885 = CBS 185.27 = IMI 75832 = NRRL 2152	NT, Belgium	AF003242
P. hirsutum Dierckx var. *hirsutum*	ATCC 10429 = CBS 135.41 = IMI 40213 = NRRL 2032	Netherlands, aphids	AF003241
P. hirsutum var. *albocoremium* Frisvad	CBS 187.88	Netherlands, *Tulipa*	AF003240
P. hirsutum var. *venetum* Frisvad	ATCC 16025	T, England, *Hyacinthus*	AF003239
P. hordei Stolk	ATCC 22053 = CBS 701.68 = IMI 151748	T, Denmark, *Hordeum*	AF003238
P. melanoconidium (Frisvad) Frisvad & Samson	IBT 3444 = IMI 321503	T, Denmark, *Triticum*	AF004160

	DAOM 216707 = IBT 3442	Denmark, *Hordeum*	n/a
	DAOM 216708 = IBT 4107	Denmark, *Triticum*	AF003237
P. nalgiovense Laxa	ATCC 10472 = CBS 352.48 = IMI 39804 = NRRL 911	T, Czechoslovakia, cheese	AF003236
P. neoechinulatum (Frisvad *et al.*) Frisvad & Samson	CBS 169.87I = MI 296937 = NRRL 13486	T, USA, *Dipodomys*	AF001207
P. polonicum Zaleski	IBT 6583 = NRRL 995	T, Poland	AF001004
	DAOM 216709 = IBT 3109		AF001206
	DAOM 216710 = IBT 3449	bird seed	AF0012105
P. roqueforti Thom	ATCC 10110 = CBS 221.30 = IMI 24313 = NRRL 849	T, USA, blue cheese	U97185
P. solitum Westling	ATCC 9923 = CBS 288.36 = IMI 39810 = NRRL 937	T?, Germany	AF000303
	DAOM 187007	Canada, *Malus*	n/a
	DAOM 214781= IBT 6181	Denmark, bacon	n/a
	DAOM 215390 = ATCC 60539	New Zealand, lamb	n/a
P. verrucosum Dierckx	ATCC 48957 = CBS 603.74 = IMI 200310 = NRRL 965	NT, Belgium	AF00934
	DAOM 213195	Canada, *Triticum*	n/a
P. viridicatum Westling	ATCC 10515 = CBS 390.48 = IMI 39758 = NRRL 963	NT, USA, air	AF000933
	DAOM 213197	Canada, *Hordeum*	U97186
	DAOM 216713 = IBT 3110		n/a
	DAOM 216714 = IBT 5112		n/a

Single enzyme restriction digests were performed on raw PCR products using *AluI, BamH1, CloI, EcoRI, HaeIII, HindIII, HinfI, HpaII, MboI, MseI, RsaI* and *TaqI*. Restriction products were separated on 5% HydroLink (BDH Chemicals) at 100 volts for about 2 hours, then soaked in 10 mg/mL ethidium bromide for one hour to visualize bands. Some gels were also prepared using NuSieve and MetaPhor XR with ethidium bromide incorporated in the matrix.

In addition, two ribosomal regions that had been used previously in examining infrageneric relationships in *Penicillium* were screened for their usefulness in delimiting species in the *P. aurantiogriseum* group. An approximately 400 bp stretch of the mitochondrial small ribosomal subunit (mtSSU) was amplified using primers MS2 (White *et al.*, 1990) and U3 (Bruns, pers. comm.) and sequenced using the same primers. The following strains were sequenced: *P. aurantiogriseum* DAOM 213189 (GenBank accession number AF003360), 216724, *P. cyclopium* DAOM 216704 (AF003364), *P. freii* DAOM 216705 (AF003361), 216706, *P. melanoconidium* DAOM 216707 (AF003367), 216708, *P. polonicum* DAOM 216709, 216710 (AF003362), *P. verrucosum* DAOM 213195 (AF003365), *P. viridicatum* DAOM 213197 (AF003366), 216713, 216714 and *P. citrinum* DAOM 216702 (AF003369) as the outgroup.

Similarly, an approximately 1770 bp stretch of the nuclear ribosomal large subunit (28S) (LROR to LR8) was amplified and sequenced using primers LROR, LR8, LR3R, LR16 and LR17R (Rehner and Samuels, 1995; Vilgalys, pers. comm.) The following strains were sequenced: *P. aurantiogriseum* DAOM 213189 (GenBank accession number AF003355), *P. cyclopium* DAOM 216703 (AF003356), *P. freii* DAOM 216705, *P. melanoconidium* DAOM 216708, *P. polonicum* DAOM 216709, *P. verrucosum* DAOM

213195 (AF003357) and *P. viridicatum* DAOM 213197 (AF003358) with *P. expansum* DAOM 215351 (AF003359) as the outgroup.

Phylogenetic analysis. Sequences were edited and aligned using the EditSeq and MegAlign modules in the Lasergene sequence software package (DNAStar, Inc.). The final alignments were prepared using the Joint Clustal method, with minor modifications afterwards. Phylogenetic analysis was performed using PAUP 3.1.1 (Swofford, 1991). Ten replicate heuristic searches were performed of each set. Five hundred bootstrap replications were performed using heuristic searches on the entire 28S data set, and on a pruned β-tubulin data set with identical sequences and species of incidental interest removed.

RESULTS

PCR and RFLP analysis of histone-3, histone-4 and β-tubulin amplicons. Both histone primer pairs yielded multiple products under the PCR conditions employed, and hence were not investigated further. RFLP analysis was confined to amplicons from the Bt1 and Bt2 primer pairs, each of which consisted of a single PCR product. The approximate sizes of the Bt1 fragments were 550 bp, and the Bt2 fragments about 425 bp. The Bt2 fragment had more restriction sites than the Bt1 fragment, resulting in digestion products for about 75% of the enzymes (AluI, CloI, HaeIII, HinfI, HpaII, MboI, MseI, RsaI and TaqI), all of which were polymorphic among the strains. The Bt1 fragment yielded digestion products for about 50% of the enzymes (HaeIII, RsaI, MboI, AluI and BamHI) but polymorphisms were detected among the strains only with HaeIII (data not shown). Restriction patterns of the Bt2 fragment were relatively constant within species, and showed sufficient differences among species to indicate that sequencing of this fragment might provide phylogenetically interesting data.

Phylogenetic analysis of Bt2 sequences. The Bt2 sequences for the *Penicillium* strains studied varied between 374 and 421 readable bp, with the longest being *P. citrinum*, with numerous short insertions. Prior to phylogenetic analysis, up to 34 bases were trimmed from 5' end of the sequences to accommodate for differences in the first readable bases, and likewise about 40 bases were eliminated from the 3' end. This left an alignment of 360 bp for phylogenetic analysis, with approximately 85 phylogenetically significant sites (indels included) if the outgroup was included once in the alignment. Heuristic analysis of the entire data set yielded 1080 equally parsimonious trees of 176 steps (Fig. 1). The reduced data set, with redundant sequences and species of incidental interest excluded, included 54 phylogenetically significant characters and yielded 21 equally parsimonious trees of 82 steps (Fig. 2).The bootstrap support for the topology of the tree was generally rather low, reflecting the relatively high amount of homoplasy in the data, and the short branch lengths separating many of the clades.

The analyses suggest the existence of five major clades within the trimmed data set, but the phylogenetic relationships among these five clades are equivocal. The five major clades correspond to

1) *P. viridicatum, P. freii* and *P. aurantiogriseum* (although *P. aurantiogriseum* falls out of this clade in some permutations of the data)

2) *P. melanoconidium*

3) *P. polonicum* and *P. cyclopium*

4) *P. verrucosum* and

5) the species with large conidia that produce alkali on creatine-sucrose agar, namely *P. commune, P. crustosum* and *P. solitum*.

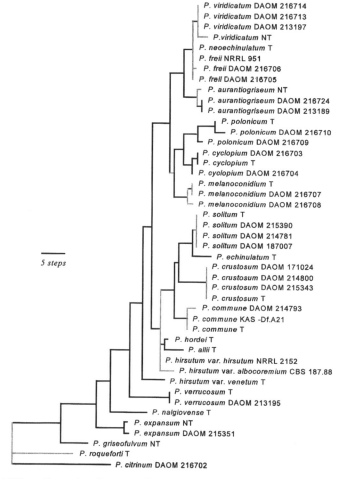

Figure 1. One of 1080 equally parsimonious trees of 176 steps based on a heuristic search of partial β-tubulin sequences using the entire data set, with *P. citrinum* as the outgroup The branches in grey are polytomies in the strict consensus tree of the equally parsimonious trees. (CI = 0.659, HI = 0.341, RI = 0.826, RC = 0.544).

With the exception of the first clade, the species within the groups are resolved into monophyletic groups, although the bootstrap values supporting the species are often only in the 70-75% range. The permutations of possible relationships between the five major clades, and of the individual strains within the clades with more than one species, are responsible for the relatively large number of equally parsimonious trees in the heuristic searches of the entire data set. The branch length between species was generally rather short, confirming the close relationship of the taxa.

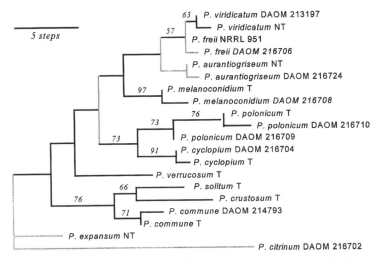

Figure 2. One of 21 equally parsimonious trees of 82 steps based on a heuristic search of partial β-tubulin sequences using the trimmed data set, with *P. citrinum* as the outgroup. The branches in grey are polytomies in the strict consensus tree of the equally parsimonious trees. The numbers represent bootstrap percentages above 50% supporting individual particular branches (CI =.732, HI = 0.268, RI = 0.763, RC = 0.559).

Of the species with more than one strain sequenced, *P. crustosum* (4 strains), *P. solitum* (4 strains) and *P. verrucosum* (2 strains) had identical sequences (i.e. the only differences being unreadable bases in 1-3 positions). For *P. aurantiogriseum*, *P. freii* and *P. cyclopium,* the sequences for two strains of each species were identical, differing in a single base from the third strains. Two of the strains of *P. melanoconidium*, the ex-type culture and DAOM 216707, had identical sequences, differing in two bases from DAOM 216708. The three strains of *P. polonicum* had sequences differing by 1-3 base pairs. The sequences of the ex-type strain of *P. viridicatum* differed from the three DAOM strains by two bases. In *P. commune*, the sequence of DAOM 214793 differed from the ex-type strain and the second DAOM strain by a single base.

Despite these minor differences in sequence among strains of some species, each species formed a monophyletic group in the heuristic analyses. The exception to this was *P. freii*, the ex-type strain of which had an identical sequence to the ex-type of *P. neoechinulatum*. This group of species formed a polytomy 1 step removed from the clade representing *P. viridicatum*. Although these species concepts are not unequivocally confirmed by this data, there is no evidence that any of these taxa may be para- or polyphyletic.

Phylogenetic analysis of mtSSU and 28S sequences. Neither of these regions provided phylogenetically useful data for the closely related group of strains under study. The 28S alignment was trimmed up to 57 bp from the 5' end and up to 20 bp from the 3', leaving an alignment of approximately 1710 bp. Only nine phylogenetically informative sites existed in the data set. Heuristic searches yield 10 equally parsimonious trees, 10 steps long. The bootstrap support was very weak for the 28S gene tree (Fig. 3), providing minimal information on the relationships between the species. Because only single strains of each species were sequenced for the 28S region, it was not possible to assess infraspecific variation.

The mtSSU primers had poor efficacy for sequencing, resulting in faint or blurred sequencing gels that were difficult to read. The resulting alignment was trimmed up to 75 bp from the 5' end, and up to 150 bp from the 3' end, leaving a stretch of 280 bp clearly readable bases. This included no phylogenetically informative sites, although there were 6 variable sites in the outgroup, *P. citrinum*. There were no unequivocal sequence differences between strains of individual species.

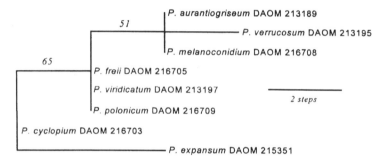

Figure 3. One of 10 equally parsimonious trees of 10 steps based on a heuristic search of partial 28S sequences, with *P. expansum* as the outgroup. The numbers represent bootstrap percentages above 50% supporting individual particular branches (CI = 0.800, HI = 0.200, RI = 0.714, RC = 0.571).

DISCUSSION

The gene trees derived from the sequences of Bt2 region of the β-tubulin correlate well with the species delimited by Frisvad and Samson (see Samson *et al.*, 1995) in the *Penicillium aurantiogriseum* group. Despite small variations in sequences from different strains of some species, the variation among species was sufficient to maintain the grouping of most of the strains into monophyletic clades. The one exception to this was *P. freii*, which was not clearly delimited from *P. neoechinulatum* by its Bt2 sequence, and forms a polytomy basal to the *P. viridicatum* clade. In general, these results demonstrate the utility of this particular gene region as a taxonomic tool. However, a longer sequence with a similar or slightly greater amount of variation would be helpful to resolve the relationships between the five major clades, and the relationships of the species in the *P. viridicatum/freii/aurantiogriseum* clade. The β-tubulin primers used by O'Donnell and Cigelnik (1995) were used to sequence about 622 bases in *Fusarium* strains. To our knowledge, these primers have not been tried in *Penicillium*. The amount of base-pair substitutions reported for this longer fragment in *Fusarium* was 55%, much higher than the 24% substitution observed here.

The β-tubulin data supports the recognition of five major clades, but the relationships between the clades are equivocal. A sister group relationship between *P. viridicatum* and *P. freii* is indicated by both the β-tubulin and 28S sequences, and these are the only two species that are not clearly deliniated into monophyletic lineages in this analysis. *Penicillium verrucosum* occupies an isolated position in both the β-tubulin and 28S analyses. *Penicillium commune, P. crustosum* and *P. solitum,* three species with large conidia that produce alkali on creatine-sucrose agar, form a relatively well-supported clade in the β-tubulin gene tree, with *P. commune* the basal species. These three taxa were not included in the 28S experiments. *Penicillium polonicum* and *P. cyclopium* form a relatively well-supported clade in the β-tubulin analyses, with each species forming a monophyletic lineage. This relationship was not evident in the 28S data. Likewise, *P. melanoconidium* represents a strongly supported monophyletic lineage in the β-tubulin analysis, although it cannot be distinguished from *P. aurantiogriseum* in the 28S analysis.

The β-tubulin gene provides an attractive addition to the ribosomal DNA gene sub-units that are more widely used for phylogenetic analysis in fungi. The amount of variation is suitable for studying relationships among closely related species, as has also been noted in *Fusarium* (O'Donnell and Cigelnik, 1995; O'Donnell, 1997). In designing the primers used here, Glass and Donaldson (1995) noted that they were useful for a broad range of filamentous ascomycetes, but yielded products of quite variable lengths among the taxa they studied. We have also used the Bt2 primers to amplify and sequence several species of *Fusarium* and have found little sequence homology between the *Penicillium* and *Fusarium* sequences (Seifert and Louis-Seize, unpublished). This indicates that the taxonomic range that can be inferred from analyses of sequences of these introns may be limited.

More evidence to support the species concepts tested here with the β-tubulin partial sequences might be obtained from sequencing other genes, but most of the regions of the ribosomal gene can be excluded. Sequences of the mitochondrial small subunit ribosomal DNA (mtSSU) are useful for delimiting species in *Fusarium* (O'Donnell, 1997), and were useful in investigations of infrageneric classification (e.g. LoBuglio *et al.,* 1994). However, our work on the mtSSU of the *P. aurantiogriseum* group reported above was abandoned when the amount of sequence variation was found to be minimal, a problem made worse by the poor quality sequencing gels resulting from the primers available at that time. Sequences of the internal transcribed spacer (ITS) do not exhibit sufficient variation in the *P. aurantiogriseum* group to allow the differentiation of the species originally delimited by mycotoxin profiles (Peterson, 1993; Frisvad, pers. comm.). Despite this, sufficient ITS divergence was found to delimit *P. roqueforti* from *P. carneum* (Frisvad) Frisvad, which had previously been considered a variety, and the new species *P. paneum* Frisvad (Boysen *et al.,* 1996). Sequences from the 28S region of the ribosomal DNA are apparently useful for delimiting species in some fungi, such as yeasts (Kurtzman and Robnett, 1995), but the variation observed here (about 0.6% phylogenetically significant substitutions) was not useful for delimiting the closely related species of the *P. aurantiogriseum* group. Similar conclusions were drawn by Peterson (1993). The remaining part of the ribosomal repeat, the intergenic spacer (IGS) is generally considered the most variable region, and is a logical candidate for future research.

ACKNOWLEDGEMENTS

We are grateful to Drs J.C. Frisvad, J.T. Mills and D. Malloch and the Canadian Collection of Fungal Cultures for providing the cultures used in this study, to Dr. L. Glass and P. Axelrood for providing information on primer sequences prior to their publication, and to Drs G. Klassen, S. Rehner and T. Ouellet for advice on DNA sequencing. Helpful presubmission reviews were provided by T. Ouellet and J. Bissett. Ms. A. Siegfried provided technical assistance in the early part of the study. Financial support from the Food Safety Program of Agriculture and Agri-Food Canada is gratefully acknowledged.

REFERENCES

Boysen, M., Skouboe, P., Frisvad, J. & Rossen, L. (1996). Reclassification of the *Penicillium roqueforti* group into three species on the basis of molecular genetic and biochemical profiles. Microbiology 142: 541-549.

Frisvad, J. C. & Filtenborg, O. (1983). Classification of terverticillate Penicillia based on profiles of mycotoxins and other secondary metabolites. Appl. Env. Microbiol. 46: 1301-1310.

Frisvad, J. C. & Filtenborg, O. (1989). Terverticillate penicillia: chemotaxonomy and mycotoxin production. Mycologia 81: 837-861.

Glass, N. L. & Donaldson, G. (1995). Development of primer sets designed for use with the PCR to amplify conserved from filamentous ascomycetes. Appl. Env. Microbiol. 61: 1323-1330.

Kurtzman, C. P. & Robnett, C. J. 1995. Molecular relationships among hyphal ascomycetous yeasts and yeastlike taxa. Can. J. Bot. 73 (Suppl. 1): S824-S830.

LoBuglio K.F., Pitt, J. I. & Taylor, J. W.(1994). Independent origins of the synnematous *Penicillium* species, *P. duclauxii, P. clavigerum,* and *P. vulpinum,* as assessed by two ribosomal DNA regions. Mycol. Res. 98:250-56

Mills, J. T., Seifert, K. A., Frisvad, J. C. & Abramson, D. (1995). Nephrotoxigenic *Penicillium* species occurring on farm-stored cereal grains in western Canada. Mycopathologia 130: 23-28.

O'Donnell, K. & Cigelnik, E.(1995). Two divergent intragenomic rDNA ITS 2 types within a monophyletic lineage of the fungus *Fusarium* are nonhomologous. Mol. Phylog. Evol. 7: 103-116.

O'Donnell, K. (1996). Towards a phylogenetic classification of *Fusarium*. Sydowia 48: 57-70.

O'Donnell,, K. (1997). Phylogenetic evidence indicates the important mycotoxigenic strains Fn-2, Fn-2B and Fn-M represent a new species of *Fusarium*. Mycotoxins (Tokyo): in press.

Peterson, S. W. (1993). Molecular genetic assessment of relatedness of *Penicillium* subgenus *Penicillium*. In: *The Fungal Holomorph: Mitotic, Meiotic and Pleomorphic Speciation in Fungal Systematics* (eds. Reynolds, D. R. & Taylor, J. W.). CAB International: Wallingford.

Pitt, J. I. (1979). The genus *Penicillium* and its teleomorphic states *Eupenicillium* and *Talaromyces*. Academic Press: London, New York, Toronto, Sydney, San Francisco.

Rehner, S. A. & Samuels, G. J. (1995). Molecular systematics of the Hypocreales: a teleomorph gene phylogeny and the status of their anamorphs. Can. J. Bot. 73: S816-S823.

Samson, R. A., Hoekstra, E. S., Frisvad, J. C. & Filtenborg, O. (1995). Introduction to Food-Borne Fungi, 4th edition. Centraalbureau voor Schimmelcultures, Baarn.

Samson, R. A., Stolk, A. C. & Hadlok, R. (1976). Revision of the subsection *Fasciculata* of *Penicillium* and some allied species. Stud. Mycol. 11: 1-47.

Seifert, K. A., Louis-Seize, G. & Savard, M. E. (1997). The phylogenetic relationships of two trichothecene-producing hyphomycetes, *Spicellum roseum* and *Trichothecium roseum*. Mycologia 89: 250-257.

Swofford, D. L. (1991). PAUP: Phylogenetic analysis using parsimony, 3.1.1. Laboratory of molecular systematics, Smithsonian Institution, Washington, DC.

White, T. J., Bruns, T. Lee, D.S. & Taylor, J. W. (1990). Amplification and direct sequencing of fungal ribosomal genes for phylogenetics. In: *PCR Protocols* (Eds. Innis, M.A., Gelfand, D.H., Sninsky, J. J. & White, T. J.). Academic Press: San Diego, CA. pp. 315-322.

DEVELOPMENT OF GENETIC MARKERS FOR POPULATION STUDIES OF *PENICILLIUM* SPP.

Søren Banke and Søren Rosendahl
Department of Mycology, University of Copenhagen, Øster Farimagsgade 2D,
DK-1353, Copenhagen, Denmark.

A new approach to developing co-dominant genetic markers for population studies of *Penicillium scabrosum* is described. The polymorphic genes were obtained by PCR using primers for known gene sequences from *Penicillium* species. Possible nucleotide differences between alleles were revealed by single strand conformation polymorphism (SSCP) and the differences were confirmed by direct sequencing of the PCR product. We obtained ten polymorphic loci which had from two to 23 alleles. The markers can be used to study population genetics of Penicillia from soil.

INTRODUCTION

Several DNA protocols have been developed for identifying polymorphic genetic markers for population studies of fungi. The methods are either based on restriction site polymorphisms: restriction fragment length polymorphism (RFLP); (McDonald and Martinez 1990) and amplified fragment length polymorphism (AFLP); (Rosendahl and Taylor, 1997) or on detection of random variable parts of the genome: RAPD (Williams *et al.* 1990) or microsatellites (Queller *et al.* 1993).

Homology of the defined alleles is critical if the markers are to be used in population genetic studies. The homology of the alleles can be difficult to assure by the RAPD technique unless the nucleotide fragments are sequenced. This can be done by sequencing with arbitrary primers (SWAP) which was used to compare homologous loci in a population genetic analysis of the human pathogen *Coccidioides immitis* (Burt *et al.* 1994, 1996).

The genetic markers generated by these various techniques will often represent DNA of unknown function including non-coding regions of the genome. Polymorphic regions of known genes can be identified and applied as genetic markers after sequencing with specific primers. This approach is laborious, however, and is not applicable if several samples are compared.

The polymorphisms in known genes can be detected by single strand conformation polymorphism (SSCP; Orita *et al.* 1989). The nucleotide differences between homologous genes are detected by electrophoresis of single stranded DNA under non-denaturing condition. The secondary structure of the strands will then reflect the nucleotide sequence and will generally result in different migration patterns. SSCP patterns are easier to interpret when the compared genes have relatively few base substitutions, and the technique is often used to detect mutations in specific genes. If the compared sequences have too many nucleotide substitutions this will result in unclear migration patterns on the SSCP gels.

In the original SSCP protocol, the patterns are visualised by radioactive labelling of the single strands. The protocol has later been improved and is now based on a discontinuous electrophoresis gel system where the DNA fragments can be stained by non-radioactive methods.

Several sequences of known structural and regulatory genes of *Penicillium* species and *Aspergillus* species are available in Genbank (http://www.ncbi.nlm.nih.gov/entrez). In this study we used sequences from Genbank to design primers for PCR amplification of known areas of the *Penicillium* genome. The polymorphic loci were subsequently detected by SSCP.

MATERIALS AND METHODS

Forty isolates of *Penicillium scabrosum* were obtained from a limestone quarry in Denmark. The isolates were grown in 500 ml Erhlenmeyer flasks with 80 ml liquid medium on a rotary shaker for 3-4 days at 20⌃C. The mycelium was harvested, washed in dd H_2O and freeze dried. DNA was extracted from 500 mg ground mycelium as described by Lee *et al.* (1988).

Primers were designed from Genbank sequences by aligning sequences from *Aspergillus* species with *Penicillium chrysogenum*. The conserved regions were used as primer sites for PCR amplification of homologous sequences from *P. scabrosum* and new specific *P. scabrosum* primers were designed from these sequences. Most of the primers were designed to produce a PCR product of about 300-500 bp (Table 1). PCR was performed in a PTC-100 termocycler (MJ-Research) using 2 min at 97°C and then 30 cycles of 1 min at 96°C, 1 min at 50-55⌃C and 1-2 min at 72⌃C. The PCR conditions were modified slightly for the different primers. The *Taq*-polymerase was purchased from Pharmacia Biotech (city, country), and the PCR buffer was 67mM Tris/HCl pH 8.5 with 2 mM $MgCl_2$, 16.6 mM $(NH_4)_2SO_4$ and 10 mM β-mercaptoethanol.

The protocol for the native SSCP gels was modified from Yab and McGee (1993). The electrophoresis used the cooled Hoefer Mighty Small SE250. The gels were 0.75 mm thick and 7 cm long. The resolving gel was composed of 25% (c:v) MDE acrylamide (FMC), 5% glycerol, 0.05% ammonium persulfate, 0.0125% TEMED and 0.5x TBE buffer. The denaturing stacking gel was made with 75% (w:v) formamide, 6% acrylamide (19:1) 0.15% APS, 0.1% TEMED and 0.2x TBE buffer.

The samples were prepared by mixing 6 µl PCR reaction with 4µl loading buffer (98% formamide, 10mM EDTA, 0.025% xylene cyanol and 0.025% bromophenol blue). The samples were then heated to 95°C for 5 min and snap cooled on ice before loading. The gels were run for 30 min at 1.5W per gel and then 2-3 h at 3W per gel with 0.5x TBE as running buffer. The gels were either stained with ethidium bromide, SYBR (Molecular Probes) or a background free silver stain (Blum *et al.* 1987).

The polymorphic loci were detected on the SSCP gel (Fig.1.) and the alleles analysed by direct sequencing. The samples for sequencing were cycle sequensed using the Thermo sequenase dye terminator pre mix kit (US 79765) Amersham (city, country), using the manufacturs instructions. The samples were then run on an automatic sequencer ABI 377 (Perkin Elmer), and the data analysed using Sequenser 3.0 ^{TM} software.

The gene sequence for TEF was cloned to detect possible nucleotide differences between multiple copies of the gene. The PCR cloning was performed using the PCR Script^{TM} cloning kit (Stratagene, (city, country) following the manufacture instructions.

RESULTS

The results show that it is possible to detect single point mutations by non-isotopic SSCP (Fig. 1). The mutations could be used as genetic markers to analyse a large number of *P. scabrosum* individuals in a population. We found eight polymorphic sites among the 13 loci that were examined in the study. The loci and the nucleotide differences responsible for the polymorphisms are listed in Table 1, along with the frequency distribution of the alleles. The nucleotide sequences of the genetic markers are shown in Figs. 2-8.

Table 1. The *Penicillium scabrosum* genes that were screened for polymorphisms.

Gene (locus)	size (bp)	Polymorphism (bp)	Allele frequency
Chitin syntase	450	1	0.13:0.87
Orotidine monophos-phate decarboxylase	468	2	0.10:0.90
β- tubulin	378	7	0.03:0.20:0.77
Nitrate reductase	281	94	(23 alleles)
Intron in actin	342	1	0.33:0.66
Transcription elongation factor	342	1	0.08:0.92
ITS 2	350	1	0.13:0.87
Unknown	467	2	0.13:0.97
ATPase subunit 6	490	1	0.03:0.97
Xylanase	450	1	0.03:0.97
ITS 1	420	monomorphic	-
Thioredoxin	300	monomorphic	-
Adenosine 5`phospho-sulfate kinase	309	monomorphic	-

The frequencies of the 23 alleles of nitrate reductase were not estimated.

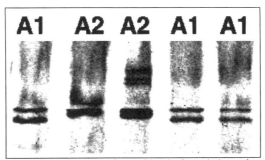

Figure 1. SSCP gel stained with EtBr showing polymorphism in the actin intron locus among five isolates of *Penicilium scabrosum*. A1 (allele 1) and A2 (allele 2) represent different alleles found on the gel, the difference in mobility of the single strands seen on the gel is due to one nucleotide substitution only.

```
AL 1   GAA CTG GTG CCC CGC AAC GCC CCC AAA TGC CCA ACT GAC CGG CCA AGT AAC
AL 2   GAA CTG GTG CCC CGC AAC GCC CCC AAA TGC CCA ACT GAC CGC CCA AGT AAC
  5´   GAA CTG GTG CCC CGC AAC GC..PRMERSITE..

AL 1   GGC CAG GTG ACG GCT CGA TGA GCC AAA GGC CCA CCT AGT CCT GGC GGC TGC
AL 2   GGC CAG GTG ACG GCT CGA TGA GCC AAA GGC CCA CCT AGT CCT GGC GGC TGC

AL 1   GCT TGG CGC CGT GAG GGT TCC AAG AAC CCT TAA TGG CGC CCG CGC CGC CGC
AL 2   GCT TGG CGC CGT GAG GGT TCC AAG AAC CCT TAA TGG CGC CCG CGC CGC CGC

AL 1   CTT GGC CGG AGC CAT GGA ATC CCC CCT GTC CGG GGT TGT AGC CTT GGG CAG
AL 2   CTT GGC CGG AGC CGT GGA ATC CCC CCT GTC CGG GGT TGT AGC CTT GGG CAG

AL 1   GGC CCC CCT GGA AGG GTG GTT GTC CTG GAG AAG AAG TTA GCT GCC GTT CGA
AL 2   GGC CCC CCT GGA AGG GTG GTT GTC CTG GAG AAG AAG TTA GCT GCC GTT CGA

AL 1   AGT CGC GGA ACA GTA TTT ATA TTG TTT ACT TAC CCT GCA TGC CCA TAT TTC
AL 2   AGT CGC GGA ACA GTA TTT ATA TTG TTT ACT TAC CCT GCA TGC CCA TAT TTC

AL 1   CTG GCA TAT TGC CCA TGC ATG GGT TGT CCC GGC ATG GGC CGA TGG CTG
AL 2   CTG GCA TAT TGC CCA TGC ATG GGT TGT CCC GGC ATG GGC CGA TGG CTG
                          PRIMERSITE..CCC GGC ATG GGC CGA TGG CTG 3´
```

Fig. 2. Partial actin intron sequences. The diagram shows the sequences of the two alleles (AL1 and AL2) found in the actin locus. One nucleotide is seen to differ between the two sequences. This position is marked on the sequences along with the primer sites.

```
  5´   GAT CTG GAA ACC CTG GAG GCA .....PRIMERSITE......
AL 1   GAT CTG GAA ACC CTG GAG GCA GTC GCA AGC CTC GGC CTC ACG GCG AAC GAC ATC
AL 2   GAT CTG GAA ACC CTG GAG GCA GTC GCA AGC CTC GGC CTC ACG GCG AAC CAC ATC
AL 3   GAT CTG GAA ACC CTG GAG GCA GTC GCA AGC TTG GGC CTC ACG GCG AAC AAC ATG

AL 1   GAT GAC CTG GTC AAC GAG CTC GGC ACC CTC AGT GTA GTG ACC CTT AGC CCA GTT
AL 2   GAT GAC CTG GTC AAC GAG CTC GGC ACC CTC AGT GTA GTG GCC CTT AGC CCA GTT
AL 3   GAT GAC CTG GTC AAC GAG CTC GGC ACC CTC AGT GTA GTG GCC CTT AGC CCA GTT

AL 1   GTT ACC AGC ACC GGA CTG ACC GAA GAC GAA GTT ATC GGG GCG GAA GAG CTT GCC
AL 2   GTT ACC AGC ACC GGA CTG ACC GAA GAC GAA GTT ATC GGG GCG GAA GAG CTT GCC
AL 3   GTT ACC AGC ACC GGA CTG ACC GAA GAT GAA GTT ATC GGG GCG GAA GAG CTT GCC

AL 1   GAA GGG ACC GGA GCG GAC AGC GTC CAT GGT ACC GGG CTC CAA ATC GAC GAG GAC
AL 2   GAA GGG ACC GGA GCG GAC AGC GTC CAT GGT ACC GGG CTC CAA ATC GAC GAG GAC
AL 3   GAA GGG ACC GGA GCG GAC AGC GTC CAT GGT ACC GGG CTC CAA ATT GAC GAG GAC

AL 1   AGC ACG GGG GAC ATA CTT GTC ACC GCT AGC CTG GGC GGT CAA AGA AAG AAG GTT
AL 2   AGC ACG GGG GAC ATA CTT GTC ACC GCT AGC CTG GGC GGT CAA AGA AAG AAG GTT
AL 3   AGC ACG GGG GAC ATA CTT GTC ACC GCT AGC CTG GGC GGT CAA AGA AAG AAG GTT

AL 1   AGA CAA CGC ACG TAA AAA GAA ATC ATA ATC ATT GTA CTC ACA TGG TTG AAG TAA
AL 2   AGA CAA CGC ACG TAA AAA GAA ATC ATA ATC ATT GTA CTC ACA TGG TTG AAG TAA
AL 3   AGA CAA CGC ACG TAA AAA GAA ATC ATA ATC ATT GTA CTC ACA TGG TTG AAG TAA

AL 1   ACG TTC ATA CGC TCC AAC TGG A GG TCG GAG GTA CCA TTG TAA
AL 2   ACG TTC ATA CGC TCC AAC TGG A GG TCG GAG GTA CCA TTG TAA
AL 3   ACG TTC ATA CGC TCC AAC TGG A GG TCG GAG GTA CCA TTG TAA
                PRIMERSITE..GG TCG GAG GTA CCA TTG TAA 3´
```

Fig. 3. Partial β- tubulin sequences. Three alleles were found in the β-tubulin locus. Seven nucleotide changes were found between allele 1 (AL1) and allele 3 (AL3), six changes between allele 2 (AL2) and AL3 and only one nucleotide difference between AL1 and AL2. The position of the nucleotide substitutions and the primer sites are marked on the sequences.

202

```
       5´ CAA GAA CAT TGA GCA CAT GT......PRIMERSITE...
AL 1      CAA GAA CAT TGA GCA CAT GTG AGA TCG TGG ACG CGA AGG TTC TAG TAG TAA AAT CAG
AL 2      CAA GAA CAT TGA GCA CAT GTG AGA TCG TGG ACG CGA AGG TTC TTG TAG TAA AAT CAG

AL 1      GGA TGT TAT CAT TGA TGG TCG TGC CAA TAT GAA CCC TCG GAC ACG CGC TGT ACT GGC
AL 2      GGA TGT TAT CAT TGA TGG TCG TGC CAA TAT GAA CCC TCG GAC ACG CGC TGT ACT GGC

AL 1      TGC TCT AGG TGT ATA CCA GGA TGG AAT TGC CAA GCA GCA AGT CGA CGG CAA GGA TGT
AL 2      TGC TCT AGG TGT ATA CCA GGA TGG AAT TGC CAA GCA GCA AGT CGA CGG CAA GGA TGT

AL 1      CAC CGC CCA TAT TTT TTA CGA GTA CAC TAC TCA AGT GGG GTT GGA AGT GAA GGG TAC
AL 2      CAC CGC CCA TAT TTT TTA CGA GTA CAC CAC TCA AGT GGG GTT GGA AGT GAA GGG TAC

AL 1      GCA GGT ACA CCT TAA GCC TAG GTC TGG ACC TCC CGT CCA AAT GAT TTT CTG GAA GGA
AL 2      GCA GGT ACA CCT TAA GCC TAG GTC TGG ACC TCC CGT CCA AAT GAT TTT CTG GAA GGA

AL 1      AAA GAA CCA AAA GAA AAT CAA CTC CCA TAG ATG GTT CTT CCA GGC TTT CGG TCG TCT
AL 2      AAA GAA CCA AAA GAA AAT CAA CTC CCA TAG ATG GTT CTT CCA CGC TAA CGG TCG TGT

AL 1      ATT GGA CCC CAA TAT CTG TGT TCT TCT TGA AGC CGG TAC CAA GCC CTG AAA GGA CCC
AL 2      ATT GGA CCC CAA TAT CTG TGT TCT TCT TGA AGC CGG TAC CAA GCC CTG AAA GGA CCC

AL 1      TAT CGA ACC CCT CGG GAA GGG TCT GAG ATT TAG CCC ATG TGT GGT GGT GCT TGT GGT
AL 2      TAT CGA ACC CCT CGG GAA GGG TCT GAG ATT TAG CCC ATG TGT GGT GGT GCT TGT GGT
                                               PRIMERSITE....GT GCT TGT GGT
AL 1      GAG ATC AAG
AL 2      GAG ATC AAG
          GAG ATC AAG 3´
```

Fig. 4. Partial chitin syntase sequences. The sequences for the two alleles, AL1 and AL2 show one base differences. The position of this difference is marked on the sequences along with the primer sites.

```
       5´ GCA TCG ATG AAG AAC GCA GC ..PRIMERSITE
AL 1      GCA TCG ATG AAG AAC GCA GCG AAA TGC GAT ACG TAA TGT GAA TTG CAG AAT TCA
AL 2      GCA TCG ATG AAG AAC GCA GCG AAA TGC GAT ACG TAA TGT GAA TTG CAG AAT TCA

AL 1      GTG AAT CAT CGA GTC TTT GAA CGC ACA TTG CGC CCT CTG GTA TTC CGG AGG GCA
AL 2      GTG AAT CAT CGA GTC TTT GAA CGC ACA TTG CGC CCT CTG GTA TTC CGG AGG GCA

AL 1      TGC CTG TCC GAG CGT CAT TGC TGC CCT CAA GCA CGG CTT GTG TGT TGG GCC CCG
AL 2      TGC CTG TCC GAG CGT CAT TGC TGC CCT CAA GCA CGG CTT GTG TGT TGG GCC CCG

AL 1      CCC CCC GAT CCC GGG GGG CGG GCC CGA AAG GCA GCG GCG GCA CCG CGT CCG GTC
AL 2      CCC CCC GAT CCC GGG GGG CGG GCC CGA AAG GCA GCG GCG GCA CCG CGT CCG GTC

AL 1      CTC GAG CGT ATG GGG CTT TGT CAC CCG CTC TGT AGG CCC GGC CGG CGC CTG CCG
AL 2      CTC GAG CGT ATG GGG CTT TGT CAC CCG CTC TGT AGG CCC GGC CGG CGC TTG CCG

AL 1      ATC AAC CAA ACT TTT TTC CAG GTT GAC CTC GGA TCA GGT AGG GAT ACC CGC TGA
AL 2      ATC AAC CAA ACT TTT TTC CAG GTT GAC CTC GGA TCA GGT AGG GAT ACC CGC TGA

AL 1      ACT TAA GCA TAT CAA TAA GCG GAG GA
AL 2      ACT TAA GCA TAT CAA TAA GCG GAG GA
PRIMERSITE.. GCA TAT CAA TAA GCG GAG GA 3´
```

Fig. 5. Partial ITS (internal transcribed spacer) sequences. Two sequences allele 1 (AL1) and allele 2 (AL2) were found at the ITS locus. The two sequences differ at only one base position. The position and the primer sites are marked on the aligned sequences.

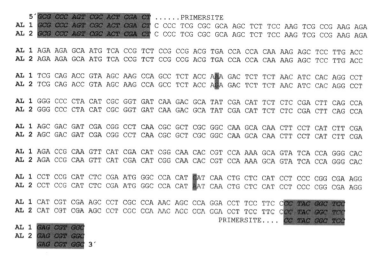

Fig. 6. Partial orotidine monophosphate decarboxylase sequences. The two sequences AL1 and AL2 represent the two alleles found at this locus. Two base pairs differs between the two sequences and these are marked on the sequences along with the primer sites.

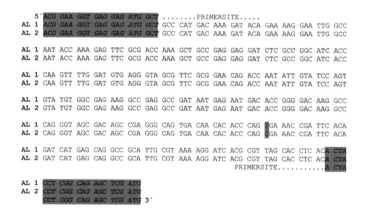

Fig. 7. Partial transcription elongation factor sequence. The two sequences represent the two alleles found (AL1 and AL2). The alleles show difference in one base position. The position of the polymorphism and the primer sites are marked on the sequences.

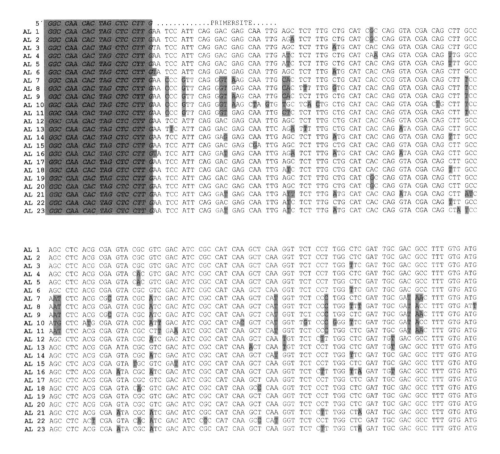

Fig. 8. Partial nitrate redutase sequences (Nitred). In this locus 23 alleles were found: AL1 to AL23. All 23 sequences are shown in the figure. The alleles can be grouped in three groups based on relatedness: AL1 to AL6 is the first group, AL7 to AL11 is the second group and AL12 to AL23 is the third group. The variable base positions and the primer sites are marked on the aligned sequences.

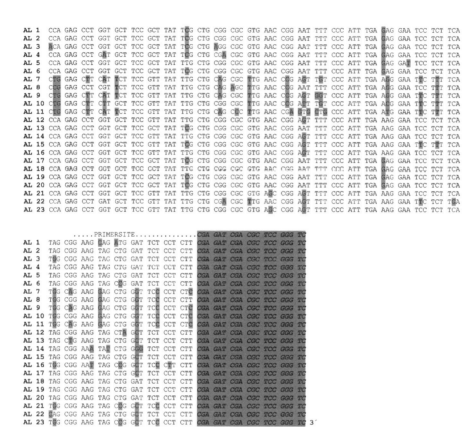

Fig 8. continuation. For legend see p. 205

For all genes except the β- tubulin gene and the Nitred promotor region we found two alleles at each locus. The alleles differed by one or two base substitutions. The β-tubulin locus was different as three alleles could be identified. Among these alleles seven nucleotide changes were found (Fig. 3).

The Nitred promotor region had several base substitutions (fig.8). It was difficult to distinguish between the alleles by SSCP and the isolates were therefore sequenced for the Nitred promotor region.

The gene for (TEF) transcription elongation factor has been reported to exist as several copies (Kambouris *et al.* 1993), The gene was then cloned to reveal possible differences in nucleotide sequences between the copies. The cloned DNA sequences showed all the same sequences indicating that all copies of the TEF gene have the same sequence. Two alleles of the TEF gene, differing by one base substitution, were seen (Fig. 7).

DISCUSSION

The use of stacking gels resulted in sharper bands and increased the sensitivity for detecting nucleotide differences compared with results using continuous gels. Single base changes may alter the mobility of one or both DNA single strands, although some changes may not result in detectable changes in conformation. The sensitivity for detection of mutations was 100% when estimated by comparing sequences of at least five isolates for each of the alleles detected on the gels. This is similar to what was found by Hayashi (1992)and better than in previous studies by Sheffield *et al.* (1993).

The SSCP technique can be used to screen for genetic markers. The polymorphic markers are easy to identify and the results can easily be reproduced. The method has some obvious advantages in comparison to other molecular markers (RAPD, AFLP). Firstly, the use of co-dominant alleles excludes null alleles from the analysis and secondly the homology of the molecular characters can be assured by direct sequencing.

The choice of molecular markers for population studies of fungi was discussed by Brown (1996) who stressed the importance of using alleles with appropriate frequencies. This can be done by excluding loci that are almost monomorphic as these will not contain sufficient phylogenetic information.

Two of the loci in the present study were nearly monomorphic. The data from the xylanase gene and the ATPase subunit 6 (Table 1) showed that only one isolate was? different from the other. Such markers have limited phylogenetic information and are not informative in population genetic analyses.

At the other extreme, a locus can also be so highly variable that each individual is represented by an unique allele. We found 23 alleles in the Nitred promotor region among the 40 isolates (Fig. 8). This locus could thus not be used as a marker in population genetic analysis of a large population as all individuals in the population would have to be sequenced for this particular gene. However, such highly variable loci could be useful in studies of microevolution in small populations. The gene showed 94 variable positions out of 281 base-pair positions and although the data were not analysed, the sequence differences indicated that *P.scabrosum* occurred as three distinct subpopulations (Fig 8).

The other loci studies had intermediate allele frequency distributions, and are useful and informative for population genetic analyses.

ACKNOWLEDGEMENTS

We would like to thank J.W. Taylor for helping with the primers, J. C. Frisvad and K. Lyhne for assisting in isolation and identification of the isolates. C.0 Hansen for sequencing of the samples and Kim B. Pedersen for critical remarks. F. Limestone quarry is thanked for allowing us to collect samples. This work was conducted within the *Centre for Identification and Characterisation of Fungi* supported by the Danish ministry of Business and Industry.

REFERENCES

Blum H., Beier H. & Gross H.J. (1987). Improved silver staining of plant protein, RNA and DNA in polyacrylamide gels. Electrophoresis 8: 93-99.

Brown J.K.M.(1996). The choice of molecular marker methods for population genetic studies of plant pathogens. New Phytologist 133: 183-193.

Burt A., Carter D.A., Koenig G.L., White T.J., & Taylor J.W. (1996) Molecular markers reveal cryptic sex in the human pathogen *Coccidioides immitis.* Proceedings of the National Academy of Science of the USA 93: 770-773.

Burt A., Carter D.A., White T.J. & Taylor J.W. (1994). DNA sequencing with arbitary primer pairs. Molecular Ecology 3: 523-525.

Hayashi K. (1992). PCR-SSCP: A method for detection of mutations. Geneti. Anal.: Techn. Appl. 90: 73-79.

Kambouris, N.G., Burke, D.J. & Creutz, C.E. 1993. Cloning and genetic characterization of a calcium/ and phospholipid/binding protein from *Saccharomyces cerevisiae* that is homologous to translation elongation factor-1 gamma. Yeast 9:151-163.

Lee S.B., Milgroom M.G. & Taylor J.W. (1988). A rapid, high yield mini-prep method for isolation of total genomic DNA from fungi. Fungal Geneti. Newsl. 35: 23-24

Lui Q. & Sommer S.S. (1994). Parameters affecting the sensitivies of Dideoxy fingerprinting and SSCP. PCR Meth. and Appli. 4: 97-104

McDonald B.A. & Martinez J.P.1990. Restriction fragment length polymorphism in Septoria tritici occur at a high frequency. Curr. Geneti. 17: 133-138.

Rosendahl S.R. & Taylor J.W. (1997). Development of multiple genetic markers for studies of genetic variation in arbuscular mycorrhizal fungi using AFLP™. Molec. Ecol. 6: 821-829

Orita M., Iwahana H., Knanzawa H., Hayashi K. & Sekiya T. (1989). Detection of polymorphisms of human DNA by gel electrophoresis as single strand conformation polymorphisms. Proc. Natl. Acad. Sci. U.S.A. 86: 2766-2770.

Queller D.C., Strassmann J.E. & Hughes C.R. (1993). Microsatellites and kinship. Trends Ecol. and Evol. 8: 285-288.

Sheffeild V.C., Beck J.S., Kwitek A.E., Sandstrom D.W. & Stone E.M. (1993). The sensitivity of single strand conformation polymorphism analysis for detection of single base substitutions. Genomics 16: .325-332.

Williams J.G.K., Kubelik A.R., Livak K.J., Rafalski J.A. & Tingey S.V. (1990). DNA polymorphism amplified by arbitary primers are useful as genetic markers Nucleic Acid Research 18: 6531-6535.

Yab E.P.H. & McGee J.O'D. (1993). Nonisotopic discontinuous phase single strand conformation polymorphism (DP-SSCP): genetic profiling of D-loop of human mitochondrial (mt) DNA. Nucl. Acids Res. 21: 4155.

A REVIEW OF CURRENT METHODS IN DNA FINGERPRINTING

James Scott and Neil Straus
Department of Botany, University of Toronto, Toronto, Ontario, CANADA, M5S 3B2

The development of the polymerase chain reaction (PCR) in the mid 1980s precipitated a series of rapid technological advancements in DNA-based diagnostics as well as DNA sequencing technology. Many such improvements have lead to the widespread acceptance and, indeed, routine use of DNA sequencing as a tool for addressing a myriad of previously untenable biological hypotheses. In particular, DNA sequencing has permitted the inferential elucidation of phylogeny in many difficult taxonomic groups, such as the Trichocomaceae.

As problems of relatedness and species concept among the anamorphs of the Trichocomaceae are resolved, questions of population biology or genetic variation below the species level become germane. To date, many molecular genetic techniques have been devised to discriminate 'individuals'. These methods include the classical hybridization approach of Restriction Fragment Length Polymorphism (RFLP) as well as several PCR-based approaches such as Random Amplified Polymorphic DNA (RAPD), Amplified Fragment Length Polymorphism (AFLP), RFLP of PCR products (PCR-RFLP), Single-Strand Conformation Polymorphism (SSCP), Microsatellite Single-Locus Fingerprinting and Heteroduplex Mobility Assay (HMA). This paper presents a critical assessment of these techniques and comments upon their appropriate uses.

INTRODUCTION

The ability to analyze and characterize genetic variation between individuals of the same species with both accuracy and precision has been a technological grail of biological science since Gregor Mendel's pioneering work on heredity in the mid-nineteenth century. Traditionally, morphological and later physiological characteristics of phenotype served as proxy measurements of genotypic variation. However, most phenotypic traits do not behave as strict Mendelian determinants and instead are regulated by multifactorial genetic expression (i.e. polygenetic inheritance) coupled with complex environmental influences. Thus, while useful as taxonomic and ecological indicators, these markers pose problems in population genetic studies. In more recent time, molecular approaches have attempted to solve these shortcomings, first with studies on protein variation and later, following the emergence of recombinant DNA technology, by analyses of the genetic material itself.

The recent introduction of the polymerase chain reaction (PCR) (Mullis, 1990; Saiki *et al.*, 1985, 1988) represents a quantum technological advance in applied molecular genetics, and has driven the development of a great diversity of ancillary methodologies

(White *et al.*, 1989). While some of these techniques represent truly novel ideas, the majority are modifications or combinations of existing procedures. Regardless that much explicit basic terminology exists for all of the fundamental techniques of molecular biology, too many authors coin new acronyms for trivial reasons. We have sought to avoid much of this redundant nomenclature by applying the nomenclatural concept of "earliest valid authority" where the core concept of a technique is decisively unique or novel.

The use of proteins to distinguish variation
Early approaches to the assessment of genetic diversity exploited polymorphisms in a variety of well characterized enzymes. Extracted proteins were separated by gel electrophoresis and allozymes (allelic variants of the same enzyme) were detected by reacting the gel with an appropriate substrate and dye. Allozymes with differing amino acid sequences appeared as bands with unique electrophoretic mobilities in native gels. With the development of allozyme analysis in the mid 1960s, it became possible to address questions about spatial and temporal distribution of genetic variability within and between populations and to investigate fundamental aspects of mating systems, recombination, gene flow and genetic drift. This early form of molecular genetic analysis remained prominent until the mid 1980s.

Hybridization-based markers
In the early 1970s, DNA-based technologies began to replace protein analyses. Initially, DNA-DNA reassociation kinetics coupled with DNA hybrid stability was used to analyze phylogenetic relationships of single-copy DNA in eukaryotic organisms (Shields and Straus, 1975; Sohn *et al.*, 1975). These pioneering techniques of the 1970s rapidly gave way to the more powerful methodologies of genetic engineering. Unlike allozyme assays, the still-widespread analysis of restriction fragment length polymorphisms (RFLP) had the advantage of examining DNA variability directly. In RFLP, genomic DNA is digested with a restriction endonuclease. The resulting DNA fragments are separated by gel electrophoresis, chemically denatured and transferred by Southern blotting to a DNA-binding membrane. The membrane is incubated in a solution containing a labelled nucleic acid probe and the hybridizing fragment bands are visualized by autoradiographic or chemiluminescent techniques. The resulting patterns may represent polymorphic forms of a structural gene with implication for breeding strategies or may be the result of more complex patterns arising from repeated elements of cryptic function.

Variable number tandem repeats (VNTRs) range from short, interspersed repeats of short nucleotide sequences to long stretches of recurrent, tandem sequence motifs. Families of VNTR sequences are known variously as "satellite" DNAs, appropriately prefixed to indicate the relative size category of the repeated element and its extent of dispersion within the genome (e.g. microsatellite, minisatellite). Variable minisatellite DNA was first investigated for DNA fingerprinting by Jefferies and co-workers (1985a, 1985b), and is discussed in detail by Jefferies (1987). RFLP technology has been widely exploited for mycological analysis and have been reviewed in depth by Weising and colleagues (1995).

Although RFLP methodologies have been powerful analytical tools, they require large amounts of genomic DNA, are difficult to automate and require substantial time to complete. The discovery of the polymerase chain reaction (PCR) (Saiki *et al.*, 1985,

1988) provided an opportunity to alleviate these constraints and in addition, offered new strategies of exploiting sequence variation.

LOW-STRINGENCY PCR -- MOLECULAR SERENDIPITY

Randomly amplified polymorphic DNA (RAPD)
Perhaps the most widespread PCR-based fingerprinting techniques currently in use are based on random-primed PCR. The use of single oligonucleotide primers of arbitrary sequence was introduced simultaneously by the independent groups of Williams and co-workers (1990) and Welsh and McClelland (1990). Williams *et al.* (1990) systematically studied single primer PCR using a set of randomly designed decamers as well as sequentially truncated versions of a selected primer. In their methodology, known as random amplified polymorphic DNA (RAPD), amplified fragments were separated using agarose gel electrophoresis and visualized by ethidium bromide staining. In contrast, Welsh and McClelland (1990) described a similar method, where single primers of arbitrary sequence of various lengths (optimally 20 bp) were used in PCR of candidate template DNAs, into which α-^{32}P dCTP was incorporated as a radiolabel. The fingerprint generated consists of a unique profile of fragment sizes separated by denaturing polyacrylamide gel electrophoresis and visualized by autoradiography. A third variation was proposed by Caetano-Anollés and colleagues (1992) where even shorter oligonucleotide primers (5 bp) were used in a single-primer PCR. Fragment profiles were separated using denaturing polyacrylamide gel electrophoresis and visualized by silver staining. Of these three related methods, RAPD (*sensu* Williams *et al.*, 1990) has enjoyed the greatest acceptance to date (Hedrick, 1992; Weising *et al.*, 1995). These techniques offer a cheaper and less time-consuming alternative to RFLPs. In addition, markers may be developed rapidly by screening a panel of candidate arbitrary-sequence oligonucleotide primers without *a priori* knowledge of target sequence (Williams *et al.*, 1990). Also, polymorphisms which are inaccessible by RFLP analysis may be accessed by these methods (Williams *et al.*, 1990). However, marker dominance is not a complication for the analysis of ascomycetous and allied fungi since their vegetative mycelia are haploid (Weising *et al.*, 1995; Williams *et al.*, 1990).

With time, numerous authors have advanced general modifications and application-specific refinements to the basic concept behind RAPD technology, such as varying the concentrations of reactants (Tommerup *et al.*, 1995), thermal cycling set-up protocols and profiles (Bielawski *et al.*, 1995; Kelly *et al.*, 1994; Yu and Pauls, 1992) and using paired primers (Micheli *et al.*, 1993; Welsh and McClelland, 1991). However, many investigators have found random primer fingerprinting methods to be hampered by problems relating to reproducibility and consequently have questioned RAPD reliability for certain types of analyses. These problems largely stem from a critical dependence of the fragment profile and relative yield on all of the basic parameters of PCR (Bielawski *et al.*, 1995), such as annealing temperature and extension time (Ellsworth *et al.*, 1993; Penner *et al.*, 1993), primer concentration (Ellsworth *et al.*, 1993; Muralidharan and Wakeland, 1993; MacPherson *et al.*, 1993), template quality (Micheli *et al.*, 1993) and concentration (Davin-Regli *et al.*, 1995; Micheli *et al.*, 1993; Muralidharan and Wakeland, 1993), reactant concentration (Ellsworth *et al.*, 1993), the particular commercial brand of DNA polymerase (Meunier and Grimont, 1993; Schierwater and

Ender, 1993; Tommerup *et al.*, 1995) and the make and model of thermal cycler (He *et al.*, 1994; MacPherson *et al.*, 1993; Meunier and Grimont, 1993). Penner and co-workers (1993) noted problems with RAPD reproducibility between seven laboratories utilizing the same primers and templates with some varied reaction conditions. Also, several workers have described the presence of non-parental bands resulting from PCR artifacts such as heteroduplexes (Ayliffe *et al.*, 1994) or other interference (Hallden *et al.*, 1996; Micheli *et al.*, 1993; Riedy *et al.*, 1992). The observation of heteroduplexed DNAs as RAPD bands, however, has been suggested as a means of identifying codominant RAPD markers (Davis *et al.*, 1995). While useful as initial screens for polymorphic loci which subsequently may be cloned and sequenced to construct site-specific primers (e.g. Groppe *et al.*, 1995), PCR-based fingerprinting techniques relying on random primers are not robust and generally unsuitable for use as population markers, particularly in critical or demanding situations such as human diagnostics or courtroom evidence (Riedy *et al.*, 1992).

Amplified fragment length polymorphism (AFLP)
A fingerprinting method which combines elements of both RFLP and random primer PCR was described by Zabeau and Vos (1993) and Vos and co-workers (1995). The authors called this technique amplification fragment length polymorphism (AFLP1, Vos *et al.*, 1995), a clearly intentional allusion to the acronym "RFLP". Although RFLPs originally examined polymorphisms at identified loci by hybridization, AFLP compares polymorphic patterns in fragments generated from the selected amplification of a subset of restriction fragments. In the first stage of this method, a template DNA is digested with restriction endonucleases. Concomitantly, one or two "adapter" molecules, consisting of 18-20 bp duplexed oligomers synthesized with 5'- and 3' sequence homology to the respective "sticky" ends produced by restriction digestion of the template DNA are ligated such that the original restriction sites are not restored. A panel of single primers (Mueller *et al.*, 1996) or primer pairs (Vos *et al.*, 1995; Zabeau and Vos, 1993) designed with homology to core sequences in the adaptor molecules, but with the addition of two arbitrary bases at the 3' termini were used in high stringency, "touchdown" PCR (Don *et al.*, 1991). In the seminal paper by Vos and colleagues (1995), primers were end-labelled using T4 polynucleotide kinase with either γ-^{32}P- or γ-^{33}P-dATP to facilitate detection of the amplified fragments by autoradiography following separation by denaturing polyacrylamide gel electrophoresis. Curiously, however, this paper specifically stated that "adapters were not phosphorylated", in which case it is apparent that the template strand of the adapters could not ligate to the 3'-OH terminus of the cleaved genomic DNA. As such, it is not clear how the primers initiate the first round of PCR.

To date, several studies have demonstrated the use of a practical variation of this methodology in resolving relatedness within bacterial (Lin *et al.*, 1996) and fungal populations (Majer *et al.*, 1996; Mueller *et al.*, 1996). Mueller and co-workers (1996) used a single adapter, and the double-stranded PCR products were resolved by agarose gel electrophoresis and visualized with ethidium bromide staining. However, the study by

---

[1] "AFLP" as "amplification fragment length polymorphism" had been applied previously by Caetano-Anollés and co-workers (1992) as a general term referring to PCR fingerprinting methods directed by single primers of arbitrary sequence.

Majer and colleagues (1996) followed more closely the original AFLP technique (Vos *et al.*, 1995). They suggest that AFLP appears to be more robust than RAPD and related methods because the longer primers and known target sequence permit a higher stringency of hybridization during the amplification procedure (Majer *et al.*, 1996). High stringency hybridizations in PCR inherently produce fewer artifactual bands due to spurious priming events (Cha and Thilly, 1993; Dieffenbach *et al.*, 1993).

HIGH-STRINGENCY PCR SITE-SPECIFIC POLYMORPHISMS

The identification of specific, polymorphic loci circumvents the problem of irreproducibility common to random-primer fingerprinting methods. However, unlike random primer techniques, locus-specific methods often require a substantial initial effort to identify and characterize suitable loci.

One category of site-specific polymorphisms exploits microsatellite DNA. Microsatellite regions are PCR-amplified using primers either based on adjacent, conserved sequences (Groppe *et al.*, 1995) or consisting of a short, repeated sequence known to be present in the test organism from probe hybridization data (Buscot, 1996; Meyer *et al.*, 1992; Morgante and Olivieri, 1993; Schönian *et al.*, 1993). Polymorphisms in microsatellite DNA consist of variation in the number of repeated elements and are detected as relative length polymorphisms following electrophoretic separation. Due to the high degree of variation in these sequences, however, individual loci may be quite taxon-specific, restricting their use as general markers. Several current reviews discuss applications of microsatellite single-locus fingerprinting (Bruford and Wayne, 1993; Weising *et al.*, 1995).

Another source of site-specific genetic variability can be found in the sequences of introns (short stretches of non-coding sequences which punctuate many genes) of single copy metabolic and structural genes. Although these loci, like microsatellite markers are difficult to develop, they are particularly useful when fingerprint data from multiple loci are needed to study patterns of genetic variability (e.g. clonality vs. mating and recombination) (Anderson and Kohn, 1995). Primers designed using highly conserved sequences flanking introns have been described in a number of genes for filamentous fungi and yeasts (Glass and Donaldson, 1995). The variability of intron sequences, however, may not necessarily involve length polymorphism and may result solely from alterations in base sequence between fragments of identical size, necessitating a more sophisticated method of detection than that used for microsatellite typing. Several of the more commonly used methods are discussed below. These techniques and others have been reviewed recently by Prosser (1993) and Cotton (1993).

Unquestionably the most widely used locus for genetic discrimination to date is the subrepeat of the nuclear and mitochondrial ribosomal RNA (rRNA) genes (Gargas and DePriest, 1996; White *et al.*, 1990). These genes provide a wide range of useful polymorphism, with highly variable regions suitable for fingerprinting (e.g. internal transcribed- and intergenic, non-transcribed spacer regions) as well as more conserved domains appropriate for several levels of phylogenetic study (e.g. small and large ribosomal subunit genes). Numerous primer sequences have been described for PCR mediated amplification of specific regions of rDNA from filamentous fungi and yeasts (Gargas and DePriest, 1996; White *et al.*, 1990). Like intron loci, polymorphisms in rRNA may be cryptic and not necessarily reflected by length polymorphism of PCR-amplified fragments, requiring a sensitive detection method.

Detection of low-level sequence variability

The determination of base sequence is the ultimate, definitive method for the discrimination of occult genetic differences and the defining of mutations. However, despite recent extraordinary technological advancements in DNA sequencing technologies that have occurred since the introduction of these methods (Maxam and Gilbert, 1980; Sanger *et al.*, 1977), the resources required for large-scale sequencing projects remain untenable for most laboratories (Chowdhury *et al.*, 1993; Cotton, 1993). Therefore, a number of techniques have been described which permit the inferential comparison of base sequence, aspiring to offer a high degree of sensitivity and reliability with low cost and ease of use.

Restriction endonuclease digestion of PCR products

The simplest method for screening sequence variability in PCR products is digestion by restriction endonucleases (PCR-RFLP) in which subject PCR products are digested individually with a panel of restriction enzymes. Typically, restriction endonucleases that cleave at quadrameric recognition sequences are chosen because the occurrence of the shorter recognition motif statistically occurs with greater frequency. Polymorphisms are recognized by different fragment profiles following electrophoretic separation. Variation detected by this method is limited to either fragment length differences or changes in base sequence which result in the loss or gain of a restriction enzyme recognition site. Several studies of fungal populations have used this method with some success (Buscot *et al.*, 1996; Donaldson and Glass, 1995; Gardes *et al.*, 1991; Glass and Donaldson, 1995). As well, PCR-RFLP has been used as a means of taxonomic positioning (Vilgalys and Hester, 1990). However, relatively little sequence information can be inferred from PCR-RFLP analysis since the probability of encountering a change in a specific site of four nucleotides is quite low. Although the inclusion of additional restriction enzymes improves the analysis, very little base sequence can be compared using this method.

Denaturing-gradient gel electrophoresis (DGGE)

Denaturing-gradient gel electrophoresis (DGGE) is a method that has long been used to identify single base mutations in DNA fragments (Fisher and Lerman, 1983). This technique relies upon the alteration of melting domains between DNA fragments differing in base composition when separated electrophoretically on gels containing an ascending gradient of chemical denaturant. Partially annealed DNA duplexes migrate differentially, modulated by the degree and relative location of melted domains. Thus, the resolution of the technique is dependent upon at least partial association of the DNA strands, and ceases to provide further useful separation upon complete denaturation. Myers and co-workers (1985) attached a segment of G/C-rich sequence (a "GC-clamp") to one end of the DNA fragment prior to DGGE, to maintain partial duplex association at higher concentrations of denaturant in order to achieve better resolution. Sheffield and colleagues (1989, 1992) described the inclusion of similar G/C-rich sequences using PCR. Several recent reviews discuss applications of DGGE (Cotton, 1993; Prosser, 1993).

214

Single-strand conformation polymorphism (SSCP)

A popular method for studying low-level sequence variability in PCR products known as single-strand conformation polymorphism (SSCP) was described by Orita and co-workers (1989a, 1989b). In this technique, radio labelled PCR products are chemically denatured, separated by electrophoresis in native polyacrylamide gels and visualized by autoradiography. The electrophoretic migration of single strands is a function of secondary structure formed due to spontaneous self-annealing upon entry into the non-denaturing gel matrix. Single base substitutions may alter this secondary structure; hence, changing the relative electrophoretic mobility of the molecule.

SSCP analysis has been used widely to identify deleterious mutations relating to human genetic diseases such as colour vision defects (Zhang and Minoda, 1996), cystic fibrosis (Ravnikglavac *et al.*, 1994), Tay-Sachs disease (Ainsworth *et al.*, 1991), phenylketonuria (Dockhorn-Dworniczak *et al.*, 1991), and a number of p53-associated carcinomas including lung cancer (Suzuki *et al.*, 1990), lymphoblastic leukemia and non-Hodgkin lymphoma (Gaidano *et al.*, 1991), breast- and colon cancer (Soto and Sukumar, 1992) and resistance to thyroid hormone (Grace *et al.*, 1995). As well, SSCP has been employed in numerous investigations of organismal variability such as human papilloma virus (HPV) (Spinardi *et al.*, 1991), major histocompatibility complex in Swedish moose (Ellegren *et al.*, 1996), and differentiation of species of *Aspergillus* section *Flavi* (Kumeda and Asao, 1996).

Sheffield and co-workers (1993) studied the sensitivity of SSCP analysis for the detection of single base substitutions using an assortment of 64 characterized mutations (e.g. murine globulin promoter, p53 and rhodopsin). Using SSCP, they were able to detect roughly 80 % of these single base mutations. This detection rate is consistent with results from other studies (Hayashi, 1992; Spinardi *et al.*, 1991). Hayashi (1991, 1992) and Sheffield and colleagues (1993) found that fragment size[2] as well as location of the substituted base are the major factors governing the sensitivity and thus the success of this technique. Dideoxy-fingerprinting (ddF) is a hybrid variant of SSCP (Orita *et al.*, 1989a, 1989b) and dideoxy sequencing (Sanger *et al.*, 1977) in which a typical Sanger reaction with one dideoxynucleotide is followed by chemical denaturation and electrophoresis on a non-denaturing polyacrylamide gel (Blaszyk *et al.*, 1995). This technique has been used to detect single base mutations with an astonishingly high level of sensitivity (Blaszyk *et al.*, 1995; Ellison *et al.*, 1994; Felmlee *et al.*, 1995; Fox *et al.*, 1995; Stratakis *et al.*, 1996).

A number of procedural modifications have been proposed to enhance the sensitivity and reproducibility of SSCP including concentrating a single strand using asymmetric PCR (Ainsworth *et al.*, 1991), decreasing dNTP concentration in initial PCR to increase incorporation of radiolabel (Dean and Gerrard, 1991), non-isotopic detection systems including silver staining (Ainsworth *et al.*, 1991; Dockhorn-Dworniczak *et al.*, 1991; Mohabeer *et al.*, 1991), fluorescence-based methods (Makino *et al.*, 1992), and ethidium bromide staining (Grace *et al.*, 1995), modifications to the gel substrate (Dean and Gerrard, 1991; Spinardi *et al.*, 1991), altering electrophoresis conditions such as temperature (Dean and Gerrard, 1991; Spinardi *et al.*, 1991), ionic strength of the electrophoresis buffer and gel matrix (Spinardi *et al.*, 1991) and changes to facilitate

---

[2] The ability to discriminate mutations diminishes with increased fragment size, e.g. > 200 bp

large-scale screening (Mashiyama *et al.*, 1990). Liu and Sommer (1995) performed multiple restriction enzyme digests on large PCR-amplified fragments and combined the products prior to SSCP, permitting this technique to be used effectively on larger amplicons. Several authors however have noted that results of SSCP analyses vary considerably between experiments under identical conditions (Dean and Gerrard, 1991; Soto and Sukumar, 1992).

Heteroduplex mobility assay (HMA)

Heteroduplex mobility assay (HMA) (Keen *et al.*, 1991; Delwart *et al.*, 1993) is a relatively new technique which is comparable in sensitivity to SSCP, but not so widely used (Cotton, 1993). In HMA, the PCR amplification products of a pair of isolates are combined in equimolar proportion, heat denatured and reannealed at lower temperature. The resulting mixture comprises duplexed molecules of all possible combinations of compatible DNAs including two populations of homoduplexes identical to the fragments of each original amplification product, as well as two hybrid DNAs (heteroduplexes) created by the cross-annealing of compatible strands originating from different "parent" duplexes. Differences in base sequence such as substitutions, insertions or deletions between the two strands of heteroduplexed DNAs produce local "bubbles" or "kinks" in these hybrid molecules. Indeed, Wang and co-workers (1992) confirmed the bent physical conformation of heteroduplexed DNA fragments by visualizing fragments containing deletional kinks using electron microscopy. These structural instabilities result in the slower migration of the heteroduplex molecules to duplexes with total base complementarity when separated by electrophoresis in native polyacrylamide gels. The rate of detection of single base mutations by this technique is comparable to that observed in SSCP (Ganguly *et al.*, 1993; Ravnikglavac *et al.*, 1994; White *et al.*, 1992). Other authors have reported significantly better discrimination of small mutations using HMA relative to SSCP (Offermans *et al.*, 1996). However, a number of studies using heteroduplexed DNA employ excessively long annealing times, or inappropriate annealing temperatures which unintentionally favour the reannealing of homoduplexes (e.g. Bachmann *et al.*, 1994; Cheng *et al.*, 1994; D'Amato and Sorrentino, 1994; Delwart *et al.*, 1993, 1994; El-Borai *et al.*, 1994; Gross and Nilsson, 1995; Soto and Sukumar, 1992; Wilson *et al.*, 1995; Winter *et al.*, 1985).

While DNA heteroduplexes were first noted as artifacts of PCR (Jensen and Straus, 1993; Nagamine *et al.*, 1989), the differential mobility of deliberate DNA heteroduplexes has since been used in the recognition of various human genetic mutations including human p53-related tumours (Soto and Sukumar, 1992), *ras* oncogenes (Winter *et al.*, 1985;), type 1 antithrombin (Chowdhury *et al.*, 1993), cystic fibrosis (Dodson and Kant, 1991; Ravnikglavac *et al.*, 1994), endometrial adenocarcinoma (Doherty *et al.*, 1995), sickle cell anaemia (Wood *et al.*, 1993) and β-thalassaemia (Cai *et al.*, 1991; Hatcher *et al.*, 1993; Law *et al.*, 1994; Savage *et al.*, 1995). As well, heteroduplex technology has been proposed as a means of human leukocyte antigen typing (HLA typing) (D'Amato and Sorrentino, 1994; El-Borai *et al.*, 1994; Martinelli *et al.*, 1996). HMA has also been used to study genetic variation in viruses such as human immunodeficiency virus (HIV) (Bachmann *et al.*, 1994; Delwart *et al.*, 1993, 1994; Louwagie *et al.*, 1994) and hepatitis C virus (Gretch *et al.*, 1996; Wilson *et al.*, 1995), as well as eukaryotes such as European populations of the basidiomycete plant pathogen *Heterobasidion annosum* (Cheng *et al.*,

1994), Swedish populations of brown trout (Gross and Nilsson, 1995) and eastern Australian rabbit populations (Fuller *et al.*, 1996).

Since the introduction of HMA as a diagnostic tool, a number of modifications have been proposed including the use of ethidium bromide staining (Bachmann *et al.*, 1994; Chowdhury *et al.*, 1993; D'Amato and Sorrentino, 1994; Delwart *et al.*, 1994; Dodson and Kant, 1991; El-Borai *et al.*, 1994; Hatcher *et al.*, 1993; Pulyaeva *et al.*, 1994; Soto and Sukumar, 1992; Wood *et al.*, 1993) or silver staining (Gross and Nilsson, 1995) for visualization, lower ionic strength of electrophoresis buffer (Chowdhury *et al.*, 1993; Soto and Sukumar, 1992), the use of buffering system other than TBE (D'Amato and Sorrentino, 1994; Ganguly *et al.*, 1993), detection of heteroduplexed fragments by capillary electrophoresis (Cheng *et al.*, 1994), electrophoresis on temperature gradient gels (Campbell *et al.*, 1995) and the use of heteroduplex generators or universal comparative DNAs (D'Amato and Sorrentino, 1995; Doherty *et al.*, 1995; El-Borai *et al.*, 1994; Gross and Nilsson, 1995; Law *et al.*, 1994; Louwagie *et al.*, 1994; Martinelli *et al.*, 1996; Savage *et al.*, 1995; Wack *et al.*, 1996; Wood *et al.*, 1993).

Much of the earlier work using DNA heteroduplex analysis used Hydrolink MDE, which forms a vinyl polymer as an electrophoresis medium (Soto and Sukumar, 1992). Zakharov and Chrambach (1994) demonstrated increased resolution of DNA heteroduplexes by electrophoresis in low-crosslinked polyacrylamide gels, noting however that the gel matrix is more prone to swelling when synthesized with lower concentrations of N,N'-methylenebisacrylamide (Bis) due to less reproducible fibre properties (Zakharov and Chrambach, 1994). Pulyaeva and co-workers (1994) used uncrosslinked polyacrylamide gels with similar success. However, they noted that uncrosslinked polyacrylamide media lacked the mechanical strength of low- or standard crosslinked gels and thus were not suitable for many applications (Pulyaeva *et al.*, 1994). Xing and colleagues (1996) amended 12 % polyacrylamide gels with 10 % glycerol and 2 % agarose for resolving DNA heteroduplexes. Several investigators have incorporated chemical denaturants such as ethylene glycol (Ganguly *et al.*, 1993), formamide (Ganguly *et al.*, 1993) and urea (White *et al.*, 1992) into polyacrylamide gels as a means of destabilizing small heteroduplexes to facilitate their detection.

A number of related techniques have developed from HMA including the enzymatic cleavage of mismatched bases in RNA/RNA or RNA/DNA heteroduplexes using RNase A as a way of facilitating the detection of heteroduplexed molecules by gel electrophoresis (Winter *et al.*, 1985), and the use of bacteriophage resolvases to cleave mismatched bases in DNA/DNA heteroduplexes (Marshal *et al.*, 1995). Gross and Nilsson (1995) performed a restriction digest of PCR-amplified growth hormone 2 (GH2) gene from brown trout prior to heteroduplex generation as a means reducing fragment size and thus enhancing electrophoretic resolution of heteroduplexes.

Oka and co-workers (1994) demonstrated that a carefully controlled thermal annealing gradient can be used to cause the preferential formation of DNA homoduplexes relative to heteroduplexes. Using double-labelled DNA fragments (one strand labelled with biotin, the other strand labelled with dinitrophenyl (DNP)) they performed a temperature-gradient annealing with a test DNA, following which the duplexes bearing a biotin-labelled strand were captured onto a streptavidin-coated microtitre plate. The treatment of these fragments with an anti-DNP conjugated alkaline phosphatase followed by the introduction of a chromogenic substrate permitted the quantification of original double-labelled homoduplexes by spectrophotometry. The population of regenerated double-labelled homoduplexes was inversely proportional to the degree of homology of the double-labelled DNA to the tester DNA. Using this method, which they named PCR-dependent preferential homoduplex formation assay (PCR-PHFA), Oka and colleagues (1994) were able to detect differences of as little as a single nucleotide substitution between the double-labelled fragment and the tester. Although proposed as an alternative to HLA typing, PCR-PHFA may hold promise for other automated diagnostic applications.

Like SSCP, HMA is a rapid technique that takes into account the entirety of the sequence variability of a PCR-amplified DNA fragment, rather than the limited amount of sequence information available from PCR-RFLP. Both SSCP and HMA require less sample manipulation than PCR-RFLP and fewer gel runs. However, SSCP requires the ability to compare lanes both within and between gel runs to assess similarity or difference between different products. This necessitates a high degree quality control and thus is demanding of hardware technology capable of precisely duplicating run conditions on an ongoing basis. On the other hand, because HMA compares two individual products in the same gel lane, there is little need for comparison of migration distances between different gels.

CONCLUSIONS

We have reviewed various approaches that could be applied to studies of fungal clonality and genetic diversity. From this we conclude that methods based on site specific polymorphisms represent a more robust approach to population analysis than RAPD or other non-specific methods. For the identification of different strains we suggest that heteroduplex mobility analysis (HMA) is superior to single-strand conformation polymorphism (SSCP) because two strains/isolates are compared in the same gel lane, eliminating difficulties in gel-to-gel comparison due to mobility variations from differences in running conditions inherent to SSCP analysis.

ACKNOWLEDGEMENTS
This work was supported by a strategic grant from the Natural Sciences and Research Council of Canada (NSERC) jointly to Dr. David Malloch and NS, and an NSERC postgraduate fellowship to JS. The authors wish to thank Laurie Ketch for proof-reading and commenting on an earlier version of this manuscript.

REFERENCES
Ainsworth, P.J., Surh, L.C. & Coulter-Mackie, M.B. (1991). Diagnostic single strand conformation polymorphism, (SSCP): a simplified non-radioisotopic method as applied to a Tay-Sachs B1 variant. Nucleic Acids Res. 19: 405-406.
Anderson, J.B. & Kohn, L.M. (1995). Clonality in soilborne, plant-pathogenic fungi. Ann. Rev. Phytopathol. 33: 369-391.
Ayliffe, M.A., Lawrence, G.J., Ellis, J.G. & Pryor, A.J. (1994). Heteroduplex molecules formed between allelic sequences cause nonparental RAPD bands. Nucleic Acids Res. 22: 1632-1636.
Bachmann, M.H., Delwart, E.L., Shpaer, E.G., Lingenfelter, P., Singal, R., Mullins, J.I. & WHO Network for HIV Isolation and Characterization. (1994). Rapid genetic characterization of HIV Type 1 strains from four World Health Organization-sponsored vaccine evaluation sites using a heteroduplex mobility assay. AIDS Research and Human Retroviruses 10: 1245-1353.

Bielawski, J.P., Noack, K. & Pumo, D.E. (1995). Reproducible amplification of RAPD markers from vertebrate DNA. BioTechniques 18: 856-857.

Blaszyk, H., Hartmann, A., Schroeder, J.J., McGovern, R.M., Sommer, S.S. & Kovach, J.S. (1995). Rapid and efficient screening for p53 gene-mutations by dideoxy fingerprinting. BioTechniques 18: 256-260.

Bruford, M.W. & Wayne, R.K. (1993). Microsatellites and their application to population genetic studies. Curr. Opin. Genet. Dev. 3: 939-943.

Buscot, F. (1996). DNA polymorphism in morels: PCR/RFLP analysis of the ribosomal DNA spacers and microsatellite-primed PCR. Mycol. Res. 100: 63-71.

Buscot, F., Wipf, D., Di Battista, C., Munch, J.-C., Botton, B. & Martin, F. (1996). DNA polymorphism in morels: PCR/RFLP analysis of the ribosomal DNA spacers and microsatellite-primed PCR. Mycol. Res. 100: 63-71.

Caetano-Anollés, G., Brassam, B.J. & Gresshoff, P.M. (1992). Primer-template interactions during DNA amplification fingerprinting with single arbitrary oligonucleotides. Mol. Gen. Genet. 235: 157-165.

Cai, S.-P., Eng, B., Kan, Y.W. & Chui, H.K. (1991). A rapid and simple electrophoretic method for the detection of mutations involving small insertion or deletion: application to β-thalassemia. Hum. Genet. 87: 728-730.

Campbell, N.J.H., Harriss, F.C., Elphinstone, M.S. & Baverstock, P.R. (1995). Outgroup heteroduplex analysis using temperature-gradient gel-electrophoresis: high resolution, large-scale screening of DNA variation in the mitochondrial control region. Molec. Ecol. 4: 407-418.

Cha, R.S. & Thilly W.G. (1993). Specificity, efficiency and fidelity of PCR. PCR Methods Appl. 3: S30-S37.

Cheng, J., Kasuga, T., Mitchelson, K.R., Lightly, E.R.T., Watson, N.D., Martin, W.J. & Atkinson, D. (1994). Polymerase chain reaction heteroduplex polymorphism analysis by entangled solution capillary electrophoresis. J. Chromatogr. 667: 169-177.

Chowdhury, V., Olds, R.J., Lane, D.A., Conard, J., Pabinger, I., Ryan, K., Bauer, K.A., Bhavnani, M., Abildgaard, U., Finazzi, G., Castaman, G., Mannucci, P.M. & Thein, S.L. (1993). Identification of nine novel mutations in type I antithrombin deficiency by heteroduplex screening. Brit. J. Haematol. 84: 656-661.

Cotton, R.G.H. (1993). Current methods of mutation detection. Mutation Res. 285: 125-144.

D'Amato, M. & Sorrentino, R. (1994). A simple and economical DRB1 typing procedure combining group-specific amplification, DNA heteroduplex and enzyme restriction analysis. Tissue Antigens 43: 295-301.

D'Amato, M. & Sorrentino, R. (1995). Short insertions in the partner strands greatly enhance the discriminating power of DNA heteroduplex analysis: resolution of HLA-DQB1 polymorphisms. Nucleic Acids Res. 23: 2078-2079.

Davin-Regli, A., Abed, Y., Charrel, R.N., Bollet, C. & de Micco, P. (1995). Variations in DNA concentrations significantly affect the reproducibility of RAPD fingerprint patterns. Res. Microbiol. 146: 561-568.

Davis, T.M., Yu, H., Haigis, K.M. & McGowan, P.J. (1995). Template mixing: a method of enhancing detection and interpretation of codominant RAPD markers. Theor. Appl. Genet. 91: 582-588.

Dean, M. & Gerrard, B. (1991). Helpful hints for the detection of single-stranded conformation polymorphisms. BioTechniques 10: 332-333.

Delwart E.L., Shpaer E.G., Louwagie J., McCutchan F.E., Grez M., Rubsamen-Waigmann H. & Mullins J.I. (1993). Genetic relationships determined by a DNA heteroduplex mobility assay: analysis of HIV-1 env genes. Science 262: 1257-1261.

Delwart, E.L., Sheppard, H.W., Walker, B.D., Goudsmit, J. & Mullins, J.I. (1994). Human immunodeficiency virus type 1 evolution in vivo tracked by DNA heteroduplex mobility assays. J. Virol. 68: 6672-6683.

Dieffenbach, C.W., Lowe, T.M.J. & Dveksler, G.S. (1993). General concepts for PCR primer design. PCR Methods Appl. 3: S30-S37.

Dockhorn-Dworniczak, B., Dworniczak, B., Brommelkamp, L., Bulles, J., Horst, J. & Bocker, W.W. (1991). Non-isotopic detection of single-strand conformation polymorphism (PCR-SSCP): a rapid and sensitive technique in the diagnosis of phenylketonuria. Nucleic Acids Res. 19: 2500.

Dodson, L.A. & Kant, J.A. (1991). Two-temperature PCR and heteroduples detection: application to rapid cystic fibrosis screening. Molec. Cell. Probes 5: 21-25.

Doherty, T., Connell, J., Stoerker, J., Markham, N., Shroyer, A.L. & Shroyer, K.R. (1995). Analysis of clonality by polymerase chain reaction for phosphoglycerate kinase-I: heteroduplex generator. Diagnostic Molec. Pathol. 4: 182-190.

Don, R.H., Cox, P.T., Wainwright, B.J., Baker, K. & Mattick, J.S. (1991). 'Touchdown' PCR to circumvent spurious priming during gene amplification. Nucleic Acids Res. 19: 4008.

219

Donaldson, G.C. & Glass, N.L. (1995). Primer sets developed to amplify conserved genes from filamentous ascomyceted are useful in differentiating *Fusarium* species associated with conifers. Appl. Environ. Microbiol. 61: 1331-1340.

El-Borai, M.H., D'Alfonso, S., Mazzola, G. & Fasano, M.E. (1994). A practical approach to HLA-DR genomic typing by heteroduplex analysis and a selective cleavage at position 86. Human Immunol. 40: 41-50.

Ellegren, H., Mikko, S., Wallin, K. & Andersson, L. (1996). Limited polymorphism at major histocompatibility complex (MHC) loci in the Swedish moose, *A. alces*. Molec. Ecol. 5: 3-9.

Ellison, J., Squires, G., Crutchfield, C. & Goldman, D. (1994). Detection of mutations and polymorphisms using fluorescence-based dideoxy fingerprinting (F-ddF). BioTechniques 17: 742.

Ellsworth, D.L., Rittenhouse, K.D., Honeycutt, R.L. (1993). Artifactual variation in randomly amplified polymorphic DNA banding patterns. BioTechniques 14: 214-217.

Felmlee, T.A., Liu, Q., Whelen, A.C., Williams, D., Sommer, S.S. & Persing, D.H. (1995). Genotypic detection of *Mycobacterium tuberculosum*-rifampin resistance: comparison of single-strand conformation polymorphism and dideoxy fingerprinting. J. Clin. Microbiol. 33: 1617-1623.

Fisher, S.G. & Lerman, L.S., (1983). DNA fragments differing by single base-pair substitutions are separated in denaturing-gradient gels: correspondence with melting theory. Proc. Natl. Acad. Sci. USA 80: 1579-1583.

Fox, S.A., Lareu, R.R. & Swanson, N.R. (1995). Rapid genotyping of hepatitis-c virus isolates by dideoxy fingerprinting. J. Virological Meth. 53: 1-9.

Fuller, S.J., Mather, P.B. & Wilson, J.C. (1996). Limited genetic differentiation among wild *Oryctolagus cuniculus* (rabbit) populations in arid eastern Australia. Heredity 77: 138-145.

Gaidano, G., Ballerini, P., Gong, J.Z., Inghirami, G., Neri, A., Newcomb, E.W., Magrath, I.T., Knowles, D.M. & Della-Favera, R. (1991). p53 mutations in human lymphoid malignancies: Association with Burkitt lymphoma and chronic lymphocytic leukemia. Proc. Natl. Acad. Sci. USA. 88: 5413-5417.

Ganguly, A., Rock, M.J. & Prockop, D.J. (1993). Conformation-sensitive gel electrophoresis for rapid detection of single-base differences in double-stranded PCR products and DNA fragments: evidence for solvent-induced bends in DNA heteroduplexes. Proc. Natl. Acad. Sci. USA 90: 10325-10329.

Gardes, M., White, T.J., Fortin, J.A., Bruns, T.D. & Taylor, J.W. (1991). Identification of indigenous and introduced symbiotic fungi in ectomycorrhizae by amplification of nuclear and mitochondrial DNA. Can. J. Bot. 69: 180-190.

Gargas, A. & DePriest, P.T. (1996). A nomenclature for fungal PCR primers with examples from intron-containing SSU rDNA. Mycologia 88: 745-748.

Glass, N.L. & Donaldson, G.C. (1995). Development of primer sets designed for use with the PCR to amplify conserved genes from filamentous ascomycetes. Appl. Environ. Microbiol. 61: 1323-1330.

Grace, M.B., Buzard, G.S. & Weintraub, B.D. (1995). Allele-specific associated polymorphism analysis: novel modification of SSCP for mutation detection in heterozygous alleles using the paradigm of resistance to thyroid hormone. Human Mutation 6: 232-242.

Gretch, D.R., Polyak, S.J., Wilson, J.J., Carithers, R.L., Perkins, J.D. & Corey, L. (1996). Tracking hepatitis-C virus quasi-species major and minor variants in symptomatic and asymptomatic liver-transplant recipients. J. Virol. 70: 7622-7631.

Groppe, K., Sanders, I., Wiemken, A. & Boller, T. (1995). A microsatellite marker for studying the ecology and diversity of fungal endophytes (*Epichloë* spp.) in grasses. Appl. Environ. Microbiol. 61: 3943-3949.

Gross, R. & Nilsson, J. (1995). Application of heteroduplex analysis for detecting variation within the growth hormone 2 gene in *Salmo trutta* L. (brown trout). Heredity 74: 286-295.

Hallden, C., Hansen, M., Nilsson, N.O., Hjerdin, A. & Sall, T. (1996). Competition as a source of errors in RAPD analysis. Theor. Appl. Genet. 93: 1185-1192.

Hatcher, S.L., Lambert, Q.T., Teplitz, R.L. & Carlson, J.R. (1993). Heteroduplex formation: a potential source of genotyping error from PCR products. Prenatal Diagnosis 13: 171-177.

Hayashi, K. (1991). PCR-SSCP: a simple and sensitive method for detection of mutations in the genomic DNA. PCR Methods Appl. 1: 34-38.

Hayashi, K. (1992). PCR-SSCP: a method for detection of mutations. Genet. Analysis: Techniques Applic. 9: 73-79.

He, Q., Viljanen, M. & Mertsola, J. (1994). Effects of thermocyclers and primers on the reproducibility of banding patterns in randomly amplified polymorphic DNA analysis. Mol. Cell Probes 8: 155-160.

Hedrick, P. (1992). Population genetics: shooting the RAPDs. Nature 355: 679-680.

Jefferies, A.J. (1987). Highly variable minisatellites and DNA fingerprints. Biochem Soc. Trans. 15: 309-317.

Jefferies, A.J., Wilson, V. & Thein, S.L. (1985a). Individual-specific 'fingerprints' of human DNA. Nature 316: 76-79.

Jefferies, A.J., Wilson, V. & Thien, S.L. (1985b). Hypervariable 'minisatellite' regions in human DNA. Nature 314: 67-73.

Jensen, M.A. & Straus, N. (1993). Effect of PCR conditions on the formation of heteroduplex and single-stranded DNA products in the amplification of bacterial ribosomal DNA spacer regions. PCR Meth. Appl. 3: 186-194.

Keen, J., Lester, D., Inglehearn, C., Curtis, A. & Bhattacharya, S. (1991). Rapid detection of single base mismatches in heteroduplexes on hydrolink gels. Trends Genet. 7: 5.

Kelly, A., Alcalá-Jiménez, A.R., Bainbridge, B.W., Heale, J.B., Pérez-Artés, E. & Jiménez-Diaz, R.M. (1994). Use of genetic fingerprinting and random amplified polymorphic DNA to characterize pathotypes of *Fusarium oxysporum* f. sp. *ciceris* infecting chickpea. Phytopathology 84: 1293-1298.

Kumeda, Y. & Asao, T. (1996). Single-strand conformation polymorphism analysis of PCR-amplified ribosomal DNA internal transcribed spacers to differentiate species of *Aspergillus* section *Flavi.* Appl. Environ. Microbiol. 62: 2947-2952.

Law, H.-Y., Ong, J., Yoon, C.-S., Cheng, H., Tan, C.-L. & Ng, I. (1994). Rapid antenatal diagnosis of β-thalassemia in Chinese caused by the common 4-bp deletion in codons 41/42 using high resolution agarose electrophoresis and heteroduplex detection. Biochem. Med. Metabol. Biol. 53: 149-151.

Lin, J-J., Kuo, J. & Ma, J. (1996). A PCR-based DNA fingerprinting technique: AFLP for molecular typing of bacteria. Nucleic Acids Res. 24: 3649-3650.

Liu, Q. & Sommer, S.S. (1995). Restriction-endonuclease fingerprinting: a sensitive method for screening mutations in long, contiguous segments of DNA. BioTechniques 18: 470-477.

Louwagie, J., Delwart, E.L., Mullins, J.I., McCutchan, F.E., Eddy, G. & Burke, D.S. (1994). Genetic analysis of HIV-1 isolates from Brazil reveals presence of two distinct genetic subtypes. AIDS Research and Human Retroviruses 10: 561-567.

MacPherson, J.M., Eckstein, P.E., Scoles, G.J. & Gajadhar, A.A. (1993). Variability of the random amplified polymorphic DNA assay among thermocyclers, and effects of primer and DNA concentration. Mol. Cell Probes 7: 293-299.

Majer, D., Mithen, R., Lewis, B.G., Vos, P. & Oliver, R.P. (1996). The use of AFLP fingerprinting for the detection of genetic-variation in fungi. Mycol. Res. 100: 1107-1111.

Makino, R., Sekiya, T. & Hayashi, K. (1992). F-SSCP: Fluorescence-based polymerase chain reaction-single-strand conformation polymorphism (PCR-SSCP) analysis. PCR Methods Appl. 2: 10-13.

Marshal, R.D., Koontz, J. & Sklar, J. (1995). Detection of mutations by cleavage of DNA heteroduplexes with bacteriophage resolvases. Nature Genet. 9: 177-183.

Martinelli, G., Trabetti, E., Farabegoli, P., Buzzi, M., Zaccaria, A., Testoni, N., Amabile, M., Casartelli, A., Devivo, A., Pignatti, P.F. & Tura, S. (1996). Fingerprinting of HLA-DQA by polymerase chain reaction and heteroduplex analysis. Molec. Cell. Probes 10: 123-127.

Mashiyama, S., Sekiya, T. & Hayashi, K. (1990). Screening of multiple DNA samples for detection of sequence changes. Technique 2: 304-306.

Maxam, A.M. & Gilbert, W. (1980). Sequencing end-labelled DNA with base-specific chemical cleavages. Meth. Enzymol. 65: 499-560.

Meunier, J.-R. & Grimont, P.A.D. (1993). Factors affecting reproducibility of random amplified polymorphic DNA fingerprinting. Res. Microbiol. 144: 373-379.

Meyer, W., Liecktfeldt, T., Kayser, T., Nürnberg, P., Epplen, J.T. & Börner, T. (1992). Fingerprinting fungal genomes with phage M13 DNA and oligonucleotide probes specific for simple repetitive DNA sequences. Adv. Mol. Gen. 5: 241-253.

Micheli, M.R., Bove, R., Calissano, P. & D'Ambrosio, E. (1993). Random amplified polymorphic DNA fingerprinting using combinations of oligonucleotide primers. BioTechniques 15: 388-390.

Mohabeer, J.A., Hiti, A.L. & Martin, W.J. (1991). Non-radioactive single strand conformation polymorphism (SSCP) using the Pharmacia 'PhastSystem'. Nucleic Acids Res. 19: 3154.

Morgante, M. & Olivieri, A.M. (1993). PCR-amplified microsatellites as markers in plant genetics. Plant J. 3: 175-182.

Mueller, U.G., Lipari, S.E., Milgroom, M.G. (1996). Amplified fragment length polymorphism (AFLP) fingerprinting of symbiotic fungi cultured by the fungus-growing ant *Cyphomyrmyx minutus.* Molecular Ecology 5: 119-122.

Mullis, K.B. (1990). Process for amplifying, detecting, and/or cloning nucleic acid sequences using a thermostable enzyme. US Patent 4,965,188.

221

Muralidharan, K. & Wakeland, E.K. (1993). Concentration of primer and template qualitatively affects products in randomly-amplified polymorphic DNA PCR. BioTechniques 14: 362-364.

Myers, R.M., Fischer, S.G., Maniatis, T. & Lerman, L.S. (1985). Nearly all single base substitutions in DNA fragments joined to a GC-clamp can be detected by denaturing gradient gel electrophoresis. Nucleic Acids Res. 13: 3111-3129.

Nagamine, C.M., Chan, K. & Lau, Y-F.C. (1989). A PCR artifact: generation of heteroduplexes. Am. J. Hum. Genet. 45: 337-339.

Offermans, M.T.C., Struyk, L., Degeus, B., Breedveld, F.C., Vandenelsen, P.J. & Rozing, J. (1996). Direct assessment of junctional diversity in rearranged t-cell receptor-beta chain encoding genes by combined heteroduplex and single-strand conformation polymorphism (SSCP) analysis. J. Immunol. Meth. 191: 21-31.

Oka, T., Matsunaga, H., Tokunaga, K., Mitsunaga, S., Juji, T. & Yamane, A. (1994). A simple method for detecting single base substitutions and its application to HLA-DPB1 typing. Nucleic Acids Res. 22: 1541-1547.

Orita, M., Iwahana, H., Kanazawa, H., Hayashi, K. & Sekiya, T. (1989a) Detection of polymorphisms of human DNA by gel electrophoresis as single-strand conformation polymorphisms. Proc. Natl. Acad. Sci. USA 86: 2766-2770.

Orita, M., Suzuki, Y., Sekiya, T. & Hayashi, K. (1989b) Rapid and sensitive detection of point mutations and DNA polymorphisms using the polymerase chain reaction. Genomics 5: 874-879.

Penner, G.A., Bush, A., Wise, R., Kim, W., Domier, L., Kasha, K., Laroche, A., Scoles, G., Molnar, S.J. & Fedak, G. (1993). Reproducibility of random amplified polymorphic DNA (RAPD) analysis among laboratories. PCR Methods Appl. 2: 341-345.

Prosser, J. (1993). Detecting single base mutations. Trends. Biol. Technol. 11: 238-246.

Pulyaeva, H., Zakharov, S.F., Garner, M.M. & Chrambach, A. (1994). Detection of a single base mismatch in double-stranded DNA by electrophoresis on uncrosslinked polyacrylamide gel. Electrophoresis 15: 1095-1100.

Ravnikglavac, M., Glavac, D. & Dean, M. (1994). Sensitivity of single-strand conformationa polymorphism and heteroduplex method for mutation detection in the cystic fibrosis gene. Human Molec. Genet. 3: 801-807.

Riedy, M.F., Hamilton W.J. III & Aquadro, C.F. (1992). Excess of non-parental bands in offspring from known primate pedigrees assayed using RAPD PCR. Nucleic Acids Res. 20: 918.

Saiki, R.K., Gelfand, D.H., Stoffel, S., Scharf, S.J., Higuchi, R., Hory, G.T., Mullis, K.B. & Erlich, H.A. (1988). Primer directed enzymatic amplification of DNA with a thermostable DNA polymerase. Science 239: 487-491.

Saiki, R.K., Scharf, S., Faloona, F., Mullis, K.B., Horn, G.T., Erlich, H.A. & Arnhein, N. (1985). Enzymatic amplification of beta-globulin genomic sequences and restriction site analysis for diagnosis of sickle cell anemia. Science 230: 1350-1354.

Sanger, F., Nicklen, S. & Coulson, A.R. (1977). DNA sequencing with chain terminating inhibitors. Proc. Natl. Acad. Sci. USA 74: 5463-5467.

Savage, D.A., Wood, N.A.P., Bidwell, J.L., Fitches, A., Old, J.M. & Hui, K.M. (1995). Detection of beta-thalassemia mutations using DNA heteroduplex generator molecules. Brit. J. Haematol. 90: 564-571.

Schierwater, B. & Ender, A. (1993). Different DNA polymerases may amplify different RAPD products. Nucleic Acids Res. 21: 4647-4648.

Schönian, G., Meunsel, O., Tietz, H.-J., Meyer, W., Gräser, Y, Tausch, I., Presber, W. & Mitchell, T.G. (1993). Identification of clinical strains of Candida albicans by DNA fingerprinting with polymerase chain reaction. Mycoses 36: 171-179.

Sheffield, V.C., Beck, J.S., Kwitek, A.E., Sandstrom, D.W. & Stone, E.M. (1993). The sensitivity of single-strand conformation polymorphism analysis for the detection of single-base substitutions. Genomics 16: 325-332.

Sheffield, V.C., Beck, J.S. & Stone, E.M. (1992). A simple and efficient method for attachment of a 40-base pair GC-rich sequence to PCR-amplified DNA. Biofeedback 12: 386-387.

Sheffield, V.C., Cox, D.R., Lerman, L.S. & Myers, R.M. (1989). Attachment of a 40-base-pair G+C-rich sequence (GC-clamp) to genomic DNA fragments by the polymerase chain reaction results in improved detection of single-base changes. Proc. Natl. Acad. Sci. USA 86: 232-236.

Shields, G.F. & Straus, N.A. (1975). DNA-DNA hybridization studies of birds. Evolution 29: 159-166.

Sohn, U.-I., Rothfels, K.H. & Straus, N.A. (1975). DNA:DNA hybridization studies in black flies. J. Molec. Evol. 5: 75-85.

Soto, D. & Sukumar, S. (1992). Improved detection of mutations in the p53 gene human tumors as single-stranded conformation polymorphisms and double-stranded heteroduplex DNA. PCR Methods Appl. 2: 96-98.

Spinardi, L., Mazars, R. & Theillet, C. (1991). Protocols for an improved detection of point mutations by SSCP. Nucleic Acids Res. 19: 4009.

Stratakis, C.A., Orban, Z., Burns, A.L., Vottero, A., Mitsiades, C.S., Marx, S.J., Abbassi, V. & Chrousos, G.P. (1996). Dideoxyfingerprinting (ddF) analysis of the type-x collagen gene (col10a1) and identification of a novel mutation (s671p) in a kindred with schmid metaphyseal chondrodysplasia. Biochem. Molec. Med. 59: 112-117.

Suzuki, Y., Orita, M., Shiraishi, M., Hayashi, K. & Sekiya, T. (1990). Detection of *ras* gene mutations in human lung cancers by single-strand conformation polymorphism analysis of polymerase chain reaction products. Oncogene 5: 1037-1043.

Tommerup, I.C., Barton, J.E., O'Brien, P.A. (1995). Reliability of RAPD fingerprinting of three basidiomycete fungi *Laccaria*, *Hydnangium*, and *Rhizoctonia*. Mycol. Res. 99: 179-186.

Vilgalys, R., & Hester, M. (1990). Rapid generic identification and mapping of enzymatically amplified ribosomal DNA from several *Cryptococcus* species. J. Bacteriol. 172: 4238-4246.

Vos, P., Hojers, R., Bleeker, M., Reijans, M., van de Lee, T., Hornes, M., Frijters, A., Pot, J., Peleman, J., Kuiper, M. & Zabeau, M. (1995). AFLP: a new technique for DNA fingerprinting. Nucleic Acids Res. 23: 4407-4414.

Wack, A., Montagna, D., Dellabona, P. & Casorati, G. (1996). An improved PCR-heteroduplex method permits high-sensitivity detection of clonal expansions in complex t-cell populations. J. Immunol. Meth. 196: 181-192.

Wang, Y.-H., Barker, P. & Griffith, J. (1992). Visualization of diagnostic heteroduplex DNAs from cystic fibrosis deletion heterozygotes provides an estimate of the kinking of DNA by bulged bases. J. Biol. Chem. 267: 4911-4915.

Weising, K., Nybom, H., Wolff, K. & Meyer, W. (1995). DNA fingerprinting in plants and fungi. CRC Press, Boca Raton. 322 pp.

Welsh, J. & McClelland, M. (1990). Fingerprinting genomes using PCR with arbitrary primers. Nucleic Acids Res. 18: 7213-7218.

Welsh, J. & McClelland, M. (1991). Genomic fingerprinting using arbitrary primes PCR and a matrix of pairwise combinations of primers. Nucleic Acids Res. 19: 5275-5279.

White, M.B., Carvalho, M., Derse, D., O'Brien, S.J. & Dean, M. (1992). Detecting single-base substitutions as heteroduplex polymorphisms. Genomics 12: 301-306.

White, T.J., Arnheim, N. & Erlich, H.A. (1989). The polymerase chain reaction. Trends Genet. 5: 185-189.

White, T.J., Bruns, T., Lee, S. & Taylor, J. (1990). Amplification and direct sequencing of fungal ribosomal RNA genes for phylogenetics. In: PCR protocols, A guide to methods and applications. Innis, M.A., Gelfand, D.H., Sninsky, J.J. & White, T.J. (eds), Academic Press, San Diego, pp. 315-322.

Williams, J.G.K., Kubelik, A.R., Livak, K.J., Rafalski, J.A. & Tingey, S.V. (1990). DNA polymorphisms amplified by arbitrary primers are useful as genetic markers. Nucleic Acids Res. 18: 6531-6535.

Wilson, J.J., Polyak, S.J., Day, T.D. & Gretch, D.R. (1995). Characterization of simple and complex hepatitis C virus quasispecies by heteroduplex gel shift analysis: correlation with nucleotide sequencing. J. Gen. Virol. 76: 1763-1771.

Winter, E., Yamamoto, F., Almoguera, C. & Perucho, M. (1985). A method to detect and characterize point mutations in transcribed genes: Amplification and overexpression of the mutant c-Ki-*ras* allele in human tumor cells. Proc. Natl. Acad. Sci. USA. 82: 7575-7579.

Wood, N., Standen, G., Hows, J., Bradley, B. & Bidwell, J. (1993). Diagnosis of sickle-cell disease with a universal heteroduplex generator. Lancet 342: 1519-1520.

Xing, Y., Wells., R.L. & Elking, M.M. (1996). Nonradioisotopic PCR heteroduplex analysis: a rapid, reliable method of detecting minor gene mutations. BioTechniques 21: 186-187.

Yu, K. & Pauls, K.P. (1992). Optimization of the PCR program for RAPD analysis. Nucleic Acids Res. 20: 2606.

Zabeau, M. & Vos, P. (1993). Selective restriction fragment amplification: a general method for DNA fingerprinting. European Patent Office, publication 0 534 858 A1.

Zakharov, S.F. & Chrambach, A. (1994). The relative separation efficiencies of highly concentrated, uncrosslinked or low-crosslinked polyacrylamide gels compared to conventional gels of moderate concentration and crosslinking. Electrophoresis 15: 1101-1103.

Zhang, Q.J. & Minoda, K. (1996). Detection of congenital color-vision defects using heteroduplex-SSCP analysis. Japanese J. Ophthalmol. 40: 79-85.

DNA HETERODUPLEX FINGERPRINTING IN *PENICILLIUM*

James Scott, David Malloch, Bess Wong, Takashi Sawa and Neil Straus
Department of Botany, University of Toronto, Toronto, Ontario, CANADA, M5S
3B2

Viable fungal spores occur in great numbers in household dust and indoor air. A large proportion of these spores typically arise outdoors in the phylloplane and are carried indoors
by mechanical means (e.g. air currents, footwear). In contrast, many species of *Aspergillus*
and *Penicillium*, while extremely common and abundant in the indoor environment, often
lack obvious outdoor sources. Thus, it is widely thought that these fungi commonly prolif-
erate indoors on various substrata, including dust itself. Empirical support for this hypothesis is lacking, however, particularly in the absence of moisture.

We have sought to examine this hypothesis by using molecular genetic techniques to investigate the population structure of common indoor asexual Penicillia, assuming that cryptic indoor amplification would, over time, lead to the establishment of resident, dominant
clones. In the present study, we have adapted PCR-based multilocus heteroduplex assay
(HMA) to screen for genotypic variability and clonality between isolates of *P. brevicompactum* from household dust. In this technique, homologous loci are amplified using the
polymerase chain reaction, and the products of an arbitrary pair of isolates are combined in
equal proportion, heat denatured and cooled to permit reannealing. The resulting mixture
comprises double-stranded DNAs (dsDNAs) of all possible combinations of compatible
DNA strands including two homoduplexes identical to each original amplification product,
along with two hybrid dsDNAs (heteroduplexes) created from the cross-annealing of compatible strands originating from different "parent" isolates. Sequence differences including
base substitutions, insertions or deletions produce local base pairing anomalies in the hybrid molecules. These non-complementary domains typically result in differentially slower
electrophoretic migration of hybrid fragments relative to duplexes in which strands bear
100 % base complementarity. Thus, small dissimilarities in sequence between two isolates
at a given locus may be detected readily.

INTRODUCTION

Filamentous fungi are among the most abundant microorganisms in household dust and
indoor air (van Bronswijk, 1981; Miller, 1992). While environmental fungal reservoirs
are rarely implicated in the etiology of human mycoses (Summerbell *et al.*, 1992; Miller,
1992), it has long been known that human exposure to environmental fungi is an important risk factor for the development of allergy and asthma (American Industrial Hygiene
Association, 1996; Flemming and Schwartz, 1946; Nilsby, 1949). Two recent studies in
this regard have independently shown strongly-supported positive correlations between
the presence of fungi indoors and the incidence of allergy, asthma and other, primarily

225

respiratory diseases (Brunekreef *et al.*, 1989; Dales *et al.*, 1991a, 1991b). In fact, work by Dales and colleagues (1991a) identified the occurrence of moulds in residential housing to be at least as significant to the development of childhood asthma as parental smoking.

While many common, building-associated fungi may be found in other habitats besides buildings, a subset of species appears to be restricted largely to human-associated settings and while abundant in indoor environments are comparatively poorly represented in outdoor air and soil. The latter group consists mainly of asexual fungi, many of which are in the genera *Aspergillus* and *Penicillium*. Little is known about the indoor ecology of this latter dust bound group despite an increasingly clear association of these fungi with building-related health complaints, typically associated with "sick building syndrome" and "building-related illness" (American Industrial Hygiene Association, 1996). These fungi are among the most common agents of indoor contamination associated with indoor moisture problems. For instance, in addition to *Stachybotrys chartarum*, we have observed *Aspergillus versicolor* and *Penicillium brevicompactum* commonly on water-damaged gypsum wallboard and decorative, vinyl wall coverings. However, where clear indoor fungal amplifiers are lacking, these fungi remain amongst the most prevalent in the indoor environment. In such situations, it is not apparent whether the major locus of fungal amplification of these dust borne fungi is cryptic in the indoor environment or if it lies outside. Furthermore, the level of clonal diversity among strictly asexual, putatively dust borne species has not been investigated.

A study to examine the fungal species structure of *Penicillium brevicompactum* and *P. chrysogenum* in household dust using a selected set of houses in Wallaceburg, a small community in southwestern Ontario, Canada was carried out. Principally, we are interested in establishing whether sources of amplification of these fungi are internal to the houses, and, if so, whether they are primarily autochthonic (within the dust substrate) or allochthonic (located elsewhere), or if they enter passively from the outdoors. In addressing these and other questions, we examine the degree of clonal diversity of these species, testing the hypothesis that individual houses are islands of low clonal diversity, indicating internal amplification. By this we expect a relatively greater amount of variability from isolates of the same species compared from a large number of houses. In this paper, we have investigated the technique of heteroduplex mobility analysis as a means of addressing these and other questions relating to the population structure of domicile associated Penicillia.

MATERIALS AND METHODS

Collection and analysis of dust samples: Vacuum cleaner bag samples of carpet dust were collected from 369 houses in Wallaceburg, Ontario by a private company under contract from Canada Mortgage and Housing Corporation over a period of five months starting in January 1994. The samples were supplied dry in sealed, 10 mL polypropylene vials and stored at room temperature until analysis.

Two subsamples of approximately 50 mg (the actual mass was recorded for use in subsequent calculations) were added individually to 10 mL of sterile 2% peptone broth and suspended by vortexing at medium speed for several minutes. Two serial dilutions of these stock suspensions were made subsequently using an adaptation of the standard technique reviewed by Malloch (1981); the first was made by diluting 1 mL of stock suspension in 9 mL of 2% peptone broth and the second was made by diluting 1 mL of the first serial dilution in 9 mL of 2% peptone broth. Four aliquots of 1 mL each were taken from each of the two sets of dilutions and dispensed individually into polystyrene 90 mm Petri dishes (Fisher). Two Petri plates were set up in this manner for each of the two stock suspensions. Molten sterile Rose Bengal agar (RBA) (Malloch, 1981) and RBA containing 25 % glycerol (RBGA), both amended with 60 ppm of chlortetracycline hydrochloride (Sigma), streptomycin sulphate (Sigma) and benzylpenicillin (penicillin-G, Sigma), were cooled to 45 °C and dispensed aseptically each into half of the Petri plate replicates. The medium was mixed with the dilution aliquot by gently swirling the Petri plates prior to solidification. After 12 to 18 hr the plates wrapped with Parafilm (Alcan) and inverted in stacks of 20 plates each. The plates were incubated under 12 hr artificial daylight at room temperature (ca. 24°C) for 7 days. One of the three sets of dilutions averaging between 15 and 60 colonies per plate was selected for identification and enumeration.

Isolation, identification and storage of cultures: Where possible, fungi were identified to the species level directly from colonies on the Rose Bengal isolation media using well-established techniques of macroscopic and microscopic examination and standard reference works for the identification of moulds (e.g. Domsch *et al.* 1980; Barron, 1968; Carmichael *et al.*, 1980; Hanlin, 1990; Barnett and Hunter, 1986, Ellis, 1971, 1976; Malloch, 1981; v. Arx, 1970), as well as numerous other monographic treatments and individual descriptions.

Species of *Penicillium* were grouped according to macroscopic and microscopic similarity, averaging five groups per house. Representatives of each of the groups of similar Penicillia were subcultured on four diagnostic media for further identification using a central-point inoculation technique modified from Pitt (1979). The Petri plates were wrapped with Parafilm, inverted and incubated under 12 hr artificial daylight at room temperature for from 7 to 14 days prior to examination. The following media were employed in the identification of cultures of *Penicillium* to species level: Czapek's yeast-autolysate agar (CYA) (Pitt, 1979), Creatine agar (CREA) (Frisvad, 1985), 25% Glycerol-nitrate agar (G25N) (Pitt, 1979) and Modified Leonian's agar (MLA) (Malloch, 1981). Cultures were identified according to colonial and microscopic morphologies produced on these media as compared with species descriptions given by Pitt (1979, 1988).

Each representative isolate of *Penicillium* was subcultured in duplicate into 2 mL screw cap (with rubber o-ring) flat-bottom microcentrifuge (microculture) tubes (Sarstedt) containing 1 mL per tube of 2 % MLA. Additional representatives of *Penicillium* and other genera were subcultured in a similar manner pending future need. The tubes were capped and incubated under 12 hr artificial daylight at room temperature for 7 to 10 days prior to transfer to 5°C for short-term storage. Cultures requiring long-term preservation (e.g. for use in fingerprinting) were subcultured subsequently and checked for purity. Axenic cultures were subcultured in triplicate in microculture tubes and incubated as outlined above. After colonies had grown out, cultures were aseptically overlaid with 1 mL sterile 20 % glycerol combined with 17 % skim milk as a cryoprotectant and stored at -70 °C (McGinnis and Pasarell, 1992).

Isolation of DNA from Penicillium conidia: Fungal isolates were inoculated centrally on a Petri plate of Weitzman and Silva-Hutner's agar (WSHA) (Weitzman and Silva Hutner, 1967), and grown for 7 days at room temperature under 12 hr artificial daylight. The plates were flooded with 2 mL of 95 % ethanol and the conidia and mycelium were suspended by gently scraping the surface of the colonies with a sterile bent glass rod. The conidial suspensions were collected in microcentrifuge tubes, centrifuged at 12 Krpm and the supernatant ethanol was discarded. The pellets were dried for 30 min in a vacuum concentrator centrifuge. This protocol yielded approximately 15 mg pelleted conidia per vial. Each vial was sufficient for a single DNA isolation.

Approximately 15 mg of sterile, acid-cleaned Dicalite 1400 (Grefco Inc., Torrance, California), was added to a tube containing a roughly equal volume of dry, pelleted, ethanol-killed conidia. Following the addition of 10 µL of 70 % EtOH, the mixture was ground with a sterile glass rod for 1 min and suspended in 600 µL of lysis buffer containing 1.4 M NaCl, 2 % w/v CTAB, 200 mM Tris·HCl pH 8.0 and 20 mM EDTA (adapted from Weising *et al.*, 1995). Tubes were incubated at 65 °C for 1 hr, during which they were mixed by inversion at 30 min intervals.

After extraction, the tubes were cooled to room temperature and centrifuged at 10 K rpm for 1 min to pellet the Dicalite and cellular debris. The supernatant liquid was extracted twice with chloroform:isoamyl alcohol (24:1) and the DNA was precipitated with 100 % isopropanol for 10 min at -80°C. The pellets were rinsed with 70 % ethanol and dried. The DNA was resuspended in 200 µL Tris-EDTA (10 mM and 1 mM, respectively) (TE) pH 8.0 (Sambrook *et al.*, 1989). Ribonuclease A was added to the DNA at a final concentration of 0.2 µg/µL, and incubated for 30 min at 37 °C. The DNA was subsequently extracted with chloroform:isoamyl alcohol as above, and following the addition of sodium acetate to a concentration of 0.3 M, the DNA was precipitated with 250 µL of 100 % ethanol at -80 °C. The DNA solution was pelleted and pellets were rinsed with 70 % ethanol, dried and resuspended in 100 µL TE pH 8.0. The concentration of DNA in solution was determined spectrophotometrically.

DNA amplification: An identified polymorphic region near the 5' end of the β-tubulin gene was amplified using primer sequences Bt2a and Bt2b described by Glass and Donaldson (1995). PCR was carried out in a 100 µL reaction volume containing 4 units of Taq DNA polymerase (Boeringher Mannheim), 50 mM KCl, 2.0 mM $MgCl_2$, 250 µM each of dATP, dTTP, dCTP and dGTP, 0.2 mM of each primer and approximately 200 ng high molecular weight template DNA in 1 x Promega Taq DNA polymerase buffer overlaid with a drop of sterile mineral oil to prevent evaporation. A template-free reaction was included in each batch to control against extraneous template contaminating reagents. Reactions were carried out in a PTC-100 thermocycler (MJ Research). The typical PCR profile used consisted of 94 °C for 30 s to denature, 58°C for 30 s to anneal primers and 72°C for 30 s to extend. This profile was repeated for 30 cycles, followed by a final extension at 72 °C for 2 min.

Cloning and sequencing of PCR product: Products for cloning were amplified directly from genomic template DNA using 5'-phosphorylated primers. Primers were phosphory- lated by combining 1 μL of 500 μM oligonucleotide primer, 0.2 μL 10 mM ATP, 1 μL 10x polynucleotide kinase buffer (Promega) and 0.5 μL T4 polynucleotide kinase (Promega) made to 10 μL total volume with sterile distilled deionized water and incu- bated at 37 °C for 30 min then transferred to 65°C for 20 min to inactivate the enzyme.

Following amplification and soft gel purification of products (Sambrook *et al.*, 1989), 42 μL aliquots of products were blunt-ended with 5 μL 10x T4 DNA polymerase buffer, 0.25 μL acetylated bovine serum albumin, 100 μM deoxynucleotide triphosphates (dNTPs) and 0.25 μL T4 DNA polymerase (New England Biologicals). The resulting products were extracted once with phenol:chloroform (1:1) and once with chloroform and precipitated at –80 °C by the addition of 0.1 M NaCl and two volumes of cold 100 % ethanol. Prior to ligation, plasmid vector was digested to completion with *Hin*dIII and treated with shrimp alkaline phosphatase according to suppliers instructions (US Bio- chemical). Ligation was performed using 45 ng of linearized pUC19 vector, 45 ng insert DNA, 2.4% polyethylene glycol 8000, 2 μL 5x DNA ligase buffer and 0.6 μL DNA li- gase (BRL) in a reaction volume made to 10 μL with sterile distilled deionized water and incubated at room temperature for 3-4 hr. Following ligation, reactions were diluted to 40 μL total volume with 1x DNA ligase buffer and transformed into *E. coli* strains DH5α and JM109 using the method described by Hanahan (1985). Cells were plated in 3 repli- cates of 100 μL aliquots of transformation reaction onto LB agar containing 60 μg/mL ampicillin. Isopropylthio-β-D-galactoside (IPTG) (80 μg/mL) and 5-bromo-4-chloro-3- indolyl-β-D-galactoside (XGAL) (20 ng/mL) were also incorporated to permit blue/white selection. Recombinant plasmids were isolated by alkaline lysis (Sambrook *et al.*, 1989). Cloned DNA fragments were sequenced by the method of Sanger and co-workers (1977), using universal and reverse primers for pUC19.

Preparation and analysis of DNA heteroduplexes: Cloned, sequenced DNA fragments were amplified as described above using the Bt2 primer set. PCRs were diluted to one half of the original concentration, with 4 mM EDTA and 50 mM KCl. Diluted PCR products were combined in equimolar proportion in a total volume of 10 μL, and over- laid with a drop of sterile mineral oil. Reactions were heated to boiling for 2 min and immediately annealed at 65°C for 2 min.

Electrophoresis and imaging: The quality and yield of PCRs was assessed by electropho- resis on 1.5 % agarose gels in 1 x TBE. Gels were stained in 250 ng/mL ethidium bro- mide for 10 min and destained in distilled water for 10 min. For resolving heterodu- plexed DNAs, agarose gel casting trays (BioRad minisubmarine) were modified by glu- ing a spacer strip of 6 mm square extruded poly(methylmethacrylate) (PMMA) rod (General Electric), equal in length to the casting tray, into each corner of the tray using a glue that consisted of 1 % (w/v) PMMA and 1 % (v/v) acetic acid in dichloromethane. The volume of the gel tray was recalculated based on the decreased width. A removable cover plate designed to rest on top of the square rods was cut from 1/8" thick PMMA (Plexiglas, General Electric). The cover plate was cut to approximately 1 cm shorter than the total length of the gel tray to accommodate insertion of the comb. The use of a cover plate ensured a uniform thickness for the gel. In addition, the cover plate prevented contact with air during polymerization. The ends of the tray were sealed with masking tape (3M). The cover plate and comb were put in place prior to pouring. A solution of 7.7 % acrylamide, 0.3 % bisacrylamide and 0.04 % ammonium persulfate was prepared in 1 x TAE (Sambrook *et al.*, 1989), to which 0.2 % N,N,N',N'-tetramethylethylenediamine (TEMED) was added immediately prior to casting after degassing of the solution. The cover plate was kept in position on the gel during electrophoresis to prevent contact between the upper surface of the gel and the running buffer. Exclusion of the running buffer from contacting the upper surface of the gel prevents ionic migration across the gel-buffer interface, thus preventing the formation of a vertical, ionic gradient from developing within the gel.

Duplicate sets of samples were electrophoresed for comparison in 1.5 % agarose (BioRad) as well as the proprietary media, Clearose® (Elchrom) and 9 % poly N-acryloyltris-(hydroxymethylaminomethane) (poly-NAT®) (Elchrom). Proprietary gel media electrophoresed in 0.75 x TAE (30 mM tris-acetate, 0.75 mM EDTA adjusted to pH 8.0) (Sambrook, 1989) using a temperature controlled, buffer-recirculating electrophoresis system (SEA 2000, Elchrom) running at 40 °C with a field strength of 5 Vcm^{-1}. Polyacrylamide and poly-NAT gels were stained for 1 hr in 250 ng/mL ethidium bromide and destained in distilled water for 3-4 hr. All gels were visualized on an ultraviolet light transilluminator at 300 nm (Fotodyne). Concentrations of amplification products were standardized based on band intensity as compared to a quantitative standard.

RESULTS AND DISCUSSION

In order to study DNA variability in *Penicillium* isolates we developed a method of DNA isolation that was suitable for use with PCR-based protocols. Many of the existing techniques for isolating fungal DNA were designed for methods requiring milligram amounts of total DNA (Weising *et al.*, 1995). These methods typically involved preparation of broth cultures, mycelial harvest and grinding. All of these procedures contained the risk of generating viable aerosols which present a potential biohazard for pathogens and opportunists as well as a contamination hazard for heavily sporulating fungi. Other risks independent of the viability of conidial or cellular aerosols, such as mycotoxins and allergenic beta-glucans, are greater with large mycelial isolations. In our procedure, ethanol serves as the initial harvesting medium. Ethanol is an efficient wetting agent for hydrophobic conidia, which prevents conidial aerosolization during the early stages of preparation. In addition, ethanol treatment kills *Penicillium* conidia thereby eliminating the potential of cross culture contamination during isolation procedures.

Dicalite was selected as a grinding agent since it is commercially available and inexpensive. It acts both as an abrasive to disrupt conidial walls, and as a bulking agent to ensure surface contact between the abrasive and the biological material. Although tight binding between DNA and glass powder in the presence of high concentrations of sodium iodide forms the basis of some proprietary technologies for DNA purification (e.g. Geneclean® II, Bio 100 Inc.), controlled experiments using herring sperm DNA indicated that negligible DNA bound to the Dicalite under the conditions used in our procedure (unpubl.). Typically, a single Petri plate of *P. brevicompactum* grown for 14 days yielded 10 to 100 mg, dry weight, of harvestable conidia. From this, 10 to 100 µg of high molecular weight DNA was isolated.

A set of morphologically and physiologically indistinguishable isolates of *P. brevicompactum* was retained for molecular analysis. Three of these isolates, 114, 132 and 244 were used in an initial study of heteroduplex variability in the Bt2 locus, which contains three introns (Figure 1). This analysis revealed a different allelic form of Bt2 for each of these isolates. An additional isolate, 112, which did not form a heteroduplex with 114 was selected as a control. All of these isolates were cloned and sequenced to determine the variability responsible for electrophoretic shifts observed in the heteroduplexed structures. Figure 2 shows the aligned sequences of the cloned fragments for the Bt2 locus of isolates 114, 132 and 244. Isolates 114 and 244 have the greatest sequence similarity (15 base substitutions and two non-contiguous insertions/deletions). Isolates 114 and 132 differ by 42 base substitutions, 9 non-contiguous insertions/deletions and a four-base contiguous insertion/deletion. Isolates 132 and 244 differ by 44 base substitutions, 8 non-contiguous insertions/deletions, and the same four-base contiguous insertion/deletion. Sequence obtained for the Bt2 amplicon from isolate 112 was identical to that of isolate 114 (data not shown).

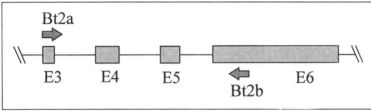

Figure 1. Position of the Bt2 primer sequences (Glass and Donaldson, 1995) on the BenA gene of *Emericella nidulans* relative to exon sequences, indicated with "E".

Each of these clones was amplified by PCR and heteroduplexed as described in Materials and Methods. The resulting reassociated structures were compared under three different conditions of electrophoresis (Figures 3-5). These conditions included Clearose BG (Elchrom) (Fig. 3), 9 % poly-NAT (Elchrom) (Fig. 4) and low-crosslinked 8 % polyacrylamide (crosslinked with 0.2 % bisacrylamide) (Fig. 5). Clearly, the data show that the low-crosslinked polyacrylamide gel provides the greatest level discriminate both in the differential mobilities between lanes and in the separation of complementary heteroduplexed structures within lanes. Of the pre-cast gels, 9 % poly-NAT would represent an acceptable degree of resolution for situations where the volume of analysis precluded in-house gel preparation. All of these analyses were completed on a temperature-regulated horizontal gel apparatus that is amenable to polyacrylamide or other polymeric gel substrates (SEA 2000, Elchrom). Horizontal gel electrophoresis systems have the advantages of rapid loading, and easy gel casting and manipulation inherent in this type of configuration (Bellomy and Record, 1989).

```
114    GGTAACCAAATCGGTGCTGCTTTCTGGTATGTATCGCACCAT GTCTTTTTCTTTTCCCG 59
132    ..................................C....T..CG........T.C ....
244    ...................................G...............

114    CTATGGCTGGGTATCAATTGACAATTTGCTAACTGGCTTAAAGGCAAACTATCTCCGGCG 119
132    .A........C...........TC..........A.A.C.................
244    ....................T..............C...............  ....

114    AGCACGGTCTCGATGGCGATGGACAGTAAGTGGA CG ACTGTGTTCGAATTATGCGTGG 177
132    ..............................G..T....G.      .TCCCAT ....
244    ...................................      ..  .............C......

114    ATGGGGTCTGAGATCTTGTTAGGTACAATGGCACCTCCGACCTCCAGCTCGAGCGTATGA 237
132    ..T..T.....T..A.............C..T................G.........
244    ..T..............................................

114    ACGTCTACTTCAACCATGTGAGTACAATG ATGTGAAGAACTCTTGTTGTGCATTTTCTC 296
132    ............................CC.TG..TA.. .T........TG...G...
244    ........................C.CTCCG...A...A...G............

114    ACCTCATATTCTT GACCGCCCAGGCTAGTGGTGACAAGTACGTTCCCCGTGCCGTCCTC 355
132    ...CGG ......T....T..........C...........T.................
244    .TT........  .........................................

114    GTCGACTTGGAGCCCGGTACCATGGACGCTGTCCGCTCCGGTCCCTTCGGCAAGCTTTTC 415
132    .....T................................................
244    ...................................................

114    CGCCCCGACAACTTCGTCTTCGGTCAGTCCGGTGCTGGTAACAACTGGGCCAAGGGTCAC 475
132    ....................................................
244    ....................................................

114    TACACTGAGGGT                                             487
132    ............
244    ............
```

Figure 2. Comparison of aligned sequences of cloned Bt2 fragments from isolates of *Penicillium brevicompactum*. Dots indicate a nucleotide that is identical to that in the first line. The absence of a dot indicates a deletion.

Figures 3-5. PCR amplified products from Bt2 clones of *Penicillium brevicompactum* isolates. Lanes (from left to right) 1, 6 and 12 size marker (100-base ladder, Pharmacia); lanes 2-4 contain individual PCR products from isolates 112, 114, 132 and 244, respectively; lane 7 contains iso-lates 112 and 114 combined (as per procedure for generation of heteroduplexes in Materials and Methods section); similarly, lane 8 contains isolates 114+244, lane 9 contains isolates 114+132, and lane 10 contains isolates 132+244. Lane 11 is empty. All gels were run at a field strength of 5 V/cm at 40 °C in 0.75 x TAE. The gel in figure 3 is Clearose®, run for 1 hr at 5 V/cm, figure 4 shows a 9 % poly-NAT® gel run for 3 hr 45 min and figure 5 shows a 8 % polyacrylamide gel run for 2 hr.

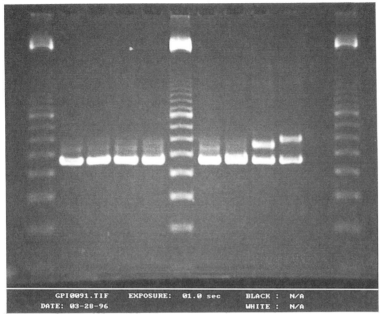

Figure 3

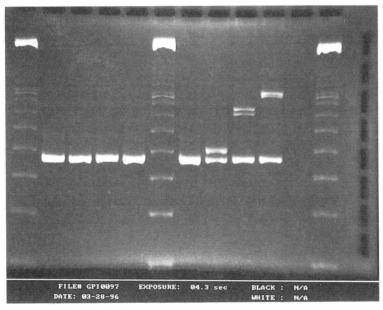

Figure 4

233

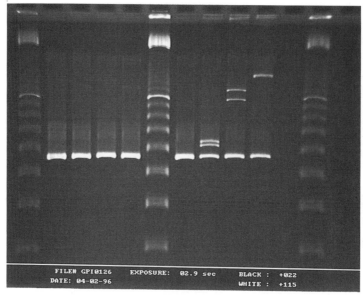

FILE# GPI0126   EXPOSURE: 02.9 sec   BLACK : +022
DATE: 04-02-96   WHITE : +115

Figure 5

The heteroduplex patterns that appeared in the poly-NAT and polyacrylamide electrophoretic gels (Figs. 6 and 7, respectively) correspond to the mobilities expected based on the differences in sequence of the fragments such that the order of retardation of heteroduplex mobility corresponds to the degree of sequence dissimilarity. However, as pointed out by White *et al.* (1992) and Ganguly *et al.* (1993), the actual gel mobility retardation in heteroduplex structures is related both to the type and distribution of sequence differences in a complex manner. Clustered base sequence changes that are centrally located produce local destabilization domains and have a much larger effect on mobility than non-contiguous differences dispersed throughout the molecule. However, the most pronounced heteroduplex mobility shifts result from molecular kinks caused by insertions or deletions between paired strands (Ganguly *et al.*, 1993; Wang *et al.*, 1992). This is seen in the substantial difference between the heteroduplex mobilities of the 114/132 and 132/244 heteroduplexes. Although the structures produced from 114/132 and 132/244 pairings show substantial difference in electrophoretic mobilities the cumulative dissimilarities are only 55 and 56, respectively. The comparative differences between these mobilities relative to both the homoduplexed DNAs and the 114/244 heteroduplexes seem unduly large since this total cumulative sequence difference between the 114/132 and 132/244 heteroduplexes is only one additional single base insertion/deletion in the latter.

Very low sequence variation between fragments, however, may require a greater running distance to achieve adequate resolution. We have found the technique described by Xing and colleagues (1996) useful in such circumstances. In this method, a hybrid low crosslinked polyacrylamide / agarose / glycerol gel is cast between large glass plates using a vertical protein-type electrophoresis apparatus (e.g. Protean II, BioRad, La Jolla, California. While this method permits the detection of very small differences, it is somewhat time-consuming and not necessary where a coarser degree of discrimination is acceptable or desirable.

CONCLUSIONS

In this study we have developed a rapid, safe method for DNA isolation from sporulating cultures grown on semi-solid substrates. Also, we have also demonstrated that heteroduplex mobility assays are useful in the discrimination of sequence differences between isolates of *Penicillium*.

ACKNOWLEDGEMENTS

The authors wish to thank B. Koster, L. Hutchison, B. Couch, E. Taylor, M. Wiebe, W. Malloch and C. McGee for assistance during various stages of this project. Clearose and poly-NAT gels were kindly provided by T. Kung, Helixx Technologies Inc. L. Ketch is thanked for proof-reading and offering comments on an earlier version of this manuscript. This work was supported by a strategic grant from the Natural Sciences and Engineering Research Council of Canada (NSERC), jointly held by DM and NS, and an NSERC postgraduate fellowship to JS.

REFERENCES

American Industrial Hygiene Association (AIHA). (1996). Field guide for the determination of biological contamination in environmental samples. Dillon, H.K., Heinsohn, P.A. and Miller, J.D. (eds). Fairfax, Virginia: AIHA Publications. 174 pp.

Arx, J.A. v. (1970). The genera of fungi sporulating in pure culture. J. Cramer, Vaduz, 2nd ed. 315 pp.

Barnett, H.L. & Hunter, B.B. (1986). Illustrated genera of fungi imperfecti. MacMillan Co., New York. 218 pp.

Barron, G.L. (1968). The genera of hyphomycetes from soil. Williams and Wilkins, Baltimore. 364 pp.

Bellomy, G.R. & Record, T.M. Jr. (1989). A method for horizontal polyacrylamide slab gel electrophoresis. BioFeedback 7: 16, 19-21.

Bruford, M.W. & Wayne, R.K. (1993). Microsatellites and their application to population genetic studies. Curr. Opin. Genet. Dev. 3: 939-943.

Brunekreef, B., Drockery, D.W., Speozer, F.E., Ware, J.H., Spengler, J.D. & Ferris, B.G. (1989). Home dampness and respiratory morbidity in children. Am. Rev. Respir. Dis. 140: 1363-1367.

Carmichael, J.W., Kendrick, W.B., Connors, I.L. Sigler, L. (1980). Genera of hyphomycetes. University of Alberta Press, Edmonton. 386 pp.

Dales, R., Zwanenburg, H., Burnett, R. & Franklin, C.A. (1991a). Respiratory health effects of home dampness and molds among Canadian children. Am. J. Epidemiol. 134: 196-203.

Dales, R., Burnett, R. & Zwanenburg, H. (1991b). Adverse health effects in adults exposed to home dampness and molds. Am. Rev. Respir. Dis. 143: 505-509.

Domsch, K. H., Gams, W. & Anderson, T.H. (1980). Compendium of soil fungi. Vol. 1. Academic Press, London. 859 pp.

Ellegren, H., Mikko, S., Wallin, K. & Andersson, L. (1996). Limited polymorphism at major histocompatibility complex (MHC) loci in the Swedish moose, *A. alces*. Molec. Ecol. 5: 3-9.

Ellis, M.B. (1971). Dematiaceous Hyphomycetes. Commonwealth Mycological Institute, C.A.B. Kew, Surrey. 608 pp.

Ellis, M.B. (1976). More Dematiaceous Hyphomycetes. Commonwealth Mycological Institute, C.A.B. Kew, Surrey. 507 pp.

235

Frisvad, J.C. (1985). Creatine-sucrose agar, a differential medium for mycotoxin producing *Penicillium* species. Letters in Appl. Microbiol. 1: 109-113.

Ganguly, A., Rock, M.J. & Prockop, D.J. (1993). Conformation-sensitive gel electrophoresis for rapid detection of single-base differences in double-stranded PCR products and DNA fragments: evidence for solvent-induced bends in DNA heteroduplexes. Proc. Natl. Acad. Sci. USA 90: 10325-10329.

Glass, N.L. & Donaldson, G.C. (1995). Development of primer sets designed for use with the PCR to amplify conserved genes from filamentous ascomycetes. Appl. Environ. Microbiol. 61: 1323-1330.

Hanahan, D. (1985). Techniques for transformation of *E. coli*. In: DNA cloning volume 1 -- a practical approach. Glover, D.M. (ed). IRL Press, Oxford. pp. 109-135.

Hanlin, R.T. (1990). Illustrated genera of Ascomycetes. APS Press, St. Paul, Minnesota. 263 pp.

Malloch, D.W. (1981). Moulds: their isolation, cultivation and identification. University of Toronto Press, Toronto, Canada. 97 pp.

McGinnis, M.C. & L. Pasarell. (1992). Viability of fungal cultures maintained at -70C. J. Clin. Microbiol. 30: 1000-1004.

Miller, J.D. (1992). Fungi as contaminants in indoor air. Atmospheric Environment 26: 2163-2172.

Nilsby, I. (1949). Allergy to moulds in Sweden. Acta Allergogica 2: 57-90.

Pitt, J.I. (1979). The genus *Penicillium* and its teleomorphic states *Eupenicillium* and *Talaromyces*. Academic Press, New York. 634 pp.

Pitt, J.I. (1988). A laboratory guide to common *Penicillium* species. North Ryde, N.S.W.: Commonwealth Scientific and Industrial Research Organization, Division of Food Processing. 187 pp.

Reymann, F. and M. Schwartz. (1946). House dust and fungus allergy. Acta Pathol. et Microbiol. Scand. 24: 76-85.

Sambrook, J., Fritsch, E.F. & Maniatis, T. (1989). Molecular cloning. A laboratory manual. 2nd ed. Cold Spring Harbor Laboratory Press, New York. Vols 1-3.

Sanger, F., Nicklen, S. & Coulson, A.R. (1977). DNA sequencing with chain terminating inhibitors. Proc. Natl. Acad. Sci. USA 74: 5463-5467.

Summerbell., R.C., Staib, F., Dales, R., Nolard, N., Kane, J., Zwanenburg, H., Burnett, R., Krajden, S., Fung D. & Leong, D. (1992). Ecology of fungi in human dwellings. J. Med. Veterin. Mycol. 30 (suppl. 1): 279-285.

Van Bronswijk, J.E.M.H. (1981). House dust biology; for allergists, acarologists and mycologists. Zoelmond: Published by the author. 316 pp.

Wang, Y.-H., Barker, P. & Griffith, J. (1992). Visualization of diagnostic heteroduplex DNAs from cystic fibrosis deletion heterozygotes provides an estimate of the kinking of DNA by bulged bases. J. Biol. Chem. 267: 4911-4915.

Weising, K., Nybom, H., Wolff, K. & Meyer, W. (1995). DNA fingerprinting in plants and fungi. CRC Press, Boca Raton. 322 pp.

Weitzman, I. & Silva-Hutner, M. (1967). Non-keratinous agar media as substrates for the ascigerous state in certain members of the Gymnoascaceae pathogenic for man and animals. Sabouraudia 5: 335-340.

White, M.B., Carvalho, M., Derse, D., O'Brien, S.J. & Dean, M. (1992). Detecting single-base substitutions as heteroduplex polymorphisms. Genomics 12: 301-306.

Xing, Y., Wells., R.L. & Elking, M.M. (1996). Nonradioisotopic PCR heteroduplex analysis: a rapid, reliable method of detecting minor gene mutations. BioTechniques 21: 186-187.

236

Chapter 5

CLASSIFICATION AND
IDENTIFICATION OF *PENICILLIUM*

SUBSTRATE UTILIZATION PATTERNS AS IDENTIFICATION AIDS IN *PENICILLIUM*

Keith A. Seifert, John Bissett, Sabina Giuseppin and Gerry Louis-Seize
Eastern Cereal and Oilseed Research Centre, Agriculture and Agri-Food Canada, Research Branch, Ottawa, Ontario, Canada K1A 0C6

The taxonomic usefulness of carbon substrate utilization patterns was assessed for 81 strains (including 13 duplicate strains) of *Penicillium* species representing 18 morphologically defined species. Standardized conidial suspensions were inoculated into Biolog™YT= plates, and growth for each substrate was assessed after 4 days incubation using a microplate reader. The resulting dendrograms were inconsistent with accepted infrageneric classifications, demonstrating that physiological similarities do not necessarily reflect phylogenetic relationships. Responses for replicates of any strain are consistent at a 90-100% level of similarity. Some species are well-defined at an approximately 80% similarity level, but strains of other species exhibit more variation and do not cluster together. In the *Penicillium aurantiogriseum* group, the amount of variation between strains of any one species seems to overshadow differences among species, although some species, such as *P. melanoconidium*, appear well-defined. Cultures tentatively assigned to the *Penicillium glabrum/spinulosum* appear to be divisible into at least four distinct clusters, with the majority of the variation occurring in *P. spinulosum*. The sensitivity of this technique makes it a useful adjunct to morphological analysis, by rapidly providing a completely independent data set that can be used as an aid for interpreting colony or micromorphological variation. The probability that the technique could be adapted as an efficient identification system for *Penicillium* is discussed.

INTRODUCTION

Penicillium taxonomy has traditionally emphasized morphological and cultural characters (Pitt, 1979), but in recent years new types of characters have been explored. Secondary metabolite profiles have become an important part of species concepts, particularly in subgenus *Penicillium* (Frisvad and Filtenborg, 1989). Although DNA based methods have so far been used mostly at the generic and subgeneric levels (eg. LoBuglio *et al.*, 1994), there is now evidence that they may be useful at the species level (Seifert & Louis-Seize, 1997). Physiological methods have also been explored. Cruikshank and Pitt (1987) correlated pectic zymogram patterns with several morphological species in subgenus *Penicillium*, noting that other species (in particular *P. aurantiogriseum* and *P. viridicatum*) could be further subdivided on this basis. Bridge (1985) assessed a variety of physiological characters, including carbon and nitrogen source assimilation on agar media, as part of a larger, multidisciplinary taxonomic study (Bridge *et al.*, 1989a, b).

239

These new technologies provide important data that can be used to further our understanding of the taxonomy of *Penicillium*. However, it is difficult to employ any of the methodologies routinely in a diagnostic mycological laboratory without the investment of a large amount of time and material. Automated identification systems have clear advantages for laboratories lacking advanced taxonomic expertise, or with a large volume of identifications. Presently, there are three types of microbial automated identification systems commercially available, based on substrate utilization profiles (eg. Biolog, BCCM Yeast Identification System), fatty acid profiles (eg. the MIDI, Inc., system), and RFLPs of genomic DNA (eg. the Qualicon RiboprinterJ). To our knowledge, none of these have been extensively tested with filamentous fungi.

In our laboratory, we are exploring the usefulness of the Biolog identification system for rapid, automated identification of filamentous fungi. The Biolog system was originally marketed for identifying bacteria or yeasts, and consists of microplates with dried nutrients, a microplate reader, and a computer with identification software. The YT microplate comprises 94 test wells containing different carbon substrates to test which are utilized by the test organism, plus two control wells with no carbon source. Our early experiments with *Trichoderma* and *Beauveria* (Bissett, unpublished) have indicated that the system is very sensitive, gives reproducible results, and can give reliable identifications at or below the species level. Compared to other automated systems, it is relatively inexpensive. All necessary hardware and software for automated operation are presently sold for about US$ 30,000, and the cost for disposables is currently about US $6.00 per identification. Modifications have been necessary to use the system for filamentous fungi.

In this paper, we present the results from experiments to assess the utility of Biolog for identifying *Penicillium* species. Two model groups were selected for study. The *Penicillium aurantiogriseum* group is discussed at length elsewhere in this volume, and substrate utilization data provides useful extra data for these discussions. The second group includes species of *Penicillium* subgenus *Aspergilloides*, particularly cultures identified as *P. spinulosum* and *P. glabrum*. This group was the focus of a previous study by ICPA members (Pitt *et al.*, 1990) and those conclusions provide an interesting basis of comparison for the substrate utilization data.

MATERIALS AND METHODS

Cultures. Cultures selected for the comparisons within the *P. aurantiogriseum* group were for the most part the same as those employed for beta-tubulin sequencing by Seifert and Louis-Seize (1997). Cultures of subgenus *Aspergilloides* were derived from several sources, and included a number of isolates from wood (Seifert and Frisvad, 1997).

Inoculation, incubation and reading of microplates. Isolates were cultured on 2% malt agar slants under ambient laboratory conditions of daylight and temperature (about 21°C) for seven days or longer to develop sufficient conidiation. To prepare inoculum, a sterile cotton swab moistened with phytagel-SLS solution was rolled across the sporulating areas of the colonies. The collected conidia were deposited into a 20 x 150 mm disposable, borosilicate test tube containing 16 mL sterile solution of 0.25% phytagel and 0.01% sodium lauryl sulphate (surfactant). The suspension was mixed in a vortex mixer for 30 sec, and adjusted to 70-80% transmittance at 590 nm by harvesting additional conidia or by dilution with additional phytagel-SLS solution, as required. YT Biolog plates were inoculated with 100 FL of conidial suspension per well using a multichannel pipette. Plates were incubated in the dark at 25°C, and absorbance readings at 590 nm were taken after four and seven days using the Biolog MicroStation™ system microplate reader. Data were recorded as % absorbance and as +/- utilization of the test substrate using the Biolog MicroLog™ system software. Every tenth strain was run in duplicate to assess interstrain variability.

Statistical analyses. Absorbance readings for each well were corrected by subtracting the absorbance in the respective control wells (A1 and D1- not containing a carbon source). Data for the 4- and 7-day readings were analysed separately. Similarity for absorbance readings was calculated with the NTSYS program SIMINT using the product-moment correlation coefficient (Rohlf, 1992). Cluster analysis was performed with the NTSYS program SAHN using the unweighted pair group method arithmetic average (UPGMA). Canonical discriminant analyses were performed using the SAS program CANDISC, and discriminant functions and identifications were determined with SAS program DISCRIM (SAS Institute Inc., 1990). Linear discriminant functions were computed based on the pooled covariance matrix. Variables in the ordination analyses were reduced in number to the minimum required to perform multivariate significance tests for the eigenvalues - i.e. one less than the total number of samples minus the number of classes (species). This was accomplished in an entirely arbitrary fashion, by retaining variables in order from the first wells (A2, A3...) and deleting the necessary number of the last-read wells (H12, H11...). Wilk's Lambda and Pillai's trace were employed to test significance in multivariate tests.

RESULTS

For the majority of cultures, the results of the 4 day incubation were essentially similar to those obtained after 7 days (data not shown). For that reason, only the 4-day results are shown here. For most of the strains that were duplicated in the experiments, the variation between replicates was low. In the cluster analyses (Figs. 1, 3), duplicated strains generally had 95% or higher similarities, the type strain of *P. glabrum* being the only exception. This strain sporulated very poorly and we suspect that the inoculum may not have been uniform.

Penicillium aurantiogriseum complex.
The cluster analysis for strains from the *P. aurantiogriseum* complex is shown in Fig. 1. Relatively high correlations reflect similar patterns of substrate utilization among the species. The amount of variation between species is often similar to the amount of observed variation within species, with the result that few species formed discrete clusters. Only two species formed clusters that included all tested strains of the species, *P. melanoconidium* at about 80% similarity (4 strains) and *P. expansum* at about 90%

241

similarity (2 strains). *P.aurantiogriseum* and *P. verrucosum* formed relatively well defined clusters near 80% similarity, but single strains of each species occurred elsewhere on the dendrogram. The relationships among species postulated by the cluster analysis do not correlate with the relationships hypothesized for these species using beta-tubulin sequences (Fig. 1 in Seifert and Louis-Seize, 2000).

Cluster analyses represent relationships in a compressed 2-dimensional space. This may be a disadvantage when applied to some types of phenetic data. Ordination analyses were performed to retain and analyse the relationships of species in multidimensional space. Canonical variates redefine the multivariate space to minimize the variation within a class (i.e. species), and to maximize the distance among the classes in the same space, resulting in a reduced dimensional space expressed by the statistically significant canonical vectors.

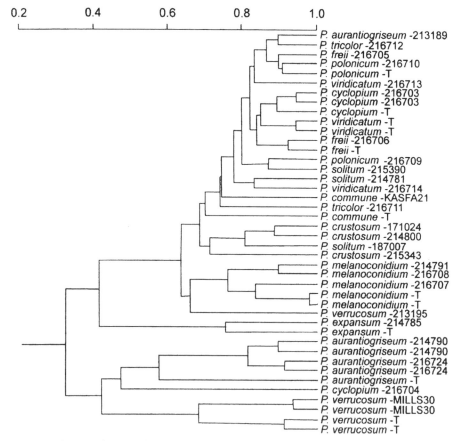

Fig 1. Cluster analysis of substrate utilization data for strains in *Penicillium* subgenus *Penicillium*.

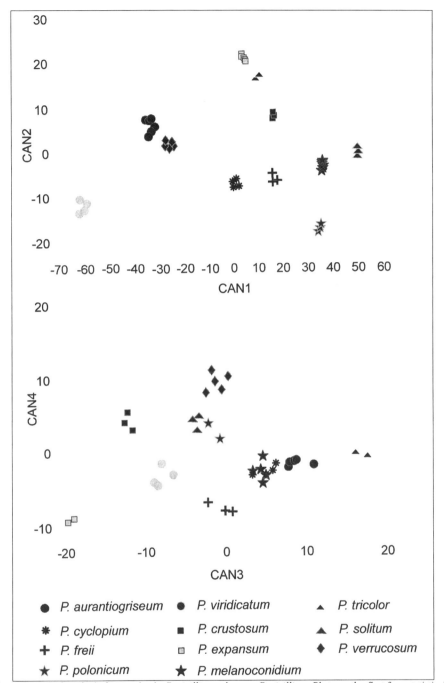

Fig 2. Canonical analysis for species in *Penicillium* subgenus *Penicillium*. Plots on the first four statistically significant canonical variates.

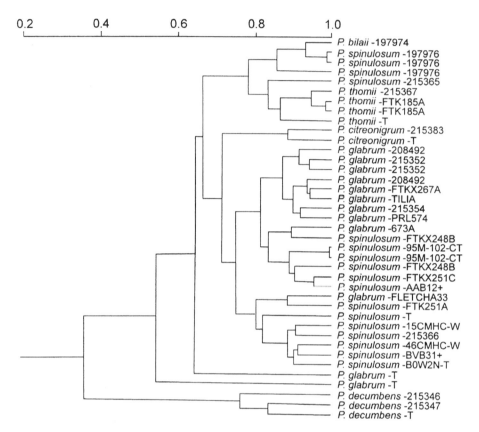

Fig 3. Cluster analysis of substrate utilization data for strains in *Penicillium* subgenus *Aspergilloides*

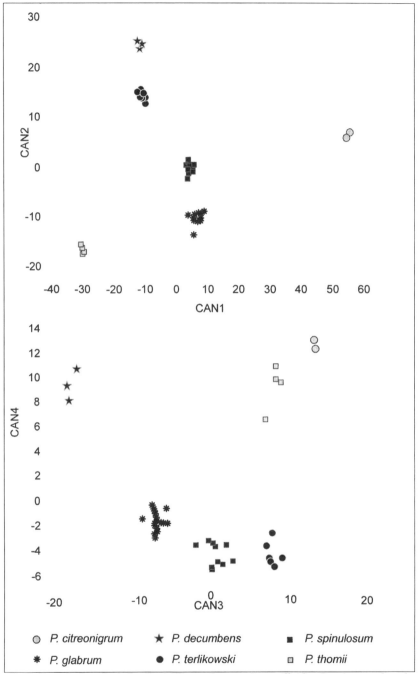

Fig 4. Canonical analysis for *Penicillium* subgenus *Aspergilloides*. Species are plotted on the four statistically significant canonical variates.

Seven statistically significant (type I error < 0.05) canonical correlations were found in the analysis of strains in subgenus *Penicillium* (Table 1). The small sample size allowed the inclusion of only the first 28 test wells as included variables. The first four canonical correlations were significant at P < 0.0001 and are plotted in Fig. 2. These four vectors summarize 96% of the total variation distinguishing the species. The plot of the strains on the first four canonical vectors clearly indicates the potential of this method to differentiate among the species, with all strains of each species tightly clustered and separated from the other species, especially on the first two vectors. In the plot of the third and fourth canonical variants, strains of *P. cyclopium* and *P. melanoconidium* were intermingled, and *P. polonicum* strains were very close strains of *P. solitum*. However, the relationships among the taxa in this analysis are dissimilar to those derived from beta-tubulin sequences (Seifert and Louis-Seize, 1997).

Table 1. *Penicillium* subgenus *Penicillium*, **multivariate test results for the significant canonical correlations.**

	Eigenvalue	Proportion	Approx. F	Num DF	Den DF	Prob > F
1	1533.5	.83	5.32	280	45.2	.0001
2	124.6	.07	3.78	243	46.5	.0001
3	93.6	.05	3.30	208	46.6	.0001
4	35.0	.02	2.75	175	45.4	.0001
5	26.8	.01	2.46	144	43.0	.0005
6	19.1	.01	2.11	115	39.2	.0045
7	9.2	.01	1.71	88	34.1	.0400

Penicillium subgenus Aspergilloides.
The cluster analysis for strains in subgenus *Aspergilloides* is shown in Fig. 3. Excluding the strains identified as *P. spinulosum* or *P. glabrum* for the moment, species are delimited in the 75-85% similarity range. The two species from section *Exilicaulis*, *P. decumbens* and *P. citreonigrum* are both well-defined. *P. decumbens* is rather distant from the rest of the data set, but *P. citreonigrum* is among species classified in section *Aspergilloides* by Pitt (1979). *P.thomii* is also well-defined in the cluster analysis. Within the *P. glabrum/P. spinulosum* complex, the strains that produce a characteristic orange brown reverse on CA, here called *P. glabrum*, form a fairly well-defined group at about 85% similarity that can be considered a species. Only two strains with an orange brown reverse grouped outside this main cluster. If the level for species similarity is assumed as 80-85%, the data support the idea of at least two and possibly four taxa in the cultures identified as *P. spinulosum*. No morphological or cultural characters were noted that could be used to distinguish these taxa, except for the clusters including *P. glabrum* FletchA33 and *P. spinulosum* FTK x251A, both of which had distinctive greyish conidia.

The cluster analysis suggests that some of the strains were misidentified: eg. *P. spinulosum* DAOM 197976 appears to be a strain of *P. bilaii* and *P. spinulosum* DAOM 215365 appears to be a sclerotium deficient strain of *P. thomii*. Unfortunately, the ex-type strains of *P. glabrum* and *P. spinulosum* available to us were badly degenerated, produced no colony pigmentation on CA, and scarcely sporulated. Therefore, their behaviour in these assays is suspect.

Four canonical correlations were significant for the analysis of strains in subgenus *Aspergilloides* (Table 2). For this analysis, cultures originally labelled *P. terlikowskii*

246

(DAOM 215365, 215366, AAB12t, 15CMHCW, 46CMHC-W, BVB31t, BOW2N-t) were considered distinct from *P. spinulosum*. Data for the first 30 test wells were included in the analysis. Plots for the 38 strains tested on the first four canonical vectors (Fig. 4) illustrate that the six species included can be clearly differentiated. The physiological similarities between the strains identified as *P. glabrum* and *P. spinulosum* are evident, because these taxa lie adjacent to one another on all four vectors.

Table 2. *Penicillium* subgenus *Aspergilloides*, multivariate test results for significant canonical correlations

	Eigenvalue	Proportion	Approx. F	Num DF	Den DF	Prob > F
1	361.9	.54	8.77	150	20.0	.0001
2	183.0	.27	6.61	116	18.5	.0001
3	81.6	.12	4.61	84	15.8	.0007
4	39.7	.06	2.87	54	12.0	.0247

Evaluation of discriminant functions as identification tools for physiological data. Discriminant functions were computed to determine their effectiveness for identification of species of *Penicillium*. The test set comprised one strain of each of 14 species. Types were not included in the test set, otherwise the strains were selected randomly for each species. Five species were removed from the test set because only one replicate had been included in the experiments. The remaining data set comprised the remaining 61 strains representing 18 species - 12 species in subgenus *Penicillium* and six in *Aspergilloides*. Eleven species were represented by as few as two strains in the data set. Only the first 24 physiological variables were included in the analysis, and seven significant canonical correlations were obtained (data not shown). Identification results are shown in Table 3. Despite the small sample size, the technique correctly identified 9 of the 14 test strains. The correct identification was obtained as the second or third choice for a further three strains. However, the procedure failed in the identification of two strains (*P. polonicum* and *P. solitum*).

Table 3. Identification of *Penicillium* strains using discriminant functions.

Strain number	Species	Primary identification	Second identification	Third identification
216724	*aurantiogriseum*	*aurantiogriseum*	none	none
214800	*crustosum*	*solitum*	*crustosum*	*glabrum*
216703	*cyclopium*	*cyclopium*	*aurantiogriseum*	*freii*
215347	*decumbens*	*decumbens*	none	none
216706	*freii*	*freii*	*cyclopium*	none
FTKX267A	*glabrum*	*spinulosum*	*glabrum*	*terlikowskii*
216707	*melanoconidium*	*freii*	*cyclopium*	*melanoconidium*
216710	*polonicum*	*melanoconidium*	*freii*	none
214781	*solitum*	*commune*	*polonicum*	none
FTKX248B	*spinulosum*	*spinulosum*	none	none
46CMHC-W	*terlikowskii*	*terlikowskii*	none	none
FTK185A	*thomii*	*thomii*	*glabrum*	none
MILLS30	*verrucosum*	*verrucosum*	none	none
216714	*viridicatum*	*viridicatum*	none	none

DISCUSSION

The use of microplates without redox dyes and the preparation of the conidial inoculum in a solution polymerizing into a transparent gel allows the Biolog system to be used as an automated system for identification of filamentous fungi. The technique is rapid and comparatively inexpensive, allowing large databases to be built that can be used for identification systems. We have demonstrated that the system can provide useful taxonomic data in *Penicillium*, although the physiological similarity among closely related strains is quite high. Although the data here are from only one of the microplates produced by Biolog, the YT plate, we obtained similar results with the SFP plate, which contains more esoteric carbon sources (data not shown).

The data presented from the two subgenera suggests that species are delimited at 80-85(-90%) similarity. Given that inter-replicate variation may be 5% or more, the detectable differences between species may be rather small. In subgenus *Penicillium*, the amount of variation among the strains of a single species overlapped with the amount of observed variation between species. This complicates the interpretation of the dendrograms resulting from cluster analysis. Some of the relationships between the species suggested by the phenetic analysis do not correlate with the relationships suggested by the cladistic analysis of beta-tubulin sequences (Seifert and Louis-Seize, 1997). For example, beta-tubulin sequences could scarcely distinguish between strains of *P. viridicatum, P. freii* and *P. aurantiogriseum*. In the Biolog data, most strains of *P. aurantiogriseum* are rather distant from *P. viridicatum* and *P. freii*, which more or less form sister clusters. In contrast, the group of species that produce base on creatine agar, with relatively large conidia (*P. solitum, P. crustosum* and *P. commune*), form a well defined clade with discrete species in the beta-tubulin analysis. In the physiological cluster analysis, the strains of these taxa are intermingled in the central section of the dendrogram.

The results for subgenus *Aspergilloides* were somewhat more satisfying. The "outgroup" species to the *P. spinulosum/P. glabrum* complex were well-defined, and a few presumably misidentified cultures were clearly evident. Within the *P. spinulosum/P. glabrum* complex, cultures with an orange-brown reverse on CYA, identified as *P. glabrum,* formed a discrete cluster. The remaining strains, received under the names *P. spinulosum* and *P. terlikowskii*, form at least two, and possibly as many as four clusters that may represent species. The usefulness of Biolog as a taxonomic tool is thus demonstrated, and these strains can now be examined by other methods, such as DNA sequencing, for evidence to support the different species hypotheses. The results obtained here support some of the conclusions made by Pitt *et al.* (1990) in their multidisciplinary study of this complex. The differentiation of *P. glabrum* from *P. spinulosum* is supported although the colony character we used to distinguish the two was not considered significant in that study. In contrast to the observations in Pitt *et al.* (1990), we detected variation within *P. spinulosum* that remains to be explained.

The Biolog MicroLog™ identification software utilizes binary data based on interpretation of growth in the test wells as positive or negative using a statistically derived threshold value. For *Penicillium*, the pattern of +/- utilization of the test substrates did not vary much among the species. In addition, determination of an appropriate threshold absorbance was unreliable, attributable to erratic germination, sporulation responses and intermediate levels of growth. The results from cluster analyses using % absorbance data reflect the high degree of similarity in patterns of utilization of C-substrates among different species in *Penicillium*, and the relatively high variation among strains of some species. These results suggest that a relatively large number of strains would need to be surveyed for each species to ensure that physiological variation within a species was represented, and to allow reliable differentiation of species in *Penicillium*. However, cluster analysis has some disadvantages for this type of phenetic data. Standardizing the variables (e.g. correlation coefficients), allows small variations in wells with low absorbance values (unused substrates) to have a disproportionate influence. An even more disproportionate effect occurs if the variables are not standardized, because of fluctuations and high absorbance values in wells containing substrates that support abundant growth and conidiation. Finally, the cluster analysis is a severe simplification of the multivariate data, compressing the 94-dimensional data into a 2-dimensional space.

Canonical variate analysis redefines the multidimensional space by sequentially extracting orthogonal eigenvectors that summarize variation between classes (e.g. species). The resulting vectors can be statistically tested resulting in a reduced number of significant vectors (canonical variates) that distinguish the classes. In the current study, the results of canonical variate analysis demonstrate the potential for the Biolog technique to accommodate physiological variation within species, and allow differentiation of closely related species in *Penicillium* subgenera *Penicillium* and *Aspergilloides*.

Discriminant functions are rearrangements of the variables in the data set summarizing the location of the classes in the original multivariate space. Identifications can be performed by evaluating the discriminant functions for unknown strains. In this study, nine of fourteen strains extracted from the data set were correctly identified using discriminant functions. The correct identity of only two strains were not suggested using this technique. These results are encouraging considering that 70 of the 94 variables (C-substrates) in the data set were arbitrarily excluded from the analysis because of the small size of the data set. We conclude that substrate utilization patterns, in combination with an effective classification protocol such as discriminant functions, is a strong candidate to become a rapid and efficient procedure for identification of species in the genus *Penicillium*.

ACKNOWLEDGEMENTS

We thank C. Babcock, Canadian Collection of Fungal Cultures, and Dr. D.Malloch for providing cultures used in this study, and Drs. S. Redhead and T. Ouellet for their critical reviews of the manuscript. Contribution number 971177.1248 from the Eastern Cereal and Oilseed Research Centre

REFERENCES

Bridge, P. D. (1985). An evaluation of some physiological and biochemical methods as an aid to the characterization of species of *Penicillium* subsection *Fasciculata*. J. Gen. Microbiol. 131: 1887-1895.

Bridge, P. D., Hawksworth, D. L., Kozakiewicz, Z., Onions, A. H. S., Paterson, R. R. M. Sackin, M. J. & Sneath, P. H. A. (1989a). A reappraisal of the terverticillate Penicillia using biochemical, physiological and morphological features I. Numerical taxonomy. J. Gen. Microbiol. 135: 2941-2966.

Bridge, P. D., Hawksworth, D. L., Kozakiewicz, Z., Onions, A. H. S., Paterson R. R. M. & Sackin, M. J. (1989b). A reappraisal of the terverticillate Penicillia using biochemical, physiological and morphological features II. Identification. J. Gen. Microbiol. 135: 2967-2978.

Cruikshank, R. H. & Pitt, J. I. (1987). Identification of species in *Penicillium* subgenus *Penicillium* by enzyme electrophoresis. Mycologia 79: 614-620.

Frisvad, J. C. & Filtenborg, O. (1989). Terverticillate penicillia: chemotaxonomy and mycotoxin production. Mycologia 81: 837-861.

LoBuglio K.F., Pitt, J. I. & Taylor, J. W. (1994). Independent origins of the synnematous *Penicillium* species, *P. duclauxii, P. clavigerum,* and *P. vulpinum*, as assessed by two ribosomal DNA regions. Mycol. Res. 98:250-56

Pitt, J. I. (1979). The genus *Penicillium* and its teleomorphic states *Eupenicillium* and *Talaromyces*. Academic Press: London, New York, Toronto, Sydney, San Francisco.

Pitt, J. I., Klich, M. A. Shaffer, G. P., Cruikshank, R. H., Frisvad, J. C., Mullaney, E. J., Onions, A. H. S., Samson R. A. & Williams, A. P. (1990). Differentiation of *Penicillium glabrum* from *Penicillium spinulosum* and other closely related species: An integrated taxonomic approach. System. Appl. Microbiol. 13: 304-309.

Rohlf, E. J. (1992). NTSYS-pc. Numerical taxonomy and multivariate analysis system, version 1.70. Exeter Software, New York.

SAS Institute, Inc. (1990). SAS/STAT7User"s Guide, Version 6, Fourth Edition, Volume 1. SAS Institute, Inc. Cary, NC.

Seifert, K. A. & Louis-Seize, G. (2000). Phylogeny and species concepts in the *Penicillium aurantiogriseum* complex as inferred from partial beta-tubulin gene DNA sequences In. Integration of modern taxonomic methods for *Penicillium* and *Aspergillus* classification. (Eds. Samson, R.A. & Pitt, J.I). Harwood Publishers, Amsterdam, pp 189-198

Seifert, K. A. & Frisvad, J. C. (2000). *Penicillium* on solid wood products. In. Integration of modern taxonomic methods for *Penicillium* and *Aspergillus* classification (Eds. Samson, R.A. & Pitt, J.I). Harwood Publishers, Amsterdam, pp. 285-298

CHARACTERIZATION OF *PENICILLIUM* BY THE USE OF BIOLOG

Helene Kiil and Mikako Sasa
Screening Biotechnology, Enzyme Research, Novo Nordisk A/S, Novo Allé, DK-2880 Bagsvaerd, DENMARK

INTRODUCTION

During an investigation of the microfungal flora in an alkaline soil 67 isolates of *Penicillium* were obtained. The BIOLOG system was tested to see whether it was able to detect physiologically distinct isolates prior to enzyme screening and how these results correlated with subsequent identification. All of the *Penicillium* isolates were *P. atramentosum* except for one isolate each of *P. citreonigrum*, *P. primulinum* and *Eupenicillium ochrosalmoneum*. BIOLOG distinguished between these four species and furthermore, with one exception, all of the isolates of *P. atramentosum* clustered together. Based on these results, further investigations were performed to determine whether the BIOLOG technique could distinguish between three species which belonged to the same subgenus as *P. atramentosum* (subgenus *Penicillium*) and are thus presumed to be closer related than the species investigated previously.

MATERIALS AND METHODS

Fungi: The *Penicillium* isolates used in the two studies came from two different sources: the first group was isolated from alkaline soil on alkaline CYA plates using the soil wash method and the soil dilution method (Parkinson, 1982). The second group of *Penicillium*, *P. chrysogenum*, *P. commune* and *P. expansum* was obtained from culture collections. *P. chrysogenum:* NN 009386, ATCC 9480, IBT 10025; *P. commune:* NN 006475, IBT 12714; *P. expansum:* ATCC 24692, ATCC 24692. (Novo Nordisk: Novo Nordisk A/S Culture Collection, Bagsvaerd; Denmark; ATCC: American Type Culture Collection; IBT: Department of Biotechnology, Technical University of Denmark, Lyngby, Denmark). An isolate of *Doratomyces stemonitis* originally presumptively identified, as *Penicillium* was also included.

Medium: Alkaline CYA was made from the standard formulation (3g $NaNO_3$, 1 g KH_2PO_4, 0.5 g $MgSO_4,7H_2O$, 0.5 g KCl, 1 ml $FeSO_4,7H_2O$ 1%, 30 g sucrose, 5 g yeast extract (Difco), 25 g agar, 1000 ml distilled water). The pH of the plates was adjusted to pH 9 with 0.04 M $NaHCO_3$.

The BIOLOG system: The BIOLOG system is a 96 well microtiter plate with 95 different carbon sources and one reference well. A nutrient buffer is added to all of the wells. The

251

system detects whether or not an organism is capable of utilizing a specific carbon source under standardized conditions, i.e. temperature and incubation time of the MicroPlates.

SF-N MicroPlate™

1	2	3	4	5	6	7	8	9	10	11	12
A1 water	A2 α-cyclodextrin	A3 dextrin	A4 glycogen	A5 tween 40	A6 tween 80	A7 N-acetyl-D-galactosamine	A8 N-acetyl-D-glucosamine	A9 adonitol	A10 L-arabinose	A11 D-arabitol	A12 cellobiose
B1 i-erythritol	B2 D-fructose	B3 L-fucose	B4 D-galactose	B5 gentiobiose	B6 α-D-glucose	B7 m-inositol	B8 α-D-lactose	B9 lactulose	B10 maltose	B11 D-mannitol	B12 D-mannose
C1 D-melibiose	C2 β-methyl D-glucoside	C3 D-psicose	C4 D-raffinose	C5 L-rhamnose	C6 D-sorbitol	C7 sucrose	C8 D-trehalose	C9 turanose	C10 xylitol	C11 methyl pyruvate	C12 mono-methyl succinate
D1 acetic acid	D2 cis-aconitic acid	D3 citric acid	D4 formic acid	D5 D-galactonic acid lactone	D6 D-galacturonic acid	D7 D-gluconic acid	D8 D-glucosaminic acid	D9 D-glucuronic acid	D10 α-hydroxybutyric acid	D11 β-hydroxybutyric acid	D12 γ-hydroxybutyric acid
E1 p-hydroxy phenylacetic acid	E2 itaconic acid	E3 α-keto butyric acid	E4 α-keto glutaric acid	E5 α-keto valeric acid	E6 D,L-lactic acid	E7 malonic acid	E8 propionic acid	E9 quinic acid	E10 D-saccharic acid	E11 sebacic acid	E12 succinic acid
F1 bromo succinic acid	F2 succinamic acid	F3 glucuronamide	F4 alaninamide	F5 D-alanine	F6 L-alanine	F7 L-alanyl-glycine	F8 L-asparagine	F9 L-aspartic acid	F10 L-glutamic acid	F11 glycyl-L-aspartic acid	F12 glycyl-L-glutamic acid
G1 L-histidine	G2 hydroxy L-proline	G3 L-leucine	G4 L-ornithine	G5 L-phenylalanine	G6 L-proline	G7 L-pyroglutamic acid	G8 D-serine	G9 L-serine	G10 L-threonine	G11 D,L-carnitine	G12 γ-amino butyric acid
H1 urocanic acid	H2 inosine	H3 uridine	H4 thymidine	H5 phenyl ethylamine	H6 putrescine	H7 2-amino ethanol	H8 2,3-butanediol	H9 glycerol	H10 D,L-α-glycerol phosphate	H11 glucose-1-phosphate	H12 glucose-6-phosphate

The SF-N MicroPlate™ (Fig 1.) was used in this study and purchased from the BIOLOG company. One MicroPlate is inoculated with one isolate. After incubation, plates are evaluated for growth with a MicroStation Reader, which measures the turbidity in each well. The MicroStation Reader is connected to a computer, which stores the results of these readings. Using the MicroLog™ software, which is also a part of the BIOLOG system, it is possible to perform cluster analyses based on the results of the turbidity readings and to calculate dendrograms using the program "MLCLUST". MLCLUST calculates a goodness of match between the patterns of substrate utilization obtained from the investigated organisms. The goodness of match is directly related to the number of mismatched carbon utilization between the isolates.

The testing scheme used in this investigation was based on that recommended by BIO-LOG Inc. The isolates were incubated on V8 agar (200 ml V8 vegetable juice, 3 g $CaCO_3$, 25 g agar, ad 1000 ml distilled water) for four days at 25°C. The colonies were harvested by scraping conidia and fragments of mycelium from the agar plates using sterile moist cotton buds. The spores and mycelium were transferred to an autoclaved 20 mm glass tube with 13.5 ml 0.2% carageenan Type II (Sigma) and 15 glass beads (2 mm diameter). The suspension was vortexed twice for 15 seconds at high speed in order to fragment the mycelium and obtain a homogenous suspension. The turbidity was measured on the BIOLOG turbidimeter. It is recommended to achieve an OD_{590} of approximately 0.22, which corresponds to a transmittance of about 60%. The suspension was adjusted until the desired transmittance was obtained and allowed to settle for approximately two minutes. It was then diluted tenfold by transferring 1.5 ml to an autoclaved 20 mm tube containing 13.5 ml 0.2% carageenan. 100 µl of the diluted suspension was then inoculated into each well using a multichannel pipette. The MicroPlates were incubated at 25°C and read every day starting day one (24 hours after inoculation). Replicates were prepared by transferring two 1.5 ml suspensions to separate tubes with 13.5 ml 0.2% carageenan and inoculating two separate MicroPlates.

The readings from day seven are used for this paper, as the reproducibility of the results of day seven was good and the turbidity readings revealed that the isolates were still actively growing in the wells.

RESULTS

The dendrogram obtained from the readings of the MicroPlates inoculated with the first group of *Penicillium* isolates (*Eupenicillium ochrosalmoneum*, *Penicillium atramentosum*, *P. citreonigrum*, *P. primulinum* and additionally *Doratomyces stemonitis*) is shown in Figure 2. Replicates are indicated by "(1)" or "(2)" following the isolate number. The dendrogram illustrates that the three species of *Penicillium* cluster separately. The dendrogram differentiates between *Eupenicillium* and *Doratomyces* and separates them from the other isolates. A single isolate of *P. atramentosum*, no. 142.1, clustered separately.

The reproducibility is apparently good, i.e. distances between replicates of the same species are merely 2-8. The largest distance found between two replicates is 8, which is seen for *E. ochrosalmoneum*. Replicates of the other species have distances between 2-6.

The dendrogram for the second group of isolates which all belong to subgenus *Penicillium* (*P. chrysogenum*, *P. commune* and *P. expansum*) is shown in Figure 3. The previous results for some of the *P. atramentosum* isolates are included. The dendrogram shows that BIOLOG is also able to separate these four species of *Penicillium* and furthermore, that each species forms a separate cluster. It is noteworthy that all strains of *P. chrysogenum* cluster together despite the fact one isolate produced a bright yellow soluble pigment in contrast to the other two investigated isolates.

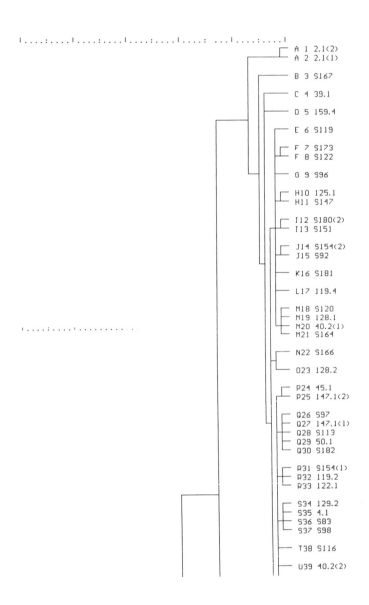

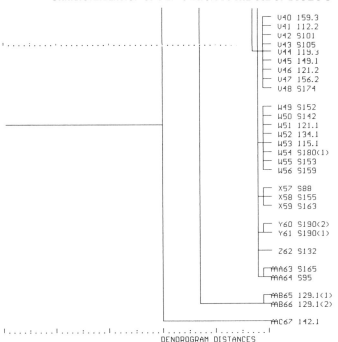

Fig. 2: Dendrogram of *Penicillium* isolates from alkaline soil. Isolate 124.1: *P. primulinum*, isolate C3 through AAB64: *Penicillium atramentosum*, isolate 2.1: *P. citreonigrum*, isolate 152.1 *Eupenicillium ochrosalmoneum*, isolate 129.1: *Doratomyces stemonitis*, isolate 142.1 *P. atramentosum*.

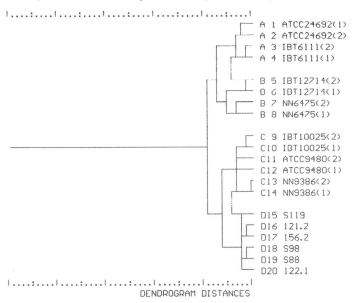

Fig. 3. Dendrogram of *P. atramentosum*, *P. chrysogenum*, *P. commune* and *P. expansum*. Cluster A: *P. expansum*; cluster B: *P. commune*; cluster C: *P. chrysogenum*; cluster D: *P. atramentosum.*. The strains of *P. atramentosum* were selected among the soil isolates presented in Figure 2.

255

DISCUSSION

Some species of the genus *Penicillium* are difficult to separate based on morphology alone, and the differentiation between especially species of *Penicillium* subgenus *Penicillium* is difficult without the aid of physiological tests (Pitt, 1988). Several approaches have been used in order to find alternative ways of differentiating between species of this genus. Bridge (1985) investigated the assimilation of carbon sources by *Penicillium* subsection *Fasciculata* and found four carbon sources and three nitrogen sources which could be used to differentiate between the species investigated. Cruickshank and Pitt (1987) investigated the use of zymograms to differentiate between species within *Penicillium* subgenus *Penicillium*.

The BIOLOG system has not been commonly used as a tool to investigate carbon utilization by filamentous fungi. Talbot *et al.* (1996) and Wildman (1995) used BIOLOG GN plates for investigating the utilization of carbon sources by *Fusarium compactum*. Talbot *et al.* (1996) determined that the ability to utilize specific carbon sources corresponded well with the secondary metabolite profile found for *F. compactum*. As secondary metabolite profiles are routinely used as a means to characterize and identify *Penicillium* (Frisvad and Filtenborg 1989), the results of Talbot *et al.* (1996) indicate that BIOLOG may be used as an alternative method to characterize species of this genus, as shown for the isolates tested in this investigation.

The turbidity readings of the MicroPlates are central to the BIOLOG system. Thus, the standardization of the testing procedures is essential to ensure that the observed patterns of carbon utilization can be attributed to physiological differences between isolates and not to differences in procedures. During the preparation of the inocula, we observed visible differences depending on the species. The required transmittance was obtained using mostly spores and small pieces of mycelium (invisible to the naked eye) when preparing the inocula of isolates with pigmented conidia (*Penicillium atramentosum, P. citreonigrum, P. expansum, P. chrysogenum* and *P. commune*). When preparing inocula of isolates with hyaline or lightly pigmented conidia (*P. primulinum* and *Eupenicillium ochrosalmoneum*) the required transmittance was obtained not only from conidia and small fragments of mycelium, but also from larger clumps of mycelium which were only partially fragmented by vortexing with glass beads. Due to this, the amount of mycelium and conidia in the inoculum which was required to obtain the same transmittance probably differed between the various species. We observed that glass beads were very useful in fragmenting the mycelium and they are recommended for this purpose. As the reproducibility of results of the replicates was good for all isolates, the testing procedure including our method of preparing the inoculum was adequate.

We found the optimal incubation time of the MicroPlates to be seven days. This was based on the growth of the isolates and results of regular readings of turbidity. The continuous increase in OD in several wells from day four to day seven indicated that the carbon sources, which were provided in the wells, were not exhausted and that the fungi were still active (results not presented here). The decision to choose the results of the turbidity readings of day seven was also based on the fact that Pitt (1988) recommends seven days as the time usually needed for species of *Penicillium* to develop characteristic morphological and physiological features on commonly used complex media.

During the incubation of the MicroPlates it was observed that most of the isolates produced coloured soluble pigments on some of the carbon sources. Although the pigments may have contributed to an increased OD reading in these wells, this did not affect the results of the readings of the three *P. chrysogenum* isolates which differed clearly in their production of pigments. The production of coloured exudates and soluble pigment on complex media is used in the identification of *Penicillium* isolates (Pitt, 1979,1988). Thus, the colour pattern produced on specific carbon sources may prove useful in the identification of species of *Penicillium*. Some carbohydrate monomers have been investigated and found useful in taxonomic distinction between species of *Penicillium* by Bridge (1985).

We found that BIOLOG could distinguish between the limited number of species of *Penicillium* investigated in this study and suggest that the system may be useful as a relatively simple and highly standardized tool for distinguishing between species of this genus. Further work is needed to support and determine the reliability of the system, but it may be a complementary method to the secondary metabolite profiles employed by some laboratories as a means of identifying and characterizing species of *Penicillium* subgenus *Penicillium*.

Our results indicate that BIOLOG may also be employed as a pre-screening tool to characterize and differentiate between a large group of unidentified isolates of microfungi in order to limit the number of isolates prior to initiating an extensive screening for e.g. secondary metabolites or enzymes. However, a correlation between carbon utilization and the production of specific enzymes needs to be established.

ACKNOWLEDGEMENTS

M.Sc. Jens Efsen Johansen and Dr. Carsten Sjøholm, Novo Nordisk A/S are acknowledged for their assistance with BIOLOG. Dr. Jens Frisvad, Technical University of Denmark is acknowledged for the identification of the *Penicillium* isolates and suggestions for the choice of *Penicillium* species which were tested by BIOLOG in the second part of this investigation. S.Å. Hansen, Novo Nordisk A/S is acknowledged for photographing the MicroPlates.

REFERENCES

Bridge, P.D. (1985). An evaluation of some physiological and biochemical methods as an aid to the characterization of species of *Penicillium* subsection *Fasciculata*. J. Gen. Microbiol. 131: 1887-1895.

Cruickshank, R.H. & Pitt, J.I. (1987). Identification of species in *Penicillium* subgenus *Penicillium* by enzyme electrophoresis. Mycologia 79: 614-620.

Frisvad, J.C. & Filtenborg, O. (1989). Terverticillate Penicillia: chemotaxonomy and mycotoxin production. Mycologia 81: 837-861.

Parkinson, D. (1982). Filamentous fungi. In Methods of Soil Analysis. Eds. Page A.L. & Klute A. Part 2: Chemical and Microbial properties. Madison, Wisconsin: American Society of Agronomy, pp.929-968.

Pitt, J.I. (1979). The Genus *Penicillium* and its Teleomorphic States *Eupenicillium* and *Talaromyces*. London, Academic Press.

Pitt, J.I. (1988). A Laboratory Guide to Common *Penicillium* Species. North Ryde, N.S.W., CSIRO Division of Food Processing.

Talbot, N.J., Vincent, P. & Wildman, H.G. (1996). The influence of genotype and environment on the physiological and metabolic diversity of *Fusarium compactum*. Fungal Gen. and Biol. 20: 254-267.

Wildman, H.G. (1995). Influence of habitat on the physiological and metabolic diversity of fungi. Can. J. Bot. 73 (Suppl. 1): s907-s916.

CYCLOHEXIMIDE TOLERANCE AS A TAXONOMIC CHARACTER IN *PENICILLIUM*

Keith A. Seifert and Sabina Giuseppin,
Eastern Cereal and Oilseed Research Centre, Agriculture and Agri-Food Canada,
Research Branch, Ottawa, Ontario, Canada K1A 0C6

Growth and conidial production response to 100 µg/ml cycloheximide of 76 strains belonging to 33 species of *Penicillium* were tested using malt extract agar. Species responses were relatively constant. The majority of the species showed some tolerance to the antibiotic at this concentration, with only four species being sensitive: *P. decumbens, P. digitatum, P. italicum* and *P. thomii*. Only *P. glabrum* included strains that were either sensitive or tolerant. The possibility of using cycloheximide tolerance as a diagnostic tool for separating some morphologically similar species is discussed.

INTRODUCTION

Cycloheximide (also known as Actidion, Naramycin A, and the active ingredient in the commercially produced Mycosel agar) is a polyene antibiotic produced by the actinomycete *Streptomyces griseus* that interferes with eukaryote protein synthesis. The best known taxonomic application of responses to this antibiotic is its use as one of the characters distinguishing *Ceratocystis* and *Ophiostoma* (Harrington, 1981). Resistance to cycloheximide is also used in the design of selective media for some medically important fungi, particularly *Coccidiodes immitis* and some dermatophytes (de Hoog and Guarro, 1995).

We became aware of cycloheximide tolerance in some species of *Penicillium* during our attempts to isolate *Ophiostoma* species from lumber. *Penicillium* species, particularly *P. brevicompactum* and *P. spinulosum,* were particularly common on these isolation plates (Seifert and Frisvad, 2000). This observation was in agreement with the findings of Marchisio *et al.* (1993), who isolated several Penicillia from air in Italy using a medium containing 400 µg/ml of cycloheximide. In this paper, we report on the ability of 33 species of *Penicillium* to grow in the presence of this antibiotic. The effect on sporulation was also noted. We attempted to determine whether the response of each species is consistent, and whether this information can be used as a supplementary diagnostic tool for problematic species.

MATERIALS & METHODS

The cultures used for this experiment are listed in Table 1. Each culture was grown on 2% malt extract agar (2% MEA; Samson *et al.*, 1995) alone and 2% MEA emended with

259

cycloheximide. A stock solution of the antibiotic in 95% ethanol was added to the medium after autoclaving to bring the final concentration to 100 µg/m. Plates were inoculated from conidial suspensions in semi-solid agar at three equidistant points (Pitt, 1979) using presterilized plastic loops. Radial growth was measured and sporulation was qualitatively determined after 7 days incubation at 25°C in the dark.

RESULTS AND DISCUSSION

The responses of the strains tested on 100 µg/ml cycloheximide are summarized in Table 1. The *Penicillium* strains tested showed a wide variation in growth responses, from complete inhibition to about 90% linear growth relative to the controls. All strains of only four species were completely inhibited: *P. decumbens, P. digitatum, P. italicum* and *P. thomii*. Responses were relatively consistent for each species where multiple strains were tested, with variation usually within +/- 15%. *Penicillium glabrum* was the only species to include some strains that were sensitive (no growth) and others that were tolerant (20-50% growth relative to the control). Tolerance to the antibiotic was found in all subgenera, and there were only a few evident patterns corresponding to infrageneric classification. For example, species classified according to Pitt (1979) in subgenus *Penicillium* section *Cylindrosporum* series *Italica* (*P. digitatum, P. italicum*) were sensitive to the antibiotic, while the closely related species *P. brevicompactum, P. griseofulvum* and *P. olsonii* all exhibited linear growth >75% relative to the controls. No clear species specific patterns were noted in the *P. aurantiogriseum* complex, which generally grew at rates 10-30% of the controls.

Sporulation was unaffected in all tested strains of *P. brevicompacum, P. citrinum, P. crustosum, P. rugulosum* and *P. variabile,* and reduced for the tested strains of *P. citreonigrum, P. expansum, P. freii, P. melanoconidium, P. oxalicum, P. solitum, P. tricolor* and *P. verrucosum.* The others species either had a variable response, or only one strain was tested. Sporulation was stimulated in single strains of *P. olsonii* and *P. steckii.*

This study demonstrates that cycloheximide tolerance is widespread in all subgenera of the genus *Penicillium.* This confirms the finding of Marchisio *et al.* (1993), who isolated several species of *Penicillium* from air in Italy using a medium containing 400 µg/ml of cycloheximide. These included species tested here (*P. brevicompactum, P. frequentans [= P. glabrum], P. funiculosum, P. roqueforti, P. rugulosum* and *P. vulpinum*) as well as some species we have not tested: *P. camemberti, P. citreoviride, P. fellutanum, P. implicatum, P. janthinellum, P. ochrochloron, P. purpurogenum, P. raistrickii,* and *P. waksmanii*. Harrington (1981) reported that linear growth of species of *Ophiostoma* is not significantly slower for cultures grown on 100 ppm cycloheximide than it is for controls, indicating a resistance to the antibiotic in the species of that genus. In contrast, species of *Penicillium* are at least somewhat inhibited, indicating a tolerant response rather than true resistance.

Table 1. Linear growth and sporulation of strains of *Penicillium* to 100 μg/ml cycloheximide in 2% MEA after 7 days incubation at 25°C, relative to control with no cycloheximide.

Species	Strain	% Growth	Effect on sporulation
Penicilium subgenus *Aspergillioides*			
P. bilaiae	DAOM 197974	76	sterile
P. citreonigrum	NRRL 761 (NT)	84	sterile
	DAOM 215383	70	sterile
	DAOM 215338	85	both sterile
P. decumbens	NRRL 741 (T)	0	na
	DAOM 215346	0	na
	DAOM 215347	0	na
P. glabrum	IMI 91944 (T)	0	na
	DAOM 208492	23	similar
	DAOM 215352	0	na
	DAOM 215354	21	reduced
	FTK 673A	0	na
	PRL 574	50	similar
P. spinulosum	IMI 24316i (T)	38	both sterile
	DAOM 215365	16	sterile
	DAOM 215366	18	similar
P. thomii	NRRL 2077 (T)	0	na
	DAOM 215367	0	na
	FTK x185a	0	na
Subgenus *Biverticillium*			
P. funiculosum	CBS 235.93	83	both sterile
	IMI 170614	92	both sterile
P. oxalicum	NRRL 787 (T)	3	both sterile
	DAOM 208491	12	sterile
	DAOM 215383B	5	sterile
	DAOM 215381	17	sterile
P. rugulosum	DAOM 215359	84	none
	DAOM 215360	74	none
	*DAOM 215361	82	none
P. variabile	DAOM 215368	76	none
	*FTK x126a	73	none
Subgenus *Furcatum*			
P. citrinum	CBS 241.85	83	none
	*94M-28 P-1	89	none
P. simplicissimum	94M-62 P-8-1	50	none
P. steckii	95M-79 DS-1	60	none
	95M-79 DS-2	46	more

Subgenus *Penicillium*

P. aurantiogriseum	*NRRL 971 (NT)	56	reduced
	*DAOM 213189	39	sterile
	*DAOM 214790	81	reduced
	*DAOM 216724	51	none
P. brevicompactum	DAOM 208489	89	none
	DAOM 214776	77	none
P. chrysogenum	DAOM 216700	46	none
P. commune	NRRL 890 (T)	29	sterile
P. expansum	NRRL 976 (NT)	21	sterile
	DAOM 214785	19	sterile
P. crustosum	DAOM 171024	28	none
	DAOM 21800	41	none
	*DAOM 215343	23	none
P. cyclopium	NRRL 1888 (T)	26	none
	*DAOM 216703	26	reduced
P. digitatum	NRRL 786	0	na
	DAOM 214780	0	na
P. freii	NRRL 951	29	reduced
	DAOM 214787	30	sterile
	DAOM 216705	28	sterile
P. griseofulvum	DAOM 216315	83	both sterile
P. italicum	NRRL 983 (NT)	0	na
	DAOM 214806	0	na
P. melanoconidium	IMI 321503 (T)	22	sterile
	DAOM 214791	23	sterile
	DAOM 216707	18	reduced
	DAOM 216708	19	sterile
P. olsonii	CBS 232.60 (NT)	61	more
P. polonicum	NRRL 995 (T)	17	sterile
	*DAOM 216709	25	reduced
	DAOM 216710	23	none
P. roquefortii	NRRL 849 (T)	39	none
P. solitum	*DAOM 187007	11	sterile
	*DAOM 215390	15	sterile
P. tricolor	DAOM 216711	22	sterile
	DAOM 216712	26	sterile
P. verrucosum	NRRL 965 (NT)	24	sterile
	*DAOM 213195	23	sterile
	Mills 30	17	sterile
P. viridicatum	NRRL 963 (NT)	22	sterile
P. vulpinum	NRRL 2031	45	reduced

Because there is a wide variation in growth and sporulation responses among different species and because the responses appear to be relatively consistent among different strains of the same species, there is the potential for incorporating cycloheximide tolerance into physiologically based identification schemes. However, the amount of work done to date is insufficient to recommend that this character be used routinely. The possibility of the occasional use of this character as an aid for distinguishing otherwise similar species, such as *P. citrinum* and *P. steckii*, is worth considering for labs where these species are frequently encountered.

ACKNOWLEDGMENTS

We thank John Bissett and Michael Corlett for their critical reviews of this manuscript. We thank Carolyn Babcock, Canadian Collection of Fungal Cultures, for providing some of the cultures used in this study.

REFERENCES

Harrington, T. C. (1981). Cycloheximide sensitivity as a taxonomic character in *Ceratocystis*. Mycologia 73: 1123-1129.

Hoog, G. S. & J. Guarro. (1995). Atlas of Clinical Fungi. Centraalbureau voor Schimmelcultures, Baarn, the Netherlands and Universitat Rovira i Virgili, Reus, Spain.

Marchisio, V. F., C. Cassinelli, V. Tullio and P. Mischiati. (1993). A preliminary survey of cycloheximide-resistant airborne fungi in Turin, Italy. Mycopathologia 123: 1-8.

Pitt, J. I. (1979). The genus *Penicillium* and its teleomorphic states *Eupenicillium* and *Talaromyces*. Academic Press: London, New York, Toronto, Sydney, San Francisco.

Samson, R. A., E. S. Hoekstra, J. C. Frisvad & O. Filtenborg. (1995). Introduction to Food-borne Fungi, 5th edition. Centraalbureau voor Schimmelcultures, Baarn.

Seifert, K. A & J. C. Frisvad. (2000). *Penicillium* on wood. In Integration of modern taxonomic methods for *Penicillium* and *Aspergillus* classification (Eds. Samson, R.A. & Pitt, J.I). Harwood Publishers, Amsterdam, pp 285-290

THE HOMOGENEOUS SPECIES AND SERIES IN SUBGENUS *PENICILLIUM* ARE RELATED TO MAMMAL NUTRITION AND EXCRETION

Jens C. Frisvad, Ole Filtenborg, Flemming Lund, and Robert A. Samson[1]
Dept. of Biotechnology, Building 221, Technical University of Denmark, DK-2800 Lyngby, Denmark and [1]Centraalbureau voor Schimmelcultures, 3740 AG Baarn, The Netherlands

Species in subgenus *Penicillium* have been regarded as difficult to classify by all experts in the genus. It is suggested here to exclude soil forms from subgenus *Penicillium* (*P. lanosum*, *P. fennelliae*, *P. scabrosum*), but include *P. sclerotigenum*, a plant associated species hitherto included in subgenus *Furcatum* Pitt. These changes leave 50 species and 15 natural series that are closely related phylogenetically and ecologically, as they are all primarily associated to mammal metabolism (feeding and dung). Based on a phylogenetic species model these species can be discovered by using the principle of the Aristotelian essence of a species, or the unique profile of extrovert expressions of differentiation (the aristotype). The species and series discovered this way are, however, also homogeneous in nutritional and physiological characters. The terverticillate Penicillia have been re-evaluated by recording their profiles of extrovert expressions of differentiation and the species and series accepted are listed. Several former varieties have been raised to species level based on their unique aristotypes and few new series are described. A list of synonyms and types are listed for each species. As some of the types are iconotypes or herbarium specimens without a surviving ex type culture a list of typical cultures for each species is provided.

INTRODUCTION

The taxonomic treatment of the asymmetric terverticillate Penicillia has been quite different in major monographs and revisions (Thom, 1930; Raper and Thom, 1949; Samson, Stolk and Hadlok, 1976; Pitt, 1979; Frisvad and Filtenborg, 1989; Pitt and Cruickshank, 1990; Stolk *et al.*, 1990). Most *Penicillium* monographic works have been based on a Linnaean principle with an emphasis on identification ease and practicability. The principles of the three different systematic schools, numerical phenetics, cladistics and evolutionary classification have practically not been used to any great extent (Frisvad, 1996), whereas scientific papers on subgenus *Penicillium* have included numerical phenetic methods, for example the work of Bridge *et al.* (1989), Svendsen and Frisvad (1994), Larsen and Frisvad, 1995, and Smedsgaard and Frisvad (1997) using numerical phenetic methods and studies on *Penicillium* (Skouboe *et al.*, 1996) using gene sequences and cladistic methods. Thus it is now possible to suggest series and genera that are holophyletic in a cladistic sense but also natural in the sense of Gower (1974). It is the aim of this paper to summarize, which natural species and series that belongs to a slightly

265

emended subgenus *Penicillium*, based on published taxonomic works on these fungi, with an emphasis on chemotaxonomical results.

Subgenera in the genus *Penicillium* and teleomorph connections

Pitt (1979) suggested four subgenera in the genus *Penicillium* and these have been used often in later taxonomic studies. Both secondary metabolite data (Frisvad and Filtenborg, 1989; 1990a,b; Frisvad *et al.*, 1990; Samson *et al.*, 1989) and DNA sequence data (Peterson, 1993; LoBuglio *et al.*, 1993; 1994; Barbee *et al.*, 1995) indicates that the subgenera *Aspergilloides* and *Furcatum* are related to *Eupenicillium javanicum* and related soilborne forms, while *E. egyptiacum* and *P. gladioli* are related to subgenus *Penicillium*. Subgenus *Biverticillium* is very different from all those supraspecific taxa and belong to the ascomycete genus *Talaromyces* (Pitt, 1979; Frisvad *et al.*, 1990; Peterson, 1993; LoBuglio *et al.*, 1993, Barbee *et al.*, 1995). The ascomycete genera and their associated anamorphic subgenera also seem to be ecologically homogeneous (Table 1).

Table 1. Phylogenetically and ecologically distinct groups in *Penicillium* and its associated teleomorphs.

Species	Primary habitats

1. Genus *Eupenicillium* pro parte
    + subgenus *Aspergilloides*
    + subgenus *Furcatum*
	Soil, plant rhizosphere

2. Subgenus *Penicillium* + some Eupenicillia	Mammal feed and dung
3. Genus *Talaromyces* + Subgenus *Biverticillium*	Wood and related products (paper) and textiles, bird feather

Species in subgenus *Penicillium*

Nearly all species included in the terverticillate Penicillia or subgenus *Penicillium* sensu Pitt (1979) are apparently both phylogenetically and ecologically related. The species associated to mammal nutrition (mammal feed and faeces) are the core species and rarely occur in soil with any human or other vertebrate activity. There are, however, important exceptions. Species, like *P. arenicola*, *P. scabrosum*, *P. lanosum*, *P. fennelliae*, *E. tularense* and *E. shearii* have biverticillate and terverticillate (often "twice biverticillate") structures, but those species have a velutinous colony texture and have soil as their primary habitat and thus can be excluded from subgenus *Penicillium*. *P. arenicola* produce canadensolides, chlorogentisylalcohol and unknown secondary metabolites not found in any other *Penicillium* species and thus may be related to another genus than *Penicillium*. It does produce asperphenamate in common with *P. brevicompactum* and *Aspergillus flavipes* however (Turner and Aldridge, 1983). *P. scabrosum*, *P. fennelliae* and *P. lanosum* produce secondary metabolites in common with certain species in both subgenus *Penicillium* and *Furcatum*, while *Eupenicillium tularense* and *E. shearii* only produce secondary metabolites (janthitrems, shearinins, paxillin and others) in common with certain *Furcatum* species (Frisvad et al., 1990c; Frisvad and Filtenborg, 1990a; Belofsky *et al.*, 1995).

266

One species that have been regarded as belonging to subgenus *Furcatum: P. scleroti-genum*, should be included in subgenus *Penicillium*. This species is predominantly biver-ticillate, but the penicilli are typically appressed and it is associated to rotting of plants: *P. sclerotigenum* has only been found in connection with yams (Filtenborg *et al.*, 1996). Recently isolated strains of *P. sclerotigenum* had several terverticillate penicilli (Frisvad, unpubl.). These few suggested alterations of subgenus *Penicillium* is also supported by the DNA sequence data that are available (Lobuglio and Taylor, 1994; Skouboe *et al.*, 1996). Another feature that unites most species in subgenus *Penicillium* as emended here is the tendency for fasciculation of the conidiophore stipes. This feature is only absent in the series centred around *P. brevicompactum, P. roqueforti, P. chrysogenum, P. mononematosum, P. digitatum* and *P. camemberti* (Raper and Thom, 1949; Samson *et al.*, 1976; Pitt, 1979).

One can then consider subgenus *Penicillium* as a polythetic class with no single defin-ing characteristics, but with a number of the following attributes present: Terverticillate structures with one or more appressed rami, fasciculate colony texture, strong sporulation on most media, association to mammal food, feed or dung, good growth at 5°C, no growth at 37°C, good growth on creatine-sucrose agar, and production of one or more of the following secondary metabolites: roquefortine C, patulin, xanthomegnin, cyclopenin, verrucosidin, puberuline, secalonic acid D, cyclopaldic acid, asteltoxin, geosmin, 2-methylisoborneol, cyclopiazonic acid, chaetoglobosin A, penitrem A, brevianamide A, aurantiamin, anacine or penicillin. These secondary metabolites are either present in many species or are exclusive produced by species in subgenus *Penicillium*. No species grow more than poorly at 37°C in this subgenus (some isolates of *P. aethiopicum, P. chrysogenum, P. mononematosum* and *P. flavigenum* may attain colony diameters of 2-4 mm after one week on CYA at 37°C). One species, *P. oxalicum* produce asymmetric biverticillate penicilli, may produce a rot on cucumber stems (Menzies *et al.*, 1995) and it also produces oxaline, thus this species have some resemblance with subgenus *Penicil-lium* species. It is consistently biverticillate and grows very fast at 37°C, so it most probably belongs outside subgenus *Penicillium sensu stricto*.

Series in subgenus *Penicillium*

The series accepted here have been based on former treatments by Raper and Thom (1949) and Pitt (1979) with necessary modifications based on new data, especially growth on creatine sucrose agar, secondary metabolite production, ecological data and other characteristics (Frisvad and Filtenborg, 1983; 1989; Filtenborg *et al.*, 1996). These data are summarized in Table 3.

The relation to *Eupenicillium* is strengthened by the fact that some or all isolates in 5 species are able to produce hard sclerotia (*P. gladioli, P. sclerotigenum, P. olsonii, P. coprobium* and *P. roqueforti*). Representative isolates of all 50 species recognized in sub-genus *Penicillium* are listed in table 3. They should not replace cultures ex type, but are handy for studies where authentic typical isolates are needed. Furthermore some species are only represented by iconotypes (*P. digitatum* and *P. coprophilum*). These typical iso-lates produce all the secondary metabolites associated with the species name (Frisvad and Filtenborg, 1989) and are in good cultural condition at present. Results from studies of secondary metabolites and morphological similarities indicate that *Eupenicillium crusta ceum, E. egyptiacum, E. molle*, and *E. osmophilum* are related to the terverticillate Penicil lia (Frisvad, unpublished; Wang *et al.*, 1995; 1998) and this should be confirmed using DNA sequence data. For example the mollenins produced by *Eupenicillium molle* are chemically closely related to the fructigenins produced by *P. polonicum* and *P. aurantio-candidum*.

Table 2. Characteristics for series in subgenus *Penicillium*. Series *Oxalica* has been included for comparison, but is not considered to belong to subgenus *Penicillium*.

Series	Growth on CREA	Primary habitat	Stipes/ conidia	Secondary metabolites	Special characteristics
0. Oxalica	Poor	Cucumber, corn	Sm/Sm, el	Secalonic acid D (1/1)*	Heavy growth at 37°C, conidial crusts
1. Mononematosa	Poor	Desert seeds	Sm/Sm	Asteltoxin & met. A (2/2)	
2. Chrysogena	Poor	Desert seeds	Sm -finely ro/ Sm	Penicillin (4/4)	Xerophilic
3. Italica	Poor	Citrus fruits	Sm/Sm, cy	Deoxybrevianamide E (2/2)	Citrus rot, conidial crusts, fasciculate
4. Digitata	Poor	Citrus fruits	Sm/Sm, large, cy	Tryptoquivalins (1/1)	Olive conidia, citrus rot
5. Olsonii	Poor	Tropical plants	Sm, broad/ finely rougened, el	Met. A, met. f (2/2)	Superficially aspergilloid capitula, occasionally sclerotia
6. Gladiolii	Poor	Bulbs	Sm/Sm	Patulin (2/2)	Sclerotia
7. Viridicata	Poor	Cereals	Ro/Sm or ro	Xanthomegnin (5/9), Penicillic acid (8/9), Cyclopenol (4/9)	Blue green conidia (7/9), brown halo on Raulin-Thom** (7/9)
8. Verrucosa	Poor	Cereals, cheese and meat	Ro/Sm	Ochratoxin A & arabenoic acid (1/1)	
9. Corymbifera	Poor	Bulbs or cereals	Ro/Sm	Roquefortine C (5/5)	Loose synnemata
10 Urticicola	Poor	Seeds	Sm/Sm, el	Cyclopiazonic acid (2/2), patulin (2/2), griseofulvin (2/2)	Short phialides
11. Expansa	Good	Pomaceous fruits and nuts	Sm or ro/ Sm	Roquefortine C & geosmin (2/2)	Apple rot, conidial crusts (1/2), fasciculate
12. Claviformae	Good	Dung	Sm or ro/ el	Patulin (6/8), roquefortine C (5/8)	Distinct synnemata (7/8)
13. Roqueforti	Good	Lactic acid ferments	Ro/Sm, globose	Roquefortine C & met. A (3/3)	growth on 0.5 % acetic acid
14. Camemberti	Good	Cheese and meat	Ro (or Sm)/ Sm	Cyclopiazonic acid (3/5), rugulovasine A (3/5)	
15. Solita	Good	Cheese, meat, nuts	Ro/ro or Sm	Cyclopenol & palitantin (3/3)	Dark conidia

* Number of species in the series that have the property/total number of species in the group; ** Growth on Raulin-Thom agar (see Raper and Thom, 194) Sm=smooth, ro= rough, el=ellipsoidal, cy=cylindrical.

Table 4. Typical isolate for each species in subgenus *Penicillium*

P. aethiopicum CBS 484.84= IBT 21501
P. albocoremium CBS 472.84 = IBT 21502
P. allii CBS 131.89= IBT 21503
P. atramentosum CBS 194.88 = IBT 21504
P. aurantiogriseum CBS 792.95 = IBT 21505
P. aurantiocandidum CBS 294.48 = IBT 21506
P. brevicompactum CBS 480.84 = IBT 21507
P. camemberti CBS 299.48 = IBT 21508
P. carneum CBS 449.78 = IBT 21509
P. caseifulvum CBS 101134 = IBT 21510 = IBT
18282
P. chrysogenum CBS 478.84 = IBT 21511
P. clavigerum CBS 255.94 = IBT 21512
P. commune CBS 468.84 = IBT 21513
P. concentricum CBS 101024?= IBT 21514 =
IBT 20230
P. confertum CBS 171.87 = IBT 21515
P. coprobium CBS 561.90 = IBT 21516
P. coprophilum CBS 186.89 = IBT 21517
P. crustosum CBS 101025 = IBT 21518 = IBT
17747
P. cyclopium CBS 101136 = IBT 11415 = IBT
21519
P digitatum CBS 101026= IBT 21520 = IBT
15179
P. dipodomyicola CBS 173.87= IBT 21521
P. dipodomyis CBS 170.87 = IBT 21522
P. discolor CBS 474.84 = IBT 21523
P. echinulatum CBS 101027 = IBT 21524 = IBT
12879
P. expansum CBS 481.84 = IBT 21525
P. flavigenum CBS 419.89 = IBT 21526

P. formosanum CBS 101028 =IBT 21527= IBT
19748
P. gladioli CBS 101029= IBT 21528 = IBT
14769
P. glandicola CBS 498.75 = IBT 21529
P. griseofulvum CBS 485.84 = IBT 21530
P. hirsutum CBS 135.41 = IBT 21531
P. hordei CBS 560.90 = IBT 21532
P. italicum CBS 489.84 = IBT 21533
P. melanoconidium CBS 641.95 = IBT 21534
P. mononematosum CBS 172.87 = IBT 21535
P. nalgiovense CBS 101030 = IBT 21536 = IBT
3800
P. neoechinulatum CBS 101135 = IBT 3493=
IBT 21537
P. olsonii CBS 833.88 = IBT 21538
P. palitans CBS 101031= IBT 21540 = IBT
14740
P. paneum CBS 101032 = IBT 12407 = IBT
21541
P. polonicum CBS 793.95 = IBT 21542
P. roqueforti CBS 479.84 = IBT 21543
P. sclerotigenum CBS 101033 = IBT 21544 =
IBT 14346
P. solitum CBS 147.86 = IBT 21545
P. tricolor CBS 637.93 = IBT 21547
P. ulaiense CBS 262.94 = IBT 21548
P. venetum CBS 405.92 = IBT 21549
P. verrucosum CBS 223.71 = IBT 21550
P. viridicatum CBS 101034 = IBT 21551 = IBT
15053
P. vulpinum CBS 101133 = IBT 11932 = IBT
21552

The 15 series recognized are discussed below mostly in the order recommended by Raper and Thom (1949), but with the creatine reaction as a major criterion for changing that order:

1. SERIES *Mononematosa* Frisvad, ser. nov.

Series in sectione *Penicillium*, conidiophora mononematica, divaricata, multiramulata, conidia globosa. Asteltoxinum et metaboliticum A producuntur.

Type species: *P. mononematosum*

Accepted species

P. mononematosum (Frisvad, Filt. & Wicklow) Frisvad, Mycologia 81: 857, 1989.
  P. glandicola var. *mononematosum* Frisvad, Filt. & Wicklow, Can. J. Bot. 65: 767, 1987.
  P. granulatum var. *mononematosum* (Frisvad, Filt. & Wicklow) Bridge, Kozak & R.R.M. Paterson, Myc. Pap. 165: 38, 1992.
  Holotype: IMI 296925
P. confertum (Frisvad, Filt. & Wicklow) Frisvad, Mycologia 81: 852, 1989
  P. glandicola var. *confertum*. Frisvad, Filt. & Wicklow, Can. J. Bot. 65: 769, 1987.
  Holotype: IMI 296930

The non-fasciculate slow-growing species have complicated structures different from the usual appressed two-ramus structures seen in most other species in subgenus *Penicillium*. Both species produce asteltoxin and metabolite A. *P. mononematosum* and *P. confertum* have been found in desertes and usually in connection with burrows of rodents. These species were characteristic in the two numerical taxonomical studies where they have been included (Bridge *et al.*, 1989; Svendsen and Frisvad, 1994). The production of verrucologen and other fumitremorgins, cyclopaldic acid, isochromantoxin, asteltoxin, viriditoxin and metabolite A (Frisvad and Filtenborg, 1989; Svendsen and Frisvad, 1994) by *P. mononematosum* and the production of meleagrin, asteltoxin, secalonic acid D & F and metabolite A clearly sets these species apart from other terverticillate Penicillia and even indicates that *P. mononematoseum* is more closely related to subgenus *Furcatum*. *P. confertum* has the habitat, growth rates and some metabolites in common with *P. mononematosum* and is not obviously related to *Claviformae* or *Chrysogena* as originally thought (Frisvad *et al.*, 1987; Banke *et al.*, 1997).

2. SERIES *Chrysogena* Raper & Thom ex Stolk & Samson, Adv. Pen. Asp. Syst.: 180, 1985

Series *P. chrysogenum* Raper & Thom, Man. Penicillia: 355, 1949 (nom. inval., arts 21,36)
Type species: *P. chrysogenum*

Accepted species:

P. chrysogenum Thom, Bull. Bur. Anim. Ind. USDA 118: 58, 1910.
  P. griseoroseum Dierckx, Ann. Soc. Scient. Brux. 25: 86, 1901.
  P. brunneorubrum Dierckx, Ann. Soc. Scient. Brux. 25: 88, 1901.
  P. citreoroseum Dierckx, Ann. Soc. Scient. Brux. 25: 89, 1901.
  P. baculatum Westling, Svensk Bot. Tidskr. 14: 139, 1910.
  P. notatum Westling, Ark. Bot. 11: 95, 1911.
  P. meleagrinum Biourge, Cellule 33: 147, 1923.
  P. flavidomarginatum Biourge, Cellule 33: 150, 1923.
  P. cyaneofulvum Biourge, Cellule 33: 174, 1923.

P. roseocitreum Biourge, Cellule 33: 184, 1923.

P. rubens Biourge, Cellule 33: 265, 1923.

P. chlorophaeum Biourge, Cellule 33: 271, 1923.

P. camerunense Heim apud Heim, Nouvel & Saccas, Bull. Acad. R. Belg. Cl. Sci. 35: 42, 1949.

P. chrysogenum var. *brevisterigma* Forster, Brit. Pat. 691: 242, 1953.

P. aromaticum f. *microsporum* Romankova, Uchen. Zap. Leningr. Gos. Univ. (Ser.Biol. Nauk. 40:) 191: 102, 1955.

P. harmonense Baghdadi, Nov. Sist. Niz. rast. 5: 102, 1968.

P. verrucosum var. *cyclopium* strain *ananas-olens* Ramírez, Man. Atlas. Penicil.: 457, 1982.

Neotype: IMI 24314

P. flavigenum Frisvad & Samson, Mycological Research 101: 620, 1997.

Holotype: CBS 419.89

P. dipodomyis (Frisvad, Filt. & Wicklow) Banke, Frisvad and S. Rosendahl, comb. nov., basionym

P. chrysogenum var. *dipodomyis* Frisvad, Filt. & Wicklow, Can. J. Bot. 65: 766, 1987.

P. dipodomyis (Frisvad, Filt. & Wicklow) Banke, Frisvad & S. Rosendahl, Mycol. Res. 101: 622, 1997 (nom. inval.).

Holotype: IMI 296926

P. nalgiovense Laxa, Zentbl. Bakt. ParasitKde, Abt. II 86: 162, 1932.

Neotype: CBS 352.48

These species (Banke *et al.*, 1987) are united by their production of penicillin, their dry habitats, salt tolerance, strictly velutinous colony texture, divaricate structure and phialide shape, fast growth rates and production of yellow and orange pigments. Secondary metabolite data and isozyme data show that *P. chrysogenum* is most close to *P. flavigenum*, while *P. nalgiovense* (the starter culture strains) is close to *P. dipodomyis*.

3. SERIES *Italica* Raper & Thom ex Pitt, Gen. Penicil.: 381, 1979

series *P. italicum* Raper & Thom, Man. Penicillia: 523, 1949 (nom. inval., arts 21,36)

Type species: *P. italicum*

Accepted species

P. italicum Wehmer, Hedwigia 33: 211, 1894.

Oospora fasciculata (Grev.) Sacc. & Vogl. apud Sacc., Syll. Fung. 4: 11, 1886.

P. aeruginosum Dierckx, Ann. Soc. Scient. Brux. 25: 87, 1901.

P. ventruosum Westling, Ark. Bot. 11: 112, 1911.

P. digitatum var. *latum* Abe, J. Gen. Appl. Microbiol. 2: 97, 1956.

P. japonicum G. Smith, Trans. Br. Mycol. Soc. 46: 333, 1963.

P. italicum var. *avellaneum* Samson, Stolk & Hadlok, Stud. Mycol. (Baarn) 11: 30, 1976.

Neotype: CBS 339.48

P. ulaiense Hsieh, Su & Tzean, Trans. Mycol. Soc. R.O.C. 2: 161, 1987.

Holotype: PPEH 29001.87

This is a natural series adapted to citrus fruits (Holmes *et al.*, 1994). The species share several secondary metabolites, yet each species both produce species specific secondary metabolites. The series has been emended to only include *P. italicum* and *P. ulaiense*, as the other species included in it were *P. fennelliae, P. digitatum* and *P. expansum* (as *P. resticulosum*) (Pitt, 1979) and these latter species are very different from the core species *P. italicum*.

4. SERIES *Digitata* Raper & Thom ex Samson & Stolk, Adv. Pen. Asp. Syst.: 183, 1985
Series *P. digitatum* Raper & Thom, Man. Penicillia: 385, 1949 (nom. inval., arts 21,36)
Type species: *P. digitatum*

P. digitatum (Pers.:Fr.) Sacc., Fung. Ital.: 894, 1881.
   Monilia digitata Pers. ex Fr., Syst. Mycol. 3: 411, 1832.
   Monilia digitata Pers., Syn. Meth. Fung.: 693, 1801.
   Aspergillus albus tenuissimus, graminis dactyloidis facie, seminibus rotundis Mich., Nova Pl. Gen.:
      213, 1729.
   Mucor caespitosus L., Sp. Pl. 2: 1186, 1753.
   Penicillium olivaceum Wehmer, Beitr. Kennt. Einh. Pilze 2: 73, 1895.
   P. olivaceum Sopp, Skr. Vidensk. Selsk. Christiana 11: 176, 1912.
   P. olivaceum var. *norvegicum* Sopp, Skr. Vidensk. Selsk. Christiana 11: 177, 1912.
   P. olivaceum var. *italicum* Sopp, Skr. Vidensk. Selsk. Christiana 11: 179, 1912.
   P. digitatoides Peyronel, Germi Atmosferici Fung. Micel.: 22, 1913.
   P. lanosogrisellum Biourge, Cellule 33: 196, 1923.
   P. terraconense Ramírez & Martínez, Mycopathologia 72: 187, 1980.
   Lectotype: icon in Saccardo, Fung. Ital.: tab. 894 Jul. 1881

This series comprises only the very distinct species *P. digitatum*. The olive coloured conidia and large micromorphological structures and cylindrical phialides and conidia are
unique in *Penicillium*. No obvious relatives are known and its placement in subgenus
Penicillium has been questioned (Stolk and Samson, 1985). There are no secondary metabolites in common between *P. digitatum* and the two members of *Italica*.

5. SERIES *Olsonii* Pitt, Gen. Penicil.: 392, 1979
Series *P. brevicompactum* Raper & Thom, Man. Penicillia: 404, 1949 (nom. inval., arts 21,36)
Type species: *P. olsonii*

Accepted species:
P. brevicompactum Dieckx, Ann. Soc. Scient. Brux. 25: 88, 1901.
   P. griseobrunneum Dierckx, Ann. Soc. Scient. Bruix. 25: 88, 1901.
   P. stoloniferum Thom, Bull. Bur. Anim. Ind. US Dept. Agric. 118: 68, 1910.
   P. tabescens Westling, Ark. Bot. 11: 100, 1911.
   P. szaferi K.M. Zalessky, Bull. Int. Acad. Pol. Sci. Lett., Sér. B, 1927: 447, 1927.
   P. hagemii K.M. Zalessky, Bull. Int. Acad. Pol. Sci. Lett., Sér. B, 1927: 448, 1927.
   P. bialowiezense K.M. Zalessky, Bull. Int. Acad. Pol. Sci. Lett., Sér. B, 1927: 462, 1927.
   P. patris-mei K.M. Zalessky, Bull. Int. Acad. Pol. Sci. Lett., Sér. B, 1927: 496,1927.
   P. brunneostoloniferum Abe, J. Gen. Appl. Microbiol. 2: 104, 1956.
   P. brunneostoloniferum Abe ex Ramírez, Man. Atlas Pen.: 412, 1982.
   Neotype: IMI 40225
P. olsonii Bain. & Sartory, Ann. Mycol. 10: 398, 1912.
   P. monstrosum Sopp, Skr. Vidensk. Selsk. Christiana 11: 150, 1912.
   P. volgaense, Beljakova & Mil'ko, Mikol. Fitopatol. 6: 147, 1972.
   P. brevicompactum var. *magnum* Ramírez, Man. Atlas Penicil.: 398, 1982.
   Neotype: IMI 192502

The two widely distributed species in series *Olsonii* have large often multiramulate penicilli on broad stipes with a superficial *Aspergillus* like appearance, ellipsoidal finely
roughened conidia, and they produce the unknown metabolites O and f in common
(Svendsen and Frisvad, 1994; Frisvad *et al.*, 1990a). They are common in pot plants,
greenhouses, tropical soil, on basidiocarps, tomatoes, bananas and several other sub-

strates. They are able to grow at lower water activities that most other terverticillate Penicillia.

6. SERIES *Gladiolii* Raper & Thom ex Stolk & Samson, Adv. Pen. Asp. Syst.: 183, 1985
Series *P. gladioli*, Man. Penicillia: 471, 1949 (nom. inval., arts 21,36)
Type species: *P. gladioli*

Accepted species
    P. gladioli McCulloch & Thom, Science, N.Y. 67: 217, 1928.
        P. gladioli Machacek, Rerp. Queb. Soc. Prot. Pl. 19: 77, 1928
        Neotype: IMI 34911
    P. sclerotigenum Yamamoto, Scient. Rep. Hyogo Univ. Agric., Agric. Biol. Ser. 2, 1: 69, 1955.
        Neotype: IMI 68616

These two sclerotium producing plant associated species are associated to *Gladiolus* corms and Yams tubers respectively. Isolates in both species are able to produce patulin, but they have no other secondary metabolites in common. It should be tested whether they are phylogenetically as closely related as their habitat and sclerotium producing ability suggest.

7. SERIES *Viridicata* Raper & Thom ex Pitt, Gen. Penicil.: 334, 1979
    Series *P. viridicatum* Raper & Thom, Man. Penicillia: 481, 1949 (nom. inval., arts 21,36)
    Series *P. cyclopium* Raper & Thom, Man. Penicillia: 490, 1949 (nom. inval., arts 21,36)
    Series *P. ochraceum* Raper & Thom, Man. Penicillia: 475, 1949 (nom. inval., arts 21,36)
    Ochracea Fassatiová, Acta Univ. Carol. Biol. 12: 324, 1977

Accepted species
    P. aurantiogriseum Dierckx, Ann. Soc. Scient. Brux. 25: 88, 1901
        P. aurantiogriseum var. *poznaniense* K.M. Zalesky, Bull. Int. Acad. Pol. Sci. Lett., Sér. B 1927: 444, 1927.
        Neotype: IMI 195050
    P. aurantiocandidum Dierckx, Ann. Soc. Scient. Brux. 25: 88, 1901.
        P. aurantiovirens Biourge, Cellule 33: 119, 1923.
        P. janthogenum Biourge, Cellule 33: 143, 1923.
        P. brunneoviolaceum Biourge, Cellule 33: 145, 1923.
        P. martensii Biourge, Cellule 33: 152, 1923.
        P. aurantio-albidum Biourge, Cellule 33: 197, 1923.
        P. johanniolii K.M. Zalessky, Bull. Int. Acad. Pol. Sci. Lett., Sér. B 1927: 453,1927.
        P. cyclopium var. *aurantiovirens* (Biourge) Fassatiová, Acta Univ. Carol. Biol. 12:326, 1977.
        P. cordubense Ramírez & Martínez, Mycopathologia 74: 164, 1981.
        Neotype: IMI 39814
    P. cyclopium Westling, Ark. Bot. 11: 90, 1911.
        ? *P. puberulum* Bain., Bull. Trimest. Soc. Mycol. Fr. 23: 16, 1907.
        P. porraceum Biourge, Cellule 33: 188, 1923.
        P. viridicyclopium Abe, J. gen. Appl. Microbiol. 2: 107, 1956.
        Neotype: IMI 89372
    P.freii Frisvad & Samson, in press.
        Holotype: IMI 285513
    P. melanoconidium (Frisvad) Frisvad and Samson, in press.

P. aurantiogriseum var. *melanoconidium* Frisvad, Mycologia 81:849.
Holotype:IMI 321503

P. neoechinulatum (Frisvad, Filt. & Wicklow) Frisvad & Samson, in press.
P. aurantiogriseum var. *neoechinulatum* Frisvad, Filt. & Wicklow, Can. J. Bot. 65: 767, 1987.
Holotype: IMI 296937

P. polonicum K.M. Zalessky, Bull. Int. Acad. Pol. Sci. Lett., Sér. B 1927: 445, 1927.
P. aurantiogriseum var. *polonicum* (K.M. Zalessky) Frisvad, Mycologia 81: 850.
P. carneolutescens G. Smith, Trans. Br. Mycol. Soc. 22: 252, 1939.
Neotype: CBS 222.28

P. tricolor Frisvad, Seifert, Samson & Mills, Can. J. Bot. 72: 937.
Holotype: DAOM 216240

P. viridicatum Westling, Ark. Bot. 11: 88, 1911.
P. olivinoviride Biourge, Cellule 33: 132, 1923.
P. blakesleei Zaleski, Bull. Int. Acad. Pol. Sci. Lett., Sér. B, 1927: 441, 1927.
P. stephaniae Zaleksi, Bull. Int. Acad. Pol. Sci. Lett., Sér. B, 1927: 451, 1927.
P. ochraceum Bain. apud Thom, Penicillia: 309, 1930.
P. verrucosum var. *ochraceum* (Bain.) Samson, Stolk & Hadlok, Stud. Mycol. (Baarn) 11: 42, 1976.
P. olivicolor Pitt, Gen. Penicil.: 368, 1979.
P. aurantiogriseum var. *viridicatum* (Westling) Frisvad & Filt., Mycologia 81: 850, 1989.
Neotype: IMI 39758ii

This series has been discussed by Frisvad and Lund (1993) and Lund and Frisvad (1994). It is characterized by the seed and cereal habitats (good amylase production) and production of several secondary metabolites. It is clearly a polythetic series concerning secondary metabolites as no metabolites are common to all nine species (Svendsen and Frisvad, 1994; Larsen and Frisvad, 1995; Smedsgaard and Frisvad, 1997).

8. SERIES *Verrucosa* Frisvad ser. nov.

Series in sectione *Penicillium*, conidiophora verrucosa, conidia globosa vel subglobosa, levibus, in CREA composito non crescit. Producta secondaria accumulata: ochratoxinum A et acidum arabenoicum.

Type species: *P. verrucosum*

P. verrucosum Dierckx, Ann. Soc. Scient. Brux. 25: 88, 1901.
P. casei Staub, Zentralbl. Bakt. ParasitKde. Abt. II, 31: 454, 1911.
P. mediolanense Dragoni & Cantoni, Ind. Aliment 155: 281, 1979.
P. nordicum Dragoni & Cantoni, Ind. Aliment 155: 283, 1979.
P. nordicum Dragoni & Cantoni ex Ramírez, Adv. Pen. Asp. Syst.: 139, 1985.
P. mongoliae Beljakova et al., in press.
Neotype: IMI 200310

A number of differences set this species apart from any other member of the subgenus. Most of the secondary metabolites produced by *P. verrucosum* (arabenoic acid, ochratoxins, the verrucins and a red brown pigment) are autapomorphic and only citrinin is shared with other *Penicillium* species. *P. verrucosum* has been found on two completely different kinds of habitats: cereals from temperate zones and salted meat products and cheese from Northern and Southern Europe. It has often been referred to species in *Viridicata*, but differs in a large number of features from those species, including growth on nitrite-sucrose agar and no acid production of creatine-sucrose agar.

9. SERIES *Corymbifera* Frisvad, ser. nov.

Series in sectione *Penicillium*, conidiophora fasciculata et verrucosa, synnematibus flavi vel albi, conidia
globosa vel subglobosa.

Type species: *P. hirsutum*

Accepted species

P. hirsutum Dierckx, Ann. Soc. Scient. Brux. 25: 89, 1901.
  P. corymbiferum Westling, Ark. Bot. 11: 92, 1911.
  P. verrucosum var. *corymbiferum* (Westling) Samson, Stolk & hadlok, Stud. Mycol.(Baarn) 11: 36,
    1976.
  ? *P. hispalense* Ramírez & Martínez, Mycopathologia 74: 169, 1981.
  Neotype: IMI 40213
P. hordei Stolk, Ant. van Leeuwenhoek 35: 270, 1969.
  P. hirsutum var. *hordei* (Stolk) Frisvad, Mycologia 81: 856, 1989.
  Holotype: CBS 701.68
P. venetum (Frisvad) Frisvad, comb. nov. Basionym:.
  P. hirsutum var. *venetum* Frisvad, Mycologia 81: 856, 1989.
  Holotype: IMI 321520
P. albocoremium (Frisvad) Frisvad, comb.nov. Basionym:
  P. hirsutum var. *albocoremium* Frisvad, Mycologia 81: 856, 1989.
  Holotype: IMI 285511
P. allii Vincent & Pitt, Mycologia 81: 300, 1989.
  P. hirsutum var. *allii* (Vincent & Pitt) Frisvad, Mycologia 81: 856, 1989.
  Holotype: MU Vincent 114

These five distinct fasciculate species are all associated with onions and flower bulbs
except *P. hordei* which is associated with barley and other cereals. *P. albocoremium*
sensu lato may contain two taxa as some isolates clustered with *P. hordei* while others
clustered with *P. allii* and *P. hirsutum* and *P. venetum* clustered separately when analyzed
for volatile secondary metabolites (Larsen and Frisvad, 1995, Smedsgaard and Frisvad,
1997). In the secondary metabolite study of Svendsen and Frisvad (1994) all these species
clustered with *P. crustosum* , except *P. hordei* which clustered with *P. aurantiogriseum*,
another cereal-borne species. The five species in this series seems to closest related to
Viridicata and *P. crustosum* in *Expansa*.

10. SERIES *Urticicola* Fassatiová, Acta Univ. Carol. Biol. 12: 324, 1977
Series P. urticae, Raper & Thom, Man. Penicillia: 531, 1949 (nom. inval., arts 21,36)
Type species: *P. griseofulvum*

Accepted species:

P. griseofulvum Dierckx, Ann. Soc. Scient. Brux. 25: 88, 1901.
  P. patulum Bain., Bull. Trimest. Soc. Mycol. Fr. 22: 208, 1906.
  P. urticae Bain., Bull. Trimest. Soc. Mycol. Fr. 23: 15, 1907.
  P. flexuosum Dale apud Biourge, Cellule 33: 264, 1923.
  P. maltum Hori & Yamamoto, Jap. J. Bacteriol. 9: 1105, 1954.
  P. duninii Sidibe, Mikol. Fitopatol. 8: 371, 1974.
  Neotype: IMI 75832
P. dipodomyicola (Frisvad, Filt. & Wicklow) Frisvad, comb. nov., basionym:
  P. griseofulvum var. *dipodomyicola* Frisvad, Filt. & Wicklow, Can. J. Bot. 65: 767, 1987.
  Holotype: IMI 296935

P. griseofulvum and *P. dipodomyicola* are very closely related. They share the characteristic very small phialides, divaricate structure, poor growth on creatine, and production of griseofulvin, cyclopiazonic acid and patulin. They have most often been found on dry cereals and seeds. Both species are distinct, however, as *P. dipodomyicola* produce predominantly bi- to rarely ter-verticillate structures while *P. griseofulvum* has ter- to quarter-verticillate structures. Each species consistently produce other species specific secondary metabolites (Svendsen and Frisvad, 1994; Smedsgaard and Frisvad, 1997)).

11. SERIES *Expansa* Raper & Thom ex Fassatiová, Acta Univ. Carol. Biol 12: 324, 1977
Series *P. expansum* Raper & Thom, Man. Penicillia: 508, 1949 (nom. inval., arts 21,36)
Series *P. terrestre* Raper & Thom, Man. Penicillia: 446, 1949 (nom. inval., arts 21,36)
Type species: *P. expansum*

Accepted species:
   P. expansum Link, Obs. Mycol. 1: 16, 1809.
      Coremium leucopus Pers., Mycol. Eur. 1: 42, 1822.
      Coremium glaucum Link ex Pers., Mycol. Eu. 1: 42, 1822.
      Floccaria glauca Grev., Scot. Crypt. Fl. 6: 301, 1828.
      Coremium alphitobus Secr., Mycol. Suisse 3: 539, 1833.
      Coremium vulgare Corda, Pracht-Fl.: 54, 1839.
      P. glaucum var. *coremium* Sacc., Syll. Fung. 4: 78, 1886.
      P. elongatum Dierckx,Corda, Pracht-Fl.: 54, 1839.
      P. glaucum var. *coremium* Sacc., Syll. Fung. 4: 78, 1886.
      P. elongatum Dierckx, Ann. Soc. Scient. Brux. 25: 87, 1901.
      P. musae Weidemann, Zentralbl. Bakt. ParasitKde., Abt. II, 19: 687, 1907.
      P. variabile Wehmer, Mykol. Zentralbl. 2: 195, 1913.
      P. plumiferum Demelius, Verh. Zool.-Bot. Ges. Wien 72: 76, 1922.
      P. aeruginosum Demelius, Verh. Zool.-Bot. Ges. Wien 72: 76, 1922.
      P. leucopus (Pers.) Biourge, C.R. Séanc. Soc. Biol. 82: 877, 1919.
      P. kap-laboratorium Sopp apud Biourge, Cellule 36: 454, 1925.
      P. resticulosum Birkinshaw, Raistrick & G. Smith, Biochem. J. 36: 830, 1942.
      P. martensii var. *moldavicum* Beljakova et al., in press.
      Neotype: CBS 325.48
   P. crustosum Thom, Penicillia: 399, 1930.
      P. pseudocasei Abe, J. Gen. Appl. Microbiol. 2: 102, 1956.
      P. pseudocasei Abe ex G. Smith, Trans. Brit. Mycol. Soc. 46: 335, 1963.
      P. terrestre sensu Raper & Thom, Man. Penicil.: 450, 1049.
      P. farinosum Novobranova, Nov. Sist. Niz. Rast. 11: 232, 1974.
      P. expansum var. *crustosum* (Thom) Fassatiová, Acta Univ. Carol.- Biol. 12: 329, 1977.
      P. solitum var. *crustosum* (Thom) Bridge, D. Hawksw., Kozak., Onions, R.R.M. Paterson, Sackin &
         Sneath, J. Gen. Microbiol. 135: 2957, 1989.
      Neotype: IMI 91917

These species share the ability to produce apple rot, to produce roquefortine C and geosmin, fast growth rates, smooth-walled conidia, and general appearance. Raper and Thom (1949) placed them in the same series and Fassatiová (1977) even combined *P. crustosum* as a variety of *P. expansum*. Pitt (1979) placed *P. chrysogenum* and *P. atramentosum* in this series and *P. crustosum* in series *Viridicata*, but his series were based more on facilitating identification than phylogenetic or overall phenetic similarity. The two species do not cluster in any of the chemotaxonomical studies performed, however (Svendsen and Frisvad, 1994; Larsen and Frisvad, 1995; Smedsgaard and Frisvad, 1997).

12. SERIES *Claviformae* Raper & Thom ex Stolk, Samson & Frisvad, Mod. Con. Pen. Asp. Clas.: 132, 1990
Type species: *P. vulpinum*

Accepted species
   ***P. clavigerum* Demelius**, Verh. Zool.-Bot. Ges. Wien 72: 74, 1922.
      Neotype: IMI 39807
   ***P. vulpinum* (Cooke & Massee) Seifert & Samson**, Adv. Pen. Asp. Syst.: 144, 1985.
      Coremium claviforme (Bain.) Peck, Bull. N.Y. St. Mus. 131: 16, 1909.
      Coremium silvaticum Wehmer, Ber. Dt. Bot. Ges. 31: 373, 1914.
      Penicillium silvaticum (Wehmer) Biourge, Celule 33: 1056, 1923.
      P. silvaticum (Wehmer) Gäumann, Vergl. Morph. Pilze: 177, 1926.
      Holotype: "on dung", *s.coll.*, in herb. Cooke (K)
   ***P. concentricum*, Samson, Stolk & Hadlok**, Stud. Mycol. (Baarn) 11: 17, 1976.
      P. glandicola var. *glaucovenetum* Frisvad, Mycologia 81: 855, 1989.
      Holotype: CBS 477.75
   ***P. coprobium* Frisvad**, Mycologia 81: 853, 1989.
      Holotype: IMI 293209
   ***P. coprophilum* (Berk. & Curt.) Seifert & Samson**, Adv. Pen. Asp. Syst.: 145, 1985.
      Holotype: Cuba, Wright 666 (K)
   ***P. aethiopicum* Frisvad**, Mycologia 81: 848, 1989.
      Holotype: IMI 285524
   ***P. glandicola* (Oud.) Seifert & Samson**, Adv. Pen. Asp. Syst.: 147, 1985.
      P. granulatum Bain., Bull. Trimest Soc. Mycol. Fr. 21: 126, 1905.
      P. divergens Bain. & Sartory, Bull. Trimest. Soc. Mycol. Fr. 28: 270, 1912.
      P. schneggii Boas, Mykol. Zentralbl. 5: 73, 1914.
      P. granulatum var. *globosum* Bridge, D. Hawksw., Kozak., Onions, R.R.M. Paterson, Sackin & Sneath, J. Gen. Microbiol. 135: 2957, 1989.
      Holotype: Netherlands, Valkenburg, Jul 1901, *Rick* in herb. Oudemans (L)
   ***P. formosanum* Hsieh, Su & Tzean**, Trans. Mycol. Soc. R.O.C. 2: 159, 1987.
      Holotype: PPEH 10001

This series is here amended to include all known synnematous coprophilic species of *Penicillium*.. They all have predominantly smooth walled ellipsoidal conidia.. All species produce patulin, except the two producers of griseofulvin in the series (*P. aethiopicum* and *P. coprophilum*). The placement of *P. aethiopicum* in this series is a provisional, but this species has several features in common with *P. coprophilum*. They all clustered in the HPLC analysis of Svendsen and Frisvad (1994), except *P. aethiopicum* which clustered with *P. digitatum* because of production of tryptoquivalins and *P. clavigerum* which was loosely connected to *P. griseofulvum*. The members of this group have several features in common with *P. expansum* in *Expansa* and members of *Urticicola*, including production of patulin and/or griseofulvin and roquefortine C, smooth walled stipes and smooth-walled ellipsoidal conidia and the production of synnemata.

13. SERIES *Roqueforti* Raper & Thom ex Frisvad, ser. nov.
      Series *P. roqueforti* Raper & Thom, Man. Penicillia, 392, 1949 (nom. inval., arts 21,36)
      Series in sectione *Penicillium*, conidiophora mononematica et tuberculati, conidia globosa et levibus, in CREA composito crescit, in 0.5 % acido acetico crescit. Producta secondaria accumulata: roquefortinum C.
Type species: *P. roqueforti*

Accepted species:

P. roqueforti Thom, Bull. Bur. Anim. Ind. US Dept. Agric. 82: 35, 1906.
  P. aromaticum casei Sopp, Zentbl. Bakt. ParasitKde., Abt. II: 4: 164, 1898.
  P. vesiculosum Bain., Bull. Trimest. Soc. Mycol. Fr. 23: 10, 1907.
  P. roqueforti var. *weidemannii* Westling, Ark. Bot. 11: 71, 1911.
  P. atroviride Sopp, Skr. Vidensk. Selsk. Christiana 11: 149, 1912.
  P. roqueforti Sopp, Skr. Vidensk. Selsk. Christiana 11: 156, 1912.
  P. virescens Sopp, Skr. Vidensk. Selsk. Christiana 11: 157, 1912.
  P. aromaticum Sopp, Skr. Vidensk. Selsk. Christiana 11: 159, 1912.
  P. aromaticum-casei Sopp ex Sacc., Syll. Fung. 22: 1278, 1913.
  P. suavolens Biourge, Cellule 33: 200, 1923.
  P. gorgonzolae Weidemann apud Biourge, Cellule 33: 204, 1923.
  P. weidemannii (Westling) Biourge, Cellule 33: 204, 1923.
  P. stilton Biourge, Cellule 33: 206, 1923.
  P. weidemannii var. *fuscum* Arnaudi, Boll. Ist. Sieroter. Milan. 6: 27 (1928).
  P. biourgei Arnaudi, Boll. Ist. Sieroter. Milan. 6: 27 (1928).
  P. roqueforti var. *viride* Dattilo-Rubbo, Trans. Br. Mycol. Soc. 22: 178, 1938.
  P. conservandi Novobranova, Nov. Sist. Niz. Rast. 11: 233, 1974.
  Neotype: IMI 24313
P. carneum (Frisvad) Frisvad, Microbiology, UK, 142: 546, 1996.
  P. roqueforti var. *carneum* Frisvad, Mycologia 81: 858, 1989.
  Type: IMI 293204
P. paneum Frisvad, Microbiology (UK) 142: 546, 1996.
  Holotype: C 25000

Species in series *Roquefortii* are closely related (Boysen *et al.*, 1996). They all grow fast, they have large smooth-walled globose conidia and tuberculate conidiophore stipes, they grow well on creatine-sucrose agar and on media with 0.5 % acetic acid, and they all produce roquefortine C. The species differ by their profiles of volatile and non-volatile secondary metabolites, by their rDNA sequences and their conidium and colony reverse colours. Members of the series have a symbiotic relationship with lactic acid bacteria and certain acid-tolerant yeasts.

14. SERIES *Camemberti* Raper & Thom ex Pitt, Gen. Penicil.: 358, 1979
Series *P. camemberti*, Raper & Thom, Man. Penicillia: 421, 1949 (nom. inval., arts 21,36) Series *P. commune*, Raper & Thom, Man. Penicillia: 429, 1949 (nom. inval., arts 21,36)
Type species: *P. camemberti*
Accepted species

P. commune Thom, Bull. Bur. Anim. Ind. USDA 118: 56, 1910.
  P. fuscoglaucum Biourge, Cellule 33: 128, 1923.
  P. flavoglaucum Biourge, Cellule 33: 130, 1923.
  P. lanosoviride Thom, Penicillia: 314, 1930.
  P. ochraceum Thom var. *macrosporum* Thom, Penicillia: 310, 1930.
  P. lanosoviride Thom, Penicillia: 314, 1930.
  P. lanosogriseum Thom, Penicillia: 327, 1930.
  P. psittacinum Thom, Penicillia: 369, 1930.
  P. australicum Sopp ex van Beyma, Ant. van Leeuwenhoek 10: 53, 1944.
  P. cyclopium var. *album* G. Smith, Trans Brit. Mycol. Soc. 34: 18, 1951.
  P. roqueforti var. *punctatum* Abe, J. Gen. Appl. Microbiol. 2: 99, 1956.
  P. caseiperdens Frank, Beitr. Tax. Gat. Pen.: 91, 1966.
  P. verrucosum var. *album* (G,. Smith) Samson, Stolk & Hadlok, Stud. Mycol. (Baarn) 11:35, 1976.
  P. album (G. Smith) Stolk & Samson, Adv. Pen. Asp. Syst.: 185, 1985.

278

Neotype: IMI 39812

P. camemberti Thom, Bull. Bur. Anim. Ind. USDA 82: 33, 1906. domesticated form of *P. commune*)

    P. album Epstein, Ark. Hyg. Bakt. 45: 360, 1902.
    P. epsteinii Lindau, Rabenh. Krypt.-Fl. 1, Abt. 8: 166, 1904
    P. rogeri Wehmer apud Lafar, Handb. Tech. Mykol. 4: 226, 1906.
    P. caseicola Bain., Bull. Trimest. Soc. Mycol. Fr. 23: 94, 1907.
    P. camemberti var. *rogeri* Thom, Bull. Bur. Anim. Ind. US Dept. Agric. 118: 52, 1910.
    P. biforme Thom, Bull. Bur. Anim. Ind. US Dept. Agric. 118: 54, 1910.
    P. camemberti Sopp, Skr. Vidensk. Selsk. Christiana 11: 179, 1912.
    P. candidum Roger apud Biourge, Cellule 33: 193, 1923.
    P. paecilomyceforme Szilvinyi, Zentralbl. Bakt. ParasitKde., Abt. II, 103: 156, 1941.
    Lectotype: IMI 27831

P. caseifulvum Lund, Filt. & Frisvad, J. Food Mycol 1: 97, 1998.
    Type: C 24999

P. palitans Westling, Ark. Bot. 11: 83, 1911.
    Neotype: IMI 40215

P. atramentosum Thom, Bull. Bur. Anim. Ind. US Dept. Agric. 118: 65, 1910.
    Neotype: IMI 39752

These creatine positive species are found on cheese and other substrates with a high content of lipid and protein. Isolates in four of the species produce rugulovasine A and three of the species produce cyclopiazonic acid.

15. SERIES *Solita* Frisvad, ser. nov.

Series in sectione *Penicillium*, conidiophora verrucosa, conidia globosa vel subglobosa et atroviridis, in CREA composito crescit, in 0.5 % acido acetico non crescit.

Type species: *P. solitum*

Accepted species

P. discolor Frisvad & Samson, Ant. van Leeuwenhoek, 72: 120, 1997.
    Holotype: IMI 285513

P. echinulatum Fassatiová, Acta Univ. Carol. Biol. 12: 326, 1977.
    P. cyclopium var. *echinulatum* Raper & Thom, Man. Penicil.: 497, 1949.
    P. palitans var. *echinoconidium* Abe, J. Gen. Appl. Microbiol. 2: 111, 1956.
    Holotype: PRM 778523

P. solitum Westling, Ark. Bot. 11: 65: 1911.
    P. majusculum Westling, Ark. Bot. 11: 60, 1911.
    P. conditaneum Westling, Ark. Bot. 11: 63, 1911.
    P. casei Staub var. *compactum* Abe, J. Gen. Appl. Microbiol. 2: 101, 1956.
    P. mali Novobr., Biol. Nauki 10: 105, 1972.
    P. verrucosum var. *melanochlorum* Samson, Stolk & Hadlok, Stud. Mycol. (Baarn) 11: 41, 1976.
    P. mali Gorlenko & Novobr., Mikol. Fitopatol. 17: 464, 1983.
    P. melanochlorum (Samson, Stolk & Hadlok) Frisvad, Adv. Pen. Asp. Syst.: 330, 1985.
    Neotype: CBS 424.89

This series contains three closely related species with dark green conidia and rough walled conidiophore stipes. They all produce the viridicatin biosynthetic family. The three species were distinct yet included in the same main cluster in the HPLC analysis based on secondary metabolites reported by Svendsen and Frisvad (1994). In an electrospray mass spectrometric (ES-MS) study of secondary metabolites of the terverticillate

279

Penicillia, *P. solitum* clustered with *P. echinulatum* when grown on CYA agar and *P. echinulatum* clustered with *P. discolor* as its nearest neighbour when cultures grown on YES agar (Smedsgaard and Frisvad, 1997). The reason these species also clustered with several members of *Viridicata* and *P. crustosum* was that the viridicatin biosynthetic family was present in all these species and these dominated the ES-MS profiles. Concerning volatile secondary metabolites the three species were distinct and not very similar (Larsen and Frisvad, 1995).

DISCUSSION

Subgenus *Penicillium sensu stricto* as delimited here appears to be a natural and phylogenetically distinct subgenus related to *Eupenicillium*. The series suggested here are polythetic, but the species in them are similar in both secondary metabolite, nutritive and physiological attributes and thus certain characteristics can be predicted when a species has been allocated to a series. A more detailed numerical taxonomical study of the species should be performed to evaluate this in quantitative terms. Several secondary metabolites are produced by species in more than one series (roquefortine C, patulin, citrinin, mycophenolic acid, griseofulvin and others), while other secondary metabolites are specific to a series or one or two species. The ability to grow on creatine as sole nitrogen source seems to be correlated with the ability to grow on proteinaceous substrates. A major exception is *P. verrucosum* that grows on cheese and meat and only poorly on creatine sucrose agar. The major habitat of *P. verrucosum* is cereals, and probably it produces different proteases than those required to degrade creatine in contrast to typical species growing on cheese or meat. Most of the species in the series suggested here have been examined by molecular methods and small differences in nucleotide sequences are seen within the series (Skouboe *et al.*, 1996, unpublished).

The series in Table 2 have been ordered according to their relationships with velutinous, creatine negative species series. *Mononematosa* and *Chrysogena* species grow at low water activities (originally desert species?). *Italica* and *Digitata* species are associated to citrus fruits only, while species *Olsonii* are widely distributed in the tropics and in pot plants and greenhouses. *Gladiolii* and *Corymbifera* are mostly associated to onions and bulbs, while series *Viridicata* and *Verrucosa* species are associated to cereals. The synnematous species are clustered as number 10-13 (*Corymbifera, Urticicola, Expansa* and *Claviformae*) and most species of the latter three series share several characteristics such as ellipsoidal smooth conidia and patulin and/or griseofulvin and roquefortine C production. Series number 14-16 are all common on cheese and meat. In the future new series may have to be described and some of the series may be expanded. For example new species have recently been discovered that are obviously to be included in *Viridicata, Urticicola* and *Expansa*, respectively.

Species series were regarded by Raper and Thom (1949) as species which had several characters in common, but were not necessarily natural series. Pitt (1979) regarded series as mostly natural, occasionally artificial, but at least of help for identification purposes. Stolk and Samson (1985) regarded series as micromorphologically homogeneous groups while Stolk *et al.* (1990) also considered secondary metabolites and habitats. Some of the series erected by Pitt (1979) have been accepted here, but many have been emended based on new evidence of relationship, mainly secondary metabolites. Preliminary results based on sequencing the ribosomal DNA, especially the ITS I and II regions have until now supported the series suggested here (Skouboe *et al.*, 1996).

The species listed in table 2 all appear to be very homogeneous and distinct and this has been confirmed in several studies based on volatile and non-volatile secondary metabolites (Frisvad and Filtenborg, 1983; 1989, Svendsen and Frisvad, 1994, Larsen and Frisvad, 1995; Smedsgaard and Frisvad, 1997). Many of these species have also been accepted by Pitt and Hocking, 1997; Cruickshank and Pitt, 1989; Bridge *et al.*, 1989; Samson *et al.*, 1976; Stolk *et al.*, 1990). Thus earlier statements that these Penicillia intergraded or were difficult to classify and identify (Thom, 1930; Raper and Thom, 1949; Ciegler *et al.*, 1973; 1981; Pitt, 1973; Samson *et al.*, 1976; Onions *et al.*, 1981; 1984; Onions and Brady, 1987; Pitt and Samson, 1990) were too pessimistic.

Eventually it is necessary to use three different teleomorphic states for *Penicillium* like anamorphs: those associated to 1) Subgenus *Furcatum* and *Aspergilloides*, 2) Subgenus *Penicillium*, and 3) Subgenus *Biverticillium*. Such a system will be phylogenetically optimal and would be required for good predictions, but there will be major nomenclatural and literature retrieval problems if this is consistently applied.

REFERENCES

Banke, S., Frisvad,. J.C. & Rosendahl, S. 1997. Taxonomy of *Penicillium chrysogenum* and related xerophilic species, based on isozyme analysis. Mycol. Res. 101: 617-624.

Barbee, M.L., Yoshiomura, A.; Suguyama, J. & Taylor, J.W. (1995) Is *Penicillium* monophyletic? An evaluation of phylogeny in the family Trichocomaceae from 18S, 5.8S and ITS ribosomal data. Mycologia 87: 210-222.

Belofsky, G.N., Gloer, J.B., Wicklow, D.T. & Dowd, P.F. 1995. Antiinsectan alkaloids: shearinines A-C and a new paxillin derivative from the ascostromata of *Eupenicillium shearii*. Tetrahedron 51: 3959-3968.

Boysen, M., Skouboe, P., Frisvad, J.C. & Rossen, L. 1996. Reclassification of the *Penicillium roqueforti* group into three species on the basis of molecular genetic and biochemical profiles. Microbiology (UK) 142: 541-549.

Bridge, P.D., Hawksworth, D.L., Kozakiewicz, Z., Onions, A.H.S., Paterson, R.R.M., Sackin, M.J. & Sneath, P.H.A. 1989. A reappraisal of the terverticillate Penicillia using biochemical, physiological and morphological features. I. Numerical taxonomy. J. Gen. Microbiol. 135: 2941-2966. Ciegler, A., Fennell, D.I., Sansing, G.A., Detroy, R.W. & Bennett, G.A. 1973. Mycotoxin producing strains of *Penicillium viridicatum*: classification into subgroups. Appl. Microbiol. 26: 271-278.

Ciegler, A., Lee, L.S & Dunn, J.J. (1981). Naphthoquinone production and taxonomy of *Penicillium viridicatum*. Applied and Environmental Microbiology 42: 446-449.

Filtenborg, O., Frisvad, J.C. & Thrane, U. 1996. Moulds in food spoilage. Int. J. Food Microbiol. 33: 85-102.

Frisvad, J.C. 1996. Fungal identification as related to numerical phenetics, cladistics and practicability. In Rosen, L., Rubio, V., Dawson, M.T. & Frisvad, J.C. (eds.) Fungal identification techniques, pp. 166-171. European Commission, Brussels.

Frisvad, J.C. & Filtenborg, O. 1989. Terverticillate Penicillia: chemotaxonomy and mycotoxin production. Mycologia 81: 837-861.

Frisvad, J.C. & Filtenborg, O. 1990a. Revision of *Penicillium* subgenus *Furcatum* based on secondary metabolite and conventional characters. Samson, R.A. & Pitt, J.I. (eds.) Modern concepts in *Penicillium* and *Aspergillus* classification. pp. 159-170. Plenum Press, New York.

Frisvad, J.C. & Filtenborg, O. 1990b. Secondary metabolites as consistent criteria in *Penicillium* taxonomy and a synoptic key to *Penicillium* subgenus *Penicillium*. In Samson, R.A. & Pitt, J.I. (eds.) Modern concepts in *Penicillium* and *Aspergillus* classification. pp. 373-384. Plenum Press, New York.

Frisvad, J.C. & Lund, F. 1993. Toxin and secondary metabolite production by *Penicillium* species growing in stored cereals. In Occurrence and significance of mycotoxins. Central Science Laboratory, MAFF, Slough, pp. 146-171.

Frisvad, J.C., Samson, R.A. & Stolk, A.C. 1990a. Notes on the typification of some species of *Penicillium*. Persoonia 14: 193-202.

Frisvad, J.C., Filtenborg, O., Samson, R.A. & Stolk, A.C. 1990b. Chemotaxonomy of the genus *Talaromyces*. Antonie van Leeuwenhoek 57: 179-189.

Frisvad, J.C., Samson, R.A. & Stolk, A.C. 1990c. A new species of *Penicillium, P. scabrosum*. Persoonia 14: 177-182.

Gower, J. 1974. Maximal predictive classification. Biometrics 30: 643-654.

Holmes, G.J., Eckert, J.W. & Pitt, J.I.. 1994. Revised description of *Penicillium ulaiense* and its role a a pathogen of citrus fruits. Phytopathology 84: 719-727.

Larsen, T.O. & Frisvad, J.C. 1995. Chemosystematics of fungi in genus *Penicillium* based on profiles of volatile metabolites. Mycol. Res. 99: 1167-1174.

LoBuglio, K., Pitt, J.I. & Taylor, J.W. 1993. Phylogenetic analysis of two ribosomal DNA regions indicates multiple independent losses of a sexual *Talaromyces* state among asexual *Penicillium* species in subgenus *Biverticillium*. Mycologia 85: 592-604.

LoBuglio, K.F., Pitt, J.I. & Taylor, J.W. 1994. Independent origins of the synnematous *Penicillium* species, *P. duclauxii, P. clavigerum* and *P. vulpinum*, as assessed by two ribosomal DNA regions. Mycol. Res. 98: 250-256.

Lund, F. & Frisvad, J.C. 1994. Chemotaxonomy of *Penicillium aurantiogriseum* and related species. Mycol. Res. 98: 481-492.

Menzies, J.G., Koch, C., Elmhirst, J. & Portree, J.D. 1995. First report of *Penicillium* stem rot caused by *Penicillium oxalicum* on long English cucumber in British Columbia greenhouses. Plant Disease 79: 538.

Onions, A.H.A. Allsop, D. & Eggins, H.O.W.. 1981. Smith's introduction to industrial mycology. 7th ed. Edward Arnold, London.

Onions, A.H.S. & Brady, B.L. 1987. Taxonomy of *Penicillium* and *Acremonium*. In: *Penicillium* and *Acremonium* (J.F. Peberdy, ed.). Plenum Press, New York, pp. 1-35.

Peterson, S. 1993. Molecular genetic assessment of relatedness of *Penicillium* subgenus *Penicillium*. In: Reynolds, D.R. & Taylor, J.W. (Eds.): The fungal holomorph: mitotic, meitic and pleomorphic speciation in fungal systematics. Pp. 121-128. CAB International, Wallingford.

Pitt, J.I. 1973. An appraisal of identification methods for *Penicillium* species: novel taxonomic criteria based on temperature and water relations. Mycologia 65: 1135-1157.

Pitt, J.I. 1979. The genus *Penicillium* and its teleomorphic states *Eupenicillium* and *Talaromyces*. Academic Press, London.

Pitt, J.I. & Cruickshank, R.H. 1990. Speciation and synonymy in *Penicillium* subgenus *Penicillium* - towards a definitive taxonomy. In: Samson, R.A. & Pitt, J.I. (eds.). Modern concepts in *Penicillium* and *Aspergillus* classification. Plenum Press, New York, pp. 103-119.

Pitt, J.I & Hocking, A.D. 1997. Fungi and food spoilage II. Chapman and Hall, London..

Pitt, J.I. & Samson, R.A. 1990. Approaches to *Penicillium* and *Aspergillus* systematics. Stud. Mycol. (Baarn) 32: 77-90.

Raper, K.B. & Thom, C. 1949. A manual of the Penicillia. Williams and Wilkins, Baltimore.

Samson, R.A., Stolk, A.C. & Hadlok, R. 1976. Revision of the subsection fasciculata of *Penicillium* and some allied species. Stud. Mycol. (Baarn) 11: 1-47.

Samson, R.A., Stolk, A.C. & Frisvad, J.C. 1989. Two new synnemateous species of Penicillium. Sud. Mycol. (Baarn) 31: 133-143.

Skouboe, P., Boysen, M., Pedersen, L.H., Frisvad, J.C. & Rossen, L. 1996. Identification of *Penicillium* species using the internal transcribed spacer (ITS) regions. In Rosen, L., Rubio, V., Dawson, M.T. & Frisvad, J.C. (eds.) Fungal identification techniques, pp. 160-164. European Commission, Brussels.

Smedsgaard, J. & Frisvad, J.C. 1997. Terverticillate Penicillia studie by direct electrospray mass spectrometric profiling of crude extracts. I. Chemosystematics. Biochem. Syst. Ecol. 25: 51-64.

Stolk, A.C. & Samson, R.A. 1985. A new taxonomic scheme for *Penicillium* anamorphs. In Samson, R.A. & Pitt, J.I. Advances in *Penicillium* and *Aspergillus* systematics. Plenm Press, New York, pp. 163-191.

Stolk, A.C., Samson, R.A., Frisvad, J.C. & Filtenborg, O. 1990. The systematics of the terverticillate Penicillia. In: Samson, R.A. & Pitt, J.I. (eds.). Modern concepts in *Penicillium* and *Aspergillus* classification. Plenum Press, New York, pp. 121-136.

Svendsen, A. & Frisvad, J.C. 1994. A chemotaxonomic study of the terverticillate Penicillia based on high performance liquid chromatography of secondary metabolites. Mycol. Res. 98: 1317-1328.

Thom, C. 1930. The Penicillia. Williams and Wilkins, Baltimore.

Turner, W.B. & Aldridge, D.C. 1983. Fungal metabolites II. Academic Press, London.

Wang, H., Gloer, J.B., Wicklow, D.T. & Dowd, P.F. 1995. Aflavinines and other antiinsectan metabolites from the ascostromata of *Eupenicillium crustaceum* and related species. Appl. Environ. Microbiol. 61: 4429-4435.

Wang, H., Gloer, J.B., Wicklow, D.T. & Dowd, P.F. 1998. Mollenines A and B: New dioxomorpholines from the ascostromata of *Eupenicillium molle*. J. Nat. Prod. 61: 804-807.

PENICILLIUM ON SOLID WOOD PRODUCTS

Keith A. Seifert and Jens C. Frisvad,
Eastern Cereal and Oilseed Research Centre, Agriculture and Agri-Food Canada,
Research Branch, Ottawa, Ontario, Canada K1A 0C6 and Department of
Biotechnology, The Technical University of Denmark, 2800 Lyngby, Denmark

Although *Penicillium* species are well-known soil microbes and contaminants of many agricultural commodities, their role in the ecology of wood is often overlooked. *Penicillium* species are early colonizers of sapwood and consequently are often also present on lumber used to build houses. In a survey of fungi colonizing stored lumber across Canada, *P. brevicompactum*, *P. glabrum* and *P. spinulosum* were all commonly isolated; *P. citreonigrum*, *P. commune* (or the similar *P. palitans*), *P. crustosum*, *P. expansum* and *P. roqueforti* were also repeatedly isolated. Similar Penicillia were isolated from lumber built into experimental test huts. All isolates examined for mycotoxin production produced profiles typical for their species. A comparison with other published studies reveals a profile of common saprobic Penicillia with no particular specialization for woody substrates.

INTRODUCTION

Species of *Penicillium* are best known as soil fungi and as colonizers of food, textiles and other organic matter used by man. Although species of *Penicillium* have been recorded from wood for many years, they tend to be overshadowed in studies of wood degradation and succession by other moulds, particularly species of *Trichoderma* and sapstaining fungi such as *Ophiostoma* species. The forestry literature contains several studies in which Penicillia have been critically identified. In this paper, we review some of this literature and document our own experiences with the *Penicillium* species occurring on lumber (excluding those isolated from various stages of the pulping process). Our intention is to highlight the species that can be considered typical wood Penicillia and to speculate on their biological activities on solid wood products.

There is both a natural ecology of wood (living trees and their decay in forests; see Kubatová, 1997), and an artificial ecology of wood, which is a consequence of the exploitation of wood fibre by man. Not surprisingly, much more is known about *Penicillium* species on wood products than about their occurrence in forests. In saw mills, wood is found in many different forms. Whole logs, with or without bark, may be stored before processing. Processed lumber may be stored prior to shipping, and piles of debris, wood chips or sawdust, are also usually produced. With these diverse substrates, it is not surprising that *Penicillium* species are frequently found in saw mills. Land *et al.*

(1985) found that *Penicillium* was the dominant genus in a softwood saw mill in Sweden. They isolated seven taxa from sawn wood, wood chips, sawdust and logs in the mill, and attributed the abundance to the ability of the species to grow and sporulate, and hence infect wood, at low temperature. Large piles of wood chips are common in pulp mills, and in the lumber mills who provide them. The piles are self-heating, with temperatures reaching 50-60°C in the centre of the piles at the peak of fermentation. Thermophilic fungi are particularly common in the piles, including *P. argillaceum* (Bergman *et al.*, 1970) and several thermophilic Talaromyces species. Sawdust is the most `soil-like` woody substrate, and good substrate for Penicillia. For example, Kubiak *et al.* (1971) isolated five *Penicillium* species from pine saw dust piles in Poland (see Table 1). Composite wood products, comprised of compressed wood wafers or wood particles held together with organic adhesives, are also a good substrate for mould growth. Kerner-Gang and Nirenberg (1987) found that *Penicillium* was the most common genus occurring on phenol-formaldehyde bonded particle board in West Germany. Thirteen species were identified (Table 1). Several were tested for their abilities to colonize particle board or its components. None of the tested species could degrade beech wood, but *P. chrysogenum* and *P. waksmanii* showed some tolerance to alkali, were able to form colonies on samples of particle board, and were able to use phenol-formaldehyde resins as a carbon source.

The strength and esthetic properties of wood are two of its strongest selling points, and much research has been done in North America, and some countries of northern Europe to understand and prevent fungal degradation of lumber. Succession studies of microorganisms in wood biodeterioration suggest that the following pattern is typical: bacteria, moulds, staining and soft rot fungi, and finally basidiomycetes. Penicillia are generally considered moulds in this sequence, although some may also cause soft rot. Soft rot fungi, which are primarily Ascomycetes, utilize cellulose, eroding the secondary cell wall in a typical pattern, beginning with a diamond shape cavity that elongates and branches until much of the cell wall is destroyed. The wood surface becomes soft and easily scraped away. Seehan *et al.* (1975) compiled the literature on species known to produce soft rots up until that time, and included seventeen *Penicillium* species (see Table 1). Despite their inability to depolymerize lignin, *Penicillium* species are nevertheless among the most common fungi isolated from badly decayed conifer wood, which is composed primarily of lignin residues (Crawford *et al.*, 1990).

The effectiveness of wood preservatives is usually studied using wooden stakes driven into soil. Despite their inability to decay wood, Penicillia are often tolerant to wood preservatives, and are often isolated in soil stake tests (Table 1). For example, Morton and Eggins (1976) isolated *Penicillium* species from the above ground surfaces of pine and beech-veneered stakes after 12 weeks in the ground. *Penicillium funiculosum,* considered by the authors as a mesophilic cellulolytic fungus, was recovered from unpainted stakes in the shade in the autumn months following the June commencement of the experiment, whereas "*P. cyclopium*" considered a non-cellulolytic fungus, was isolated only from shaded, black-painted stakes. Other unidentified Penicillia were isolated in the warm summer months from plain and black-painted stakes in the sunlight.

Table 1. *Penicillium* species recorded from wood and solid wood products in other studies.

Species	Product	Reference
P. aculeatum	wood stakes in soil	Pitt, 1979
P. argillaceum	Pinus chips	Seehan et al., 1975
P. brasilianum	wooden stake in soil	Seehan et al., 1975 (as P. paraherquei)
P. brevicompactum	wood in sea	Picci, 1966
	sawn wood	Land et al., 1985
	particle board	Kerner-Gang and Nirenberg, 1987 (as P. stoloniferum)
	yed Pseudotsuga	Crawford et al., 1990 (as P. stoloniferum)
P. canescens	wood in sea	Picci, 1966
	wood stakes in soil	Gersonde and Kerner-Gang, 1968
P. charlesii	Pinus stakes in soil	Gersonde and Kerner-Gang, 1968
P. chrysogenum	wood in mine	Ioachimescu, 1972
	particle board	Kerner-Gang and Nirenberg, 1987
	Pinus sawdust	Kubiak et al., 1972 (as P. notatum)
	particle board	Kerner-Gang and Nirenberg, 1987 (as P. notatum and P. cyaneofulvum)
P. citrinum	wooden stake in soil	Seehan et al., 1975
	Pinus stakes in soil	Gersonde and Kerner-Gang, 1968
	particle board	Kerner-Gang and Nirenberg, 1987
	decayed Pseudotsuga	Crawford et al., 1990
P. commune	wood in sea	Picci, 1966
P. corylophilum	Pinus stakes in soil	Gersonde and Kerner-Gang, 1968
	decayed Pseudotsuga	Crawford et al., 1990
P. crustosum	Pinus strobus lumber	Brown, 1953
	wood stakes in soil	Pitt, 1979
	Pinus stakes in soil	Gersonde and Kerner-Gang, 1968
"P. cyclopium"	Pinus strobus lumber	Brown, 1953
	wood chips	Land et al., 1985
P. decumbens	wood stakes in soil	Pitt, 1979
P. diversum	Pinus wood in soil	Seehan et al., 1975
	Pseudotsuga poles	Wang and Zabel, 1990
P. duclauxii	mine timber	Pitt, 1979
P. expansum	wood in sea	Picci, 1966
	wood stake in soil	Seehan et al., 1975
	Pinus stakes in soil	Gersonde and Kerner-Gang, 1968
	wood chips, sawdust, logs	Land et al., 1985
	wood in sea	Picci, 1966 (as P. glaucum)
P. funiculosum	Pinus, Betula wood	Seehan et al., 1975
	Pinus poles	Wang and Zabel, 1990
	Pinus stakes in soil	Gersonde and Kerner-Gang, 1968
P. glabrum	Pinus contorta trees	Bourchier, 1961
	wood in sea	Picci, 1966 (as P. frequentans)
	CCA treated Pinus radiata	Butcher, 1971 (as P. frequentans)
	wood in mine	Ioachimescu, 1972 (as P. frequentans)
	stakes in soil	Pitt, 1979
	decayed Pseudotsuga	Crawford et al., 1990 (as P. frequentans)
P. humuli	particle board	Kerner-Gang and Nirenberg, 1987
P. implicatum	decayed Pseudotsuga	Crawford et al., 1990
P. janczewskii	wood in mine	Ioachimescu, 1972 (as P. nigricans)
P. janthinellum	wood in sea	Picci, 1966
	Pinus wood in soil	Seehan et al., 1975
	Pinus poles	Wang and Zabel, 1990
	Pinus stakes in soil	Gersonde and Kerner-Gang, 1968
	decayed Pseudotsuga	Crawford et al., 1990
P. lanosum	decayed Pseudotsuga	Crawford et al., 1990
P. lignorum	Pinus, Fagus wood	Pitt, 1979
P. lividum	decayed Pseudotsuga	Crawford et al., 1990

287

P. minioluteum	particle board	Kerner-Gang and Nirenberg, 1987
	wood chips, stakes in soil	Pitt, 1979
P. ochrochloron	wooden stake in soil	Seehan *et al.*, 1975
	Pinus stakes in soil	Gersonde and Kerner-Gang, 1968
P. palitans	particle board	Kerner-Gang and Nirenberg, 1987
	Pinus wood in soil	Seehan *et al.*, 1975
	Pinus stakes in soil	Gersonde and Kerner-Gang, 1968
P. pinophilum	old Pinus wood	Pitt, 1979
	wood chips	Pitt, 1979
	Pinus poles	Wang and Zabel, 1990 (as *P. funiculosum*)
P. primulinum	Pinus sawdust	Kubiak *et al.*, 1971 (as *P. diversum* var. *aureum*)
P. purpureogenum	wood in sea	Picci, 1966
	decayed Pseudotsuga	Crawford *et al.*, 1990
P. purpurescens	stakes in soil	Pitt, 1979
P. raciborski	wood chips, sawdust	Land *et al.*, 1985
P. restrictum	decayed Pseudotsuga	Crawford *et al.*, 1990
P. roqueforti	wood in sea	Picci, 1966
	sawn wood, logs	Land *et al.*, 1985
	wood stakes in soil	Pitt, 1979
	wood in sea	Picci, 1966 (as *P. casei*)
P. roseopurpureum	Pinus sawdust	Kubiak *et al.*, 1971
P. rubrum	Pinus sawdust	Kubiak *et al.*, 1971
	Pinus ponderosa lumber	Davidson, 1985
	particle board	Kerner-Gang and Nirenberg, 1987
P. rugulosum	Pinus wood in soil	Seehan *et al.*, 1975
	Pinus stakes in soil	Gersonde and Kerner-Gang, 1968 (as *P. tardum*)
P. simplicissimum	Pinus wood in soil	Seehan *et al.*, 1975
	wood chips	Land *et al.*, 1985
	Pinus stakes in soil	Gersonde and Kerner-Gang, 1968
	decayed Pseudotsuga	Crawford *et al.*, 1990
P. spinulosum	particle board	Kerner-Gang and Nirenberg, 1987
	stakes in soil	Pitt, 1979
	decayed Pseudotsuga	Crawford *et al.*, 1990
P. steckii	particle board	Kerner-Gang and Nirenberg, 1987
	decayed Pseudotsuga	Crawford *et al.*, 1990
P. thomii	wood	Pitt, 1979
P. variabile	wood chips	Pitt, 1979
	Pinus stakes in soil	Gersonde and Kerner-Gang, 1968
P. verruculosum	Pinus wood in soil	Seehan *et al.*, 1975
P. verrucosum	sawn wood	Land *et al.*, 1985
P. viridicatum	wood stakes in soil	Pitt, 1979
	Pinus stakes in soil	Gersonde and Kerner-Gang, 1968
	particle board	Kerner-Gang and Nirenberg, 1987 (as *P. olivino-viride*)
P. waksmanii	particle board	Kerner-Gang and Nirenberg, 1987

Butcher (1971) studied fungi invading *Pinus radiata* sapwood treated with a copper-chrome-arsenate preservative in New Zealand. *Penicillium glabrum* (as *P. frequentans*) was one of the dominant primary colonizers, and was most commonly isolated from below ground portions of the wood. The isolates were tolerant to up to 17,500 ppm copper, and up to 52,800 ppm arsenic.

The best known stains of lumber are the so-called blue stains caused by species of *Ophiostoma* and black yeasts. However, species of *Penicillium* are often implicated in other stains. The spreading green spore masses often seen on lumber are usually *Trichoderma* species, but *Penicillium* colonies are also common. These stains are usually considered trivial because they do not penetrate the wood surface and are easily planed off. Brown (1953) demonstrated that isolates of *P. cyclopium* and *P. crustosum* from *Pinus strobus* lumber caused an internal blue stain in 2-4 weeks when inoculated onto sterile wood. The epithelial tissues of the resin ducts were destroyed, but the ray parenchyma cells were not visibly affected. Typical of sapstain fungi, the fungus spread through the wood in the tracheids, passing between cells through the bordered pits. Conidia and conidiophores produced in the tracheids were apparently the cause of the stain. No effect on wood toughness was noted. Davidson (1985) reported on a reddish discoloration in the sapwood of *Pinus ponderosa* attacked by the bark beetle *Dendroctonus ponderosae*. The culprit proved to be *Penicillium purpurogenum* sensu Raper & Thom (as *P. rubrum*), which excretes a conspicuous reddish pigment into agar, and presumably also into wood.

It would be tempting to speculate that truly lignicolous species of *Penicillium* would not produce mycotoxins. Land and Hult (1987) studied mycotoxin production by twenty-five isolates of *Penicillium* spp. isolated during their study of a Swedish saw mill (Land *et al.*, 1985). They found that P. expansum was able to produce patulin when cultured on wood blocks or wood chips. Ochratoxin A was produced by a strain of *P. nordicum* grown on wood chips, even though the strain did not originate from wood. Several taxa, namely P. verrucosum var. verrucosum, P. verrucosum var. cyclopium and *P. roqueforti*, did not produce detectable mycotoxins when grown on wood. They concluded that wood associated *Penicillium* spp. generally produced mycotoxins less frequently than food isolates.

MATERIALS AND METHODS

Identifications. Purified cultures were grown on Czapek Yeast Agar (CYA) with added trace elements, at 25°C and in some cases 5°C and 37°C, Malt Extract Agar (MEA) at 25°C, and in some cases on Yeast Extract Sucrose Agar (YES) at 25°C and Creatine Sucrose Agar (CREA) at 25°C for 7 days in the dark. All formulae are those given in Samson *et al.* (1995).

1990 Sapwood survey: From March-July 1990, wood samples were requested from selected lumber mills across Canada. Each mill provided three samples cut from their stock with sapstain, and three without. Each sample was subdivided into three sections and four subsurface pieces of wood (about 5 x 5 x 1 mm) were removed from each section using ethanol-rinsed and flamed chisels and forceps. Two pieces from each section were plated two different media: 1) 2 % malt agar (2% MA: 20 g Difco malt extract, 20 g Difco Bacto Agar, 1000 mL distilled water) with 100 ppm tetracycline and 2) 2% MA with 100 ppm tetracycline and 100 ppm cycloheximide. Plates were incubated at 27°C and observed periodically for 3-28 days. Identifications were made from purified subcultures. The wood species of some samples were indicated by the mills; these identifications were considered correct.

1995-96 Hem-fir survey. Moulds were isolated from hem-fir and *Pseudotsuga menziesii* lumber in British Columbia by J. Clark and T. Byrne, Forintek Canada, Corp.,

Vancouver. Isolations were made from surface discolorations on products treated with experimental chemical preservatives. Representative isolates were sent to the senior author for identification.

Moisture huts: in 1990, the Canadian Mortgage initiated a project and Housing Corporation to determine the relationship between moisture in wooden wall studs, rates of drying, and the presence of moulds that might affect indoor air quality. Two test huts were assembled from soaked conifer wood, one in Edmonton, Alberta and the other in Waterloo, Ontario. After one year, the huts were disassembled and fungi were isolated from the studs. Sterile swabs were wiped across the studs and then inoculated onto 2% MA with 100 ppm tetracycline. In addition, small slivers were removed from the studs and plated directly on the same medium.

Mycotoxin production. Selected isolates from Canadian wood, as well as strains sent by Carl Johan Land (Land and Hult, 1985) were examined for secondary metabolites using the methods described by Filtenborg *et al.* (1983) and Frisvad and Thrane (1987, 1993). Strains were grown on optimal media for production of secondary metabolites (CYA, MEA, YES, OAT, two 11 cm Petri dishes of each medium, for medium composition, see Samson *et al.*, 1995) for two weeks at 25°C. The agar was extracted with chloroform/methanol/ethylacetate (2:1:3, vol/vol/vol) with 1 % (vol) formic acid. The solvents were then evaporated to dryness after filtering through a hydrophobic filter, redissolved in methanol, and "defatted" with petroleum ether. The extract was analysed by gradient HPLC with diode array detection and both UV spectra and retention indices compared with a library of approximately 400 authentic standards. The identity of the compounds was confirmed by the TLC methods mentioned in Filtenborg *et al.* (1983)

RESULTS

1990 Sapwood survey: *Penicillium* species were isolated at high frequency in both the east and the west (Tables 2 and 3). Twelve species were isolated in British Columbia in 1990, with three species isolated from more than 10% of the samples: *P. brevicompactum*, *P. glabrum* and *P. commune* (or the similar *P. palitans*). In eastern Canada, the incidence of colonization by *Penicillium* species was lower, with no species isolated from more than 10% of the samples. However, the species diversity was slightly higher, with sixteen taxa isolated. The most frequently isolated species was still *P. brevicompactum*, followed by *P. crustosum* and *P. spinulosum*.

Table 2. *Penicillium* species isolated from sapwood of stored lumber in eastern Canada in 1990, in order of frequency of isolation. Based on 71 samples from thirteen mills. NT indicates strains not examined for mycotoxin production. Cultures with acronyms other than DAOM are in the personal collection of the senior author. Spruce-fir is a lumber sorting category used in some lumber mills that process multiple wood species.

Species	% samples	% mills	Substrate	Representative cultures	Metabolites
P. brevicompactum	8.5	30.7	Picea Populus	DAOM 215331, 215335 SOV A11c	Raistrick phenols, mycophenolic acid"
P. brevicompactum (slow form)			Spruce-fir [1]	DAOM 215332	Raistrick phenols. mycophenolic acidmycochromenic acid, metabolites O, f " + pebrolide
P. crustosum	7.0	15.4	Picea Pinus	BLE A22c DAOM 215343, 215344 DAOM 215345	penitrem A, roquefortine C, terrestric acid, viridicatin"
P. spinulosum (P. terlikowskii)	4.6	23.1	Picea Pinus Populus	AAB12t BOW A21t SOV B31t	NTNTNT
P. sizovae	4.2	7.7	Picea Abies	LSAG	NTNT
P. citreonigrum	1.4	7.7	Picea	DAOM 215338	NT
P. thomii	1.4	7.7	Pinus	NPj	NT
P. variabile	1.4	7.7	Picea	DAOM 215368	rugulosin
P. steckii	1.4	7.7	Picea	BLS	NT
P. raistrickii	1.4	7.7	Picea	BLS	NT
P. commune (P. palitans)	1.4	7.7	Picea	COM	NT
P. carneum	1.4	7.7	Picea	COM B11t, B13t	NT
P. spinulosum (P. tannophagum)	1.4	7.7	Picea	BLS A32t	metabolite QQmany exotic metabolites
P. lanosum	1.4	7.7	Picea	BLS A11t	kojic acid, griseofulvin
P. rugulosum	1.4	7.7	Picea	DAOM 215362	rugulosin
P. smithii	1.4	7.7	Picea	DAOM 215363	citreoviridin
P. decumbens	1.4	7.7	Picea	DAOM 215346	NT

[1] Spruce-fir is a lumber sorting category used in some lumber mills that process multiple wood species.

Table 3. *Penicillium* species isolated from sapwood of stored lumber in British Columbia, Canada in 1990, in order of frequency of isolation. Based on 84 samples from fourteen mills. NT indicates strains not examined for mycotoxin production. Cultures with acronyms other than DAOM are in the personal collection of the senior author.

Species	% samples	% mills	Substrates	Representative cultures	Metabolites
P. brevicompactum	32.1	57.1	Picea Pinus Pseudotsuga Tsuga	DAOM 215333 not saved not saved DAOM 215334	Raistrick phenols, mycophenolic acidNTNTNT
P. glabrum	15.5	35.7	Picea Pseudotsusga Tsuga	DAOM 215353 DAOM 215354 not saved	NTNTNT
P. commune (P. palitans)	10.7	21.4	Pseudotsuga Tsuga	MayoDFA21 not saved	cyclopiazonic acidNT
P. spinulosum	10.7	14.2	Pseudotsuga Tsuga	not saved not saved	NTNT
P. expansum	3.6	21.4	Picea Tsuga	DAOM 215350 DAOM 215349	expansolide, patulin, roquefortine C, chaetoglobosin A, C, communesin A, BNT
P. roqueforti	2.4	7.1	Picea	DAOM 215358	NT
P. chrysogenum	1.2	7.1	Tsuga	DAOM 215337	roquefortine C, meleagrin, chrysogine
P. citreonigrum	1.2	7.1	Picea	DAOM 215339	NT
P. crustosum	1.2	7.1	Picea	DAOM 215342	penitrem A, roquefortine C, terrestric acid, viridicatin
P. decumbens	1.2	7.1	Pinus	not saved	NT
P. echinulatum	1.2	7.1	Picea	DAOM 215348	NT
P. simplicissimum	1.2	7.1	Abies	not saved	NT

1995-96 Hem-fir survey. The results from the Forintek hem-fir and Douglas fir survey are shown in Table 5. *P. brevicompactum* was again the most commonly isolated species, followed by *P. spinulosum* and *P. glabrum*. The diversity and frequency of species was similar to that obtained for British Columbian lumber during the 1990 survey.

Moisture hut isolations: Penicillium species isolated from studs in wall cavities after one year are listed in table 4. The isolations were not made in a strictly quantitative manner, and the relative frequency of isolation refers simply to the number of cultures of each species isolated. The most frequently isolated species was *P. spinulosum* (which was also isolated from the surface of the tar papered wall board as the huts were being built).

Table 4. *Penicillium* **species isolated from studs in walls of test huts one year after construction, and their mycotoxin producing abilities, in order of relative frequency. NT indicates strains not examined for mycotoxin production. Cultures with acronyms other than DAOM are in the personal collection of the senior author.**

Species	Representative cultures	Mycotoxin production
P. spinulosum	DAOM 215365, 215366	NT
P. corylophilum	DAOM 215341, 215366	NT
P. rugulosum	DAOM 215359, 215360	rugulosin
P. minioluteum sensu Pitt	DAOM 215356	NT
P. chrysogenum	DAOM 215336, 215337	roquefortine C, meleagrinchrysogine
P. smithii	DAOM 215364	citreoviridin, canescins
P. expansum	DAOM 215351	expansolide, patulin, roquefortine C, chaetoglobosin A, Ccommunesin A, B
P. hispanicum (=P. implicatum)	DAOM 215355	NT
P. aculeatum	CMHC 3T	NT
P. citrinum	DAOM 215340	citrinin
P. roqueforti	DAOM 215357	NT
`P. decumbens`'	DAOM 215347	xanthocillins, many exotic compounds

Table 5. *Penicillium* **species isolated from hem-fir[1] and Douglas fir lumber in British Columbia during 1995-1997. Courtesy of J. Clarke and T. Byrne, Forintek Canada.**

Species	Number of isolates
Penicillium brevicompactum	18
Penicillium spinulosum	14
Penicillium glabrum	11
Penicillium rugulosum	3
Penicillium thomii	3
Penicillium carneum	2
Penicillium solitum	2
Penicillium miczynskii	2
Penicillium expansum	1
Penicillium lividum	1
Penicillium minioluteum	1
Penicillium steckii	1
Penicillium variabile	1

Mycotoxin production: Mycotoxin profiles were determined for selected cultures from the sapwood survey and moisture hut experiment (Tables 2-4). Mycotoxin production on agar was consistent with strains of the same taxa isolated from other substrates. The strains isolated by Land and Hult (1985) produced all the mycotoxins known from these species. The three strains of *P. roqueforti* produced PR-toxin, mycophenolic acid, roquefortine C and the unknown metabolite A. The strains called P. verrucosum var. cyclopium and var. verrucosum from wood were reidentified as P. palitans and produced cyclopiazonic acid and isofumigaclavine A. The strains of P. expansum all produced expansolide, patulin, roquefortine C, chaetoglobosin A, and C and communesin A and B.

DISCUSSION

A comparison of the species reported from wood in other studies (Table 1), with the species isolated from the different surveys reported above (Tables 2-5), reveals the consistent isolation of several species of *Penicillium*. Although these different studies cannot be compared in any statistical way, the results suggest a profile of the species that are "typical" wood Penicillia.

The results of independent surveys of Canadian lumber in 1990 and in 1995-96 show that *P. brevicompactum* is one of the most common inhabitants of the sapwood of conifer lumber in North America. Interestingly, we did not recover this species from the moisture hut survey, although it is frequently reported in indoor air (Samson *et al.*, 1994). It has also been isolated during several studies in Europe. Many of the strains from Canada, particularly those isolated on media with cycloheximide, grew more slowly than reported by Pitt (1979), growing only 2-4 mm diam on CYA, and 0-11 mm on MEA after 7 days at 25°C. Conidiophores were often predominantly biverticillate and had swollen metulae (Fig. 1a, c), giving the fungus much in common with *P. inflatum* in micromorphology and growth rate. However, these cultures often grew better on CYA at 5°C (1-9 mm) than at 25°C, and typical terverticillate morphology can be observed at 5°C (Fig. 1 b, d). After several transfers, terverticillate morphology began to dominate cultures at 25°C, which were then more typical of *P. brevicompactum*. It is possible that these cultures may represent a distinct taxon. Secondary metabolites include the Raistrick phenols and mycophenolic acid typically produced by *P. brevicompactum*. Several additional metabolites are also produced, including mycochromenic acid and metabolites A2, O, and f, and these strains do not produce brevianamide A. Similar strains have also been recovered from Danish forest soils.

Another constant between our isolations and studies on other continents is the frequent occurrence of *P. glabrum* and *P. spinulosum*. Given the confusion that still exists surrounding the precise delimitation of these species (Pitt *et al.*, 1990), it is risky to state which species is actually most common. Both species occur at relatively high frequency on western Canadian hem-fir (a mixture of Tsuga and Abies species) and Douglas fir lumber. Other monoverticillate species also consistently appear in wood surveys, although at lower frequencies. *Penicillium citreonigrum, P. decumbens* and *P. thomii* were isolated in different aspects of our study, and in several other studies (see

294

Table 1). We have observed the orange sclerotia of *P. thomii* directly on lumber samples. Of the monoverticillate species commonly reported in other studies, only *P. fellutanum* was absent from our surveys, although the fungus is reportedly a frequent contaminant of indoor air in Canada.

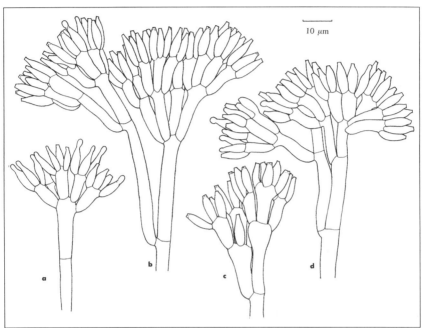

Fig. 1. *Penicillium brevicompactum* strains isolated from wood, showing repressed morphology on MEA at 25°C (a, c) and typical morphology on CYA at 5°C (b, d) for strains isolated with cycloheximide. (a. SOV A11c, 7 d on MEA, 25□C. b. Prix B12c, 11 d, CYA, 5□C. c. BLE A23c, 7 d at 25°C on MEA. d. WW B11, CYA, 11 days, 5°C)

The rough-stiped, terverticillate species with large conidia *P. roqueforti, P. carneum, P. commune* (or *P. palitans*) and *P. crustosum* were isolated with moderate frequency in our studies, and have also been recorded from studies in Europe (see Table 1). Recent western Canadian surveys also revealed the presence of the another species with a similar micromorphology, *P. solitum*. The similar species with small conidia in the *P. aurantiogriseum* complex were noticeably absent from our surveys, although *P. verrucosum* and *P. viridicatum* have been reliably reported by other authors. *Penicillium expansum* and *P. chrysogenum* were also isolated several times in our surveys, and have also been recorded several times in surveys on other continents (Table 1).

Species in subgenus *Biverticillium* isolated more than once in our studies include *P. rugulosum, P. minioluteum* (*sensu* Pitt, 1979) and *P. variabile*, all of which have also been reported in other surveys. Absent from our surveys are relatively well-known lignicolous members of this subgenus, such as *P. funiculosum*, which has been used as a test soft-rot organism, and *P. rubrum*, the cause of a red lumber stain.

Species in subgenus Furcatum were relatively infrequent in our surveys. Single isolates of *P. smithii, P. citrinum* and *P. steckii* were obtained from different surveys. The latter two species have been recorded in other studies (see Table 1). The apparent absence of *P. janthinellum* in our surveys is a surprise since it seems to be relatively frequently isolated from wood in Europe. In our experience, the species is common in many North American soils.

The *Penicillium* species most frequently isolated from wood, as discussed above, are usually considered general saprobes and there is no evidence of specialization for woody substrates. This idea is supported by the fact that the profile of species isolated from badly decayed wood in forests (Table 1, Crawford *et al.*, 1990) is similar to the species isolated from freshly exposed wood. Most of these Penicillia are relatively common in soil and food. It would be surprising if the lignicolous representatives of these species were metabolically different that those from other substrates. Land and Hult (1987) tested the ability of several Penicillia isolated from a lumber mill to produce mycotoxins when grown on YES and on wood in vitro. Their statistical comparison of toxin producing ability of lignicolous isolates and those from food suggested that Penicillia isolated from wood produced fewer secondary metabolites. However, we has tested mycotoxin production on agar by several of the strains from the Swedish study, and found them to produce typical profiles and concentrations of secondary metabolites.

No *Penicillium* species are known to degrade lignin, and thus none of the species discussed here are likely to be responsible for significant structural damage. At worst, some soft-rot may be occurring, but this is likely only in very wet wood in contact with the ground. More likely, the species are exploiting the remaining simple nutrients in the sapwood, and producing a minimal degradation of the cellulose polymers.

Penicillium colonies on lumber or other solid wood products are usually scarcely visible as diffuse greyish or greenish powdery masses, and rarely have the density of Trichoderma colonies on the same substrates. Our studies on stored lumber firmly establish the presence of Penicillia on stored lumber. Differences in experimental designs prevent the establishment of a direct link between the *P. spinulosum* that is apparently common on lumber, and the *P. spinulosum* that was isolated from test huts one year after they were built with wet lumber. However, this possible coincidence suggests at least one mechanism for the introduction of moulds into human dwellings.

ACKNOWLEDGMENTS
We are grateful to B. Grylls and B. Bilmer, who isolated many of the strains from stored conifer lumber, and Industry Science and Technology Canada, who provided financial support for the survey of sapwood inhabiting fungi in 1990. A portion of this work was undertaken when KAS was an employee of Forintek Canada, Corp. We are extremely grateful to the Forintek member companies that participated in the sapstain survey. The moisture hut experiments were funded by Canadian Mortgage and Home Corporation: some of the isolations from that study were made by L. Sigler. J. Clark and T. Byrne, Forintek Canada, Corp., Vancouver, kindly allowed us to use some of the data from their studies of hem-fir and Douglas fir lumber. We thank Dr. C. J. Land for providing strains

from his studies of Swedish lumber mills. Drs S. Redhead and J. Ginns are thanked for their critical reviews of the manuscript.

REFERENCES

Bergman, O., T. Nilsson & P. Jerkeman. (1970). Reduction of microbial deterioration in outside chip storage by alkali treatment. Svensk Papperstidning 73: 653-666.

Brown, F. L. (1953). Mercury tolerant penicillia causing discoloration in northern white pine lumber. J. For. Prod. Res. Soc., Nov. 1953: 67-69.

Bourchier, R. J. (1961). Laboratory studies on microfungi isolated from the stems of living lodgepole pine, Pinus contorta Dougl. Can. J. Bot. 39: 1373-1385.

Butcher, J. A. (1971). Colonisation by fungi of Pinus radiata sapwood treated with a copper-chrome-arsenate preservative. J. Inst. Wood Sci. 28: 16-25.

Crawford, R. H., S. E. Carpenter & M. E. Harmon. (1990). Communities of filamentous fungi and yeast in decomposing logs of Pseudotsuga menziesii. Mycologia 82: 759-765.

Davidson, R. W. (1985). Brown and red stain in Ponderosa pine lumber. Mycologia 77: 494-496.

Frisvad, J. C. & Thrane, U. (1987). Standardized High-Performance Liquid Chromatography of 182 mycotoxins and other fungal metabolites based on alkylphenone indices and UV-VIS spectra (diode-array detection). J. Chromatogr. 404: 195-214.

Frisvad, J. C. & Thrane, U. (1993). Application of high performance liquid chromatography. In: Betina, V. (ed.): Chromatography of mycotoxins: techniques and applications. Journal of Chromatography Library 54. Elsevier, Amsterdam. pp. 253-372.

Filtenborg, O., J. C. Frisvad & J. A. Svendsen. (1983). Simple screening method for molds producing intracellular mycotoxins in pure cultures. Appl. Environ. Microbiol. 45: 581-585.

Gersonde, M. & W. Kerner-Gang. (1968). Untersuchungen an Moderf□ule-Pilzen aus Holzst□ben nach Freilandversuchen. Mat. und Org. 3: 199-212.

Ioachimescu, M. (1972). Contributii la cunoayterea micoflorei de pe lemnul din mine. St. Si. Cerc. Biol., Ser. Bot., 24: 507-510 (in Romanian).

Kerner-Gang, W. & H. I. Nirenberg. (1987). Identifizierung von Schimmelpizen aus Spanplatten und deren matrixbezongenes Verhalten in vitro. Mat. und Ord. 20: 265-276.

Kubatová, A. (1977). Neglected *Penicillium* spp. associated with declining trees. In Integration of modern taxonomic methods for *Penicillium* and *Aspergillus* classification (Eds. Samson, R.A. & Pitt, J.I). Harwood Publishers, Amsterdam, pp. 299-307.

Kubiak, M., E. Dymalski & S. Balazy. (1971). Wst□pne obserwacje nad sk□adem mikoflory stos□lw trocin sosnowych. Foria Forest. Polonica, Ser. B, 10: 97-105.

Land, C. J., Z. G. Banhidi & A. C. Albertson. (1985). Surface discoloring and blue stain in by cold-tolerant filamentous fungi on outdoor softwood in Sweden. Mat. und Org. 20: 133-156.

Land, C. J. & K. Hult. (1987). Mycotoxin production by some wood-associated *Penicillium* spp. Letters Appl. Microbiol. 4: 41-44.

Morton, L. H. G. & H. O. W. Eggins. (1976). The influence of insolation on the pattern of fungal succession onto wood. Int. Biodeter. Bull. 12: 100-105.

Picci, G. (1966). Sulla micoflora presente nelle strutture in legno soggette all'azione dell'acqua de mare. La Ricerca Scient. 36: 153-157.

Pitt, J. I. (1979). The genus *Penicillium* and its teleomorphic states Eu*penicillium* and Talaromyces. Academic Press: London, New York, Toronto, Sydney, San Francisco.

Pitt, J. I., M. A. Klich, G. P. Shaffer, R. H. Cruikshank, J. C. Frisvad, E. J. Mullaney, A. H. S. Onions, R. A. Samson & A. P. Williams. (1990). Differentiation of *Penicillium* glabrum from *Penicillium* spinulosum and other closely related species: An integrated taxonomic approach. System. Appl. Microbiol. 13: 304-309.

Samson, R. A., B. Flannigan, M. E. Flannigan, A. P. Verhoeff, O. C. G. Adan & E. S. Hoekstra. (1994). Health implications of fungi in indoor environments. Elsevier: Amsterdam, Lausanne, New York, Oxford, Shannon, Tokyo.

Samson, R. A., E. S. Hoekstra, J. C. Frisvad & O. Filtenborg. (1995). Introduction to Food-borne Fungi, 4th edition. Centraalbureau voor Schimmelcultures, Baarn.

Seehan, F., W. Liese & B. Kesa. (1975). List of fungi in soft-rot tests. International Working Group on Wood Preservation, Document no. 1RG/WP/105.

Wang, C. J. K. & R. A. Zabel. (1990). Identification manual for fungi from utility poles in the eastern United States. American Type Culture Collection: Rockville, MD.

NEGLECTED *PENICILLIUM* SPP. ASSOCIATED WITH DECLINING TREES

Alena Kubátová
Culture Collection of Fungi (CCF), Department of Botany, Faculty of Science, Charles University, Benátská 2, 128 01 Prague, Czech Republic

Microfungi associated with declining trees were studied in the period 1991-1995 in the Czech Republic. Samples from roots, stems and branches were cut out from declining, sound and dead trees and incubated in moist chambers. Besides Ophiostomatoid fungi, *Trichoderma* spp. and other Hypocrealean anamorphs, *Penicillium* species were very frequently associated with trees. They were isolated from 70 % of all samples. Altogether 27 *Penicillium* and *Geosmithia* species were recovered from eight wood species (*Quercus petraea, Q. robur, Q. pubescens, Fagus sylvatica, Picea abies, Larix decidua, Pinus sylvestris, P. nigra*). *Penicillium minioluteum, P. glabrum, Geosmithia* spp., *P. spinulosum* and *P. glandicola* were the most frequent. The highest number of *Penicillium* species was observed on *Quercus petraea* samples. *Penicillium* spp. were present only in a limited degree on *P. nigra, L. decidua* and *F. sylvatica*. *Penicillium minioluteum* was markedly dominant on *Picea abies* samples, and *Geosmithia* spp. were strongly dominant on *Quercus pubescens*. *Penicillium* species were found on healthy, damaged and dead trees, too. Penicillia were isolated both from bark and cambium, and from wood. Affinity to one of these substrates was recorded only in four frequent species. *Penicillium minioluteum* and *Geosmithia* spp. were frequently found in the cambial zone, *P. glabrum* and *P. citreonigrum* were predominantly observed on wood. During this study three Penicillia were found for the first time in the Czech Republic: *P. minioluteum, P. smithii* and *P. purpurogenum var. rubrisclerotium*. It seems that some of the observed Penicillia cannot be considered as contaminants but they have affinity to this substrate. Their role in connection with living trees is not clear.

INTRODUCTION

Penicillium species are known to be distributed world wide. They are found on almost all substrata. In the current literature, many authors deal with declining trees, especially in connection with Ophiostomatoid fungi. In some of these articles Penicillia represent a major part of the observed fungi (e.g. Shigo, 1958; Amos and True, 1967; Kowalski, 1991; Przybyl, 1995, 1996; Kaus *et al.*, 1996, etc.). However, only a small part of these articles give species determination of Penicillia (e.g. Maňka and Truszkowska, 1958; Bills and Polishook, 1990).

This contribution is one of the results of a major project primarily centered on occurrence and isolation of Ophiostomatoid fungi associated with damaged and dead forest wood species in the Czech Republic, particularly with oaks and Norway spruce.

Table 1. List of tree species, numbers of trees, samples and localities [a]

Tree species	healthy	damaged	dead	total	Type of samples	Localities and data of collection
Quercus petraea	-	7/24	4/14	11/38	A B C D K	5: Okrouhlo, central Bohemia; XI.91; Poněšice, near Hluboká n.V., south Bohemia; V.92; Dřevič-Skalka, Křivoklátsko reg., central Bohemia; X.94; Pláně, Český kras reg., central Bohemia; VII.93; Lipí, Plzensko reg., west Bohemia; VII.95
	3/13	4/16	1/5	8/34	A B C D K	3: Dešov, south Moravia; IV.93; Kuntínov, south Moravia; VI.94; Valtice, south Moravia; VI.94
Q. pubescens	1/2	7/15	1/3	9/20	A B C D	2: Pláně, Český kras reg., central Bohemia; IX.92; Třebotov, Český kras reg., central Bohemia; X.92
Fagus sylvatica	-	5/19	-	5/19	A B C D K	2: Borová Lada, Šumava Mts., south Bohemia; IV.92; Jelenec, Křivoklátsko reg., central Bohemia; XI.94
Picea abies	1/3	18/35	-	19/38	A B C	2: Okrouhlo, central Bohemia; XI.91; Františkov, Šumava Mts., south Bohemia; I.92
Larix decidua	-	6/18	1/3	7/21	A B C D K	4: Slapy, central Bohemia; IX.92; Čížová, near Milevsko, south Bohemia; X.92; Kostelec N/Vlt, south Bohemia; X.92; Dešov, south Moravia; VII.95
Pinus sylvestris	1/4	-	1/4	2/8	A B C K	1: Hudlice-Diboř, Křivoklátsko reg., central Bohemia; III.94
Pinus nigra	-	1/4	1/4	2/8	A B C K	1: Hudlice-Diboř, Křivoklátsko reg., central Bohemia; III.94
total number	**6/22**	**48/131**	**9/33**	**63/186**		

[a] A, base of stem; B, middle part of stem; C, upper part of stem; D, branch, K, root

300

The main aim of this study was to discover the spectrum of microfungi occurring on declining trees comparing to healthy and dead trees. As well as *Ophiostoma*, *Trichoderma* and Hypocrealean anamorphs, many Penicillia were found. Due to scant information on *Penicillium* diversity in connection with bark or wood of living trees attention was focused on this problem, too. This paper presents differences in *Penicillium* diversity on several wood species showing different level of damage.

MATERIALS AND METHODS

The study was carried out between November 1991 and December 1995. Collection data are given in Table 1. Over 180 samples of 63 trees of eight species from 16 localities were examined. The age of trees varied from 50 to 150 years. The tree species selected for this study are forest tree species declining in numbers in the Czech Republic. For comparison, mycoflora of several sound-appearing and dead trees was studied, too. Localities included central, south and west Bohemia and south Moravia.

The trees were cut down and discs about 1 - 2 cm thick were taken from cross sections from the basal, middle and upper part of stem, from branches and roots. The discs were washed under tap water and incubated in Petri dishes or plastic bags under moist chamber condition at room temperature.

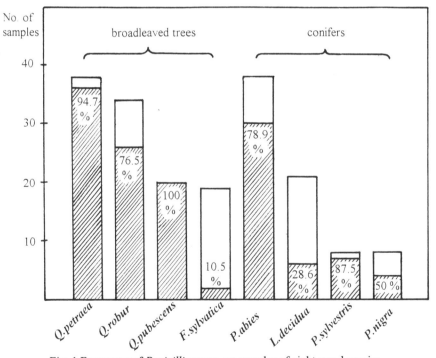

Fig. 1.Frequency of *Penicillium* spp. on samples of eight wood species

After two to six weeks microfungi were observed under dissecting microscope, isolated and identified according their microscopic features and morphology on CYA and MEA (after Pitt 1979 a,b).

Identification was made according to Raper and Thom (1949), Pitt (1979 a,b), Ramírez (1982), Quintanilla (1985), van Reenen-Hoekstra *et al.*. (1990), and Stolk *et al.*. (1990).

RESULTS

A characteristic association of *Penicillium* species was recovered from the wood and bark of the trees studied. *Penicillium* species were the most frequently isolated fungi, as well as Ophiostomatoid fungi, *Trichoderma* spp. and other Hypocrealean anamorphs. In total, *Penicillium* species were found on 70 % of all samples. The data on individual wood species are given on Figure 1. Penicillia were recorded on all *Quercus pubescens* samples (100 %) and on the majority of *Q. petraea* (94.7 %), *Pinus sylvestris* (87.5 %), *Picea abies* (78.9 %), and *Q. robur* (76.5 %). On other trees, Penicillia were found in a limited degree (*Pinus nigra*: 50% and *Larix decidua*: 28.6 %). Interestingly, only a few Penicillia were found on samples of *Fagus sylvatica* (10.5 %), although a comparable number of samples were studied as for *Q. pubescens*.

The list of all *Penicillium* species identified is given in Table 2. Altogether 27 *Penicillium* and *Geosmithia* species were isolated on eight wood species. *Penicillium minioluteum, P. glabrum, Geosmithia* spp., *P. spinulosum* and *P. glandicola* were found to be the most frequent species with the total frequency of 18.8 %, 17.2 %, 11.3 %, 9.1 %, and 7.5 %, respectively. Some of the frequent Penicillia are noteworthy because of their yellow or red pigmentation in mycelium, exudate or in the agar medium, e.g. *P. minioluteum, P. glandicola, P. citreonigrum, P. miczynskii, P. purpurogenum, P. atrosanguineum.*

In Table 2 differences between individual wood species are apparent. The highest number of *Penicillium* species was observed on *Quercus petraea, Q. robur* and *Picea abies* samples (19, 13 and 12 species, respectively). *Penicillium glabrum* and *P. spinulosum* were the dominant Penicillia on *Quercus petraea* samples. *Quercus robur* had three dominants: *P. glabrum, P. glandicola* and *P. minioluteum. Geosmithia* spp. were typical species of *Q. pubescens*. On *Picea abies* high numbers of *P. minioluteum* characteristically occurred. In other trees Penicillia occurred in a lesser extent and no dominant *Penicillium* species was detected.

Penicillium species were found on healthy, damaged and dead trees, too. Twice as many species of *Penicillium* were isolated from samples of damaged trees than from healthy and dead ones. This figure is probably influenced by the higher number of damaged trees studied.

Penicillia were isolated both from bark and cambium, and from wood. Affinity to one of these substrates was recorded only in three frequent species. *Penicillium minioluteum* was frequently observed in the cambial zone as elongate colonies, *P. glabrum* and *P. citreonigrum* was recorded predominantly on wood.

Penicillia were present on all types of samples studied (Table 3). The total frequency of Penicillia on roots, stem bases, middle and upper parts of stems and branches was 90 %, 63.4 %, 75.9 %, 63.5 % and 68.4 %, respectively. In Table 3 the most frequent Penicillia on these parts are given, too. During this study three *Penicillium* species were found for the first time in the Czech Republic: *P. minioluteum* (Kubátová, 1995), *P. smithii* and *P. purpurogenum* var. *rubrisclerotium*.

Table 2. List of *Penicillium spp.* and their frequency on eight wood species (numbers indicate the quantity of samples from which the species was isolated).

Penicillium species	Quercus petraea	Quercus robur	Quercus pubescens	Fagus sylvtica	Picea abies	Larix decidua	Pinus nigra	Pinus sylvestris	TF (%)*
P. minioluteum	5	10	-	-	17	3	-	-	18.8
P. glabrum	13	11	1	1	-	-	2	4	17.2
Geosmithia spp.	3	-	18	-	-	-	-	-	11.3
P. spinulosum	12	2	-	-	2	-	-	1	9.1
Penicillium spp.	5	1	-	-	6	2	1	1	8.6
P. glandicola	2	11	-	1	-	-	-	-	7.5
P. brevicompactum	1	1	-	-	5	-	-	3	5.4
P. citreonigrum	6	1	-	-	1	1	-	-	4.8
P. miczynskii s.l.	2	-	3	-	4	-	-	-	4.8
P. fellutanum	2	-	4	-	2	-	-	-	4.3
P. purpurogenum var. rubrisclerotium	2	2	-	-	1	1	-	-	3.2
P. simplicissimum s.l.	2	2	-	1	-	1	-	-	3.2
P. expansum	1	-	-	-	2	-	-	2	2.7
P. cf. manginii	1	4	-	-	-	-	-	-	2.7
P.roqueforti group	-	4	-	-	-	-	-	-	2.2
P. corylophilum	-	-	-	-	-	-	1	2	1.6
P. verruculosum	1	2	-	-	-	-	-	-	1.6
P. citrinum	-	-	-	-	2	-	-	-	1.1
P. daleae	-	2	-	-	-	-	-	-	1.1
P. smithii	2	-	-	-	-	-	-	-	1.1
P. atramentosum	1	-	-	-	-	-	-	-	0.5
P. canescens	1	-	-	-	-	-	-	-	0.5
P. crustosum	-	-	-	-	1	-	-	-	0.5
P. griseofulvum	-	-	1	-	-	-	-	-	0.5
P. roseopurpureum	-	-	1	-	-	-	-	-	0.5
P. solitum	-	-	-	-	1	-	-	-	0.5
P. thomii	1	-	-	-	-	-	-	-	0.5
total No. of samples	38	34	20	19	38	21	8	8	186
total No. of species	19	13	6	3	12	5	3	6	27

* TF = Total frequency was assessed in % by the number of samples in which a species occurred in relation to the total number of samples.

DISCUSSION

The list of *Penicillium* species observed on the wood and bark of the trees studied was very characteristic and dissimilar from those found on foods and feeds, soils, air etc. Many *Penicillium* species are undoubtedly living on dead wood and bark (Stewart *et al.*, 1979; Davidson, 1985; Land and Hult, 1987). A detailed review is given by Seifert and Frisvad (2000).

Table 3. Occurrence of *Penicillium* spp. on different parts of trees

Part of tree	No. of samples	Frequency of Penicillia %	Number of Penicillium sp.	Most frequent species
Roots	20	90	9	P.minioluteum
				P.glandicola
				P.glabrum
				P.simplicissimum
Stem bases	41	63.4	18	P.glabrum
Middle part of stems	54	75.9	16	P.minioluteum
				P.glabrum
				P.spinulosum
Upper part of stems	52	63.5	16	P.minioluteum
				Geosmithia spp.
				P.glabrum
				P.miczynskii
Branches	19	68.4	7	Geosmithia spp.

Penicillia isolated from living trees have been identified in several studies. Mañka and Truszkowska (1958) recorded *Penicillium commune*, *P. glabrum* (as *P. frequentans*), *P. janthinellum*, *P. chrysogenum* (as *P. notatum*), *P. raciborskii* and *P. waksmanii* on roots of *Picea excelsa*. Bourchier (1961) isolated *P. glabrum* (as *P. frequentans)* infrequently from living *Pinus contorta* where the fungus was capable of growing through the tracheids. Bills and Polishook (1990) found *P. janthinellum*, *P. thomii*, *P. spinulosum*, *P. pseudostromaticum* and other *Penicillium* spp. on the bark of *Carpinus* trees.

However, the role of the Penicillia associated with declining trees is under question. Our material was incubated in moist chambers without surface sterilization. This method was favourable for growth of Penicillia. On the other hand, the surface and internal mycoflora could not be distinguished. Many Penicillia originated very probably from surface contamination of samples in forest. For example, *P. glabrum* and *P. spinulosum* are often occurring in soils in the Czech Republic (Nováková and Kubátová, 1995) and they were presumably present in soil of our localities. On the other hand, other frequent species *Penicillium minioluteum* and *Geosmithia spp.* are not known from Czech soils. Therefore, some affinity of these species to local trees is supposed. It is not excluded that they may grow in tissues weakened by other factors.

A similar study was carried out by Novotný (1995). His work was focused on roots of *Quercus* spp. in one locality in the Czech Republic. Although he used surface sterilization, he found similar species of *Penicillium* on roots as presented in this study, with the difference that the dominant *Penicillium* species were *P. glandicola*, *P. glabrum*, *P. simplicissimum*, *P. spinulosum*, *P. daleae* and *P. minioluteum*.

Researchers from many parts of the world have recorded *Penicillium* spp. after surface sterilization of samples (e.g. Mañka and Truszkowska, 1958; Shigo, 1958; Amos and True, 1967; Przybyl, 1995, etc.). Other authors studied endophytic fungi of trees after sterilization.

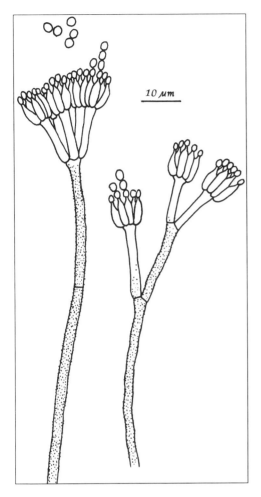

Fig. 2. *Penicillium atrosanguineum* (AK 63/93).
Conidiophores and conidia.

The majority of these authors have not listed Penicillia (Carroll, 1990; Halmschlager *et al.*, 1993; Barklund and Kowalski, 1996, etc.). However, Fisher *et al.* (1991) observed *Penicillium spp.* during their study on endophytic fungi of *Pinus sylvestris* roots. They admitted the possibility that these fungi live in plant tissues or, in some cases, the structure of the root surface may protect them against the surface sterilization agents. Kehr and Wulf (1993) isolated *Penicillium* spp. in low numbers from aboveground parts of declining *Q. robur* in Germany. Bettucci and Saravay (1993) recorded *Penicillium dendriticum* in the xylem of *Eucalyptus globulus* with other endophytic fungi. In studying endophytes from *Opuntia stricta*, Fisher *et al.* (1994) did not consider the Penicillia observed to be true endophytes. Dix and Webster (1995) concluded that records of soil fungi such as *Penicillium* and *Trichoderma* species in lists of endophytes should be treated with caution.

In some cases Penicillia were isolated from necrotic tissue of trees (e.g. Gibbs, 1982). Kowalski (1991), in agreement with Gibbs (1982), concluded that *Penicillium* spp. are secondary colonizators of tissues. *Penicillium* species were commonly found by Kaus *et al.* (1996) in vessels of healthy and damaged trees.

A high proportion of the Penicillia could be produced by spores which penetrated into the vessels during cutting. The influence of Penicillia in tree decline is not clear, according to Kaus *et al.* (1996).

Identification of some species was very difficult. *P. glabrum* and *P. spinulosum, P. simplicissimum* and *P. miczynskii* appear to be problematic taxa. Two distinct *Geosmithia* species are included which could not be identified from current keys (Raper and Thom, 1949; Pitt, 1979; Ramírez, 1982). Another species identified as *P. atrosanguineum* is a very interesting fungus. Its pigmentation resembled that of *P. manginii*. In agreement with the description of *P. manginii* of Stolk and Samson (1983), it has strongly pigmented colonies (on CYA yellow, on MEA red) and rough conidiophores. It differs by its conidia and colony growth. Conidia are globose to subglobose rather then ellipsoidal. Conidiophores are terminally branched as well as somewhat irregular, metulae of inequal length sometimes occurred (Fig. 2). The growth of colonies is faster both on CYA and MEA than that of *P. manginii*. No growth at 37 °C was observed.

ACKNOWLEDGEMENTS

I sincerely thank Olga Fassatiová for her advice and collaboration on the project. I wish to thank the workers of the Forestry and Game Management Research Institute in Strnady, Czech Republic, for providing tree samples. I would like to thank Ellen S. Hoekstra for her help in identification of *P. purpurogenum* var. *rubrisclerotium* and Jens C. Frisvad for help in identification of *P. smithii* and *P. atrosanguineum*. This work was supported by the project of Ministry of Agriculture of the Czech Republic No. 29-91-9106.

REFERENCES

Amos, R.E. & True, R.P. (1967). Longevity of *Ceratocystis fagacearum* in roots of deep-girdled oak-wilt trees in West Virginia. Phytopathology 57: 1012-1015.

Barklund, P. & Kowalski, T. (1996). Endophytic fungi in branches of Norway spruce with particular reference to *Tryblidiopsis pinastri*. Can. J. Bot. 74: 673-678.

Bettucci, L. & Saravay, M. (1993). Endophytic fungi of *Eucalyptus globulus*: a preliminary study. Mycol. Res. 97: 679-682.

Bills, G.F. & Polishook J.D. (1990). Microfungi from *Carpinus caroliniana*. Can. J. Bot. 69: 1477-1482.

Bourchier, R. J. (1961). Laboratory studies on microfungi isolated from the stems of living lodgepole pine, *Pinus contorta* Dougl. Can. J. Bot. 39: 1373-1385.

Carroll, G. C. (1990). Fungal endophytes in vascular plants: mycological research opportunities in Japan. Trans. Mycol. Soc. Japan 31: 103-116.

Davidson, R.W. (1985). Brown and red stain in Ponderosa pine lumber. Mycologia 77: 494-496.

Dix, N. J. & Webster, J. (1995). Fungal ecology. London, Chapman & Hall.

Fisher, P.J., Petrini, O. & Petrini, L.E. (1991). Endophytic Ascomycetes and Deuteromycetes in roots of *Pinus sylvestris*. Nova Hedwigia 52: 11-15.

Fisher, P.J., Sutton, B.C., Petrini, L.E. & Petrini, O. (1994). Fungal endophytes from *Opuntia stricta*: a first report. Nova Hedwigia 59: 195-200.

Gibbs, J.N. (1982). An oak cancer caused by a gall midge. Forestry 55: 69-78.

Halmschlager, von E., Butin, H. & Donaubauer, E. (1993). Endophytische Pilze in Blättern und Zweigen von *Quercus petraea*. Eur. J. For. Path. 23: 51-63.

Kaus, A., Schmitt, V., Simon, A. & Wild, A. (1996). Microscopical and mycological investigations on wood of pedunculate oak (*Quercus robur* L.) relative to the occurrence of oak decline. J. Plant Physiol. 148: 302-308.

Kehr, R. D. & Wulf, A. (1993). Fungi associated with above-ground portions of declining oaks (*Quercus robur*) in Germany. Eur. J. For. Path. 23: 18-27.

Kowalski, T. (1991). Oak decline: I. Fungi associated with various disease symptoms on overground portions of middle-aged and old oak (*Quercus robur* L.). Eur. J. For. Path. 21: 136-151.

Kubátová, A. (1995). New records of Penicillia from the Czech and Slovak Republics: *Penicillium coprophilum, P. minioluteum*, and *P. rubefaciens*. Novit. Bot. Univ. Carol. 8: 7-19.

Land, C. J. & Hult, K. (1987). Mycotoxin production by some wood-associated *Penicillium* spp. Lett. Appl. Microbiol. 4: 41-44.

Maňka, K. & Truszkowska, W. (1958). Próba mykologicznej analizy korzeni swierka (*Picea excelsa* Lk.). Acta Soc. Bot. Pol. 27: 45-73.

Nováková, A. & Kubátová, A. (1995). Studium rodu *Penicillium* v Èeské a Slovenské republice a pøehled zástupcù zjištìných na tomto území (Study of the genus *Penicillium* in Czech and Slovak Republics and survey of reported species). In Souèasný stav, využití moderních metod a perspektivy studia rodu *Penicillium* (Present State, Modern Methods and Perspectives in *Penicillium* Study) (Eds Kubátová, A. & Prášil, K.). Praha, ÈVSM. pp. 31-88.

Novotný, D. (1995). Pøíspìvek k mykoflóøe koøenù dubù s tracheomykózními pøíznaky (Contribution to mycoflora fo the oak roots with tracheomycotics symptoms). In Sborník referátù (II) z odborného semináøe Aktuální problémy ochrany døevin (Proceedings (II) of the workshop Present problems of the tree species protection) (Eds Ĉížková, D. & Švecová, M.). Praha, PøF UK. pp. 52-64.

Pitt, J. I. (1979a). *Geosmithia* gen. nov. for *Penicillium lavendulum* and related species. Can. J. Bot. 57: 2021-2023.

Pitt, J. I. (1979b). The Genus *Penicillium* and its Teleomorphic States *Eupenicillium* and *Talaromyces*. London, Academic Press.

Przybyl, K. (1995). Zamieranie dębów w Polsce. Idee Ekologiczne 8: 1-85.

Przybyl, K. (1996). Disease symptoms and fungi occurring on overground organs of *Quercus petraea*. Acta Mycologica 31: 163-170.

Quintanilla, J.A. (1985). Three new species of *Penicillium* belonging to subgenus *Biverticillium* Dierckx, isolated from different substrates. Mycopathologia 91: 69-78.

Ramírez, C. (1982). Manual and Atlas of the Penicillia. Amsterdam, Elsevier Biomedical Press.

Raper, K. B. & Thom, C. (1949). A Manual of the Penicillia. Baltimore, Williams & Wilkins.

Reenen-Hoekstra, E.S.van, Frisvad, J.C., Samson, R.A. & Stolk, A.C. (1990). The *Penicillium funiculosum* complex - well defined species and problematic taxa. In Modern concepts in *Penicillium* and *Aspergillus* classification (Eds Samson, R.A. & Pitt, J.I.). New York, Plenum Press. pp. 173-191.

Seifert, K. A. & Frisvad, J. C. (2000): *Penicillium* on wood. In. Integration of modern taxonomic methods for *Penicillium* and *Aspergillus* classification (Eds. Samson, R.A. & Pitt, J.I) Harwood Publishers, Amsterdam pp. 285-298

Shigo, A.L. (1958). Fungi isolated from oak-wilt trees and their effects on *Ceratocystis fagacearum*. Mycologia 50: 758-769.

Stewart, E. L., Palm, M. E., Palmer, J. G. & Eslyn, W. E. (1979). Deuteromycetes and selected Ascomycetes that occur on or in wood: indexed bibliography. Madison, Wisconsin, US Department of Agriculture.

Stolk, A.C. & Samson, R.A. (1983). The Ascomycete genus *Eupenicillium* and related *Penicillium* anamorphs. Stud. Mycol., Baarn 23: 1-149.

Stolk, A.C., Samson, R.A., Frisvad, J. C. & Filtenborg, O. (1990). The systematics of the terverticillate Penicillia. In Modern Concepts in *Penicillium* and *Aspergillus* Classification (Eds Samson, R.A. & Pitt, J.I.). New York, Plenum Press. pp. 121-136.

307

PENICILLIUM SPECIES DIVERSITY IN SOIL AND SOME TAXONOMIC AND ECOLOGICAL NOTES

M. Christensen¹, J.C. Frisvad² and D.E. Tuthill¹
¹Department of Botany, University of Wyoming, Laramie, WY 82071, USA and
²Department of Biotechnology, The Technical University of Denmark, 2800 Lyngby, Denmark.

To assess the prominence of Penicillia in soil in relation to latitude and community type, we examined species lists in 74 soil microfungal surveys. Penicillia were major contributors to soil microfungal biodiversity, accounting for 3 and 4% (legume crop soils, India and Israel) to 48 and 49% (forest soil, Mexico; various soils, Syria) of the total species reported per survey. The average proportion of Penicillia was 21% among an average of 90 microfungal species per survey. In contrast to *Aspergillus*, which peaks in prominence in the subtropics, *Penicillium* is prominent across a broad range of latitudes, from up to 24-29% of total species at low latitude (3-9°) to 27-49% at mid-latitudes and 21-32% at high latitude sites in the northern hemisphere (Alaska and Sweden; 58-71°). Monoverticillate species were proportionally most abundant in relation to total Penicillia in conifer forest and bog soils, *Penicillium* subgenus *Furcatum* was well represented and common in grassland and hardwood forest soils, and terverticillate species were proportionally most numerous in agricultural soils.

Current taxonomic uncertainty at the species level in *Penicillium*, however, has made geographic and ecological understanding for *Penicillium* species equally uncertain. As an aid to taxonomic clarification and consensus, we strongly recommend integrative taxonomic studies that involve 1) many isolates, recently obtained from nature, and 2) comparisons using 100 or more characters from complementary data sets. Our study of 40 isolates in the *Penicillium miczynskii* complex using 53 morphological characters and 42 secondary metabolite characters, separately and in combination, will be reviewed briefly to illustrate a system now in use in our laboratories.

INTRODUCTION

The early work on *Penicillium* revealed that members in the genus are extremely common in nature and delicately beautiful microscopically (Zaleski, 1927; Raper and Thom, 1949; Pitt, 1979). More recent students of *Penicillium* have discovered that they also are chemical factories, with amazing capacities for specific biosyntheses (Raper, 1957; Frisvad, 1994b; Samson *et al.*, 1995).

In what follows, we will consider some geographical and ecological aspects of soil-dwelling Penicillia, based upon a selection of microfungal surveys and so comparable to our earlier analysis for *Aspergillus* (Christensen and Tuthill, 1985). The middle section is a brief review of habitat selectivity in general, and the final section will introduce you to our current research and to our suggestions for future studies in the fascinating area of *Penicillium* ecology and taxonomy.

Table 1. Occurrence of *Penicillium* species in soil in relation to latitude country and vegetation*

Citation	Country	Vegetation	Latitude	# Pen spp.	Percent of total spp.
Vargese 1972	Malaysia	Forest	3-8	8	14
Bettucci 1995	Malaysia	Rain forest	4	24	24
Maggi 1983	Ivory Coast	Native	5-6	41	21
Maggi 1983	Ivory Coast	Cultivated	5-6	39	22
Mueller-Dombois 1971	Ceylon	Grassland	6-9	18	29
Ogbonna 1982	Nigeria	Savanna	10	10	8
Gochenaur 1970	Peru	Low desert	12-15	2	7
Gochenaur 1970	Peru	High desert	11-17	37	30
Nour 1956	Sudan	Mixed	16	2	6
Rao 1970	India	Mixed	18	12	30
Rodriguez 1990	Mexico	Conifer forest	19	11	48
Dutta 1965	India	Cereal crop	21	40	24
Lee 1975	Hawaii, USA	Swamp	21	15	27
Rai 1971	India	Desert	22-28	13	9
Moubasher 1970	Egypt	Mixed	22-32	33	32
Moura Sarquis 1996	Brazil	Beach	23	35	16
Saksena 1963	India	Cereal crop	25	8	11
Dayal 1971	India	Legume crop	25	2	3
Dwivedi 1966	India	Grassland	25	9	11
Eicker 1974	South Africa	Savanna	25	18	14
Gochenaur 1975	Bahamas	Tree crop	25	14	23
Dwivedi 1971	India	Sal forest	25-27	8	16
Mehrotra 1972	India	Cultivated	26	14	25
Kanaujia 1977	India	Mixed	26-28	4	6
Ranzoni 1968	SW USA	Desert	28-34	22	10
Moustafa 1975	Kuwait	Saline	29	3	5
Eicker 1969	South Africa	Tree crop	29	15	19
Eicker 1969	South Africa	Savanna	29	12	18
Yousseff 1974	Libya	Coastal	30-33	10	16
Borut 1960	Israel	Desert	31	21	26
Steiman 1995	Israel	Desert	31-33	25	12
Joffe 1966	Israel	Peanut crop	31-33	4	4
Joffe 1967	Israel	Citrus crop	32	48	27
Miller 1957	Georgia, USA	Mixed	31-35	42	25
Sizova 1967	Syria	Unknown	33	59	49
Al-Doory 1959	Iraq	Unknown	32-35	23	15
Bettucci 1995	Uruguay	Tree crop	34	16	47
Horie 1977	Japan	Peanut crop	35	14	18
Warcup 1957	Australia	Cereal crop	35	16	13
England 1957	Oklahoma, USA	Grassland	35	22	23
Durrell 1960	Nevada, USA	Desert	36	3	7
States 1978	Arizona, USA	Native	38-39	31	14
Huang 1975	Ohio, USA	Native	39-40	40	22
McLennan 1954	Australia	Heathland	38	37	41
Vardavakis 1991	Greece	Shrub	40	8	26
Scarborough 1970	Colorado, USA	Grassland	41	8	15
Gochenaur 1978	New York, USA	Deciduous forest	41	23	26

Gochenaur 1974	New York ,USA	Mixed forest	41	28	32
Wacha 1979	Iowa, USA	Cultivated	42	12	21
Tuthill 1985	Iowa, USA	Grassland	43	9	24
Gochenaur 1967	Wisconsin, USA	Deciduous forest	43	29	19
Orpurt 1957	Wisconsin, USA	Grassland	43	15	16
Christensen 1962	Wisconsin, USA	Deciduous forest	43	38	40
Ruscoe 1973	New Zealand	Wet pasture	44	8	13
Clarke 1981	S. Dakota, USA	Grassland	44	8	20
Wicklow 1978	Wisconsin, USA	Mixed forest	44-46	7	13
Christensen 1969	Wisconsin, USA	Mixed forest	44-46	23	26
Christensen 1965	Wisconsin, USA	Bog	45	9	24
Wicklow 1974	Oregon, USA	Mixed forest	45	19	31
Montemartini 1975	Italy	Meadow	45	21	33
Widden 1986	Canada	Mixed forest	45	5	17
Luppi-Mosca 1960	Italy	Alpine grass	46	12	16
Singh 1976	Canada	Conifer forest	48-49	8	13
Steiman 1995	Kerguelen Island	Grassland	49	25	27
Morrall 1974	Canada	Aspen forest	49-57	37	18
Bissett 1979	Canada	Alpine	51	16	13
Brown 1958	Britain	Dune system	51	44	26
Sewell 1959	Britain	Heathland	51-53	31	29
Apinis 1958	Britain	Grassland	53	13	20
Morrall 1968	Canada	Conifer forest	54	11	18
Soderstrom 1978	Sweden	Conifer forest	56-61	20	15
Arnebrant 1987	Sweden	Conifer forest	58	7	32
Nilsson 1992	Sweden	Bog	61	2	11
Flannigan 1974	Alaska	Tundra	71	14	21

*Complete citations available on request from the authors.

SPECIES DIVERSITY IN SOIL-INHABITING PENICILLIA

Many workers have shown that *Penicillium* species are abundant in soil, in stored grains and other food and feed materials, and in dispersed plant and animal debris (Domsch *et al.*, 1980; Farr *et al.*, 1989; Samson *et al.*, 1995). According to an earlier review, Penicillia can account for 0-67% (average 35%) of the prevalent species in any soil microfungal community beneath native vegetation, where diversity commonly reaches 50-75 or more species per gram of soil (Christensen, 1981, 1989). Proportionately, *Penicillium* species were most abundant in heathland and flood plain forest soils and least common in desert and tundra soils (Bissett and Parkinson, 1979; Christensen, 1981). Overall, biodiversity in *Penicillium* greatly exceeded that in any other single genus.

In an effort to compile a current consensus of *Penicillium* geographical distribution and habitat selectivity (if any) as a soil form, we examined for this workshop 74 surveys and calculated the contribution of *Penicillium* to microfungal diversity on the basis of total reported species.

Specifically, we recorded number of species obtained, number of *Penicillium* species, *Penicillium* species as a percentage of total species, and ordered the surveys by increasing latitude (Table 1). In 12 of the 74 surveys, the number of *Penicillium* species was 37 or more. Although six of the studies reported recovery of only two or three species, no survey was without a member in the genus. The highest number of recovered Penicillia was 59 species (49% of total species) in a study of the microfungi in various Syrian soils (Sizova *et al.*,

311

1967). Proportionately, *Penicillium* species accounted for 3-49% (average 21%) of 18 - 229 total species (average 90). The genus contributed only 3 and 4% of total species in two legume crop soils in India and Israel, but 48 and 49% of total species in a Mexican forest soil and in the Syrian soils (Table 1).As can be seen in Figure 1, *Penicillium* is prominent across a broad range of latitudes, from up to 24-29% of total species at low latitude (3-9°) to 27-49% at mid-latitudes and 21-32% at high-latitude sites in the northern hemisphere (Alaska and Sweden; 58-71°). In contrast, our earlier calculations for Aspergilli revealed a clear peak in *Aspergillus* species diversity and prominence in the subtropics (Christensen and Tuthill, 1985).

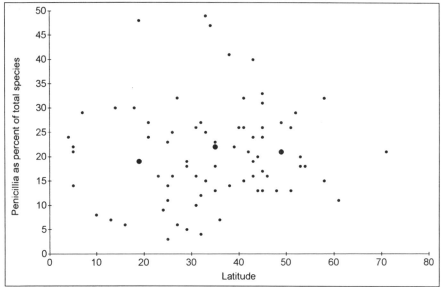

Fig. 1. Prominence of *Penicillium* species in relation to latitude, based on an analysis of 74 soil surveys at latitudes ranging from 3° to 7°(see Table 1). Large dots represent averages in three successive latitude segments, embracing 25, 25, and 24 surveys, respectively

Numbers of undescribed species in *Penicillium* probably also is a random variable against latitude. In a survey of desert soils in Wyoming, we encountered 12 *Penicillium* species among 3600 isolates. Six of those 12 match species treated in one or both of the current monographs (Pitt, 1979; Ramirez, 1982) or described since 1982 and six appear to be undescribed species.

The analysis of *Penicillium* species diversity in relation to habitat confirmed our earlier impression of a diverse and proportionately prominent assemblage of Penicillia in heathland soils (31 and 37 species, 29 and 41% of total species in British and Australian heaths) and certain forest soils (38 species, 40% of total species in the soils of wet-mesic hardwood flood-plain forests in Wisconsin, USA; Christensen *et al.*, 1962). In general, differences in prominence among habitats at the genus level were not evident because of high variability within each habitat category (Fig. 2).

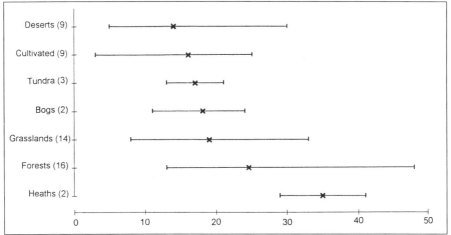

Fig. 2. Range and average of *Penicillium* species as percent of total species isolated, for several community types. Numbers in parentheses are the numbers of surveys included

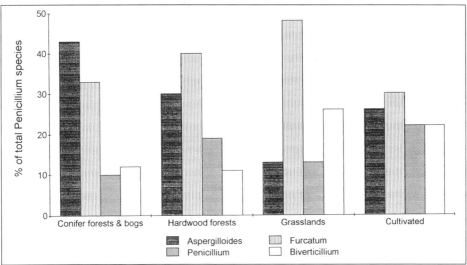

Fig. 3. Soil Penicillia: apportionment of species among the four subgenera. Data shown are averages from a selection of five surveys for each community type.

Our careful examination of a subset of the surveys, however, indicated a possible habitat selectivity at the subgeneric level (Fig. 3). Thus, in our selection of surveys, monoverticillate species were proportionately most abundant in relation to total Penicillia in conifer forest and bog soils, *Penicillium* subgenus *Furcatum* was well-represented and common in grassland and hardwood forest soils, and terverticillate species were proportionately most numerous in agricultural soils.

313

HABITAT SELECTIVITY AT THE SPECIES LEVEL

Habitat selectivity at the species level is apparent in several species in *Aspergillus* and in other genera as well. Among Aspergilli, six species described by workers at the University of Wyoming from soil are known only from communities in the western United States -- primarily sagebrush and grassland communities of the intermountain basins or the western fringe of the Great Plains (*A. bicolor, A. bridgeri, A. campestris, A. leporis, A. navahoensis, A. spectabilis*) (Christensen and Tuthill, 1985). Similarly, *A. elongatus, A. ochraceoroseus, A. sparsus, Emericella desertorum* and many others may be restricted ecologically (Christensen and Tuthill, 1985). *Aspergillus terreus, A. candidus* and *A. flavipes* are prevalent species in New World grasslands (Christensen, 1981).

In , all known isolates of a beautiful species in subgenus *Biverticillium*, P. coalescens, are from pine forest soil in Spain and a pine cone in Japan (CBS, 1996; Frisvad, unpublished). *P. confertum* and *P. dipodomyicola,* obtained originally from the cheek pouches of kangaroo rats (*Dipodomys spectabilis*), have not to date been isolated elsewhere (Frisvad and Filtenborg, 1989; CBS, 1996). And there are several examples of *Penicillium* species that are cosmopolitan geographically, but exhibit well-defined habitat selectivity. Thus, *P. restrictum* along with *Aspergillus fumigatus, Paecilomyces lilacinus* and some *Fusarium* species are reliable indicators of grassland soils world wide. An alliance in the soil of *P. pinetorum, P. radulatum, P. montanense* and *Torulomyces lagena* along with two or three distinctive *Mortierella* species and *Oidiodendron* are as definitive for conifer and conifer-hardwood communities as are the trees (Christensen and Backus, 1964; Christensen, 1981; CBS, 1996).

Of the Penicillia occurring in foods and aboveground organic debris above ground, a high percentage are subgenus , apparently as many as 70-90% of the total Penicillia isolated (Williams, 1990; Samson *et al.,* 1995). In contrast, numbers of terverticillate species in soil is in the range of approximately 5-25% of total Penicillia isolated, and subgenus Furcatum subsection Asymmetrica-Divaricata of Ramirez, 1982), is more commonly encountered. Here it can account for 40-60% of the total species isolated (Christensen *et al.,* 1962; Christensen and Whittingham, 1965; Christensen, 1969, 1981; Gochenaur, 1978, 1984; Gochenaur and Woodwell, 1974; see Fig. 3).

Since it will require taxonomic consensus and major additional exploration to add to our knowledge of what Penicillia occur where and how commonly a more complete understanding of habitat selectivity in Penicillia is not an immediately attainable goal. But it is without doubt a goal we should be working toward! In 1985, we strongly advocated Plant Hunting (and by that we meant and Aspergillus hunting) (Wilson, 1927; Christensen and Tuthill, 1985). Now, 12 years later, quantitative surveys and examination of isolates fresh from nature remains as essential to contemporary and Aspergillus systematists as were massive plant collections to botanical-systematists at the turn of the century!

NOTES AND COMMENTARY -- TAXONOMIC AND ECOLOGICAL

Background
At the species level, *Penicillium* has been extremely difficult taxonomically. A well-written account of the early contributors to taxonomy and their publication was provided by Pitt, in

1979. Considering just the monographs since 1930: Thom (1930) accepted approximately 290 strictly asexual species in *Penicillium*; Raper and Thom (1949) accepted 118 species as clearly valid (excluding 137 "probably synonymous" species and 19 species that now are in *Eupenicillium* or *Talaromyces*); Pitt, (1979) accepted 97 species (also excluding *Eupenicillium, Talaromyces* and their culturally-connected anamorphs); while Ramirez in (1982), accepted 227 species (excluding *Eupenicillium, Talaromyces* and their undoubted anamorphs). Ramirez (1982) accepted approximately 126 (68%) of the 186 species described as new in the 30 years following 1949 in contrast to the acceptance of exactly 17 (9%) of those 186 species by Pitt (1979).

Since both of the recent monographers worked primarily through the 1970's, consideration of biochemical and molecular aspects of species was not an option. But neither, at that time, recognized the absolute obligation--axiomatic in plant and animal systematics--of obtaining large numbers of isolates from nature (Christensen, 1989). Scrutiny of Ramirez, (1982) has revealed that the author's descriptions of exactly 91 (61%) of the 148 species in his Monoverticillata, Asymmetrica-divaricata and Asymetrica-velutina sections are based on examinations of a single culture!

Interestingly, the Eighth Edition of the Ainsworth and Bisby Dictionary of Fungi offered an estimate of 223 species in in 1995, after publication of 95 additional species in 1980-1994 (Hawksworth *et al.*, 1995; CMI index of Fungi). If Hawksworth (1991) is correct, that fewer than 5% of fungal species are known, the actual number of Penicillia may be over 3000 species.

The taxonomic state of affairs in *Penicillium*, then, has been unsettling to fungal ecologists, physiologists, industrial mycologists and others through the last 10-15 years. As others have pointed out, a reliable and widely-accepted taxonomy is fundamental to all other disciplines in plant, animal and fungal biology (Cronquist, 1988; Christensen, 1989; Stanton and Lattin, 1989; Wilson, 1994), as well as in applied mycological research (Dreyfuss and Chapela, 1994).

In the long term, through the next 50 years or so, what sorts of efforts are likely to lead to an improved understanding of , biologically and taxonomically? That question is a useful one and the wording indicates our philosophy, that is, that taxonomy must be based on as much information as is possible. In the words of Arthur Cronquist, a well-known plant taxonomist, "... sound taxonomy proceeds by the use of multiple correlations." (Cronquist, 1988). Further, individual representatives of the given taxa must serve as the basis for discovery of multiple correlations and integrative taxonomy. Gleason's individualistic concept is valid and fundamental in both taxonomy and community ecology (Gleason, 1939; Cronquist, 1988).

The authors' response to taxonomic uncertainty in Penicillium
Our own response to the taxonomic uncertainty in has involved.
1. Assemblage of a large collection of recent isolates. The Wisconsin Soil Fungi (WSF) and Rocky Mountain Fungi (RMF) collections, all isolates from soil, contain 1084 strains in *Penicillium*, and there are approximately 8000 Penicillia in the department of Biotechnology (IBT) collection (Technical University of Denmark).

2. Preservation of those isolates by lyophilization or on silica gel, following assignment of permanent culture numbers accompanied by geographical and habitat data. Estimates of frequency of occurrence in specific communities and community types are available for all RMF and WSF isolates. No survey should be sanctioned which does not preserve all or at minimum the major isolated taxa under permanent culture numbers.

3. Comparisons with type cultures (or neotypes or "primary basis" cultures) (Raper and Thom, 1949; Pitt, 1979; Pitt and Samson, 1993).

4. An integrative taxonomic analysis, using the isolates as Operational Taxonomic Units (OTUs) and thus conforming to Gleason's individualistic concept (Gleason, 1939), and using characters in complementary suites of features -- biochemical and morphological characters for example.

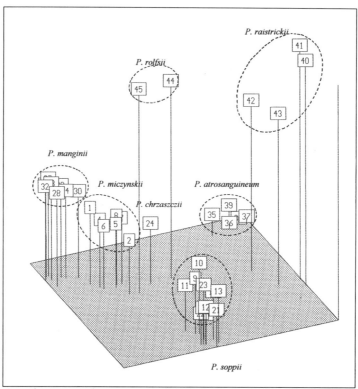

Fig. 4. An integrative taxonomic assessment of OTUs in the *Penicillum miczynskii* complex of species plus isolates in *P. rolfsii* and *P. raistrickii* for comparison. The correspondence analysis, generated from 95 characters, support recognition of five species in *P. miczynskii sensu* Pitt (Christensen *et al.*, 1999).

At least two other teams in research have conducted integrated taxonomic analyses (Bridge *et al.*,; 1989 a,b; Patterson *et al.*, and Cruickshank, 1990; Pitt *et al.*, 1990). Such studies with other genera of fungi include a classical integrative analysis by Whalley and Greenhalgh (1973) and more recently, studies by Sheard (1978), Wolfe (1984), Mueller (1985), Mordue

316

et al. (1989) and Jun *et al.*, (1991). Integrative taxonomic studies with plants, which flourished through the 1970's and 1980's, have influenced the effort in mycology (Cronquist, 1988; Taylor *et al.*, 1994).

Our own first attempt at integrative taxonomy involved calculations of similarity among 71 isolates (Christensen, 1994; Frisvad, 1994a). Twenty-two of the 71 cultures were ex-types, ex-neotypes, (Pitt and Samson, 1993) or strains listed by Raper and Thom (1949) as "primary basis" cultures. Character presence or absence, scored for each OTU, resulted in a binary matrix, and subsequently, matrices of similarity coefficients were calculated for all pairs of OTUs (NTSYS-pc version 1.80; Rohlf, 1993). The characters used were: 1) 89 morphological characters, 2) 162 secondary metabolite characters, 3) 54 cultural characters, thus 305 characters in all. Despite some high variability in cultural features, the phenogram generated using the 305 characters in combination (by unweighted pair-group method, arithmetic average: UPGMA; Rohlf, 1993 and see Frisvad and Thrane, 1993) contained clusters centered on nine ex-type or primary basis cultures (*P. brasilianum - P. paraherquei, P. canescens, P. cremeogriseum, P. glaucoroseum, P. janthinellum, P. javanicum, P. miczynskii, P. rolfsii*); eight ex-type or "primary basis" cultures were unrelated to one another and to all clusters; and there were separate but mixed alliances of *P. chermisinum - P. indicum - P. bilaiae* and three soil isolates, and *P. pulvillorum - P. simplicissimum* and 11 soil isolates.

More recently, we used a similar integrative taxonomic method in an examination of 40 isolates in the *P. miczynskii* complex of species (Christensen *et al.*, 1999). The OTUs were compared in respect to 38 micromorphological characters, 15 cultural characters and 42 secondary metabolites (a total of 95 characters; Table 2), again using the NYSYS-pc program (Rohlf, 1993) for cluster analysis (UPGMA) and correspondence analysis. Constellations of OTUs in the cluster phenograms were clearly defined, and the discovery that our separate analyses by secondary metabolites and morphological-cultural features allocated 39 of the 40 OTUs to the same clusters justified our combining the data sets (Fig. 4).

Table 2. Features used in the calculations of similarity among 40 OTUs in the *Penicillium miczynskii* complex of species (Christensen *et al.*, 1999)*.

1.	Condium size, shape, surface (5)
2.	Conidial aggregation (3)
3.	Phialide morphology (3)
4.	Metula features (11)
5.	Rami presence/absence (2)
6.	Stipe surface, length, origin (10)
7.	Sclerotia: presence/absence, colour (4)
8.	Colony features, MEA (6)
9.	Colony features, Cz, CYA, YES; acid production on CREA; growth at 37°C (9)
10.	Secondary metabolites (42)

*Figures in parentheses indicate number of characters in each class. Total characters used in the combined analysis (Fig. 4) was 95 (38 micromorphological, 15 cultural, 42 secondary metabolite characters).

Biogeographical notes and some further evidence of habitat selectivity.
As soon as there is reasonable confidence in species circumscription, it becomes exciting then to consider the geographical and ecological data.

Three of the species for which we have assembled biogeographical and ecological notes (Table 3) are our own circumscriptions (*P. brasilianum, P. soppii, P. pulvillorum*) (Frisvad, 1994c; Christensen *et al.*, 1999), and three (*P. islandicum, P. montanense* and *P. rolfsii*) are valid species by general consensus (Pitt, 1979; Ramirez, 1982; Pitt and Samson, 1993).

Finally, we want to emphasize again the point that biological understanding at the species level in any genus requires taxonomic clarification and consensus. In, it is our strong opinion that: 1) multiple isolates and 2) attention to complementary kinds of characterization are the essential prerequisites to taxonomic clarification.

Table 3. Geographic and habitat distributions for selected species of *Penicillium*

Species	Geographic Distribution	Habitat
P. brasilianum (28)	Brazil, Canada, Costa Rica, Czechoslovakia, Japan, Germany, Netherlands, South Africa, USA, USSR, Zimbabwe.	Grassland soil (7), cultivated soil (5), forest soil (5), uniden. soil (4), unknown or non-soil (7)
P. islandicum (47)	Australia, Canada, Ethiopia, Ghana, India, Island of Skyr, Isracl, Italy, Japan, Nepal, Pakistan, Solomon Islands, South Africa, Spain, Switzerland, United Kingdom, USA, USSR, Zaire.	Seeds-grains-flour-feeds- fruit (21), feathers (1), kapok (1), animal (2), cultivated soil (7), heath soil or peat (4), forest soil (3), grassland soil (1), uniden. soil (3), unknown (4).
P. montanense (26)	Australia, Austria, Canada, Colombia, Germany, India, Spain, Sweden, Switzerland, USA.	Forest soil or litter (20), peat (1), uniden. soil (4), unknown (1).
P. pulvillorum (49)	Canada, Germany, India, Italy, Nepal, South Africa, Tanzania, United Kingdom, USA.	Grassland soil (34), cultivated soil (2), forest soil (3), uniden. soil (5), unknown (5).
P. rolfsii (7)	Australia, USA (Florida, Wisconsin).	Ericaceous bog peat (5), Epacridaceae litter and soil (1), pineapple fruit (1).
P. soppii (16)	Denmark, Poland, Spain, USA, USSR	Conifer forest soil or litter (9), heathland soil (2), hardwood forest soil (3), unknown or non-soil (2).

Figures in brackets are numbers of records in this compilation.

ACKNOWLEDGMENTS

This study has been supported in part by two grants from the United States National Science Foundation (RII-8610680 and DEB 9632880). We thank Ellen Kirstine Lyhne for her valuable technical assistance.

318

REFERENCES

Bills, G.F., Holtzman, G.I. & Miller, O.K. Jr., (1986). Comparison of ectomycorrhizal basidiomycote communities in red spruce versus northern hardwood forests of West Virginia. Can. J. Bot. 64: 760-768.

Bissett, J. & Parkinson, D. (1979). The distribution of fungi in some alpine soils. Can. J. Bot. 57:1609-1629.

Bridge, P.D., Hawksworth, D.L., Kozakiewicz, Z., Onions, A.H.S., Paterson, R.R.M., Sackin, M.J. & Sneath, P.H.A. (1989a). A reappraisal of the terverticillate Pencillia using biochemical, physiological and morphological features. I. Numerical taxonomy. J. Gen. Microbiol. 135: 2941-2966.

Bridge, P.D., Hawksworth, D.L., Kozakiewicz, Z., Onions, A.H.S., Paterson, R.R.M. & Sackin, M.J. (1989b). A reappraisal of the terrerticillate Pencillia using biochemical, physiological and morphological features. II. Identification. J. Gen. Microbiol. 135: 2967-2978.

CBS (1996). Centraalbureau voor Schimmelcultures. List of Cultures: Fungi and Yeasts. 34th Ed. Baarn, Netherlands, Centraalbureau voor Schimmelcultures.

Christensen, M. (1969). Soil microfungi of dry to mesic conifer-hardwood forests in northern Wisconsin. Ecology 50: 9-27.

Christensen, M. (1981). Species diversity and dominance in fungal communities. In The fungal community, its organization and role in the ecosystem (Eds Wicklow, D.T. & Carroll, G.C.). New York, Marcel Dekker, Inc., pp. 208-232.

Christensen, M. (1989). A view of fungal ecology. Mycologia 81: 1-19.

Christensen, M. (1994). Morphological analysis and introduction to our comparative study of isolates in *Penicillium janthinellum* and related species. Abstract, Proceedings of the Fifth international Mycological Congress, Vancouver, British Columbia.

Christensen, M. & Backus, M.P. (1964). Two varieties of *Monocillium humicola* in Wisconsin forest soils. Mycologia 61: 498-504.

Christensen, M. & Tuthill, D.E. (1985). *Aspergillus*: an overview. In Advances in *Penicillium* and *Aspergillus* systematics (Eds Samson, R.A. & Pitt, J.I.). New York, Plenum Press. pp. 195-209.

Christensen, M. & Whittingham, W.F. (1965). The soil microfungi of open bogs and conifer swamps in Wisconsin. Mycologia 57: 882-896.

Christensen, M., Whittingham, W.F. & Novak, R.O. (1962). The soil microfungi of wet-mesic forests in southern Wisconsin. Mycologia 54: 374-388.

Christensen, M., Frisvad, J.C. & Tuthill, D.E. (1999). Zaleski's *Penicillium miczynskii* and its synonyms *sensu* Pitt, 1979. Mycological Research 103: 527-541.

Cronquist, A. (1988). The evolution and classification of flowering plants. 2nd Ed. Bronx, New York, New York Botanical Garden.

Domsch, K.H., Gams, W. & Anderson, T.H. (1980). Compendium of soil fungi. London, Academic press.

Dreyfuss, M.M. & Chapela, I.H. (1994). Potential of fungi in the discovery of novel, low-molecular weight pharmaceuticals. In The discovery of natural products with therapeutic potential (Ed Gullo, V.P.). Boston, Butterworth-Heinemann. pp. 49-80.

Farr, D.F., Bills, G.F., Chamuris, G.P. & Rossman, A.Y. (1989). Fungi on plants and plant products. St. Paul, Minnesota, American Phytopathological Society.

Frisvad, J.C. (1994a). Correspondence, principal coordinate and redundancy analysis used on mixed chemotaxonomical qualitative and quantitative data. Chemomet. Intell. Lab. Syst. 23: 213-222.

Frisvad, J.C. (1994b). Classification of organisms by secondary metabolites. In The identification and characterization of pest organisms (Ed. Hawksworth, D.L.). Wallingford, UK, CAB International. pp. 303-320.

Frisvad, J.C. (1994c). Secondary metabolites and consensus analysis in *Penicillium janthinellum* and related species. Abstract, Fifth International Mycological Congress, Vancouver, British Columbia.

Frisvad, J.C. & Filtenborg, O. (1989). Terverticillate Penicillia: Chemotaxonomy of and mycotoxin production. Mycologia 81: 837-861.

Frisvad, J.C. & Thrane, U. (1993). Liquid column chromatography of mycotoxins. In Chromatography of mycotoxins (Ed. Betina, V.). Amsterdam, Elsevier. Pp. 253-372.

Gleason, H.A. (1939). The individualistic concept of the plant association. Am. Midland Natural. 21: 92-110.

Gochenaur, S.E. (1978). Fungi of a Long Island oak-birch forest. I. Community organization and seasonal occurrence of the opportunistic decomposers of the A horizon. Mycologia 70: 975-994.

Gochenaur, S.E. (1984). Fungi of a Long Island oak-birch forest. II. Population dynamics and hydrolase patterns for the soil Penicillia. Mycologia 76: 218-231.

319

Gochenaur, S.E. & Woodwell, G.M. (1974). The soil microfungi of a chronically irradiated oak-pine forest. Ecology 55: 1004-1016.

Hawksworth, D.L. (1991). The fungal dimension of biodiversity: magnitude, significance, and conservation. Mycol. Res. 95: 641-655.

Hawksworth, D.L., Kirk, P.M., Sutton, B.C. & Pegler, D.N. (1995). Ainsworth & Bisby'sDictionary of the Fungi. Wallingford, UK, CAB International.

Jun, Y., Bridge, P.D. & Evans, H.C. (1991). An integrated approach to the taxonomy of the genus *Verticillium*. J. Gen. Microbiol. 137: 1437-1444.

Mordue, J.E.M., Currah, R.S. & Bridge, P.D. (1989). An integrated approach to *Rhizoctonia* taxonomy: cultural, biochemical and numerical techniques. Mycol. Res. 92: 78-90.

Mueller, G.M. (1985). Numerical taxonomic analyses on *Laccaria* (Agricales). Mycologia 77: 121-129.

Paterson, R.R.M., Bridge, P.D., Crosswaite, M.J. & Hawksworth, D.L. (1989). A reappraisal of the terrterticillate Pencillia using biochemical, physiological and morphological features III. An evaluation of pectinase and amylase isoenzymes for species characterization. J. Gen. Microbiol. 135: 2979-2991.

Pitt, J.I. (1979). The Genus *Penicillium* and it Teleomorphic States *Eupenicillium* and *Talaromyces*. London, Academic Press.

Pitt, J.I. & Cruickshank, R.H. (1990). Speciation and synonymy in *Penicillium* subgenus *Penicillium* -- towards a definitive taxonomy. In Modern concepts in *Pencillium* and *Aspergillus* Classification (Eds Samson, R.A. & Pitt, J.I.). New York, Plenum Press. pp. 103-119.

Pitt, J.I., Klich, M.A., Shaffer, G.P., Cruickshank, R.H., Frisvad, J.C., Mullaney, E.J., Onions, A.H.S., Samson, R.A. & Williams, A.P. (1990). Differentiation of *Penicillium glabrum* from *Penicillium spinulosum* and other closely related species: an integrated taxonomic approach. System. Appl. Microbiol. 13: 304-309.

Pitt, J.I. & Samson, R.A. (1993). Species names in current usc in the *Trichocomaceae* (Fungi, Eurotiales). In Names in Current Use in the Families *Trichocomaceae, Cladoniaceae, Pinaceae,* and *Lemnaceae* (Ed Greuter, W.). Königstein, Germany, Koeltz Scientific Books. pp. 13-57.

Ramirez, C. (1982). Manual and Atlas of the Penicillia. Amsterdam, Elsevier Biomedical Press.

Raper, K.B. (1957). Microbes -- man's mighty midgets. Am. J. of Bot. 44: 56-65.

Raper, K.B. & Thom, C. (1949). A Manual of the Penicillia. Baltimore, Williams and Wilkins.

Rohlf, F.J. (1993). NYSYS-pc: Numerical Taxonomy and Multivariate Analysis System. Version 1.80. Setauket, New York, Exeter Software.

Samson, R.A., Hoekstra, E.S., Frisvad, J.C. & Filtenborg, O. (1995). Introduction to Food-borne Fungi. Baarn, Netherlands, Centraalbureau voor Schimmelcultures..

Sheard, J.W. (1978). The taxonomy of the *Ramalina siliquosa* species aggregate (lichenized Ascomycetes). Can. J. Bot. 56: 916-938.

Sizova, T.P., Baghdadi, J.K. & Gorlenko, M.V. (1967). Mycoflora of Mukhafez of Damascus and Es-Suveida (Syria). Mikol. Fitopatol. 1: 286-294.

Stanton, N.L. & Lattin, J.D. (1989). In defense of species. BioScience 39: 67.

Taylor, R.J., Patterson, T.F. & Harrod, R.J. (1994). Systematics of Mexican spruce -- revisited. Syst. Bot. 19: 47-59.

Thom, C. (1930). The Penicillia. Baltimore, Wlliams and Wilkins.

Whalley, A.J.S. & Greenhalgh, G.N. (1973). Numerical taxonomy of *Hypoxylon* I. Comparison of classifications of the cultural and the perfect states. Trans. Br. Mycol. Soc. 61: 435-454.

Williams, A.P. (1990). *Penicillium* and *Aspergillus* in the food microbiology laboratory. In Modern Concepts in *Penicillium* and *Aspergillus* Classification (Eds Samson, R.A. & Pitt, J.I.). New York, Plenum Press. pp. 67-71.

Wilson, E.H. (1927). Plant hunting. Boston, Stratford Company.

Wilson, E.O. (1994). Naturalist. Washington, D.C., Island Press.

Wolfe, C.B., Jr. (1984). A numerical taxonomic analysis of the tribe Ixechineae (Boletaceae). Mycologia 76: 140-147.

Zaleski, K. (1927). Über die in Polen gefundenen Arten der Gruppe *Penicillium* Link. I, II and III Teil. Bulletin international de l'académie Polonaise des sciences et de lettres. Classe des sciences mathematiques et naturelles Series B: Sciences naturelles, 417-563.

Chapter 6

MOLECULAR TAXONOMY OF *ASPERGILLUS*

PHYLOGENETIC RELATIONSHIPS IN *ASPERGIL-LUS* BASED ON RDNA SEQUENCE ANALYSIS

Stephen W. Peterson
National Center for Agricultural Utilization Research, Agricultural Research Service, U. S. Department of Agriculture, Peoria, IL 61604, USA

Aspergillus taxonomy is based on morphological and physiological similarities. To test the current phenotype-based classification systems, I have sequenced the D1 and D2 regions of the large subunit ribosomal RNA (lsu-rDNA) genes from 215 named taxa in *Aspergillus*. Data were analyzed by parsimony and maximum likelihood methods. Apical regions of the trees were generally well supported in bootstrap analysis, but basal regions of the resulting trees are not. The maximum parsimony trees are incompatible with the current subgeneric taxonomy of *Aspergillus*. Accordingly, the three major lineages are given subgeneric rank as subgenera *Aspergillus*, *Nidulantes* and *Fumigati*. A few of the sections (Gams *et al.*, 1985) are monophyletic, but most are revised to reflect the phylogeny of the species. The revised subgenus *Fumigati* contains sections *Fumigati* and *Clavati*, in contrast to their current placement in separate subgenera. *Hemicarpenteles acanthosporus* (section *Ornati*) also belongs in subgenus *Fumigati*, section *Clavati*. Subgenus *Nidulantes* contains the *Emericella* species, some species from sections *Versicolores*, *Usti*, *Sparsi* and *Ornati*. Subgenus *Aspergillus* contains sections *Restricti*, *Aspergillus*, *Flavi*, *Nigri*, *Circumdati*, *Terrei*, *Flavipedes*, and *Cervini*. Some species synonymies are suggested by identical lsu-rDNA sequences.

INTRODUCTION

Aspergillus is a form genus composed of species whose teleomorphs have been described in eight different genera. However, the anamorphic species have been viewed as naturally related forms, and for this reason they have been dealt with monographically as a unit (Thom and Church, 1926; Thom and Raper, 1945; Raper and Fennell, 1965; Klich and Pitt, 1988a, Kozakiewicz, 1989). Additional data from different research techniques have appeared since the last comprehensive monograph (Raper and Fennell, 1965) including morphological (Samson, 1979; Kozakiewicz, 1989); biochemical (Samson, *et al.*, 1990; Kuraishi *et al.*, 1990); ecological and genetic (Yamatoya et al, 1990; Cruickshank and Pitt, 1990; Egel *et al.*, 1994; Cotty *et al.*, 1994); and molecular genetic (Dupont *et al.*, 1990; Chang *et al.*, 1991; Berbee and Taylor, 1992; Peterson 1992; Berbee *et al.*, 1995).

Placement of species into groups on the basis of morphology has sometimes been problematical (Raper and Fennell, 1965), and continues to be so (Samson, 1979; Christensen, 1981). Various phenotypic characters have been used to formulate keys to species and to derive sub-generic classification (Gams *et al.*, 1985). Colony colours, growth rate

323

on media with low water activity, production of particular metabolites, and microscopic morphology are all used in currently assessing taxonomy.

The current study was designed to test the validity of the phenotypic characters with the phylogeny of *Aspergillus* from analysis of molecular data. Because eight teleomorphic genera are associated with *Aspergillus* teleomorphs, the question whether *Aspergillus* is monophyletic, was also addressed.

MATERIALS AND METHODS

Isolates used in this study included an ex type culture for each species studied, and when possible, two or more additional isolates assigned to the species. The isolates examined are listed in Table 1, and may be accessed electronically at http://nrrl.ncaur.usda.gov. The lsu-rDNA sequences cited here are available from Genbank or at ftp://nrrl.ncaur. usda.gov.

Table 1. Isolates examined.

Aspergillus niger v. Tiegh.

A. 326 = ATCC 16888 = CBS 554.65 = IMI 50566 = Thom 2766 = WB 326. Ex neotype of *A. niger*. Received from Hollander, 1913. Isolated from tannin-gallic acid fermentation, Connecticut.

A. 3 = ATCC 9029 = CBS 120.49 = IMI 41876 = WB 3, 566. Received from A. J. Moyer.

A. 348 = ATCC 1027 = CBS 103.12 = IMI 16148 =Thom 3534B = WB 348. Received as *Aspergillus cinnamomeus* from Schiemann, 1914. Chemically induced cinnamon coloured mutant.

A. 363 = ATCC 10698 = CBS 126.49 = Thom 5135.16. Received as *Aspergillus batatae* from Yakoyama, 1930.

Aspergillus phoenicis (Corda) Thom

B. 365 = CBS 629.78 = Thom 4640.484. Received from Bainier collection via Da Fonseca 1922.

B. 4750 = CBS 128.52 = QM 8183. Received as *Aspergillus foetidus* var. *acidus* WB 4750, 1969.

B. 4851= ATCC 16879 = CBS 115.48, 558.65 = WB 4851. Received as ex type culture of *Aspergillus pulverulentus* WB 4851, 1965. Type of *Aspergillus elatior*.

Aspergillus carbonarius (Bainier) Thom

C. 369 = ATCC 1025 = CBS 111.26 = IMI 16136 = Thom 4030.1 = WB 369. Ex neotype of *A. carbonarius*. Received from Blakeslee, 1915.

C. 67 = ATCC 8740 = CBS 420.64 = IMI 41875 = WB 67. Received as *Aspergillus niger* from Da Fonseca, 1923.

C. 346 = ATCC 6277 = Thom 5373.16 = WB 346. Received from M. M. Mannis, Honduras, 1933, from paper.

C. 4849 = CBS 114.29 = WB 4849. Received 1967. Type of *Sterigmatocystis aciniuvae*.

Aspergillus japonicus Saito

D. 360 = ATCC 1042 = CBS 123.27 = Thom 3522.30 = WB 360. Ex lectotype of *A. japonicus*. Received from J. R. Johnson, 1914. Isolated from Puerto Rican soil.

D. 359 = Thom 4030.5. Received as *Aspergillus niger* from Blakeslee, 1915.

D. 1782 = ATCC 16873 = CBS 568.65 = WB 1782. Isolated from Panamanian soil by J. T. Bonner, 1941.

D. 2053 = CBS 620.78. Received as J745, 1946. Isolated by W. L White from tent material, New Guinea.

Aspergillus aculeatus Iizuka

E. 5094 = ATCC 16872 = CBS 172.66. Ex type of *A. aculeatus*. Received as WB 5094, 1965. Isolated from tropical soil.

Aspergillus ellipticus Raper & Fennell

F. 5120 = ATCC 16876 = CBS 482.65 = IMI 172283. Ex lectotype of *A. ellipticus*. Received as WB 5120, 1965. Isolated by K. J. Kwon from Costa Rican soil.

Aspergillus flavus Link

G. 1957 = ATCC 16883 = CBS 569.65 = IMI 124930 = WB 1957. Ex neotype of *A. flavus*, received from Chemical Warfare Service, 1944. Isolated from cellophane diaphragm of optical mask, S. Pacific.

G. 20521. Received as *Aspergillus flavus* from N. Zummo, USDA, ARS, Miss. St. Univ. 1990. Isolated from corn. Naturally occurring albino mutant.

G. 3751 = CBS 542.69 = QM 9326. Received as *Aspergillus kambarensis* IMI 141553, 1970. Isolated from stratigraphic drilling core Japan. Type of *Aspergillus kambarensis*.

G. 3518. Received as *Aspergillus flavus* var. *columnaris* F55-36 from R. Graves, 1966. Isolated from wheat flour.

G. 4818 = ATCC 16870 = CBS 485.65 = IMI 124932 = WB 4818. Ex lectotype. Received as *Aspergillus flavus* var. *columnaris* WB 4818, 1965. Isolated from butter. Type of *Aspergillus flavus* var. *asper*.

G. 4823. Received as *Aspergillus flavus* var. *columnaris* WB 4823, 1969. Type of *Aspergillus oryzae* var. *wehmeri*.

G. 4822. Received as *Aspergillus flavus* var. *columnaris* WB 4822, 1969. Type of *Aspergillus oryzae* var. *variablis*.

G. 4998 = ATCC 16862 = CBS 501.65 = IMI 44882. Ex lectotype of *Aspergillus subolivaceus*. Received as WB 4998, 1965. Isolated from lintafelt cotton, England.

G. 2097 = ATCC 16859 = CBS 120.51 = IMI 45644 = WB 2097. Ex lectotype of *Aspergillus thomii*. Received as BB.213 from G. Smith, LSHTM, 1947. Isolated as culture contaminant.

Aspergillus oryzae (Ahlburg) Cohn

G. 447 = ATCC 1011; 4814; 7561; 9102; 12891 = CBS 102.07 = IMI 16266; 44242 = Thom 113 = WB 447. Ex neotype of *Aspergillus oryzae*. Received from Westerdijk, CBS, 1909.

G. 458 = ATCC 9376; 10063 = IMI 51983. Received from Oshima, from Takamine's plant.

G. 506 = ATCC 1010 = CBS 574.65 = IMI 16142, 124935 = Thom 130 = WB 506. Received as *Aspergillus oryzae* var. *effusus* from B. F. Lutman, 1910. Isolated from milk.

G. 1958. Received as *Aspergillus oryzae* var. *effusus* from Chemical Warfare Service, 1944. Isolated from optical mask, S. Pacific.

Aspergillus parasiticus Speare

G. 502 = ATCC 1018, 6474, 7865 = CBS 103.13 = IMI 15957 = Thom 3509 = WB 502. Ex lectotype of *Aspergillus parasiticus*. From Speare, Hawaii, 1913. Isolated from mealy bug on sugarcane.

G. 4123. Received as *Aspergillus parasiticus* WB 4123, 1966. Isolated from toxic grain.

G. 6433. Received as RB 274 from D. T. Wicklow, 1978. Isolated from corn, N. Carolina.

G. 1988 = ATCC 9362 = CBS 133.52 = IMI 87159 = WB 1988. Received as *A. oryzae* for sauce, from King, Chungking, China. Used for making soy sauce. Identified as *A. sojae*.

G. 425 = Thom 4640.478. Received as *Aspergillus terricola* , 1922.

G. 426 = ATCC 16860 = CBS 579.65 = IMI 172294 = WB 426. Ex neotype of *Aspergillus terricola*. Origin unknown.

G. 424 = ATCC 1014; 16863 = CBS 580.65= IMI 16127 = Thom 4838 = WB 424. Ex lectotype of *Aspergillus terricola* var. *americana*. Received 1925. Isolated from soil, Georgia.

Aspergillus tamarii Kita

H. 20818 = CBS 104.13. Ex lectotype of *A. tamarii*. Received as *Aspergillus tamarii* QM 9374, 1991. Isolated from activated carbon.

H. 4911 = ATCC 16864 = CBS 484.65 = IMI 124938. Received as WB 4911, 1965. Isolated as culture contaminant. Ex type culture of *A. flavofurcatis*.

Aspergillus caelatus B. W. Horn

I. 25528. Ex type. Isolated from soil, Georgia, by B. W. Horn, 1982.

Aspergillus nomius Kurtzman, Hesseltine & Horn

J. 13137. Ex type of *A. nomius*. Isolated from moldy wheat, 1964.

J. 3161. Received as *Aspergillus flavus* from A. C. Keyl, 1965. Isolated from *Cycas circinalis*, Guam.

J. 3353. Received as #126 from D. Shemanuki, Univ. of Wyoming, 1965. Isolated from diseased alkali bees.

J. 6552 = ATCC 96015. Received as C-641 from C. R. Benjamin, Beltsville, MD, 1967. Isolated from pine sawfly, Wisconsin.

Petromyces albertensis Tewari

K. 20602 = UAMH 2976. Received as ATCC 58745, 1990. Ex type. Isolated from human ear, Alberta, Canada.

Petromyces alliaceus Malloch & Cain

L. 4181 = ATCC 16891 = CBS 542.65 = IMI 126711. Ex type. Received as WB 4181, 1965. Isolated from Australian soil by J. H. Warcup as SA 117.

L. 315 = ATCC 10060 = CBS 536.65 = IMI 51982 = Thom 4656 = WB 315. Received from M. M. High, 1922. Isolated from blister beetle.

L. 5108 = CBS 612.78. Received as WB 5108, 1969.

L. 3648 = CBS 650.74 = WB 5347. Ex lectotype of *Aspergillus lanosus*. Received as IMI 130727, 1969. Isolated from teak forest soil, India.

Aspergillus leporis States & M. Christensen

M. 3216 = ATCC 16490 = CBS 151.66 = WB 5188. Ex type of *A. leporis*. Received as RMF 99 from M. Christensen, Univ. of Wyoming, 1966. Isolated from dung of *Lepus townsendii*, Wyoming.

M. 6599 = ATCC 44565. Received as O-168 from M. Christensen, Univ. of Wyoming, 1981. Isolated from *Artemisia* grasslands soil, Wyoming. Variant producing "blond" sclerotia.

Aspergillus avenaceus G. Smith

N. 4517. Ex type of *A. avenaceus*. Received as WB 4517, 1969. Isolated from California soil, by J. Cavender.

N. 517 = Thom 5725. Received from G. Smith, as isolate BB155, 1940.

Aspergillus ochraceus Wilhelm

O. 398 = ATCC 1008 = CBS 108.08 = IMI 16247 = Thom 112 = WB 398. Ex type of *A. ochraceus*. Received from Westerdijk, CBS, 1909.

O. 419 = CBS 624.78 = IMI 16265 = Thom 4640.476. Received from the Bainier Collection via Da Fonseca, 1922.

O. 4752 = ATCC 12066 = CBS 123.55 = IMI 211804. Received as WB 4752, 1969. Isolated from scalp lesion. Type of *Aspergillus ochraceopetaliformis*, Batista's # 270.

O. 420 = ATCC 16887 = CBS 103.07 = IMI 15960 = Thom 4724.35 = WB 420. Ex neotype of *Aspergillus ostianus*. Received from H. Raistrick, 1924.

O. 422 = CBS 627.78 = Thom 4640.471 = WB 422. Received as *A. ostianus* from Bainier collection via Da Fonseca, 1922.

O. 423 = CBS 101.23 = Thom 4876.1 = WB 423; 4762. Received as *A. ostianus* from Westerdijk, CBS, 1926.

O. 416 = ATCC 12337 = CBS 628.78 = Thom 4291.6 = WB 416. Received as *A. petrakii* from Hanzawa, 1918. Isolated from katsuobushi (fermented fish).

O. 4369 = ATCC 16885 = CBS 105.57 = IMI 172291 =WB 4777. Ex lectotype of *Aspergillus petrakii*. Received as WB 4369, 1965. Isolated from *Leptinotarsa* moths, Hungary.

O. 4748 = CBS 115.51. Received as *A. petrakii* WB 4748, 1969.

O. 4789 = CBS 640.78. Received as *A. petrakii* WB 4789, 1969.

O. 394 = CBS 622.78 = Thom 5667.446 = WB 394. Received from Bliss, 1939. Isolated from dates, CA.

O. 5103 = ATCC 16889 = CBS 546.65. Ex neotype of *A. melleus*. Received as WB 5103, 1965. Isolated from soil, India.

Aspergillus elegans Gasperini

P. 4850 = ATCC 13829, 16886 = CBS 102.14, 543.65 = IMI 133962. Ex neotype of *A. elegans*. Received as WB 4850, 1965.

P. 407 = CBS 310.70; 614.78 =Thom 5400.1 = WB 4813. Received as *Aspergillus rehmii* from Westerdijk, CBS, 1933.

P. 4820 = CBS 615.78. Received as WB 4820, 1969.

325

Aspergillus sclerotiorum Huber
Q. 415 = ATCC 16892 = CBS 549.65 = IMI 56673 = Thom 5351 = WB 415. Ex lectotype of *A. sclerotiorum*. Isolated from apple.
Q. 4482 = CBS 631.78. Received as WB 4482, 1965.
Q. 4901 = CBS 632.78. Received as WB 4901, 1965.
Q. 5584 = CBS 385.75 = WB 5280. Received as WB 5280, 1973. Isolated from soil, India. Ex type of *Aspergillus sulphureus* var. *crassus*.
Q. 13077 = ATCC 46856. Received as RMF 7132 from M. Christensen, Univ. of Wyoming, 1982. Isolated from soil, Nebraska.

Aspergillus auricomus (Guégen) Saito
R. 388 = CBS 638.78 = Thom 4754.C76. Received from D. H. Linder, 1925.
R. 389 = CBS 639.78 = Thom 5402.3 = WB 389. Received from Biourge, 1933.
R. 391 = ATCC 16890 = CBS 467.65 = IMI 172277 = Thom 5479.A41= WB 391. Received from Biourge, 1935.
R. 397 = CBS 613.78 = Thom 5479.A42 = WB 397. Received as *Aspergillus vitellinus* from G. Smith, LSHTM, 1935.

Aspergillus insulicola Montemayor & Santiago
S. 6138 = ATCC 26220 = CBS 382.75. Ex lectotype of *A. insulicola*. Received as T1 from I.. Montemayor, Caracas, Venezuela 1975. Isolated from Venezuelan soil.

Aspergillus bridgeri Christensen
T. 4565 = CBS 625.78. Received as *A. ochraceus* WB 4565, 1969. Isolated from desert grass, Haiti, by W. Scott.
T. 13078 = ATCC 46854. Received as RMF 7127 from M. Christensen, Univ. of Wyoming, 1982. Isolated from soil, Nebraska.
U. 13000 = ATCC 44562 = CBS 350.81 = IMI 259098. Ex type. Received as JB 26-2 from M. Christensen, Univ. of Wyoming, 1981. Isolated from soil under *Atriplex*, WY.

Aspergillus sulphureus (Fres.) Thom & Church
V. 4077 = ATCC 16893 = CBS 550.65 = IMI 211397. Ex neotype of *A. sulphureus*. Received as WB 4077, 1965. Isolated from soil, India.
V. 6161. Received as *A. fresenii* ATCC 18413, 1969. Isolated from shelled Brazil nuts, Canada.

Aspergillus robustus Christensen & Raper
W. 6362= ATCC 36106 = CBS 428.77 = IMI 216610. Ex type. Received as WB 5286 from M. Christensen, Univ. of Wyoming, 1977. Isolated from thorn forest soil, Kenya.

Fennellia flavipes Wiley & Simmons
X. 302 = ATCC 24487 = IMI 171885 = Thom 4640.474 = WB 302. Ex lectotype of *F. flavipes*. Received as *Sterigmatocystis flavipes* from Bainier Collection via Da Fonseca 1922.
Y. 295 = ATCC 16814 = CBS 585.65 = IMI 135422 = Thom 5335.205 = WB 295. From H. Macy, St. Paul, MN, 1933.
Y. 4578 = ATCC 16805 = CBS 586.65 = IMI 135423. Received as *A. flavipes* WB 4578, 1965. Isolated from Haitian soil by Wm. Scott.
Z. 4263. Received as *A. flavipes* WB 4263, 1968. Isolated from Indian soil by B. K. Bakshi.
Z. 3750 = CBS 541.69. Received as *Aspergillus iizukae* IMI 141552, 1970. Isolated from stratigraphic drilling core Japan. Type of *Aspergillus iizukae*.

Fennellia nivea (Wiley & Simmons) Samson
AA. 515 = ATCC 12276 = CBS 114.33 = IMI 135425 = Thom 5402.1 = WB 515. Received as *A. eburneus* from Biourge 1933.
AA. 1955 = ATCC 56745 = WB 1955 from M. Timonin, Ottawa, CA.
AA. 4751. Received as *A. niveus*, WB 4751, 1969. Type of *Aspergillus niveus* var. *bifida*
AB. 6134. Received as *A. niveus* var. *indicus* CBS 444.75, 1975. Isolated from Indian soil. Type of *Aspergillus niveus* var. *indicus*.
AB.1923 = ATCC 16793 = CBS 503.65 = IMI 82431 =WB 1923. Ex lectotype of *A. terreus* var. *aureus* from soil, Greeneville, TX.

Aspergillus carneus Blochwitz
AC. 298 = Thom 5714.111. Received as *Aspergillus flavipes* from H. Fellows, Manhattan, KS, 1940. Isolated from soil, Kansas.
AC. 527 = ATCC 16798 = CBS 494.65 = IMI 135818 = Thom 5740.4 =WB 527. Ex neotype of *A. carneus*. NRRL isolate, 1940. Isolated from air contaminant on agar plate.
AD. 1928. Received as ARK A3, 1942. Isolated from soil from Fayettville, AR.
AE. 4610. Received as *A. carneus* WB 4610, 1969. Isolated from Haitian soil by Wm. Scott.
AF. 6326. Received as *A. aureofulgens* CBS 653.74, 1977. Isolated from truffle soil, France. Type of *Aspergillus aureofulgens*.

Aspergillus terreus Thom
AC. 260 = Thom 5323.5042. Received from J. J. Taubenhaus, College St., TX, 1933. Isolated from Texas soil.
AC. 680 = ATCC 16794 = CBS 594.65 = IMI 135817 = WB 680. Received from G. A. Ledingham.
AC. 1913 = ATCC 46533. Received from C. W. Emmons NIH, Bethesda MD, 1941. Isolated from lung of pocket mouse, AZ.
AC. 2399 = ATCC 16792 = CBS 130.55 = IMI 61457 = QM 1913 = WB 2399. Ex lectotype of *Aspergillus terreus* var. *africanus*. Received 1950. Isolated from soil, Gold Coast.
AC. 4609. Received as *Aspergillus terreus* var. *africanus* WB 4609, 1969. Isolated from Panamanian pasture soil by W. Scott as 6A.
AG. 4017 = WB 4017, Received from D. Fennell, WBC, 1969. Isolated from Argentine soil.

Aspergillus janus Raper & Thom
AH. 1787 = ATCC 16835 = CBS 118.45 = IMI 16065 = WB 1787. Ex lectotype of *A. janus*. Received 1941. Isolated from soil, Panama.
AI. 1936 = WB 1936. Received 1942. Isolated from soil, Panama.

Aspergillus janus var. *brevis* Raper & Thom
AJ. 1935 = ATCC 16828 = CBS 111.46 = IMI 16066 = WB 1935. Ex lectotype. Received 1942. Isolated from soil, Mexico City, Mexico.

Aspergillus allahabadii Mehrotra & Agnihotrti
AK. 4101 = WB 4101. Received 1969. Isolated from soil, San Salvador.
AK. 4539 = ATCC 15055 = CBS 164.63 = IMI 139273. Ex lectotype. Received as WB 4539, 1965. Isolated byB. S. Mehrotra from soil, India.

Aspergillus ambiguus Sappa

AL. 4737 = ATCC 16827 = CBS 117.58 = IMI 139274. Ex lectotype. Received as WB 4737, 1965. Isolated from savannah soil, Somalia.

Aspergillus microcysticus Sappa

AM. 4749 = ATCC 16826 = CBS 120.58 = IMI 139275. Ex lectotype. Received as WB 4749, 1965. Isolated from savannah soil, Somalia.

Chaetosartorya chrysella (Kwon & Fennell) Subramanian

AN. 5084 = ATCC 16852 = CBS 472.65 = IMI 238612. Ex type. Isolated from forest soil, Costa Rica, 1963.

AN. 5085. Received as WB 5085, 1969. Isolated from forest soil, Costa Rica.

Chaetosartorya cremea (Kwon & Fennell) Subramanian

AO. 4518. Received as WB 4518, 1969. Isolated from soil, Panama, by J. H. Warcup.

AO. 5081 = ATCC 16857 = CBS 477.65 = IMI 123749. Ex lectotype. Received as WB 5081, 1965. Isolated from Costa Rican forest soil.

Chaetosartorya stromatoides Wiley & Simmons

AP. 5501 = ATCC 24480 = CBS 265.73 = IMI 171880. Ex type of as *A. stromatoides*. Received as QM 8944, 1972. Isolated from forest soil, Thailand.

Aspergillus stromatoides Raper & Fennell

AQ. 4519 = ATCC 16854; 24485 = CBS 500.65 = IMI 123750. Ex lectotype culture of *A. stromatoides*. Received as WB 4519, 1965. Isolated from soil, Panama.

Aspergillus itaconicus Kinoshita

AR. 161 = ATCC 10021 = CBS 115.32 = IMI 16119 = Thom 5660.48 =WB 161. Ex lectotype. Received from G. Smith, LSHTM, 1939.

Aspergillus flaschentraegeri Stolk

AS. 5042 = ATCC 15535 = CBS 108.63 = IMI 101651. Ex lectotype. Received as *A. flaschentraegeri* WB 5042, 1965. Isolated from gut of *Prodenia litura* (cutworm).

Aspergillus pulvinus Kwon & Fennell

AT. 5078 = ATCC 16842 = CBS 578.65 = IMI 139628. Ex lectotype. Received as WB 5078, 1965. Isolated from forest soil, Costa Rica.

Aspergillus wentii Wehmer

AU. 375 = ATCC 1023 = CBS 104.07 = IMI 17295 = Thom 116 = WB 375. Ex neotype. Received 1909.

AU. 377 = Thom 4230. Received 1917. Isolated from palm nut, Brazil.

AU. 378 = Thom 4291.16. Received 1918. Isolated from katsuobushi (fermented fish), Japan.

AU. 382 = CBS 121.32 = Thom 5346. Received 1933.

AU. 3650 = CBS 649.74. Ex lectotype of *Aspergillus dimorphicus*. Received as IMI 131553, 1969. Isolated from garden soil, India.

Aspergillus gorakhpurensis Kamal & Bhargava

AV. 3649 = CBS 648.74 = WB 5346. Ex type of *A. gorakhpurensis*. Received as IMI 130728, 1969. Isolated from teak forest soil, India.

Aspergillus candidus Link

AW. 303 = ATCC 1002 = CBS 566.65 = IMI 91889 = Thom 106 = WB 303. Ex neotype of *A. candidus*. Received 1909, from Westerdijk, CBS.

AX. 313 = IMI 15962 = Thom 4337 = WB 313. Received as *Aspergillus okazaki*, 1919.

Aspergillus campestris Christensen

AY. 13001 = ATCC 44563 = CBS 348.81 = IMI 259099. Ex type. Received as ST 2-1 from M. Christensen, Univ. of Wyoming, 1981. Isolated from native mixed prairie soil, North Dakota.

AY. 312 = ATCC 16871 = CBS 567.65 = Thom 5695.481D = WB 312. From Reis, Instituto Biologico, Brazil, 1939, as *A. candidus*.

AY. 4646. Received as *A. candidus* WB 4646, 1969. Isolated from barn litter.

AY. 4809 = ATCC 11380 = IFO 4310. Received as *A. candidus* WB 4809, 1969.

Aspergillus peyronelii Sappa

AZ. 4899 = ATCC 16831 = CBS 572.65 = IMI 139272 = WB 4207. Received as WB 4899, 1965. Isolated from paint, West Indies.

AZ. 4754. Ex lectotype of *A. peyronelli.*

Aspergillus arenarius Raper & Fennell

BA. 5012 = ATCC 16830 = CBS 463.65 = IMI 55632 = WB 4429. Ex lectotype. Received as WB 5012, 1965. Isolated from soil, India.

Eurotium herbariorum (Wiggers : Fries) Link

BB. 71 = Thom 5600A. Received from V. K. Charles, Maryland, 1937. Isolated from leafhoppers. Conidial and cleistothecial variant, misidentified as *E. rubrum.*

BB. 114 = ATCC 10076 = IMI 211808 = Thom 5619.19. Received from O. W. Richards, NY, 1937. Isolated at Woods Hole Biological Station, Massachusetts.

BB. 116 = ATCC 16469 = CBS 516.65 = Thom 5629.C = WB 116. Ex type culture of *Eurotium herbariorum*. Received as *Aspergillus minor*, 1937, isolated from unpainted boards, K. B. Raper's basement.

BB. 117 = Thom 5629.D = WB 117. Received as *Aspergillus minor* from 1937. Isolated from unpainted boards, K. B. Raper's basement.

Eurotium repens de Bary

BC. 13 = ATCC 9294 = CBS 529.65 = IMI 16114 = Thom 4640.404; 5612.AC61; and 5632.3776 = WB 13. Ex neotype. Received as *Aspergillus mollis* from the Bainier collection via Da Fonseca 1922.

BC. 17 = ATCC 10079 = Thom 5305.5 = WB 17. Received from U. S. Army Medical School, 1932. Isolated from human wrist.

BC. 40 = ATCC 10066 = CBS 123.28 = IMI 16122 = Thom 5343 = WB 40. Ex lectotype culture of *Eurotium pseudoglaucum*. Received from CBS 1933.

Eurotium tonophilum Ohtsuki

BD. 5124 = ATCC 16440, 36504 = CBS 405.65 = IMI 108299. Received as WB 5124, 1965. Ex lectotype. Isolated from binocular lens.

Eurotium rubrum König *et al.*

BE. 52 = ATCC 16441 = CBS 530.65 = Thom 5599B = WB 52. Ex neotype. Received from U.S. Army Medical School, 1937.

BE. 76 = IMI 91868 = Thom 5479.A31. Received from G. Smith LSHTM, 1935. Ex type culture of *Aspergillus lovainensis.*

BE. 5000 = ATCC 16923 = CBS 464.65 = IMI 32048 = WB 5000. Received as WB 5000, 1965. Isolated from incubated coffee beans. Ex lectotype of *Edyuillia athecia.*

Eurotium chevalieri Mangin

BF. 78 = ATCC 16443 = CBS 522.65 = IMI 211382 = Thom 4125.3 = WB 78. Ex neotype. Received 1916.

BF. 79 = Thom 5061. Received, 1929. Isolated in Indiana.

BF. 82 = ATCC 16444 = CBS 523.65 = IMI 89278 = Thom 5612.107 = WB 82. Received from G. Smith, LSHTM, 1937. Ex type culture of *Aspergillus chevalieri* var. *intermedius.*

BF. 4755. Received as WB 4755, 1969. Isolated as culture contaminant.

BF. 4817. Received as WB 4817, 1969. Ex type of *Aspergillus chevalieri* var. *ruber*.

Eurotium amstelodami Mangin

BG. 90 = ATCC 16464 = CBS 518.65 = Thom 126 = WB 90. Ex neotype.

BG. 89 = ATCC 10065 = IMI 211806 = Thom 109 = WB 89

BG. 4716. Received as WB 4716, 1968. Isolated from candied grapefruit rind.

BG. 108 = ATCC 10077 = CBS 491.65 = IMI 172290 = Thom 5290 = Thom 5633.24 = WB 108. Ex type culture of *A. montevidense*. Received from Talice and MacKinnon, 1932. Isolated from tympanic membrane of human ear.

BG. 4222 = ATCC 16468 = CBS 123.53 = IMI 172280 = WB 4222. Ex lectotype culture of *Eurotium cristatum*. Received as WB 4222, 1965. Isolated from S. Africa by H. Swart as #168.

Eurotium umbrosum Bain. & Sart.

BH. 120 = ATCC 16925 = CBS 117.46; 532.65 = Thom 4803.1914; 5633.9 = WB 120. Received from F. A. McCormick, 1925.

BH. 121 = Thom 5612.A37. Received from G. Smith LSHTM, 1937.

BH. 126 = ATCC 16924 = CBS 471.65 = IMI 172279 =Thom 5612.A32 = WB 126. Ex lectotype culture of *Eurotium carnoyi*. Received from G. Smith, LSHTM, 1937.

Eurotium echinulatum Delacroix

BI. 131 = ATCC 1021 = CBS 112.26; 524.65 = IMI 211378 = Thom 4481, 5633.4 = WB 131. Ex neotype. Received as *Eurotium verruculosum* from E. H. Phillips, Fresno, CA, 1921. Isolated from figs.

BI. 133 = Thom 5612.A28. Received from G. Smith LSHTM, 1937. Produces no ascomata.

BI. 124 = ATCC 1036 = CBS 113.27 = IMI 29188 =Thom 4724.45; 5612.AC45; 5633.8 = WB 124. Ex type culture of *E. medium*. Received G. Smith LSHTM, 1937.

Eurotium halophilicum C. M. Christensen *et al.*,

BJ. 2739 = ATCC 16401 = CBS 122.62 = IMI 211802 = WB 4679. Ex lectotype. Received from C. M. Christensen, Univ. of Minnesota, 1957. Isolated from stored wheat seeds.

Eurotium niveoglaucum (C. Thom & Raper) Malloch & Cain

BK. 127 = ATCC 10075 = CBS 114.27, 517.65 =IMI 32050 = Thom 5612.A16; 5633.7; 7053.2 = WB 127, 130. Ex lectotype. Received from M. B. Church. Ex type culture of *Aspergillus glaucus* mut. *albus*.

BK. 128 = Thom 5479.A35. Received from G. Smith LSHTM, 1935.

BK. 136 = Thom 5633.6. Received from CBS, 1938. .

BK. 137 = Thom 5479.A36. Received as *Aspergillus mongolicus*, from G. Smith, LSHTM, 1935. Biourge's type of *Aspergillus mongolicus*.

Aspergillus proliferans Smith

BL. 1908 = ATCC 16922 = CBS 121.45 = IMI 16105 = WB 1908. Ex lectotype. Received from G. Smith LSHTM, 1943. Isolated from cotton fabric.

Eurotium leucocarpum Hadlok & Stolk

BM. 3497 = CBS 353.68. Ex type. Received from CBS, 1969. Isolated from dried raw sausage, Germany.

Eurotium xerophilum Samson & Mouchacca

BN. 6131. Ex type. Received as CBS 938.73, 1975. Isolated from desert sand, Egypt.

BN. 6132. Received as CBS 755.74, 1975. Isolated from desert soil, Egypt.

Aspergillus caesiellus Saito

BO. 5061 = ATCC 11905 = CBS 470.65 = IMI 172278. Ex lectotype. Received as WB 5061, 1965.

Aspergillus conicus Blochwitz

BP. 149 = ATCC 16908 = CBS 475.65 = IMI 172281 = Thom 4733.701= WB 149. Ex neotype. Received from Biourge, 1924.

** 4661 = WB 4661. Received from D. Fennell, 1968.

Aspergillus restrictus Smith

BQ. 145 = Thom 4246. Received from Piper, BPI, USDA, MD, 1917.

BQ. 148 = CBS 118.33 = IMI 16268 = Thom 5367.2, 5660.83 −WB 148. Received from G. Smith, LSHTM, 1933. Type strain *A. restrictus* type B.

BQ. 151 = Thom N.O.7. Received as *Aspergillus penicillioides*. Isolated in Louisiana.

BQ. 154 = ATCC 16912 = CBS 117.33, 541.65 = IMI 16267 = Thom 5660.93 = WB 154. Ex lectotype. Received as *Aspergillus restrictus* from G. Smith, LSHTM, 1939. Clinical isolate.

BQ. 4783 = CBS 331.59 = IMI 68226. Received as WB 4783, 1968. Type of *Penicillium fuscoflavum*.

Aspergillus gracilis Bainier

BR. 4962 = ATCC 16906 = CBS 539.65 = IMI 211393. Ex neotype. Received as WB 4962, 1965. Isolated from gun-firing mechanism.

Aspergillus penicillioides Spegazzini

BS. 4548 = ATCC 16910 = CBS 540.65 = IMI 211342. Ex neotype. Received as WB 4548, 1965. Clinical isolate from a case of lobomycosis.

BS. 4550. Received as WB 4550, 1968.

BT. 5125 = ATCC 16905 = 36505 = IMI 108298. Received as WB 5125, 1968. Isolated from binocular lens. Type of *Aspergillus vitricolae*.

Aspergillus cervinus Massee

BU. 3157=ATCC 15508 = CBS 196.64 = IMI 107684 = WB 5026. Received as WT 540 from M. Christensen, Univ. of Wyoming, 1964. Isolated from Malaysian soil by M. Christensen.

BU. 5025=ATCC 16915=CBS 537.65=IMI 126542. Ex type. Received as *A. cervinus* WB 5025, 1965. Isolated from Malaysian soil.

BU. 4220. Received as *A. kanagawaënsis* WB 4220, 1969. Isolated from, coniferous forest soil, Wisconsin.

BU. 5023 = CBS 412.64. Received as *A. kanagawaënsis* WB 5023, 1969. Isolated from soil under *Pinus banksiana*, Wisconsin.

BU. 2667 = WB 5028. Received as *A. parvulus* 1957. Isolated from Georgia soil.

BU. 4753. Ex type of *A. parvulus*. Received 1965. Isolated from pine and sweetgum forest soil, S. Carolina.

BU. 4994 = WB 4994. Received as *A. parvulus* 1969. Isolated from scotch pine forest soil, U.K.

BU. 4364. Ex type of *A. nutans*. Received 1965. Isolated from Australian soil.

Aspergillus kanagawaënsis Nehira

BV. 2161. Received as 17-15 from J. W. Warcup, Cambridge Univ., 1948. Isolated from beech forest soil, New Zealand.

BV. 5027. Received as *A. kanagawaënsis* WB 5027, 1969. Isolated from Wisconsin pine soil by M. Christensen as WT 813.

BW. 4774. Received as *A. kanagawaënsis* WB 4774, 1965. Isolated from soil under *Pinus banksiana*, Wisconsin.

Aspergillus nutans McLennan & Ducker

BX. 4897 = CBS 122.56 = WB 4201; 4897. Received 1969 as *A. nutans*. Isolated from soil, South Africa.

Aspergillus zonatus Kwon & Fennell

BY. 5079 = ATCC 16867 = CBS 506.65 = IMI 124936. Ex lectotype. Received as *Aspergillus zonatus* WB 5079, 1965. Isolated from forest soil, Costa Rica.

Aspergillus clavatoflavus Raper & Fennell

BZ. 5113 = ATCC 16866 = CBS 473.65 = IMI 124937. Ex lectotype. Received as *Aspergillus clavatoflavus* WB 5113, 1965. Isolated from forest soil by J. H. Warcup as a 186/1, Australia.

Warcupiella spinulosa (Warcup) Subramanian

CA. 4376 = ATCC 16919 = CBS 512.65 = IMI 75885 = IMI 238611. Ex lectotype. Received as WB 4376, 1965. Isolated from jungle soil, Borneo, by J. H. Warcup as A 41/4.

Penicilliopsis clavariiformis Solms-Lubach

CB. 2482. Received as from CBS 1955. Isolated from *Diospyros celebica* seed, Indonesia.

Sclerocleista ornata (Raper, Fennell & Tresner) Subramanian

CC. 2256 = ATCC 16921 = CBS 124.53 = IMI 55295 = WB 2256. Ex lectotype. Received as 437 from M. P. Backus, Univ. of Wisconsin, 1951. Isolated by H. Tresner from soil in oak woods, Wisconsin.

CC. 2291 = CBS 385.53 = IMI 63918 = WB 2291. Received from M. P. Backus, Univ. of Wisconsin, 1951. Isolated from woodland soil by H. Tresner.

CD. 4225. Received as WB 4225, 1965. Isolated from forest soil, MA.

CD. 4735. Received as WB 4735, 1969. Isolated frompeat soil, Wisconsin.

Aspergillus speluneus Raper & Fennell

CE. 4990. Received as WB 4990, 1969. Isolated from cave soil, W. Virginia.

Aspergillus asperescens Stolk

CF. 4738. Isolated from bat dung, Poland.

CF. 4770. Ex type. Received as WB 4770, 1965. Isolated from cave soil, U.K.

CF. 5036. Received as WB 5036, 1969. Isolated from cave soil, as Brian 517, U. K.

Aspergillus aureolatus Munt.-Cvet. & Bata

CG. 5126. Ex type. Received as WB 5126, 1965. Isolated from air, Yugoslavia.

Emericella desertorum Samson & Mouchacca

CH. 5921. Ex type. Received as CBS 653.73, 1974. Isolated from grey soil near Kharga Oasis, Egypt.

Aspergillus egyptiacus Moubasher & Moustafa

CI. 5920 = ATCC 32114 = CBS 656.73 = IMI 141415. Ex lectotype. Received as CBS 656.73, 1974. Isolated from sandy soil, olive tree plantation, Egypt.

Emericella nidulans (Eidam.) Vuill.

CJ. 187 = ATCC 10074 = CBS 589.65 = IMI 86806 =Thom 4640.5 = WB 187. Ex neotype. Received from Bainier Collection via Da Fonseca, 1922.

CJ. 2241. Isolated from soil, Gold Coast, 1950.

CJ. 4266. Received as WB 4266, 1969.

CJ. 206 = ATCC 16820 = CBS 133.60 = IMI 136775 = Thom 4138.T11= WB 206. Ex lectotype of *Emericella*

rugulosa. Received from S. Waksman, 1916. Isolated from New Jersey soil.

CJ. 4581. Received as *Emericella rugulosa* WB 4581from D. Fennell, WBC 1969.

CJ. 2394 = ATCC 16839 = CBS 119.55 = IMI 61453 = WB 2394. Ex lectotype of *Emericella nidulans* var. *acristata*. Received as 62X from M. E. Sorte, Ohio, 1952. Clinical isolate, New Mexico.

CJ. 4904 = ATCC 16822 = CBS 493.65 = IMI 139280. Ex lectotype of *Emericella parvathecia*. Received as WB 4904, 1965. Isolated from skin by G F ORR as O-326.

CJ. A-9919. Received as *Emericella nidulans* var. *acristata* BR38, from S. A. Lohhi, Punjab Univ., Lahore, Pakistan, 1960. Isolated from soil.

CJ. 201 = ATCC 16816 = CBS 591.65 = IMI 89351 = Thom 4138.N8 = WB 201. Ex lectotype of *Emericella quadrilineata*. Received from S. Waksman, 1916. Isolated from New Jersey soil.

CJ. 4992. Received as *Emericella quadrilineata* WB 4992, 1969. Isolated from soil, NSW, Australia.

CJ. 2240 = ATCC 16813 = CBS 138.55 = IFO 8106 = WB 2240. Ex lectotype of *Emericella violacea*. Isolated from Gold Coast soil, 1950.

CJ. 4178. Received as *Emericella violacea* WB 4178, 1969. Isolated from Australian soil by J. H. Warcup as SA 116.

CJ. 4908 = ATCC 16829 = CBS 114.63 = IMI 126693. Ex lectotype of *Emericella dentata*. Received as WB 4908, 1965. Isolated from fingernail, India.

Emericella echinulata (Fennell & Raper) Hori

CK. 2395 = ATCC 16825 = CBS 120.55 = IMI 61454 = WB 2395. Ex lectotype. Received as M-354 from J. Winitzky, Argentina, 1950. Isolated from soil, Argentina.

CK. 200 = ATCC 1000, 16848 = CBS 492.65 = IMI 74181 = Thom 110 = WB 200. Received as *E. nidulans* var. *lata* from Westerdijk, CBS, 1909.

Emericella striata (Rai *et al.*,) Malloch & Cain

CL. 4699 = ATCC 16815 = CBS 283.67; 592.65 = IMI 96679. Ex lectotype. Received as WB 4699, 1965. Isolated from mangrove mud, India.

Emericella fruticulosa (Raper & Fennell) Malloch & Cain

CM. 4903 = ATCC 16823 = CBS 486.65 = IMI 139279. Ex lectotype. Received as WB 4903, 1965. Isolated from California soil by G. F. Orr as O-1077.

Aspergillus caespitosus Raper & Thom

CN. 1929 = ATCC 11256 = CBS 103.45 = IMI 16034 = WB 1929. Ex lectotype. Received 1942. Isolated from soil, Fayetteville, AR.

Aspergillus unguis (Emile-Weil & Gaudin) Thom & Raper

CO. 216 = ATCC 10073 = CBS 595.65 = IMI 136525 = Thom 5076.1 = WB 216. Ex lectotype. Received from R. Ottenberg, NY, 1929.

CO. 5041 = CBS 118.37. Received as WB 5041, 1969. Type of *Aspergillus laokiashanensis*.

CO. 6328 = ATCC 24715. Received as CBS 652.74, 1977. Isolated from apple fruit and leaves, USSR.

Emericella heterothallica (Kwon, Fennell & Raper) Malloch & Cain

CP. 5096 = ATCC 16847 = CBS 488.65 = IMI 139277. Ex type. Received as WB 5096, 1965. Isolated from soil, Costa Rica. Mating type A (orange).

CP. 5097 = ATCC 16824 = CBS 489.65 = IMI 139278. Ex type. Received as WB 5097, 1965. Isolated from soil, Costa Rica. Mating type alpha (yellow).

Aspergillus aeneus Sappa
CQ. 4649 = IMI 86833. Received as WB 4649, 1969. Isolated from soil, India.
CR. 4769 =ATCC 16803 = CBS 128.54 = IMI 69855 = WB 4279. Ex lectotype. Received as WB 4769, 1965. Isolated from forest soil, Somalia.

Aspergillus eburneocremeus Sappa
CS. 4773 = ATCC 16802 = CBS 130.54 = IMI 69856. Ex lectotype. Received as WB 4773, 1965. Isolated from forest soil, Somalia.

Aspergillus crustosus Raper & Fennell
CT. 4988 = ATCC 16806 = CBS 478.65 = IMI 135819. Ex lectotype. Received as WB 4988, 1965. Isolated from skin scrapings, IL.

Aspergillus multicolor Sappa
CU. 4775 = ATCC 16804 = CBS 133.54 = IMI 69857 = WB 4281. Ex lectotype. Received as WB 4775, 1965. Isolated from forest soil, Somalia.

Emericella spectabilis Christensen
CV. 6363 = ATCC 36105 = CBS 429.77 = IMI 216611. Ex type. Received as RMFH 429 from M. Christensen, Univ of Wyoming, 1977. Isolated from coal mine spoils, WY.

Emericella bicolor M. Christensen & States
CW. 6364 = ATCC 36104 = CBS 425.77 = IMI 216612. Ex type. Received as RMF 2058 from M. Christensen, Univ. of Wyoming, 1977. Isolated from grassland-*Artemisia* soil, WY.

Emericella variecolor Berkeley & Broome
CX. 212 = ATCC 10067 = CBS 597.65 = IMI 136777 = Thom 5602.3 = WB 212. Received from Verona, Italy,1937. Isolated from olive fruits.
CY. 1858 = ATCC 16819 = CBS 598.65 = IMI 136778 = WB 1858. Ex type. Received as Coghill C2, 1942. Isolated from soil, Panama.
CZ. 1954 = ATCC 14883 = WB 1954. Isolated from soil, Tucson, AZ.
CZ. 4736. Received as WB 4736, 1969. Isolated by M. Christensen from baobob forest soil. Rhodesia.

Emericella astellata (Fennell & Raper) Horie
DA. 2396 = ATCC 16817 = CBS 134.55 = IMI 61455 = WB 2396. Ex lectotype. Received as 6273 from G. W. Martin, Univ. of Iowa, 1947. Isolated from leaf, Baltra Island, Galapagos.
DA. 2397 = CBS 135.55 = WB 2397. Received as 6368 from G. W. Martin, Univ. of Iowa, 1948. Isolated from leaf, S. Seymour Island, Galapagos.

Aspergillus versicolor (Vuill.) Tiraboschi
DB. 227 = ATCC 16853 = CBS 599.65 = Thom 3555.21 = WB 227. Received 1915. Isolated from soil, New Jersey.
DC. 238 = ATCC 9577 = CBS 583.65 = Thom 5519.57 = WB 238. Ex neotype. Received 1935.
DC. 239 = ATCC 16856 = CBS 584.65 = Thom 5667.506 = WB 239. Received 1939. Isolated from dates, California.
DB. 4642. Received as WB 4642, 1969.
DC. 4791. Received as WB 4791, 1969. Isolated from tobacco. Type of *Aspergillus tabacinus*.
DC. 4838. Received 1965. Isolated from *Berberis* sp. fruits. Type of *Aspergillus amoenus*.

Aspergillus sydowii (Bain. & Sart.) Thom & Church
DD. 254 = ATCC 16844 = CBS 593.65 = IMI 211384 = Thom 5706.45590 = WB 254. Ex neotype. Received 1940. Clinical isolate.

DE. 250 = Thom 5119. Received 1930.
DE. 4768. Received as WB 4768, 1969. Isolated from soil by G. F. Orr, California.

Aspergillus granulosus Raper & Thom
DF. 1931 = CBS 119.58 = WB 1931. Received 1942. Isolated from soil, Greenville, TX.
DF. 1932 = ATCC 16837 = CBS 588.65 = IMI 17278 = WB 1932. Ex lectotype. Received 1942. Isolated from soil, Fayetteville, AR.

Aspergillus ustus (Bain.) Thom & Church
DG. 275 = ATCC 1041, 16818 = CBS 261.67 = IMI 211805 = Thom 3556 = WB 275. Ex neotype. Received 1914. Isolated as a culture contaminant.
DG. 1852 = CBS 128.62 = WB 1852. Received 1942. Isolated from soil, Louisiana.
DG. 4688 = WB 4688. Received 1969. Isolated from Panamanian soil by W. Scott.
DH. 4991 = WB 4991 Received 1969. Isolated from Brazilian soil.
DG. 5077 = ATCC 16800 = CBS 495.65 = IMI 126692. Ex lectotype of *Aspergillus puniceus*. Received as WB 5077, 1965. Isolated from soil, Costa Rica.

Aspergillus pseudodeflectus Samson & Mouchacca
DI. 6135. Ex type. Received as CBS 756.74, 1975. Isolated from desert soil, Egypt.
DI. 1974 = ATCC 16801 = CBS 561.65 = IMI 127148 = WB 1974. Received as *A. ustus* 1942. Isolated from soil, Panama.
DI. 4876. Received 1969. Isolated from Iowa soil. Type of *Aspergillus minutus*.

Aspergillus cavernicola Lorinczi
DJ. 6327 = CBS 117.76. Received 1977. Isolated from soil on cave wall, Romania. Type of *Aspergillus cavernicola*.

Aspergillus subsessilis Raper & Fennell
DK. 4905 = ATCC 16808 = CBS 502.65 = IMI 135820. Ex lectotype. Received as WB 4905, 1965. Isolated from desert soil by G F ORR as O 325.
DK. 4906 = CBS 987.72 = IMI 335781. Received as WB 4906, 1969. Isolated from desert soil by G. F. Orr as Orr 322.
DK. 4907 = CBS 988.72 = IMI 335782. Received as WB 4907, 1969. Isolated from desert soil by G. F. Orr as Orr A2.
DK. 3752 = BKM F 1080 = CBS 419.69. Received as IMI 140345, 1970. Isolated from soil, Syria. Type of *Aspergillus kassunensis*.

Aspergillus ivoriensis Rambelli et al.,
DL. 22883. Ex type. Received as CBS 551.77. Isolated from forest soil, Ivory Coast.

Aspergillus raperi Stolk
DM. 2640 = CBS 124.56 = IMI 70948 = WB 5040. Isolated from grassland soil, Zaire.
DM. 5039 = CBS 125.56 = IMI 70947 = WB 5039. Received as WB 5039, 1969. Isolated from grassland soil, Zaire.

Aspergillus elongatus Rai & Agarwal
DN. 5176 = CBS 387.75. Ex lectotype. Received as *A. elongatus* WB 5495. Isolated from alkaline soil, India. Produces numerous initials but no cleistothecia.

Aspergillus deflectus Fennell & Raper
DO. 2206 = ATCC 16807 = CBS 109.55 = IMI 61448 = WB 2206. Ex lectotype. Received 1949. Isolated by A. Cury, from Brazilian soil.

DO. 4235 = WB 4235. Received 1969. Isolated from potting soil.

DP. 4993 = ATCC 16799 = CBS 504.65 = IMI 135420. Received as WB 4993, 1965. Isolated from soil, Turkey.

Aspergillus lucknowensis Rai, Tewari & Agarwal

DQ. 3491 = ATCC 18607 = CBS 449.75. Ex type. Received as WB 5377, 1969. Isolated from alkaline soil, India.

Emericella aurantiobrunnea (Atkins, Hindson, & Russell) Malloch & Cain

DR. 4545. Ex type. Received as WB 4545, 1965. Isolated from canvas haversack, Australia.

Aspergillus silvaticus Fennell & Raper

DS. 2398 . = ATCC 16843; 46904 = CBS 128.55 =IMI 61456 = WB 2398. Ex lectotype. Received 1950. Isolated from soil, Gold Coast.

Aspergillus bisporus Kwon-Chung & Fennell

DT. 3690. Received 1970. Isolated from pine plantation soil, Georgia. Pathogenic for mice.

DT. 3692. Received 1969. Isolated from soil stored 10 yr at 30C, MD. Pathogenic for mice, isolated by mouse passage technique.

DU. 3693 = ATCC 22527 = CBS 707.71. Ex type. Received 1969. Pathogenic for mice, isolated by mouse passage technique.

Aspergillus anthodesmis Rambelli *et al.*,

DV. 22884. Ex type. Received as CBS 552.77. Isolated from forest soil, Ivory Coast.

Aspergillus panamensis Raper & Thom

DW. 1785 = ATCC 16797 = CBS 120.45 = IMI 19393 = QM 8897 = WB 1785. Ex lectotype. Received 1941. Isolated by J. T. Bonner from soil, Panama.

DX. 1786. Received 1941. Isolated by J. T. Bonner from soil Panama.

Aspergillus conjunctus Kwon & Fennell

DY. 5080 = ATCC 16796 = CBS 476.65 = IMI 135421. Ex lectotype. Received as WB 5080, 1965. Isolated from forest soil, Costa Rica.

Aspergillus funiculosus Smith

DZ. 4744. Ex type. Received as WB 4744, 1965. Isolated from soil, Nigerian.

Aspergillus biplanus Raper & Fennell

EA. 5071= ATCC 16858 = CBS 468.65 = IMI 235602. Ex lectotype of *A. biplanus*. Received as WB, 1965. Isolated from soil, Costa Rica.

EA. 5072. Received as WB 5072, 1969. Isolated from soil, Costa Rica, by K. J. Kwon, as *A. biplanus* TRI-11.

EA. 5073 = ATCC 16850 = CBS 469.65. Received as *A. biplanus* WB 5073, 1965. Isolated from soil, Costa Rica.

EA. 5074 = ATCC 16849 = CBS 480.65 = IMI 232882. Ex lectotype of *Aspergillus diversus*. Received as WB 5074, 1965. Isolated from soil, Costa Rica.

EA. 5075. Received as WB 5075, 1969. Isolated from soil, Costa Rica by K. J. Kwon as *A. diversus* E -9.

Aspergillus sparsus Raper & C. Thom

EB. 1933. Ex type of *A. sparsus*. Isolated from Costa Rican soil, 1943, by K. B. Raper.

EB. 1937. Received as Clare M1, 1943. Isolated from soil, San Antonio, TX.

EB. 4568. Received as WB 4568, 1969. Isolated from desert soil, Haiti, by W. Scott, as #103B.

EB. 4569. Received as WB 4569, 1969. Isolated from soil under sage and cactus, Haiti, by W. Scott as #113A.

Aspergillus amylovorus Panasenko

EC. 5813 = ATCC 18351 = CBS 600.67 = IMI 129961. Ex type. Received as BKM F-906, 1969. Isolated from wheat starch, USSR.

Aspergillus varians Wehmer

ED. 4793 = ATCC 16836 = CBS 505.65 = IMI 172297 = Thom 115. Ex lectotype. Received as WB 4793, 1965.

Aspergillus recurvatus Raper & Fennell

EE. 4902= ATCC 16809 = CBS 496.65 = IMI 136528. Ex lectotype. Received as WB 4902, 1965. Isolated from lizard dung by G. F. Orr as O-566, California.

Aspergillus clavatus Dezm.

EF. 1. Ex type. Received from Westerdijk, 1909.

EF. 2 = Thom 4074.C1. Received 1916.

EF. 8 = Thom 5626. Received 1937.

EF. 2254. Received 1946. Isolated from dung found in British Guiana.

EF. 4097. Received as WB 4097, 1969. Isolated from toxic feed pellets.

EF. 5811 = ATCC 18327 = CBS 344.67 = IMI 129967. Received as BKM F 1136 1969. Isolated from soil, Moldavia, USSR. Ex type of *Aspergillus pallidus.*

Aspergillus clavatonanicus Batista, Maia & Alecrim

EG. 4741 = ATCC 12413 = CBS 474.65 = IMI 235352. Ex lectotype. Received as WB 4741, 1965. Isolated from nail lesion.

Aspergillus giganteus Wehmer

EH. 10 = ATCC 10059 = CBS 526.65 = Thom 5581.13A =WB 10. Ex neotype. Received 1936. Isolated from bat dung, Yucatan, Mexico.

EH. 4560. Received as WB 4560, 1969. Isolated from clay in a patch of *Selaginella*, Panama.

EH. 4763 = ATCC 16439 = CBS 515.65. Received as WB 4763, 1965. Isolated from mouse dung, by G. F. Orr as O-823, California.

EI. 6136. Received as type of *A. rhizopodus*, CBS 450.75. Isolated from alkaline soil, India.

Aspergillus longivesica Huang & Raper

EJ. 5215 = ATCC 22434 = CBS 530.71 = IMI 156966. Ex type. Received as *A. longavesica* N 1129, L.S. Huang, Univ of Wisconsin, 1971. Isolated from Nigerian soil.

Hemicarpenteles acanthosporus Udagawa & Takada

EK. 5293 = ATCC 22931 = CBS 558.71 = IMI 164621. Ex type. Received as NHL 2462 from S. Udagawa 1972. Isolated from Solomon Islands soil.

Aspergillus fumigatus Fresenius

EL. 163 = ATCC 1022; 4813 = CBS 133.61 = IMI 16152 = WB 163. Ex neotype. Received from C. Thom. Isolated 1909 from chicken lung.

EL. 164 = ATCC 1028 = CBS 113.26 = WB 164. Received as *Aspergillus cellulosae* from C. Neuberg, Berlin, 1921. Isolated from soil, Germany.

EL. 165 = Thom 4640.4. Received as *Aspergillus fumigatus* from Bainier Collection via Da Fonseca, 1922.

EL. 166 = Thom 4967.57. Received as *Aspergillus* sp. from Melin, 1927. Isolated from soil, Adirondack Mts., New York.

EL. 5587 = ATCC 36962 = CBS 457.75 = ATCC 2588. Received as *Aspergillus fumigatus* var. *acolumnaris* WB 5452, 1973. Isolated from soil, India. Ex type of *Aspergillus fumigatus* var. *acolumnaris.*

EL. 5517 = ATCC 22268 = CBS 158.71. Received as BKM F 1488, 1972. Isolated from soil, USSR. Ex type of *Aspergillus anomalus* Pidoplichko & Kirilenko,

EL. 6113 = ATCC 26606 = CBS 542.75. Received from K. J. Kwon-Chung, NIAID NIH, Bethesda, MD, 1973. Isolated from clinical material, patient with sinusitis. Ex type isolate of *Aspergillus phialiseptus*.

EL. 5109 = ATCC 16903 = CBS 487.65 = IMI 172286. Ex type culture of *A. fumigatus* var. *ellipticus*. Received as WB 5109, 1965. Isolated from pus, case of chronic emphysema, Illinois.

Aspergillus unilateralis Thrower

EM. 577 = ATCC 16902 = CBS 126.56 =IMI 62876. Ex type. Received as 2600/27 from McLennan, Univ. of Melbourne, 1955. Isolated from rhizosphere of *Epacris & Hibbertia*, Australia.

Neosartorya fischeri (Wehmer) Malloch & Cain

EN. 181 = ATCC 1020 = CBS 544.65 = IMI 211391 = Thom 4651.2 =WB 181. Ex lectotype. Received as *Aspergillus fischeri* from Wehmer 1922.

EN. 4075 = ATCC 13831 = IMI 16143 Isolated from garden soil, U.K.

Neosartorya glabra (Fennell & Raper) Kozakiewicz

EO. 183 = Thom 5047. From Beauverie, Lyons, France 1929. Isolated from beerwort, France.

EO. 2163 = ATCC 16909 = CBS 111.55 = IMI 61447 = WB 2163. Ex lectotype. Received as 6354 from G. W. Martin, Univ. of Iowa 1947. Isolated from rubber tire scrap.

EO. 3434 = CBS 448.75. Received as *Sartorya fumigata* var. *verrucosa* NHL 5083 from Y. Sato 1968. Isolated from soil Japan. Ex type of *Aspergillus fischeri* var. *verrucosus*.

EO. 4179. Received as WB 4179, 1969. Isolated from soil by J. H. Warcup as SA 57, Australia.

EO. 2392 = CBS 112.55 = IMI 61450 = WB 2392. Received as SA 14 from J. H. Warcup, 1952. Isolated from garden soil, Australia.

Neosartorya spinosa (Raper & Fennell) Kozakiewicz

EP. 5034 = ATCC 16898 = CBS 483.65 = IMI 211390 Ex lectotype. Received as WB 5034, 1965. Isolated from Nicaraguan soil, K. B. Raper.

EP. 185 = Thom 5136.11 = WB 185. From Gadd, 1930. Isolated from teakwood, Sri Lanka.

EP. 3435. Received as NHL 5084, from Y. Sato, 1968

EQ. 4076 = IMI 16061. Received as WB 4076, 1969. Isolated from hay, U.K.

EQ. A-1914. Received as *Aspergillus malignus* from G. Smith, LSHTM 1946.

Neosartorya quadricincta (Yuill) Malloch & Cain

ER. 2154 = ATCC 16897 = CBS 135.52 = IMI 48583 = WB 2154. Ex lectotype. From E. Yuill, 1947.

ER. 4175. Received as WB 4175, 1969. Isolated from soil, Australia.

Neosartorya aureola (Fennell & Raper) Malloch & Cain

ES. 2244 = ATCC 16896 = CBS 105.55 = IMI 61451 = WB 2244. Ex lectotype. Isolated from agricultural soil, Gold Coast. Received from C. F. Charter, 1950.

ES. 2391 = CBS 106.55. 1953. Isolated from Liberian soil by J. T. Baldwin.

Neosartorya stramenia (Novak & Raper) Malloch & Cain

ET. 4652 = ATCC 16895 = CBS 498.65 = IMI 172293. Ex lectotype. Received as WB 4652, 1965. Isolated from forest soil, Wisconsin.

Neosartorya aurata (Warcup) Malloch & Cain

EU. 4378 = ATCC 16894 = CBS 466.65 = IMI 75886. Ex type. Received as WB 4378, 1965. Isolated from jungle soil, Indonesia, by J. H. Warcup as A 13/1.

EV. 4379 = WB 4771. Received as WB 4379, 1969. Isolated from jungle soil, Indonesia, by J. H. Warcup as A 52/3.

Neosartorya fennelliae Kwon-Chung & Kim

EW. 5534 = ATCC 24325 = CBS 598.74. Received as AF 5 from K. J. Kwon-Chung, NIAID, NIH, Bethesda, MD, 1972, isolated from rabbit eyeball. Ex type culture of mating type A.

EW. 5535 = ATCC 24326 = CBS 599.74. Received as AF 4 from K. J. Kwon-Chung, NIAID, NIH, Bethesda, MD, 1972, isolated from rabbit eyeball. Ex type of mating type alpha.

Neosartorya pseudofischeri Peterson

EX. 20748. Ex type. Received as B-4812 from A.A. Padhye, CDC, Atlanta, GA, 1989. Isolated from vertebrae (human infection).

EX. 180 = Thom 4188.21. From Levine, Iowa, 1917.

EX. 3496 = ATCC 18618 = CBS 404.67. Received from J. W. Paden, Univ. of Victoria, B. C., Canada, 1969. Isolated from moldy cardboard in Victoria. Ex type culture of *Aspergillus fischeri* var. *thermomutatus*.

Neosartorya spathulata Takada & Udagawa

EY. 20549. Ex type. Received as ATCC 64222, 1990. Isolated from cultivated soil, Taiwan. Ex type culture of mating type A.

EY. 20550. Ex type. Received as ATCC 64223, 1990. Isolated from cultivated soil, Taiwan. Ex type culture of mating type alpha.

Aspergillus brunneouniseriatus S. Singh & B. K. Bakshi

EZ. 4273 = ATCC 16916 = CBS 127.61 = IMI 227677 = WB 4273. Ex lectotype. Received as WB 4273, 1965. Isolated from soil under *Dalbergia sissoo*.

Monascus purpureus Went

FA. 1596 = ATCC 16365; 16426 = CBS 109.07 = IMI 210765. Ex type. Received as 143 from Harvard Univ., 1940.

Hemicarpenteles paradoxus Sarbhoy & Elphick

FB. 2162= ATCC 16918 = CBS 527.65 = IMI 61466 = WB 2162. Ex lectotype. Received as A28 from J. H. Warcup, Cambridge Univ., U.K., 1948. Isolated from opossum dung, New Zealand.

FB. 4695 = IMI 86829. Received as WB 4695, 1969.

Aspergillus malodoratus Kwon & Fennell.

FC. 5083 = ATCC 16834 = CBS 490.65 = IMI 172289. Ex lectotype. Received as WB 5083, 1965. Isolated from forest soil, Costa Rica.

Aspergillus crystallinus Kwon & Fennell

FD. 5082 = ATCC 16833 = CBS 479.65 = IMI 139270. Ex lectotype. Received as WB 5082, 1965. Isolated from forest soil, Costa Rica.

Isolates were revived from lyophilized storage and grown on Czapek's and malt extract agars (Raper and Fennell, 1965) to confirm the morphology and purity of the strains. Conidia or ascospores from cultures on agar slants were used to inoculate 50 ml of steril-ized broth medium [usually Czapek's broth, amended with 20% sucrose when needed, or M40Y broth (Raper and Fennell, 1965) when that was the most satisfactory medium] contained in a 300-ml Erlenmeyer flask stoppered with a tightly fitting cotton plug. Flasks were incubated at 25°C on a rotary shaker at 200 rpm. Mycelium was harvested from the flasks after 2-3 days growth, or when ca. 1 g biomass had accumulated. Myce-lium and medium were filtered over four layers of cheesecloth and the resulting mycelial mat was pressed dry by hand. Ca. 0.3 g portions of mycelium were placed in 2.0 ml snap-cap tubes, frozen at -20°C, and then freeze-dried. The dried mycelium was ground to a powder in the snap-cap tubes using a disposable pipette tip as pestle. The powdered my-celium was suspended in 500 µl DNA extraction buffer (50 mM Tris, 50 mM KCl, 10 mM disodium EDTA, 1% sodium lauryl sarcosinate, pH 8.0), and proteins were extracted with 500 µl of phenol:chloroform (1:1, w:v). Organic and aqueous phases were separated at room temperature by centrifugation in a bench top micro-centrifuge operated at full speed for 5 min (ca. 10,000 x g). The aqueous phase was pipetted to a clean tube and nucleic acids were precipitated by the addition of 1.3 volumes of 95% ethanol. The pre-cipitate was pelleted by centrifugation in a micro centrifuge (as above). Supernatant liq-uid was removed and the pellet was dissolved in 500 µl TE (10 mM Tris, 1 mM disodium EDTA, pH 8.0). The DNA solution was further purified by adsorption to glassmilk (Bio101, La Jolla, CA) using the manufacturer's protocol, and desorption in 500 µl 1/10th strength TE. These DNA preparations were stored at 20°C until used for DNA amplifica-tions

Ribosomal DNA fragments were amplified using the technique and conditions out-lined by White *et al.* (1990). Primers used to amplify the large subunit rDNA fragment (lsu DNA) are listed in Table 2. Thermal profiles used to amplify nuclear DNA were 96°C, 30 sec; 60°C, 45 sec; 72°C, 90 sec, for 30 cycles followed by 7 min at 72°C. Am-plified fragments were cleaned by adsorption to glassmilk using the manufacturer's in-structions, and were eluted into 50 µl 1/10th strength TE. Sequence reactions contained ca. 3 µl purified DNA template (200-400 ng), 1.5 pmol of primer (listed in Table 2) and the reaction mix contained in the DyeDeoxy sequencing kit (Applied Biosystems, Inc). Sequence reactions were incubated in a Perkin Elmer thermal cycler using the Applied Biosystems recommended thermal profile. Completed reactions were cleaned using spun-column chromatography over Sephadex (G 50 ultra fine, Pharmacia). Eluate was dried in a vacuum centrifuge (Savant) and the residue was dissolved in 3.5 µl ultrapure forma-mide. Sequences were determined by electrophoresis on a 373a DNA sequencer (Applied Biosystems).

Sequence ambiguities were corrected by comparison of opposite strand sequences and by multiple sequencing reactions of the regions in question. Sequences were initially aligned using CLUSTAL V (Higgins and Sharp, 1988; 1989), and these aligned se-quences were further checked using a text editor. Sequences were deposited in GenBank.

Aligned data sets were analyzed using parsimony programs contained in the phyloge-netic inference package of programs PHYLIP (Felsenstein, 1993) and PAUP 3.1.1 (Swof-ford, 1993). Bootstrap analysis was performed using PAUP 3.1.1. Bootstrap values below 70% are not indicated in the results as they lack 95% confidence (Hillis and Bull, 1993;

sequences were further checked using a text editor. Sequences were deposited in Gen-Bank.

Aligned data sets were analyzed using parsimony programs contained in the phylogenetic inference package of programs PHYLIP (Felsenstein, 1993) and PAUP 3.1.1 (Swofford, 1993). Bootstrap analysis was performed using PAUP 3.1.1. Bootstrap values below 70% are not indicated in the results as they lack 95% confidence (Hillis and Bull, 1993; Felsenstein and Kishino, 1993) General searches for a consensus tree were performed using DNAPARS in multiple analyses with random input order.

RESULTS AND DISCUSSION

The D1 and D2 regions of large subunit rDNA are considered highly variable (Guého *et al.*, 1990; Peterson and Kurtzman, 1991), and have sufficient variability that most species of yeasts can be distinguished by sequence differences in those regions. The sequences are also sufficiently conserved that alignment, and analysis of the relationships of species in a family, or even in an order (Guého *et al.*, 1990) is possible. In the context of *Aspergillus*, the majority of the species accepted by Raper & Fennell (1965) can be distinguished by differences in the DNA sequence from these regions. In addition, little or no intraspecific variability has been noted (Peterson *et al.*, 2000). Therefore, for the purpose of interpreting this set of data, strains with identical lsu-rDNA sequences will be considered conspecific. Additional data (e.g., DNA complementarity measurements, isozyme studies, studies of additional genes and patterns of heteromorphism in those genes) are needed to conclude that two strains are conspecific.

Figure 1 shows in overview the relationships of some *Eupenicillium* and *Aspergillus* species based on the lsu DNA data set. Gams *et al.*(1985) divided *Aspergillus* and its teleomorphs into six subgenera and 18 sections, which largely correspond to the 18 groups of Raper and Fennell(1965). The phylogenetic analysis of *Aspergillus* and its teleomorphs differs however, from that subgeneric taxonomy, indicating the need to revise the subgeneric and sectional classification if phylogeny is reflected. Table 2 lists the disposition of current subgenera and sections based on the consensus tree of the lsu-rDNA data.

There are three major branches in the phylogram (Fig 1) of *Aspergillus* and its teleomorphs that could be used to define subgenera. The first includes the teleomorphs *Eurotium, Fennellia, Chaetosartorya*, and *Warcupiella* and the sections *Aspergillus, Flavi, Nigri, Circumdati, Flavipedes, Candidi, Terrei*, parts of *Versicolores, Restricti* and *Cerv ini*. Because this group includes the type species of *Aspergillus*, it must be assigned as subgenus *Aspergillus*.

Formulated in this way, subgenus *Aspergillus* includes both uniseriate and biseriate species. The presence or absence of metulae is not a character that can be used to define monopheletic taxa as recognized by Raper and Fennell (1965) and more recent authors. The subgenus also includes species with quite different colony colour and appearance. Peterson (1995) showed that colour was not a reliable predictor of phylogenetic placement, although it is useful in many cases for identifying cultures.

334

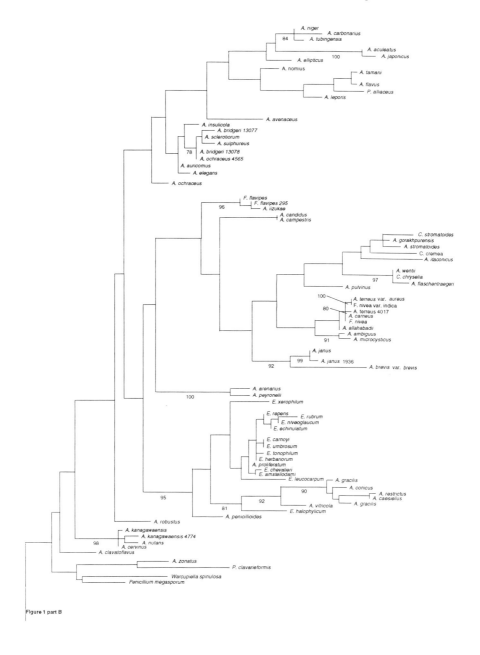

Figure 1 part B

335

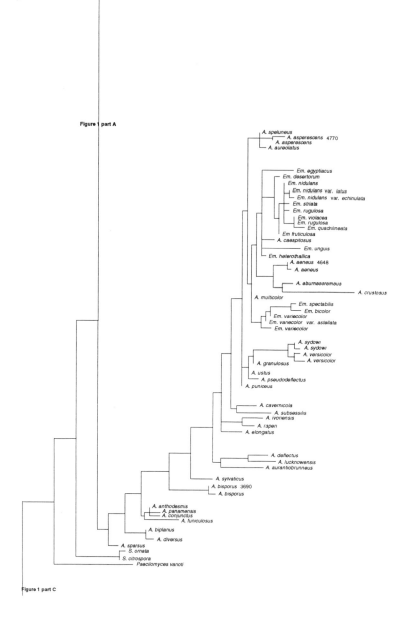

Figure 1 part A

Figure 1 part C

Figure 1, part B

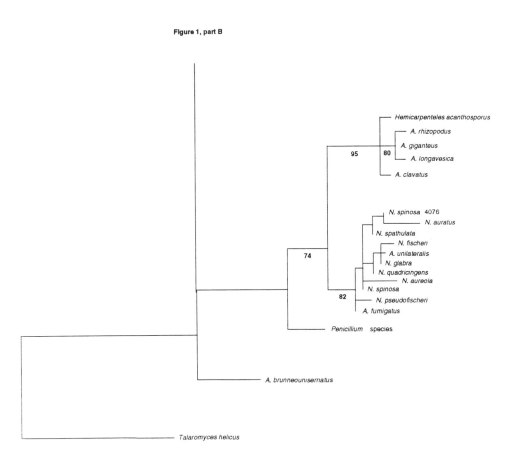

Fig. 1. a, b and c. One of over 1500 equally parsimonious trees of all *Aspergillus* species. The tree was run with the lsu-rDNA data set using only *Aspergillus* species with sequences different from all others, and 88 strains representing the diversity of *Penicillium*. The *Penicillium* species occur mostly on a single branch, which has been removed to simplify this tree. The tree was subjected to bootstrap analysis (PAUP 3.1.1) and bootstrap values above 70% are placed on the branches. The tree has four main subbranches that are used to define three subgenera in *Aspergillus*. The fourth branch contains *Penicillium* species. It should be noted that in the basal parts of the tree bootstrap values are generally low, and the relationships portrayed in this figure may shift as more informative data are obtained.

Table 2. Revisions to *Aspergillus* subgeneric taxonomy based on monophyletic groups determined from rDNA sequences

Gams *et al.*, 1985	Disposition in this study
Subgenus *Aspergillus*	retained
Section *Aspergillus*	retained
Section *Restricti*	retained
Subgenus *Fumigati*	retained
Section *Fumigati*	retained
sect *Cervini*	retained section, and place in subgenus *Aspergillus*
Subgenus *Ornati*	delete subgenus
Section *Ornati*	retained section, and place in subgenus *Nidulantes*
Subgenus *Clavati*	delete subgenus *Clavati*
Section *Clavati*	retained section, and place in subgenus *Fumigati*
Subgenus *Nidulantes*	retained
Section *Nidulantes*	retained
Section *Versicolores*	delete section *Versicolores*, species included in section *Nidulantes*
Section *Usti*	delete section *Usti*, species included in section *Nidulantes*
Section *Terrei*	modify section, and place in subgenus *Aspergillus*
Section *Flavipedes*	modify section, and place in subgenus *Aspergillus*
Subgenus *Circumdati*	delete subgenus
Section *Wentii*	delete section
Section *Flavi*	retained and modify in subgenus *Aspergillus*
Section *Nigri*	retained and place in subgenus *Aspergillus*
Section *Circumdati*	retained and place in subgenus *Aspergillus*
Section *Candidi*	retained and place in subgenus *Aspergillus*
Section *Cremei*	retained and place in subgenus *Aspergillus*
Section *Sparsi*	modify and place in subgenus *Nidulantes*

The second subgenus appropriately called subgenus *Nidulantes* includes the teleomorph genera *Emericella* and *Sclerocleista*, and species from section *Nidulantes*, parts of *Versicolores*, *Usti*, *Ornati* and *Sparsi*. As in subgenus *Aspergillus*, this subgenus is heterogeneous in terms of the uni- or biseriate pattern of the conidial heads and colony colour. The third subgenus, assigned here as subgenus *Fumigati* includes the teleomorph genus *Neosartorya*, *Hemicarpenteles acanthosporus, A. brunneouniseriatus* and species from section *Clavati* and *Fumigati*. The species in this subgenus are uniseriate, but have some colour and colony differences.

The sections within each subgenus require some revisions to make them monophyletic. The accommodation of each species into appropriate sections is discussed below.

Subgenus *Aspergillus*, sections *Aspergillus* and *Restricti*

The relationships of species within the sections *Aspergillus* and *Restricti* are shown in Fig. 2. Section *Aspergillus* contains the teleomorph species: *Eurotium herbariorum, E. repens , E. tonophilum, E. rubrum, E. chevalieri, E. amstellodami, E. umbrosum, E. echinulatum, E. halophilicum, E. niveoglaucum, E. leucocarpum, E. xerophilum*, and the anamorph *A. proliferans*.

Blaser (1975) reduced *E. umbrosum* to synonymy with *E. herbariorum* because the phenotypic differences were too slight to *justify E. umbrosum*. Pitt (1985) noted that cultures labeled *E. herbariorum* generally had smaller ascospores and the cultures had less brown and more orange colour than *E. umbrosum*, but on balance agreed with Blaser (1975). In this study, two strains considered typical of *E. umbrosum* by Raper and Fennell (1965) showed a single nucleotide difference from strains of *E. herbariorum*. Kozakiewicz (1989) placed *E. umbrosum* strains in synonymy with E. *rubrum* on the basis of the SEM morphology of conidia and ascospores, but *E. rubrum* and *E. umbrosum* differ at seven nucleotide positions in the lsu-rDNA. On the basis of the DNA sequence differences, and the phenotypic differences noted by Blaser (1975) and Pitt (1985), *E. umbrosum* has a sibling species relationship with *E. herbariorum*. Strain NRRL 71 identified by Thom and Raper (1945) as a cleistothecial and conidial variant of E. *rubrum*. The DNA sequence shows that it is a strain of *E. herbariorum*.

Raper and Fennell (1965) maintained *A. montevidensis* as a species distinct from *E. amstelodami*, in part because the former species was isolated from the tympanic membrane of a human ear. Pitt (1985) regarded *E. montevidense* as a synonym of *E. amstellodami*, and Kozakiewicz (1989) reduced it to varietal status. The rDNA sequence of *E. montevidense* is identical to that of *E.* amstellodami and these taxa can be considered conspecific. A third species E. *cristatum* also have the same rDNA sequence as *E. amstelodami*. Although E. *cristatum* produces larger and differently ornamented ascospores (Pitt, 1985; Kozakiewicz, 1989), the DNA sequences suggest that the differences are not sufficient to justify a separate species.

A. athecius produces naked asci rather than the cleistothecia characteristic for *Eurotium* and it was therefore raised to generic rank by Subramanian (1972) as *Edyuillia athecia*. Von Arx (1974) placed the species in *Eurotium* which was accepted by Pitt (1985). However, Kozakiewicz (1989) maintained *Edyuillia* while classifying the species with the *Eurotium* species. The DNA sequence of *E. rubrum* and *E. aethecia* are identical showing that this is merely an atypical strain of *E. rubrum*.

Raper & Fennell (1965) accepted *A. pseudoglaucum* as a distinct species, while Pitt (1985) regarded it as a synonym of *E. repens* and Kozakiewicz (1989) reduced it as a variety of *E. repens*. The lsu-rDNA sequences of strains of the species are identical, indicating that these species are synonyms.

Pitt (1985) regarded *E. medium* as a synonym of *E. echinulatum,* while Kozakiewicz (1989) retained them as distinct species. The rDNA sequences showed that both species are synonyms

Among strains from section *Restricti*, four *A. restrictus* strains had identical lsu-rDNA sequences; two strains of *A. gracilis* differed at six nucleotide positions; two strains of *A. penicillioides* were identical, while a third (ex type culture of *A. vitricola*) differed from the other two strains at 20 nucleotide positions. *Aspergillus conicus* and *A. caesiellus* had lsu-rDNA sequences different from other species in the section. The lsu-rDNA sequence of strain *A. conicus* NRRL 4661 was identical with an ex type strain of *A. fumigatus* and upon re-examination it was re-identified as *A. fumigatus*.

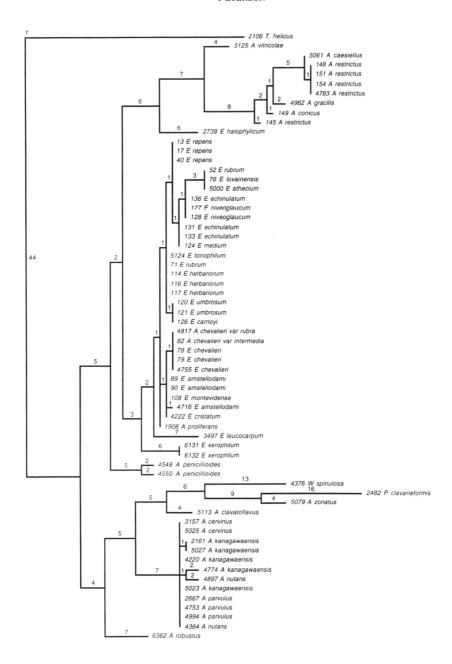

This analysis of section *Restricti* contrasts with the morphological study by Pitt & Samson (1990) who reduced the species in section *Restricti* to three: *A. restrictus, A. caesiellus* and *A. penicillioides*. Considerable genetic variation is present in the 12 strains from this section that were sequenced. The DNA sequence differences found among those strains strongly suggests that at least six anamorphic species in section *Restricti* should be retained.

Species of the teleomorph *Eurotium* are all placed in subgenus *Aspergillus*, while the highly xerophilic anamorphs with similar micromorphology are placed in section Restrici. Both sections occur on one well supported main branch (95% bootstrap value; Fig.2) of the phylogenetic tree, but are separated at a lower level (Fig. 2). However the branch that includes section *Restricti* species also includes *E. halophilicum*, and this has strong bootstrap support (81%). Topological constraints were placed on the tree and tested using the log likelihood test (Kishino and Hasegawa, 1989). *E. halophilicum* was forced into the *Eurotium* clade and tested against the tree where it was included with section *Restricti* species. The tree with all *Eurotium* species on the same branch was significantly less likely in the log likelihood test (Kishino & Hasegawa, 1989), than when *E. halophilicum* was grouped with the section *Restricti*. The presence of *E. halophilicum* on the branch with the anamorph species of section *Restricti*; and the presence of *A. proliferans* among the holomorphic section *Aspergillus* species, demonstrates that these sections are artificial distinctions based on the presence of ascomata. Several species, including *E. spiculosum, E. appendiculatum*, and *E. glabrum*, that were described by Blaser (1975) were not examined in this study.

Subgenus Aspergillus, section Cremei

Section *Cremei* (Figs.1,3) has recently been studied by Peterson (1995). In that analysis based on the lsu-rDNA, *A. gorakpurensis* (section *Versicolores*), *A. dimorphicus* (section *Circumdati*) and *A. wentii* (section *Wentii*) were transferred into section *Cremei*. The remaining species in section *Wentii* were transferred to section *Flavi*. Because the type species of section *Wentii* is very closely related to *Chaetosartorya chrysella*, it was transferred into section *Cremei*, and on the basis of phylogenetic analysis, there is no reason to maintain section *Wentii* (Peterson, 1995). Species accepted in section *Cremei* are *Chaetosartorya cremea, C. chrysella, C. stromatoides, A. wentii, A. flaschentraegeri, A. itaconicus, , A. gorahkpurensis, A. dimorphicus*, and *A. pulvinus*. The DNA sequence differences between strains assigned as the anamorph *Aspergillus stromatoides* and those of the teleomorph *Chaetosartorya stromatoides* are large enough to suggest that this species is not monopheletic (Peterson, 1995). *Aspergillus wentii* and *A. dimorphicus* have identical rDNA sequences and might be synonyms; both differ from *C. chrysella* at a single base position. Raper and Fennell (1965) acknowledged the close morphological relationship of *C. chrysella* and *A. wentii* but kept them in different groups because colony colour was a fundamental part of their taxonomic system.

Fig. 2. Phylogram of species from sections *Aspergillus, Restricti, Cervini* and an unnamed group centred around *Warcupiella* spinulosa. The clear taxonomic distinctions of sections *Aspergillus* and *Restricti* are not reflected in the phylogram, where *E. halophilicum* is on the branch with *A. restrictus*, and *A. penicillioides* is basal to all species of both sections. *A. robustus* is figured here and with section *Ochraceus* but its association with either section is clarified by this data. Strains from section *Cervini*, without *A bisporus* form a clade. In Fig. 1, the species branching with *W. spinulosa* are basal to sections *Cervini, Aspergillus* and *Restricti*.

Subgenus *Aspergillus*, sections *Terrei* and *Flavipedes*

Opinions on the taxonomic relationships of species from sections *Terrei* and *Flavi*pedes have differed and overlapped over the years. Thom and Raper (1945) placed *A. terreus* and its varieties in the *A. terreus* group along with *A. carneus* and *A. nivea*, while Raper and Fennell (1965) placed *A. carneus* and *A. nivea* in the A. *flavi*pes group, keeping only *A. terreus* in the *A. terreus* group. Phylogenetically (Figs 1,3), *A. carneus*, *A. terreus* and *F. nivea* are closely related, but they are not closely related to *Fennellia flavi*pes. Because of this, *A. carneus* and *F. nivea* are transferred to section *Terrei*. The *A. janus* series of section *Versicolores* (Raper and Fennell, 1965; Klich, 1993) is phylogenetically related to *A. terreus*. Species in the *A. janus* series are unusual because the species develop a white colony appearance when grown at temperatures near 20°C, and a green colony appearance when grown at 30°C. Although Raper and Fennell (1965) placed them in the *A. versicolor* group, the species in this series are phylogenetically related to the white or light-coloured species in section *Terrei* and are transferred there. *Aspergillus aureofulgens* was placed in section *Flavipedes* (Samson, 1979) as a probable synonym of *A. carneus*. It is phylogenetically distinct from other species, is related to the species in section *Terrei* and is transferred to that section.

Fennellia nivea and *F. flavi*pes differ from each other at 20 nucleotide sites, and appear on distinct branches. Wiley and Simmons (1973) originally placed the teleomorph of *A. niveus* in *Emericella*, and Samson (1979) transferred it to *Fennellia* as *F. nivea*. The evidence presented here indicates that *F. nivea* is correctly placed in *Fennellia*, but the species in *Fennellia* are phylogenetically relatively distant.

Subgenus *Aspergillus*, Section *Flavipedes* contains *Fennellia flavipes*, *A. iizukae*, and sequence variants of *F. flavipes*. These species occur on a branch with 96% bootstrap support (Fig 3). *Fennellia flavi*pes strains do not have identical rDNA sequences, suggesting that there is more than one species contained in the current species concept. *Aspergillus iizukae* is phylogenetically close to *F. flavipes*. Although Samson (1979) reduced *A. iizukae* to synonymy with *F. flavipes*, it is a distinct species according to my analysis. It is interesting to note that no teleomorph in the type isolate was found, but the rDNA sequence is identical with several strains of *Fennellia*. It is likely that *A. iizukae* is an anamorph isolate of *Fennellia*.

Aspergillus peyronelii and *A. arenarius* (section *Versicolores*) show DNA similarity to species related to section *Flavipedes,* although the relationship is not close, they are transferred to section *Flavipedes*. In a consensus tree, *Fennellia*, *A. janus* and the section *Terrei* are monophyletic.

Fig. 3. Phylogram showing sections *Cremei, Candidi, Terrei,* and *Flavi*pedes. Section *Cremei* includes *A. pulvinus* and the species branching with the *Chaetosartorya* species. The section is supported by 73% bootstrap value. Section *Terrei* includes *Fennellia* nivea, *A. terreus, A. carneus* and *F. nivea* var. indica, along with A. allahabadii, A. microcysticus, and A. ambiguus transferred from section *Versicolores*. The branch containing section *Terrei* is supported in 80% of the bootstrap samples. All of the species in the clade have colony colours in light shades, from creamy to cinnamon tones. Section *Flavipedes* contains *Fennellia flavi*pes (but not *F. nivea*), A. iizukae, A. aureofulgens and a variant strain of *A. carneus* placed in that species by Raper and Fennell (1965). All of the light tan to white species occur in a single branch. The teleomorphic species *Fennellia flavipes* and *F. nivea* are placed in the sections *Flavipedes* and *Terrei* respectively. A. campestris is transferred from section *Circumdati* to section *Candidi* on the basis of the lsu-rDNA data. The *A. janus* series (section *Versicolores*) is transferred to section *Flavipedes*, along with *A. arenarius* and *A. peyronelli.*

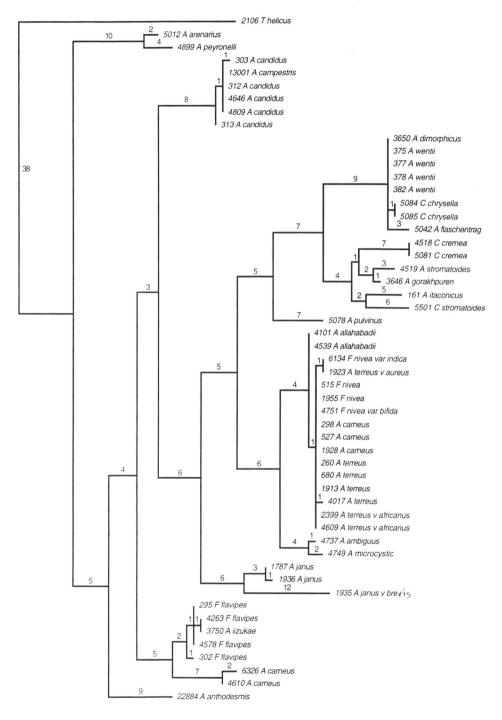

Subgenus *Aspergillus*, section *Candidi*

A. candidus (Fig. 3) branches between sections *Terrei* and *Flavipedes*, sharing a common ancestor with those species. Closely related to *A. candidus* is *A. campestris*, which was placed in section *Circumdati* (Christensen, 1982). However this analysis indicates that it is phylogenetically related to *A. candidus* so it is transferred to section *Candidi*.

Subgenus *Aspergillus* Section *Circumdati*

Christensen (1982) reduced the *A. ochraceus* group to 15 species by removing *A. ochraceoroseus* and *A. coremiiformis*, and accepting the synonymy of *A. sulphureus* and *A. fresenii*. Peterson (1995) removed three additional species from the section i.e. *A. dimorphicus*, *Petromyces alliaceus*, and *P. albertensis*. Some strains from the remaining 12 species (Fig. 4) have identical sequences (*Aspergillus ochraceus, A. melleus, A. ostianus,* and *A. petrakii*). *Aspergillus lanosus* appears to be synonymous with *Petromyces alliaceus* and is moved to section *Flavi* (Peterson, 1995). *Aspergillus robustus* has an unresolved relationship to other species in section *Circumdati*. The species that have distinct lsu-rDNA sequences are *A. ochraceus, A. sulphureus, A. sclerotiorum, A. insulicola, A. bridgeri, A. auricomus,* and *A. elegans.* These seven species have similar sequences and form a monophyletic group, section *Circumdati*.

A. robustus is not shown in the consensus tree but is present near the base of section *Aspergillus* and is of uncertain position in the bootstrapped tree (Fig. 1). *A. robustus* was originally assigned to section *Circumdati*, on morphological basis. In the Kishino Hasagawa test, *A. robustus* was forced into section *Circumdati* and compared with a consensus tree that placed it with the *Eurotium* species. There was no significant difference between the trees. Because *A. robustus* is morphologically distinct from *Eurotium*, it is retained in section *Circumdati*.

Subgenus *Aspergillus*, section *Flavi*

Seventeen ex type strains representing *A. flavus, A. oryzae, A. parasiticus, A. sojae, A. terricola* var. *americana, A. subolivaceus, A. kambarensis, A. flavus* var. *columnaris* and *A. thomii* (Fig. 5) showed identical sequences in the lsu-rDNA, indicating they are very closely related taxa. *A. tamarii, A. flavofurcatis,* and *A. terricola* were reduced to synonymy (Peterson *et al.*, 2000). *A. caelatus, A. leporis, A. nomius* and *Petromyces alliaceus* have similar sequences and are closely related to *A. flavus. A. avenaceus* is more distantly related, but appears in the phylogram as part of section *Flavi. A. zonatus* and *A. clavato-flavus* are not phylogenetically part of section *Flavi*, and will be dealt with later in this chapter. Peterson (1995) showed that *Petromyces alliaceus* belongs in the *A. flavus* clade and that *A. thomii, A. terricola,* and *A. terricola* var. *americana*, formerly assigned to section *Wentii* belong in section *Flavi*.

Fig. 4. Phylogram showing relationship in section *Circumdati*. Three species are synonymous with *A. ochraceus* based on the lsu-rDNA data; *Petromyces* species and *A. lanosus* are transferred to section *Flavi*; *A. campestris* is transferred to section *Candidi*, and the placement of *A. robustus* is unclear although it is illustrated with species of section *Circumdati*. Eight species shown in this figure comprise the revised section *Ochraceus*.

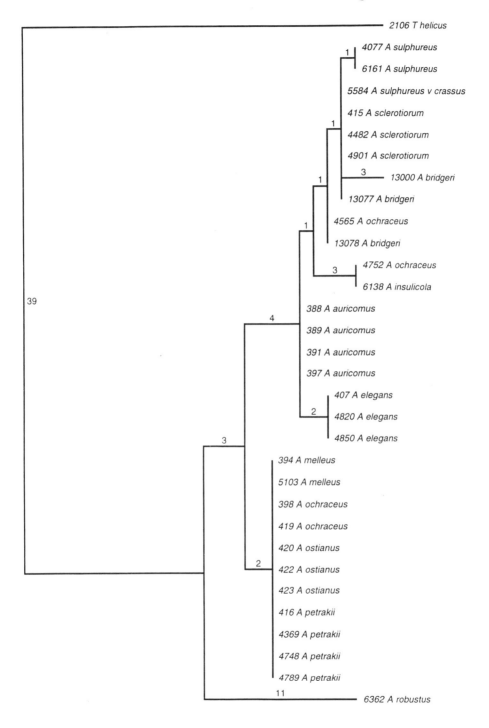

The current study places *A. nomius* near the base of the section *Flavi* clade. When ITS region data are added to the tree (Peterson *et al.*, 2000), *A. leporis* takes the basal position in the section. *A. flavus, A. parasiticus, A. nomius* and *A. caelatus* (Goto *et al.*, 1996; Peterson *et al.*, 2000) occur on the same branch and are the only species that produce aflatoxin, but other species in the tree lie between these species and do not make aflatoxin, suggesting multiple losses of aflatoxigenesis.

Subgenus *Aspergillus*, section *Nigri*

Raper and Fennell (1965) accepted 12 species and two varieties. Al-Musallam (1980) reduced many species to synonymy on the basis of morphology. Kusters-Van Someren *et al.*, (1991) in an RFLP study of ribosomal DNA were largely in agreement with the findings of Al-Musallam (1980), accepting only six species. The current finding are also in agreement with the taxonomy of Al- Musallam (1980) and provide a phylogenetic framework for understanding the species relationships. The uniseriate species form one branch in the section while the biseriate species are on a separate branch. Sections *Nigri* and *Flavi* are sister groups in this phylogram, already apparent from morphological data.

Subgenus *Aspergillus*, section *Cervini*

Section *Cervini* (Figs. 1,2) currently consists of five species *A. cervinus, A. bisporus, A. kanagawaensis, A. parvulus* and *A. nutans*. Sequence differences between strains indicate that *A. kanagawaensis*, may contain more than a single species(Table 1). *A. parvulus* and *A. nutans* may be synonyms on the basis of identical rDNA sequences. The species of section *Cervini*, except *A. bisporus*, occur as a single branch on the consensus tree with 98% bootstrap support. *A. bisporus* is not phylogenetically related to other species in section *Cervini*. It is most closely related to some species in subgenus *Nidulantes* and is transferred there in section *Sparsi*.

Subgenus *Aspergillus*, Genus *Warcupiella*

Five species that do not fit into existing groups branch together near the base of subgenus *Aspergillus* (Figs. 1,2). Those species are *A. clavatoflavus, A. zonatus, Penicilliopsis clavariiformis, Warcupiella spinulosa* and *Penicillium megasporum*. There is no current section in *Aspergillus* that accommodate these species and they are referred here as the "*Warcupiella* group" until a formal taxon is proposed. The phylogenetic relatedness of a *Penicillium* species to members of the *Warcupiella* group is surprising because *Penicillium* species mostly are distinct from *Aspergillus* species in this tree.

Fig 5. Maximum parsimony tree of the isolates from the revised section *Nigri* and *Flavi. Petromyces alliaceus* in treated as a member of section *Flavi* (Peterson, 1995). The section *Nigri* species with uniseriate vesicles are closely related and in the larger tree, their relationship to other species in the section is not proved by high bootstrap values. The strict consensus tree however, places these species together as a sister group to section *Flavi.* Section *Flavi* species other than *A.avenaceus* and *A. leporis* are on a branch showing 81% bootstrap support. The relationship of these two species to the other section *Flavi* taxa is not proven by the bootstrap statistic, and as additional data are added, the hypothesized relationship presented here can be tested. In this tree A. avenaceus branches with section *Nigri* species, but in Fig. 1a, it branches with section *Flavi.* This effect is from the addition of taxa, and there is no high bootstrap value in Fig. 1a validating the position of this species in the tree. Seventeen strains in this tree have sequence identical to *A. flavus* including species that were placed in section *Wentii* by Raper and Fennell (1965). *Aspergillus zonatus* and A. *clavatoflavus* are not closely related to *A. flavus*, branching most closely with *Warcupiella* spinulosa (Fig. 1, 4).

346

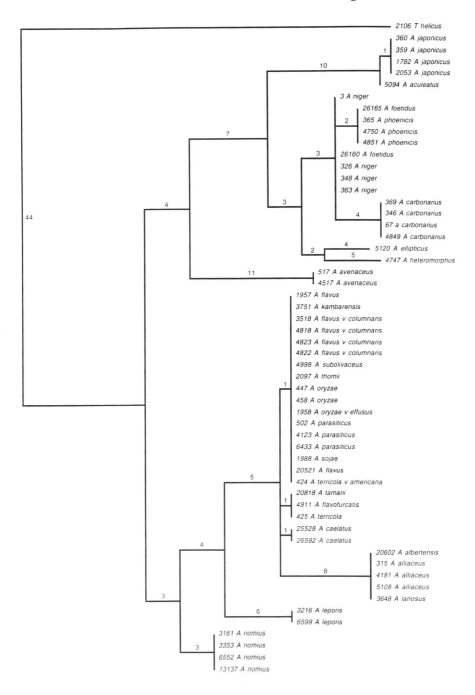

347

Both morphology and sequence data of the *P. megasporum* isolate were rechecked and confirmed. Pitt (1979) stated that this species had little affinity with other species of *Penicillium* because of the distinctive size and ornamentation of the conidia. Samson and Seifert (1985) considered *Penicilliopsis clavariiformis* to be similar to hypothetical ancestral species that gave rise to the mononematous *Aspergillus* and *Penicillium* species. The basal position in the tree of the *Warcupiella* group lends support to that hypothesis. Raper and Fennell (1965) stated their uncertainty about the placement of *A. zonatus* and *A. clavatoflavus* in the *A. flavus* group and this has been confirmed by our phylogenetic analysis.

Subgenus *Nidulantes*, section *Nidulantes*

All examined *Emericella* species lie on a single branch that has 75% bootstrap support (Fig. 6). Most of the species in this group are so closely related that the lsu-rDNA does not resolve the relationships of the taxa with any statistical support. On the same branch with the *Emericella* species are some species from sections *Versicolores*, *Usti*, and *Ornati*. *A. ustus*, type species of section *Usti*, branches well inside of the *Emericella* species and is transferred to section *Nidulantes*, invalidating section *Usti*. *A. deflectus*, also placed in section *Usti* occurs on the *Emericella* branch and is transferred to section *Nidulantes*. *A. raperi* (Raper and Fennell, 1965) and *A. ivoriensis* (Samson, 1979) have been placed in section *Ornati*, but both species occur in the phylogram between species of *Emericella*. These species belong in section *Nidulantes* and are transferred there.

   A. cavernicola differs from *A. varians* at eight nucleotide positions indicating that they are distinct species although (Samson, 1979; Klich 1993) suggested that these species are synonyms. *A. amylovorus* was placed in section *Versicolores* (Samson, 1979), and is closely related to *A. cavernicola*. It is transferred to section *Nidulantes*. *A. pulvinus* was transferred to section *Cremei* (Peterson, 1995).

   A. crystallinus, *A. malodoratus* (section *Versicolores*) and the teleomorph species *Hemicarpenteles paradoxus* all branch with species from *Penicillium* subgenus *Penicillium*, both in the bootstrap tree and in the consensus tree and phylogenetically belong in this genus. *H. paradoxus* is very reminiscent of *Eupenicillium* (Samson, 1979) but according to the original description produces an *Aspergillus* anamorph. Examination of the holotype and ex-type culture of *H. paradoxus* have however not revealed that this species produces an ascigerous state (R.A. Samson, pers. communication), indicating that the teleomorph connection between *Aspergillus* with *Hemicarpenteles* is doubtful.

   All remaining species assigned to section *Versicolores*, except *A. sylvaticus* occur on the *Emericella* branch, between species of *Emericella*, and are transferred to section *Nidulantes*. Species of Section *Versicolores* all belong in other sections.

Fig. 6. One of the equally parsimonious trees of the species in revised section *Nidulantes*. This branch is supported in 75% of the bootstrap samples. Most of the relationships between the species in this clade are not proven statistically. *A. sydowi* and *A. versicolor* form a strongly supported clade. Species included are all *Emericella* species, along with anamorphic taxa from sections *Versicolores*, *Usti* and *Ornati*. The type species of section *Versicolores* branches among *Emericella* species and is transferred to section *Nidulantes*, invalidating section *Versicolores*. All other species from section *Versicolores* fit here or in other sections. Similarly, the type species of section *Usti* is transferred to *Nidulantes* invalidating that section, and the species from section *Usti* all fit in section *Nidulantes*.

Subgenus *Nidulantes*, section *Sparsi*

Section *Sparsi* contains five species (Samson, 1979). Peterson (1995) transferred *A. gorakhpurensis to* section *Cremei*. The remaining species occur on a branch of the phylogram ancestral to the *Emericella* branch. Also included in this branch are *A. bisporus, A. sylvaticus, A. panamensis, A. anthodesmis* and *A. conjunctus. A. bisporus* was placed in section *Cervini* (Samson, 1979) but has no close phylogenetic relationship to that section and the species is transferred to section *Sparsi. A. sylvaticus* was placed in section *Versicolores* (Raper and Fennell, 1965) but branches basal to all *Emericella* species and, because it is phylogenetically related to *A. sparsus*, it is placed in section *Sparsi. A. panamensis* and *A. conjunctus* (section *Usti*) are closely related to *A. sparsus* and they are transferred to section *Sparsi. A. anthodesmis* also belongs in section *Sparsi*.

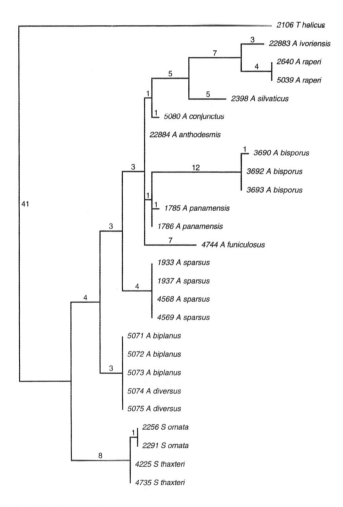

Fig. 7. Revised section *Sparsi* and *Ornati. A. sylvaticus, A. raperi, A. anthodesmis, A. bisporus* and A. ivoriensis are transferred into section *Sparsi. Sclerocleista ornata* and *S. thaxteri* comprise section *Ornati*.

Subgenus *Nidulantes,* section *Ornati*

The two *Sclerocleista* teleomorphs are deeply rooted in the subgenus *Nidulantes* branch (Figs 1,7). These two species are placed in section *Ornati* as the only species of the section.

Subgenus *Fumigati,* section *Clavati*

Section *Clavati* contained three species in the Raper and Fennell (1965) treatment. Samson (1979) accepted one additional species. In the phylogenetic tree (Fig. 8), the species in the *Clavati* clade include *A. clavatus, A. giganteus, A. rhizopodus, A. clavatonanica, A. longivesica* and *Hemicarpenteles acanthosporus* (95% bootstrap value). Samson (1979) considered *A. pallidus* a white variant of *A. clavatus* and the identical sequences of the two species support this. He also considered *A. rhizopodus* as a synonym of *A. giganteus,* but there are lsu-rDNA sequence substitutions compared with *A. giganteus* and it may be a valid species. The *Aspergillus* morphology of *H. acanthosporus* (section *Ornati*) is similar to other species in section *Clavati*. It differs from *A. clavatus* at two nucleotide positions and therefore it is transferred to section *Clavati. H. acanthosporus* has similar teleomorph characteristics as that of *Warcupiella spinulosa* and *Neosartorya spinosa,* but rDNA sequence data (Fig 1,8) show that these taxa are not closely related.

Subgenus *Fumigati,* section *Fumigati*

Section *Fumigati* (Fig. 8) contains holomorphic species from the genus *Neosartorya* and anamorphic species that are uniseriate, and have blue-green to gray-green colony colour (Raper and Fennell, 1965). Few changes to their taxonomic treatment are presented her. Kozakiewicz (1989) gave species status to *N. spinosa* and *N. glabra,* taxa that had previously been varieties of *N. fischeri.* Peterson (1992) agreed but also suggested that *N. spinosa* might contain more species. The rDNA sequences of several strains of *N. spinosa* fall into two categories, those with sequences identical to that of the ex type isolate, and Peterson (1992) showed that *A. fischeri* var. *thermomutatus* is a distinct species and named this taxon *N. pseudofischeri,* a conclusion supported by the lsu-rDNA data. These two genetically distinct groups within the current morphological concept of *N. spinosa* appear to warrant species status. *N. pseudofischeri* branches at the deepest level of section *Fumigati* and *A. fumigatus* branches next on the tree (Fig. 3). The five strains of *N. glabra* examined have identical DNA sequences, but SEM pictures showed that there is some variation in the size and ornamentation of their ascospores (Kozakiewicz, 1989; Peterson, 1992). DNA complementarity of these strains (Peterson, 1992; unpublished data) is above 70%. For that reason, Peterson (1992) considered the variation in the ascospore ornamentation to be subspecific.

Two species from section *Fumigati, A. fumigatus* and *N. pseudofischeri,* are up to now reported to cause human disease (Rippon, 1975; Peterson, 1992; Padhye, *et al.,*1996). Perhaps pathogenicity arose only once in species from this section

A. brunneouniseriatus is basal on the branch that includes subgenus *Fumigati* (Figs 1,8). Raper and Fennell (1965) placed the species in section *Ornati* with the caveat that a close relationship to other species in the section was not implied. This species fits well phenotypically in section *Fumigati*. It is possible that *A. brunneouniseriatus* may have an undiscovered teleomorph.

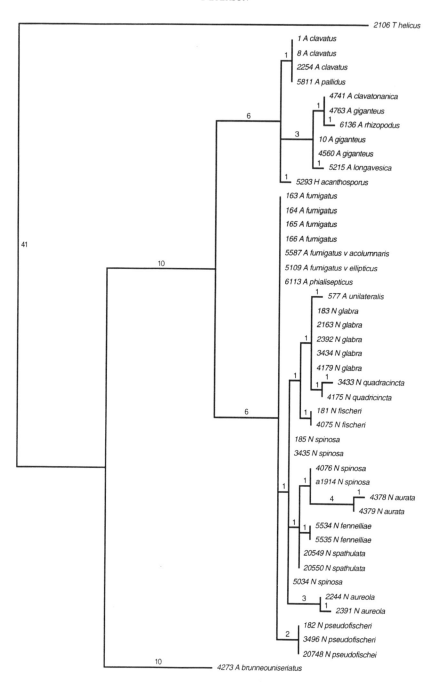

ACKNOWLEDGMENTS

The author is indebted to Gordon A. Adams and Paul A. Bonneau for technical assistance in this project.

REFERENCES

Al- Musallam, A. 1980. Revision of the black *Aspergillus* species. Thesis, Rijksuniversiteit, Utrecht.

Berbee, M. L. & Taylor, J. W. (1992). Two ascomycete classes based on fruiting body characters and ribosomal DNA sequences. Mol. Biol. Evol. 9:278-284.

Berbee, M. L., Yoshimura, J., Sugiyama, J., & Taylor, J. W. (1995). Is *Penicillium* monophyletic? An evaluation of phylogeny in the family Trichocomaceae from 18S, 5.8S and ITS ribosomal DNA sequence data. Mycologia 87:210-222.

Blaser, P. (1975). Taxonomische und physiologische untersuchungen uber die gettung *Eurotium* Link ex Fries. Sydowia 28:1-49.

Chang, J.-M., Oyaizu, H. & Sugiyama, J. (1991). Phylogenetic relationships among eleven selected species of *Aspergillus* and associated teleomorphic genera estimated from 18S ribosomal RNA partial sequences. J. Gen. Appl. Microbiol. 37:289-308.

Chang, P.K., Bhatnagar, D, Cleveland, T. E., and Bennett, J. W. (1995). Sequence variability in homologs of the aflatoxin pathway gene aflR distinguishes species in *Aspergillus* section *Flavi*. Applied and Environmental Microbiology 61:40-43.

Christensen, M. (1981). A synoptic key and evaluation of species in the *Aspergillus flavus* group. Mycologia 72:1056 - 1084.

Christensen, M. (1982). The *Aspergillus ochraceus* group: two new species from western soil and a synoptic key. Mycologia 74:210-225.

Cotty, P. J., Bayman, P., Egel, D. S., & Elias, K. S. (1994). Agriculture, Aflatoxins and *Aspergillus*, p1-28. In: The Genus *Aspergillus*, from taxonomy and genetics to industrial applications, K. A. Powell, A. Renwick & J. F. Perberdy (eds). Plenum Press, New York.

Cruickshank, R. H. & Pitt, J. I. (1990). Isoenzyme patterns in *Aspergillus flavus* and closely related species. In: Advances in *Penicillium* and *Aspergillus* systematics. (Eds. R. A. Samson & J. I. Pitt). Pp.259-265. Plenum Press, NY.

Dupont, J., Dutertre, M., Lafay, J.F., Roquebert, M.F., & Brygoo, Y. (1990). A molecular assessment of the position of Stilbothamnium in the genus *Aspergillus*. Pp. 335-342. In: Modern Concepts in *Penicillium* and *Aspergillus* Classification. (Eds. R. A. Samson & J. I. Pitt). Plenum Press, New York.

Egel, D. S., Cotty, P. J. & Elias, K. S. (1994) Relationships among isolates of *Aspergillus* sect. *Flavi* that vary in aflatoxin production. Phytopathology 84:906-912.

Fig 8. Phylogram of species assigned to sections *Clavati* and *Fumigati*. In the bootstrapped parsimony tree of all species (Fig. 1), these two sections occur on a single branch with a 74% bootstrap value. In the subgeneric treatment of *Aspergillus* (Gams *et al.,* 1985), section *Clavati* is the only section in subgenus *Clavati*, and section *Fumigati* is placed in subgenus *Fumigati* along with section *Cervini*. In the bootstrapped parsimony tree sections *Clavati* and *Fumigati* are sister groups, while section *Cervini* occurs on a distinct branch of the generic tree (Fig. 1). Revised subgenus *Fumigati* includes the sections *Fumigati* and *Clavati* and excludes section *Cervini* and includes *A. brunneouniseriatus*. The species placed in section *Clavati* by Raper and Fennell (1965) and subsequent workers occur on a single branch with 95% bootstrap support. In this tree, A. pallidus appears to be a synonym of *A. clavatus*. Other distinct species in the section are *A. giganteus, A. clavatonanica, A. rhizopodus, A. longavesica* and *Hemicarpenteles acanthosporus. H. acanthosporus* has been placed in section *Ornati* (Samson, 1979). The section is revised to accommodate this teleomorphic species. The other main branch of the tree contains isolates assigned to section *Fumigati. A. fumigatus* strains have identical lsu-rDNA sequences that are also the same as the sequences of varieties ellipticus and a*columnaris* and the species *A. phialisepticus*. These strains appear to be synonymous. *Neosartorya spinosa* isolates appear on two distinct branches, in agreement with the DNA complementarity data of Peterson (1992) and the SEM data of Kozakiewicz ((989). Although there is some morphological divergence among N. glabra strains (Samson et al, 1990; Peterson, 1992) they have identical rDNA sequences. *A. brunneouniseriatus* was placed in section *Ornati* with reservations, by Raper and Fennell (1965) but phylogenetically belongs in subgenus *Fumigati*

Felsenstein, J. (1993). PHYLIP version 3.5, University of Washington, Seattle.

Felsenstein, J. & J. Kishino. (1993). Is there something wrong with the bootstrap on Phylogenies? A reply to Hillis and Bull. Systematic Biology 42:193-200.

Gams, W., Christensen, M., Onions, A. H., Pitt, J. I. & Samson, R. A. (1985). Infrageneric taxa of *Aspergillus*. Pp. 55 62 In: Advances in *Penicillium* and *Aspergillus* systematics. (Eds. R. A. Samson & J. I. Pitt). Plenum Press, NY.

Goto, T., D. T. Wicklow, & Y. Ito. (1996). Aflatoxin and cyclopiazonic acid production by a sclerotium-producing *Aspergillus tamarii* strain. Appl. Environ. Microbiol.: 62:4036-4038.

GuJho, E., Kurtzman, C. P. & Peterson, S. W. (1990). Phylogenetic relationships among species of Sterigmatomyces and Fellomyces as determined from partial rRNA sequences. International Journal of Systematic Bacteriology 40:60-65.

Higgins, D. G. & Sharp, P. M. (1988). CLUSTAL: a package for performing multiple sequence alignments on a microcomputer. Gene 73, 237-244.

Higgins, D. G. & Sharp, P.M. (1989). Fast and sensitive multiple sequence alignments on a microcomputer. CABIOS 5, 151-153.

Hillis, D. M. & J. J. Bull. (1993). An empirical test of bootstrapping as a method for assessing confidence in phylogenetic analysis. Systematic Biology 42:182-192.

Kishino, H. & M. Hasegawa. (1989). Evaluation of the maximum likelihood estimate of the evolutionary tree topologies from DNA sequence data, and the branching order in Hominoidea. Journal of Molecular Evolution 29:170-179.

Klich, M. A. (1993). Morphological studies of *Aspergillus* section *Versicolores* and related species. Mycologia 85:100-107.

Klich, M. A. & Pitt, J. I. (1988a). A Laboratory guide to common *Aspergillus* species and their teleomorphs. CSIRO, Division of Food Processing, North Ryde, Australia. 116 pp.

Klich, M. A. & Pitt. J. I. (1988b). Differentiation of *Aspergillus flavus* from A. *parasiticus* and other closely related species. Transactions of the British Mycological Society 91:99-108.

Kozakiewicz, Z. (1989). *Aspergillus* Species on Stored Products. Mycological Papers, No. 161:1-188.

Kuraishi, H., Itoh, M., Tsuzaki, N., Katayama, Y., Yokoyama, T. & Sugiyama, J. (1990). The ubiquinone system as a taxonomic aid in *Aspergillus* and its teleomorphs, p 395-406. In: Modern Concepts in *Penicillium* and *Aspergillus* Classification. (Eds. R. A. Samson & J. I. Pitt).Plenum Press, New York.

Kurtzman, C.P., Smiley, M. J., Robnett, C. J., & Wicklow, D. T. (1986). DNA relatedness among wild and domesticated species in the *Aspergillus flavus* group. Mycologia 78:955-959.

Kusters-van Someren, M., Samson, R. A., & Visser, J. (1991). The use of RFLP analysis in classification of the *Aspergillus niger* aggregate. Current Genetics 19:21-26.

Malloch, D. & Cain, R. F. (1972). The Trichocomataceae: Ascomycetes with *Aspergillus*, Paecilomyces and *Penicillium* imperfect states. Canadian Journal of Botany 50:2613-2628.

McAlpin, C. E. & Mannarelli, B. (1995). Construction and characterization of a DNA probe for distinguishing strains of *Aspergillus flavus*. Applied and Environmental Microbiology 61:1068-1072.

Padhye, A. A., Godfrey, J. H., Chandler, F. W. & Peterson, S. W. (1994). Osteomyelitis caused by *Neosartorya pseudofischeri*. J. Clinical Microbiol. 32:2832-2836.

Peterson, S. W. (1992). *Neosartorya pseudofischeri* sp. nov. and it relationship to other species in *Aspergillus* section *Fumigati*. Mycological Research 91:547-554.

Peterson, S. W. (1995). Phylogenetic analysis of *Aspergillus* sections *Cremei* and *Wentii*, based on ribosomal DNA sequences. Mycological Research 99:1349-1355.

Peterson, S. W., Horn, B. W. Itoh, Y., & Goto, T. (1997). Genetic variation and aflatoxin production in *Aspergillus tamarii* and A. *caelatus*. In Integration of modern taxonomic methods for *Penicillium* and *Aspergillus* classification (Eds. Samson, R.A. & Pitt, J.I). Harwood Publishers, Amsterdam, pp 447-458.

Peterson, S. W. & Kurtzman, C. P. (1991). Ribosomal RNA sequence divergence among sibling species of yeasts. Systematic and Applied Microbiology 14:124-129.

Pitt, J. I. 1979. The genus *Penicillium* and its teleomorphic states *Eupenicillium* and *Talaromyces*. Academic Press, New York. 634p.

Pitt, J. I. (1985). Nomenclatorial and taxonomic problems in the genus *Eurotium*. In: Advances in *Penicillium* and *Aspergillus* systematics. (eds. R. A. Samson & J. I. Pitt). Pp. 383-396. Plenum Press, New York.

Pitt, J. I. & Samson, R. A. (1990). Taxonomy of *Aspergillus* section *Restricti*. In: Modern Concepts in *Penicillium* and *Aspergillus* Classification. (Eds. R. A. Samson & J. I. Pitt). Pp. 249-258. Plenum Press, New York.

Raper, K. B. & Fennell, D. I. (1965). The Genus *Aspergillus*. Williams and Wilkins, Baltimore, 686 pp.

Rippon, J. W. (1974). Meidcal Mycology. W.B. Saunders Company, Philadelphia. 587pp.

Samson, R. A. (1979). A compilation of the Aspergilli described since 1965. Studies in Mycology 18:1-38.

Samson, R. A. Nielsen, P. V., & Frisvad, J. C. (1990). The genus *Neosartorya*, differentiation by scanning electron microscopy and mycotoxin profiles. In: Modern Concepts in *Penicillium* and *Aspergillus* Classification. (Eds. R. A. Samson & J. I. Pitt). pp. 455-467. Plenum Press, New York.

Samson, R. A. & K. A. Seiffert. (1985). The ascomycete genus *Penicilliopsis* and its anamorphs. In. Advances in *Penicillium* and *Aspergillus* systematics. (eds. R. A. Samson & J. I. Pitt). Pp 397-428. Plenum Press, New York.

Subramanian, C. V. (1972). The perfect states of *Aspergillus*. Curr. Sci. 41:755-761.

Swofford, D. L. (1993). PAUP: Phylogenetic Analysis using Parsimony, Version 3.1.1 Computer program distributed by the Illinois Natural History Survey, Champaign, IL.

Thom, C. & Church, M. B. (1926). The Aspergilli. Williams and Wilkins, Baltimore. pp272.

Thom, C. & Raper, K. B. (1945). A Manual of the Aspergilli. Williams and Wilkins, Baltimore. pp373.

White, T. J., Bruns, T. D., Lee, S. B., & Taylor, J. W. (1990). Amplification and direct sequencing of fungal ribosomal DNA for phylogenetics. In PCR Protocols: A guide to the methods and applications. (eds. Innis, M. A., Gelfand, D. H., Sninsky, J. J., & White, T. J.). pp.315- 322. Academic Press, New York

Yamatoya, K., Sugiyama, J. & Kuraisi, H. (1990). Electrophoretic comparison of enzymes as a chemotaxonomic aid among *Aspergillus* taxa: (2) *Aspergillus* sect. *Flavi*. In Modern Concepts in *Penicillium* and *Aspergillus* Classification. (Eds. R. A. Samson & J. I. Pitt). pp. 395-405. Plenum Press, New York.

Yuan, G. F., Liu, C.S., & Chen, C.C. (1995). Differentiation of *Aspergillus parasiticus* from *Aspergillus sojae* by random amplification of polymorphic DNA. Applied and Environmental Microbiology 61:2384

MOLECULAR PHYLOGENY OF *ASPERGILLUS* AND ASSOCIATED TELEOMORPHS IN THE TRICHOCOMACEAE (EUROTIALES)

Miki Tamura[1], Kazuyoshi Kawahara[2] and Junta Sugiyama[1]
[1] Institute of Molecular and Cellular Biosciences, The University of Tokyo, 1-, Yayoi 1-chome, Bunkyo-ku, Tokyo 113-0032, and [2]Department of Bacteriology, The Kitasato Institute, 5-9-1, Shirokane, Minato-ku, Tokyo 108-8642, Japan.

Phylogenetic relationships among *Aspergillus*, its associated teleomorphs, and relatives within the Trichocomaceae were investigated using ribosomal DNA sequence divergence. We determined the 18S rDNA sequences for 26 species of *Aspergillus* and related teleomorphs and the 28S rDNA partial sequences for three species of *Aspergillus* subgenus *Ornati*. The sequences were analysed phylogenetically by the neighbour-joining, maximum parsimony, and maximum likelihood methods. The phylogeny based on 18S rDNA sequence suggests that *Aspergillus* is basically monophyletic. All examined taxa including eight teleomorphs except for two type species of the teleomorphic genera *Hemicarpenteles* and *Warcupiella*, form a monophyletic group although the bootstrap confidence level is comparatively low. However, *Hemicarpenteles paradoxus* and *Warcupiella spinulosa* aligned with section *Ornati* together *Eupenicillium crustaceum*, *Penicillium chrysogenum*, and *Talaromyces avellaneus* (*Hamigera avellanea*), respectively. All xerophilic species tested of *Aspergillus* sections *Aspergillus* and *Restricti* grouped together, whereas taxa included in subgenera *Circumdati*, *Ornati*, and *Nidulantes* are spread across the genus. *Hemicarpenteles acanthosporus* in section *Ornati* and *A. clavatus* in section *Clavati* form a monophyletic group with strong bootstrap support. Our 18S full and 28S rDNA partial sequence (a total of 2223 aligned sites) analyses support the major topologies in the tree based on 18S rDNA sequences. The impact of molecular evidence on phylogenetic hypotheses concerning *Aspergillus* and associated teleomorphs is discussed.

INTRODUCTION

Aspergillus is one of most economically important fungal genera in fermentation industry, food microbiology, biodeteriolation, and human health. Therefore, the establishment of the taxonomic system reflecting phylogeny and evolution has been requested from various aspects (Bennett and Klich, 1992; Bossche *et al.*, 1988; Powell *et al.*, 1994; Sugiyama *et al.*, 1991).

Traditionally, the systematics of *Aspergillus* and its associated teleomorphs have been based primarily on differences in morphological and cultural characteristics (Raper and Fennell, 1965; Samson, 1979; Klich and Pitt, 1988). Raper and Fennell (1965) recognized 132 species and 18 varieties, and accommodated these in the 18 groups, which were a subgeneric category with no nomenclatural standing. Gams *et al.* (1985) subsequently classified the genus into six subgenera and 18 sections, which were nomenclaturally formalized.

On the other hand, the root of modern systematics of trichocomaceous genera with emphasis on the ascoma ontogeny and anamorph associations goes back to the work of C. R. Benjamin (1955), Fennell (1973, 1977), Malloch and Cain (1972), Subramanian (1972), and Benny and

Kimbrough (1980). Subsequently Malloch (1981) classified 20 genera into five groups based on stages of reduction in ascoma complexity, of which eleven teleomorph genera with an *Aspergillus* anamorph are scattered in all groups (Fig. 1; cf. Fig. 2). Taxonomically, furthermore, Malloch (1985) proposed two new subfamilies, i.e., Trichocomoideae and Dichlaenoideae, within the Trichocomaceae. The former subfamily is characterized by ascomata never produced within a stromatic tissue, prolate ascospores, and a *Penicillium* (biverticillate), *Paecilomyces* or *Polypaecilum* anamorph, whereas the latter subfamily is characterized by various ascomata within stromata or not, oblate, bivalved ascospores, and an *Aspergillus*, *Penicillium, Paecilomyces* or *Polypaecilum* anamorph (cf., Berbee *et al.*, 1995).

In addition to the ascoma diversifications, chemotaxonomically *Aspergillus* and associated teleomorphs are characterized by the heterogeneity of three major ubiquinone systems, i.e., Q-9, Q-10, and Q-10 (H_2) (Fig. 1; Kuraishi *et al.*, 1985, 1990; Yamatoya *et al.*, 1990). Ubiquinone distribution in teleomorph genera producing an *Aspergillus* anamorph is uniform as a whole, except for a few sections, where taxa have been based on the form and structure (Kuraishi *et al.*, 1985, 1990; Yamatoya *et al.*, 1990); for *Penicillium* and associated teleomorphs (see Kuraishi *et al.*, 1991; Ogawa *et al.*, 1997; Yaguchi *et al.*, 1996). One of representatives for exceptional taxa is subgenus *Ornati* section *Ornati* (Sugiyama and Yamatoya, 1990; cf. Kuraishi *et al.*, 1990). The heterogeneity of ubiquinone systems suggests that the taxa of *Aspergillus* and associated teleomorphs should be revised taxonomically as well as in *Penicillium* and associated teleomorphs (Kuraishi *et al.*, 1991). To date the anamorph genus *Aspergillus* contains 114 recognized species and is associated with 72 species of eleven teleomorph genera (Samson, 1992; Pitt and Samson, 1993).

Nuclear ribosomal RNA gene (rDNA) sequence is considered to be very useful for estimating phylogeny and evolution of fungi (e.g., Bruns *et al.*, 1991; Hibbett, 1992; Hill *et al.*, 1990; Hillis *et al.*, 1996). In recent years, full or partial sequence data of small (16S to 18S) and large (23S to 28S) subunit rDNA have exercised great impact on phylogenetic and evolutionary aspects of fungal systematics (e.g., Berbee and Taylor, 1993; Bruns *et al.*, 1992; Nishida and Sugiyama, 1993; Nishida *et al.* 1995; Ogawa *et al.*, 1997).

Dupont *et al.* (1990) sequenced about 130 base pairs of the 28S rRNA in several *Stilbothamnium* and *Aspergillus* strains. Chang *et al.* (1991; cf. Berbee *et al.*, 1995) demonstrated phylogenetic relationships among eleven selected species of *Aspergillus* and associated teleomorphs using the 18S rRNA partial (558 aligned sites) sequence divergence. On the other hand, Berbee and Taylor (1993) suggested closed, cleistothecial ascomata originated more recently from molecular clock evidence (18S rDNA) and fossil records; the family Trichocomaceae, including the *Penicillium*-producing *Talaromyces flavus* and the *Aspergillus*-producing *Eurotium rubrum* is less than 100 Ma old (Berbee and Taylor, 1993). Using 18S, 5.8S and ITS (internal transcribed spacer) rDNA sequence data, Berbee *et al.* (1995) evaluated about whether *Penicillium* is monophyletic. Unfortunately their phylogenetic analysis included only three species of *Aspergillus* and associated teleomorphs (i.e., *Neosartorya fischeri, A. fumigatus*, and *Eurotium rubrum*).

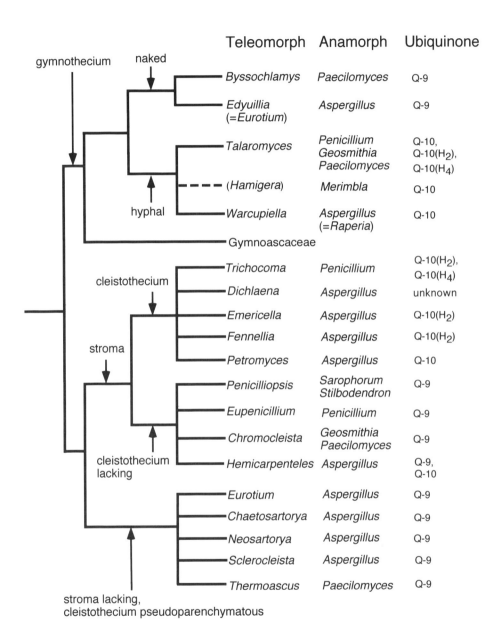

Fig. 1. Morphology-based phylogenetic scheme for genera related to the Trichocomaceae (Eurotiales) with teleomorph-anamorph connections (Pitt, 1995; Pitt and Samson, 1993; Samson, 1992) and the major ubiquinone system (Kuraishi *et al.*,1985, 1990, 1991, unpublished data; Ogawa *et al.*, 1997; Yaguchi *et al.*, 1996).

In the paper, they suggested that *Aspergillus* may be monophyletic. On the other hand, Peterson (1995) demonstrated a phylogenetic hypothesis of *Aspergillus* sections *Cremei* and *Wentii* by a parsimony analysis of large subunit rDNA sequence data; i.e., all species formerly assigned to section *Wentii* are phylogenetically related to species in other sections of subgenus *Circumdati*. Geiser *et al.* (1996) tested phylogenetic relationships among closely related meiotic and strictly mitotic taxa with *Aspergillus* anamorphs using mitochondrial and nuclear DNA sequences. Their analyses support the hypothesis that strictly mitotic lineages arise frequently from more ancient meiotic lineages with *Aspergillus* anamorphs.

Using 18S full and 28S partial rDNA sequences this paper analyses whether *Aspergillus* is monophyletic and discusses phylogenetic relationships in *Aspergillus* and its associated teleomorphs.

MATERIALS AND METHODS

Fungal strains and rDNA sequences. Fungal species and DNA data bank accession numbers of *Aspergillus* species, representing both teleomorphs and anamorphs, are provided in Table 1, together with their major ubiquinone systems. Authentic strains, in most cases derived from the nomenclatural types, were used for DNA sequencing.

Cultivation, harvest of mycelia, and nucleic acid extraction. Cultures were prepared by incubation for 24-48h on a reciprocal shaker at 26°C in 100 ml Czapek yeast extract or Czapek Yeast extract with 20% sucrose liquid medium (Klich and Pitt, 1988). Mycelial cells were harvested by filtration and washed with 0.05 M phosphate buffer (pH 7.8) containing 0.25 M sucrose, then frozen at −80°C and lyophilized. Total DNA was isolated from lyophilized mycelium using a modification of the methods of Ogawa *et al.* (1997).

PCR and DNA sequencing. Methods used for PCR (polymerase chain reaction) and rDNA sequencing were fully described by Ogawa *et al.* (1997). PCR products spanning about 1700 bp of the 18S rDNA were amplified with the primer pairs N1 and N2 (Nishida and Sugiyama, 1993). The PCR reaction (100 µL) included 0.1 µg of fungal genomic DNA as template, 1x PCR buffer, 1.5 mM $MgCl_2$ and 0.2m M dNTPs, 0.8 µM of each primer and 2.5 U Takara *Taq* DNA polymerase (Takara Shuzo Co., Ltd., Shiga, Japan). Thermal cycling parameters used were an initial denaturation at 94°C for 2 min, followed by 35 cycles consisting of denaturation at 94°C for 1 min, annealing at 52°C for 1 min, and extension at 72°C for 2.5 min. A final extension at 72°C for 7.5 min was done at the end of the amplification. The 5' end of the 28S rDNA partial fragments for *Hemicarpenteles acanthosporus*, *H. paradoxus*, and *Warcupiella spinulosa*, spanning approximately 1400 bp was amplified with the primer pairs LROR and LR7 (Ogawa *et al.,* 1997). The 18S rDNA fragments from *Eurotium repens*, *Sclerocleista ornata*, and *Chaetosartorya cremea* were purified with Sephaglas[a] BandPrep Kit (Pharmacia Biotech Co., Ltd., California, USA) to remove unincorporated primers and sequences were obtained for both strands using direct sequencing performed with Sequenase 2.0 (US Biochemical Corp., Ohio, USA) and ^{35}S-labeled dATP (Amersham Life Science Corp., Illinois, USA) following the manufacturer's recommendations.

Table 1. Strains and rDNA sequences determined in this study.

Species[a]	Strain[b]	Accession number 18S	Accession number 28S	Major ubiquinone system
Subgenus *Aspergillus*				
Section *Aspergillus*				
A. manginii (Margin) Thom & Church[c]	NRRL 116 (**T**)	AB002069		Q-9[f]
A. proliferans G. Smith[d]	WB 1908 (**T**)	AB002083		Q-9[f]
Eurotium amstelodami L. Mangin	FRR 21792 (**NT**)	AB002076		Q-9[f]
Eurotium cristatum (Raper & Fennell) Malloch & Cain	WB 4222 (**T**)	AB002073		Q-9[f]
Eurotium repens de Bary	WB 13 (**NT**)	AB002084		Q-9[f]
Edyuillia athecia (Raper & Fennell) Subramanian[e]	NRRL 5000 (**T**)	AB002082		Q-9[f]
Sect. *Restricti*				
A. restrictus G. Smith	FRR 2176	AB002079		Q-9
A. penicillioides Spegazzini	NRRL 4548 (**NT**)	AB002060		Q-9
	IMI 144121	AB002077		Q-9
	IFO 8155	AB002078		Q-9
Subgenus *Fumigati*				
Sect. *Cervini*				
A. cervinus Massee	NRRL 5025 (**T**)	AB003808		Q-9[f]
Subgenus *Ornati*				
Hemicarpenteles paradoxus A.K. Sarbhoy & Elphick	IFO 8172 (**LT**)	AB002080	AB003807	Q-10[f]
Hemicarpenteles acanthosporus Udagawa & Takada	IFO 9490(**T**)	AB002075	AB003809	Q-10[f]
Sclerocleista ornata (Raper et al.) Subramanian	NRRL 2256(**T**)	AB002067		Q-10[f]
Warcupiella spinulosa (Warcup) Subramanian	IFO 31800(**T**)	AB002081	AB003806	Q-10[f]
Subgenus *Clavati*				
A. clavatus Desmazières	NRRL1 (**T**)	AB002070		Q-10[f]
Subgenus *Nidulantes*				
Sect. *Usti*				
A. ustus (Bainier) Thom & Church	NRRL 275 (**T**)	AB002072		Q-10(H$_2$)[f]

361

Table 1(continued). Strains and rDNA sequences determined in this study.

Species[a]	Strain[b]	Accession number		Major ubiquinone[e] system
		18S	28S	
Sect. *Flavipedes*				
A. flavipes (Bainier & Sartory) Thom & Church	NRRL 302 (**NT**)	AB002062		Q-10(H$_2$)[f]
Fennellia flavipes Wiley & Simmons	NRRL 5504T	AB002061		Q-10(H$_2$)[f]
Sect. *Versicolores*				
A. versicolor (Vuillemin) Tiraboschi	NRRL 238 (**T**)	AB0020640		Q-10(H$_2$)[f]
Subgenus *Circumdati*				
Sect. *Circumdati*				
A. ochraceus K. Wilhelm	NRRL 1642 (**NT**)	AB002068		Q-10(H$_2$)[f]
Petromyces alliaceus Malloch & Cain	NRRL 4181 (**T**)	AB002071		Q-10[f]
Sect. *Wentii*				
A. wentii Wehmer	JCM 2724 (**NT**)	AB002063		Q-9[f]
Sect. *Candidi*				
A. candidus Link : Fries	IAM 13850 (**T**)	AB002065		Q-10(H$_2$)[f]
Sect. *Cremei*				
Chaetosartorya cremea (Kwon-Chung & Fennell) Subramanian	NRRL 5081L (**T**)	AB002074		Q-9[f]
Sect. *Sparsi*				
A. sparsus Raper & Thom	IAM 13904 (**T**)	AB002066		Q-10(H$_2$)[f]

[a] For anamorph names, see Samson (1992), and Pitt and Samson (1993). [b] (**T**), strain derived holotype; (**LT**), strain derived from lectotype; (**NT**), strain derived from neotype. Abbreviations of culture collections: IMI, International Mycological Institute, Bakeham Lane, Surrey, U.K.; FRR, CSIRO Division of Food Science, North Ryde, NSW Australia; IAM, Institute of Molecular and Cellular Biosciences, University of Tokyo, Tokyo, Japan; IFO, Institute for Fermentation, Osaka, Japan; JCM, Japan Collection of Microorganisms, RIKEN, Wako, Saitama, Japan; NRRL, ARS Culture Collection, National Center for Agricultural Utilization Research, USDA, Peoria, IL, USA; WB, Department of Bacteriology and Botany, University of Wisconsin, Madison, WI, USA; [c] Isolate WB 116 has been received as *Aspergillus mangini* (Mangin) Thom & Church from M. Christensen in 1989. Pitt (1990) placed *A. mangini* [sic] in synonym with *Eurotium herbariorum* (Wiggers) Link. [d] This species has not been reated as a separate species (Pitt and Samson, 1993), because it is an aberrant, abortive form of an *Eurotium* (see Samson, 1993)[e]. For the taxonomic status see Samson (1993)[,]. Data from Kurashi *et al.,* (1990)

362

Four sequencing primers were used in each direction; N1, NS 3, NS 5 and NS 7 for sequences in the 5' to 3' direction, and NS 2, NS 4, and NS 6 and NS 2 for sequences in the 3' to 5' directions (Ogawa *et al.*, 1997; White *et al.*, 1990). Sequencing reactions were performed by a modification of Ogawa *et al.* (1997). For the remaining of 18S and 28S rDNAs sequence determination, each of the PCR products was cloned in the TA vector pCR2.1 (Invitrogen Corp., San Diego, California, USA). After purifying TA vectors with QIAprep spin miniprep (QIAGEN GmBH, Hilden, Germany), sequence reactions were performed. For every sequence determination, an AmpliTaq FS DyePrimer cycle sequencing kit (Perkin-Elmer Corp., California, USA) was used for sequencing the 5' and 3' ends of inserted DNA fragments. The remaining regions of the 18S and 28S rRNA genes were cycle sequenced using Dye Primer cycle sequencing kit (Applied Biosystems, Foster City, California, USA). The data collections were performed on an Applied Biosystems 373S automated DNA sequencer.

Sequence and phylogenetic analysis. The accession numbers of DNA sequences determined in this study are given in Table 1. Other published rDNA sequences used in our phylogenetic analyses are as follows: *Hypocrea lutea* (Tode) Petch (D14407); *Neurospora crassa* Shear & B. Dodge (X04971, U40124); *Geosmithia putterillii* (Thom) Pitt (D88318, D88326); *Ascosphaera apis* (Maasen ex Claussen) Olive & Spiltoir (M83264); *Eremascus albus* Eidam (M83258); *Coccidioides immitis* Rixford & Gilchrist (X58571); *Onygena equina* (Willdenow: Fries) Persoon (U45442); *Elaphomyces maculatus* Vittadini (U45440); *E. leveillei* Tulasne (U45441); *Thermoascus crustaceus* (Apinis & Chesters) Stolk (M83263); *Byssochlamys nivea* Westling (M83256); *Monascus purpureus* Went (M83260); *Talaromyces bacillisporus* (Swift) C. R. Benjamin (D14409); *T. macrosporus* (Stolk & Samson) Frisvad *et al.*, (M83262); *T. emersonii* Stolk (D88321, 28S rDNA; Ogawa *et al.*, 1997); *T. avellaneus* (Thom & Turesson) C. R. Benjamin [*Hamigera avellanea* Stolk & Samson] (D14406, AB000620); *Merimbla ingelheimensis* (van Beyma) Pitt (D14408); *Eupenicillium crustaceum* Ludwig (D88324, AB000486); *Penicillium chrysogenum* Thom (M55628); *Chromocleista malachitea* Yaguchi & Udagawa (D88323, AB000621); *Geosmithia namyslowskii* (Zaleski) Pitt (D88319, AB000487); *G. cylindrospora* (G. Smith) Pitt (D88320, 28S rDNA; Ogawa *et al.*, 1997); *Aspergillus clavatus* Desmazieres (U28889); *A. flavus* Link (X78537, D63696 for 18S rRNA gene; U15487 for 28S rRNA gene); *A. sojae* Sakaguchi & Yamada (D63700); *A. niger* van Tieghem (D63697); *A. awamorii* Nakazawa (D63695); *A. tamarii* Kita (D63701); *A. oryzae* (Ahlburg) Cohn (D63698); *A. parasiticus* Speare (D63699); *A. fumigatus* Fresenius (M55626); *Fennellia flavipes* B.J. Wiley & E.G. Simmons (U28853); *A. penicillioides* Spegazzini (U29640, U29646); *Emericella nidulans* (Eidam) Vuillemin (X78539, U29841); *Neosartorya fischeri* (Wehmer) Malloch & Cain (U21299) and *Eurotium rubrum* König *et al.* (U00970, U29543).

Methods used for phylogenetic analyses were also fully described by Ogawa *et al.* (1997). DNA sequences were aligned with Clustal W (Version 1.60, an updated version of Clustal W Version 1.4, Thompson *et al.*, 1994) and visually corrected. Phylogenetic relationships were estimated from the aligned sequences for each data set using Clustal W 1.6 (the neighbor-joining (NJ) method developed by Saitou and Nei (1987) with Kimura's (1980) two-parameter model; transition/transversion = 2.0), for distance analyses, or PAUP 3.1.1 (method developed by Swofford, 1993), using heuristic search option with 10 random-addition sequences, run on a PowerMacintosh 8500/120 computer. The neighbor-joining tree was constructed excluding gaps, while the maximum parsimony (MP) tree was constructed using PAUP 3.1.1 with gaps treated as missing data. The DNAML program in the PHYLIP 3.572c package (Felsenstein, 1993) was used for the maximum likelihood (ML) analysis. Support for the topologies was

363

obtained with bootstrap analysis using 1000 replications (Felsenstein, 1985).

RESULTS AND DISCUSSION

In this study we determined full 18S rDNA sequences of 26 species of *Aspergillus* and associated teleomorphs, and partial 28S rDNA sequences (549 aligned sites) for three teleomorph taxa (Table 1).

Phylogenetic relationships among plectomycete orders and families within the Ascomycota
Sequences from 26 fungal taxa, including ten species of *Aspergillus* determined in this study, were aligned and yielded 1633 sites. The maximum parsimony tree is shown in **Fig. 2** with groupings based on stages of reduction in ascoma complexity (Malloch, 1981), the major ubiquinone systems (Kuraishi *et al.,* 1990, 1991, unpublished data; Ogawa *et al.,* 1997), and teleomorph-anamorph connection (e.g., Samson, 1992; Pitt and Samson, 1993). The major topologies in this tree agreed well with those in neighbor-joining and maximum likelihood trees (not illustrated). In Fig. 2, the taxa of the families Trichocomaceae, Monascaceae, and Elaphomycetales appear as a monophyletic group in 100% of bootstrap replications. The 15 taxa in the Trichocomaceae also cluster in 74% of the parsimony bootstrap replications. Nine representative species of *Aspergillus* and associated teleomorphs have 56% bootstrap support. Outlieres include *Hemicarpenteles paradoxus* and *Warcupiella spinulosa.. H. paradoxus*, the type species of the genus, forms a monophyletic clade at 91 % confidence level, with the anamorph *Penicillium chrysogenum* and *Eupenicillium crustaceum*. On the other hand, *W. spinulosa*, grouped with the *Merimbla*-producing *Talaromyces* (*Hamigera*) *avellaneus* with 75% bootstrap support. The phylogenetic position of *H. acanthosporus* is remarkable. This species classified in section *Ornati* and *Aspergillus clavatus* from section *Clavati* form a well supported clade (98%). The relationship of both species is discussed later.

In the Trichocomaceae the genera *Byssochlamys*, *Talaromyces*, and *Warcupiella*, which are characterized by the production of naked or hyphal cleistothecia, are basal to *Aspergillus*, *Penicillium*, and *Merimbla* species and their teleomorphs.

In this cladogram, the position of the hypogeous genus *Elaphomyces* represented by *E. leveillei* and *E. maculatus* is closer to the Eurotiales (Trichocomaceae) rather than the Onygenales. This genus has previously been placed in the discomycete family Elaphomycetaceae in the Tuberales with orther hypogeous taxa (Korf, 1973). As shown in Fig. 2, the Elaphomycetales. Our placement for *Elaphomyces* is in agreement with those of Landvik *et al.* (1996) and LoBuglio *et al.* (1996).

Is *Aspergillus* monophyletic ?
We analysed whether *Aspergillus* is monophyletic or not from the bootstrapped neighbor-joining and maximum parsimony trees (Fig. 3) were constructed based on a total of 2223 aligned sites from 18S (1674 bp) and 28S partial (549 bp) of rDNA using two pyrenomycetous species, i.e., *Geosmithia putterilli* (Ogawa *et al.,* 1997) and *Neurospora crassa*. In the neighbouring joining tree, nine *Aspergillus* species, three anamorph species and six teleomorphs, grouped together, but the bootstrap confidence level is low (37%).

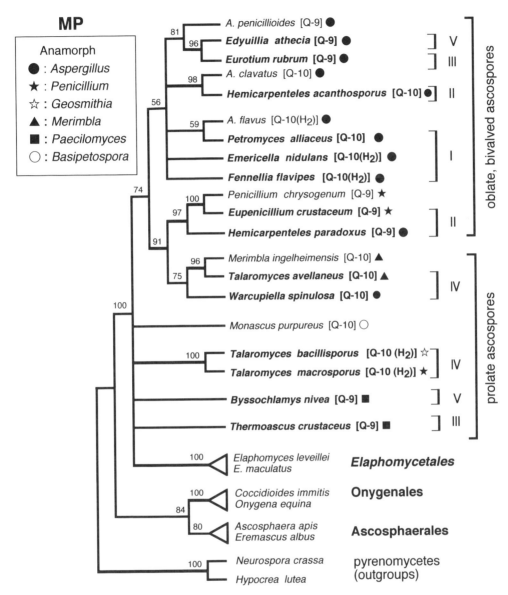

Fig. 2. Phylogenetic relationships of the Eurotiales and related taxa inferred from 1663 alignable sites of 18S rDNA sequences. The bootstrapped 50% majority-rule consensus tree was constructed by the heuristic unweighted parsimony analysis (Tree length=516, CI=0.688, HI=0.312, RI=0.728, RC=0.501). The bootstrap percentages derived from 1000 replications were indicated at the respective nodes of the tree. The grouping of ascoma form and development is based on Malloch (1981). I: Stroma with cleistothecia. II: Stroma with cleistothecia lacking. III: Stroma lacking, cleistothecia pesudoparenchymatous. IV Stroma lacking, cleistothecia (gymnothecia) hyphal. V: Stroma and cleistothecia lacking. Major ubiquinone systems (Kuraishi *et al.*, 1990; 1991, unpublished data; Ogawa *et al.*, 1997) are also shown. Bold letters indicate teleomorph species.

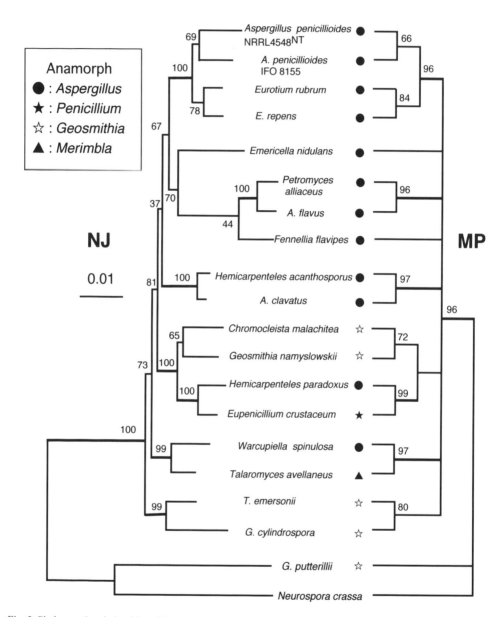

Fig. 3. Phylogenetic relationships of the Eurotiales inferred from 2223 aligned sites of 18S & 28S rDNA sequences. The neighbor-joining tree (NJ) was constructed by the NJ method. The scale bar indicates one base change per 100 nucleotide positions. The bootstrap 50% majority-rule consensus tree (MP) was constructed by the maximum parsimony method. It was made by the heuristic unweighted parsimony analysis (tree length=772, CI=0.710, HI=0.290, RI=0.652, RC=0.463). The bootstrap percentages derived from 1000 replications were indicated at the respective nodes of both trees.

H. paradoxus and *Eupenicillium crustaceum*, both species characterized by sclerotioid cleistothecia and the identical major ubiquinone system Q-9, form a monophyletic group in 100% of bootstrap replications (cf. Fig. 1). Pitt's speculation (Pitt, 1995) based on comparative morphology of cleistothecia supports our molecular phylogenetic trees (Figs. 2 and 3). Actually *Hemicarpenteles* is distinguished primarily by production of an aspergillum, not a penicillus.

On the other hand, *Warcupiella spinulosa* and the *Merimbla*-producing *Talaromyces avellaneus*, also form a monophyletic group at a high bootstrap confidence level (99% in the NJ tree, 97% in the MP tree). The two are characterized by hyphal ascomata which lack a pseudoparenchymatous peridium and the identical major ubiquinone system Q-10 (Figs. 1 and 2). *Hamigera* has been proposed by Stolk and Samson (1971) for *T. avellaneus*, which forms asci singly from croziers. *Talaromyces* differs in having asci borne in chains and *Byssochlamys* differs in lacking ascomata (Benny and Kimbrough,1980; Malloch and Cain, 1972). Our molecular phylogeny clearly indicates *Warcupiella spinulosa* is closely related to *Hamigera avellanea* Stolk & Samson (*T. avellaneus*) rather than other *Talaromyces* species. Therefore, the placement of *Warcupiella* in section *Ornati* is questionable at least. By Subramanian and Rajendram (1975), *Raperia* was proposed to accommodate an anamorph species *Aspergillus warcupii* Samson & Gams (*A. spinulosus* Warcup) associated with a teleomorph *Warcupiella spinulosa* (Pitt and Hocking, 1985; Subramanian, 1972). Our molecular phylogeny may suggest a revival of *Raperia* which is distantly related to the majority of *Aspergillus* species. *Warcupiella spinulosa*, *Chaetosartorya chrysella*, *Eurotium amstelodami*, and *A. restrictus* clustered together in DNA sequence-based trees from the mitochondrial small subunit, the nuclear ribosomal ITS, and the nuclear 5.8S ribosomal gene (Geiser *et al.,* 1996). Our placement for *W. spinulosa* in the 18S rRNA gene phylogeny may conflict with the results of phylogenetic analysis of mitochondrial and nuclear data sets by Geiser *et al.* (1996). However, no account was given for *W. spinulosa* in their paper. We look forward to future studies with other genes which resolve this conflict and evolutionary relationships among *Warcupiella*, *Talaromyces*, and *Hamigera*, and their anamorphic taxa. For the moment, the results of our phylogenetic analyses suggest that *Aspergillus* is not monophyletic but has three different evolutionary lines within the Trichocomaceae.

As stated above, the majority of *Aspergillus* taxa and associated with teleomoprhs, which are characterized by various grades of cleistothecial ascoma ontogeny (Figs.1 -3), cluster within the Trichocomaceae. These results may suggest that the diversity of the ascoma form and structure (Malloch, 1981; Malloch and Cain, 1972), in addition to the heterogeneity of the major ubiquinone systems, occurred rapidly during the last ca. 100 Ma years in the light of molecular clock of the 18S rRNA gene (Berbee and Taylor, 1993).

Phylogenetic relationships among the representative taxa of *Aspergillus* and associated teleomorphs.

Further we extensively analysed phylogenetic or genealogical relationships among 34 selected species of *Aspergillus* and associated teleomorphs, including the 18 type species of six subgenera and 18 sections, using the ML method. The ML tree (Fig. 4) was constructed based on 1614 aligned sites of 18S rDNA sequence, using two *Talaromyces* species as outgroups.

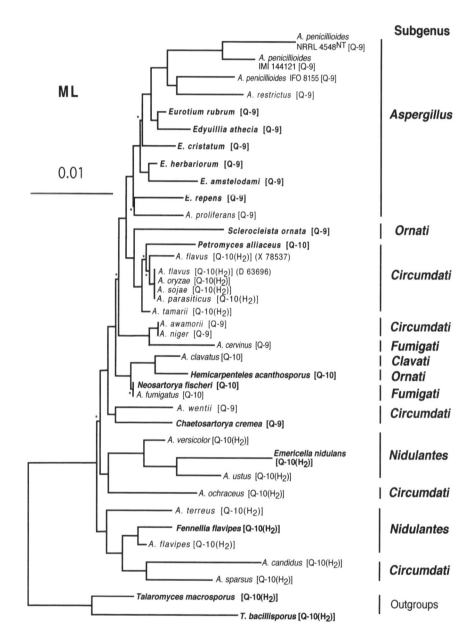

Fig. 4. Phylogenetic relationships of *Aspergillus* and related teleomorph genera inferred from 1614 aligned sites of 18S rDNA sequences by the maximum likelihood (ML) method. The scale bar indicates one base change per 100 nucleotide positions. The asterisks are not significantly positive probability error (P>0.05) and the others are significantly positive probability error (P>0.01). Bold letters indicate teleomorph species.

As a result, there is a good correlation between the major clusters and the major ubiquinone systems. In addition, the phylogeny indicates the diversity of subgenera *Ornati, Circumdati, Fumigati,* and *Nidulantes.* In contrast, the subgenus *Aspergillus,* accommodating xerophilic species, with *Eurotium* and *Edyuillia* teleomorphs, may be monophyletic. In the tree the strictly anamorph species *A. penicillioides, A. restrictus,* and *A. proliferans,* five *Eurotium* teleomorphs represented by *E. herbariorum,* and the monotypic teleomorph species *Edyuillia athecia,* group together. All species have Q-9 as the major ubiquinone system (Kuraishi *et al.,* 1990; Tamura *et al.,* 1996). In the 18S rDNA sequence-based phylogeny both genera *Eurotium* and *Edyuillia* show a close relationship. It supports Samson's opinion (1992) that *Edyuillia* is an aberrant form of *Eurotium.* Further accounts and discussion concerning this problem and the phylogenetic relationships among xerophilic taxa of *Aspergillus* and associated teleomorphs will be published elsewhere.

Petromyces alliaceus, the teleomorph of *A. alliaceus* Thom & Church, and *A. flavus* and phenotypically similar four species of section *Flavi,* all assigned to the subgenus *Circumdati,* grouped together. Taxa of section *Flavi* strictly lack their teleomorphs. The phylogenetic position of *P. alliaceus* is worthy of notice. Our placement of *P. alliaceus* into section *Flavi* is supported by large subunit ribosomal DNA sequence divergence (Peterson, 1995). In the major ubiquinone system, *P. alliaceus* has Q-10, whereas *A. flavus* and its relatives posses Q-10(H_2). Therefore, the case of *P. alliaceus* is similar to *A. flavus* var. *asper* having Q-10 in section *Flavi* (Kuraishi *et al.,* 1990; Yamatoya *et al.,* 1990). In the *A. flavus* cluster, an alignment comparison between CBS 108.30 (X78537, Melchers *et al.,* 1994) and NFRI 1212 (= NRRL 11612; D63696, Nikkuni *et al.,* 1996) of *A. flavus* showed nine base changes, of which four were assigned to gaps. Verification of strain identity for CBS 108.30 may be required in the light of nuclear DNA relatedness among wild and domesticated species in section *Flavi* (Kurtzman *et al.,* 1986).

The molecular phylogeny indicates two black *Aspergillus* species, *A. awamori* and *A. niger* having the Q-9 system in section *Nigri,* related to *A. cervinus* having the Q-9 system in section *Cervini.* It suggests that the subgeneric assignment of section *Cervini* should be reconsidered. *Aspergillus clavatus,* the type species of section *Clavati,* subgenus *Clavati,* groups with *Hemicarpenteles acanthosporus,* the second species of section *Ornati,* both having the identical major ubiquinone system Q-10. Our ML tree also shows *H. acanthosporus* related to *Neosartorya fischeri* having Q-10 as the major ubiquinone in section *Fumigati.* This relationship is attractive in the light of ascoma ontogeny and the major ubiquinone system.

In conclusion, the 18S and 28S rRNA gene phylogeny indicates that *Aspergillus* is not monophyletic within the Trichocomaceae. However, the majority of *Aspergillus* and associated teleomorphs groups together, except for *Hemicarpenteles paradoxus* and *Warcupiella spinulosa.* The *Aspergillus*-producing *H. paradoxus,* the type species of the genus in section *Ornati* is related to the *Penicillium*-producing *Eupenicillium crustaceum* and the strictly anamorph species *Penicillium chrysogenum,* whereas the *Aspergillus*-producing *W. spinulosa* in section *Ornati* is related to the *Merimbla*-producing *Talaromyces avellaneus* (*Hamigera avellanea*) rather than other *Talaromyces* species. The existing taxonomies of *Aspergillus* and associated teleomorphs at the subgeneric and sectional levels shculd be reconstructed based on their phylogeny and evolutionary relationships.

ACKNOWLEDGEMENTS

We thank Dr. H. Kuraishi for providing unpublished ubiquinone data on *Paecilomyces* spp. We also thank Prof. Emeritus M. Christensen (WB strains), Prof. D. L. Hawksworth (IMI, now

369

MycoNova), Dr. C. P. Kurtzman (NRRL), Dr. T. Nakase (JCM), Dr. S. P. Peterson (NRRL), Dr. J. I. Pitt (FRR), and Dr. M. Takeuchi (IFO) for providing fungal cultures. Our thanks to Mr. H. Ogawa for assistance in computer work.

REFERENCES

Benjamin, C. R. (1955). Ascocarps of *Aspergillus* and *Penicillium*. Mycologia 47: 669- 687.

Bennett, J.W. & Klich, M.A. (Eds) (1992). *Aspergillus*, biology and industrial applications. Boston, Butterworth-Heinemann.

Benny, G.L. & Kimbrough, J.W. (1980). A synopsis of the order and families of plectomycetes with keys to genera. Mycotaxon 12: 1-91.

Berbee, M.L. & Taylor, J.W. (1993). Dating the evolutionary radiations of the true fungi. Can. J. Bot. 71: 1114-1127.

Berbee, M.L., Yoshimura, A., Sugiyama, J. & Taylor, J.W. (1995). Is *Penicillium* monophyletic? An evaluation of phylogeny in the family Trichocomaceae from 18S, 5.8S and ITS ribosomal DNA sequence data. Mycologia 87: 210- 222.

Bossche, H. V., Mackenzie, D. W. R. & Cauwenbergh, G. (Eds) (1988). *Aspergillus* and aspergillosis. New York, Plenum Press.

Bruns, T.D., Fogel, Vilgalys, R., Barns, S.M., Gonzalez, D., Hibbett, D.S., Lane, D. J., Simon, L., Stickel, S., Szaro, T.M., Weisburg, N.G. & Sogin, M.L. (1992). Evolutionary relationships within the Fungi: Analyses of molecular small subunit rRNA gene. Mol. Phylogenet. Evol. 1: 231-241.

Bruns, T.D., White, T.J .& Taylor, J.W. (1991). Fungal molecular systematics. Annu. Rev. Ecol. Syst. 22: 525-564.

Chang, J.M., Oyaizu, H. & Sugiyama, J. (1991). Phylogenetic relationships among eleven selected species of *Aspergillus* and associated teleomorphic genera estimated from 18S ribosomal RNA partial sequences. J. Gen. Appl. Microbiol. 37: 289-308.

Dupont, J., Dutertre, M., Lafay, J.-F., Roquebert, M.-F. & Bryoo, Y.(1990). A molecular assessment of the position of *Stilbothamnium* in the genus *Aspergillus*. In Modern concepts in *Penicillium* and *Aspergillus* classification (Eds Samson, R. A. & Pitt, J. I.). New York, Plenum Press. pp. 335-342.

Felsenstein, J. (1985). Confidence limits on phylogentics: an approach using the bootstrap. Evolution 39: 783-791.

Felsenstein, J. (1993). PHYLIP (Phylogeny Inference Package) version 3.5c. Distributed by the author. Seattle, Department of Genetics, University of Washington.

Fennell, D.I. (1973). Plectomycetes; Eurotiales. In The fungi, an advanced treatise, Vol. 4A (Eds Ainsworth, G.C, Sparrow, F.K. & Sussman, A.S.). New York, Academic Press. pp. 45-68.

Fennell, D.I. (1977). *Aspergillus* taxonomy. In Genetics and physiology of *Aspergillus* (Eds Smith, J.E. & Paterman, J.A.). London, Academic Press. pp. 1-21.

Gams, W., Christensen, M., Onions, A.H., Pitt, J.I. & Samson, R.A. (1985). Infrageneric taxa of *Aspergillus*. In Advances in *Penicillium* and *Aspergillus* systematics (Eds Samson, R.A. & Pitt, J.I.). New York, Plenum Press. pp. 55-62.

Geiser, D.M., Timberlake, W.E. & Arnold, M.L. (1996). Loss of meiosis in *Aspergillus*. Mol. Biol. Evol. 13: 809-817.

Hibbett, D.S. (1992). Ribosomal RNA and fungal systematics. Trans. Mycol. Soc. Japan 33: 533-556.

Hill, W.E., Dahlberg, A., Garrett, R.A., Moore,P.B., Shlessinger, D. & Warner. J.R. (Eds) (1990). The ribosome: structure, function, & evolution. Washington, D.C., American Society of Microbiology.

Hillis, D.M., Moritz, C. & Mable, B.K.(Eds) (1996). Molecular systematics, 2nd edition. Sunderland, Massachusetts, Sinauer Assoc.

Kimura, M. (1980). A simple method for estimating evolutionary rate of base subsittutions through comparative studies of nucelotide sequence. J. Mol. Evol. 16: 111-120

Klich, M.A. & Pitt, J.I. (1988). A laboratory guide to common *Aspergillus* species and their teleomorphs. North Ryde, CSIRO Division of Food Processing.

Korf, R.P. (1973). Discomycetes and Tuberales. In The fungi, an advanced treatise, Vol. 4A (Eds Ainsworth, G.C., Sparrow, F.K. & Sussman, A.S.). New York, Academic Press. pp. 249-319.

Kuraishi, H., Itoh, M., Katayama, Y., Sugiyama, J. & Pitt, J.I. (1991). Distribution of ubiquinones in *Penicillium* and related genera. Mycol. Res. 95: 705-711.

Kuraishi, H., Itoh, M., Tsuzaki, N., Katayama, Y., Yokoyama, T. & Sugiyama, J. (1990). The ubiquinone system as a taxonomic aid in *Aspergillus* and its teleomorphs. In Modern concepts in *Penicillium* and *Aspergillus* classification (Eds Samson, R.A. & Pitt, J.I.). New York, Plenum Press. pp. 407-421.

370

Kuraishi, H., Katayama- Fujimura, Y., Sugiyama, J. & Yokoyama, T. (1985). Ubiquinone systems in fungi I. Distribution of fungi in the major families of ascomycetes, basidiomycetes, and deuteromycetes, and their taxonomic implications. Trans. Mycol. Soc. Japan 26: 383-395.

Kurtzman, C. P., Smiley, M. J., Robnett, C. J. & Wicklow, D. T. (1986). DNA relatedness among wild and domesticated species in the *Aspergillus flavus* group. Mycologia 78: 955-959.

Landvik, S., Shailer, Neil, F. & Eriksson, O.E. (1996). SSU rDNA sequence support for a close relationship between the Elaphomycetales and the Eurotiales and Onygenales. Mycoscience 37: 237-241.

LoBuglio, K.F., Berbee, M.L. & Taylor, J.W. (1996). Phylogenetic origins of the asexual mycorrhizal symbiont *Cenococcum geophilum* Fr. and other mycorrhizal fungi among the ascomycetes. Mol. Phylogenet. Evol. 6: 287-294.

Malloch, D. (1981). The plectomycete centrum. In Ascomycete systematics, the Luttrellian concept (Ed Reynolds, D.R.). New York, Springer-Verlag. pp.73-91.

Malloch, D. (1985). Taxonomy of the Trichocomaceae. In Filamentous microorganisms, Biomedical aspects (Ed Arai, T.). Tokyo, Japan Scientific Societies Press. pp.37-45.

Malloch, D. & Cain, R.F. (1972). The Trichocomataceae: Ascomycetes with *Aspergillus*, *Paecilomyces* and *Penicillium* imperfect states. Can. J. Bot. 50: 2613-2628.

Melchers, W. J. G., Verweij, P. E., van den Hurk, P., van Belkum, A., de Pauw, B. E., Hoogkamp-Korstanje, J. A. A. & Meis, J. F. G. (1994) General primer-mediated PCR detection of *Aspergillus* species. J. Clin. Microbiol. 32: 1710-1717.

Nikkuni, S., Kosaka, N., Suzuki, C. & Mori, K. (1996) Comparative sequence analysis on the 18S rRNA gene of *Aspergillus oryzae, A. sojae, A. flavus, A. parasiticus, A. niger, A. awamorii* and *A. tamarii*. J. Gen. Appl. Microbiol. 42: 181-187.

Nishida, H. & Sugiyama, J. (1993). Phylogenetic relationships among *Taphrina, Saitoella,* and other higher fungi. Mole. Biol. Evol. 10: 431-436.

Nishida, H., Ando, K., Ando, Y., Hirata, A. & Sugiyama, J. (1995). *Mixia osmundae*: transfer from the Ascomycota to the Basidiomycota based on evidence from molecules and morphology. Can. J. Bot. 73: S660-S666.

Ogawa, H., Yoshimura, A. & Sugiyama, J. (1997). Polyphyletic origins of species of tanamorphic genus *Geosmithia* and the relationships of cleistothecial genera: Evidence from 18S, 5S and 28S rDNA sequence analyses. Mycologia 89: 756-771.

Peterson, S.W. (1995). Phylogenetic analysis of *Aspergillus* sections *Cremei* and *Wentii*, based on ribosomal DNA sequences. Mycol. Res. 99: 1349-1355.

Pitt, J.I. (1985). Nomenclatorial and taxonomic problems in the genus *Eurotium*. In Advances in *Penicillium* and *Aspergillus* systematics (Eds Samson, R.A. & Pitt J.I.). New York, Plenum Press, pp. 393-396.

Pitt, J. I. (1995). Phylogeny in the genus *Penicillium*: a morphologist's perspective. Can. J. Bot. 73: S768-S777.

Pitt, J. I. & Hocking, A. D. (1985) Interfaces among genera related to *Aspergillus* and *Penicillium*. Mycologia 77: 810-824.

Pitt, J.I. & Samson, R.A. (1993). Species names in current in the Trichocomaceae (Fungi, Eurotiales). In Names in current use in the families Trichocomaceae, Cladoniaceae, Pinaceae, and Lemnaceae (Ed Greuter, W.). Königstein, Germany, Koeltz Scientific Books. pp. 13-57.

Powell, K.A., Renwick, A. & Peberdy, J.F.(Eds) (1994). The genus *Aspergillus* from taxonomy and genetics to industrial application. New York, Plenum Press.

Raper, K.B. & Fennell, D.I. (1965). The genus *Aspergillus*. Baltimore, Williams & Wilkins.

Saitou, N. & Nei, M. (1987). The neighbor-joining method: A new method for reconstructing phylogenetic trees. Mol. Biol. Evol. 4: 406-425.

Samson, R.A. (1979). A compilation of the Aspergilli described since 1965. Stud. Mycol. 18: 1-38.

Samson, R.A. (1992). Current taxonomic schemes of the genus *Aspergillus* and its teleomorphs. In *Aspergillus*, biology and industrial aspects (Eds Bennett, J.W. and Klich, M.A.). Boston, Butterworth-Heinemann. pp. 355-390.

Stolk, A. C. & Samson, R. A. (1971). Studies on *Talaromyces* and related genera. I. *Hamigera* gen. nov. and *Byssochlamys*. Persoonia 6: 341-357.

Subramanian, C.V. (1972). The perfect states of *Aspergillus*. Curr. Sci. 41: 755- 761.

Subramanian, C.V. & Rajendran, C. (1975). *Raperia*, a new genus of the Hyphomycetes. Kavaka 3: 129-133.

Sugiyama, J. & Yamatoya, K. (1990). Electrophoretic comparison of enzymes as a chemotaxonomic aid among *Aspergillus* taxa : (1) *Aspergillus* sects. *Ornati* and *Cremei*. In Modern concepts in *Penicillium* and *Aspergillus* classification (Eds Samson, R.A. & Pitt, J.I.). New York, Plenum Press. pp. 385-393.

Sugiyama, J., Rahayu, E.S., Chang, J.M. & Oyaizu, H. (1991). Chemotaxonomy of *Aspergillus* and associated teleomorphs. Jpn. J. Med. Mycol. 32: S39-S60.

371

Swofford, D.L. (1993). PAUP : Phylogenetic analysis using parsimony, version 3.1.1. Illinois. Illinois Natural History Survey.

Tamura, M., Rahayu, E. S., Gibas, C. & Sugiyama, J. (1996). Electrophoretic comparison of enzymes as a chemotaxonomic aid among *Aspergillus* taxa:(3) The identity of the xerophilic species *Aspergillus penicillioides* in subgen. *Aspergillus* sect. *Restricti*. J. Gen. Appl. Microbiol. 42 : 235-248.

Thompson, J.D., Higgins, D.G. & Gibbson, T.D. (1994). CLUSTAL W: Improving the sensitivity of progressive multiple sequence alignment through sequence weighting, position specific gap penalties, and weight matrix choice. Nucl. Acids Res. 22: 2673-4680.

White, T. J., Bruns, T. D. , Lee, S. & Taylor, J. W. (1990). Amplification and direct sequencing of fungal ribosomal RNA genes for phylogenetics. In PCR protocols (Eds Innis, M.A. & Gelfand, D.H., Sninsky, J.J. & White, T.J.). San Diego, Academic Press. pp. 315-322.

Yaguchi, T., Someya, A. & Udagawa, S. (1996). A reappraisal of intrageneric classification of *Talaromyces* based on the ubiquinone systems. Mycoscience 37: 55-60.

Yamatoya, K., Sugiyama, J. & Kuraishi, H. (1990). Electrophretic comparison of enzymes as a chemotaxonomic aid among *Aspergillus* taxa: (2) *Aspergillus* section *Flavi*. In Modern concepts in *Penicillium* and *Aspergillus* classification (Eds Samson, R A & Pitt, J.I.). New York, Plenum Press. pp. 395-405

Present address of the authors:

Dr **M. Tamura,** Research Center for Pathogenic and Microbial Toxicoses, Chiba University, 8-1, Inohana 1 chome Chuo-ku, Chiba 260-8673, Japan

Prof. Emeritus **J. Sugiyama,** Department of Botany, The University Museum, The University of Tokyo, 3-1, Hongo 7 chome, Bunkyo-ku Tokyo 113-0033 and NCIMB, Kaminakazato Office, 9-2, Sakae-cho, Kita-ku, Tokyo 114-0005 Japan

FACTORS AFFECTING THE USE OF SEQUENCE DIVERSITY OF THE RIBOSOMAL RNA GENE COMPLEX IN THE TAXONOMY OF *ASPERGILLUS*

Brian W. Bainbridge,
Molecular Mycology Group, Division of Life Sciences, King's College, Campden Hill Road, London W8 7AH, U.K

Sequence analysis of DNA or RNA is increasingly used as a tool for the analysis of diversity in the genus *Aspergillus*. Base sequence differences may be analyzed directly by computer programmes or indirectly by techniques depending on hybridization / annealing and/or restriction enzyme digestion. The information accumulated is frequently used as an aid to taxonomy and the assumption is made that sequence diversity is related to evolutionary history. The choice of target molecule is obviously important and over the last 10 years bacterial taxonomists have successfully argued for the use of the small ribosomal RNA subunit (16S-like) (Pace, 1997). This molecule has also been enthusiastically analyzed by fungal taxonomists and useful information has been obtained (Chang *et al.,*1991; Verweij *et al.,*1995). However some caution should be exercised as there are differences between the prokaryotic and eukaryotic systems, for example, in the size and number of copies of the ribosomal RNA molecules. In the fungi there are effectively two different versions of the ribosomal RNA subunits, nuclear and mitochondrial. In addition the genetic systems generating diversity in the fungi may be different from those in the prokaryotes. In some cases the sequence differences in the ribosomal subunits have not been sufficiently discriminating and spacer regions of the ribosomal RNA gene complex (rDNA) have been analyzed. The aim of this paper is to describe the molecular and genetic mechanisms which generate sequence diversity in rDNA and to discuss potential problems in the use of this region in taxonomy and phylogeny.

CHOICE OF RDNA REGION FOR ANALYSIS

There have been several recent reviews on the use of rDNA sequences for fungal taxonomic purposes (Bruns *et al.,* 1991; Seifert *et al.,*1995; Samuels and Seifert 1995). To discuss these, it is first necessary to identify the various regions of the tandemly arranged copies in the rDNA complex. The first 5' region is the intergenic spacer (IGS) which contains a non-transcribed spacer (NTS) and an external transcribed spacer (ETS). The two regions are separated by a single promoter from which a polygenic mRNA spanning a single unit of the complex is transcribed. The second region is the 18S, which produces a ribosomal RNA subunit, and this is followed by an internal transcribed spacer (ITS1), the 5.8S ribosomal RNA submit and the second ITS region, ITS2. The complex is completed by the 26-28S ribosomal RNA submit. The 5S subunit is usually found at a different locus in the genome of filamentous fungi. Metzenberg (1991) drew attention to the invariant and highly conserved sequences in the genes coding for ribosomal RNA subunits. These were separated by variable transcribed sequences, subject to rapid evolutionary change, and non transcribed sequences of most recent origin. Regions could be selected depending on the level of taxonomic discrimination required e.g. genus, species, variety or isolate. Seifert *et al.* (1995) were more specific and ranked the regions in order of decreasing taxonomic resolution as follows 18S, 28S, ITS and IGS but concluded that a unified classification of the Ascomycetes using, for example, the ITS regions, was not possible. For *Aspergillus* and *Penicillium*, Samuels & Seifert (1995) quoted the use of 18S and 28S for genus and species level discrimination, and mitochondrial rDNA, ITS and 5.8S for more detailed discrimination. Little information is available on the use of the IGS region (Bainbridge, 1994) in spite of evidence showing species level discrimination in *Drosophila* (Tautz *et al.*, 1987), rice (Cordesse *et al.*, 1992) and *Fusarium* (Appel & Gordon, 1996).

DIFFERENCES BETWEEN PROKARYOTIC AND EUKARYOTIC RIBOSOMAL SYSTEMS

Perhaps the most obvious difference is in the size of the subunits of the rDNA complex. For example, the nucleotide sedimentation coefficient of the small subunit is 16S in prokaryotes and 18S in eukaryotes. In spite of this, a comparison of a recent phylogenetic trees shows that there are far fewer base pair differences between the members of the eukaryotic domain containing the fungi, plants and animals compared with the striking differences between the two domains now know as the bacteria and the archaea (Pace, 1997). Additionally there are significantly different numbers of tandem copies: seven per nucleoid in *Escherichia coli* compared to 50 up to several hundreds per nucleus in the fungi. Furthermore there are two types of genes for the ribosomal subunits in most eukaryotes: mitochondrial subunit genes and nuclear subunit genes. Although there are multiple copies of the mitochondrial genome, each individual DNA molecule has only one copy of each of the genes for the mitochondrial ribosome. The constraints on the sequences of these two types of ribosomal genes will therefore be very different and the molecules are likely to evolve at different rates. Indeed variation will depend on the life cycle of the fungus involved. Uniparental inheritance of mitochondrial DNA will preclude the occurrence of recombination between different mitochondrial genomes whereas vegetative compatibility and anastomosis of hyphae will increase the probability. Likewise the scope for recombination between nuclear genes will be more limited in asexual genera than in genera with sexual and/or parasexual cycles (Clutterbuck, 1992).

VARIATION IN THE SEQUENCE AND ORGANIZATION OF THE RDNA COMPLEX

Historically, sequence variation has been analyzed indirectly by restriction enzyme analysis followed ideally, by hybridization to homologous rDNA probe. Although there are limitations to restriction fragment length polymorphism (RFLP) analysis (Bainbridge, 1994), it has the advantage of simplicity and many isolates can be screened relatively simply. Species specific variation in RFLPs has been detected in *Aspergillus* (Bainbridge *et al.*, 1990). *Neurospora*, strain specific restriction enzymes sites have been reported, particularly in the NTS region (Russell *et al.*, 1984). Direct sequencing of units within the rDNA has now become technically much easier and more data is now accumulating. Verweij *et al.*, (1995) compared five species of *Aspergillus* by comparing 18S sequences. Peterson *et al.*, (1995) compared sections *Aspergillus Cremei* and *Wentii* using the large ribosomal RNA subunit. Sequencing of the ITS1/5.8S/ITS2 region, after amplification by the polymerase chain reaction (PCR), has become very popular due to the length of the region, only about 600 bp. Kumeda & Asao (1996) have recently reduced the need for sequencing by comparing sequence differences in ITS1 and ITS2 DNA amplified by PCR by using single strand conformation polymorphism (SSCP), in the presence of formamide, to analyze member of *Aspergillus* section *Flavi*.

It is possible that the number of copies of the rDNA complex may vary between different isolates within a species as has been demonstrated for *Saccharomyces cerevisiae* (Chindamporn *et al.*, 1993). The occurrence of many repeats of the rDNA unit in one genome has raised the possibility that they may not be identical even within a clonal population. Indeed it is a mystery why there should not be more extensive variation within different copies as the selective pressures on any one copy must be relatively low. Variation in RFLPs, using an IGS probe, have been detected in *A. fumigatus* both within a clonal isolate and between isolates (Spreadbury *et al.*, 1990). This was interpreted to mean that there polymorphisms in length existed between individual subunits of the same tandem array. Similar evidence is available for a number of different fungi (Bainbridge, 1994) and more recently direct evidence has been found for a number of different sequences amplified from the same spore of a mycorrhizal fungus (Sanders *et al.*, 1996). However it should be realized that a spore in these fungi may have up to a thousand nuclei.

It is generally assumed that in a set of tandem copies most of the copies will have identical sequences although there is evidence for a variable minority of copies which have different sequences (see above). The process which ensures that the majority of copies are identical is known as concerted evolution and Dover (1986) has suggested a number of mechanisms which would ensure standardization of sequences. These will be discussed later but it is difficult to imagine how one copy with a particular sequence can be used as template for up to one hundred further copies. What is required is a mechanism equivalent to the discarded master-slave hypothesis suggested several decades ago for producing uniformly repetitive DNA. Evidence for a molecular version of this theory is not extensive but papers on yeasts suggest that a topoisomerase may be implicated in the production of circular intermediates containing a single rDNA unit. This is then replicated by a rolling circle method producing identical concatenates which are then reinserted into the chromosome by recombination (Kim & Wang, 1989; Christman *et al.*, 1993; Moss & Stevanofsky, 1995). Further evidence will be needed to elucidate the mechanism(s) by which concerted evolution occurs.

It is of course possible that, for example, the 18S regions are conserved in duplicate copies while the spacer regions differ in sequence. It has been argued that the selective pressure for a functional ribosome will maintain the conserved sequences for the ribosomal genes whereas the IGS region is able to show co-evolution between the DNA sequence and RNA polymerase I giving rise to promoters specific for species (Dover & Flavell, 1984; Moss & Stefanofsky, 1995).

GENETIC MECHANISMS FOR THE GENERATION OF SEQUENCE DIVERSITY IN RDNA

A number of the mechanisms the generation of sequence diversity in rDNA are well known and need no treatment here. These include mutation and recombination during meiosis. The genus *Aspergillus* contains both meiotic and anamorphic species. Sexual recombination may not be present in some of these species (Geiser *et al.*, 1996). However it should be remembered that if diploid strains occur recombination can also happen during mitosis. *Aspergillus nidulans* provides the model system for this and for other parasexual processes including nondisjunction and haploidisation (Clutterbuck, 1992). Gene conversion is a consequence of hybrid DNA during recombination and it can result in the loss of one allele (sequence) and the gain of another alternative allele (sequence). The molecular nature of the initial mutant lesion can result in a unidirectional gene conversion resulting in the increase in the frequency of the favoured allele. Unequal crossing over can also occur particularly in regions of repetitive DNA and this will result in duplications and deletions producing the length polymorphisms mentioned in the previous section. Duplications artificially introduced into *Neurospora* have been observed to undergo premeitoic deletion and /or repeatedly induced point mutations (reviewed in Kistler & Miao, 1992) but this phenomenon has not been reported for *Aspergillus*. Inversions or translocations between non homologous chromosomes can also occur resulting in significant changes in chromosome length, detectable by pulsed field gel electrophoresis as well as by altered RFLPs. Kistler & Miao (1992) have suggested that these effects are inversely related to the presence of meiosis which normally acts as a filter for chromosome abnormalities.

MOLECULAR MECHANISMS FOR THE GENERATION OF SEQUENCE DIVERSITY IN RDNA

The distinction between genetic and molecular mechanisms is not clear cut; an alternative distinction would be nuclear and cytoplasmic mechanisms for the generation of sequence diversity. An important process here is horizontal transfer which would include the acquisition of mitochondria by the evolving eukaryotes. The general importance of horizontal transfer is difficult to assess but occurrence should be born in mind and sequence comparisons made with both prokaryotic and eukaryotic molecules. Novel methods of genetic change, including mycoviruses, transposons, mitochondrial plasmids and RNA-mediated transfer, have been reviewed by Kistler & Miao (1992). A number of these phenomena have been detected in *Aspergillus,* including transposons (Glayzer *et al.,* 1995; Amutan *et al.,* 1996; Nyssonen *et al.,* 1996) and retrotransposons (Neuveglise *et al.,* 1996). All of these elements can cause deletions and sequence alterations although their contribution to overall variation is unknown. Mitochondrial polymorphisms have been used as an aid to taxonomy in *Aspergillus niger* (Varga *et al.* 1994) although there appears to be considerable variation in the usefulness of mitochondrial data. Polymorphism can be easily detected by the use of GC-rich restriction enzymes four bases long used directly on genomic DNA. However, in the long term it is likely that sequencing of the small subunit of the mtDNA will prove to be more reliable (Bruns *et al.,* 1991). Larger mitochondrial genomes have been reported to more variable than smaller molecules. Variation can also occur due to the presence or absence of introns (Yamamoto *et al.,* 1995).

MOLECULAR DRIVE AND THE RDNA COMPLEX

On the basis of research on plants and *Drosophila*, Dover (1986) proposed a theory to explain variation in repetitive gene families located on homologous or non homologous chromosomes. This process is considered to be distinguishable from natural selection and genetic drift . It results in the homogenization of the preferred copy of a multigene family and its fixation in the population. The ribosomal RNA gene complex is subject to molecular drive which is considered to be a property of the molecules themselves and of molecular processes not directly influenced by selection. Reference has already been made to concerted evolution

376

which can be a consequence of molecular drive. Dover (1986) has listed a number of processes involved and these include gene conversion, slippage replication, transposition, unequal recombination and RNA-mediated transfers. Methods for detection and quantification
of concerted evolution of promoters in *Drosophila* have also been published (Dover *et al.*,
1993) but it should be remembered that these analyses are based on an obligately sexually
reproducing organism. As indicated above other processes such as the modified master and
slave theory may also aid the production of homogenized sequences. A number of these
processes are known to occur in *Aspergillus* but their relative importance is unknown.

SEQUENCE DIVERSITY IN THE RDNA AND THE TAXONOMY OF *ASPERGILLUS*

So far I have discussed the source of sequence diversity and the mechanisms which produce
it. The assumption made in the use of this diversity for taxonomic purposes is that present
day sequences have been derived from earlier sequences in a regular fashion and in such a
way as to reflect natural or phylogenetic groupings. Analysis of sequence differences should
therefore give rise to clades which reflect evolutionary history. An alternative approach is to
treat the data in ways which make no previous assumptions about origins. This is known as a
numerical taxonomic approach and does not weight characters in any way. A variety of
computer programmes are available for phylogenetic analysis of sequence differences,
RFLP and other fingerprint based systems (Swofford & Olsen, 1990). Inevitably data will be
incomplete and inaccurate and some allowance can be made for this (States, 1992). Hillis *et
al.* (1994) have made a comparative study of different methods by using simulation studies
and laboratory experiments. They conclude that the methods were reliable but needed to
make allowance for substitution bias.

The changes described above will not always occur at a constant rate and this is reflected
in the two theories of evolution by gradual change with small steps or by punctuated evolution with major changes caused by horizontal transfer or other molecular events involving
more than one base pair change. The fixation of genetic change can occur either by natural
selection, neutral effects (Brookfield & Sharp, 1994) or molecular drive as already discussed. In many cases we simply do not know the mechanisms giving rise to particular sequences. The occurrence of deletion of bases followed by additions followed by further deletions makes interpretation very difficult and may cause misleading results. Furthermore the
groupings which result will depend on the evolutionary state of a particular fungal group.
There may be strong incompatibility barriers, genetic and cytoplasmic, between two groups
so that there is no exchange of genetic material and the two gene pools are distinct. Alternatively there may be complete compatibility within a diverse group with continuous variation
in sequence diversity. The number of taxa identified will then depend on the detail of sampling as well as on the methods used to analyze the data.

A major problem is therefore the collecting of a sample of sequences for phylogenetic
analysis. The occurrence of different sequences in the spacer regions of the same clonal
material has raised potential problems. How many independent PCR amplification should be
made? How many of these should be checked by sequence analysis? How many different
isolates of the same taxon or species should be analyzed? Should both the teleomorph and
the anamorphic forms be analyzed? Geiser *et al.* (1995) have recently compared ribosomal
sequences from meiotic and strictly mitotic asexual taxa of *Aspergillus*. They found four

clades with both sexual and asexual forms and concluded that their results supported the hypothesis that asexual lineages were frequently derived from more ancient meiotic forms.

CONCLUSIONS

Bearing in mind the diverse mechanisms which can give rise to sequence variation and its fixation, it becomes clear that it is advisable to analyze as many isolates as possible from a particular species or taxon. In addition more than one target molecule should be used. Initially it may be advisable to screen a large number of morphologically or biochemically similar isolates by Randomly Amplified Polymorphic DNA (RAPD) techniques to group and detect any unusual isolates. However, Thorman *et al.*, (1994) compared RFLP and RAPD techniques in the crucifers and showed that RFLP techniques are more useful than RAPDs for detecting inter-specific variation. The next stage could be the PCR amplification of rDNA followed by RFLP to group the strains further. Alternatively PCR-SSCP could be used to detect sequence differences indirectly (Kumeda *et al.*, 1996). This should then be followed by sequence analysis of representative isolates. The approach described should avoid the analysis of a single abnormal isolate even if it is derived from the type specimen. Furthermore a better understanding of the mechanisms which generate and fix sequence variation may allow us to select regions less prone to excessive or misleading variation thus providing most reliable sequence data for taxonomic or phylogenetic purposes.

REFERENCES

Amutan, M., Nyyssonen, E., Stubbs, J., Diaztorres, M.R. & Dunncoleman, N. (1996). Identification and cloning of a mobile transposon from *Aspergillus niger* var *awamori*. Curr. Genet. 29: 468-473.

Appel, D.J. & Gordon, T.R. (1996). Relationships among pathogenic and nonpathogenic isolates of *Fusarium oxysporum* based on the partial sequence of the internal spacer region of the ribosomal RNA. Mol. Plant Microbe Interact. 9: 125-138.

Bainbridge, B.W., Spreadbury, C.L., Scalise, F.G. & Cohen, J. (1990). Improved methods for the preparation of high molecular weight DNA from large and small scale cultures of filamentous fungi. FEMS Micro. Let. 66: 113-118.

Bainbridge, B.W. (1994). Modern approaches to the taxonomy of Aspergillus. In The genus *Aspergillus* : from taxonomy and genetics to industrial application, (Eds Powell, K. A., Renwick, A. and Peberdy, J..F.). New York, Plenum Press. pp 291-301.

Brookfield, J.F.Y. & Sharp, P.M. (1994). Neutralism and selectionism face up to DNA data. Trends Genet.10: 109-111.

Bruns, T., White, T.J. & Taylor, J.W. (1991). Fungal molecular systematics. Annu. Rev Ecol. Syst. 22: 525-64.

Chang, J., Oyaizu, H. & Sugiyama, J. (1991). Phylogenetic relationships among eleven selected species of *Aspergillus* and associated teleomorphic genera estimated from 18S ribosomal RNA partial sequences. J. Gen. Microbiol. 37:289-308.

Chindamporn, A., Iwaguchi, S.I., Nakagawa, Y., Homma, M. & Tanaka, K. (1993). Clonal size variation of rDNA cluster region of chromosome XII of *Saccharomyces cerevisiae*. J. Gen. Microbiol. 139: 1409-1415.

Christman, M.F., Dietrich, F.S., Levin, N.A., Sadofff, B.U. and Fink, G.R. (1993). The rRNA-encoding DNA array has an altered structure in topoisomerase I mutants of *Saccharomyces cerevisiae*. Proc. Nate. Acad. Sci. USA 90: 7637-7641.

Clutterbuck, A.J. (1992). Sexual and parasexual genetics of *Aspergillus* species. In *Aspergillus*: Biology and industrial application (Eds Bennett, J.W. & Klich, M.A.). Boston, Butterworth-Heinemann. pp. 3-18.

Cordesse, F., Grellet, F., Reddy, A.S. & Delseny, M. (1992). Genome specificity of rDNA spacer fragments from *Oryza sativa* L. Theor. Appl. Genet. 83: 864-870.

Dover, G.A. (1986). Molecular drive in multigene families: how biological novelties arise, spread and are assimilated. Trends Gent. 2: 159-165 .

Dover, G.A. & Flavell, R.B. (1984) Molecular coevolution: DNA divergence and the maintenance of function. Cell 38: 622-623.

Dover, G.A., Linares, A.R., Bowen, T. & Hancock, J.M. (1993). Detection and quantification of concerted evolution and molecular drive. Meth. Enzymol. 224: 525-541.

Geiser, D.M., Timberlake, W.E. & Arnold, M.L. (1996). Loss of meiosis in *Aspergillus*. Mol. Biol. Evol. 13: 809-817.

Glayzer, D.C., Roberts, I.N., Archer, D.B. & Oliver, R.P. (1995). The isolation of Ant1, a transposable element from *Aspergillus niger*. Mol. Gen. Genet. 249: 432-438.

Hillis, D.M., Huelsenbeck, J.P. & Cunningham, C.W. (1994). Application and accuracy of molecular phylogenies. Science 264: 671-677.

Kim, R.A. & Wang, J.C. (1989) A subthreshold level of DNA topoisomerase leads to the excision of yeast rDNA as extrachromosomal rings. Cell 57: 975-985.

Kistler, H.C. & Miao, V.P.W. (1992). New modes of genetic change in filamentous fungi. Annu. Rev. Phytopathol. 30: 131-152.

Kumeda, Y. & Asao, T. (1996). Single strand confirmation polymorphism analysis of PCR-amplified ribosomal DNA internal transcribed spacers to differentiate species of *Aspergillus* section *Flavi*. Appl. Environ. Microbiol. 62: 2947-2952.

Metzenberg, R.L. (1991). Benefactors' lecture: the impact of molecular biology on mycology. Mycol. Res. 95: 9-13.

Moss, T. and Stefanovsky ,V.Y. (1995). Promotion and regulation of ribosomal transcription in eukaryotes by RNA polymerase I. Prog. Nucl. Acid. ERs Mol. Biol. 50: 25-66.

Neuveglise, C., Sarfati, J., Latgé, J.P. & Paris, S. (1996). Afut1, a retrotransposon-like element from *Aspergillus fumigatus*. Nucl. Acid Res. 24: 1428-1434.

Nyysonen, E., Amutan, M., Enfield, L. Stubbs, J. & Dunncoleman, N.S. (1996). The transposable element tan1 of *Aspergillus niger* var. *awamori,* a new member of the fot1 family. Mol Gen. Genet. 253: 50-56.

Pace, N.R. (1997). A molecular view of microbial diversity and the biosphere. Science 276: 734-740.

Peterson, S.W. (1995). Phylogenetic analysis of *Aspergillus* Sections *Cremei* and *Wentii* based on ribosomal DNA sequences. Mycol. Res. 99: 1349-1355.

Russell, P. J. Wagner, S., Rodland, K.D., Feinbaum, R.L., Russell, J.P., Bret-Harte, M.S., Free, S.J. & Metzenberg, R.L. (1984). Organization of the ribosomal ribonucleic acid genes in various wild-type strains and wild-collected strains of *Neurospora*. Mol. Gen. Genet. 196: 275-282.

Samuels, G.J. & Seifert, K.A. (1995). The impact of molecular characters on systematics of filamentous Ascomycetes. Annu. Rev. Phytopathol. 33: 37-67.

Sanders, I.R., Clapp, J.P. & Wiemken, A. (1996). The genetic diversity of arbuscular mycorrhizal fungi in natural ecosystems - a key to understanding the ecology and functioning of the mycorrhizal symbiosis. New Phytol. 133: 123-134.

Seifert, K.A., Wingfield, B.D. & Wingfield, M.J. (1995). A critique of DNA sequence analysis in the taxonomy of filamentous Ascomycetes and ascomycetous anamorphs. Can. J. Bot. 73(Suppl.1): S760-S767.

Spreadbury, C.L., Bainbridge, B.W. & Cohen, J. (1990). Restriction fragment length polymorphisms in isolates of *Aspergillus fumigatus* probed with part of the intergenic spacer region from the ribosomal RNA gene complex of *Aspergillus nidulans*. J. Gen. Microbiol. 136: 1991-1994.

States, D.J. (1992). Molecular sequence accuracy: analyzing imperfect data. Trends Genet. 8: 52-55.

Swofford, D.L. & Olsen, G.J. (1990) Phylogeny reconstruction. In Molecular Systematics (Eds Hillis, D.M. & Moritz, C.). Sunderland UK, Sinhauer Associates Inc. pp. 411-501.

Tautz, D., Tautz, C.. Webb, D. & Dover, G.A. (1987). Evolutionary divergence of promoters and spacers in the rDNA family of four *Drosophila* species. J. Mol. Biol. 195: 525-542.

Thorman, C.E., Ferreira, M.E., Camargo, L.E.A.., Tivang, J.G & Osborn, T.C. (1994). Comparison of RFLP and RAPD markers for estimating genetic relationships within and among cruciferous species. Theor. Appl. Genet. 88: 973-980.

Varga, J., Kevei, F., Vriesema, A., Debets, F., Kozakiewicz, Z. & Croft, J.H. (1994). Mitochondrial - DNA restriction fragment length polymorphisms in field isolates of the *Aspergillus niger* aggregate. Can. J. Microbiol. 40: 612-621.

Verweij, P.E., Meis, J.F.G.M., van den Hurk, P., Zoll, J., Samson, R.A. & Melchers, W.J.G. (1995). Phylogenetic relationships of 5 species of *Aspergillus* and related taxa as deduced by comparison of sequences of small unit ribosomal RNA. J. Med. Vet. Mycol. 33: 185-190.

Yamamoto, H., Naruse, A., Ohsaki, T. & Sekiguchi, J. (1995). Nucleotide sequence and characterization of the large mitochondrial ribosomal RNA gene of *Penicillium urticae,* and its comparison with those of other fungi. J. Biochem. 117: 888-896.

MOLECULAR AND ANALYTICAL TOOLS FOR CHARACTERIZING *ASPERGILLUS* AND *PENICILLIUM* SPECIES AT THE INTRA- AND INTERSPECIFIC LEVELS

David M. Geiser, Frederick M. Harbinski and John W. Taylor
Fusarium Research Center , Department of Plant Pathology, The Pennsylvania State University University Park, PA 16802 and Department of Plant and Microbial Biology, 111 Koshland Hall, University of California, Berkeley, CA 94720-3102, USA

Molecular genetic tools are useful for answering a variety of evolutionary questions about fungi at the intra- and interspecific levels, including questions about population structure, clonal versus recombinant modes of propagation, biological species, gene flow, and phylogenetic relationships among close relatives. Multiple independent sources of molecular sequence data are desirable or necessary for these purposes. Ideally, molecular markers are present in a single copy in the genome, and are easy to assay and interpret. In this paper, we describe how to isolate molecular markers for these purposes, based on PCR amplification and sequence characterization of portions of genes encoding proteins. These markers are useful for population genetic analyses, and for comparing both closely and more distantly related species. In addition, sequence data from multiple genes can be used to identify cryptic biological species, even in cases where mating tests are not possible.

INTRODUCTION

At the Second International *Aspergillus* and *Penicillium* Workshop (Samson and Pitt, 1990), the point was made that molecular phylogenetic techniques could determine the genetic distances between mitosporic fungi, but could not necessarily provide a reliable means for defining species (pp. 354-355). As reasonable as that statement seemed at the time, subsequent data have proved it wrong. Research in fungal phylogenetics and population genetics, including some presented at this Third International *Aspergillus* and *Penicillium* Workshop, documents two surprises: 1) the ribosomal genes discussed at the Second Workshop evolve too slowly to resolve the phylogenetic history of some closely related *Penicillium* and *Aspergillus* species; and 2) recombination is occurring in natural populations of mitosporic fungi, which means that molecular phylogenetic techniques can provide a means of recognizing and defining species of *Aspergillus* and *Penicillium*.

Interestingly, the data that are providing better phylogenetic resolution among closely related *Penicillium* and *Aspergillus* species and the data that can be used to define species are the same, i.e. variable nucleotide positions in genes coding for protein. This report addresses the means of finding nucleotide variation in such genes and analyzing the data to reveal species and analyze their phylogenies.

Background

Molecular analyses have revolutionized fungal systematics, and most of this revolution has involved analysis of ribosomal RNA genes (rDNA). These genes provide a large number of characters for phylogenetic analysis useful at a range of taxonomic levels, from among closely related species to among divisions. A number of systematic and evolutionary analyses using rDNA sequences have been performed in recent years on *Penicillium* and *Aspergillus* species and their teleomorphs (Logrieco *et al.*, 1989; Chang *et al.*, 1991; LoBuglio *et al.*, 1993, 1994; LoBuglio and Taylor, 1993, 1995; Peterson, 1993, 1995; Berbee *et al.*, 1995; Geiser *et al.*, 1996, 1998b; Ogawa et al., 1997), along with a large scale project to sequence multiple rDNA genes from every *Penicillium* and *Aspergillus* species (Peterson, 1993). Even with these efforts incomplete, we are already getting a very good picture of relationships among sections and subgenera based on rDNA sequence data (Peterson, 2000; Sugiyama and Tamura, 2000).

However, rDNA sequences cannot answer all of our taxonomic and evolutionary questions about *Aspergillus* and *Penicillium*. First, they tend to be highly conserved, such that relationships among very closely related species cannot be inferred with strong support. The nuclear rDNA intergeneric spacer (IGS) region offers the most variation, and is likely to be useful within species and among close relatives (Spreadbury *et al.*, 1991; Appel and Gordon, 1995). However, this is variation at only a single locus, which leads to a second shortcoming of rDNA sequences. When nucleic acid sequences are used to make a phylogenetic inference, an assumption is made that the inferred gene genealogy is the same as the genealogy of the organisms from which the genes were sampled. This is not necessarily the case, and this assumption becomes particularly dangerous among closely related taxa and among conspecific individuals (Avise and Ball, 1990). One way to increase the confidence in the assumption that a given gene tree reflects the underlying organismal tree is to sample additional, independent genes. If two different genes produce the same phylogenetic inference, one can be more confident that that inference reflects the true organismal phylogeny. Unless the two genes are tightly linked, or unless there has been recent interspecific hybridization, it is highly unlikely that two different genes would have the same genealogy, different from the corresponding organismal genealogy. Conservatively speaking, fungal rDNA offers two independent sets of characters: nuclear [18S or small subunit, internal transcribed spacers (ITSs), 5.8S subunit, 28S or large subunit, intergeneric spacers (IGSs)] and mitochondrial (small and large subunits). Neither mitochondrial rDNA gene offers variation that is superior to the nuclear rDNA repeat for comparing very closely related taxa nor for making intraspecific comparisons e.g., LoBuglio *et al.*, 1993. Finally, recent studies have shown that in somewhat rare cases, single individuals can possess multiple forms of nuclear rDNA, which can lead to misleading results (O'Donnell and Cigelnik, 1997; O'Donnell et al., 1998).

In addition to bolstering phylogenetic inferences, analysis of multiple loci is a necessity for most population genetic analyses, where multilocus approaches have been used for over thirty years. Many options exist for finding polymorphic loci at the intraspecific level, including protein electrophoresis, restriction fragment length polymorphism (RFLP) analysis using either random or known segments of DNA, and analysis of anonymous PCR products [(randomly amplified polymorphic DNA (RAPD), microsatellites, etc.]. In this paper, we will concentrate on the development of a particular type of nucleic acid marker, variable nucleotide sequence in genes coding for proteins, because they can provide extensive variation that is useful at a variety of taxonomic levels, and that variation is particularly simple to assay and interpret.

Ideal markers for these analyses would fulfil the following criteria. They would be single-copy. They would be co-dominant (i.e., the genotype would be known for all individuals, with no individuals having a null or missing allele or no sequence). Therefore, in haploid Ascomycetes we would like to obtain a single DNA fragment of variable sequence via PCR using primers with high specificity. The amplification protocol should be simple and robust, so that the amplification is highly repeatable. Variation in the marker should be easily analyzed via a simple electrophoretic method, with or without restriction enzyme digestion, but direct sequencing should not be ruled out if necessary. Fortunately, several *Aspergillus* and *Penicillium* species and related teleomorphs are among the best studied genetically of all fungi, particularly *Emericella nidulans*, *Aspergillus fumigatus*, *A. flavus*, *A. parasiticus*, *A. oryzae*, *A. niger* and *Penicillium chrysogenum*. The entire genome of *Emericella nidulans* has been sequenced, and this information is available in a limited fashion (V. Gavrias, pers. comm.)

In this paper, we will describe how multiple molecular markers can be developed by using the wealth of protein-encoding sequence information available in *Aspergillus* and *Penicillium*, which is accumulating at an accelerating rate. We will then discuss how the data derived from these markers can be used to ask a number of basic questions regarding the taxonomy and evolution of *Aspergillus* and *Penicillium* species, some of which were not previously addressable.

Identifying useful gene regions

Generating such markers involves 1. selecting a DNA region to study; 2. design of PCR primers to amplify a small (~400-600 bp) region of a protein-encoding gene using known sequences as a guide, and 3. identifying variation within that region in species or strains of interest. Approaches for identifying these regions in Ascomycetes is described in Carbone and Kohn (1999).

No absolute rules exist for identifying gene regions for analysis. As different rDNA regions are expected to be variable at different taxonomic levels, so are protein-encoding genes. Furthermore, different regions within a protein-encoding gene will have different levels of variation. For example, non-coding regions (the region 5' to the first amino acid (ATG= methionine), the region 3' to the stop codon (TAA, TAG or TGA), introns) tend to be more variable than protein-encoding regions. Therefore, non-coding regions may be most useful for intraspecific comparisons, or for comparing closely related species, whereas coding regions may be more suitable for more distantly related taxa. Furthermore, different genes will harbour vastly different levels of variation. A non-essential gene (one that is not absolutely lethal if missing, such as a gene encoding a secondary metabolite) is likely to harbour more sequence polymorphism than a more globally necessary gene (say, beta-tubulin or calmodulin). These different levels of variation within and among genes make different gene regions useful for different purposes, although one cannot always predict how much variation will be identified in a particular sequence based on its assumed relative importance. Because of this uncertainty, it is worthwhile to test new genes on a subset of the taxa of interest before embarking on an exhaustive study. Members of the subset should include species thought to be closest relatives as well as those thought to be most distant.

Furthermore, these gene regions tend not to be as universally useful as the rDNA genes. Although 18S genes can be useful over a very broad range of taxa (from among genera to among different kingdoms), less conserved rDNA genes may be useful only among species or closely related genera. More conserved protein encoding genes can be useful over an ever greater range of taxa, because there is variation in non-coding regions (useful within species and among very closely related species) and coding regions (useful for comparing higher taxa). Less conserved protein-encoding genes are less widely applicable. In these cases, the high level of variation that is useful at the intraspecific level may be so extreme among species that the mutable nucleotide positions are saturated with mutations, which interferes with phylogenetic inference, and makes broadly applicable PCR primer design impossible.

Accessing *Aspergillus* and *Penicillium* sequences via the World Wide Web

The National Center for Biotechnology Information (NCBI) maintains the www-accessible search tool Entrez (http://www3.ncbi.nlm.nih.gov/Entrez/), from which all sequences in GenBank can be found and downloaded. Searching the NCBI nucleotide sequence database, all sequences can be accessed by using a variety of query fields, and detailed information obtained about the sequences. For example, as of February, 2000, one can access all of the 13,693 sequences generated from members of the Trichocomaceae by entering "Trichocomaceae" as the search term under the nucleotide search.. As another example, Entrez finds 185 sequences from *Aspergillus flavus*, most of which are rDNA sequences, or sequences related to aflatoxin production. There are a number of other genes as well, including polygalacturonases, alcohol dehydrogenases, and beta-tubulin. For studies within *A. flavus*, or perhaps comparing *A. flavus* and *A. parasiticus*, the aflatoxin genes may be good markers. However, if we are interested in comparing other non-toxigenic species, these aflatoxin genes might not be useful due to their potential absence in non-toxigenic taxa or to rapid evolution.

Designing primers in highly conserved genes

Beta-tubulin is useful for analysis at a variety of taxonomic levels. Its amino acid sequence is highly conserved, but in fungi it is also rich in introns. The highly conserved exons are expected to be good sites for placing primers that flank both exons and introns, providing a source of nucleotide characters with different levels of variation. If Entrez is queried for all *Aspergillus* beta-tubulin sequences (query "Aspergillus AND beta-tubulin", the search produces four full gene sequences (beside the dozens of parital sequnces generated for phylogenetic studies): the *A. flavus* beta-tubulin sequence, an *A. parasiticus* beta-tubulin, and two *Emericella nidulans* sequences. *E. nidulans* (and probably all *Aspergillus* species) has two beta-tubulin genes, *ben*A and *tub*C. This might be a problem, because we want to know with certainty that we are looking at one gene or the other, not both. It turns out that *ben*A and *tub*C have a number of sequence features that distinguish them (May *et al.*, 1987), and that the *A. flavus* and *A. parasiticus* sequences are most similar to *ben*A.

Figure 1 shows part of an alignment of the *A. flavus* and *E. nidulans ben*A sequences. The 5' and 3' non-coding regions of the two genes were first trimmed off, using information about the locations of the start (ATG) and stop (TAA, TAG, or TGA) codons found in the *GenBank report* from Entrez, and the remaining coding sequences were aligned using the ClustalV algorithm (Higgins et al., 1992). Further corrections to the alignment involved correctly aligning introns vs. exons (the locations of which are also given in the GenBank reports), and were made by eye. The pattern of variation is clear: most variable nucleotide sites are located in introns, and most exon variation is found in third codon positions.

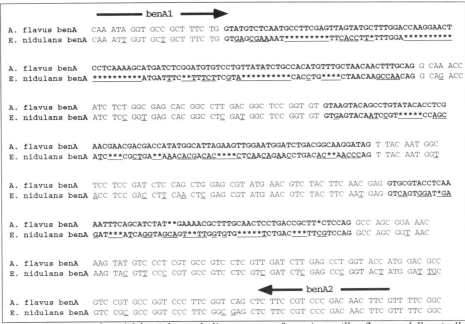

Fig. 1. Alignment of partial *ben*A beta-tubulin sequences from *Aspergillus flavus* and *Emericella nidulans*, with locations of primer sites. Bold-face regions represent introns. Underlined sites differ between the two species. Asterisks represent gaps

Primers benA1 and benA2 were designed based on the following criteria:

1. The primer sites are in regions of high sequence conservation, so we expect them to work in all isolates in a variety of taxa (*A. flavus*, *E. nidulans*, *A. niger*, etc.). In this case the primers correspond exactly to the *A. flavus* sequence. Designing degenerate primers, producing a mixture of primers representing both the *E. nidulans* and *A. flavus* sequences, would be another option to increase their breadth of applicability. However, for each degenerate nucleotide position, the effective concentration of the primer is reduced by one half.

2. The primer sites flank both introns and exons, providing variation at a variety of taxonomic levels. It appears from the alignment that the level of variation within introns is too high for comparing taxa as distantly related as *A. flavus* and *E. nidulans*, but should be useful within species or between closely related species (within sections). The apparent level of variation in exons appears more appropriate for comparing more distant relatives like *E. nidulans* and *A. flavus*.

3. The primers are predicted to produce a ~477 bp PCR product, which can be sequenced in both directions using two sequence reactions to provide twofold redundancy over all but the terminal 30 bp on each end.

4. The primers are kinetically appropriate for PCR, meaning that they are not expected to produce stable dimers or hairpin structures, and they are expected to anneal to template at reasonable temperatures (50-60°C). Software packages such as Oligo v4.0 (Rychlik, 1992) can be used to test these parameters.

We were able to design primer pairs that amplify a portion of the *ben*A gene in a variety of *Aspergillus* species and related teleomorphs (Figure 2a). We also designed a set of primers for amplifying portions of the calmodulin gene (Figure 2b), which also has highly conserved exons and variable introns.

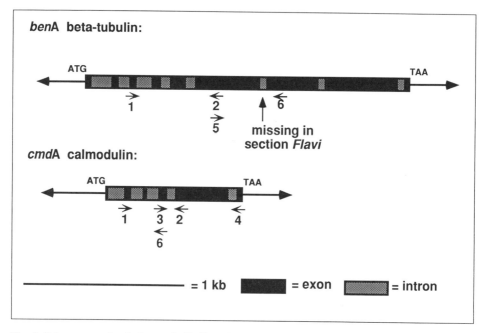

Fig. 2: Primer maps for the beta-tubulin (*ben*A) and calmodulin (*cmd*A) genes.

DESIGNING PRIMERS IN LESS CONSERVED GENES

Because of their conserved exons and variable non-coding regions, beta-tubulin and calmodulin are predicted to have variation useful for comparing isolates both within and between species. However, there are relatively few candidate genes as conserved as beta-tubulin and calmodulin, and for population genetic analyses, as many as ten to fifteen molecular markers may be needed, and they may need to be quite variable. In this case, it is more difficult to develop loci that are useful in a wide variety of species, but if there is sequence information available for the species of interest or a very close relative, it is possible to develop less conserved coding genes as molecular markers. Carbone and Kohn (1999) discuss methods for developing such markers in less-studied Ascomycete groups.

An example of a more variable gene is that encoding acetamidase (*amd*S), which catalyzes the breakdown of certain amides. *E. nidulans amd*S mutants are quite viable except in the presence of acetamide (Hynes *et al.*, 1983), which may explain why the gene sequence is less conserved than tubulin or calmodulin. A partial alignment of the *E. nidulans* and *A. oryzae amd*S genes is shown in Figure 3. Note that the *amd*S exons are less conserved than those in the *ben*A locus and the introns are even more poorly conserved. In fact, *E. nidulans* lacks two of the introns present in *A. oryzae*. Primers amdS1 and amdS2 were designed based on the *A. oryzae* sequence, and produce the expected PCR products in *A. oryzae*, *A. flavus* and *A. parasiticus*, but we did not have success in other species, including *E. nidulans*, *A. nomius*, and *A. tamarii*.

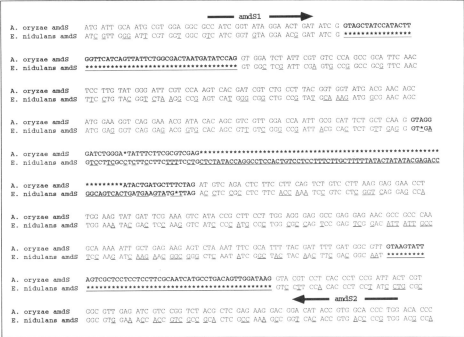

Fig. 3: Alignment of partial *amd*S acetamidase sequences from *Aspergillus oryzae* and *Emericella nidulans*, with locations of primer sites. Bold-face regions represent introns. Underlined sites differ between the two species. Asterisks represent gaps. *E. nidulans* lacks two of the introns found in *A. oryzae*.

IDENTIFYING VARIATION WITHIN LOCI

To evaluate their usefulness, newly designed PCR primers should be tested on a small subset of the taxa of interest, chosen to represent the phylogenetic range from closest relatives to most distant. For PCR amplification, ~50-200 ng of genomic template DNA works best; keep in mind that since these genes are single-copy, more template is required than with rDNA. Rather than measuring the template DNA concentration, it may be easier to simply try different dilutions of the genomic DNA preparation (e.g., 1:10, 1:100, 1:1000). For the first attempt, the following cycling parameters are recommended: 1) 1 step of 2 min at 94°C; 2) 35 steps of the following three cycles: 1 min at 94°C, 1 min at the annealing temperature (primer-pair specific, usually at or close to 56°C), 1 min at

72°C; 3) one final 5 min step at 72°C. If the primers work well, electrophoresing the product on an agarose gel should show a single band of predicted size. If the band is weak or if no product is visible, increasing the template concentration, decreasing the annealing temperature, or trying "touch-down" PCR parameters (Chou et al., 1992), may provide a better result. If multiple bands are produced, one of which is approximately the expected size of the correct product, raising the annealing temperature, touch-down PCRprimer redesign, or gel isolation and re-amplification is recommended. In these cases, one must weigh the cost of spending the time and resources to perfect a particular primer pair versus starting over with new primers or moving on to another gene region.

If the primers work perfectly, and a single band of expected size is produced in all isolates, one still needs to determine whether sequence variation exists within the PCR product (see Figure 4 for an example). One can be virtually certain that variation exists among distantly related taxa, but within species or among close relatives, it is best to check first before wasting time and resources determining its nucleotide sequence. Single-strand conformation polymorphism (SSCP) analysis is one simple means for doing this (Orita *et al.*, 1989; Yap and McGee, 1993). This method involves separating the two strands of the PCR product by heating them, snap cooling, and separating the strands on a non-denaturing gel. The cooled single-stranded DNA forms a secondary structure that depends on its sequence. In almost all cases involving relatively short fragments (<~500 bp), two PCR products that differ at even a single nucleotide will show different patterns on an SSCP gel. If all templates produce PCR products that give the same SSCP pattern, one can assume that there is no sequence variation in that gene region, and move on to something else. In cases where variation is observed, each SSCP pattern can be considered an "allele". To find the sequence differences between different alleles, representatives of each allele type can be sequenced. Once the nucleotide variation is identified, experience shows that most differences will lie in the recognition sequence of some restriction enzyme, which provides an easier and more informative means to assay nucleotide variation than SSCP.

PHYLOGENETICS WITH MULTIPLE GENES
There are two approaches to phylogenetic analysis of more than one gene. First, each gene can be analyzed separately and the resulting phylogenies can be compared to see if they support the same or conflicting phylogenies. Because gene phylogenies do not necessarily represent species phylogenies, separate analysis is needed to validate or challenge phylogenetic inferences based on single genes. Conversely, phylogenies based on single genes may not resolve important evolutionary relationships, whereas combining data from more than one gene may provide sufficient data for complete resolution.

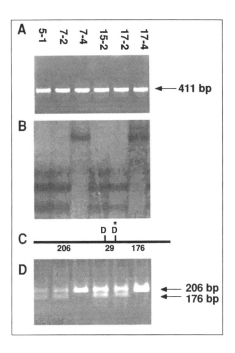

Fig. 4: Identification of polymorphism within a gene region. A. Primers benA5 and benA6 (see Figure 1) were used to amplify a ~411 bp region of the *ben*A beta-tubulin gene in six Australian *Aspergillus flavus* isolates. A single band of predicted size was produced in each isolate. B. To identify sequence polymorphisms, the PCR products were subjected to single-strand conformation polymorphism (SSCP) analysis, which uncovered two distinct patterns, representing different alleles. C. Representatives of the two alleles were sequenced, and found to possess two nucleotide differences, one of which was located in a *Dpn*II restriction site (D*). D. Digestion of the PCR products with *Dpn*II yields distinctive patterns on an agarose gel related to the two SSCP alleles. One allele produces two fragments of roughly equal size that appear as one band on the gel, whereas the other allele produces three bands, one of 206 bp, one of 179 bp, and a third 27 bp band that is too small to appear on the gel.

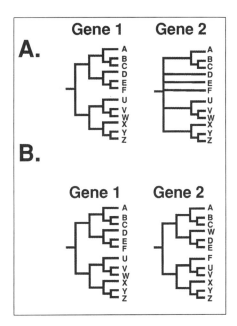

Fig.5 A. Concordant gene trees suitable for a "total evidence" approach. B. Non-concordant gene trees that would be unsuitable.

389

There are good arguments for beginning with separate analyses and, provided that the single gene trees are not in conflict, proceeding to the combined analysis to achieve the maximum resolution (Huelsenbeck *et al.*, 1996). The combined analyses take the form of comparing single gene trees to find a consensus phylogeny, or simply combining the data to create one "total evidence" phylogenetic tree. In Figure 5A, two gene trees are shown that are different, but not in conflict; the data used to make these trees could be combined to make a more fully resolved phylogeny. In Figure 5B, two gene trees are shown that are in topological conflict. Combined analysis is not advised in this case because both phylogenies cannot be correct. A re-examination of the data, or acquisition of new data would be necessary to resolve the conflict.

USE OF SEQUENCE DATA TO IDENTIFY SPECIES

Until recently, phylogenetics and species concepts have not been considered together in mycology. Most phylogenetic studies of fungi have been concerned with relationships among species and not with the limits of species; they have relied upon a phenotypic species concept. In cases where more than one individual from a species has been included in a phylogenetic study, conspecificity is assumed from phenotypic characters. Should multiple individuals from a species not be found as closest neighbours on the resulting tree, re-evaluation of the phenotype is undertaken, often uncovering misidentification of the individual.

In fungi, current species concepts are overwhelmingly phenotypic and morphological, although, with *Aspergillus* and *Penicillium*, biochemistry in the form of secondary compounds has influenced species definitions (Frisvad and Filtenborg, 1989). In a handful of cases, beginning with the description of *Neurospora* species by Shear and Dodge (1927), mating tests have been used to define fungal species biologically. Mayr (1963) defined biological species as "groups of actually or potentially interbreeding natural populations which are reproductively isolated from other such groups." This is a particularly useful definition for population genetics, where defining interbreeding groups is central. With fungi, morphological species typically have been found to contain more than one biological species, and in general, members of different biological species will not produce viable offspring. With *Aspergillus* and *Penicillium*, most species are not known to engage in sexual reproduction, and those that do are usually homothallic, making a biological species concept (BSC) unattainable.

Phylogenetics can also provide a species concept: known as the phylogenetic species concept (PSC; Cracraft, 1983)). Although there is controversy over precisely how the PSC should be defined, a working definition is the smallest clade of organisms sharing a derived character. As noted above, for fungi and morphological characters, the PSC is likely to be broader than the BSC. For nucleic acid characters from single regions, the breadth of the PSC will depend on the variation in the DNA region. Slowly evolving regions could lump morphologically distinct species, but regions having more than one sequence in a species would split the species into as many groups as there are different sequences.

For recombining fungi, phylogenetic analysis of several different nucleic acid regions can bring phylogenetics and species concepts together by combining the PSC and the BSC. Whereas phylogenetic trees based on single DNA regions (single locus) could subdivide biological species by the different sequences (or alleles) found in the region, when the tree is compared to trees based on other DNA regions, branches leading to biological species should be common to all gene trees, but branches due to allelism within biological species should be in conflict. For example, in species of *Gibberella* and its anamorph *Fusarium* O'Donnell and Cigelnik (1997) found phylogenetic species based on rDNA repeat sequences and beta tubulin sequences to be identical to biological species that Leslie (1995) proposed from mating tests. For truly clonal fungi, the BSC will not be applicable, and the PSC might be meaningless because the smallest clade define by a shared derived character could be the progeny from one individual and the individual.

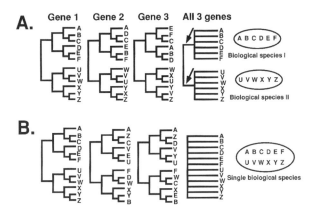

Fig. 6: Identification of biological species by phylogenetic analysis (see Dykhuizen and Green, 1991). Three genes are sequenced in twelve different individuals, represented by the letters A-F and U-Z. A. A-F and U-Z represent two reproductively isolated biological species. The phylogenies of each gene are significantly different in topology with respect to individuals A-F and U-Z, but each phylogeny supports the distinction between the two groups. When all three data sets are analyzed together, conflicts between the gene trees cause a lack of resolution within the two biological species, but the resolution between the two biological species remains (arrowed branches). B. All twelve individuals are members of the same biological species. In this case, the different genes are completely non-concordant and do not agree in any groupings. When all three data sets are analyzed together, none of the groupings apparent in the individual gene trees remain.

Molecular phylogenetic methods provide a means by which biological species can be identified without mating tests (Dykhuizen and Green, 1991; Koufopanou *et al.*, 1997). Figure 6 demonstrates this. The first example (Fig. 6A) includes two biological species, and three genes are sequenced in six individuals in each species. Members of the same biological species interbreed and recombine, so that different genes have different phylogenies. When different genes are taken together in a phylogenetic analysis, inconsistencies between the different genes cause conflict (homoplasy), the result of which is a lack of resolution *within* species. However, there is no interbreeding *between* biological species, so all genes support the distinction between them. Accordingly, each of the three gene trees generated by separate analysis supports the distinction between members of the two biological species as separate clades. By combined analysis, when all three genes are analyzed together, the branches distinguishing the two clades remain intact, whereas there is no resolution of clades within biological species. In the second example (Fig. 5B), all twelve individuals are members of the same biological species. By separate analysis of each gene individually, there is no agreement between genes with respect to clades. By combined analysis, there is no resolution at all, with no support for distinguishing any clades among the twelve individuals.

Only a few cases have been reported where such analyses have been performed in microorganisms (one word). But, with the decreasing cost and increasing efficiency of DNA sequencing, along with the increased availability of molecular markers like those described in this paper, the utility of this method for determining biological species is quite realistic. Interestingly, as was the case for biological species defined based on mating tests, more biological species are apparent based on phylogenetic tests than were apparent previously. Dykhuizen and Green (1991) sequenced three genes, and found evidence for at least four biological species among 15 strains of *Escherichia coli*. With fungi, Koufopanou *et al.* (1997) found two distinct biological species of the human pathogen *Coccidioides immitis* based on the sequences of five genes. Geiser *et al.* (1998b) found two distinct biological species of *A. flavus* among Australian isolates, with no obvious geographical or morphological means to distinguish them. These data strongly suggest that there are more biological species present in nature than we currently recognize, even within species we have already described.

Reproductive isolation, either due to intrinsic or extrinsic barriers, has an essential role in speciation and biological diversification (Dobzhansky, 1937). Certainly, there are important yet unrecognized biological differences among the cryptic biological species identified in phylogenetic analyses. At this point we do not know what biological differences exist between the two biological species of *A. flavus*, but they may include differences in colonization potential or pathogenicity, secondary metabolite (including aflatoxin) production, and susceptibility to control methods. Therefore, recognition of reproductively isolated subtaxa is expected to make an important contribution to our ability to understand and control *Penicillium* and *Aspergillus* species.

REFERENCES

Appel, D.J. & Gordon, T.R. (1996). Relationships among pathogenic and non-pathogenic isolates of *Fusarium oxysporum* based on the partial sequence of the intergenic spacer region of ribosomal DNA. Mol. Plant Microbe Int. 9: 125-138.

Avise, J.C. & Ball, R.M. (1990). Principles of genealogical concordance in species concepts and biological taxonomy. Oxford Surv. Evol. Biol. 7: 45-67.

Berbee, M. L., Yoshimura, A., Sugiyama, J. & Taylor, J.W. (1995). Is *Penicillium* monophyletic? An evaluation of phylogeny in the family Trichocomaceae from 18s, 5.8s and ITS ribosomal DNA sequence data. Mycologia 87: 210-222.

Carbone, I., and Kohn, L.M. (1999) A method for designing primer sets for speciation studies in filamentous ascomycetes. Mycologia 91 : 553-556.

Chang, J.M., Oyzaizu, H. & Sugiyama, J. (1991). Phylogenetic relationships among 11 selected species of Aspergillus and associated teleomorphic genera estimated from 18S ribosomal RNA partial sequences. J. Gen. Appl. Microbiol. 37: 289-308.

Chou, Q., Russell, M. Birch, D.E., Raymond, J. and Bloch, W. 1992. Prevention of pre-PCR mis-priming and primer dimerization improves low-copy-number amplificatin. Nuc. Acids Res. 20:1717-1723.?

Cracraft, J. (1983). Species concepts and speciation analysis. In Current Ornithology (Ed. Johnston, R.F.) New York, Plenum Press. pp. 159-187.

Dobzhansky, T. (1937). Genetics and the Origin of Species. Columbia University Press, New York.

Dykhuizen, D.E. & Green, L. (1991). Recombination in *Escherichia coli* and the defintion of biological species. J. Bacteriol. 173: 7257-7268.

Frisvad, J.C. & Filtenborg, O. 1989. Terverticillate Penicillia: chemotaxonomy and mycotoxin production. Mycologia 81: 837-861.

Geiser, D.M., Timberlake, W.E. & Arnold, M. L. (1996). Loss of meiosis in *Aspergillus*. Mol. Biol. Evol. 13: 809-817.

Geiser, D.M., Frisvad, J.C., and Taylor, J.W. (1998a) Evolutionary relationships in Aspergillus section Fumigati inferred from partial beta-tubulin and hydrophobin DNA sequences. Mycologia 90: 831-845.

Geiser, D. M., Pitt, J.I. & Taylor, J.W. (1998b). Cryptic speciation and recombination in the aflatoxin producing fungus *Aspergillus flavus*. Proc. Natl. Acad. Sci. USA. 05: 388-393..

Higgins, D.G., Beasby, A.J. & Fuchs, R. (1992). CLUSTAL V: improved software for multiple sequence alignment. CABIOS 8: 189-191.

Huelsenbeck, J.P., Bull, J.J. & Cunningham, C.W. (1996). Combining data in phylogenetic analysis. Trends in Ecol. Evol. 11: 152-158.

Hynes, M.J., Corrick, C.M. & J.A. King. (1983) Isolation of genomic clones containing the *amd*S gene of *Aspergillus nidulans* and their use in the analysis of structural and regulatory mutations. Mol. Cell. Biol. 3:1430-1439.

Koufopanou, V., Burt, A. & Taylor, J.W. (1997). Concordance of gene genealogies reveals reproductive isolation in the pathogenic fungus *Coccidioides immitis*. Proc. Natl Acad. Sci. USA 94: 5478-5482.

Leslie, J.F. 1995 *Gibberella fujikuroi*: available populations and variable traits. Can. J. Bot. 73 (Suppl. 1): S282-S291.

LoBuglio, K. F., Pitt, J.I. & Taylor, J.W. (1993). Phylogenetic analysis of two ribosomal DNA regions indicates multiple independent losses of a sexual *Talaromyces* state among asexual *Penicillium* species in subgenus *Biverticillium*. Mycologia 85: 592-604

LoBuglio, K. F., & Taylor, J.W. (1993). Molecular phylogeny of *Talaromyces* and *Penicillium* species in subgenus *Biverticillium*. In The Fungal Holomorph: Mitotic, Meiotic and Pleomorphic Speciation in Fungal Systematics (Eds Reynolds, D.R. & Taylor, J.W.). Wellingford, UK, CAB International. pp. 115-119.

LoBuglio, K. F., Pitt, J.I. & TaylorJ.W. (1994). Independent origins of the synnematous *Penicillium* species, *P. duclauxii*, *P. clavigerum*, and *P. vulpinum*, as assessed by two ribosomal DNA regions. Mycol. Res. 98: 250-256.

LoBuglio, K. F. & TaylorJ.W. (1995). Phylogeny and PCR detection of the human pathogenic fungus *Penicillium marneffei*. J. Clin. Microbiol. 33: 85-89.

Logrieco, A., Peterson, S.W. & Wicklow, D.T. (1989). Ribosomal RNA comparisons among taxa of the terverticilliate Penicillia. In Modern Concepts in *Penicillium* and *Aspergillus* Classification (Eds Samson, R.A. & Pitt, J.I.). New York, Plenum Press. pp. 343-355.

May, G.S., Tsang, M.L.-S., Smith, H., Fidel, S. & Morris, N.R. (1987). *Aspergillus nidulans* beta-tubulin genes are unusually divergent. Gene 55: 231-243.

Mayr, (1963) Animal species and evolution. Bellknap Press, Cambridge, MA, USA.

O'Donnell, K. & Cigelnik, E. (1997). Two divergent intragenomic rDNA ITS2 types within a monophyletic lineage of the fungus *Fusarium* are non-orthologous. Mol. Phyog. Evol. 7: 103-116.

O'Donnell, K., Cigelnik, E., and Nirenberg, H.I. (1998) Molecular systematics and phylogeography of the *Gibberella fujikuroi* species complex. Mycologia 90: 465-493.

Ogawa, H., Yoshimura, A. and Sugiyama, J. (1997) Polyphyletic origins of the anamorphic genus Geosmithia and the relationships of the cleistothecial genera: Evidence from 18S, 5S and 28S rDNA sequence analyses. Mycologia 89: 756-771.

Orita, M., Iwahana, H., Kanazawa, H., Hayashi, K. & Sekiya, T. (1989). Detection of polymorphisms of human DNA by gel electrophoresis as single-strand conformation polymorphisms. Proc. Matl Acad. Sci. USA 86: 2766-2770.

Peterson, S. W. (1993). Molecular genetic assessment of relatedness of *Penicillium* subgenus *Penicillium*. In The Fungal Holomorph: Mitotic, Meiotic and Pleomorphic Speciation in Fungal Systematics (Eds Reynolds, D.R. & Taylor, J.W.). Wellingford, U.K., CAB, International. pp. 121-128

Peterson, S.W. (1995). Phylogenetic analysis of *Aspergillus* sections *Cremei* and *Wentii*, based on ribosomal DNA sequences. Mycol. Res. 99: 1349-1355.

Peterson, S.W. (2000) Phylogenetic relationships in *Aspergillus* based on rDNA sequence analysis In Integration of modern taxonomic methods for *Penicillium* and *Aspergillus* classification (Eds. Samson, R.A. & Pitt, J.I). Harwood Publishers, Amsterdam, pp 323-355

Rychlik, W. (1992). Oligo v4.04 Primer Analysis Software. Plymouth, Minnesota, National Biosciences, Inc.

Samson, R.A. & Pitt, J.I. eds. (1990). Modern Concepts in *Penicillium* and *Aspergillus* Classification. New York, Plenum Press.

Shear, C.L. & Dodge, B.O. (1927). Life histories and heterothallism of the red bread-mold fungus of the *Monilia sitophila* group. J. Agric. Res. 34: 1014-1042.

Spreadbury, C.L., Bainbridge, B.W. & Cohen, J. (1990). Restriction fragment length polymorphisms in isolates of *Aspergillus fumigatus* probed with part of the intergeneric spacer from the ribosomal RNA gene complex of *Aspergillus nidulans*. J. Gen. Microbiol. 136: 1991-1994.

Tamura, M., Kawahara, K. & Sugiyama, J. (2000). Molecular phylogeny of *Aspergillus* and associated teleomorphs in the Trichocomaceae (Eurotiales). In Integration of modern taxonomic methods for Penicillium and Aspergillus classification (Eds. Samson, R.A. & Pitt, J.I). Harwood Publishers, Amsterdam, pp. 357-372

Yap, E. P. H., and McGee, J.O.D. 1993. Nonisotopic discontinuous phase single strand conformation polymorphism (DP-SSCP): genetic profiling of D-loop of human mitochondrial (mt) DNA. Nucleic Acids Research 21:4155.?].

Chapter 7

TAXONOMY OF *ASPERGILLUS* SECTION *NIGRI* AND SECTION *FLAVI*

GENOTYPIC AND PHENOTYPIC VARIABILITY AMONG BLACK ASPERGILLI

János Varga[1], Ferenc Kevei[1], Zsuzsanna Hamari[1], Beáta Tóth[1], József Téren[2], James H. Croft[3] and Zofia Kozakiewicz[4]
[1]Department of Microbiology, Attila József University, H-6701 Szeged, Hungary, [2]Animal Health and Food Control Station,H-6701 Szeged, Hungary, [3]School of Biological Sciences, University of Birmingham, Birmingham B15 2TT, UK, [4]CABI Bioscience, Egham, Surrey TW20 9TY, UK

Aspergillus section *Nigri* is industrially one of the most important taxa of filamentous fungi. Several strains belonging to this section are used in the fermentation industry as producers of different organic acids and hydrolytic enzymes. A taxonomic evaluation of this group was carried out by using different methods. Among the genotypic approaches, electrophoretic karyotypes and the mycovirus content of the strains had little taxonomic value. Nuclear DNA polymorphisms with special regard to the variability of the ribosomal RNA gene cluster, mitochondrial DNA polymorphisms and amplified fragment length polymorphisms were useful from the taxonomic point of view. Among the phenotypic methods, carbon source utilization spectra were applicable for distinguishing the *A. niger* species complex from other more distantly related black *Aspergillus* species. The other phenotypic approaches including scanning electronmicroscopic examination of the conidial ornamentations of the strains, isoenzyme analysis and thin layer chromatography of the secondary metabolites together with the genotypic approaches led to the recognition of eight species (*A. heteromorphus, A. ellipticus, A. carbonarius, A. japonicus, A. aculeatus, A. niger, A. tubingensis* and "*A. brasiliensis*"), and one subspecies ("*A. carbonarius* var. *indicus*") within *Aspergillus* section *Nigri*. In this scheme, some well-known species names such as *A. foetidus, A. awamori, A. phoenicis* or *A. ficuum* are treated as synonims. Formal taxonomic descriptions of "*A. carbonarius* var. *indicus*" and "*A. brasiliensis*" are in progress.

INTRODUCTION

Black Aspergilli (*Aspergillus niger* species group, Raper and Fennell, 1965; *Aspergillus* section *Nigri*, Gams *et al.*, 1985) have a significant impact on modern society. Many species belonging to this group cause food spoilage, and several of them are used in the fermentation industry to produce different hydrolytic enzymes such as amylases or lipases, and organic acids like citric acid and gluconic acid (for references, see Raper and Fennell, 1965; Smith and Pateman, 1977; Bennett and Klich, 1992; Kozakiewicz, 1989). They are also candidates for genetic manipulation in the biotechnological industries since *A. niger* has been granted the GRAS (generally regarded as safe) status by the Food and Drug Administration of the US government. Black Aspergilli have long had one of the better taxonomic descriptions among the fungi. Mosseray (1934; cf. Raper and Fennell, better taxonomic descriptions among the fungi. Mosseray (1934; cf. Raper and Fennell, 1965) described 35 species of the black Aspergilli. Raper and Fennell (1965) reduced the number of species accepted within the *A. niger* group to twelve. Al-Musallam (1980) revised the taxonomy of the *A. niger* group by taking mainly morphological features into account. She recognized seven species within this group (*A. japonicus, A. carbonarius, A. ellipticus, A. helicothrix, A. heteromorphus, A. foetidus, A. niger*), and described *A. niger* itself as an aggregate consisting of seven varieties and two formae. Kozakiewicz (1989) distinguished *A. ellipticus, A. heteromorphus, A. japonicus, A. helicothrix, A. atroviolaceus* (treated as *A. aculeatus* or *A. japonicus* var. *aculeatus* in other classifications) and *A. carbonarius* species exhibiting echinulate conidial ornamentations from the rest of black *Aspergillus* strains, which displayed verrucose conidia. Within the verrucose category, *A. fonsecaeus, A. acidus* (*A. foetidus* var. *acidus*), *A. niger* var. *niger*, *A. niger* var. *phoenicis, A. niger* var. *ficuum, A. niger* var. *tubingensis, A. niger* var. *pulverulentus, A. niger* var. *awamori, A. citricus* (*A. foetidus*) and *A. citricus* var. *pallidus* (*A. foetidus* var. *pallidus*) taxa were distinguished [for a more detailed evolution of the taxonomic schemes within *Aspergillus* section *Nigri*, see Table 16-4 in Samson (1992)]. In the recent years, several publications have dealt with the application of different phenotypic and genotypic markers for clarifying the taxonomy of black Aspergilli. In this paper, we wish to give a short summary of these approaches, relying heavily on our results in this field.

MATERIALS AND METHODS

For the black *Aspergillus* strains tested, see Varga *et al.* (1993, 1994a, 1994b), Kevei *et al.* (1996), and Hamari *et al.* (1997). The methods applied are also referred to in the literature (Varga *et al.*, 1993, 1994a, 1994b, 1996; Téren *et al.*, 1996; Kevei *et al.*, 1997). Scanning electron micrographs of the conidia of black *Aspergillus* strains were made as described earlier (Kozakiewicz, 1989).

RESULTS

GENOTYPIC APPROACHES

1. Nuclear DNA polymorphisms
a. Restriction fragment length polymorphisms (RFLPs)
Kusters-van Someren *et al.* (1990) used Western blotting and DNA hybridization with a pectin lyase (*pelD*) gene to ascertain whether these methods could be used for rapid strain identification. The DNA hybridization experiments showed that the *pelD* gene is conserved in all isolates belonging to the *A. niger* aggregate. Hybridization was also observed in DNAs of all *A. foetidus* strains. The authors established three groups within the *A. niger* aggregate on the basis of presence or absence of three other bands which hybridized strongly to the *pelD* gene. Other species of black Aspergilli did not show homology to *pelD*. As a continuation of this work, Kusters-van Someren *et al.* (1991) carried out a more extensive study on nuclear DNA RFLPs of several black *Aspergillus* collection strains. Two groups of strains were distinguished according to their *Sma*I-

generated ribosomal DNA (rDNA) patterns. The two groups were also clearly distinguishable by their hybridization patterns when pectin lyase genes (*pelA, pelB*) and the pyruvate kinase (*pki*) gene were used as probes in DNA hybridization experiments. The two groups found were proposed to represent different species, namely *A. niger* and *A. tubingensis*. Examination of other species not belonging to the *A. niger* aggregate was also carried out. *A. foetidus* strains, classified into a different species by Al-Musallam (1980), showed the same nuclear DNA RFLPs as *A. niger. A. helicothrix* was found to represent only a morphological variant of *A. ellipticus*, and *A. aculeatus* should only be ranked to subspecies status as it showed the same *Sma*I-digested rDNA pattern as the *A. japonicus* strains examined.

Bussink *et al.*, (1991), Graaff *et al.*, (1994) and Gielkens *et al.*(1997) detected further differences in the nuclear genes encoding polygalacturonase II, arabinoxylan-arabinofuranohydrolase and xylanase enzymes of *A. niger* and *A. tubingensis* strains.

According to our results, the *Sma*I digested repetitive DNA profiles hybridized with the ribosomal repeat unit of *A. nidulans* have distinctive value among black Aspergilli. *A. ellipticus, A. heteromorphus, A. japonicus* and *A. carbonarius* exhibited species specific hybridization patterns, with the exception of *A. carbonarius* strain IN7, which revealed a slightly different profile than the other *A. carbonarius* strains examined (Fig. 1). *A. fonsecaeus* displayed the same *Sma*I digested rDNA profile as the majority of the *A. carbonarius* strains (Kevei *et al.*, 1996). Among the strains of the *A. niger* species complex, four profiles were observed, among which rDNA types I and III were shown by *A. niger* and by some Brazilian black *Aspergillus* strains, respectively, while rDNA types II and II' were characteristic of the *A. tubingensis* strains (Fig. 1; Varga *et al.*, 1994b).

b. Amplified fragment length polymorphisms (AFLPs)

Megnegneau *et al.* (1993) applied random amplified polymorphic DNA (RAPD) technique for examining variability among black Aspergilli. By applying six random primers, they could differentiate *A. carbonarius, A. japonicus, A. aculeatus, A. heteromorphus* and *A. ellipticus* from each other, and could divide the *A. niger* species complex into two groups corresponding to the *A. niger* and *A. tubingensis* species.

We also successfully used the RAPD technique for the examination of genetic variability within *A. carbonarius* and *A. japonicus* species. *A. carbonarius* strain IN7 could readily be distinguished from the other *A. carbonarius* strains examined (Kevei *et al.*, 1996). The strains representing the *A. japonicus* var. *aculeatus* subspecies could also be distinguished from the other *A. japonicus* strains by using 4 random primers (Hamari *et al.*, 1997). The results obtained were in agreement with those obtained by other approaches (detailed below). The Brazilian strains which exhibited type III rDNA profiles could also be distinguished from the other strains of the *A. niger* species complex by RAPD (Fig. 2).

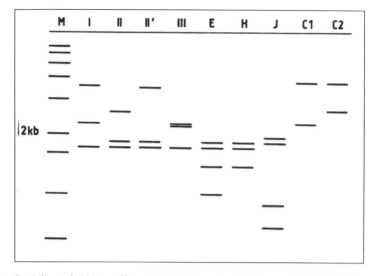

Fig. 1. *Sma*I digested rDNA profiles of the black *Aspergillus* strains examined. M, 1 kb DNA ladder.

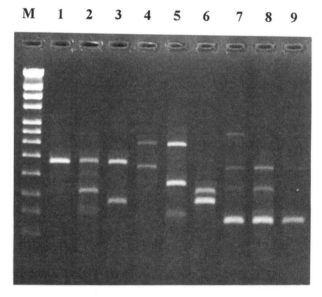

Fig. 2. RAPD profiles of some strains belonging to the *A. niger* species aggregate by using OPC-19 as primer. M, lambda-pUC mix (#SM0291, Fermentas); lane 1, *A. niger* CBS 120.49 (mtDNA type 1a); lane 2, *A. niger* SZMC 0851 (mtDNA type 1b); lane 3, *A. niger* 1.8A.9 (mtDNA type 1d); lane 4, *A. tubingensis* CBS 117.32 (mtDNA type 2a); lane 5, *A. tubingensis* SZMC 0932 (mtDNA type 2b); lane 6, *A. tubingensis* JHC 806 (mtDNA type 2c); lane 7, "*A. brasiliensis*" JHC 607 (mtDNA type 3a); lane 8, "*A. brasiliensis*" JHC 614 (mtDNA type 3a); lane 9, "*A. brasiliensis*" JHC 602 (mtDNA type 3b).

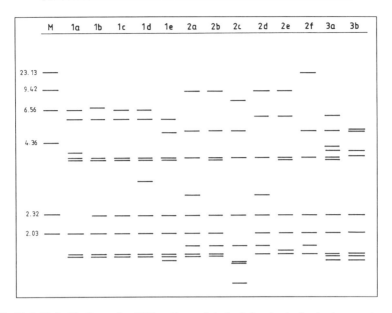

Fig. 3. HaeIII-BglII double digested mtDNA patterns of strains belonging to the A. niger species aggregate. M, HindIII digested lambda DNA.

c. Electrophoretic karyotyping

Analysis of electrophoretic karyotypes among black Aspergilli revealed the presence of high levels of intraspecific variability of the banding patterns observed (Megnegneau et al., 1993; Swart et al., 1994). However, the estimated total genome sizes did not differ significantly, ranging from 35.9 Mb in an A. niger strain to 43.8 Mb in an A. ellipticus strain. The average genome size of strains belonging to the A. niger species complex was 38.3 Mb, which is much larger than that found in A. nidulans strains (31.3 Mb; Brody and Carbon, 1989). In general, electrophoretic karyotyping seems to be of little taxonomic value in such a variable group as black Aspergilli.

2. Mitochondrial DNA polymorphisms

Wide-ranging mitochondrial DNA (mtDNA) variation was observed both among collection strains and in natural populations of the A. niger species complex (Varga et al., 1993, 1994b). Within the A. niger species complex, mtDNA types could be differentiated from each other when total DNA samples were double-digested with HaeIII and BglII restriction enzymes (Fig. 3). Using this enzyme combination the nuclear DNA digests to small fragments which migrate to the anodic end of the gel leaving the mtDNA fragments clearly isolated towards the cathodic end. These fragments were shown to be of mitochondrial origin by comparison with HaeIII-BglII digests of purified mtDNA, and by hybridization experiments. This method was used for fast grouping of black Aspergillus isolates. Most isolates were classifiable as A. niger or A. tubingensis according to their HaeIII-BglII digested mtDNA patterns. The mtDNA variation was distributed unevenly in the populations studied; some local populations displayed very little polymorphism in their mtDNA and rDNA patterns, while Australian populations exhibited very high degree of variability. Hybridization experiments in which cloned *A. niger* and *A. nidulans* mtDNA fragments were used revealed that the two main mtDNA groups corresponding to *A. niger* and *A. tubingensis* are more distantly related than concluded earlier (Kusters-van Someren *et al.*, 1991). The *A. niger* and *A. tubingensis* species could be grouped into 5 and 6 mtDNA types, respectively. Six of the 13 Brazilian isolates examined exhibited mtDNA and rDNA types different from those of all the other strains; these strains were proposed to represent a new subspecies or a new species within *Aspergillus* section *Nigri* (Varga *et al.*, 1994b). A physical map of an *A. tubingensis* strain harbouring type 2a mtDNAs is available (Kirimura *et al.*, 1992); physical mapping of the mtDNAs of other strains is in progress.

The sizes of the mtDNAs of the black *Aspergillus* strains examined were highly variable. The mtDNA of type 3 was the largest (35 kb) followed by those of types 2f and 2e (34 kb and 32.5 kb, respectively). The smallest mtDNA molecule (26 kb) was that of type 2c. All the other mtDNA types had sizes in the range 28-31 kb, which is in agreement with the mtDNA sizes of other *Aspergillus* species (Hudspeth, 1992; Moody and Tyler, 1990).

MtDNA transfers between different RFLP groups of black Aspergilli were attempted by polyethylene glycol induced protoplast fusion, using a strain carrying mitochondrial oligomycin-resistance. The transfers resulted in recombination of the mitochondrial genomes of the donor and acceptor strains in most cases. Recombination readily occurred even between mtDNAs of strains representing different species like *A. niger* and *A. tubingensis*, but we could not transfer the oligomycin resistance marker to more distantly elated black Aspergilli (*A. carbonarius, A. japonicus*; Kevei *et al.*, 1997).

For *A. japonicus* isolates, analysis of RFLPs of the rDNA and mtDNA gave different results. The rDNAs proved to be invariable, even strains of the subspecies *A. japonicus* var. *aculeatus* exhibited the same restriction profile (Fig. 1), while the strains could be classified into seven different mtDNA RFLP groups based on their *Hae*III-digested mtDNA profiles (Fig. 4). Hybridization data suggest that six of these mtDNA types have certain common features in their organization, while mtDNA type 7, which was exhibited by the *A. japonicus* var. *aculeatus* type strain and two other strains, probably have quite different mtDNA structure (Hamari *et al.*, 1997). The sizes of *A. japonicus* mtDNAs were in the range of 43-50 kb. Recent results indicate that there is even more variability within the mtDNAs of the *A. japonicus* strains (Hamari *et al.*, unpublished results).

Among the 16 collection strains and field isolates of *Aspergillus carbonarius* examined, the *Hae*III-digested mtDNA profiles revealed only slight variations, except for one field isolate (IN7), which exhibited completely different mtDNA patterns (Fig. 4). The mitochondrial DNAs of these strains were found to be much larger (45 to 57 kb) than those found earlier in the *A. niger* aggregate. The physical maps of the mtDNAs of *A. carbonarius* strain IN7 and the other *A. carbonarius* strains are quite different from each other, however, the order of the genes on these molecules seems to be conserved (Hamari *et al.*, unpublished results).

In Table 1, a short summary of the mtDNA and rDNA variability within *Aspergillus* section *Nigri* is given. For the rDNA and mtDNA patterns observed, see Figs. 1, 3 and 4.

3. Mycoviruses

Altogether, 493 black *Aspergillus* isolates were checked for the presence of double-stranded RNA (dsRNA) genomes. About 10% of the field isolates examined proved to be infected with dsRNA elements indicative of mycovirus infection.

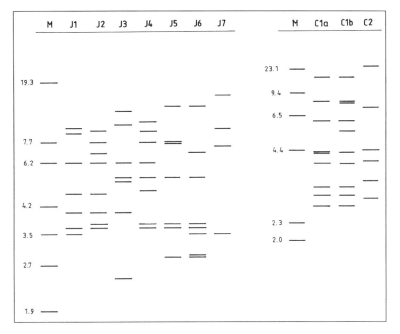

Fig. 4. *Hae*III digested mtDNA patterns of the *A. japonicus* and *A. carbonarius* strains. M (for *A. japonicus* strains), *Hind*III-*Eco*RI double digested lambda DNA, M (for *A. carbonarius* strains), *Hind*III digested lambda DNA.

In addition, fourteen of the 53 collection strains belonging in *Aspergillus* section *Nigri* also carried dsRNA genomes (Varga *et al.*, 1994a). The identity of these bands was proved by S1 nuclease and RNase treatments. The dsRNA patterns observed were highly variable, representing 20 different profiles. Both *A. heteromorphus* strains revealed the same dsRNA profiles, while dsRNA elements were only rarely observed in *A. tubingensis* and *A. carbonarius* strains. Most of the virus-containing natural isolates came from Indonesia.

Electronmicroscopic examination of the strains revealed the presence of virus-like particles (VLPs) in the mycelia of the strains examined; all the VLPs were isometric and their size was around 30-35 nm, while some Indonesian isolates also contained VLPs in the size range 23-25 nm (Varga *et al.*, 1994a).

1. Morphological examinations
Morphologically distinct species such as *A. ellipticus*, *A. heteromorphus*, *A. japonicus* and *A. carbonarius* also show individually distinctive patterns using the molecular and physiological techniques employed here. Furthermore, conidial ornamentation as revealed using the scanning electron microscope (SEM) is also unique (Fig. 5).

Table 1. Representatives of the different mtDNA types among black Aspergilli.

MtDNA type [a]	RDNA type [b]	Number of isolates	Representative strains[c]
A. niger		252	
1a	I	105	CBS 554.65 (A. niger)
1b	I	76	JHC 411 (from soil, Tunisia)
1c	I	65	IMI 211394 (A. awamori)
1d	I	1	1.8A.9 (from soil, Indonesia)
1e	I	1	JHC 245 (from soil, Hungary)
A. tubingensis		90	
2a	II-II'	45	IMI 172296 (A. tubingensis)
2b	II-II'	21	IMI 091881 (A. ficuum)
2c	II-II'	9	IMI 104688 (A. foetidus var. acidus)
2d	II-II'	8	JHC 550 (from soil, Australia)
2e	II-II'	3	JHC 554 (from soil, Australia)
2f	II	3	JHC 565 (from soil, Australia)
"*A. brasiliensis*"		7	
3a	III	6	JHC 607 (from soil, Brazil)
3b	III	1	JHC 602 (from soil, Brazil)
A. japonicus	.	51	
J1	J	27	IMI 119894
J2	J	1	IN10 (from soil, India)
J3	J	10	CBS 114.51
J4	J	7	Fr.1.2.1 (from soil, France)
J5	J	2	JHC 557 (from soil, Australia)
J6	J	1	JHC 564 (from soil, Australia)
J7	J	3	CBS 172.66 (A. aculeatus)
A. carbonarius		16	
C1a	C1	11	IMI 016136 (A. carbonarius)
C1b	C1	4	ETH 8969
C2	C2	1	IN7 (from soil, India)
A. ellipticus		2	
E	E	2	IMI 172283
A. heteromorphus		2	
H	H	2	IMI 172288

In the case of the *A. niger* species complex, *Hae*III-*Bgl*II double digestions were taken into account, while for the other species, *Hae*III-digested mtDNA profiles were examined. *Sma*I-digested repetitive DNA profiles hybridized with the *A. nidulans* ribosomal repeat unit were examined. Abbreviations: CBS, Centraalbureau voor Schimmelcultures, Baarn, The Netherlands; ETH, Eidgenössische Technische Hochschule, Zürich, Switzerland; IMI, International Mycological Institute, Egham, UK; JHC, J. H. Croft's Culture Collection, Birmingham, UK.II. Phenotypic approaches

A. japonicus and *A. aculeatus* are both distinct members of section *Nigri*, being the only two species in this section which are strictly uniseriate (Gams *et al.*, 1985; Raper and Fennell, 1965; Al-Musallam,1980). Like *A. niger*, *A. japonicus* is one of the most commonly isolated of the black Aspergilli, occurring on a range of different substrates such as soil, plant materials or green coffee beans (Kozakiewicz, 1989). Despite the fact that these two species gave similar rDNA patterns, their morphological characteristics are singularly different. Using light microscopy (LM), conidia of *A. japonicus* are echinulate and globose, whilst those of *A. aculeatus* are also echinulate but distinctly elliptical.

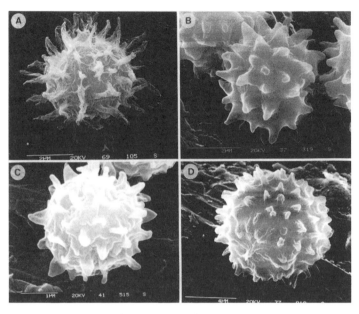

Fig. 5. Scanning electron micrographs illustrating conidial ornamentation in morphologically distinct species a. *A. ellipticus*, b. *A. heteromorphus*, c. *A. japonicus*, d. *A. carbonarius*.

Fig. 6. Scanning electron micrographs illustrating conidial ornamentation in a. *A. niger*, b. *A. tubingensis*, c. *A. japonicus*, d. *A. aculeatus.*

Fig. 7. Scanning electron micrographs illustrating distinctive conidial ornamentation in five Brazilian isolates
a. 602, b. 603, c. 606, d. 607, e. 614.

Conidial ornamentation using SEM is also subtly different. The echinulations on *A. aculeatus* are more numerous and robust in appearance, whilst those on *A. japonicus* are more delicate and less numerous (Fig. 6c, d). It is interesting to note that both AFLPs and mtDNA techniques could also distinguish between these two species.

A. carbonarius is possibly the most distinct member of this section, exhibiting a multi-nucleated conidial system (Raper and Fennell, 1965; Al-Musallam, 1980). Strains of this species can easily be recognised using LM since conidia are much larger than those of other black Aspergilli (Yuill, 1950). Conidial ornamentation in the SEM is also unique (Fig. 5d). Despite the fact that strain IN7 had conidia which were somewhat smaller in size to the other isolates of *A. carbonarius*, using LM and SEM they were identical and unequivocal for *A. carbonarius*. In this case AFLPs and mtDNA characteristics gave a different result for IN7 compared to the other *A. carbonarius* isolates.

Examination of the Brazilian isolates belonging to the *A. niger* species aggregate using LM indicated that whilst most characters were consistent with *A. niger*, conidia were much rougher than those of a typical *A. niger* conidium. This was confirmed using SEM (compare Fig. 6a to Fig. 7). Previous work on the ontogenic development of conidial ornamentation in *A. niger* (Kozakiewicz, 1989) suggests that the Brazilian isolates possibly represent a new taxon.

2. Carbon source utilization

During carbon source utilization tests, more than 100 compounds (sugars and sugar alcohols, oligo- and polysaccharides, organic acids, amino acids, nucleotides, etc.) were examined (Kevei *et al.*, 1996). Seven of these (melezitose, xylitol, galactitol, vanillic acid, cis-aconitic acid, L-serine and L-tyrosine) revealed variable utilisation patterns among black Aspergilli. Almost every strain of the *A. niger* species aggregate utilized all these 7 compounds, with the exception of L-tyrosine (Table 2); less than 5% of the strains examined were not able to utilize either melezitose, xylitol or galactitol as sole carbon source. As regards the *A. carbonarius* strains, vanillic acid, cis-aconitic acid, L-serine and L-tyrosine were not utilised, melezitose and xylitol were accepted as single carbon source by all strains, while assimilation of galactitol was variable (Table 2). For *A. japonicus* strains, only L-tyrosine was accepted as sole carbon source by most of the 51 strains tested. *A. japonicus* IN9 behaved differently from the other *A. japonicus* strains, and was able to utilise all seven compounds as sole carbon sources, while IN11 could not utilize any of these compounds (Table 2; Hamari *et al.*, 1997).

Table 2. Carbon source utilization of black *Aspergillus* strains. In the lines with double border the characteristic carbon source utilization spectra of the different black *Aspergillus* species are shown.

Strains	Glu	Mel	Xyl	Gal	Cis	Van	Ser	Tyr
A. niger sp. complex (more than 90% of the examined 300 strains)	++	+	+	+	+	+	+	-
A. foetidus var. *acidus* IMI 104688	++	(+)	-	+	(+)	-	-	-
A. foetidus var. *pallidus* IMI 175963	++	+	-	+	+	-	-	-
10V10 (J.L. Azevedo)	++	-	(+)	+	(+)	+	(+)	-
A. japonicus (51)	++	-	-	-	-	-	-	+
IN9	++	+	+	+	+	+	+	+
IN11	++	-	-	-	-	-	-	-
A. carbonarius (12)	++	+	+	(±)	-	-	-	-
IMI 016136	++	+	+	++	-	-	-	-
N/1	++	+	+	-	-	-	-	-
N/2	++	+	+	-	-	-	-	-
IN7	++	++	+	(+)	-	-	-	-
ETH 8969	++	+	+	+	-	-	-	-
Isr1.1.1	++	+	+	++	-	-	-	-
M1.4.1	++	+	+	+	-	-	-	-
1.4.29	++	+	+	++	-	-	-	-
NRRL 67	++	+	+	-	-	-	-	-
Haren	++	+	+	+	-	-	-	-
CBS 111.26	++	+	+	++	-	-	-	-
C-P01	++	+	+	+	-	-	(+)	-
A. ellipticus (2)	++	+	-	-	(+)	-	-	-
A. heteromorphus (2)	++	+	+	-	(+)	-	-	-

Abbreviations: Glu, D-glucose; Mel, melezitose; Xyl, xylitol; Gal, galactitol; Cis, cis-aconitic acid; Van, vanillic acid; Ser, L-serine; Tyr, L-tyrosine.

A. ellipticus and *A. heteromorphus* strains could also be differentiated from other black Aspergilli based on their carbon source utilization profiles (Table 2).

This approach could not be used to group the strains of the *A. niger* species complex into distinct slots.

3. Isoenzyme analysis

Isoenzyme analysis was successfully applied to distinguish *Aspergillus* strains by Megnegneau *et al.* (1993). The results obtained from these data were in agreement with those obtained from RAPD analysis, i.e. the *A. niger* species complex could be grouped into two slots corresponding to the *A. niger* and *A. tubingensis* species.

We successfully used this technique to reveal intraspecific variability within *A. carbonarius* and *A. japonicus*. The NADP-dependent glutamate dehydrogenase patterns made it possible to distinguish one strain, IN7 from the other *A. carbonarius* strains (data not shown; Kevei *et al.*, 1996). Similarly to the observations of Megnegneau *et al.* (1993), the highest variability was observed in the cases of arylesterase and acid phosphatase isoenzymes (data not shown). Our results are also in agreement with those obtained by other approaches including RAPD and mtDNA RFLP analyses (Kevei *et al.*, 1996; Hamari *et al.*, 1997).

4. Secondary metabolites

About 160 black *Aspergillus* collection strains and field isolates were tested for ochratoxin A production by an immunochemical method, and by thin layer chromatography. The strains examined include 12 *A. carbonarius* and 45 *A. japonicus* collection strains and isolates from all over the world, and about 100 strains belonging to the *A. niger* species complex. Ochratoxin production was observed in about 6% of the strains representing the *A. niger* species complex (Abarca *et al.*, 1994; Téren *et al.*, 1996). Among the 13 *A. carbonarius* strains tested, 6 produced both ochratoxins A and B (Fig. 8; Téren *et al.*, 1996; Wicklow *et al.*, 1996). *A. ellipticus*, *A. heteromorphus*, *A. japonicus* and *A. tubingensis* strains were found not to produce detectable amounts of ochratoxins. This method was also successfully applied to detect ochratoxins in *A. albertensis*, *A. auricomus* and *A. wentii* strains (Fig. 8; Varga *et al.*, 1996).

The black *Aspergillus* strains displayed complex secondary metabolite profiles. The Brazilian strains which were suggested to represent a new black *Aspergillus* species or subspecies based on their unique mitochondrial DNA and nuclear DNA profiles in our earlier studies could readily be distinguished from strains belonging to the *A. niger* or *A. tubingensis* species, based on their secondary metabolite profiles (data not shown). Similarly, *A. carbonarius* strains also exhibited species-specific patterns (data not shown).

CONCLUSIONS

Altogether, nearly 500 black *Aspergillus* strains were examined in our laboratory by most of the methods described above. 349 of these were found to belong to the *A. niger* species complex. 252 strains belong to the *A. niger* species as delimited by Kusters-van Someren *et al.* (1991), 90 represent the *A. tubingensis* species, and 7 Brazilian isolates are proposed to belong to a new species, "*A. brasiliensis*", based on their specific nuclear and mitochondrial DNA patterns, amplified DNA patterns, conidial ornamentations and secondary metabolite profiles. The strains belonging to both the *A. niger* and *A. tubingensis* species could be further divided to 5 and 6 groups, respectively, based on their mtDNA patterns, while those of "*A. brasiliensis*" exhibit two slightly different mtDNA profiles.

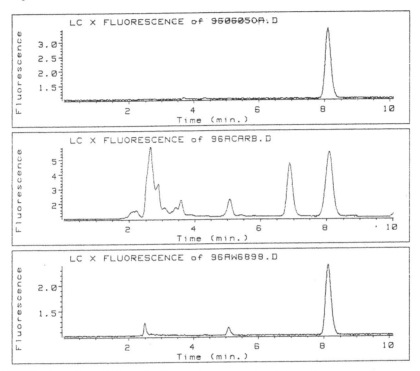

Fig. 8. HPLCs of an ochratoxin A standard (upper chromatogram), and cleaned, concentrated extracts of *A. carbonarius* N-1 (middle chromatogram) and *A. wentii* IMI 017295 (bottom chromatogram).

Fifty-one *A. japonicus* strains were examined, three of which are proposed to represent *A. aculeatus* based on their mtDNA profiles, and morphology. The other *A. japonicus* strains were highly variable, harbouring 6 different mtDNA molecules (Fig. 4; Hamari *et al.*, 1997). However, the rDNA organization, and secondary metabolite profiles of these strains were quite similar. We should mention, that the rDNA regions of *A. japonicus* and *A. aculeatus* strains were different when *Eco*RI was used to digest their nuclear DNA instead of *Sma*I (Megnegneau *et al.*, 1993).

Sixteen *A. carbonarius* strains could easily be recognized based on the size and nuclear content of their conidia. All except one strain revealed the same rDNA patterns, and similar mtDNA and isoenzyme profiles. *A. carbonarius* IN7 was found to have the smallest conidia, and mtDNAs which are different in structure from those of the other *A. carbonarius* strains. The rDNA, RAPD and isoenzyme patterns of this strain were also different from the other examined isolates. These observations led us to propose that this strain represents a new subspecies, "*A. carbonarius* var. *indicus*" of the *A. carbonarius* species. The formal description of "*A. brasiliensis*" and "*A. carbonarius* var. *indicus*" is in progress.

The two *A. ellipticus* and *A. heteromorphus* strains could easily be differentiated from the other black *Aspergillus* species based on their morphological and physiological features.

Proposed species and subspecies within *Aspergillus* section *Nigri* based on the results detailed above:

Conidia echinulate:	*A. heteromorphus*
	A. ellipticus
	A. carbonarius
	"*A. carbonarius* var. *indicus*" (IN7)
	A. japonicus
	A. aculeatus
	"*A. brasiliensis*" (rDNA type III)
	A. niger species complex:
Conidia verrucose:	*A. niger* (rDNA type I)
	A. tubingensis (rDNA type II)

According to this scheme, eight species and one subspecies are included in *Aspergillus* section *Nigri*. This taxonomic scheme treats some well-known species names such as *A. foetidus, A. awamori, A. phoenicis, A. pulverulentus, A. ficuum* or *A. usami* as synonyms. We suspect that this taxonomic scheme is quite disturbing from the industrial point of view. However, we believe that taxonomy based on such a broad spectrum of techniques outlined above should not lag behind industrial applications just because of convenience.

ACKNOWLEDGEMENTS

Part of the study referred to in this review was carried out while J. Varga was in receipt of a joint one-year postdoctoral fellowship of the Royal Society and the Hungarian Academy of Sciences, partly sponsored by the Soros Foundation. This study was also supported financially by Hungarian Scientific Research Fund (OTKA) grants #F014641, #F023062 and #T013044.

REFERENCES

Abarca, M.L., Bragulat, M.R., Castella, G. & Cabanes, F.J. (1994). Ochratoxin A production by strains of *Aspergillus niger* var. *niger*. Appl. Environ. Microbiol. 60, 2650-2652.

Al-Musallam, A. (1980). Revision of the black *Aspergillus* species. Ph.D. Thesis, Utrecht.

Bennett, J.W. & Klich, M.A. (1992). *Aspergillus*: Biology and industrial applications. Butterworth-Heinemann, Boston.

Brody, H. & Carbon, J. (1989). Electrophoretic karyotype of *Aspergillus nidulans*. Proc. Natl. Acad. Sci. USA 86, 6260-6263.

Bussink, H.J.D., Buxton, F.P. & Visser, J. (1991). Expression and sequence comparison of the *Aspergillus niger* and *Aspergillus tubingensis* genes encoding polygalacturonase II. Curr. Genet. 19, 467-474.

Croft, J.H. & Varga, J. (1994). Application of RFLPs in systematics and population genetics of Aspergilli. In The genus *Aspergillus*: from taxonomy and genetics to industrial applications (Eds Powell, K.A.,

Renwick, A. & Peberdy, J.F.). New York, Plenum Press. pp. 277-289.

Gams, W., Christensen, M., Onions, A.H.S., Pitt, J.I. & Samson, R.A. (1985). Infrageneric taxa of *Aspergillus.* In Advances in *Penicillium* and *Aspergillus* systematics (Eds Samson, R.A. & Pitt, J.I.). New York, Plenum Press. pp. 55-61.

Gielkens, M.M.C., Visser, J. & de Graaff, L.H. (1997). Arabinoxylan degradation by fungi: characterization of the arabinoxylan-arabinofuranohydrolase encoding genes from *Aspergillus niger* and *Aspergillus tubingensis.* Curr. Genet. 31, 22-29.

Graaff, L.H.de, Broeck, H.C. van den, Ooijen, A.J.J. & Visser, J. (1994). Regulation of the xylanase-encoding *xlnA* gene of *Aspergillus tubingensis.* Mol. Microbiol. 12, 479-490.

Hamari, Z., Kevei, F., Kovács, É., Varga, J., Kozakiewicz, Z. & Croft, J.H. (1997). Molecular and phenotypic characterization of *Aspergillus japonicus* strains with special regard to their mitochondrial DNA polymorphisms. Antonie van Leeuwenhoek (submitted).

Hudspeth, M.E.S. (1992). The fungal mitochondrial genome - a broader perspective. In: Handbook of applied mycology Vol. 4. Fungal biotechnology (Eds Arora, D.K., Elander, R.P. & Murekji, K.G.). New York, Marcel Dekker, Inc. pp. 213-241.

Kevei, F., Hamari, Z., Varga, J., Kozakiewicz, Z. & Croft, J.H. (1996). Molecular polymorphism and phenotypic variation in *Aspergillus carbonarius.* Antonie van Leeuwenhoek 70, 59-66.

Kevei, F., Tóth, B., Coenen, A., Hamari, Z., Varga, J. & Croft, J.H. (1997). Recombination of mitochondrial DNA following transmission of mitochondria among incompatible strains of black Aspergilli. Mol. Gen. Genet. (in press).

Kirimura, K., Fukuda, S., Abe, H., Kanayama, S. & Usami, S. (1992). Physical mapping of the mitochondrial DNA from *Aspergillus niger.* FEMS Microbiol. Letters 90, 235-238.

Kozakiewicz, Z. (1989). *Aspergillus* species on stored products. Mycol. Papers 161, 1-188.

Kusters-van Someren, M.A., Kester, H.C.M., Samson, R.A. & Visser, J. (1990). Variation in pectinolytic enzymes in black Aspergilli: a biochemical and genetic approach. In Modern concepts in *Penicillium* and *Aspergillus* classification (Eds Samson, R.A. & Pitt, J.I.). New York, Plenum Press. pp.321-334.

Kusters-van Someren, M.A., Samson, R.A. & Visser, J. (1991). The use of RFLP analysis in classification of the black Aspergilli: reinterpretation of *Aspergillus niger* aggregate. Curr. Genet. 19, 21-26.

Megnegneau, B., Debets, F. & Hoekstra, R.F. (1993). Genetic variability and relatedness in the complex group of black Aspergilli based on random amplification of polymorphic DNA. Curr. Genet. 23, 323-329.

Moody, S.F. & Tyler, B.M. (1990). Restriction enzyme analysis of mitochondrial DNA of the *Aspergillus flavus* group: *A. flavus, A. parasiticus,* and *A. nomius.* Appl. Environ. Microbiol. 56, 2441-2452.

Raper, K.B. & Fennell, D.I. (1965). The genus *Aspergillus.* Williams & Wilkins, Baltimore.

Samson, R.A. (1992). Current taxonomic schemes of the genus *Aspergillus* and its teleomorphs. In *Aspergillus:* Biology and industrial applications (Eds Bennett, J.W. & Klich, M.A.). Boston, Butterworth-Heinemann. pp. 355-390.

Smith, J.E. & Pateman, J.A. (1977). Genetics and physiology of *Aspergillus.* Academic Press, London.

Swart, K., Debets, A.J.M., Holub, E.F., Bos, C.J. & Hoekstra, R.F. (1994). Physical karyotyping: genetic and taxonomic applications in Aspergilli. In The genus *Aspergillus*: from taxonomy and genetics to industrial applications (Eds Powell, K.A., Renwick, A. & Peberdy, J.F.). New York, Plenum Press. pp. 233-240.

Téren, J., Varga, J., Hamari, Z., Rinyu, E. & Kevei, F. (1996). Immunochemical detection of ochratoxin A in black *Aspergillus* strains. Mycopathologia 134, 171-176.

Varga, J., Kevei, F., Fekete, C., Coenen, A., Kozakiewicz, Z. & Croft J.H. (1993). Restriction fragment length polymorphisms in the mitochondrial DNAs of the *Aspergillus niger* aggregate. Mycol. Res. 97, 1207-1212.

Varga, J., Kevei, F., Vágvölgyi, C., Vriesema, A. & Croft, J.H. (1994a). Double-stranded RNA mycoviruses in section *Nigri* of the *Aspergillus* genus. Can. J. Microbiol. 40, 325-329.

Varga, J., Kevei, F., Debets, F., Kozakiewicz, Z. & Croft, J.H. (1994b). Mitochondrial DNA restriction fragment length polymorphisms in field isolates of the *Aspergillus niger* aggregate. Can. J. Microbiol. 40, 612-621.

Varga, J., Kevei, É., Rinyu, E., Téren, J. & Kozakiewicz, Z. (1996). Ochratoxin production by *Aspergillus* species. Appl. Environ. Microbiol. 62, 4461-4464.

Wicklow, D.T., Dowd, P.F., Alfatafta, A.A. & Gloer, J.B. (1996). Ochratoxin A: an antiinsectan metabolite from the sclerotia of *Aspergillus carbonarius* NRRL 369. Can. J. Microbiol. 42, 1100-1103.

Yuill, E. (1950). The numbers of nuclei in conidia of *Aspergilli.* Trans. Br. Mycol. Soc. 33, 324-331.

MOLECULAR TOOLS FOR THE CLASSIFICATION OF BLACK ASPERGILLI

Lucie Parenicova[1], Pernille Skouboe[2], Robert A. Samson[3], Lone Rossen[2] and Jaap Visser[1]

[1] Wageningen Agricultural University, Section MGIM, NL-6703 HA, Wageningen, The Netherlands, [2]Biotechnological Institute, DK-2970 Hørsholm, Denmark and [3]Centraalbureau voor Schimmelcultures, 3740 AG Baarn, The Netherlands

Fifty five isolates of black Aspergilli (*Aspergillus* section *Nigri*) morphologically classified into six species were re-examined by molecular methods including restriction fragment length polymorphism analysis and DNA sequencing. The banding patterns of the genomic DNA obtained after restriction with the enzymes *Sma*I, *Pst*I and *Sal*I or *Kpn*I and *Xho*I were subsequently visualised in ethidium bromide by UV or radioactively after hybridisation with well defined probes. This analysis and the sequence data of the internal transcribed spacers of the rDNA unit allowed us to distinguish several clusters. It was possible to unambiguously separate the strains morphologically classified as *A. carbonarius* into one group, the *A. niger* aggregate and *A. foetidus* into three groups and *A. aculeatus* and *A. japonicus* into two groups. By using evolutionarily less conserved genes as probes, for instance the gene *pel* A that encodes pectin lyase A in *A. niger*, differences were also found between *A. heteromorphus* and *A. ellipticus*. The molecular methods used proved to be fast, reliable and reproducible and will be useful for applications requiring rapid identification of any black *Aspergillus* isolate.

INTRODUCTION

The black Aspergilli, classified in *Aspergillus* section *Nigri*, have been extensively studied during the past seventy years by methods of classical morphology. The number of species accepted within the section has been reduced from 13 (Thom and Church, 1926) to six (Samson, 1992, 1994a): *A. japonicus, A. niger, A. tubingensis, A. carbonarius, A. heteromorphus* and *A. ellipticus*. However, differences in the numbers of species recognized by different authors demonstrate the difficulties in establishing sound morphological criteria for species identification. Moreover, the external conditions affecting the growth of a fungus (Samson, 1994b) may result in phenotypic changes and subsequently to misidentification of an *Aspergillus* isolate (Kusters-van Someren *et al.*, 1991; Parenicova *et al.*, 1996; ten Hoor-Suykerbuyk, 1997).

The black Aspergilli find wide application in the food and feed industry. They are commonly used for the production of a large number of extracellular enzymes, including pectinases, xylanases, proteinases and cellulases.

413

The most important black biseriate species are *A. niger* and *A. tubingensis* and uniseriate species *A. japonicus* and *A. aculeatus*. Unambiguous identification of industrially important strains is very important before they are used in biotechnological processes. Various biochemical and molecular approaches have been used to develop rapid and reliable methods for identification of isolates which belong to the black Aspergilli. Kusters-van Someren *et al.* (1990, 1991) analysed more then twenty strains belonging to the *A. niger* aggregate and some representatives of other species of the black Aspergilli. For restriction fragment length polymorphism analysis these authors used mainly genes encoding different pectin lyases in *A. niger* CBS 120.49 as probes while imunological detection relied on Western blotting with specific antibodies against PLI and PLII, two major pectin lyases isolated from Ultrazym^R - a commercially available *A. niger* pectinolytic preparation. In combination with the pattern obtained from digestion of rDNA with the restriction endonuclease *Sma*I, the *A. niger* aggregate and several *A. foetidus* isolates including the type strain were divided into two distinct species, *A. niger* and *A. tubingensis*. However, the data suggested that a further division could be made, and this was supported by Varga *et al.* (1994). After an extended analysis of mitochondrial RFLP of a large number of isolates of the black Aspergilli, they separated a third group consisting of a limited number of Brazilian isolates. Also, recent data published by Parenicova *et al.* (1997) of the RFLP patterns of twenty one *A. foetidus, A. awamori* and *A. phoenicis* isolates, the latter two considered as varieties of *A. niger* shows that the combining of *A. niger* and *A. foetidus* into one species (Samson, 1992) is not supported. RFLP analysis demonstrated three rather than two separate groups.

Use of molecular techniques such as DNA sequencing, RFLP analysis of mitochondrial and chromosomal DNA, and RAPD analysis appear to be powerful tools in the recognition of intra- and inter-species variation among the black Aspergilli. Kevei *et al.* (1996) showed that intraspecific variations in the morphologically uniform species *A. carbonarius* can be clearly identified from mt DNA RFLP and RAPD patterns while examination of phenotypical features like carbon source utilisation patterns or isoenzyme analysis showed only small differences. Another RFLP study carried out with the chromosomal DNA of ten *A. carbonarius* strains (Parenicova *et al.*, 1996) showed a high degree of identity between the isolates except for the *pki*A locus for which three different patterns were obtained. These results again confirmed that the described molecular methods can be used for the fast classification of isolates which belong to the black Aspergilli and that the combination of several molecular approaches can furthermore reveal diversity within a single species.

Several taxa of the black Aspergilli are difficult to distinguish by morphological criteria, including *A. japonicus, A. aculeatus, A. niger, A. tubingensis* and *A. foetidus*. This paper looks at molecular criteria based on which a clear distinction between the *Aspergillus* species can be made.

MATERIALS AND METHODS

Strains and plasmids: Aspergillus strains (Table 1) were obtained from the Centraalbureau voor Schimmelcultures in Baarn, the Netherlands and from the Agricultural Research Service of the United States Department of Agriculture (USDA, Peoria, IL, USA). *Aspergillus tubingensis* NW 756 is an industrial strain which belonged to the *A.niger* aggregate (Kusters-van Someren *et al.*, 1991).

Plasmids carrying the pyruvate kinase encoding gene, *pki*A, of *Aspergillus nidulans* (de Graaff and Visser, 1988b), the pectin lyase A encoding gene, *pel*A, of *A. niger* (Harmsen *et al.*, 1990), a 0.9 kb fragment of the 28S ribosomal gene of *Agaricus bisporus* (EMBL,acc.nr. X91812) as well as the whole rDNA unit (Schaap *et al.*, unpublished results) were provided by our own laboratory.

Growth conditions and isolation of chromosomal DNA: Cultures were grown by inoculating minimum medium (Pontecorvo *et al.*, 1953) with approximately 10_6 conidia/ml using 1 % (w/v) of glucose as a carbon source and 0.85% (w/v) yeast extract in 50 ml of medium in 250 ml flasks. The cultures were grown at 30°C on a rotary shaker (250 rpm). Mycelium was harvested by filtration after 20 h of culturing. DNA isolation was carried out as described by de Graaff *et al.* (1988a).

DNA digestion and gel electrophoresis: The concentration of DNA was established from an agarose gel by comparison with molecular weight lambda markers. Digestions of 5 µg of DNA were carried out overnight with 40 units of enzyme in 200 µl volume at the temperature and using buffers as recommended by the manufacturer of the restriction endonucleases (Gibco BRL, Life Technologies Inc., Gaithersburg). Gel electrophoresis in 0.8% agarose gel was carried out at low field strength to obtain optimal separation, applying 1.5V/cm for 16 to 20 h in TAE (0.04 M Tris-acetate, 0.001M EDTA, pH 8.0) buffer, using a horizontal gel kit according to Sambrook *et al.* (1989). Bacteriophage lambda DNA digested with restriction enzymes *Hind*III and *Eco*RI was used as a size marker. Band size was judged precisely by using restricted DNA from two previously analysed strains as reference. The separated DNA was visualised by UV translumination.

Southern blotting and hybridisation: The DNA was transferred to a Hybond^{TM}-N membrane (Amersham, Life Science, Little Chalfont) by vacuum blotting (VacuGene^{TM}XL unit, Pharmacia, Uppsala) using a standard protocol for transfer of high molecular weight DNA as described in the instruction manual. The DNA was fixed to the membrane by UV crosslinking for 3 min. Probes were made according to Kusters-van Someren *et al.* (1990). Blots were incubated for 2 h in prehybridisation buffer described by Sambrook *et al.* (1989) containing 5mg/ml heat-denatured herring sperm DNA, hybridized overnight at 60°C and subsequently washed twice for 30 min with 2 portions of SSC (0.3 M NaCl, 0.03 M tri-sodium citrate dihydrate) and 0.5% sodium dodecyl sulphate (SDS) at the same temperature. After exposure of the first probe (the *pki*A, or the *pel*A probes) signals were stripped by rinsing the membranes for 1 min in 0.1% (w/v) SDS at 100°C and the blots were subsequently used for hybridisation with the 28S probe. The membranes were exposed either to Konica X-ray film using Kodak intensifying screens or to a Phosphor Screen (Molecular Dynamics, Sunny Vale).

Polymerase chain reaction (PCR) and DNA Sequencing: Approximately 610 bp spanning the ITS spacer regions and the 5.8S gene was amplified by PCR using the ITS4 and

ITS5 primers essentially as described by White *et al.* (1990). PCR mixtures contained 5 μl of appropriately diluted genomic DNA (5-10 ng), 1 μM each of primers ITS4 and ITS5, reaction buffer (50 mM KCl, 50 mM Tris-HCl, pH 8.3, 0.1 mg/ml bovine serum albumin), 3 mM MgCl$_2$, 200 μM of each dNTP, 10 % dimethylsulfoxide (DMSO), 0.25 % Tween 20, and 2.5 U *AmpliTaq* DNA polymerase (Perkin Elmer Cetus, Norwalk) in a total volume of 100 μl. Amplification was performed on a Perkin Elmer Cetus thermal cycler model 9600 using an initial denaturation at 94 °C for 1 min followed by 45 cycles of 94°C for 15 sec, 53 °C for 1 min and 72 °C for 1 min, followed by a final extension step of 72°C for 10 min. PCR products were visualised in 1.5% (w/v) agarose gels stained with ethidium bromide.

The ITS1, ITS2, ITS3 and ITS4 primers were used for direct cycle sequencing of the amplified PCR products (Murray, 1989). Cycle sequencing reactions were performed using the Thermo Sequenase fluorescent labelled primer cycle sequencing kit (Amersham, Life Science, Little Chalfont) according to the manufacturer's instructions, and analysed on an automated DNA sequencer (A.L.F. express, Pharmacia Biotech, Uppsala). DNA sequencing was carried out in both directions, and the sequences were aligned using the CLUSTAL V (IntelliGenetics, Inc., Mountain View) package and manual corrections.

Initially, nine black *Aspergillus* strains, some of them listed in Table 1, were selected for DNA sequencing. The selected strains included *A. japonicus* CBS 114.51 and NRRL 360, *A. niger* CBS 120.49, *A. tubingensis* CBS 127.49 and NW 756, *A. carbonarius* CBS 111.26, *A. heteromorphus* CBS 117.55, *A. ellipticus* CBS 707.79 and *A. foetidus* var. *acidus* CBS 564.65.

RESULTS AND DISCUSSION

Table 1 list the isolates studied which belong to the following taxa: *A. japonicus*, *A. aculeatus*, *A. carbonarius*, *A. niger*, *A. tubingensis*, *A. foetidus*, *A. ellipticus* and *A. heteromorphus*. These isolates were selected on the basis of extensive molecular studies (Parenicova *et al.*, 1996; 1997; ten Hoor-Suykerbuyk, 1997) as typical representatives of black *Aspergillus* species.

For RFLP analysis we used the following restriction enzymes: *Sma*I; *Kpn*I/*Xho*I and *Sal*I/*Pst*I. As probes we used (1) a 0.85 kb internal *Sal*I fragment of the *A. nidulans pki*A gene, (2) a 0.9 kb *Eco*RI fragment comprising the 3'end and some downstream sequence of the 28S rDNA gene from *Agaricus bisporus* and (3) a 1.6 kb *Cla*I fragment containing the C-terminal part and a downstream sequence of the *pel*A gene encoding pectin lyase A in *A. niger*.

Division of the black Aspergilli based on Sma*I restriction patterns*
*Sma*I restriction of chromosomal DNA leads to a division of the black Aspergilli into five distinct groups based on readily recognisable bands obtained after separation and UV-visualisation of ethidium bromide agarose gels. A typical banding pattern is: *A. carbonarius* (pattern A, 4.1 and 2.3 kb), *A. niger* (pattern B, 3.2, 2.3 and 1.8 kb), *A. tubingensis* and *A. foetidus* (pattern C, 2.6, 1.9 and 1.8 kb), *A. japonicus* and *A. aculeatus* (pat-

tern D, 1.9 and 1.8 kb) and *A. heteromorphus* and *A. ellipticus* (pattern E, 1.9, 1.8 and 1.6 kb).

Table 1. Typical representatives of the black Aspergilli groups used in the RFLP analysis and their CBS, NRRL and ATCC numbers.

Original name	Strain number	Source	
A. aculeatus	CBS 101.43	Pterocarpus santalinus	
A. aculeatus	CBS 172.66	tropical soil	K.B. Raper
A. aculeatus	CBS 610.78	tropical soil	Blakeslee
A. aculeatus	CBS 313.89	soil Figi	J.C. Frisvad
A. atroviolaceus (T)	CBS 522.89	air Amsterdam	R. Mosseray
A. atroviolaceo-fuscus mut.	CBS 122.35	unknown	A. Blochwitz
grisea (T)			
A. awamori	ATCC 22342	unknown	J. van Lanen
A. awamori	CBS 113.33	unknown	A. Blochwitz
A. awamori	CBS 121.48	unknown	R. Mosseray
A. awamori	CBS 115.52	Kuro-koji, Japan	K. Sakaguchi
A. carbonarius (NT)	CBS 111.26	paper	A.F. Blakeslee
A. carbonarius (T) of	CBS 114.29	Spain	A. Caballero
Sterigmatocystis acini-uvae			
A. carbonarius	CBS 113.80	*Theobroma cacao* Nigeria	J.A. Broadbent
A. ellipticus (T)	CBS 707.79	soil, Costa Rica	Al-Musallam
A. foetidus	CBS 103.14	unknown	A. Blochwitz
A. foetidus (T) of A. citricus	CBS 618.78	unknown	C. Wehmer
A. foetidus var acidus (T)	CBS 564.65	Japan	R. Nakazawa
A. foetidus var pallidus (T)	CBS 565.65	Japan	R. Nakazawa
A. heteromorphus (T)	CBS 117.55	culture contaminant	A. C. Batista
A. japonicus (T)	CBS 114.51	unknown	K. Kominami
A. japonicus	CBS 568.65	Panama soil	K.B. Raper
A. japonicus	NRRL 2053	New Guinea tent cloth	White
A. japonicus	CBS 312.80	unknown	
A. japonicus var. capillatus	CBS 114.34	man skin	R. Nakazawa
A. niger	CBS 120.49	unknown	Anthony
A. phoenicis	ATCC 13156	whole shelled corn	A.E. Staley
A. phoenicis	CBS 126.49	Japan	Yakoyama
A. phoenicis	CBS 139.48	unknown	R.Mosseray
A. phoenicis	CBS 629.78	France	F. de Fonseca
A. phoenicis	CBS 114.37	unknown	Y.K. Shih
Sterigmatocystis fusca *	CBS 420.64	unknown	F. de Fonseca
A. tubingensis	NW756	unknown	M.A. Kusters-van Someren

T, type strain; NT, neotype strain. * probably T of *S. fusca*

These intensely staining bands are expected to represent the restriction fragments of the highly repetitive rDNA unit (100-300 copies per haploid genome). In order to prove this some of the digested DNA`s were hybridised with the 10 kb *Bam*HI rDNA probe of *A. bisporus*. It was observed that not all visualised fragments hybridise with the DNA probe. Only in case of the *A. carbonarius* strain both bands (4.1 and 2.3 kb) appeared on the autoradiograph. For the remaining patterns the 3.2 and 1.8 kb bands of pattern B, the

2.6 and 1.8 kb bands of pattern C and the 1.8 kb band of pattern D were hybridising. In the latter case also a 0.8 kb band, not visualised by UV, was hybridising. A likely explanation for this observation is that the ITS and NTS of the rDNA unit may in the black Aspergilli be considerably different from those in *A. bisporus* sequences while the 28S, 18S and 5.8S rDNA genes present in the cluster are expected to be highly conserved among different fungi.

Division of the black Aspergilli based on polymorphisms in the pki *A, 28 S and* pel *loci*
Tables 2 and 3 show the typical banding patterns obtained for particular combinations of restriction enzymes and probes. Two combinations of restriction enzymes, *PstI/SalI* and *KpnI/XhoI*, were found to be suitable for revealing restriction polymorphisms among closely related taxa . The complete RFLP banding patterns for the individual isolates together with the proposed division of the black Aspergilli into eight taxa is presented in the Table 4. These are conveniently discussed in four groupings below.

Aspergillus foetidus, A. tubingensis and A. niger
Based on the presented data the morphologically closely related species of *A. niger*, including *A. phoenicis* and *A. awamori*, *A. tubingensis* and *A. foetidus* can be separated into three taxa each having its own distinctive banding pattern. Using the *Sma*I, *pki*A (*PstI/SalI*), 28S (*PstI/SalI*) and *pel*A (*PstI/SalI*) patterns in that order, *A. foetidus* is characterised by C-B-D and C patterns, *A. tubingensis* by C-B-B and B and *A. niger* by B-B-C and E patterns (Table 4). Using the *Sma*I digestion the isolates belonging to *A. niger* can be easily separated from the other black Aspergilli by the unique pattern B. In case of the *pki*A locus the isolates of all three groups had pattern B, with only one exception: *A. tubingensis* NW 756 has pattern E, which most probably arises from an introduction of one extra restriction site, as the 2.8 and 0.65 kb fragments add up to the approximate size observed in pattern B. However, the banding profiles of the other two combinations examined (*PstI/SalI* digests of the 28S and *pel*A loci) allow a clear distinction among these three taxa. It shows the usefulness of the less conserved *pel*A gene and the adjacent regions (NTS downstream) of the 28 S gene when examining these closely related species by RFLP analysis.

When the *pel*A gene is used as a probe more intra-specific variations can be expected (Kusters-van Someren, 1991). This can be seen in case of the isolates separated in *A. niger* and *A. tubingensis* clusters. An exception is the 28 S pattern D of *A. awamori* ATCC 22342 which indicates this isolate should belong to *A. foetidus*. However, it was placed in *A. tubingensis* based on the *pel*A pattern B. This placement is supported by the observation that the 28S patterns D and B share the 5.0 kb band and differ only in the small hybridising bands, 1.8 and 0.9 kb, respectively, which could originate from each other by a single point mutation, while in the case of the *pel*A patterns no similarity can be detected.

Aspergillus carbonarius
The RFLP analysis of the *A. carbonarius* isolates shows that they have distinct patterns from the other black Aspergilli for all the loci examined with the exception of the *pki* A locus for which in this group a restriction polymorphism was detected (Table 2 and 4).

The diversity might be explained by a mutation of one *Sal*I restriction site within the gene and differences in the non-coding region of the gene since the pattern F (0.9 kb) of *A. carbonarius* CBS 114.29 is similar in size to the internal *Sal*I fragment of the *A. nidulans pki*A gene used as a probe and is therefore expected to encode the corresponding part of the *pki*A gene of *A. carbonarius*. However, in spite of distinct morphological features of the *A. carbonarius* taxon from the other black Aspergilli, like the large biseriate conidial heads or large verrucose to tuberculate conidia (Raper & Fennell, 1965; Al-Musallam, 1980), it was shown by the RFLP analysis that a morphological misclassification of some isolates had occurred (Parenicova *et al.*, 1996). As the result of RFLP analysis, two *A. carbonarius* isolates, CBS 101.14 and CBS 127.49, had to be reclassified as *A. japonicus* and *A. tubingensis*, respectively.

Table 2. RFLP patterns observed in *Pst*I/*Sal*I and *Kpn*I/*Xho*I digests of chromosomal DNA of various black Aspergilli using a 0.8 kb *Sal*I fragment of *A. nidulans pki*A and a 0.9 kb *Eco*RI fragment of *A. bisporus* 28S rDNA as probes. Fragment lengths are indicated in kb.

Digest and probe	PstI/SalI; pkiA	PstI/SalI; 28S	KpnI/XhoI; 28S
Pattern A	6.0 [a]	5 & <0.3 [a]	7 [c]
Pattern B	3.4 [a]	5 & 0.9 [a]	4 [c]
Pattern C	3.0 [a]	5 & 1.5 [a]	7.5 [c]
Pattern D	2.8 [a]	5 & 1.8 [a]	
Pattern E	2.8 & 0.65 [a]	6 & 1.3 [c]	
Pattern F	0.9 [b]	6 & 1.32 [c]	
Pattern G	4.8 [b]	7.5 & <0.5 [a]	
Pattern H		6 & 1.35 [c]	
Pattern I		6 & 1.47 [c]	
Pattern J		6 & 1.5 [c]	
Pattern K		6 & 1.43 [c]	
Pattern L		6 & 1.1 [c]	

[a] Parenicova *et al.* (1997) ; [b] Parenicova *et al.* (1996); [c] ten Hoor-Suykerbuyk (1997)

Aspergillus japonicus and A. aculeatus

A. japonicus and *A. aculeatus* are the only uniseriate species among the black Aspergilli and are therefore morphologically readily distinguishable from the biseriate species. However, the further division of these taxa into separate species is arguable.

Using RFLP analysis, *A. japonicus* can be easily distinguished from biseriate black Aspergilli isolates by the typical *Sma*I DNA pattern D or the conserved pattern C of the *pki*A locus. Based on the banding pattern A or B, obtained after *Kpn*I/*Xho*I digestion of DNA and probing with the 28S fragment, further division into two groups is posssible (Table 4). A high degree of DNA polymorphism among the isolates was observed when another restriction enzyme combination - *Pst*I and *Sal*I - was used in combination with the 28S probe or the *pel* A probe. The first phenomenon can be explained by the position of the restriction sites within the rDNA cluster. From the known restriction map of the *A. niger* rDNA unit (O`Connell *et al.*, 1990) it can be seen that the *Xho*I restriction sites are placed within the 28S or 5.8S genes which are both expected to be highly conserved while the *Pst*I and *Sal*I restriction sites are located within the nonconserved NTS region. The *pel*A banding profiles reflect the intra-specific variability in the *pel* loci among the *A. japonicus* isolates which is more pronounced in this case than in case of the biseriate black Aspergilli. However, it can be seen in Table 3 that the *pel*A patterns belonging to the isolates of one cluster share some common bands. For instance, the I and L patterns share three out of four hybridising bands and are typical for the isolates separated by the 28 S (*Kpn*I/*Xho*I) pattern A into the *A. japonicus*.

From the RFLP analysis presented here it is clear that the *A. japonicus* and *A. aculeatus* isolates have a very variable banding profile. Other independent analyses should be used to clearly differentiate of these two taxa.

Table 3. RFLP patterns observed in *Pst*I/*Sal*I digests of chromosomal DNA of various black Aspergilli using a 1.6 kb *Cla*I fragment of the *A. niger pel*A gene as a probe. Fragment lengths are indicated in kb ([a] Parenicova *et al.* 1997; [b] ten Hoor-Suykerbuyk, 1997)

Pattern	
A	6.0 & 4.5 [a]
B	5.0 & 1.2 [a]
C	4.8, 2.8 & 2.2 [a]
D	4.8, 2.2 & 1.2 [a]
E	4.5, 1.2 & 1.0 [a]
F	4.5 & 0.9 [a]
G	3.8 & 2.0 [a]
H	3.5, 1.7 & 0.7 [b]
I	3.2, 2.3, 1.6 & 0.6 [b]
J	2.9 & 1.2 [a]
K	2.4 [a]
L	3.4, 2.3, 1.6 & 0.6 [b]
M	3, 1.7, 1.4 & 0.7 [b]
N	5, 4.5, 2.1 & 0.7 [b]
O	5, 3.5, 1.7 & 0.7 [b]
P	5, 4.5, 2.5 & 0.7 [b]
Q	2.9, 2.3, 1.7 & 0.7 [b]
R	3.5, 2.6, 1.7 & 0.7 [b]

Table 4. Clusters of strains belonging to the black *Aspergillus* species.

Cluster	Species	strain	rDNA *Sma*I	*pki*A *Pst*I/ *Sal*I	28S *Kpn*I/ *Xho*I	28S *Pst*I/ *Sal*I	*pel*A *Pst*I/ *Sal*I
A. foetidus	*A. awamori*	CBS 115.52	C	B	n.d.[a]	D	C
	A. foetidus	CBS 103.14	C	B	n.d.	D	C
	A. foetidus	CBS 564.65	C	B	n.d.	D	C
	A. foetidus	CBS 565.65	C	B	n.d.	D	C
	A. phoenicis	CBS 139.48	C	B	n.d	D	C
A. tubingensis	*A. awamori*	ATCC 22342	C	B	n.d.	D	B
	A. phoenicis	CBS 114.37	C	B	n.d.	B	B
	A. tubingensis	NW 756	C	E	A	B	B
	A. phoenicis	CBS 629.78	C	B	n.d.	B	A
A. niger	*A. awamori*	CBS 113.33	B	B	n.d.	C	E
	A. foetidus	CBS 618.78	B	B	n.d.	C	E
	A. niger	CBS 120.49	B	B	B	C	E
	A. phoenicis	CBS 126.49	B	B	n.d.	C	E
	A. awamori	CBS 121.48	B	B	n.d.	C	F
	A. phoenicis	ATCC 13156	B	B	n.d	C	D
A. carbonarius	*A. carbonarius*	CBS 111.26	A	A	C	A	J
	A. carbonarius	CBS 114.29	A	F	n.d.	A	J
	A. carbonarius	CBS 113.80	A	G	C	A	J
	S. fusca	CBS 420.64	A	G	C	A	J
A. japonicus	*A. japonicus*	CBS 114.51	D	C	A	E	I
	A. japonicus	CBS 568.65	D	n.d.	A	F	L
	A. atroviolaceus	CBS 522.89	D	n.d.	A	H	L
	A. atroviolaceo- fuscus mut. grisea	CBS 122.35	D	C	A	H	I
A. aculeatus	*A. japonicus*	NRRL 2053	D	n.d.	B	I	M
	A. japonicus	CBS 312.80	D	C	B	J	N
	A. japonicus	CBS 114.34	D	C	B	K	O
	A. aculeatus	CBS 101.43	D	C	B	I	H
	A. aculeatus	CBS 172.66	D	C	B	J	P
	A. aculeatus	CBS 610.78	D	C	B	J	Q
	A. aculeatus	CBS 313.89	D	C	B	L	R
A. heteromorphus	*A. heteromorphus*	CBS 117.55	E	D	n.d	G	K
A. ellipticus	*A. ellipticus*	CBS 707.79	E	D	n.d	G	G

[a]not determined

Aspergillus heteromorphus and A. ellipticus

The molecular analysis of these species is based on the analysis of only two available isolates. *A. heteromorphus* and *A. ellipticus* appeared to be closely related morphologically (Raper & Fennell, 1965; Al-Musallam, 1980). These two taxa share the same banding patterns for the *Sma*I restriction analysis and for the *Pst*I/*Sal*I digests using the *pki*A and 28S probes. The patterns are uniquely different from those of the other black Aspergilli and can be separated from each other by their *pel* A pattern. *A. heteromorphus* is characterised by pattern K (only one hybridising band) while pattern G of *A. ellipticus* consists of two hybridising bands.

Sequence analysis of the ITS regions

Examination of sequences for the nine strains studied showed that no differences were found between strains belonging to the same taxon. Alignment of the ITS sequences showed that the two *A. japonicus* strains had a deletion of 6 bp in the ITS I region and three deletions in the ITS II region compared to the other *Aspergillus* strains sequenced. Sequence differences among the eight clusters separated by RFLP analysis (see Table 4) were found in approximately 60 positions. Half of these sequence differences were located in the ITS I region and the rest in the ITS II region. The number of ITS sequence differences between the individual clusters have been calculated and a matrix was generated (Table 5). The matrix illustrates that the highest number of sequence differences was seen between *A. japonicus* and the rest of the taxa. *A. niger* and *A. tubingensis* could be separated on the basis of three base differences, and there were five differences between *A. niger* and *A. foetidus*.

Table 5. Numbers of nucleotide differences in ITS sequences among the eight clusters of black Aspergilli.

	japonicus	niger	tubingensis	foetidus	carbonarius	heteromorphus	ellipticus
japonicus	X	46	48	48	51	43	38
niger		X	3	5	18	23	13
tubingensis			X	2	19	25	15
foetidus				X	21	25	17
carbonarius					X	26	19
heteromorphus						X	6

ITS sequencing of an additional six strains from *A. japonicus*, eight strains of *A. aculeatus* and seven others of closely related strains were then performed. Preliminary results showed that it was not possible to distinguish between *A. japonicus* and *A. aculeatus* based on the ITS sequences.

In accordance with the RFLP analysis, *A. carbonarius* could easily be separated from the other black Aspergilli using ITS sequencing, i.e. there were 18-51 base differences separating *A. carbonarius* from the other taxa.. The RFLP analysis showed that *A. heteromorphus* and *A. ellipticus* are very closely related, which is supported by the ITS sequence data (only 6 base differences separated these two taxa).

In conclusion, the grouping of the black Aspergilli shown by RFLP analysis is supported by the ITS sequence analysis with the reservation, that the ITS sequence analysis was based on only 9 strains.

CONCLUSIONS

The molecular analysis of the black Aspergilli using RFLP analysis of chromosomal DNA and sequencing of the highly conserved ITS region in the rDNA repeat showed to be a powerful tool for fast and relatively easy identification of individual isolates. Patterns from *Sma*I RFLP analysis and sequencing of the ITS region permitted division of the strains into several groups, while using different restriction enzymes in combination with well defined probes provided further separation. Three groups were clearly distinguishable *A. niger*, *A. tubingensis* and *A. foetidus* within the *A. niger* aggregate. These probes also allowed separation of *A. carbonarius,* plus *A. japonicus* and *A. aculeatus* based on the hybridisation patterns of *XhoI/KpnI* digested DNA. *A. heteromorphus* and *A. ellipticus* showed only small differences, the sequence of the ITS and in the *pel* A pattern. Clearer molecular distinction of these species has to await further analysis.

REFERENCES

Al-Musallam, A. (1980). Revision of the black *Aspergillus* species. PhD. thesis, Rijksuniversiteit Utrecht, Utrecht, Netherlands.

de Graaff, L.H., van den Broeck, H. & Visser, J. (1988a). Isolation and expression of the *Aspergillus nidulans* pyruvate kinase gene. Curr. Genet. 13: 315-321.

de Graaff, L.H. & Visser, J. (1988b). Structure of the *Aspergillus nidulans* pyruvate kinase gene. Curr. Genet. 14: 553-560.

Harmsen, J.A.M., Kusters-van Someren, M.A. & Visser, J. (1990). Cloning and expression of a second *Aspergillus niger* pectin lyase gene (*pel*A): indications of a pectin lyase gene family in *Aspergillus niger.* Curr. Genet. 18: 161-166.

Kevei, F., Hamari, Z., Varga, J., Kozakiewicz, Z. & Croft, J.H. (1996). Molecular polymorphism and phenotypic variation in *Aspergillus carbonarius.* Antonie van Leeuwenhoek 70: 59-66.

Kusters-van Someren, M.A., Samson, R.A. & Visser, J. (1990). Variation in pectinolytic enzymes of the black Aspergilli: a biochemical and genetic approach. In Modern concepts in *Penicillium* and *Aspergillus* classification (Eds. Samson, R.A. & Pitt, J.I.). New York, Plenum Press. pp. 321-334

Kusters-van Someren, M.A., Samson, R.A. & Visser, J. (1991). The use of RFLP analysis in classification of the black Aspergilli: reinterpretation of the *Aspergillus niger* aggregate. Curr. Genet. 19: 21-26.

Megnegneau, B., Debets, F. & Hoekstra, R.F. (1993). Genetic variability and relatedness in the complex group of black Aspergilli based on random amplification of polymorphic DNA. Curr. Genet. 23: 323-329.

Murray, V. (1989). Improved double-stranded DNA sequencing using the linear polymerase chain reaction. Nucleic Acids Res. 17: 88-89.

O'Connell, M.J., Dowzer, C.E.A. & Kelley, J.M. (1990). The ribosomal repeat of *Aspergillus niger* and its effects on transformation frequency. Fungal Genet. Newsl. 37: 29-30.

Parenicova, L., Suykerbuyk, M.E.G., Samson, R.A. & Visser, J. (1996). Evaluation of restriction fragment length polymorphism for the classification of *Aspergillus carbonarius.* African J. Mycol. Biotechnol. 4: 13-19.

Parenicova, L., Benen, J.A.E., Samson, R.A. & Visser, J. (1997). Evaluation of RFLP analysis for the classification of selected black Aspergilli. Mycol. Res. 101 (7): 810-814.

Pontecorvo, G., Roper, J.A., Hemmons, L.J., MacDonald, K.D. and Bufton, A.W.J. (1953). The genetics of *Aspergillus nidulans.* Adv. Genet. 5: 141-239.

Raper, K.B. & Fennell, D.I. (1965). The genus *Aspergillus.* Baltimore, MD, Williams and Wilkins Company.

Samson, R.A. (1992). Current taxonomic schemes of the genus *Aspergillus* and its teleomorphs. In *Aspergillus*: Biology and Industrial Applications (Eds Bennet, J.W. & Klich, M.A). Reed Publishing, USA. pp. 355-390.

Samson, R.A. (1994a). Taxonomy - current concepts of *Aspergillus* systematics. In *Aspergillus* (Ed. Smith, J.E.). New York, Plenum Press. pp. 1-22.

Samson, R.A. (1994b). Current systematics of the genus *Aspergillus.* In The Genus *Aspergillus*: From Taxonomy and Genetics to Industrial Application (Eds Powell, K.A., Renwick, A. & Peberdy, J.F.). London, Plenum Press. pp 261-276

Sambrook, J., Fritsch, E. & Maniatis, T. (1989). Molecular cloning. A laboratory manual. New York, Cold Spring Harbor Laboratory.

ten Hoor-Suykerbuyk, M.E.G. (1997). Molecular analysis of endo-rhamnogalacturonan hydrolases in *Aspergillus.* PhD thesis. Landbouwuniversiteit Wageningen, Wageningen, Netherlands.

Thom, C. & Church, M.B. (1926). The Aspergilli. Baltimore, MD, Williams and Wilkins. pp. 1-272

Varga, J., Kevei, F., Fekete, C., Coenen, A., F., Kozakiewicz, Z. & Croft, J.H. (1993). Restriction fragment length polymorphisms in the mitochondrial DNAs of the *Aspergillus niger* aggregate. Mycol. Res. 97: 1207-1212.

Varga, J., Kevei, F., Vriesema, A., Debets, F., Kozakiewicz, Z. & Croft, J.H. (1994). Mitochondrial DNA restriction fragment polymorphisms in field isolates of the *Aspergillus niger* aggregate. Can. J. of Microbiol. 40: 612-621.

White T.J., Bruns T., Lee S. & Taylor, J. (1990). Amplification and direct sequencing of fungal ribosomal RNA genes for phylogenetics. In PCR Protocols: A Guide to Methods and Applications (Eds Innis, M.A., Gelfand, D.H., Sninsky, J.J. & White, T.J.). London, Academic Press. pp 315-322.

ASPERGILLUS SYSTEMATICS AND THE MOLECULAR GENETICS OF MYCOTOXIN BIOSYNTHESIS

Maren A. Klich and Thomas E. Cleveland.
USDA, ARS, Southern Regional Research Center,. New Orleans LA 70124.

In this study we review our current state of understanding of the molecular biology of aflatoxin biosynthesis and that of a closely related polyketide mycotoxin, sterigmatocystin. The genes from these pathways have proven to be useful tools in taxonomic studies of the sections of *Aspergillus* that produce these metabolites. Increased understanding of the genetics of the secondary metabolite biosynthetic pathways is helping to elucidate the nature of the apparent links between secondary metabolism and morphological characteristics.

INTRODUCTION

Aflatoxins and sterigmatocystin are secondary metabolites produced by closely related polyketide biosynthetic pathways. Indeed, sterigmatocystin is an intermediate compound in the aflatoxin biosynthetic pathway. Both sterigmatocystin and the aflatoxin family of compounds are known to be toxic and carcinogenic to animals, but the aflatoxins are considered to be a much higher risk to food and feed safety due to the high incidence of *A. flavus* and *A. parasiticus* infection of field crops such as maize, cottonseed, peanuts and tree nuts.

Aflatoxin contamination of agricultural commodities poses a potential threat to the health of human beings (Bennett and Goldblatt, 1973; Jelinek *et al.*, 1989; Cleveland and Bhatnagar, 1992), so it is regulated in many countries at levels from 5-50 µg/kg. A major emphasis of our laboratory work since 1985 has been to gain an understanding of the basic genetic and molecular mechanisms by which aflatoxin is synthesized so that new strategies may be developed to interrupt this process.

Aflatoxin and Sterigmatocystin biosynthesis

The aflatoxin and sterigmatocystin biosynthetic pathway is a series of enzymatically catalyzed conversion reactions initiated by the synthesis of polyketide from acetate, a process similar to fatty acid synthesis (Hopwood & Sherman, 1990). The process of aflatoxin synthesis is known to involve 17 or more enzymatic steps (Dutton, 1988; Bhatnagar *et al.*, 1993; Yabe *et al.*, 1993; Yu *et al.*, 1995b) and the sterigmatocystin pathway is estimated to have 15 enzymatic steps (Brown *et al.* 1996a; Keller and Hohn, 1997). Some of the enzymes involved in aflatoxin biosynthesis have been characterized and their respective genes cloned. The genes for both of these biosynthetic pathways are

425

clustered, a phenomenon common in prokaryotes, but not prevalent in eukaryotes. There is evidence that another clustered pathway, that for penicillin production, has allowed for transfer of the whole cluster between species. We have no direct evidence for this kind of transfer in the aflatoxin or sterigmatocystin pathways, but the recent finding of an aflatoxin-producing strain of *A. tamarii* leads one to consider the idea that such transfer is possible (Keller *et al.*, 1992; Goto *et al.*, 1997).

The entire aflatoxin/sterigmatocystin biosynthetic pathway is transcriptionally regulated by the activation of the *aflR* gene that encodes protein containing a zinc finger (AFLR) required for activation of the structural pathway genes (Payne *et al.*, 1993; Woloshuk *et al.*, 1994; Chang *et al.*, 1995b). The biosynthetic pathway and genes involved, as we currently understand them, are summarized in Figure 1. The regulatory gene, *aflR*, coding for the pathway regulatory factor (AFLR protein), controls the expression of the structural genes at the transcriptional level. The *fas-1, fas-2* and the *pksA* gene products, fatty acid synthase and polyketde synthase, respectively, are involved in the conversion steps between the initial acetate unit to the synthesis of the decaketide backbone in aflatoxin synthesis. The *nor-1* gene encodes a reductase for the conversion of NOR to AVN. The *avnA* gene encodes a P450 monooxygenase for the conversion of AVN to HAVN. The *adhA* (homology to an alcohol dehydrogenase), *ver1* (encoding a deydrogenase, *ord-2 cyp*450 and *avf1* gene products have been demonstrated to be functioning at various stages of the pathway, but their exact enzymatic role has not been fully characterized and is under investigation. The *omtA* gene encodes an O-mehtyltransferase for the conversion of ST to OMST and DHST to DHOMST. The *vbs* gene encodes a Ver B synthase (cyclase), which has been reported to be involved in the conversion of VHA to VER B. The oxidoreductase and esterase have been characterized to be involved in the aflatoxin biosynthesis pathway, however, their corresponding genes have not been confirmed.

The initial steps in aflatoxin and sterigmatocystin biosynthetic pathways have been shown recently to require both a polyketide synthase (PKS) and fatty acid synthases (FAS). The genes involved are *pksA* or *pksL1* for the PKS enzyme and *fas-1* and *fas-2* for the FAS enzyme. An FAS composed of an alpha and beta subunit is probably responsible for synthesis of a six carbon fatty acid (hexanoate) which serves as the starter unit for the PKS in forming the first stable intermediate in the pathway, norsolorinic acid (NOR)

Fig. 1. Summary of the generally accepted cluster of aflatoxin pathway genes, corresponding biosynthetic enzymes, and precursor intermediates involved in the aflatoxin B_1 and B_2 synthesis (adapted from Brown *et al.*, 1997). Names of the individual genes are labelled next to the open boxes and unnamed transcripts are labelled by a question mark. Arrows inside the open boxes indicate the direction of transcription. Arrows indicate the relationships from the genes to the enzymes they encode; from the enzymes to the bioconversion steps they are involved in; and from the intermediates to products in the aflatoxin bioconversion steps.

Abbreviations: NOR, norsolorinic acid; AVN, averantin; HAVN, 5'-hydroxyaverantin; AVNN, averufanin; AVF, averufin; VHA, versiconal hemiacetal acetate; VAl, versiconal; Ver B, versicolorin B; Ver A, versicolorin A; ST, sterigmatocystin; DHST, dihydrosterigmatocystin; OMST, O-methylsterigmatocystin; DHOMST, dihydro-O-methyl-sterigmatocystin; AFB_1, aflatoxin B_1; AFB_2, alflatoxin B_2. In addition, 1-hydroxyversicolorone is an intermediate between AVF, and VHA and demethysterigmatocystin is an intermeidate between Ver A and ST.

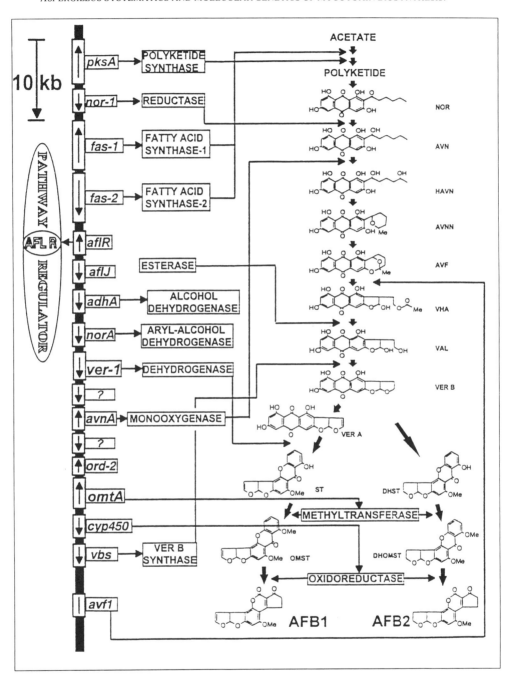

(Brown *et al.*, 1996b; Mahanti *et al.*, 1996; Watanabe *et al.*, 1996). The *nor*-1 gene encodes a reductase (Trail *et al.*, 1994) required for the conversion of NOR to averantin (AVN), and the *nor*A gene encodes an aryl-alcohol dehydrogenase which can also participate in the NOR to AVN conversion step (Cary *et al.*, 1996). The *avn*A gene (Yu *et al.* 1997), previously named *ord*-1, encodes a P450 monooxygenase responsible for the conversion of averantin to averufin (AVF).

Another gene (*avf*1) located on the gene cluster was verified to be involved in toxin biosynthesis by complementation to aflatoxin production of an AVF accumulating strain of *A. flavus* (Prieto *et al.*, 1996). The *vbs* gene encodes versicolorin B sythase which is required for the conversion of versiconal hemiacetal (VAL) to versicolorin B (VER B) (Silva *et al.*, 1997). The *ver*-1 gene product is involved in the conversion of versicolorin A (VER A), since disruption of this gene led to VER A accumulation (Skory *et al.*, 1992). The *stc*P gene in *A. nidulans* was recently reported to encode an O-methyltransferase catalyzing sterigmatocysin biosynthesis from the demethyl-sterigmatocystin precursor (Kelkar *et al.*, 1996). The *omt*A gene (previous name, *omt*1 - Yu *et al.*, 1995b) encodes an S-adenosylmethionine-dependent O-methyltransferase involved in both the aflatoxin B1 and aflatoxin B2 (AFB_1) and (AFB_2) pathways which catalyze conversion of sterigmatocystin to O-methylsterigmatocystin (OMST) in the AFB_1 pathway and dihydrosterigmatocystin (DHST) to dihydro-O-methylsterigmatocystin (DHOMST) in the AFB_2 pathway (Bhatnagar *et al.*, 1988). Several other genes in the cluster have been identified whose function is not yet known. These include: *afl*J (Payne, G.A. personal communication); *ord*-2 (Yu *et al.*, 1995a); *adh*A Chang, P.K. unpublished data); *adh*1 (Woloshuk & Payne, 1994); and CYP450 (Silva *et al.*, 1996).

The gene cluster for aflatoxin is about 75 kilobases (kb) in length and that for sterigmatocystin about 60 kb (Keller & Hohn, 1997). Although functionally very similar, a comparison of the steigmatocystin and aflatoxin gene clusters has shown that the order and direction of transcription have not been perfectly conserved (reviewed in Keller & Hohn, 1997).

Genes of the aflatoxin/sterigmatocystin pathways produce very similar gene products, but these are not highly conserved at the amino acid level. The *afl*R gene in *A. flavus*, *A. parasiticus* and *A. nidulans* have a zinc finger binding domain that gives them DNA binding specificity. This region is virtually identical in amino acid sequence in *A. flavus* and *A. parasiticus*, but only 71% identical to the *A. nidulans* zinc finger region. Comparison of the open reading frames yielded only 33% amino acid sequence similarity between the *A. flavus* - *A. parasiticus* *afl*R and the *A. nidulans* *afl*R (Yu *et al.*, 1996).

Species that produce aflatoxin or sterigmatocystin.
Only four species have been confirmed to produce aflatoxin. These are *Aspergillus flavus*, *A. parasiticus*, *A. nomius* and, very recently, a single strain similar to *A. tamarii* (Goto *et al.*, 1997). All of these species are classified in *Aspergillus* Subgenus *Circumdati* Section *Flavi*. Section *Flavi* is unusual in that it holds both blessings and curses for humankind. While *A. flavus* and *A. parasiticus* cause millions of dollars of crop losses every year from aflatoxin production, their domesticated forms, *A. oryzae* and *A. sojae,* are critical for the production of enormous quantities of certain fermented foods and beverages in some Asian countries.

Sterigmatocystin production occurs in a broader range of fungi than does aflatoxin., including *Aspergillus, Bipolaris, Chaetomium, Farrowia* and *Monocillium* (Cole and

428

Cox, 1981; Barnes *et al.*, 1994). Among the Aspergilli, sterigmatocystin production has been reported from *A. ustus, A. versicolor, A. sydowii, A. caespitosus, A. multicolor, Emericella nidulans, Em. quadrilineata, Em. rugulosa, Em. variecolor, Em. unguis* and several other *Emericella* species, *Eurotium amstelodami, Eu. chevalieri, Eu. repens, and Eu. rubrum* Bennett & Deutsch, 1986; Smith & Ross, 1991; Klich, unpublished).

Interactions of the aflatoxin-sterigmatocystin biosynthetic pathways with other biosynthetic pathways.
Early workers in the field of aflatoxin biosynthesis noted that the beginning of aflatoxin production was associated with the formation of conidiophores. Until recently, there have been no good explanations for this phenomenon. We are just beginning to understand how the mycotoxin metabolic pathways interact with pathways for other morphological and physiological characteristics of these fungi. Some of these interactions are discussed below.

Strain degeneration and mycotoxin production
Strain degeneration is a problem associated with subculturing of many fungi and the aflatoxigenic fungi are no exception. Many isolates will lose their ability to produce aflatoxin with repeated transfer. Although is no major morphological change is associated with this loss, sometimes it is associated with morphological degeneration. Colonies become floccose, with a 'fan' or 'fluff' morphology and sporulation is reduced (Bennett *et al.*, 1986). Working with *A. parasiticus* strains, Kale *et al.* (1994, 1996) conducted repeated mycelial transfers, until aflatoxin and/or aflatoxin precursor production was lost, and made detailed observations of colony and micromorphological features. Colonies no longer able to produce aflatoxins were floccose, sporulation was reduced, and conidiophores had smaller vesicles with fewer phialides, bearing fewer conidia than the wild type. The conidial morphology was not affected. In *A. nidulans*, Keller *et al.* (1995) observed that a wild type strain with normal sporulation produced sterigmatocystin whereas developmental mutants with a floccose morphology and no conidia did not.

Morphological effects of Gene deletion, disruption or insertion
What are the morphological effects of deleting certain genes or blocking biosynthesis? Disruption of some of the genes in the aflatoxin-sterigmatocystin biosynthetic pathway has no morphological effect. For instance, disruption of the two FAS genes involved in sterigmatocystin production produces strains that were morphologically identical to the wild type. With these genes, only sterigmatocystin production was blocked. When the fatty acid sythase genes involved in primary metabolism were disrupted, however, colonies did not grow unless supplemented with myristic acid (a C_{14} straight chain fatty acid) Such colonies did not sporulate (Brown *et al.*, 1996b). When interference with aflatoxin or sterigmatocystin biosynthesis does have morphological effects, these effects are seen in sporulation, pigment formation and, in the case of *A. flavus* and *A. parasiticus*, sclerotial production.

Pigments
Investigations have been conducted on the nature of the spore pigments in *A. nidulans*. The major ascospore pigment is ascoquinone A, probably a product of a polyketide synthase (Brown & Salvo, 1994). This pigment is structurally very similar to norsolorinic

acid, the first stable intermediate in the aflatoxin-sterigmatocystin biosynthetic pathway, suggesting an evolutionary relationship between the two polyketide synthase systems.

In molecular studies on the *pksA* gene, a polyketide synthase involved in the early steps of aflatoxin biosynthesis, Chang *et al.* (1995a) found that *pksA* was a homolog of the *A. nidulans wA* gene, a polyketide synthase gene involved in conidial wall pigment biosynthesis. When they disrupted the *pksA* gene in *A. parasiticus* SRRC 2043, the yellow pigment usually produced in the medium disappeared, but the conidial colors did not change. Conidial wall pigments in *A. parasiticus* and *A. nidulans* have not been fully characterized, although evidence indicates that they are polyketide derivatives (Brown, *et al.*, 1993). Indirect evidence has indicated that conidial colors in *A. flavus* and *A. parasiticus* have a different biosynthetic pathway than that of many other green-spored Aspergilli and Penicillia (Wheeler and Klich, 1995).

An important aspect of early research to elucidate the genetic and chemical mechanisms of aflatoxin biosynthesis included the use of UV mutagenesis of *A. parasiticus* to disrupt aflatoxin production (Bennett and Goldblatt, 1973). Aflatoxin Mutants which did not produce aflatoxin generated in these studies were useful in: 1) investigating genetic inheritance and establishing linkage groups of traits governing aflatoxin production; and 2) in chemical characterization of aflatoxin biosynthetic pathway intermediates which accumulated in fungal mycelia when genes encoding critical biosynthetic enzymes were inactivated. Inactivation of individual genes governing steps in aflatoxin biosynthesis in the anthraquinone sequence of pathway reactions (intermediates NOR to VER A in the aflatoxin biosynthetic pathway, see Figure 1) led to accumulation of an orange-red pigmentation in the fungal mycelium (e.g. Lee *et al.*, 1971). The pigmentation is readily visible in the reverse of mutant colonies. Such mutant colonies, when observed under long wave UV light, showed relatively less fluorescence associated with aflatoxin production than the original wild type *A. parasiticus* strain (Bennett and Goldblatt, 1973). It is noteworthy to the fungal taxonomist that, with minimal genetic change (such as as a single gene disruption), very dramatic phenotypic changes (such as in colony pigmentation and UV fluorescence) can occur.

Sclerotia
The relationship between aflatoxin production and sclerotial production has long been a topic of interest to taxonomists and mycotoxicologists. Many factors influence production of both aflatoxin and sclerotia, which has generated much of the discussion on the relationship between the two. Serial transfer of a toxigenic sclerotial strain of *A. parasiticus* resulted in loss of both sclerotial and aflatoxin production (Bennett *et al.*, 1986). In *A. flavus*, strains that produce small sclerotia have been associated with higher aflatoxin production than those that produce large sclerotia (for review see Cotty *et al.*, 1994). Although mycotoxin production is inhibited by light, there is contradictory evidence in the literature regarding the effects of light on aflatoxin production (Bennett *et al.*, 1978), indicating that some strains may be more sensitive to light than others. Production of sclerotia and ascomata by some fungi are inhibited by light (Marsh *et al.*, 1959, Bennett *et al.*, 1978).

The pH of the media is important in the production of aflatoxin and sterigmatocystin. Generally, lower pH yields higher levels of these mycotoxins and their intermediates and recent studies indicate that pH may override the effects of nitrogen and carbon source

(Cotty, 1988; Keller *et al.*, 1997). Cotty, Using a single strain of *A. flavus* Cotty, 1988), reported that sclerotial production was higher at higher pH levels.

The few molecular studies a Hemting to relate aflatoxin production and sclerotial production (Skory *et al.*, 1992; Trail *et al.*, 1995) suggest that, in *A. parasiticus*, genetically blocked mutants that accumulate intermediate compounds in the aflatoxin biosynthetic pathway produce fewer sclerotia than the wild type. If no aflatoxin or intermediates are allowed to accumulate, there is actually an enhancement of sclerotial production.

In a recent study of the relationship between aflatoxin biosynthesis and sclerotial development, extra copies of the regulatory gene *aflR* and/or an associated gene of unknown function, *aflJ*, were introduced into *Aspergillus parasiticus* SRRC 2043. This strain accumulates O-methylsterigmatocystin. Additional copies of *aflJ* had no effect on the levels of O-methylsterigmatocystin or other aflatoxin precursors. Extra copies of *aflR* yielded an approximately 10-fold increase in aflatoxin intermediates, and colonies with elongate sclerotia rather than the globose sclerotia characteristic of the untransformed isolate. When extra copies of both *aflR* and *aflJ* were introduced, a 20-25-fold increase in aflatoxin precursor levels occurred, and the colonies had elongate apiculate sclerotia (Chang *et al.*, submitted). Elongate sclerotia are not common in wild type *A. parasiticus* strains, however, they are characteristic of *A. nomius* (Kurtzman *et al.*, 1987). Goto *et al.* (1996) described an aflatoxigenic *A. tamarii* isolate, which produces small black pear-shaped sclerotia. Further study of this apparent relationship between these elongate sclerotia and the aflatoxin biosynthetic pathway may further elucidate the relationships between members of this interesting group of fungi.

Use of mycotoxin biosynthetic pathway genes as probes to study species relationships.

With the genes in the biosynthetic pathways of aflatoxin-sterigmatocystin available, we have a new tool for examining species relationships in sections of the genus that produce these two secondary metabolites. One obvious question was - do the domesticated forms of *A. flavus* and *A. parasiticus* that do not produce aflatoxin (*A. oryzae* and *A. sojae*) have the genes for aflatoxin production? When we probed DNA isolates of this section with the regulatory gene *aflR* and the structural gene *omt*-1, all of the *A. parasiticus*, *A. sojae* and *A. flavus* isolates hybridized to the probes. None of the *A. oryzae* or *A. tamarii* isolates hybridized with the *aflR* probe. Two of the three *A. oryzae* isolates hybridized with the *omt*-1 probe, but none of the *A. tamarii* isolates did (Klich *et al.*, 1995). Apparently, at least the *A. oryzae* and *A. tamarii* isolates examined in this study are missing some of the genes necessary for aflatoxin production. Further examination of the DNA of *A. parasiticus* and *A. sojae* using five other aflatoxin biosynthesis gene probes (*nor*-1, *pks*A, *uvm*8, *norA* and *ver-1*), revealed that all of the isolates of both species hybridized to all of the probes. The only probe that revealed a polymorphism was *ver-1*, which had two bands for two of the *A. parasiticus* isolates and only one band for the third *A. parasiticus* and the three *A. sojae* isolates. This indicated that *A. sojae* probably possesses the genes necessary for aflatoxin production. To determine whether or not these genes were being transcribed, Northern blots were performed using RNA from the same six *A. parasiticus* and *A. sojae* isolates and gene probes from the same seven genes used in the DNA study. Results demonstrated that some of the genes are not transcribed in *A. sojae* or in *A. parasiticus* isolates which do not produce aflatoxins (Klich *et al.*, 1997).

REFERENCES

Barnes, S.E., Dola, T.P., Bennett, J.W. & Bhatnagar, D. (1994). Synthesis of sterigmatocystin on a chemically defined medium by species of *Aspergillus* and *Chaetomium*. Mycopathologia 126: 173-178.

Bennett, J.W. & Goldblatt, L.A. (1973). The isolation of mutants of *Aspergillus flavus* and *A. parasiticus* with altered aflatoxin producing ability. Sabouraudia 11: 235-241.

Bennett, J.W., Fernholz, F.A. & Lee, L.S. (1978). Effect of light on aflatoxins, anthraquinones and sclerotia in *Aspergillus flavus* and *A. parasiticus*. Mycologia 70: 104-116.

Bennett, J.W. & Deutsch, E. (1986). Genetics of mycotoxin biosynthesis. In Mycotoxins and Phycotoxins (Eds Steyn, P.S. & Vleggaar R.). Amsterdam, Elsevier Science Publishers. pp. 51-64.

Bennett, J.W. Leong, P.-M., Kruger, S. & Keyes, D. (1986). Sclerotial and low aflatoxigenic morphological variants from haploid and diploid *Aspergillus parasiticus*. Experientia 42: 848-851.

Bhatnagar, D., Ullah, A.H.J. & Cleveland, T.E. (1988). Purification and characterization of a methyltransferase from *Aspergillus parasiticus* SRRC 163 involved in aflatoxin biosynthetic pathway. Prep. Biochem. 18: 321-349.

Bhatnagar, D., Ehrlich, K.C. & Cleveland, T.E. (1992). Oxidation-reduction reactions in biosynthesis of secondary metabolites, In Handbook of Applied Mycology, vol. 5. Mycotoxins in Ecological Systems (Eds. Bhatnagar, D., Lillehoj, E.J. & Arora, D.K.), New York, Marcel Dekker, Inc. pp 255-286.

Brown, D.W., Hauser, F.M., Tommasi, R., Corlett, S. & Salvo, J.J.. (1993). Structural elucidation of a putative conidial pigment intermediate in *Aspergillus parasiticus*. Tetrahedron Let.s 34: 419-422.

Brown, D.W. & Salvo, J.J. (1994). Isolation and characterization of sexual spore pigments from *Aspergillus nidulans*. Appl. Environ. Microbiol. 60: 979-983.

Brown, D.W., Yu, J-H., Kelkar, H.S., Fernandes, M., Nesbitt, T.C., Keller, N.P., Adams, T.H. & Leonard, T.J. (1996a). Twenty-five coregulated transcripts define a stergimatocystin gene cluster in *Aspergillus nidulans*. Proc. Natl Acad. Sci. USA 93: 1418-1422.

Brown, D.W., Adams, T.H. & Keller, N.P. (1996b). *Aspergillus* has distinct fatty acid synthases for primary and secondary metabolism. Proc. Natl Acad. Sci. USA 93: 12873-14877.

Brown, R.L., Cleveland, T.E., Bhatnagar, D. & Cary, J.E. Recent advances in preventing mycotoxin contamination. In Mycotoxins in Agriculture and Food Safety (Eds. Sinha K.K. & Bhatnagar, D.). New York, Marcel Dekker. (in press) 1977.

Cary, J.W., Wright, M., Bhatnagar, D., Lee, R. & Chu, F.S. (1996). Molecular characterization of an *Aspergillus parasiticus* dehydrogenase gene, *norA*, located on the aflatoxin biosynthesis gene cluster. Appl. Environ. Microbiol. 62: 360-366.

Chang, P-K. Cary, J.W., Yu, J., Bhatnagar, D. & Cleveland, T.E. (1995a). The *Aspergillus parasiticus* polyketide synthase gene *pksA*, a homolog of *Aspergillus nidulans WA*, is required for aflatoxin B₁ biosynthesis. Mol. Gen. Genet. 248: 270-277.

Chang, P-K., Ehrlich, K.C., Yu, J-J., Bhatnagar, D. & Cleveland, T.E. 1995b. Increased expression of *Aspergillus parasiticus aflR*, encoding a sequence-specific DNA-binding protein, relieves nitrate inhibition of aflatoxin biosynthesis. Appl. Environ. Microbiol. 61: 2372-2377.

Chang, P-K., Cotty, P.J., Bhatnagar, D., Bennett, J.W. & Cleveland, T.E. 1977. Overproduction of aflatoxin precursors affects sclerotial development and production in *Aspergillus parasiticus* SRRC 2043. Fungal Genet. and Biol. (submitted).

Cleveland, T. E., & Bhatnagar, D. (1992). Molecular strategies for reducing aflatoxin levels in crops before harvest. In Molecular Approaches to Improving Food Quality and Safety (Eds Bhatnagar, D. & Cleveland, T.E.), New York, Von Nostrand Reinhold. pp. 205-228.

Cole, R.J. & Cox. R.H. (1981). Sterigmatocystins. In Toxic Fungal Metabolites (Eds Cole, R.J. & Cox, R.H.). New York, Academic Press. pp. 67-93.

Cotty, P.J. (1988). Aflatoxin and sclerotial production by *Aspergillus flavus*: influence of pH. Phytopathology 78: 1250-1253.

Dutton, M. F. 1988. Enzymes and aflatoxin biosynthesis. Microbiol. Rev. 52: 274-295.

Goto, T., Wicklow, D.T. & Ito, Y. (1996). Aflatoxin and cyclopiazonic acid production by a sclerotium-producing *Aspergillus tamarii* strain. Appl. Environ. Microbiol. 62: 4036-4038.

Hopwood, D.A. & Sherman, D.H. (1990). Molecular genetics of polyketides and its comparison to fatty acid biosynthesis. Annu. Rev. Genet. 24: 37-66.

Jelinek, C.F., Pohland, A.E. & Wood, G.E. (1989). Worldwide occurrence of mycotoxins in foods and feeds -- an update. J. Assoc. Off. Anal. Chem. 72: 223-230.

Kale, S.P., Bhatnagar, D. & Bennett, J.W. (1994). Isolation and characterization of morphological variants of *Aspergillus parasiticus* deficient in secondary metabolite production. Mycol. Res. 98: 645-652.

Kale, S.P., Cary, J.W., Bhatnagar, D. & Bennett, J.W. (1996). Characterization of experimentally induced, nonaflatoxigenic variant strains of *Aspergillus parasiticus*. Appl. Environ. Microbiol. 62: 3399-3404.

Kelkar, H.S., Keller, N.P. & Adams, T.H. (1996). *Aspergillus nidulans stcP* encodes an O-methyltransferase that is required for sterigmatocystin biosynthesis. Appl. Environ. Microbiol. 62:4296-4298.

Keller, N.P., Cleveland, T.E. & Bhatangar, D. (1992). A molecular approach towards understanding aflatoxin production. In Handbood of Applied Mycology vol 5: Mycotoxins in Ecological Systems (Eds Bhatnagar, D., Lillehoj, E.B. & Arora, D.K.) New York, Marcel Dekker, pp. 287-310.

Keller, N.P., Brown, D., Butchko, R.A.E., Fernandes, M., Kelkar, H., Nesbitt, C., Segner, S., Bhatnagar, D., Cleveland, T.E. & Adams, T.H. (1995). In Molecular Approaches to Food Safety Issues Involving Toxic Microorganisms (Eds Eklund, M., Richard, J.L. & Mise, K.) Fort Collins, Colorado, Alaken Inc. pp. 263-277.

Keller, N.P. & Hohn, T.M. (1997) Metabolic pathway gene clusters in filamentous fungi. Fungal Genet. and Biol. 21: 17-29.

Keller, N.P., Nesbitt, C., Saar, B., Phillips, T.D. & Burow, G.B. (1997). pH regulation of sterigmatocystin and aflatoxin biosynthesis in *Aspergillus* spp. Phytopathology 87 (in press).

Klich, M.A., Yu, J., Chang, P-K., Mullaney, E.J., Bhatnagar, D. & Cleveland, T.E. (1995). Hybridization of genes involved in aflatoxin biosynthesis to DNA of aflatoxigenic and non-aflatoxigenic Aspergilli. Appl. Microbiol. Biotechnol. 44: 439-443.

Klich, M.A., Montalbano, B. & Ehrlich, E. (1997). Northern analysis of aflatoxin biosynthesis genes in *Aspergillus parasiticus* and *Aspergillus sojae*. Appl. Microbiol. Biotechnol. 47:246-249.

Kurtzman, C.P., Horn, B.W. & Hesseltine, C.W. (1987). *Aspergillus nomius*, a new aflatoxin-producing species related to *Aspergillus flavus* and *Aspergillus tamarii*. Antonie van Leeuwenhoek 53: 147-158.

Lee, L.S., Bennett, J.W., Goldblatt, L.A. & Lundin, R.E. (1971). Norsolorinic acid from a mutant strain of *Aspergillus parasiticus*. J. Am. Oil Chem. Soc. 48: 93-94.

Mahanti, N., Bhatnagar, D., Cary, J.W., Joubran, J. & Linz, J.E. (1996) Structure and function of *fas*-1A, a gene encoding a putative fatty acid synthetase directly involved in aflatoxin biosynthesis in *Aspergillus parasiticus*. Appl. Environ. Micorbiol. 62: 191-195.

Marsh, P.B., Taylor, E.E. & Bassler, L.M. (1959) A guide to the literature oncertain effects of light on fungi: reproduction, morphology, pigmentation and phototropic phenomena. Plant Dis. Rep. 261: 251-312.

Payne, G.A., Nystorm, G.J., Bhatnagar, D., Cleveland, T.E. & Woloshuk, C.P. (1993). Cloning of the afl-2 gene involved in aflatoxin biosynthesis from *Aspergillus flavus*. Appl. Environ. Microbiol. 59:156-162.

Prieto, R., Yousibova, G.L. & Woloshuk, C.P. (1996). Identification of aflatoxin biosynthesis genes by genetic complementation in an *Aspergillus flavus* mutant lacking the aflatoxin gene cluster. Appl. Environ. Microbiol. 62: 3567-3571.

Silva, J.C., Minto, R.E., Barry, C.E., III, Holland, K.A. & Townsend, C.A. (1996). Isolation and characterization of the versicolorin B synthase gene from *Aspergillus parasiticus*. J. Biol. Chem. 273: 13600-13608.

Skory, C.D., Chang, P-K., Cary, J. & Linz, J.E. (1992). Isolation and characterization of a gene from *Aspergillus parasiticus* associated with the conversion of versicolorin A to sterigmatocystin in aflatoxin biosynthesis. Appl. Environ. Microbiol. 58: 3527-3537.

Smith, J.E. & Ross, K. (1991). The toxigenic Aspergilli. In Mycotoxins and Animal Foods (Eds Smith, J.E & Henderson R.S.) Boca Raton, Florida, CRC Press. pp. 102-118.

Trail, F., Mahanti, N., Chang, P-K., Cary, J.W. & Linz, J.E. (1994). Structural and functional analysis of the *nor*-1 gene involved in the biosynthesis of aflatoxins by *Aspergillus parasiticus*. Appl. Environ. Microbiol. 60: 4078-4085.

Trail, F., Mahanti, N., Rarick, M., Mehigh, R., Liang, S-H., Zhou, R. & Linz, J.E. (1995). Physical and transcriptional map of an aflatoxin gene cluster in *Aspergillus parasiticus* and functional disruption of a gene involved early in the aflatoxin pathway. Appl. Environ. Microbiol. 61: 2665-2673.

Watanabe, C.M.H., Wilson, D., Linz, J.E. & Townsend, C.A. (1996). Demonstration of the catalytic roles and evidence for the physical association of tyep 1 fatty acid synthases and a polyketide synthase in the biosynthesis of aflatoxin B_1. Chem. and Biol. 3: 463-469.

Wheeler, M.H. & Klich, M.A. (1995) The effects of tricyclazole, pyroquilon, phthalide, and related fungicides on the production of conidial wall pigments by *Penicillium* and *Aspergillus* species. Pesticide Biochem. Physiol. 52: 125-136.

Woloshuk, C. P. & Payne, G.A. (1994). The alcohol dehydrogenase gene *adh1* is induced in *Aspergillus flavus* grown on medium conducive to aflatoxin biosynthesis. Appl. Environ. Microbiol. 60: 670-676.

Woloshuk, C. P., Foutz, K.R., Brewer, J.F., Bhatnagar, D., Cleveland, T.E. & Payne, G.A. (1994). Molecular characterization of *aflR*, a regulatory locus for aflatoxin biosynthesis. Appl. Environ. Microbiol. 60: 2408-2414.

Yabe, K., Matsuyama, Y., Ando, Y., Nakajima, H. & Hamasaki, T. (1993). Stereochemistry during aflatoxin biosynthesis: conversion of norsolorinic acid to averufin. Appl. Environ. Microbiol. 59: 2486-2492.

Yabe, K., Nakamura, Y., Nakajima, H., Ando, Y. & Hamasaki, T. (1991). Enzymatic conversion of norsolorinic acid to averufin in aflatoxin biosynthesis. Appl. Environ. Microbiol. 57: 1340-1345.

Yu, J., Chang, P-K., Cary, J.W., Wright, M., Bhatnagar, D., Cleveland, T.E., Payne, G.A. & Linz, J.E. (1995a). Comparative mapping of aflatoxin pathway gene clusters in *Aspergillus parasiticus* and *Aspergillus flavus*. Appl. environ. Microbiol. 61:2365-2371.

Yu, J., Chang, P-K., Payne, G.A., Cary, J.W., Bhatnagar, D & Cleveland, T.E. (1995b). Comparison of the *omtA* genes encoding O-methyltransferases involved in aflatoxin biosynthesis from *Aspergillus parasiticus* and *Aspergillus flavus*. Gene 153:121-125.

Yu, J., Chang, P-K., Cary, J.W., Bhatnagar, D. & Cleveland, T.E. (1997). AvnA, a gene encoding a cytochrome P-450 monooxgenase, is involved in the conversion of averantin to averufin in aflatoxin biosynthesis in *Aspergillus parasiticus*. Appl. Environ. Microbiol. 63: 1349-1356.

Yu, J-H., Butchko, R.A.E., Fernandes, M., Keller, N.P., Leonard, T.J. & Adams, T.H. (1996). Conservation of structure and function of the aflatoxin regulatory gene *aflR* from *Aspergillus nidulans* and *A. flavus*. Curr. Genet. 29: 549-555.

434

ANALYSIS OF THE MOLECULAR AND EVOLUTIONARY BASIS OF TOXIGENICITY AND NON-TOXIGENICITY IN *ASPERGILLUS FLAVUS* AND *A. PARASITICUS*

N. Tran-Dinh[1], S. Kumar[1] J.I. Pitt[2] and D.A. Carter[1]
[1]Department of Microbiology, University of Sydney, NSW 2006, Australia and
[2]CSIRO Division of Food Science and Technology, North Ryde, NSW 2113, Australia

INTRODUCTION

Aflatoxins are potent hepatocarcinogens produced exclusively by three species of *Aspergillus*: *Aspergillus flavus, Aspergillus parasiticus,* and *Aspergillus nomius*. These fungi are frequently found associated with a number of human and animal food crops, including sorghum, maize, nuts (especially peanuts) and cottonseed, making aflatoxins an important food safety issue in endemic regions.

Economic and health concerns have prompted numerous studies aimed at understanding and controlling aflatoxin biosynthesis. The aflatoxin pathway is now well characterised, and involves as many as 17 different enzymatic steps, which convert a polyketide precursor to aflatoxins B_1 and B_2 in *A. flavus*, and to B_1, B_2, G_1 and G_2 in *A. parasiticus* (Bhatnagar, *et al.*, 1992; Klich and Cleveland, 1998). The pathway is governed by *aflR*, a regulatory locus, whose gene product contains a putative zinc finger DNA-binding motif (Woloshuk, *et al.*, 1994). This appears to act at several steps in the biosynthetic pathway, in particular the conversion of acetate to the first biosynthetic intermediate, norsoloronic acid (Payne, *et al.*, 1993).

Other studies have focused on controlling aflatoxin contamination by preventing the aflatoxigenic fungi from becoming established on susceptible crop plants. Naturally occurring, nontoxigenic strains of *A. flavus* are commonly found and can be used to displace their toxigenic counterparts. Studies on maize (Brown, *et al.*, 1991) and cottonseed (Cotty, 1990) found over 90% reduction in aflatoxin contamination using this strategy of competitive exclusion, although the ability to suppress aflatoxin contamination varied according to the nontoxigenic strain employed (Cotty and Bhatnagar, 1994). Natural, nontoxigenic strains of *A. parasiticus* are much less common, however, and no data are yet available using these as biocompetitive agents, although some studies have been done using experimentally induced nontoxigenic mutants (Dorner, *et al.*, 1992; Erlich, 1987).

In Australia, peanut crops are commonly contaminated by aflatoxins produced by *A. flavus* and *A. parasiticus*. We are therefore interested in developing a biocontrol strategy to prevent colonisation by both of these species. To help predict the value of this biocontrol approach, and to aid our choice of suitable nontoxigenic strains, we have begun by examining the evolutionary history and molecular genetics of naturally occurring, non-

435

toxigenic strains of *A. flavus* and *parasiticus*. We have reported the genetic relatedness of a population of toxigenic and nontoxigenic isolates of both species by analysing Randomly Amplified Polymorphic DNA (RAPD) profiles by neighbour joining (Tran Dinh, *et al.*, 1999). This study showed *A. flavus* and *A. parasiticus* to be genetically separate from one another, with *A. flavus* isolates also separating into two distinct groups. Toxigenic and nontoxigenic isolates of *A. flavus* were interspersed in these two groups, suggesting that either toxigenicity has been lost multiple times and independently by different strains, or that it has been redistributed by genetic recombination onto a variety of different genetic backgrounds, The latter result is supported by Geiser *et al.*, who also found *A. flavus* strains to separate into two distinct groups. They showed that recombination occurs in one of these groups by analysing the distribution of sequence polymorphisms using a phylogenetic approach (Geiser, *et al.*, 1997). Our study also found that the five nontoxigenic strains of *A. parasiticus* separated into two groups, and these were interspersed among toxigenic *A. parasiticus* strains. In this paper we review our previous work and extend our analysis to look at the molecular lesion responsible for nontoxigenicity in *A. parasiticus*. Results indicate that toxigenicity has been lost at least twice in these isolates of *A. parasiticus*, and that they may be diverging from one another by a process of microevolution.

MATERIALS AND METHODS

Isolates: All isolates of *A. flavus* and *A. parasiticus* were obtained from peanut plants and soil samples taken from the Kingaroy region of Queensland, Australia, from 1979-1995, and are maintained in the FRR culture collection of the CSIRO Division of Food Science and Technology, North Ryde, NSW. Strain designations, toxigenicity, sources, and years of isolation are given in Table 1. The five nontoxigenic isolates of *A. parasiticus* are the only nontoxigenic ones to have been isolated to date from Australian soils. These produce neither B nor G aflatoxins, and only isolate FRR 4468 produces cyclopiazonic acid. Otherwise, these appear to be identical in morphology and growth characteristics to toxigenic strains of *A. parasiticus*.

Toxin assay: The production of aflatoxin was initially assessed by growing strains on coconut cream agar (Dyer and McCammon, 1994). The presence of aflatoxins was detected by intense fluorescence in the reverse of colonies examined under long wavelength UV light. Nontoxigenicity of presumptively nontoxigenic isolates was subsequently confirmed by growth on solid substrates, extraction and HPLC.

DNA isolation and standardisation: Small scale DNA extractions for PCR followed the method of Lee and Taylor (1990). Fresh mycelium (1-2 g wet weight) was ground in liquid nitrogen, suspended in lysis buffer (50 mM Tris-HCl; 100 mM NaCl; 5 mM EDTA; 1% SDS) and incubated at 80°C for 10 min. This was cooled to 40°C, 10 ml of proteinase K (10 mg/ml) was added and the lysate incubated at 37°C for a minimum of 2.5 hr. One phenol:chloroform:isoamyl and one chloroform:isoamyl extraction was performed, the DNA precipitated using 10 ml 3M sodium acetate (pH 5.2) and 0.54 vol isopropanol, washed once with 500 µl of 70% ethanol, and resuspended in 100 µl sterile water. Extracted DNA (5 µl) was electrophoresed in a 1% agarose gel and visualised with ethidium bromide staining and UV transillumination to assess yield. All DNA samples were standardised to approximately 20 ng/µl.

Table 1. Isolates of *Aspergillus flavus* and *A. parasiticus* used in this study

Non-toxigenic *A. flavus*			Non-toxigenic *A. parasiticus*		
FRR 4351	1991	peanut	FRR 4471	1993	soil
FRR 4288	1991	peanut	FRR 4470	1993	soil
FRR 4086	1990	peanut	FRR 4469	1993	soil
FRR 5314	1995	soil	FRR 4468	1993	soil
FRR 5313	1995	soil	FRR 4467	1993	soil
FRR 5310	1995	soil			
FRR 5306	1995	soil	Toxigenic *A. parasiticus*		
FRR 5311	1995	soil	FRR 2756	1984	peanut
FRR 5312	1995	soil	FRR 2753	1984	soil
FRR 5309	1995	soil	FRR 2752	1984	soil
			FRR 2749	1984	soil
Toxigenic *A. flavus*			FRR 2745	1984	peanut
FRR4474	1990	soil	FRR 2744	1984	peanut
FRR4473	1990	peanut	FRR 2503	1982	peanut
FRR4472	1990	soil	FRR 2502	1982	peanut
FRR2748	1984	soil	FRR 2501	1982	peanut
FRR2746	1984	peanut	FRR 2242	1979	soil
FRR 5316	1995	soil			
FRR 5307	1995	soil			
FRR 5305	1995	soil			
FRR 5308	1995	soil			
FRR 5315	1995	soil			

Large scale extractions for restriction digestion and Southern blotting were based on the method of Moody and Tyler (1990a). Fresh mycelium (20 g) was ground under liquid nitrogen with a mortar and pestle and resuspended in extraction buffer (20 mM Tris-HCl, pH 8.5; 250 mM NaCl; 25 mM EDTA, pH 8.0; 0.5% SDS). One volume of phenol:chloroform (3:7) was added and the mixture was homogenised by orbital shaking at 75 rpm for 30 min. The phases were separated by centrifugation at 9,000 rpm for 1 hr. The DNA was extracted 4 more times with extraction buffer mixed with an equal volume of phenol:chloroform (3:7), and once with chroroform:isoamyl. It was then precipitated overnight with 0.54 vol isopropanol, pelleted by centrifugation at 4,000 rpm for 30 min, washed once with 70% ethanol, and resuspended in 10 ml sterile TE buffer (10 mM Tris-HCl, 1 mM EDTA, pH 8.0).

RAPD-PCR: Ten base RAPD primers were purchased from the University of British Columbia Nucleic Acids - Protein Service (NAPS) Unit. These were used both singly and in combination with the M13-40 sequencing primer. RAPD-PCR reactions (50 µl) contained 1X PCR buffer (Perkin-Elmer), 5% glycerol, 0.5 mM each dNTP, 2.5 U Amplitaq polymerase (Perkin-Elmer) 10 p mol primer(s) and approx 20 ng DNA, and were overlaid with sterile mineral oil. Amplifications were carried out in a Perkin Elmer 480 thermocycler with the following cycling parameters: 2 cycles of 5 min each at 94˚C, 36˚C and

72˚C, followed by 40 cycles of 1 min at 94˚C, 1 min at 36˚C and 2 min at 72˚C, with a final elongation step of 7 min at 72˚C. Amplified DNA was electrophoresed in 2% agarose and visualised by ethidium bromide staining and UV transillumination. Each amplification with a given RAPD primer or primer pair was performed simultaneously on all of the isolates included in this analysis, and the amplified DNAs were electrophoresed in a single gel. RAPD primer sequences included in this analysis and the M13-40 primer sequence are shown in Table 2.

Analysis of RAPD profiles: Each RAPD profile was scored as a single character for neighbour joining analysis. Scoring was by visually comparing profiles and assigning each profile two numbers. The first number represented the major type, with each distinctly different profile (sharing less than 50% of bands with any other profile) assigned a different first number. Major types were then divided into subtypes representing minor variants in the RAPD profile, and differentiated by a second number. Values were weighted such that the first number (major type) was given twice the value of the second number (subtype). The data were analysed by neighbour joining using PHYLIP 3.0 (Felsenstein, 1982).

Restriction digestion, Southern blotting and hybridisation: BglII was chosen for DNA digestion as it was predicted to cut only once within the *aflR* gene region. Approximately 3 µg of DNA from each nontoxigenic strain of *A. parasiticus* and the toxigenic strain FRR 2503 was digested with 10 U of BglII for 4 hr at 37˚C. DNA was electrophoresed in 1% agarose, stained with 10 µg/ml ethidium bromide and visualised by UV transillumination. DNA was transferred by capillary blotting onto a nylon membrane (Qiabrane, QIAGEN GmbH, Hilden, Germany) following the recommendations of the manufacturer, and fixed by UV cross-linking using a Stratalinker (Stratagene, La Jolla, CA) on the autofix setting. A probe for the upper region of the *aflR* gene was made by amplifying this region from DNA from isolate FRR 2503 using PCR primers aflRU1 and aflRL1, which were designed from the published *aflR* sequence (Woloshuk, *et al.*, 1994). The aflR primer sequences are listed in Table 2. PCR reactions were carried out in 100 µl reaction volumes, using 1 X PCR buffer, 5% glycerol, 0.5 mM each dNTP, 2.5 U Amplitaq polymerase, 10 pmol each primer, and 20 ng DNA. Amplifications were carried out in a Perkin-Elmer 480 thermocycler and involved 30 cycles of 1 min at 94˚C, 1.5 min at 60˚C and 1 min at 72˚C, followed by a final 7 min elongation step at 72˚C. The PCR product was labelled by DIG incorporation (Boehringer Mannheim GmbH, Mannheim, Germany) following the recommendations of the manufacturer. Labelled probe (12.5 µg) was used in the hybridisation reaction. Prehybridisation, hybridisation and all washes were performed in a Hybaid oven at 42˚C. Hybridising DNA was detected using the chemiluminescent substrate Lumigen-PPD (Lifecodes Corp, Stamford, CT) and autoradiography.

RESULTS

RAPD-PCR amplification
Of over 90 amplifications tested using RAPD primers both singly and with primer M13-40, 24 gave amplification profiles that were suitable for analysis. Two representative gels are shown in Figure 1. From these it can be seen that the majority of isolates of *A. flavus*

and *A. parasiticus* are distinctly different from one another, and also that a number of differences exist between isolates from each species. Two isolates, FRR 5316 and FRR 2749, which had been identified as *A. flavus* and *A. parasiticus* respectively using morphological criteria, consistently gave amplification profiles that were characteristic of the other species, and were judged to have been misidentified previously.

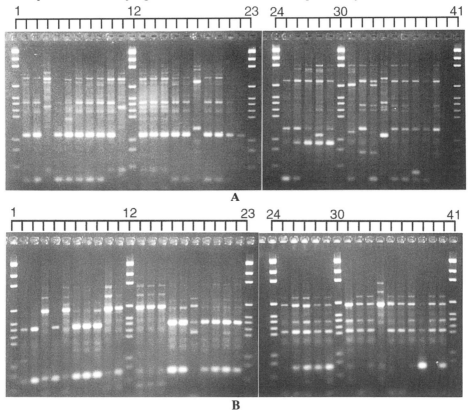

Fig. 1. Agarose gel electrophoresis of two representative RAPD amplifications. A. amplification using primers 626 + M13-40; B. amplification using primer 631 + M13-40. Lanes: 1, 12, 23, 24, 30 and 41: size standard (pGEM), 2-11: nontoxigenic isolates of *A. flavus*; 13-22: toxigenic isolates of *A. flavus*; 25-29: nontoxigenic isolates of *A. parasiticus*; 31-40: toxigenic isolates of *A. parasiticus*. The order of the isolates within each group is the same as that given in Table 1.

Profiles were scored rather than individual bands to reduce problems of co-dependence of bands within a single RAPD profile. Thus each RAPD profile represented one character for neighbour joining analysis. Band intensities were generally ignored when assigning scores for different profiles, as the genetic basis behind intensity is unclear, and profiles were differentiated according to the presence and absence of bands. The reproducibility of RAPD profiles was tested for some of the primers by amplifying DNA from a number of isolates on separate occasions, and from different extractions of a single isolate. Amplifications were reproducible when all components of the PCR reaction (eg. the thermocycler used, DNA concentration, reaction buffer, etc) were standardised. Problems with reproducibility were further reduced by ensuring that all isolates were amplified with a given primer at one time, and all amplifications were electrophoresed and scored on a single agarose gel.

Analysis by Neighbour Joining
The dendogram resulting from neighbour joining analysis using PHYLIP 3.0 is shown in Fig. 2. *A. parasiticus* and *A. flavus* were separated from one another by long branches, and two distinct clusters were seen within the *A. flavus* isolates. Toxigenic and nontoxigenic isolates of *A. flavus* were found in both clusters and were interspersed. The five nontoxigenic isolates of *A. parasiticus* separated into two groups, and within each group the individual nontoxigenic isolates were distinct from one another.

Hybridisation with the upper region of the aflR gene
PCR primers aflRU1 and aflRL1 successfully amplified the predicted fragment of approximately 900 bp from the upper region of the *aflR* gene from FRR 2503. This was labelled by DIG incorporation and used as a probe against digested DNA from the five nontoxigenic isolates of *A. parasiticus,* by Southern blotting and hybridisation. Digested DNA from FRR 2503 was included on the blot as a positive control. The results of hybridisation are shown in Fig. 3. *Bgl*II cuts once within the amplified *aflR* gene region, thus the probe should hybridise with two bands in fully digested *A. parasiticus* DNA. The toxigenic isolate FRR 2503 contained two bands of approximately 0.8 bp and 3.3 kb which hybridised strongly with the labelled probe, as well as with two larger, cross-hybridising bands. Identical hybridising bands of 0.8 bp and 3.3 kb were seen in nontoxigenic isolates FRR 4467, FRR 4468 and FRR 4469. Isolates FRR 4470 and FRR 4471 completely lacked the 3.3 kb band and showed only very faint hybridisation with the 800 bp band, suggesting that most of this region of the *aflR* gene is deleted or substantially rearranged in these two strains. The nontoxigenic isolates possessed cross-hybridising bands of the same size as those seen in FRR 2503, but in all of these hybridisation with the smaller of the two bands was less intense than in the toxigenic isolate.

DISCUSSION

RAPD-PCR is a simple means of generating many characters from around the genome of any organism (Welsh and McClelland, 1990; Williams, *et al.*, 1990). Most analyses of RAPD profiles score individual bands for their presence or absence and use a limited number of different amplifications. However, some studies suggest that the bands in a singe RAPD profile may not all be independent of one another but may be co-amplified, with small bands amplified from within larger amplicons (Welsh, *et al.*, 1995). In this case a single mutational event may cause the simultaneous loss or gain of a number of different bands, which could influence estimates of strain relatedness. To avoid this problem we performed 24 separate RAPD amplifications, scoring each amplification as a single, multi-state character for neighbour joining analysis. Using a large number of different RAPD PCR amplifications also helped to reduce potential problems with nonreproducible, artefactual bands that sometimes occur in RAPD amplifications (Levitan and Grosberg, 1993; Hadrys, *et al.*, 1992).

A.flavus and *A. parasiticus* are morphologically and functionally similar species, separated on the basis of spore morphology and the ability of *A. parasiticus* to produce G as well as B aflatoxins, (Klich and Pitt, 1988). Our results found most isolates of *A. flavus* and *A. parasiticus* to fall into two clearly separated clusters. This is in agreement with other molecular studies which have indicated that these species are genetically distinct (Moody and Tyler, 1990a) (Moody and Tyler, 1990b), and supports their classification as separate species. A single isolate of *A. flavus*, FRR 5316, consistently appeared to give amplification profiles that were characteristic of *A. parasiticus* isolates, and grouped with the *A. parasiticus* isolates on the dendogram (Figure 2). Likewise, an *A. parasiticus* isolate, FRR 2749 grouped with the *A. flavus* isolates. The morphology of these isolates was re-examined, and it was found that their original identifications had been made in error. This ability to reconfirm species identification clearly demonstrates the value of integrating molecular and morphological techniques in fungal taxonomy.

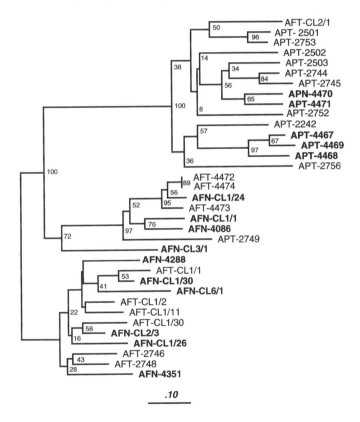

Fig. 2. Dendogram constructed using RAPD profile data by PHYLIP 3.0. Nontoxigenic isolates of *A. flavus* and *A. parasiticus* are shown in bold.

The 20 *A. flavus* isolates included in this study separated into two distinct groups which were joined by a branch that was almost as long as that joining *A. flavus* and *A. parasiticus*. This suggested that these groups may be diverging into separate species. A similar division was seen by Geiser *et al.* (1997), who analysed the population structure of *A. flavus* isolates taken from a single field using DNA sequence polymorphisms as genetic markers. No obvious differences in morphology were apparent between the two *A. flavus* groups, and it is not known if these represent the "L" and "S" types reported by Cotty (1989). The *A. parasiticus* isolates also fell into two groups supported by a bootstrap value of 100, but it will be necessary to analyse more isolates to determine whether this division is significant.

Two toxigenic isolates, FRR 4472 and FRR 4474, could not be separated by RAPD analysis. These were obtained from the same region at the same time and are probably clonemates. The ability of the RAPD analysis to distinguish clonemates is encouraging, as it indicates that identical RAPD profiles can be produced from DNA obtained from separately cultivated and extracted fungi with identical genotypes. Therefore, the observed differences in isolates should be based on genuine differences in DNA sequence, and not the presence of non-reproducible artefact bands, which can sometimes occur in RAPD amplifications (Levitan and Grosberg, 1993; Hadrys, *et al.*, 1992).

Toxigenic and nontoxigenic isolates were interspersed in the two groups of *A. flavus*, and there did not appear to be any clear genetic division between these two phenotypes. This could be explained by multiple, independent losses of toxigenicity by different isolates, or by genetic recombination reassorting this phenotype onto a variety of different genetic backgrounds. Geiser et al (1997) suggested that genetic recombination may occur in *A. flavus*, despite the absence of a known sexual form for this species. In a similar study, cryptic sex was revealed in *Coccidioides immitis,* another fungus previously believed to be strictly asexual (Burt, *et al.*, 1996). It may therefore be that genetic exchange sometimes occurs in fungi in the absence of obvious sexual structures.

The five nontoxigenic isolates of *A. parasiticus* included in this study are the only nontoxigenic ones that have been isolated from Australian soils to date, and are among the few to have been found world-wide. FRR 4467, FRR 4469, FRR 4470 and FRR 4471 do not produce B or G aflatoxins or cyclopiazonic acid (CPA), whereas FRR 4468 produces CPA only. Rigorous analysis of these isolates on different media and employing different growth conditions has failed to detect any aflatoxin, indicating that the nontoxigenic phenotype is stable. All five strains were obtained from the same area at the same time, and it was therefore suspected that these could be clonemates or at least share a very recent evolutionary history. However, RAPD analysis indicated that no two of these isolates are completely identical. In particular, FRR 4467, FRR 4468 and FRR 4469 occur as a cluster which is a considerable genetic distance from FRR 4470 and FRR 4471, which form a second cluster.

To determine the genetic basis behind nontoxigenicity in these isolates, we examined their ability to hybridise with the upper region of the *aflR* gene (Fig. 3). This region contains a zinc finger-like motif thought to be responsible for DNA binding and gene regulation, and appears to regulate the first steps in the aflatoxin biosynthetic pathway (Woloshuk, *et al.*, 1994). As none of the nontoxigenic strains accumulates biosynthetic pathway intermediates, it is likely that failure to make aflatoxin is due to a mutation in the *aflR* gene. Fig. 3 clearly shows that FRR 4470 and FRR 4471 are missing most or all of the upper region of this gene, whereas FRR 4467, FRR 4468 and FRR 4469 have strongly hybridising bands of 0.8 and 3.3 kb which are indistinguishable from the toxigenic isolate, FRR 2503. It is highly likely that the nontoxigenic phenotype observed in FRR 4470 and FRR 4471 is due to the absence of the upper region of the *aflR* gene, but nontoxigenicity must have a different cause in FRR 4467, FRR 4468 and FRR 4469. All of the nontoxigenic isolates of *A. parasiticus* show a significantly lower level of hybridisation than FRR 2503 with a larger, cross-hybridising band, but the significance of this hybridisation in not known. It thus appears that nontoxigenicity has been acquired independently at least twice in *A. parasiticus*. We will now amplify and sequence the *aflR* gene from isolates FRR 4467, FRR 4468 and FRR 4469 to determine whether additional lesions have occurred in this gene. The *aflR* gene will also be analysed in nontoxigenic isolates of *A. flavus*.

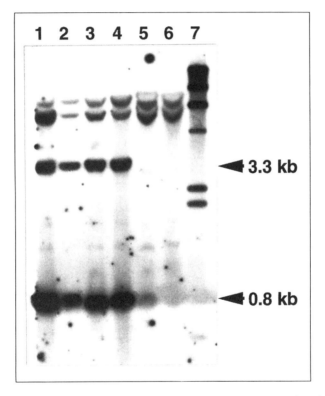

Fig. 3. Autoradiograph of *Bgl*II-digested DNA hybridised with the upper region of the *aflR* gene. Lanes: 1. FRR 2503; 2. FRR 4467; 3. FRR 4468; 4. FRR 4469; 5. FRR 4470; 6. FRR 4471; 7. size standard (λ-*Hind*III).

Within the two groups of nontoxigenic isolates of *A. parasiticus,* the isolates are not identical to one another but are separated by short branches indicating genetic differentiation. FRR 4468 is the most genetically distinct, and is also phenotypically distinguished by its ability to produce CPA. The nontoxigenic isolates within each group were only ever distinguished by minor bands in their RAPD profiles, causing them to be scored as different subtypes. FRR 4470 and FRR 4471 were scored as different subtypes for seven of the 24 RAPD amplifications. FRR 4467 and FRR 4469 differed in only three of the amplifications, whereas FRR 4468 differed from these in eight amplification profiles.

Changes in amplified bands reflect polymorphisms in some of the primer binding sites, which may occur over time through a process of microevolution. Microevolution appears to be common in a number of different fungal species, and has been demonstrated by Pulsed Field Gel Electrophoresis (PFGE) in *Magnaporthe grisea* (Talbot, *et al.*, 1993) by RAPD-PCR in *Candida albicans* (Lockhart, *et al.*, 1995) and by both techniques in *Cryptococcus neoformans* (Fries, *et al.*, 1996; Sullivan, *et al.*, 1996). Microevolution has been suggested as a means by which predominantly asexual fungi can generate genetic variation, and may be very important in fungal evolution (Talbot, *et al.*, 1993; Sullivan, *et al.*, 1996).

If the large deletion of the upper region of the *aflR* gene is responsible for nontoxigenicity as speculated, isolates FRR 4470 and FRR 4471 could make ideal biological control agents as they would be extremely unlikely to revert to toxigenicity. These isolates are also genetically distinct from other isolates included in this analysis, which should enable them to be tracked in the field following their application. Greenhouse and field trials are now necessary to test the efficacy of these isolates in suppressing toxin formation. Further studies will determine whether any of the *A. flavus* isolates possess similar lesions causing nontoxigenicity that are unlikely to revert to the toxigenic phenotype. If so, these may likewise be useful in competitive biocontrol strategies.

ACKNOWLEDGMENTS
We thank Barbara Stephens and Mike Williams for providing the isolates, for their assistance in *Aspergillus* cultivation, and for information on morphological identification and toxin production. Ruiting Lan and Stephen Radajewski helped with data analysis and provided useful feedback on the interpretation of results. We gratefully acknowledge Roche Molecular Systems for their gift of a Perkin Elmer TC480, and for providing the Amplitaq polymerase used this work.

REFERENCES
Bhatnagar, D., Ehrlich, K.C. & Cleveland, T.E. (1992) Oxidation-reduction reactions in biosynthesis of secondary metabolites. In Handbook of applied mycology. (Eds Bhatnagar, D., Lillehoj, E.B. & Arora, D.K.). Marcel Dekker, New York, pp 255-286.
Brown, R.L., Cotty, P.J. & Cleveland, T.E. (1991) Reduction in aflatoxin content of maize by atoxigenic strains of *Aspergillus flavus*. J. Food Prot. 54:623-626.
Burt, A.C., Carter, D.A., White, T.J. & Taylor, J.W. (1996) Molecular markers reveal cryptic sex in the human pathogen *Coccidioides immitis*. Proc. Natl Acad. Sci., USA 93:770-773.
Cotty, P.J. (1989) Virulence and cultural characteristics of two *Aspergillus flavus* strains pathogenic on cotton. Phytopathol. 79:808-814.
Cotty, P.J. (1990) Effect of atoxigenic strains of *Aspergillus flavus* on aflatoxin contamination of developing cottonseed. Plant Dis. 74:233-235.
Cotty, P.J. & Bhatnagar, D. (1994) Variability among atoxigenic *Aspergillus flavus* strains in ability to prevent aflatoxin contamination and production of aflatoxin biosynthetic pathway enzymes. Appl. Environ. Microbiol. 60:2248-2251.
Dorner, J.W., Cole, R.J. & Blankenship, P.D. (1992) Use of a biocompetitive agent to control preharvest aflatoxin in drought stressed peanuts. J. Food Prot. 55:888-892.
Dyer, S.K. & McCammon, S. (1994) Detection of toxigenic isolates of *Aspergillus flavus* and related species on coconut cream agar. J. Appl. Bacteriol. 76:75-78.
Erlich, K. (1987) Effect on aflatoxin production of competition between wild-type and mutant strains of *Aspergillus parasiticus*. Mycopathologia 97:93-96.
Felsenstein, J. (1982) Numerical methods for inferring evolutionary trees. Quart. Rev. Biol. 57:379-404.

Fries, B.C., Chen, F.Y., Currie, B.P. & Casadevall, A. (1996) Karyotype instability in *Cryptococcus neoformans* infection. J. Clin. Microbiol. 34:1531-1534.

Geiser, D.M., Pitt, J.I. & Taylor, J.W. (1997) Cryptic speciation and recombination in the aflatoxin producing fungus *Aspergillus flavus*. Proc. Natl. Acad. Sci. USA 95: 388-393

Hadrys, H., Balick, M. & Schierwater, B. (1992) Applications of Random Amplified Polymorphic DNA (RAPD) in molecular ecology. Molec. Ecol. 1:55-63.

Klich, M.A. & Cleveland, T.E. (1998) *Aspergillus* systematics and the molecular genetics of mycotoxin biosynthesis. In Integration of modern taxonomic methods for *Penicillium* and *Aspergillus* classification (Eds. Samson, R.A. & Pitt, J.I). Harwood Publishers, Amsterdam, pp. 425-434.

Klich, M.A. & Pitt, J.I. (1988) Differentiation of *Aspergillus flavus* from *A. parasiticus* and other closely related species. Trans. Br. Mycol. Soc. 91:99-108.

Lee, S.B. & Taylor, J.W. (1990) Isolation of DNA from fungal mycelia and single spores. In PCR Protocols. (Eds Innis, M.A., Gelfand, D.H., Sninsky, J.J. & White, T.J.). Academic Press, San Diego, pp 282-287.

Levitan, D.R. & Grosberg, R.K. (1993) The analysis of paternity and maternity in the marine hydrozoan *Hydractinia symbiolongicarpus* using Randomly Amplified Polymorphic DNA (RAPD) markers. Molec. Ecol. 2:315-326.

Lockhart, S.R., Fritch, J.J., Meier, A.S., Schroppel, K., Srikantha, T., Galask, R. & Soll, D.R. (1995) Colonizing populations of *Candida albicans* are clonal in origin but undergo microevolution through C1 fragment reorganization as demonstrated by DNA fingerprinting and C1 sequencing. J. Clin. Microbiol. 33:1501-1509.

Moody, S.F. & Tyler, B.M. (1990a) Restriction enzyme analysis of mitochondrial DNA of the *Aspergillus flavus* group: *A. flavus, A. parasiticus* and *A. nomius*. Appl. Environ. Microbiol. 56:2441-2452.

Moody, S.F. & Tyler, B.M. (1990b) Use of nuclear DNA RFLPs to analyse the diversity of the *Aspergillus flavus* group: *A. flavus, A. parasiticus* and *A. nomius*. Appl. Environ. Microbiol. 56:2453-2461.

Payne, G.A., Nystrom, G.J., Bhatnagar, D., Cleveland, T.E. & Woloshuk, C.P. (1993) Cloning of the *afl-2* gene invloved in aflatoxin biosynthesis from *Aspergillus flavus*. Appl. Environ. Microbiol. 59:156-162.

Sullivan, D., Haynes, K., Moran, G., Shanley, D. & Coleman, D. (1996) Persistence, replacement, and microevolution of *Cryptococcus neoformans* strains in recurrent meningitis in AIDS patients. J. Clin. Microbiol. 34:1739-1744.

Talbot, N.J., Salch, Y.P., Ma, M. & Hamer, J.E. (1993) Karyotypic variation within lineages of the rice blast fungus, *Magnaporthe grisea*. Appl. Environ. Microbiol. 59:585-593.

Tran Dinh, N., Pitt, J.I. & Carter, D.A. (1999) Molecular genotype analysis of natural toxigenic and nontoxigenic strains of *Aspergillus flavus* and *Aspergillus parasiticus*. Mycoll. Res. 103: 1485-1490

Welsh, J. & McClelland, M. (1990) Fingerprinting genomes using PCR with arbitrary primers. Nucl. Acids Res. 18:7213-7218.

Welsh, J., Ralph, D. & McClelland, M. (1995) DNA and RNA fingerprinting using arbitrarily primed PCR. In PCR Strategies. (Eds Innis, M.A., Gelfand, D.H. & Sninsky, J.J.). Academic Press, San Diego, pp 249-276.

Williams, J.G., Kubelik, A.R., Livak, K.J., Rafalski, J.A. & Tingey, S.V. (1990) DNA polymorphisms amplified by arbitrary primers are useful as genetic markers. Nucl. Acids Res. 18:6531-6535.

Woloshuk, C.P., Foutz, K.R., Brewer, J.F., Bhatnagar, D., Cleveland, T.E. & Payne, G.A. (1994) Molecular characterization of *aflR*, a regulatory locus for aflatoxin biosynthesis. Appl. Environ. Microbiol. 60:2408-2414

GENETIC VARIATION AND AFLATOXIN PRODUCTION IN *ASPERGILLUS TAMARII* AND *A. CAELATUS*

Stephen W. Peterson[1], Bruce W. Horn[2], Yoko Ito[3], and Tetsuhisa Goto[4]
[1]National Center for Agricultural Utilization Research, USDA, ARS, Peoria, IL 61604; [2] National Peanut Research Laboratory, USDA, ARS, Dawson, GA 31742; [3] National Research Institute for Vegetables, Ornamental Plants and Tea, Ministry of Agriculture Forestry and Fisheries, Kanaya, Shizuoka 428 Japan; [4]National Food Research Institute, Ministry of Agriculture Forestry and Fisheries, Kannondai, Tsukuba 305 Japan.

A molecular genetic study was performed to test whether *A. tamarii* is composed of different genotypes that correspond to the named species, and to determine whether *A. tamarii* strains producing aflatoxins are correctly identified. Phylogenetic analysis showed that *A. tamarii* and *A. caelatus* are distinct species. The strains of *A. tamarii* producing aflatoxins were found to be variants of *A. caelatus* rather than *A. tamarii*. *A. caelatus* strains could mistakenly be identified as *A. tamarii*

INTRODUCTION

Aflatoxins are potent carcinogenic compounds that occur in nature as metabolites of *Aspergillus flavus* Link, *A. parasiticus* Speare and *A. nomius* Kurtzman *et al.* (Cotty *et al.*, 1994; Scholl and Groopman, 1995)). Taxonomically, these three species belong in *Aspergillus* section *Flavi* (Gams *et al.*, 1985) that are phylogenetically related (Kurtzman *et al.*, 1987; Egel *et al.*, 1994). In addition to the aflatoxigenic species, section *Flavi* includes *A. tamarii*, which does not produce aflatoxin, with the exception of a single report by Goto *et al.*, (1996) who reported the production of aflatoxins B_1 and B_2, and cyclopiazonic acid (CPA). However, this single aflatoxin-producing strain of *A. tamarii*, isolated from tea-field soil in Japan, was atypical in terms of the metabolites and numerous sclerotia it produced.

Skill is required to identify isolates of *Aspergillus* in section *Flavi* solely on the basis of morphology (Klich and Pitt, 1988), and it is not always reliable for *Aspergillus* identification (Peterson, 1995). Molecular genetic data may be needed to provide unequivocal identifications of species in this section (Chang *et al.*, 1995; McAlpin and Mannarelli, 1995; Yuan *et al.*, 1995; Shapira *et al.*, 1996).

Because *A. tamarii* has not previously been shown to be aflatoxigenic, we initiated a molecular systematic study of Japanese tea-field soil isolates (Ito and Goto, 1994) and some reference isolates of *A. tamarii* to determine whether the aflatoxigenic strain was properly identified. We included isolates of *A. caelatus*, because this recently described species (Horn, 1997), isolated from agricultural soils in Georgia, U.S.A., would have previously been classified as *A. tamarii*. We also examined isolates whose vegetative compatibility group had been determined to see whether the aflatoxigenic strain might represent a different compatibility group from the non-aflatoxigenic strains.

MATERIALS AND METHODS

A total of 153 isolates of *A. tamarii* were examined. They were isolated at different times between 1913 and 1996 from locations in Africa; North, Central and South America; Southeast Asia; and the Indian subcontinent (Table 1). Forty-three isolates of *A. caelatus* from soil in Georgia (USA) and Japan were also studied.

The culture collection accession numbers, GenBank sequence numbers, isolation data, and identification of the fungal isolates examined are presented in Table 1. These isolates are maintained as lyophilized preparations in the Agricultural Research Service Culture Collection (NRRL), Peoria, Illinois, USA. In addition six strains from soil in Japanese tea fields, identified as *A. tamarii* (Goto *et al.*, 1997) were examined. Morphological comparisons were carried out on Czapek and malt extract agar after incubation in the dark at 25°C for 7 and 14 days.

DNA isolation: Fungal isolates were revived from lyophilized storage and grown on slants of malt extract agar (Raper and Fennell, 1965) for seven days to provide a source of inoculum. One ml of sterile 0.1% Triton X-100 was added to each slant, and conidia were dislodged from the agar surface with a sterile inoculating wire. The conidial suspension was pipetted into a 500-ml Erlenmeyer flask containing 100 ml of malt broth (Raper and Fennell, 1965) and incubated on a rotary shaker platform (200 rpm, 25°C) for 36-48 h, until 1-2 g of biomass was produced. The mycelium was harvested by filtration over cheesecloth or filter paper, then portions (0.2 g) were suspended in 3 ml of breaking buffer (100 mM Tris, 50 mM EDTA, 1% sarcosyl, pH 8.0) in a 15-ml screw-cap disposable centrifuge tube, and 1.5 g of glass beads (0.5 mm diam) was added. Cell walls were broken by vortexing for 30-45 seconds. Phenol-chloroform (1:1, w:v) was added to the tube (3 ml) and an emulsion was formed by gentle rocking for 15 minutes. The aqueous and organic phases were separated by low speed centrifugation (*ca.* 2000 x g) for 5 minutes. The aqueous phase was transferred to a fresh tube, and 0.1 volume of 3 M sodium acetate pH 6.0 and 1.3 volumes 95% ethanol were added. The contents of the tube were mixed, and the nucleic acids were pelleted by low speed centrifugation as above. Nucleic acids were dissolved in TE/10 (1 mM Tris, 0.1 mM EDTA, pH 8.0) and further purified by adsorption to a silica matrix (Geneclean, Bio101, LaJolla, California USA) according to the manufacturer's instructions. DNA was desorbed from the matrix in TE/10 and stored frozen (-20°C) until used for PCR amplification. DNA was prepared from some strains for use in DNA complementarity testing, using the methodology of Peterson (1992). Briefly, 1 l of malt broth in a Fernbach flask was inoculated with conidia from a slant culture and grown for 2 days (25°C, 200 rpm). Mycelium was harvested by filtration over cheese cloth, suspended in DNA breaking buffer, and passed through a French pressure cell.

Table 1. Fungal isolates (NRRL Culture Collection) examined in this study. All isolates within a sequence group have identical ITS and lsu region sequences.

Aspergillus tamarii Sequence[3] group 1

20818 ex lectotype, from activated carbon, 1913
427 from tomato, MD, 1914
428 from insect, 1941
429, 430 from soy sauce, China, 1917
441 Sao Paulo, Brazil, 1939
559 From Illinois, 1940
572 New Guinea, 1940's
1252 from soil, Panama, 1941
1654 1940
4910 Received as ex type of *A. effusus* var. *furcata*
4911 Ex type culture of *A. flavofurcatis*, from soil, Brazil
4960 Received as ex type of *A. parasiticus* var. *rugosus*
13139 from rodent cheek pouch, Arizona, 1982
25399 25400 25401 from soil, Japan, 1993
25565, 25570, 25628, 26004, 26005, 26006 from soil, Japan, 1995
25579, 25626 from soil, Japan, 1996
26011, 26012 from soil, South Texas, 1992
26013, 26014 from soil, Arizona, 1992
26016, 26018 from soil, Mississippi, 1992
26070, 26076, 26080, 26084, 26086, 26087, 26089, 26092, 26097 (all VCG 2) from soil, Georgia, 1992
26079, 26081, 26082, 26083, 26093, 26094 (all VCG 3), from soil, Georgia, 1992
26099 VCG 3 from peanut seed, Georgia, 1992
26071 VCG 4 from soil, Georgia, 1992
26077 VCG 5 from soil, Georgia, 1992
26242, 26243, 26244, 26245, from soil, Venezuela, 1944
26247 from tomato seed, New Jersey, 1945
26248 from cellophane, New Guinea, 1945
26249 from cloth, Hawaii, 1945
26250 from tentage, Hawaii, 1945
26251, 26254, 26255, 26257, 1945
26252 from shoe leather, Guadalcanal, 1945
26253 Salinas, California, 1945
26256 from soil, Nicaragua, 1945
26258 from soil, Brazil, 1945
26259 from rope, Florida, 1946
26260 from nylon hammock, Florida, 1946
26261 from cotton cloth, Florida 1946
26262 from canvas, Florida, 1946
26263 New Jersey, 1946
26264, 26265 Thailand, 1946
26266 from black ink, Pennsylvania, 1947
26267 New York, 1948
26268, 26269, 26270, 26271, 26272, from soil, Brazil, 1949
26273, 26274, 26275, 26276, 26277 England, 1949

26297, 26298 Puerto Rico, 1961
26299 Kansas, 1968
26300, 26301 Germany, 1968
26304 from cottonseed, Texas, 1969
26305 Air sample, Pacific, 1970
26306, 26307 from peanuts, Texas, 1970
26308 India, 1971
26309 from paddy rice, Malaysia, 1971
26311 India, 1972
26312, 26313 from castor seed, India, 1972
26315 from corn, Missouri, 1972
26316 from soil, Arizona, 1973
26317 SE Asia, 1973
26318 from corn, 1973
26319 from mitral valve prosthesis, South Carolina, 1974
26320 Nasal infection, Georgia, 1975
26321 Received as an ex type culture of *A. flavus* mut. *rufus*
26322 from apple, California, 1977
26323, 26324 from import dock coffee beans (Louisiana), 1977
26325 Air sample, Illinois, 1977
26326 from fermented food, Malaysia, 1979
26328 from rodent cheek pouch, Arizona, 1979
26329 from seed, Arizona, 1979
26330 Illinois, 1979
26331 from fermented food, Indonesia, 1979
26333 from corn, Georgia, 1982
26334 from rodent cheek pouch, Arizona, 1983
26337 from seed, 1984
26338 Georgia, 1988
26339, 26340 from peanut, Uganda, 1988

Aspergillus tamarii Sequence group 2

26066, 26068, 26072, 26074, 26075, 26073, 26069, 26067, 26078 26085, 26088, 26090, 26091, 26095, 26096 (all VCG 1), from soil, Georgia, 1992
26098 VCG 1 from peanut seed, Georgia, 1992

Aspergillus caelatus Sequence group 1

25528 VCG 1 Ex type culture, from soil, Georgia, 1992
25404 from soil, Japan, 1994
25566, 25568, 25571, from soil, Japan, 1995
25576, 25577 from soil, Japan, 1996
26015 from soil, Louisiana, 1993
26017 from soil, Mississippi, 1993
26100, 26104, 26105, 26106, 26107, 26109, 26110, 26111, 26112, 26113, 26114, 26120, 26121, 26122, 26123 (all VCG 1) from soil, Georgia, 1992
26115, 26116, 26118 VCG 2 from soil, Georgia,

449

26278 from soil, Gold Coast, 1950	1992
26279 Louisiana, 1950	26127 VCG 2 from peanut seed, Georgia, 1992
26280 from cassava wash water, Brazil, 1950	26124 VCG 3 from soil, Georgia, 1992
26281 from soil, Africa, 1950	26126 VCG 3 from peanut seed, Georgia, 1992
26283 from bran, Maryland, 1952	26130 VCG 3 from peanut seed, Georgia, 1992
26284 Ohio, 1952	26101 VCG 4 from soil, Georgia, 1992
26285 1952	26102 VCG 5 from soil, Georgia, 1992
26286 Sudan, 1953	26103 VCG 6 from soil, Georgia, 1992
26287 from 1% morphine solution, 1953	26108 VCG 7 from soil, Georgia, 1992
26288 from ham, Illinois, 1956	26117 VCG 8 from soil, Georgia, 1992
26289 Alabama, 1957	26119 VCG 9 from soil, Georgia, 1992
26290 from soil, Jamaica, 1958	26125 VCG 10 from peanut seed, Georgia, 1992
26291 from bamboo, Illinois, 1959	26306 from peanut, Texas, 1970
26292 from cacao beans, 1960	
26293, 26294, 26295. 26296 from soil, Pakistan, 1960	*Aspergillus caelatus* **Sequence group 2**
	25517 from tea-field soil, Japan, 1993
26128, 26129 VCG 1 from peanut seed, Georgia, 1992	443 Brazil, from De Fonseca, 1923

[1] NRRL is the Agricultural Research Service culture collection accession number. [2]VCG is vegetative compatibility group. ; [3]Sequences were deposited for one representative of each sequence group, and their GenBank accession numbers are: *Aspergillus tamarii* NRRL 20818 = AF004929; NRRL 26066 = AF004932; *Aspergillus caelatus* NRRL 25528 = AF004930; NRRL 443 = AF004931.

Proteins were extracted with phenol:chloroform (1:1 w:v) and sodium perchlorate. DNA was further purified by density gradient ultracentrifugation in CsCl, sheared by passage through a French pressure cell, and dialyzed in 0.1X SSC buffer (Peterson, 1992). DNA was melted and reassociated in the thermal cell of a Gilford Response II spectrophotometer and percent DNA complementarity calculated (Peterson, 1992).

PCR amplification and sequencing: A 1200 nucleotide fragment of DNA that includes the internal transcribed spacer regions (ITS1, ITS2), the 5.8S rDNA, and about 600 bases from the 5' end of the lsu rDNA was amplified from the genomic DNA using PCR. The PCR amplification tubes contained 5 µl of genomic DNA, 10 µl of 10X buffer (White *et al.*, 1990), 1 µl of deoxynucleotide triphosphate mix (each dNTP present in a master mix at 1 mM concentration), 1 µl of primer ITS1 (50 µM solution), 1 µl of primer D2R (50: M solution; Table 2), 0.5 µl of Taq polymerase (5 U per µl) and 81.5 µl of sterile distilled water. The solution was overlaid with *ca.* 50 µl of mineral oil and amplified during 30 thermal cycles of 96°C, 30 sec; 51°C, 30 sec; 72°C, 150 sec, followed by 10 minutes at 72°C. The amplified DNA fragments were purified by adsorption to a silica matrix (Geneclean), desorbed into TE/10, and stored frozen (-20°C) until used in sequencing reactions. DNA sequences were determined using Applied Biosystems (ABI) DyeDeoxy sequencing kits (fluorescent labelling, Taq polymerase) and the ABI 373 DNA sequencer. Sequence reaction conditions were those recommended by the manufacturer, using primers ITS1, ITS2, ITS3 and ITS4 and primers D1, D1R, D2 and D2R (Table 2). These primers allow sequencing of both strands of the entire fragment. The DNA sequences of the isolates were aligned using an ASCII text editor and compared using programs in PAUP 3.1.1 (Swofford, 1993) and from the PHYLIP package (Felsenstein, 1993).

Table 2. Primers used for amplification and sequencing. The ITS primers are described in White et al. (1990), and the lsu primers are described in Peterson (1993).

ITS 1 5'-TCCGTAGGTGAACCTGCGG
ITS 2 5'-GCTGCGTTCTTCATCGATGC
ITS 3 5'-GCATCGATGAAGAACGCAGC
ITS 4 5'-TCCTCCGCTTATTGATATGC

D1 5'-GCATATCAATAAGCGGAGGA
D1R 5'-ACTCTCTTTTCAAAGTGCTTTTC
D2 5'-GAAAAGCACTTTGAAAAGAGAGT
D2R 5'-AACCAGGCACAAAGTTCTGC

RESULTS

The majority of the isolates were correctly identified as *A. tamarii* by the original depositors, and the nucleotide sequence of most isolates (137) was identical to that obtained from NRRL 20818, an *ex type* culture of *A. tamarii*. Five vegetative compatibility groups (VCGs) were identified among the 34 *A. tamarii* strains isolated in Georgia (Horn and Greene, 1995). Strains from VCG 2,3,4, and 5 had DNA sequences identical to each other and NRRL 20818. Each of sixteen strains identified as *A. tamarii* VCG 1 had identical DNA sequences, but that sequence differed from the *ex type* DNA sequence at a single base position in the ITS-2 region. All of the strains with the minority sequence were isolated in 1992 from a single field in Georgia.

Forty-one isolates of *A. caelatus* had a DNA sequence identical to that of the *ex type* culture, NRRL 25528. The *A. caelatus* DNA sequence in the amplified fragment differs from the *A. tamarii* sequence at five nucleotide positions, two in the ITS-1 region, one in the ITS-2 regions and two substitutions in the D2 region. Six strains from Japanese tea-field soil that had initially been identified as *A. tamarii* (Goto *et al.*, 1997) or aberrant *A. parasiticus* strains had sequences identical to the *A caelatus* sequence.

Re-examination of isolates of *A. caelatus* showed that the colonies were greener (serpentine green) than those of *A. tamarii* (tawny-olive to Saccardo's umber) and numerous sclerotia were produced in most *A. caelatus* isolates. The colony reverse of the *A. caelatus* cultures (on Cz) were flame-scarlet to orange-chrome, while that of *A. tamarii* was carnelian-red to rufous or uncoloured. Colony and conidium morphology were consistent with the description of *A. caelatus*, together with the failure to produce cyclopiazonic acid.

Two isolates of *A. tamarii* examined in this study produced aflatoxin B_1 and B_2 as well as cyclopiazonic acid. One isolate was that of Goto *et al.* (1996), and the other was from the NRRL Culture Collection, obtained from Brazil in the 1920's as a culture of *A. wentii* (Goto *et al.*, 1997).

A.

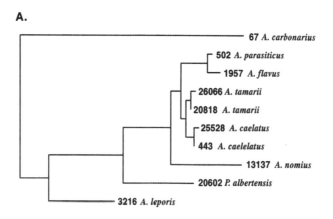

B.

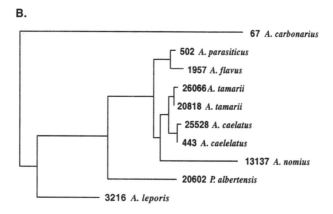

Fig. 1A and B. Phylograms produced using PAUP 3.1.1, and sequence data from ITS1, ITS2, 5.8S rDNA and 28S rDNA. An exhaustive search of the data produced two equally parsimonious trees. The primary difference between the trees is the insertion point of *A. nomius*, either ancestral *A. flavus*, *A. parasiticus*, *A. tamarii* and *A. caelatus*, or ancestral to only the last two mentioned species. *Aspergillus carbonarius* was chosen as outgroup for this analysis on the basis of a more comprehensive phylogenetic analysis of the genus *Aspergillus* (this volume). NRRL numbers precede species names.

Both of these isolates are now re-identified as *A. caelatus*. In culture, the colony colour is intermediate between that of *A. caelatus* and *A. tamarii*. The conidia are similar to those found in both *A. tamarii* and *A. caelatus*. Colony reverse colour is flame-scarlet to orange-chrome, as in *A. caelatus*. The sequences of these two isolates are identical, and this sequence differs from the sequence of the *ex type A. caelatus* culture at only two nucleotide positions in the ITS2 region. This amount of ITS2 region variability has been found in strain of *Emmonsia crescens* that form ascomata when crossed (Peterson and Sigler, unpublished). Thus we accept that *A. tamarii* strains producing aflatoxins are correctly identified as *A. caelatus* (see also note page 458).

Nucleotide sequences of *A. caelatus*, *A. tamarii*, *A. flavus*, *A. parasiticus*, *A. nomius*, *A. leporis*, *Petromyces albertensis* and *A. carbonarius* were aligned in a single data set

452

and analyzed phylogenetically using PAUP 3.1.1. Exhaustive searching of the data found two equally parsimonious trees (Fig. 1) that differ in the placement of *A. nomius*. In one tree (Fig. 1a) *A. nomius* is ancestral to *A. flavus, A. parasiticus, A. caelatus* and *A. tamarii*, while in the other tree (Fig. 1b) *A. nomius* is ancestral only to the *A. tamarii-A. caelatus* branch. *Aspergillus carbonarius* was chosen as an out-group species on the basis of a more general phylogenetic analysis of the genus *Aspergillus* (Peterson, this volume). Bootstrap analysis was performed with 100 replications, and the bootstrap values of the consensus tree along with branch lengths are shown in Fig. 2.

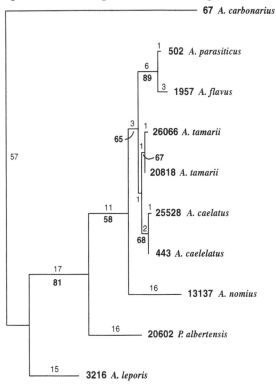

Fig. 2. Phylogram with bootstrap values below the branches and branch lengths above the branches. Bootstrapping was performed using PAUP 3.1.1 with 100 bootstrap samples. The *A. caelatus* and *A. tamarii* clades have bootstrap values of 68 and 67% and show moderately strong support for the branching. The *A. parasiticus* and *A. flavus* branching was supported in 89 bootstrap samples. The low bootstrap value (58%) near the branching of *A. nomius* shows the uncertainty in this data set, concerning the insertion point of *A. nomius* in the tree. These data do not conclusively show whether *A. nomius* is ancestral to *A. flavus, A. parasiticus, A. tamarii* and *A. caelatus*, or just to *A. caelatus* and *A. tamarii*.

Representatives of the two sequence types of *A. caelatus* form a branch found in 68% of the bootstrap samples. Support for this branch is moderately strong although the bootstrap proportion may be an underestimate of the probability that the grouping is true (Hillis and Bull, 1993; Felsenstein and Kinoshino, 1993). This data set needs to be augmented with sequences from more rDNA regions or additional genes in order to obtain complete confidence in the branching. Similarly, the two representatives of *A. tamarii* occur on a branch with a bootstrap value of 67%, and the *A. tamarii* and *A. caelatus* branches form a node supported by 72% of the bootstrap samples. The species producing brown spores form a clade joining the *A. flavus* and *A. parasiticus* clade with support in 65% of the bootstrap samples and *A. nomius* is ancestral to all these species in 58% of the bootstrap samples. The lower bootstrap values for these last two nodes reflect the lack of phylogenetic resolution afforded by these data, regarding whether *A. nomius* or *A. flavus* shares a most recent common ancestor with the brown-spored species.

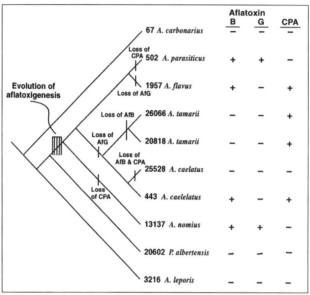

Fig. 3. Cladogram of species from the *Aspergillus* section *Flavi* and their abilities to produce aflatoxins B and G, and cyclopiazonic acid (CPA). Because of its complexity, and its restricted distribution in fungi, aflatoxigenesis evolved only once. Parsimony analysis shows which common ancestor must have produced aflatoxin. Analysis of present day species and their toxin producing abilities allows us to predict at what point in the tree various biosynthetic abilities were lost.

The cladogram in Fig. 3 has aflatoxin and cyclopiazonic acid production data appended to it along with notations about the probable loss of toxigenesis. Cyclopiazonic acid production is useful for identifying strains of *A. caelatus* from those of *A. tamarii*. Most strains of *A. caelatus* (except sequence group 2) do not produce cyclopiazonic acid, whereas most strains of *A. tamarii* do.

Six of the strains studied are *ex type* or authentic strains representing species other than *A. tamarii* (Table 1). The DNA sequence of each of these strains was identical to that of the *A. tamarii ex type* culture.

Six strains of *A. tamarii* (Table 2) displayed 94-99% DNA complementarity with NRRL 20818 (*ex type*), and had DNA nucleotides sequences identical to the *ex type* strain. The strains of *A. tamarii* identified as VCG 1 uniformly showed a single ITS-2 region base difference from the sequence of the *ex type* culture of *A. tamarii*. Because these isolates conform morphologically and physiologically to *A. tamarii* (Horn, *et al.*,

1996), and because single base changes have been noted among sexually compatible strains of *G. fujikuroi* (Peterson and Logrieco, 1991), the strains from VCG 1 are regarded as strains of *A. tamarii*.

Two strains identified morphologically as *A. caelatus* (NRRL 26015 and 26017; Table 2) had sequences identical to the *ex type* culture of *A. caelatus* and 95% intraspecific DNA complementarity. These strains were not compared to the *A. caelatus ex type* strain by DNA hybridization. Ten VCGs were identified among the 32 strains of *A. caelatus* from Georgian agricultural soils and they shared a common DNA sequence identical to the *ex type* strain. The nine additional isolates from Japan, Texas, Louisiana and Mississippi are not identified for VCG type, but they have the same DNA sequence as the other strains. DNA complementarity between *A. tamarii* and *A. caelatus* was *ca.* 65% and the DNA sequences differed from each other at five nucleotide positions.

Table 3. Nuclear DNA complementarity between selected field isolates and NRRL 20818, ex type of *Aspergillus tamarii*.

NRRL Isolate		% DNA Complementarity	NRRL Isolate		% DNA Complementarity
4911	*A. tamarii*	100	26016	*A. tamarii*	94.8
26011	*A. tamarii*	96	26018	*A. tamarii*	99.5
26012	*A. tamarii*	97.3	26015	*A. caelatus*[1]	62.8
26013	*A. tamarii*	98.3	26017	*A. caelatus*	68.1
26014	*A. tamarii*	93.7			

[1] Complementarity of the *A. caelatus* strains 26015 and 26017 with each other was 94.6%.

DISCUSSION

Raper and Fennell (1965) and Kozackiewicz (1989) retained *A. flavofurcatis* as a taxon distinct from *A. tamarii*. The DNA complementarity (100%, Table 2) and the DNA sequence identity between *A. tamarii* and *A. flavofurcatis* found in this study show that these taxa are conspecific. The DNA nucleotide sequences of NRRL 4910 *(ex lectotype strain of A. effusus var. furcata)*, NRRL 4960 (*ex type* culture of *A. parasiticus* var. *rugosus*), NRRL 441 (representative of *A. flavo-viridescens),* and NRRL 26321 (*ex type* culture of *Aspergillus flavus* mut. *rufus*) were identical with that of NRRL 20818, an *ex type* culture of *A. tamarii*. The first two taxa were placed in synonymy with *A. tamarii* by Raper and Fennell (1965) on the basis of morphology-physiology of the strains, a decision that our data supports. The other two strains can be considered synonyms of *A. tamarii* on the basis of the DNA sequence identity with the *ex type* strain of *A. tamarii*.

The two aflatoxin-producing strains of *A. tamarii* collected 70 years apart on different continents present a less clear picture. These strains have identical DNA nucleotide sequences that are most closely related to *A. caelatus,* from which they differ at two ITS-2 region nucleotide positions (Fig. 2); but colonial appearance and conidial morphology indicate, that they could be phenotypic variant strains of *A. tamarii* from which they differ at three ITS region and two D2 region nucleotide positions, or *A. caelatus*. The two strains also produce cyclopiazonic acid, a characteristic associated with *A. tamarii* (Horn *et al.*, 1996). In examining only the lsu sequences, all strains of both *A. tamarii* and *A.*

caelatus have identical sequences. The species differ at two lsu region nucleotide positions. The two aflatoxin producing species have the *A. caelatus* sequence in the lsu region. The fact that these two isolates produce aflatoxins while other *A. caelatus* isolates do not, does not justify describing them as a new taxon. On the basis of the lsu nucleotide sequence identity among the toxigenic strains and strains of *A. caelatus*, and the very few total ITS region sequence differences between them, we identify the aflatoxigenic isolates as variants of *A. caelatus*.

The most parsimonious interpretation of the secondary metabolite data (Fig. 3) is that aflatoxigenesis arose in the most recent common ancestor of *A. nomius, A. flavus, A. parasiticus, A. tamarii* and *A. caelatus*, along with the genes for production of cyclopiazonic acid. Because isolates accurately identified as *A. tamarii* have never been shown to produce aflatoxins, crucial parts of the biosynthetic pathway were probably lost from a recent common ancestor of *A. tamarii*. Strains of *A. caelatus* lack the ability to form aflatoxins, except for the two isolates discussed here. Those two strains represent a lineage that has retained the ability to produce both aflatoxins and cyclopiazonic acid. *A. tamarii* remains a non-aflatoxigenic species. However, the overall gross similarity of *A. tamarii* and the aflatoxigenic *A. caelatus* should serve as a warning that some brown-spored strains from *Aspergillus* section *Flavi* can produce aflatoxins.

REFERENCES

Blackwell, M. (1993). Phylogenetic systematics and Ascomycetes, p. 93-106. *In* D. R. Reynolds and J. W. Taylor (eds.), The Fungal Holomorph: Mitotic, Meiotic and Pleomorphic Speciation in Fungal Systematics. CAB International, Wallingford, United Kingdom.

Brown, D. W., Yu, J.H., Kelkar, H.S., Fernandes, M., Nesbitt, T.C., Keller, N.P., Adams T.H. & Leonard, T.J. (1996). Twenty-five co-regulated transcripts define a sterigmatocystin gene cluster in *Aspergillus nidulans*. Proc. Natl. Acad. Sci. (USA) 93:1418-1422.

Bruns, T. D., White, T.J. & Taylor J.W. (1991). Fungal molecular systematics. Annu. Rev. Ecol. Syst. 22:525-564.

Chang, P.K., Bhatnagar, D., Cleveland, T.E. & Bennett, J.W. (1995). Sequence variability in homologs of the aflatoxin pathway gene *aflR* distinguishes species in *Aspergillus* section *Flavi*. Appl. Environ. Microbiol. 61:40-43.

Cotty, P. J. & Bayman, P. (1993). Competitive exclusion of a toxigenic strain of *Aspergillus flavus* by an atoxigenic strain. Phytopathology 83:1283-1287.

Cotty, P. J., Bayman., P., Egel , D.S. & Elias, K.E. (1994). Agriculture, aflatoxins and *Aspergillus*, p. 1-28. In Powell, K.A., Renwick, A. & Perberdy, J.F. (ed.). The Genus Aspergillus, From Taxonomy and Genetics to Industrial Applications. Plenum Press, NY.

Egel, D. S., Cotty, P.J. & Elias, K.S. (1994). Relationships among isolates of *Aspergillus* Sect. *Flavi* that vary in aflatoxin production. Phytopathology 84:906-912.

Felsenstein, J. (1993). PHYLIP (Phylogeny Inference Package) version 3.5c. Distributed by the author. Department of Genetics, University of Washington, Seattle.

Felsenstein, J. & Kishino, J. (1993). Is there something wrong with the bootstrap on phylogenies? A reply to Hillis and Bull. Systematic Biology 42:193-200.

Gams, W., Christensen, M., Onions, A.H.S., Pitt, J.I. & Samson, R.A. (1985). Infrageneric taxa of *Aspergillus*, p. 55-64. *In*: Samson, R.A. and Pitt, J.I. (eds.) Advances in Penicillium and Aspergillus Systematics, Plenum Press, NY.

Goto, T., Ito, Y., Peterson, S.W. & Wicklow, D.T. (1997). Mycotoxin producing ability of *Aspergillus tamarii*. Mycotoxins 44:17-20.

Goto, T., Wicklow, D.T. & Ito, Y. (1996). Aflatoxin and cyclopiazonic acid production by a sclerotium-producing *Aspergillus tamarii* strain. Appl. Environ. Microbiol.: 62:4036-4038.

Hillis, D. M. & Bull., J.J. (1993). An empirical test of bootstrapping as a method for assessing confidence in phylogenetic analysis. Systematic Biology 42:182-192.

Horn, B. W. (1997). *Aspergillus caelatus*, a new species in section *Flavi*. Mycotaxon 61:185-191.

Horn, B.W. & Greene, R.L. (1995). Vegetative compatibility within populations of *Aspergillus flavus, A. parasiticus,* and *A. tamarii* from a peanut field. Mycologia 87:324-332.

Horn, B.W., Greene, R.L., Sobolev, V.S., Dorner, J.W., Powell, J.H. & Layton, R.C. (1996). Association of morphology and mycotoxin production with vegetative compatibility groups in *Aspergillus flavus, A. parasiticus,* and *A. tamarii.* Mycologia 88:574-587.

Ito, Y. & Goto, T. (1994). *Aspergillus flavus* group fungi isolated from Japanese tea fields. Mycotoxins 40:52-55.

Keller, N. P. & Hohn, T.M. (1997). Metabolic pathway gene clusters in filamentous fungi. Fungal Genetics and Biology 21:17-29.

Klich, M. A. & Pitt, J.I. (1988). Differentiation of *Aspergillus flavus* from *A. parasiticus* and other closely related species. Transactions of the British Mycological Society 91:99-108.

Kozackiewicz, Z. (1989). *Aspergillus* Species on Stored Products. Mycological Papers, No. 161:1-188.

Kurtzman, C.P. & Robnett, C.J. (1991). Phylogenetic relationships among species of *Saccharomyces, Schizosaccharomyces, Debaryomyces,* and *Schwanniomyces* determined from partial ribosomal RNA sequences. Yeast 7:61-72.

Kurtzman, C. P., Smiley, M.J., Johnson, J., Wickerham, L.J. & Fuson, G.B. (1980). Two new and closely related heterothallic species, *Pichia amylophila* and *Pichia mississippiensis*: characterization by hybridization and deoxyribonucleic acid reassociation. International Journal of Systematic Bacteriology 30:208-216.

Kurtzman, C. P., Smiley, M.J., Robnett, C.J. & Wicklow, D.T. (1986). DNA relatedness among wild and domesticated species in the *Aspergillus flavus* group. Mycologia 78:955-959.

Kurtzman, C. P., Horn, B.W. & Hesseltine, C.W. (1987). *Aspergillus nomius,* a new aflatoxin-producing species related to *Aspergillus flavus* and *Aspergillus tamarii.* Antonie van Leeuwenhoek 53:147-158.

McAlpin, C.E. & Mannarelli, B. (1995). Construction and characterization of a DNA probe for distinguishing strains of *Aspergillus flavus.* Applied and Environmental Microbiology 61:1068-1072.

Peterson, S. W. (1992). *Neosartorya pseudofischeri* sp. nov. and it relationship to other species in *Aspergillus* section *Fumigati.* Mycological Research 91:547-554.

Peterson, S.W. (1993). Molecular genetic assessment of relatedness of *Penicillium* subgenus *Penicillium,* p. 121-128. *In* R. Reynolds, D.R. & Taylor, J.W. (eds.), The Fungal Holomorph: Mitotic, Meiotic and Pleomorphic Speciation in Fungal Systematics. CAB International, Wallingford, United Kingdom.

Peterson, S.W. (1995). Phylogenetic analysis of *Aspergillus* sections *Cremei* and *Wentii* based on ribosomal DNA sequences. Mycological Research 99:1349-1355.

Peterson, S.W. & Kurtzman, C.P. (1991). Ribosomal RNA sequence divergence among sibling species of yeasts. System. Appl. Microbiol. 14:124-129.

Peterson, S.W. & Logrieco, A. (1991). Ribosomal RNA sequence variation in strains of *Gibberella pulicaris* and *Gibberella fujikuroi* varieties. Mycologia 83:397-402.

Queiroz, K. & Gauthier, J. (1992). Phylogenetic taxonomy. Annu. Rev. Ecol. Syst. 23:449-480.

Raper, K. B. & Fennell, D.I. (1965). The Genus *Aspergillus.* Williams and Wilkins, Baltimore, MD.

Scholl, P. & Groopman, J.D. (1995). Epidemiology of human aflatoxin exposures and its relationship to liver cancer, p. 169-182. *In* Eklund, M., Richard, J.L. and Mise, K. (eds.) Molecular Approaches to Food Safety: Issues Involving Toxic Microorganisms. Alaken, Inc. Fort Collins, CO.

Shapira, R., Paster, N., Eyal, O., Menasherov, M., Mertt, A. and Salomon, R. (1996). Detection of aflatoxigenic molds in grain by PCR. Appl. Environ. Microbiol. 62:3270-3273.

Swofford, D. L. (1993). PAUP: Phylogenetic Analysis Using Parsimony, Version 3.1.1 Computer program distributed by the Illinois Natural History Survey, Urbana, IL.

Taylor, J. W. (1993). A contemporary view of the holomorph: nucleic acid sequence & computer databases are changing fungal classification, p. 3-14. In Reynolds, D.R. and Taylor, J.W. (eds.), The Fungal Holomorph: Mitotic, Meiotic & Pleomorphic Speciation in Fungal Systematics. CAB International, Wallingford, United Kingdom.

Vilgalys, R. and Hibbett, D.S. (1993). Phylogenetic classification of fungi & our Linnaean heritage, p. 255-260. *In* Reynolds, D.R. and Taylor, J.W. (eds.), The Fungal Holomorph: Mitotic, Meiotic & Pleomorphic Speciation in Fungal Systematics. CAB International, Wallingford, United Kingdom.

Wayne, L.G., Brenner, D.J., Colwell, R.R., Grimont, P.A.D., Kvler, O., Krichevsky, M.L., Moore, L.H., Moore, W.E.C., Murray, R.G.E., Stackebrandt, E., Starr, M.P. & Trupper, H.G. (1987). Report of the ad hoc committee on reconciliation of approaches to bacterial systematics. International Journal of Systematic Bacteriology 36:463-464.

White T. J. , Bruns, T.D., Lee, S.B. & Taylor, J.W. (1990). Amplification and direct sequencing of fungal ribosomal DNA for phylogenetics. p. 315-322. In: PCR Protocols: A guide to the methods and applications. Eds., Innes, M.A., Gelfand, D.H., Sninsky, J.J. & White, T.J. Academic Press, New York.

Woese, C. R. (1987). Bacterial evolution. Microbiol. Rev. 51:221-271.

Yuan, G.F., Liu, C.S. & Chen, C.C. (1995). Differentiation of *Aspergillus parasiticus* from *Aspergillus sojae* by random amplification of polymorphic DNA. Appl. Environ. Microb. 61:2384-2387.

NOTE ADDED IN PROOFS: Additional genetic data collected from *A. tamarii, A. caelatus* and aflatoxigenic *A. tamarii* isolates has prompted us to propose a new taxon for *A. tamarii* isolates: *Aspergillus pseudotamarii*

TOLERANCE AND STABILITY OF MAJOR CHROMOSOMAL REARRANGEMENTS IN AN INDUSTRIAL *ASPERGILLUS ORYZAE* STRAIN

Wendy T. Yoder and Deborah C. Lin
Novo Nordisk Biotech Inc., 1445 Drew Avenue, Davis, CA 95616, USA

INTRODUCTION

At Novo Nordisk Biotech we are involved routinely in the transformation of *Aspergillus oryzae* (Ahlburg) Cohn using heterologous genes encoding industrial enzymes such as cellulases, proteases, lipases and amylases. Analysis of a set of such transformants using CHEF (contour-clamped homogeneous electric field) pulsed-field electrophoresis had revealed a high frequency of strains bearing altered chromosomal-sized band mobilities (3/11), relative to the parent strain. This observation prompted the current study to determine the frequency of such events in a larger number of transformants generated using a pyrG- recipient strain and a Lipolase™ vector harbouring the pyrG gene as the selectable marker. We were interested in determining whether such alterations were specifically related to the integration event, and their effects on Lipolase™ yield and morphology. Our observations, that 52% of transformants with resolvable karyotypes bear major alterations (compared with 0% of "mock" transformants), have implications relating to the use of electrophoretic karyotypes in strain identification, studies of parasexual recombination and genome organization and evolution.

MATERIALS AND METHODS

Fungal strains and media: A pyrG mutant of an industrial strain (20-02-09) was generated by plating 3 X 107 conidia on MM containing 1 g/l 5-fluoroorotic acid (FOA) and 10mM uridine. The mutant was spore purified and its genotype confirmed by Southern analysis (Howard Brody, unpublished). It was cultured on minimal medium MM (see below) supplemented with 10 mM uridine. Transformants were selected on regeneration minimal medium (RMM). Long-term storage of parent and transformant strains was in 10% (v/v) glycerol at -140°C.

Minimal medium (MM): per litre - Glucose 25g, 10mM urea, Noble agar 25g, salts 20ml (per liter- KCl 26g, $MgSO_4.7H_2O$ 26g, KH_2PO_4 76g, Trace metals 50 ml (per liter- $Na_2B_4O7.10H_2O$ 400mg, $CuSO_4.5H_2O$ 400mg, $FeSO_4.7H_2O$ 1.2g, $MnSO_4.H_2O$ 700mg, $Na_2MoO_2.2H_2O$ 800mg, $ZnSO_4.7H_2O$ 10g), pH6.

Regeneration Medium (RMM): As MM but 342 g sucrose replacing the 25g glucose. YPGU: per litre - Yeast extract 10g, Bacto peptone 20g, glucose 25g, 10mM urea. PDA:- per litre- Difco PDA 39g.

pJeRS6 Vector: The vector pJeRS6 was constructed by inserting the pyrG gene (Suzie Otani, unpublished), into a Lipolase™ -containing PUC19-based vector. The Lipolase gene was driven by the TAKA promoter (Christensen *et al.*, 1988).

Protoplast Generation: The procedure was a modification of that used by Brody and Carbon, (1989). Five hundred ml plastic shake flasks containing 100ml YPG (supplemented with 10mM uridine when necessary) were inoculated with spore suspensions prepared by adding 10ml sterile distilled water (SDW) to 1 week old PDA cultures and scraping the spores gently from the mycelia. Shake flasks were incubated at 37°C and shaken at 180-200 rpm for 16-18 h. Mycelia were harvested by filtration through sterile Miracloth (Calbiochem, San Diego, CA) and rinsed with 50ml 0.6M $MgSO_4.7H_2O$. The washed mycelia were transferred to 125ml glass shake flasks containing (for CHEF plug preparation) 10ml 7mg/ml Mureinase (US Biochemicals; Catalog # 19278, lot # 74849) in MgP (1.2 M $MgSO_4.7H_2O$, 10 mM $NaH_2PO_4.H_2O$, pH 5.8) or (for transformation) 10ml Novozyme 234 (lot # 5155) in MgP. These were shaken at 80 rpm at 30°C for 1-2h. Protoplast formation was monitored every 20 min until the majority of the mycelium had been converted to protoplasts. The protoplast solution was filtered through sterile Miracloth and the filtrate overlaid carefully with 5 ml ST (0.6 M sorbitol, 100 mM Tris, pH 7) and then spun at 2500 rpm for 15 min. The resulting band of protoplasts was collected with a bent glass pipette and transferred to a new tube, to which 2 volumes STC (1.2M sorbitol, 10mM Tris, 10mM $CaCl_2.2H_2O$, pH7.5) were added. After spinning at 2500 rpm for 5 min the pellet was washed with 5 ml STC, and the washing step repeated.

Transformation : Protoplasts were counted and adjusted to 2 X 107/ml. One hundred ml aliquots of protoplast suspension were transferred to 50ml plastic tubes and 4 ml (uncut) pJeRS6 DNA (2mg) added. Then 200 ml PEG solution (60% PEG 4000 (BDH catalogue # 295764W, lot # 2A1759729 543), 10mM Tris, 10mM $CaCl_2.2H_2O$, pH 8) were added and mixed gently but thoroughly. After 35 min incubation at 34°C, 1 ml STC was added and mixed well. Finally 12 - 15 ml overlay agar (RMM) were added, the tube was inverted gently 3 times and the mixture was poured over prepoured RMM plates (70 ml in 15 cm Petri dish). Plates were incubated at 37°C for 6 days, after which time transformants were picked to MM plates and spore purified 3 times. They were then assayed for Lipolase production and analyzed on CHEF gels.

CHEF Analysis: Protoplasts were resuspended in GMB (0.125M EDTA, pH 8, 0.9M Sorbitol) at a concentration of 2 X 108/ml in an Eppendorf tube and held in a waterbath at 37°C. An equal volume of 1.4% low gelling temperature agarose (FMC) in GMB, which had been pre-cooled to, 42°C was added, mixed gently and the mixture was then quickly but carefully dispensed into plastic molds (Biorad). The molds were placed on ice for 10-15 min and the plugs were then transferred to NDS (0.5M EDTA pH8, 10mM Tris HCl, pH8, 1% sodium N-lauryl sarcosine, 2mg/ml Proteinase K (Amresco, Ohio; catalogue # 0706, lot # 2435A15), (added just before washing plugs) at 50°C for 24h. After this time the plugs were washed 3 times in 50mM EDTA, pH8 at 50°C, then stored at 5°C in 50mM EDTA.

CHEF electrophoresis was performed using gels prepared using 0.8% Sea Kem Gold agarose (FMC) in 0.5X TAE buffer (prepared from a 50X stock, i.e. Tris base 242g/l, glacial acetic acid 57.1ml, 0.5M EDTA (pH8) 100ml). Plugs were sealed into wells using

0.6% agarose. Two gels were run simultaneously (one on top of the other) and the running buffer was 0.5X TAE. All runs were done using a Chef Mapper (Biorad) at 12°C, 1.4V/cm, with a 30 to 50 min linear ramp over 162 h. Gels were then stained with 2mg/ml ethidium bromide in 0.5X TAE for 40 min and destained for 1 h in 1 mM MgSO₄.

Gels were soaked in 0.25M HCl for 15min twice, followed by soaking in 0.5M NaOH, 1M NaCl for 10 min twice. They were rinsed gently in deionized water for 1-2 minutes then soaked in 0.5M Tris HCl pH 7.5 for one 10min period, then for a further 15 min. Gels were rinsed gently with deionized water and the DNA transferred to Nytran membranes (Schleicher & Schuell; Lot No. G255/5; Order No. 77409) overnight (~ 18h) with 5X SSC buffer. After the transfer membranes were rinsed with 2X SSC for 1-2 min, air-dried for 20 min and then the DNA was UV-cross linked to the membrane.

Hybridizations and detection: An internal fragment of the Lipolase gene was DIG-labelled and used as a probe. The probe was denatured by heating for 10 minutes in a boiling water bath and rapidly cooled on ice for 15 minutes. It was then mixed with fresh, pre-warmed (42°C) DIG Easy Hyb solution (Boehringer Mannheim, Cat. No. 1603558) to a final concentration of 2 ng/ml. Blots were pre-hybridized in DIG Easy Hyb solution (Boehringer Mannheim) at 42°C for 2 hours. The filters were hybridized with DIG-labelled probe at 42°C overnight. The probed membranes were washed for 5 minutes in 2 X SSC/0.1% SDS at room temperature and for 15 minutes at 0.5 X SSC/0.1% SDS at 650C twice. Detection was performed according to the Boehringer Mannheim DIG Wash and Block Buffer Set instructions (Cat. No. 1585762). Exposure to X-ray film was for 15-25 minute at room temperature. Filters were stripped, when necessary, by washing twice in 0.4N NaOH, 0.1% SDS at 370C for fifteen minutes each time.

Lipolase assays: Assays for Lipolase™ activity were made on supernatants form shake flask cultures of the parent strain and all transformants using the protocol described in Royer *et al.*, (1995).

RESULTS

CHEF analysis of the parent strain, 20-02-09, resolved six chromosomal-sized bands, two of them probably doublets on the basis of their intensity. Because of the lack of a complete genetic linkage map and telomeric probes for individual chromosomes cannot unambiguously be assigned to the different bands we observed. However, our findings correspond with the observations made by Kitamoto *et al.,* (1994), who resolved seven bands (one of them a doublet), in the same size range as ours, using a different strain (RIB40). These authors assigned at least one of twelve genes to each of the separated bands.

The overall genome size of the parent strain, 20-02-09, was difficult to determine accurately due to the unresolved larger bands (>5.7 Mb), but appears to be close to the 35 Mb estimated for RIB40 (Kitamoto *et al.*, 1994), assuming our largest and smallest bands are doublets. This is in good agreement with the estimated genome sizes of 31 Mb for A. nidulans (Brody and Crabon, 1989) and 35.5-38.5 Mb for *A. niger* van Tieghem (Debets *et al.*, 1990).

Overall genome sizes of transformants bearing CHEF-mobility alterations appeared to be close to that of the parent strain (estimated by visualization of ethidium bromide-stained CHEF gels). However, two transformants (20-10-02 and 20-13-10) may be reduced in size by 3 or 4 Mb, due to the apparent loss of band #4 in both these transformants. Unfortunately no probes specific to band #4 in the parent strain were available to use in hybridizations to determine whether part or all of the missing bands had translocated to or were co-migrating with the largest band.

A range of altered karyotypes was observed (Figs. 1 and 2) including loss of a single band with no obvious alteration in size of any other bands, a single missing band with other bands increased in size (possible translocation), a single band of altered size (smaller or larger), and appearance of new bands and loss of and size alterations in other bands (possible translocations) (Figs. 1 and 2). We assume that the mechanisms responsible for the observed rearrangements include insertions, deletions, duplications and translocations. Because the largest band observed under our parameters is probably a doublet and is of very high molecular weight we would not necessarily see small alterations in size of either of the two (putative) component bands. In addition, lethal alterations would obviously not have been selected following transformation. Our estimates of alterations and possible translocations are therefore conservative.

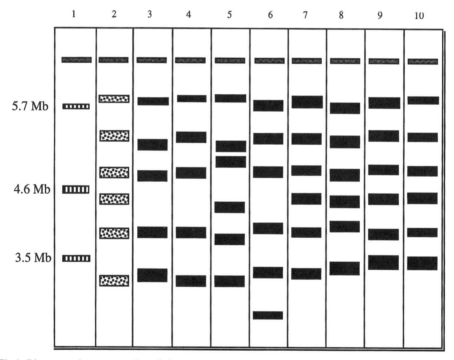

Fig.1. Diagrammatic representation of electrophoretic karyotypes of parent and selected transformants. Lane 1. *Schizosaccharomyces pombe* size markers 5.7, 4.6 and 3.5 Mb; lane 2. 20-02-09 (parent strain); lanes 3-10. examples of electrophoretic karyotypes of different transformants; lane 3. 20-10-02; lane 4. 20-13-10; lane 5. 20-10-4; lane 6. 20-10-03; lane 7. 20-13-03; lane 8. 20-13-14; lane 9. 20-13-21; lane 10. 20-13-23.

462

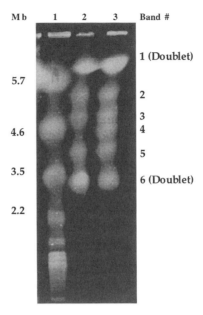

Fig. 2. Example of karyotype alteration in transformant. Lane 1, *S. pombe* and *S. cerevisiae* size markers; Lane 2, Transformant 20-10-02 (missing band #4); Lane 3, Untransformed parent 20-02-09.

No significant variation in karyotype patterns for any of six strains over repeated runs was observed, suggesting that the altered band mobilities we observed were not artifactual. The transformants used to generate the above data were obtained in two independent transformations. In both experiments the same percentages of transformants examined bore chromosome mobility changes (50% (4/8) in experiment 20-10 and 53.5% (15/28) in experiment 20-13).

Of the 36 transformants, which gave clearly resolvable karyotypes, 19 (52%) had undergone obvious rearrangements. In order to determine whether these rearrangements had been induced by the protoplasting procedure itself, mock transformants were generated (as for experimental transformations but without vector DNA) and after selection on RMM plus 10mM uridine, twenty were spore purified three times and agarose-embedded protoplasts prepared for CHEF analysis. None of the 20 mock transformants analysed were found to bear any kind of chromosomal rearrangement (data not shown).

Three selected rearranged transformants (20-10-02, 20-10-03 and 20-13-10) were mitotically stable as evidenced by unaltered electrophoretic karyotypes after four serial subcultures on selective (MM plus chlorate) and non-selective (potato dextrose agar (PDA)) media (data not shown).

All transformants were phenotypically indistinguishable from the parent strain and grew at the same linear extension rate and with the same colony morphology as the parent strain on MM and PDA.

Hybridizations to the Lipolase™ probe (Table 1) revealed that in 35% of all transformants the Lipolase gene had integrated to the largest band (band #1), in 26% of transformants it had integrated to the smallest band (band #6) and in 23% it had integrated into band #4. If only those transformants bearing clear rearrangements were included in the analysis (Table 1) the corresponding frequencies were 22.3, 27.7 and 44.5%, suggesting a strong correlation between integration to band #6 and altered karyotype. Seventy seven percent of all integrations to band #6 resulted in size alterations in this band and 44% of integrations to band #6 resulted in alterations in band # 1 and band #6. A correlation was also observed between integration to band #4 and size alterations in band #6 (62.5%) (data not shown).

Table 1. Frequency Of Integration Of pJeRS6 To Different CHEF Bands In Transformants

CHEF Band Number	Number Of Integration Events (All Transformants)	% Integration Events (All Transformants)	Number Of Integration Events (Altered Transformants)	% Integration Events (Altered Transformants)
1 (Doublet)	12[1]	35.2	4[1]	22.3
2	1	3	0	0
3	1	3	0	0
4	8	23.5	5	27.7
5	3	8.9	1	5.5
6 (Doublet)	9[1]	26.4	8[1]	44.5
n	34[1][3]	100	18[1][2]	100

(1) In one transformant (20-10-03) the probe hybridized to band #1 and band #6. (2) Two of the 19 altered transformants did not hybridize to the Lipolase probe under the conditions tested. (3) Of thirty six transformants which gave clearly resolvable karyotypes three did not hybridize to the Lipolase probe under the conditions tested.

Reprobing of one membrane with the pyrG gene revealed identical hybridization patterns to those seen using the Lipolase™ probe, suggesting the two markers (pyrG and Lipolase™) had integrated at the same locus.

Lipolase™ yields of transformants subjected to CHEF gel electrophoresis ranged from 1 to 30 relative units, with no correlation being evident between Lipolase™ yield and integration site or Lipolase™ yield and any particular type of alteration.

DISCUSSION

Our results demonstrate that integrating heterologous DNA can cause alterations in the mobility of chromosomal-sized bands of transformants, without causing any obvious deleterious effects on colony morphology, growth rate or Lipolase secretion. This phenomenon may be due to initial integration events and/or subsequent rearrangements of the integrated DNA, such as mitotic recombination between dispersed copies of transforming DNA (in transformants carrying multiple ectopic integrations).

There are several reports in the literature of chromosome rearrangements resulting from the presumably traumatic process of transformation, in both sexual and asexual fungi. Asch *et al.* (1992), compared the junctions of *N. crassa* Shear & B.O.Dodge transformants bearing ectopic integrations with the wild type strain and found that the junctions appeared to have resulted from the end joining of truncated vector sequences to breaks in chromosomes (in one case bridging sequences from two different linkage groups), without the loss of chromosomal DNA. Kistler and Benny (1992) found that during integrative transformation of *Nectria haematococca* Berk & Br. deletions of > 1Mb occurred and that these appeared to be non-essential for growth, at least in artificial culture.

One report has documented the same phenomenon we have observed in *A. oryzae*, in *Aspergillus nidulans* (Eidam)Vuill. Xuei and Skatrud (1994) observed similar variation in the electrophoretic karyotypes of transfomants generated using a vector carrying a hygromycin B phosphotransferase gene. Their data indicate that selected bands are more susceptible to alteration - one of their bands (#1 (the largest band)) is involved in 3/8 (37.5%) of rearrangements and another (#3) is involved in 5/8 (62.5%) of rearrangements. Their transforming vector contained a 1.6 kb fragment of DNA from the untransformed parent strain, which may have influenced such targeting and/or been a contributing factor to the generation of novel karyotypes. These authors presented no hybridization data so it is not known whether they saw any correlation between integration site and rearrangement.

In a rigorous, quantitative genetic study, Perkins *et al.* (1993) found chromosome rearrangements in >10% of mitotically stable transformants of *Neurospora crassa* using an assortment of markers, vectors and recipient strains. Breakpoints were randomly distributed among the seven linkage groups and segments of transforming DNA were closely linked to breakpoints in many of the rearrangements. They also determined that when homologous integration events restored the resident locus to wild type, no or very few rearrangements were observed. However, frequent rearrangements were found when multiple ectopic events had occurred.

Kistler and Miao (1992) have suggested that deletions, duplications and translocations (as well as the presence of dispensable chromosomes) may all be a source of genomic variation in filamentous fungi. Perhaps mitosporic fungi such as *A. oryzae,* which are not considered to be susceptible to processes such as RIPing (Repeat-Induced Point mutation) and MIPing (Methylation Induced Pre-meiotically), are also more tolerant than meiotically-active species, of the type of major karyotypic alterations we observed in the present study, whether such alterations are induced artificially (e.g. by transformation or UV-mutagenesis) or in response to environmental stimuli or as yet unidentified mutagenic factors.

Interestingly a high degree of targeting was observed in the current study to bands which were previously the sites of the pyrG gene in these strains. I.e. in sixty three percent (21/33) of rearranged transformants the pyrG / Lipolase cassette had integrated to either band #1 (35%) (site of the pyrG gene in the progenitor of the parent strain) or to band #6 (26%) (site of the pyrG gene in the strain from which the latter strain was ultimately derived). It should be borne in mind however that since our CHEF parameters did not resolve bands 1 and 6 into their component doublets, more convincing evidence of targetting would require resolution of bands 1 and 6 and southern hybridizations of filters made from gels run under such conditions. It is also possible that targetting to the TAKA promoter may have occurred, since it is not known whether the TAKA promoter was deleted in the generation of the progenitor of strain 20-02-09.

Our data suggest a non-random distribution with respect to the altered bands, i.e. a predominance of size alterations in bands 1 and 6 and of size alterations and/or disappearance of band 4. These are presumably related to the correlations described above (see Results section) with integration site. A correlation between bands 1 and 6 is not unexpected given that both these bands formerly harbored pyrG sequences. However the relationship between these bands and band #4 cannot be explained at the present time. We do not have a sufficiently extensive collection of chromosome-specific markers, which would allow us to identify different types of translocation or rearrangement.

Presumably endonucleases and repair and recombination enzymes will have effects on the host DNA in addition to modifying the extraneous DNA, on its introduction to the fungal cells. It would be of interest to determine whether the use of purely homologous transforming DNA (e.g. linearized and gel purified fragments bearing only homologous selectable markers) would induce rearrangements with the same frequency as did our vector, which was uncut and contained homologous and heterologous sequences. It would also be of interest to determine whether transformations involving different selectable markers, without homology to the recipient strain, would target different, marker-specific chromosomes.

Since our parent strain (20-02-09) was generated in a strain, which was itself, a translocation strain it is possible that it may be predisposed to additional translocations on further transformation and /or mutation.

It is not known whether the type and scale of alterations we observed in derivatives of a single isolate would have any significance with respect to parasexual recombination, but potential differences in electrophoretic karyotype should be considered if unexpected results are obtained in parasexual programmes. Similarly, it would probably be wise not to use electrophoretic karyotypes for identification (fingerprinting) of specific strains, at least from the lineage used in the present study, as has been suggested for *A. flavus* and *A. parasiticus* (Keller *et al.* (1992)). These authors found that six out of six mutant strains, which had been subjected to several rounds of mutagenesis and propagated for several years, exhibit identical karyotypes (Keller *et al.* (1992)).

We did not investigate the extent of intraspecific electrophoretic karyotypic variation between wild type isolates of *A. oryzae*. However, in a study of the closely related *A. flavus*, Keller *et al.* (1992) found that the majority of the twenty isolates they examined had different electrophoretic karyotypes. If the same is true for it would be interesting to determine whether the degree of intra-isolate variation we found is within or exceeds the "natural" limits of intraspecific variation.

It is not known whether the types and frequencies of rearrangements we have observed in transformants of in the laboratory are representative of the types of rearrangements which can occur in nature in response to environmental mutagens such as UV, pesticides, toxins produced by other fungi, plants and animals, and currently unidentified factors contributing to "spontaneous" muatations. If such rearrangements do occur and persist in nature they represent a mechanism for generating significant intra-specific variation and as such may play a substantial role in the speciation process, especially (but not necessarily exclusively) in those species incapable of undergoing meiosis

One means of taking advantage of the phenomenon reported here would be to deliberately induce alterations in electrophoretic karyotype by transforming with selectable markers borne on vectors with homology to specific chromosome-sized bands. (Keller *et al.* (1991) found that, in *Cochliobolus heterostrophus* (Drechsler) Drechler transforming plasmids integrate into chromosomes primarily by homologous recombination if they carry relatively large (>2 kb) regions of homologous DNA). Such an approach could form the basis for a systematic investigation of the limits of tolerance for karyotypic variability. In addition it could allow a comparison to be made of the susceptibility of individual

466

chromosomes to rearrangement. Such an approach could provide valuable information to those studying genome organization and evolution.

REFERENCES

Asch, D.K., Frederick, G., Kinsey, J.A. & Perkins, D.D. (1992). Analysis of junction sequences resulting from integration at nonhomologous loci in *Neurospora crassa*. Genetics 130: 737-48.

Brody, H. & Carbon J., (1989). Electrophoretic karyotype of *Aspergillus nidulans*. Proc. Natl. Acad. Sci. USA. 86: 6260-6263.

Christensen, T., Woeldike, H., Boel, E., Mortensen, S.B., Hjortshoej, K., Thim, L. & Hansen, M.T. (1988). High level expression of recombinant genes in *Aspergillus oryzae*. Biotechnology 6: 1419-1422.

Debets, A.J.M., Holub, E.F., Swart, K., van den Boed, H.W.J. & Bos, C.J. (1990). An electrophoretic karyotype of *Aspergillus niger*. Mol. Gen. Genet. 224: 264-268.

Keller, N.P., Bergstrom, G.C. & Yoder, O.C. (1991). Mitotic stability of transforming DNA is determined by its chromosomal configuration in the fungus *Cochliobolus heterostrophus*. Curr. Genet. 19: 227 - 233.

Keller, N.P., Cleveland, T.E. & Bhatnagar, D. (1992). Variable electrophoretic karyotypes of members of *Aspergillus* Section *Flavi*. Curr. Genet. 21: 371-375.

Kistler, H.C. & Benny, U. (1992). Autonomously replicating plasmids and chromosome rearrangement during transformation of *Nectria haematoccocca*. Gene 117: 81-9.

Kistler, H.C. & Miao, V.P.W. (1992). New modes of genetic change in filamentous fungi. Annu. Rev. Phytopath. 30: 131-152

Kitamoto,K., Kimura, K., Gomi, K. & Kumagai, C. (1994). Electrophoretic karyotype and gene assignment to chromosomes of *Aspergillus oryzae*. Biosci. Biotech. Biochem. 58(8): 1467 - 70.

Perkins, D., Kinsey, J.A., Asch, D.K & Frederick, G.D. (1993). Chromosome rearrangements recovered following transformation of *Neurospora crassa*. Genetics 134: 729-36.

Royer, J.C., Moyer, D.L., Reiwitch, S.G., Madden, M..S., Jensen, E.B., Brown, S.H., Yonker, C.C., Johnstone, J.A., Golightly, E.J., Yoder, W.T. & Shuster, J.R. (1995). *Fusarium graminearum* A3/5 as a novel host for heterologous protein production. Biotechnology 13: 1479-1483.

Xuie, X. & Skatrud, P.L. (1994). Molecular karyotype alterations induced by transformation in *Aspergillus nidulans* are mitotically stable. Current Genetics 26: 225-227.

Chapter 8

PATHOGENIC ASPERGILLI AND PENICILLIA

MOLECULAR TYPING OF *ASPERGILLUS FUMIGATUS*

J.P. Latgé[1], P. E. Verweij[2] and S. Bretagne[3]
[1]Laboratoire des Aspergillus, Institut Pasteur, Paris, France, [2] Department of Medical Microbiology, University Hospital Nijmegen, Nijmegen , The Netherlands and [3]Laboratoire de Parasitologie-Mycologie, Hôpital Henri Mondor, Créteil, France

INTRODUCTION

Aspergillus fumigatus is a saprophytic fungus playing an essential role in recycling the earth carbon and nitrogen (Pitt, 1994). Its natural ecological niche is the soil where it grows on organic debris. Although this species is not the world's most prevalent fungus, it is one of the most ubiquitous fungi present worldwide on most surfaces and aerial environments (Mullins *et al.,* 1976; Nolard, 1994). It is a very thermophilic fungus. It sporulates very abundantly and the conidia released in the atmosphere have a small diameter (2-3 μm) (Samson *et al.,* 1995). Virtually, every cubic meter of air we breathe indoor and outdoor contains spores. It can be estimated that in a normal indoor or outdoor environment every individual inhales 5-500 conidia of *A. fumigatus* per day.

This fungus load is usually harmless since these spores are normally eliminated or at least contained efficiently after inhalation by the innate immune response of the host. For this reason, *A. fumigatus* has been seen for a long time as a weak pathogen responsible for aspergilloma characterized by a limited invasion of preexisting lung cavities in tuberculosis patients. It has been also implicated in allergic disorders observed amongst patients repeatedly exposed to the fungus. This situation has changed dramatically in recent years which has seen an increase in the number of immunosuppressed patients and in the degree of severity of the immunosuppressive therapies. Due to these changes in the hospital practice, *A. fumigatus* has become today the most prevalent airborne fungal pathogen in developed countries (Bodey and Vartivarian, 1989; Dixon and Walsh, 1992). It is responsible for severe and often fatal invasive pulmonary infections amongst hematology patients and specially those undergoing bone marrow transplantation (BMT). It is also seen in AIDS patients, in solid organ transplant recipients, and patients treated with corticosteroids.

This new importance of *A. fumigatus* as a human pathogen has been at the origin of the development of molecular approaches to characterize *A. fumigatus* at an inter- or infraspecific level. This paper will review briefly the non-morphological criteria currently used to identify *A. fumigatus* and more extensively the advantages and disadvantages of the molecular methods available today to type *A. fumigatus* strains.

Non-morphological approaches to the taxonomy of the species *A. fumigatus*
Several biochemical and molecular techniques have been applied to *Aspergillus* in view to better characterize the species *fumigatus*. Although used more widely for species other

than *fumigatus*, these approaches have shown some potential for a better taxonomical definition of the species *fumigatus*. Table 1 presents the different biochemical and molecular criteria which can be used to define taxonomically *A. fumigatus*.

Table 1. Non-morphological tools available to characterize *A. fumigatus*

Methods	Discrimination at the level of		
	Species	subspecies	Strain
Biochemical			
Secondary metabolites	+		
Ubiquinone systems	+		
Multilocus enzyme electrophoresis patterns		+	
Molecular			
DNA / DNA reassociation	+		
ITS sequence	+		
Mitochondrial DNA RFLP	+		
Ribosomal DNA RFLP	+		
Total DNA RFLP		+	
IGS sequence		?	
RAPD pattern			+
Microsatellite pattern			+
Hybridization pattern with repeated DNA sequences			+

Profiles of secondary metabolites produced by *A. fumigatus* and described varieties are homogeneous (Frisvad and Samson, 1990). Isozyme electrophoretic patterns have been examined in the *A. fumigatus* complex by several groups (Lin *et al.*, 1995 ; Matsuda *et al.*, 1992 ; Rinyu *et al.*, 1995 ; Rodriguez *et al.*, 1996) with conflicting results. Mono- and polymorphic patterns have been described. Multilocus enzyme electrophoresis patterns can be used for intraspecific clustering but in no way for strain typing. In addition, isozyme patterns of closely related species have not been investigated which question the usefulness of this technique from a taxonomic point of view. Pulse field gel electrophoresis has not been investigated enough to judge of its taxonomical interest for *A. fumigatus*.

Mitochondrial and ribosomal DNA have shown a good homogeneity within the species *A. fumigatus* (Rinyu *et al.*, 1995). Analysis of the rDNA of *A. fumigatus* has been limited. As with other fungi, sequences of the 18S and 28S subunits do not show enough variability to be informative. Heterogeneity of the IGS region has been detected in *A. fumigatus*, but the absence of sequencing data does not allow to estimate the potential of this region at an infraspecific level. Sequencing of the internally transcribed spacers ITSI and ITS2 has not been completed but sequences already available are sufficiently different to distinguish Neosartorya species and *A. fumigatus* (Chrzavzez, 1995). Another very informative approach has been the analysis and sequencing of the introns of unique genes such as the ß tubulin 2 gene.

DNA/DNA reassociation values higher than 92 % have been found for strains of *A. fumigatus* whereas values <70% have been calculated for *Aspergillus fumigatus* and *Neosartorya* species (Peterson, 1992).

Strain fingerprinting in *A. fumigatus*

The rationale for typing isolates of *A. fumigatus* has a clinical relevance. Due to its presence in the atmosphere, outbreaks of nosocomial invasive aspergillosis (IA) have been often associated in the past with an increase in the concentration of air borne spores resulting, for example, from construction works or deficient ventilation systems (Arnow *et al.,* 1991; Dewhurst *et al.,* 1990; Streifel *et al.,* 1983). The occurrence of such outbreaks has led to the establishment of prophylactic measures. High efficiency air filtration has only reduced the hospital air contamination and the risk of infection during bone marrow transplantation. It has not solved the problem of late infections (Wald *et al.,* 1997). In addition, many IA cases have occurred under laminar flow. As a result, IA remains a major cause of mortality amongst transplant patients in the hospital setting and the source of contamination is still unclear.

This lack of understanding of the reservoir of the infectious inoculum has caused the supplementation of epidemiological studies with genomic typing data to investigate the nosocomiality of *Aspergillus* infections and to identify the source of the inoculum (Fridkin and Jarvis, 1996). A thorough epidemiological study of *A. fumigatus* infections will only be possible if a very precise method of strain fingerprinting is available.

Such a method must be both highly discriminative and reproducible and independent of the environment where the fungus grows. For these reasons, phenotypic methods based on protein patterns detected by antibodies or enzymatic substrates (Burnie *et al.,* 1989; Symoens *et al.,* 1993) should be discarded because they do not fulfil the criteria mentioned above: proteins produced are highly depending on the culture conditions and protein patterns can be used at the most to rank strains at a subspecies level. In contrast, genotypic methods are highly independent of the external milieu. Three methods to fingerprint the DNA of *A. fumigatus* strains have been developed up to date: two use the polymerase chain reaction (PCR) with different types of primers and principles, whereas one is based on restriction fragments length polymorphisms (RFLP) visualized after hybridization with a specific probe.

DNA-RFLP based typing method

RFLP obtained after digestion of total genomic DNA by XbaI, and better SalI and XhoI has shown some degree of discrimination. However, the complex banding patterns displayed in EtBr stained gels are difficult to interpret. In particular, the presence of a large number of faint bands makes the analysis of a high number of strains very subjective (Burnie *et al.,* 1992; Denning *et al.,* 1990). Hybridization of restriction enzyme fragments with repeated DNA sequences, a method successfully used for typing of other fungal pathogens, has been also used to type *A. fumigatus* strains and has given to date the best strain differentiation (Girardin *et al.,* 1993, 1994a, b, 1995; Neuvéglise *et al.,* 1996, 1997; Debeaupuis *et al.,* 1997). This RFLP method uses a species-specific repeated sequence λ 3.9 as a marker and provide unique Southern blot hybridization patterns for each strain tested. The repeated sequence Afut1 isolated from the repetitive DNA sequence λ3.9 used for strain fingerprinting is an inactive retro element of 6.9 kb bounded by two long terminal repeat (LTR) of 282 bp, with sequences and features characteristic of retroviruses and retrotransposons (Neuvéglise *et al.,* 1996). The 5' and 3' LTRs are not perfect direct repeats since they share only 90 % nt identity. In addition, the 5' LTR of another copy of Afut1 isolated from another phage λ4.11 which cross hybridizes with λ3.9, is 86.5 % identical with the 5'LTR of the retrotransposon isolated from the λ3.9 phage. A 5-bp duplication site was found at the border of Afut1. Afut1 encodes amino acid sequences homologous to the reverse transcriptase, RNaseH and endonuclease encoded by the pol genes of retro elements. Comparison of the peptidic sequences with other LTR retrotransposons showed that Afut1 is a member of the gipsy family identified originally in *Drosophila*. At least 10 copies of the retrotransposon element are found in the genome of *A. fumigatus*. However, Afut1 is a defective element : the putative coding domains contain multiple stop codons due exclusively to transitions from C:G to T:A. Such a pattern of nucleotide variation would recall the repeated-induced point mutation (RIP) occurring in *Neurospora* repeated sequences. However, in *A. fumigatus* no sexual reproduction is known and no methylation of cytosine, an event typically associated with mutations in sequences affected by RIP, was detected. This result would suggest that Afut1 has been subjected to RIP at a time when *A. fumigatus* possessed a functional sexual cycle and an active DNA methylation process. The copies of this repeated sequence found today could be relics of RIP consecutive to and fixed at a time where *A. fumigatus* has lost its sexual stage.

PCR-based typing methods

Although the DNA-based method described above is extremely discriminatory, it is a very time-consuming technique due to the need for DNA preparation, digestion, electrophoresis, transfer and hybridization. Due to their rapidity, PCR-based methods seem more compatible with the necessities of the routine laboratory.

Random amplification of polymorphic DNA by PCR.

Random amplification of polymorphic DNA (RAPD) by PCR relies on primers of arbitrary sequence to amplify segments of genomic DNA in a polymerase chain reaction. The primers consist of single, short oligonucleotides and differences in the distance between primer-binding sites or existence of these sites lead to synthesis of amplified DNA fragments which differ in length. Low annealing temperatures are used to permit primer binding to template DNA with one or two mismatches. The amplified DNA is usually length separated by electophoresis on agarose gel and visualized by ethidium bromide staining. Fingerprints are recorded by banding pattern and comparisons can be made by visual inspection. Genotyping by RAPD has been used succesfully to differentiate microorganisms including bacteria (Verweij *et al.*, 1995), parasites (Tannich and Burchard, 1991), viruses (Sokol *et al.*, 1992) and fungi (Van Belkum *et al.*, 1994 ; Crowhurst *et al.*, 1991). The use of RAPD to distinguish isolates of *Aspergillus fumigatus* was first described by Aufauvre-Brown *et al.* (1992). Although many primers have been evaluated only a small proportion of those examined could detect variability among *A. fumigatus* strains (Aufauvre-Brown *et al.*, 1992 ; Van Belkum *et al.*, 1993 ; Rinyu *et al.*, 1995). In one study only one of 44 oligonucleotide primers (R108, GTATTGCCCT) allowed differentiation between nine unrelated *A. fumigatus* isolates (Aufauvre-Brown, *et al.*, 1992). Comparison of the discriminatory power of primer R108 for *A. fumigatus* with that of other primers showed primer R108 to be superior in generating the greatest number of genotypes (Table 2). Besides genotyping of isolates, RAPD has been used for taxonomy and species identification of *Aspergillus* (Van Belkum *et al.*, 1993).

Several studies have compared the discriminatory power of RAPD with other typing methods such as isoenzyme analysis or RFLP. A summary of the comparison data is shown in Table 3.

474

Table 2. Summary of studies in which two or more oligonucleotide primers were evaluated for typing of *Aspergillus* species by RAPD-PCR.

References	Species	Number of isolates	Number of evaluated primers	Primer demonstrating highest degree of discrimination	Comments
Aufauvre-Brown et al., 1992	A. fumigatus	9	44	R108	Primer R108 was the only one which enables differentiation between all 9 types.
Van Belkum et al., 1993	A. fumigatus, A. flavus, A. niger, A. nidulans, A. terreus	20	6	N.D.	All primers produced species specific PCR fingerprints, 6 of 7 A. fumigatus isolates could be differentiated.
Loudon et al., 1993	A. fumigatus	19	5	primer 2 and 5	primers 2 and 5 generated 5 and 6 genotypes, respectively, while immunoblot fingerprinting generated 11 types.
Lin et al., 1995	A. fumigatus	35	5	R108	Primer R108 detected 14 types compared to 28 types if RAPD was combined with isoenzyme analysis and restriction endonuclease analysis.
Rinyu et al., 1995	A. fumigatus	62	12	OPC-10	Primer OPC-10 detected 21 different types and combined with primer OPC-7 37 types could be identified.
Anderson et al., 1996	A. fumigatus	16	4	R108	Primer R108 differentiated 8 genotypes among 16 paired isolates compared to 10 types detected by southern hybridization with bacteriophage M13.
Leenders et al., 1996	A. fumigatus, A. flavus	15/31	10	ERIC1 (A. fumigatus); primer 5-6 (A. flavus)	For A. fumigatus primer ERIC1 differentiated 12 of 13 types among 15 isolates and for A. flavus primer 5-6 19 of 22 types among 31 isolates.
Loudon et al., 1996	A. niger	15	2	both primers equally effective	4 types were distinguished by the primers.

In general RAPD performs well. Discrepancies between genotypes generated by RAPD with use of different primers are sometimes combined to create an overall genotype, which has a higher resolution than each of the primers separately (Leenders *et al.*, 1996). The interpretation of discrepancies between different typing methods, however, may be more difficult. In comparing RAPD using primer R108 with RFLP combined with hybridization with a repetitive DNA probe for the typing of A. fumigatus isolates we found that typing results were not always in agreement (Verweij *et al.*, 1996). Isolates with identical RAPD banding pattern were differentiated by RFLP, while identical isolates according to the RFLP showed unique patterns with RAPD. It remains unclear how these differences should be interpreted.

Table 3. Comparison of discriminatory power of RAPD of *Aspergillus* species with that of other typing methods.

Ref.	discriminatory power with single primer/enzyme[1]		discrimnatory power with multiple primers/probes	
	RAPD primer		RAPD primers	
Anderson *et al.*, 1996	R108	SH (10/16) > RAPD (8/16)	N.D.	
Lin *et al.*, 1995	R108	REA(*Xho*I)(17/35) > RAPD (14/35) > IEA (9/35)	R108+R151+ UBC90	REA(*Xho*I+*Sal*I)(22/35) > RAPD (21/35)
Verweij *et al.*, 1996	R108	RFLP(*Eco*RI)*(11/23) > RAPD (10/23)	N.D.	
Loudon *et al.*, 1993	2, 5	IB (10/20) > SS = RAPD(5)(6/20) > RFLP(*Xba*I)= RAPD(2)(5/20)	2 + 5	RAPD (12/20)
Loudon *et al.*, 1994	2, 5	IB (6/8) > SS (5/8) > RAPD(5)(4/8) > RAPD(2)(1/8)	N.D.	
Rinyu *et al.*, 1995	OPC-10	RAPD (21/61) > IE (7/61) > RFLP (*Sma*I, *Eco*RI) (2/61)	N.D.	

[1]Between parenthesis are the primer or enzyme used and the number of genotypes / total number of isolates typed. SH, Southern hybridization with bacteriophage M13; REA, restriction endonuclease analysis; IEA, isoenzyme activity; IB, immunoblot with rabbit antiserum to *A. fumigatus*; SS, silver stain; RFLP, restriction fragment length polymorphism. *RFLP products were hybridized with a repetitive probe.

The reproducibility of the RAPD-PCR has been found to be very good in some institutes if the reaction conditions are standardized (Van Belkum *et al.*, 1993 ; Loudon *et al.*, 1993, 1996 ; Paugam *et al.*, 1995). Others have found that differences in efficacy of amplifications may alter the banding pattern by loss of larger fragments (Anderson *et al.*, 1996). Variations in several components of the reaction may affect the results. The source of Taq DNA polymerase and the magnesium ion concentrations have been shown to be key variables in producing good discrimination (Anderson *et al.*, 1996 ; Loudon *et al.*, 1995). If the Taq DNA polymerase is not carefully titrated against the magnesium ion concentration loss of discriminatory power may occur potentially generating a "pseudocluster" (Loudon *et al.*, 1995) or variations in band intesity may occur (Anderson *et al.*, 1996). Variable or vague bands may introduce subjectivity when banding patterns are compared by visual inspection. Automated screening by densitometers will reduce this bias especially when the number of fingerprints increases. Furthermore, densitometers will yield more quantitative data since not only the peak position is recorded but also the peak intensity (Van Belkum, 1994).

Microsatellites

Microsatellites are stretches of short repetitive sequences distributed across eukaryotic genomes. The repeat units at different loci vary from one to five nucleotides in length (Tautz and Schlotterer, 1994). Because of the great variability of repeat number at most loci, microsatellites are used widely in genetic mapping (Weissenbach *et al.,* 1992) and a population markers (Field and Wills 1996). In contrast with minisatellites (or VNTR for variable number of tandem repeat) whose repetitive units may be as large as 200 bp and which are mainly studied using conventional Southern blotting techniques, microsatellites are short enough to be assayed by PCR combined with gel electrophoresis, avoiding the need for Southern blotting. Microsatellites have several advantages : (i) Variations in microsatellite loci can be assessed from minute amounts of material that might contain highly degraded DNA, avoiding laborious DNA extraction as used for RAPD techniques to assume reproducibility. (ii) Amplification of a short DNA sequence at a high annealing temperature increases reproducibility upon sequential tests and between laboratories. (iii) The pattern observed with microsatellites is easier to read than those observed after RAPD. As a specific locus is amplified with a given set of primer, the PCR consisted in one single band for haploid organisms or in two bands for heterozygous diploid organisms. Standardization can be achieved using automated procedures with fluorescent primers and analysis of the amplified fragments on a denaturing gel with an automatic sequencer. The reproducibility of the technique is consequently high and the data computerisable. (iv) Because of their high mutation rates, including high rates of reverse mutation, microsatellites might be able to fulfil a very precise molecular role in evolution. Thus, grouping of isolates according to sharing of common alleles could have an evolutionary meaning, in contrast to RAPD for which the meaning of grouping isolates depending on the absence or presence of an amplified band is questionable (Backeljau *et al.* 1995). The predominant mean by which new length alleles are generated is intra-allelic polymerase slippage during replication, followed by lack of repair in subsequent DNA replication (Strand *et al.* 1993). Such errors tend to add or delete a single repeat unit and it is the accumulation of length mutations, which renders micro satellites among the most variable classes of repetitive DNAs. For humans, the mutation rate at micro satellite loci has been estimated around 10^{-4} per generation (Weissenbach *et al.* 1992).

The main drawback to the use of micro satellites appears to be the generation of artifactual bands during some amplification reactions, which can impair the delineation of alleles (Hauge and Litt 1993). This is often observed with dinucleotide repeats but this becomes less of a problem as the size of the repeat unit increases. Another disadvantage is the ability of some polymerases to add an extrabase at the end of the amplified fragment (Ginot *et al.* 1996). When these shortcomings occur, they can be detected by including systematically a reference strain with perfectly known alleles in each PCR run.

If thousands of micro satellite loci have been cloned from the genome of higher plants and animals, micro satellites in the genome of fungi have been poorly studied. To obtain micro satellite markers, the easiest way is to screen data banks (Field and Wills 1996) and find micro satellites in coding (Field *et al.* 1996) or regulating regions (Bretagne *et al.* 1997). However, most of the micro satellites are located in non-coding regions, usually not available in data bank. To obtain such micro satellite markers, it is possible to screen a genomic library of the organism studied with a micro satellite based probe.

Two libraries of digested total DNA from *A. fumigatus* have been screened with a $(CA)_{10}$ oligonucleotide. Four clones contained long CA repeats [$(CA)_9(GA)_{25}$],

[(CA)₂C(GA)₂₃], [(CA)₈], and [(CA)₂₁]. Four sets of primers were designed to amplify these micro satellites. One primer of each set was 5'labeled with fluorescein to allow sizing of PCR products with an automatic sequencer and analysis with the GeneScan software (Applied BioSystems). Eight, 10, 10, and 17 alleles were found for the four microsatellites, respectively, in studying 50 independant isolates. Combining the four microsatellite markers, 38 different associations were obtained (Bart-Belabesse and Bretagne 1997).

Epidemiological results

The repeated DNA sequence λ3.9 mentioned above, is to date the only probe that has provided an efficient and precise computer-aided analysis of DNA fingerprints of a large population of *A. fumigatus* strains (Chazalet *et al.*, 1997 ; Debeaupuis *et al.*, 1997). Using this probe, a fingerprinting analysis of about 2000 isolates of *A. fumigatus* with different origins from all over the world has shown the extremely high diversity of this fungal species. RAPD-typing has also be used in the past to demonstrate or exclude a nosocomial acquisition of *Aspergillus* infection in patients wiht IA. However, these studies were always limited to a very small number of isolates usually obtained during a cluster of infection (Leenders *et al.*, 1996 ; Paugam *et al.*, 1995 ; Rath and Ansorg, 1997 ; Loudon *et al.*, 1994)

Analysis of hundreds of clinical and environmental isolates of *A. fumigatus* using the λ.9 probe demonstrated that no genetical discrimination can be made on the basis of the saprophytic or pathogenic origin of the isolates (Debeaupuis *et al.*, 1997). This result means that every isolate present in the environment can become pathogenic if it encounters the appropriate host. It has some practical implication since it will indicate that the prevention measures should be applied to any environmental *A. fumigatus* conidia. The absence of environmental strains of *A. fumigatus* with highly reduced aggressiveness is in agreement with previous biochemical, molecular as well as immunological studies which have been unable to identify a key-factor responsible for the pathogenicity of *A. fumigatus* (Latgé, 1997). The high genetic diversity of *A. fumigatus* is also found geographically either at the localized level of a hospital or worldwide. Similar ratios of divergence were found for comparisons between (i) strains of one Paris hospital and strains from Paris environment or (ii) strains from the same Paris hospital and a random selection of strains from the rest of the world. Inside one hospital, around 85 % of the strains was recovered only once. Only 10-15 % of the genotypes were found at least twice and some of these strains can persist for prolonged period in different buildings of the same hospital (Chazalet *et al.*, 1997). However it was never found a strain population specific of a particular location which would have resulted from the development of the fungus in this specific location.

In the hospital setting, the main purpose of fingerprinting *A. fumigatus* isolates is to investigate the nosocomial acquisition of *Aspergillus* infections. The identity found in numerous occasions between patient and environment isolates suggest that most patients have acquired their infection in hospital (Chazalet *et al.*, 1997). However, even during an aspergillosis outbreak, it is very rare to encounter several patients infected by the same strain. This result is in agreement with the typing data of the environmental studies indicating that an extremely diverse population of strains surrounds each patient. So, the chances for two patients to be infected by the same strain are very low. If IA occurs 3 months after BMT, it can be calculated from our data that every patient have inhaled during this period at least 6000 different genotypes. Accordingly, the absence of identity between genotypes found in patients and the environmental strains does not exclude the nosocomiality of an infection. It only indicates that the environmental population typed only reflects a limited portion of the fungal population actually breathed by every patient. Assessing a nosocomial infection remains difficult and requires the molecular typing of several isolates per patient as well as hundreds of strains from the environment collected from a prolonged period of time to be able to find, due to the huge diversity of the outside population, the contaminating environmental strain. Nosocomiality of IA can still be demonstrated after several months of incubation of the fungus in the patient or during the relapse of the underlying disease. Such eventuality must be carefully checked before any IA infection could be assigned to a community-acquired case.

The question of the origin of the variability encountered in this species remains open. No teleomorph has been found in this species and the possibility for *Neosartorya fischeri* to represent the sexual stage of *A. fumigatus* has been definitely discarded (Girardin *et al.* 1995 ; Peterson, 1992). Three hypotheses can be put forward (i) continuous genetic exchange can occur though parasexual cycle and *A. fumigatus* remains today a continuously evolving species. Vegetative compatible groups have been found in *A. fumigatus* (Chazalet and Latgé, unpubl) indicating the presence of a parasexual cycle in this species. However, it has never been shown that non-meiotic parasexual recombination can account for recombination in a natural population of any fungus. (ii) This variability has been fixed long time ago through meiotic exchange at a time where *A. fumigatus* had a sexual stage. This would be in agreement with the earlier occurrence of a repeated-induced point mutation process found until now only on teleomorphic species (Neu-véglise *et al.*, 1996). Continuous exchange of conidia through air currents all around the world in addition of the absence of any adaptation to a parasitic condition would explain the lack of recovery of identical multilocus genotypes from geographically and temporally unassociated hosts. **(iii)** A third hypothesis is the presence of a sexual stage in *A. fumigatus* which has gone undetected. Recent population genetic studies have suggested that such cryptic sexual stage may exist in the case of the anamorphic human pathogen A. flavus (see chapter of Geiser). The presence of such sexual stage would be correlated with variations occurring through meiotic recombination.

CONCLUSION

Although the biological and morphological characters of *A. fumigatus* are well defined, this review has shown that molecular methods can be helpful to complement the taxonomical description of this species. These methods should be specifically applied to abnormal or poorly sporulating strains which seems to occur more widely than originally described (Leslie *et al.*, 1988 ; Samson, 1994). In addition, these techniques have been essential to discard the hypothesis for Neosartorya fisheri to represent the teleomorph stage of *A. fumigatus* (Girardin *et al.*, 1995). If these non-morphological methods may seem of a limited interest on a pure taxonomical basis, molecular techniques are the only tool available to type strains of *A. fumigatus*. Three method have been developed up to date : RFLP patterns visualized by Southern blot hybridization with an inactive retrotransposon, RAPD and microsatellites. Each method has his pros and cons supporters. A comparative trial is presently undertaken with these 3 methods to select the most discriminative one. In addition, the association of several typing methods has already shown a better efficacy in the identification of strains of *A. fumigatus* than a single method (Lin *et al.*, 1995 ; Verweij *et al.*, 1996).

REFERENCES

Anderson, M. J., K. Gull, & D. W. Denning. (1996). Molecular typing by random amplification of polymorphic DNA and M13 southern hybridization of related paired isolates of *Aspergillus fumigatus*. J Clin Microbiol. 34: 87-93.

Arnow, P. M., M. Sadigh, C. Costas, D. Weil, & R. Chudy. (1991). Endemic and epidemic aspergillosis associated with in-hospital replication of *Aspergillus* organisms. J Infect Dis. 164: 998-1002.

Aufauvre-Brown, A., J. Cohen, & D. W. Holden. (1992). Use of randomly amplified polymorphic DNA markers to distinguish isolates of *Aspergillus fumigatus*. J Clin Microbiol. 30: 2991-2993.

Backeljau, T., L. De Bruyn, H. De Wolf, K. Jordaens, S. Van Donge, R. Verhagen, & B. Winnepenninckx. (1995). Random amplified polymorphic DNA (RAPD) and parsimony methods. Cladistics. 11: 119-130.

Bart-Delabesse, E., & S. Bretagne. (1997). Presented at the 13th congress of the International Society for Human and Animal Mycology, 8-13 June, Parma, Italy.

Bodey, G. P., & S. Vartivarian. (1989). Aspergillosis. Eur J Clin Microbiol Infect Dis. 8: 413-437.

Bretagne, S., J. M. Costa, C. Besmond, R. Carsique, & R. Calderone. (1997). Microsatellite polymorphism in the promoter sequence of the elongation factor 3 gene of Candida albicans as the basis for a typing system. J Clin Microbiol. 35: 1777-1780.

Burnie, J. P., A. Coke, & R. C. Matthews. (1992). Restriction endonuclease analysis of *Aspergillus fumigatus* DNA. J Clin Pathol. 45: 324-327.

Burnie, J. P., R. C. Matthews, I. Clark, & L. J. R. Milne. (1989). Immunoblot fingerprinting *Aspergillus fumigatus*. J Immunol Methods. 118: 179-186.

Chazalet, V., J. P. Debeaupuis, J. Sarfati, J. Lortholary, P. Ribaud, P. Shah, E. Gluckman, G. Brücker, & J. P. Latgé. (1997). Molecular typing of saprophytic and clinical isolates of *Aspergillus fumigatus* in the hospital environment. J Clin Microbiol. (submitted).

Chrzavzez, E. (1995). Détection de champignons contaminants de produits à base de fruits : étude du polymorphisme et détermination de sondes moléculaires. Doctorat ès Sciences de la Vie. Université Paris XI.

Crowhurst, R. N., B. T. Hawthorne, E. H. A. Rikkerink, & M. D. Templeton. (1991). Differentiation of Fusarium solani f. sp. cucurbitae races 1 and 2 by random amplification of polymorphic DNA. Curr Genet. 20: 391-396.

Debeaupuis, J. P., J. Sarfati, V. Chazalet, & J. P. Latgé. (1997). Genetic diversity among clinical and environmental isolates of *Aspergillus fumigatus*. Infect. Immun. 65: 3080-3085.

Denning, D. W., K. V. Clemons, L. H. Hanson, & D. A. Stevens. (1990). Restriction endonuclease analysis of total cellular DNA of *Aspergillus fumigatus* isolates of geographically and epidemiologically diverse origin. J Infect Dis. 162: 1151-1158.

Dewhurst, A. G., M. J. Cooper, S. M. Khan, A. P. Pallett, & J. R. E. Dathan. (1990). Invasive aspergillosis in immunocompromised patients: potential hazard of hospital building work. Brit Med J. 301: 802-805.

Dixon, D. M., & T. J. Walsh. (1992). Human pathogenesis, p. 249-267. In J. W. Bennet & Klich (ed.), *Aspergillus*, biology and industrial application. Butterworth-Heinemann, Boston, London.

Field, D., L. Eggert, D. Metzgar, R. Rose, & C. Wills. (1996). Use of polymorphic short and clustered coding-region microsatellites to distinguish strains of Candida albicans. FEMS. Immunol Med Mic. 15: 73-79.

Field, D., & C. Wills. (1996). Long, polymorphic microsatellites in simple organisms. Proc Roy Soc London. 263: 209-215.

Fridkin, S. K., & W. R. Jarvis. (1996). Epidemiology of nosocomial fungal infections. Clin Microbiol Rev. 9: 499-511.

Frisvad, J. C., & R. A. Samson. (1990). Chemotaxonomy and morphology of *Aspergillus fumigatus* and related taxa, p. 201-208. In R. A. Samson & J. I. Pitt (ed.), Modern concepts in Penicillium and *Aspergillus* classification. Plenum Press, New York.

Ginot, F., I. Bordelais, S. Nguyen, & G. Gyapay. (1996). Correction of some genotyping errors in automated fluorescent microsatellite analysis by enzymatic removal of one base overhangs. Nucleic Acids Research. 24: 540-541.

Girardin, H., J. P. Latgé, T. Srikantha, B. Morrow, & D. R. Soll. (1993). Development of DNA probes to fingerprinting *Aspergillus fumigatus*. J Clin Microbiol. 31: 1547-1554.

Girardin, H., M. Monod, & J. P. Latgé. (1995). Molecular characterization of the food-borne fungus Neosartorya fischeri (Malloch & Cain). Appl Environ Microbiol. 61: 1378-1383.

Girardin, H., J. Sarfati, H. Kobayashi, J. P. Bouchara, & J. P. Latgé. (1994a). Use of DNA moderately repetitive sequence to type *Aspergillus fumigatus* isolates from aspergilloma patients. J Infect Dis. 169: 683-685.

Girardin, H., J. Sarfati, F. Traoré, J. Dupouy-Camet, F. Derouin, & J. P. Latgé. (1994b). Molecular epidemiology of nosocomial invasive aspergillosis. J Clin Microbiol. 32: 684-690.

Hauge, X., & M. Litt. (1993). A study of the origin of shadow bands' seen when typing dinucleotide repeat polymorphisms by the PCR. Hum mol Gen. 2: 411-415.

Latgé, J. P. (1997). *Aspergillus fumigatus* and aspergillosis. Clin Microbiol Rev: (submitted).

Leenders, A., A. Van Belkum, S. Janssen, S. De Marie, J. Kluytmans, J. Wielenga, B. Lowenberg, & H. Verbrugh. (1996). Molecular epidemiology of apparent outbreak of invasive aspergillosis in a hematology ward. J Clin Microbiol. 34: 345-351.

Leslie, C. E., B. Flannigan, & L. J. R. Milne. 1988. Morphological studies on clinical isolates of *Aspergillus fumigatus*. J Med Vet Mycol. 26: 335-341.

Lin, D. M., P. F. Lehmann, B. H. Hamory, A. A. Padhye, E. Durry, R. W. Pinner, & B. A. Lasker. (1995). Comparison of three typing methods for clinical and environmental isolates of *Aspergillus fumigatus*. J Clin Microbiol. 33: 1596-1601.

Loudon, K. W., J. P. Burnie, A. P. Coke, & R. C. Matthews. (1993). Application of polymerase chain reaction to fingerprinting *Aspergillus fumigatus* by random amplification of polymorphic DNA. J Clin Microbiol. 31: 1117-1121.

Loudon, K. W., A. P. Coke, & J. P. Burnie. (1995). "Pseudoclusters" and typing by random amplification of polymorphic DNA of *Aspergillus fumigatus*. J Clin Pathol. 48: 183-184.

Loudon, K. W., A. P. Coke, J. P. Burnie, G. S. Lucas, & J. A. Liu Yin. (1994). Invasive aspergillosis: clusters and sources ? J Med Vet Mycol. 32: 217-224.

Loudon, K. W., A. P. Coke, J. P. Burnie, A. J. Shax, B. A. Oppenheim, & C. Q. Morris. (1996). Kitchens as a source of *Aspergillus* niger infection. J Hosp Infect. 32: 191-198.

Matsuda, H., S. Kohno, S. Maesaki, H. Yamada, H. Koga, M. Tamura, H. Kuraishi, & J. Sugiyama. (1992). Application of ubiquinone systems and electrophoretic comparison of enzymes to identification of clinical isolates of *Aspergillus fumigatus* and several other species of *Aspergillus*. J Clin Microbiol. 30: 1999-2005.

Mullins, J., R. Harvey, & A. Seaton. 1976. Sources and incidence of airborne *Aspergillus fumigatus* (Fres). Clin Allergy. 6: 209-217.

Neuvéglise, C., J. Sarfati, J. P. Debeaupuis, H. Vu-Thien, J. Just, G. Tournier, & J. P. Latgé. (1997). Longitudinal study of *Aspergillus fumigatus* strains isolated from cystic fibrosis patients. Eur J Clin Microbiol Infect Dis. 16: (in press).

Neuvéglise, C., J. Sarfati, J. P. Latgé, & S. Paris. (1996). Afut1, a retrotransposon-like element from *Aspergillus fumigatus*. Nucl Acids Res. 24: 1428-1434.

Nolard, N. (1994). Les liens entre les risques d'aspergillose et la contamination de l'environnement. Revue de la littérature. Pathol Biol. 43: 706-710.

Paugam, A., M. E. Bougnoux, F. Robert, J. Dupouy-Camet, L. Fierobe, J. F. Dhainaut, & H. Girardin. (1995). Use of randomly amplified polymorphic DNA markers (RAPD) to demonstrate nosocomial contamination in a case of lethal invasive aspergillosis. J Hosp Infect. 29: 158-161.

Peterson, S. W. (1992). Neosartorya pseudofischeri sp. nov. and its relationship to other species in *Aspergillus* section Fumigati. Mycol Res. 96: 547-554.

Pitt, J. I. (1994). The current role of *Aspergillus* and Penicillium in human and animal health. J Med Vet Mycol. 32(S1): 17-32.

Rath, P. M., & R. Ansorg. (1997). Value of environmental sampling and molecular typing of aspergilli to assess nosocomial sources of aspergillosis. J Hosp Infect. 37: 47-53.

Rinyu, E., J. Varga, & L. Ferenczy. (1995). Phenotypic and genotypic analysis of variability in *Aspergillus fumigatus*. J Clin Microbiol. 33: 2567-2575.

Rodriguez, E., T. De Meeus, M. Mallié, F. Renaud, F. Symoens, P. Mondon, M. A. Piens, B. Lebeau, M. A. Viviani, R. Grillot, N. Nolard, F. Chapuis, A. M. Tortorano, & J. M. Bastide. (1996). Multicentric epidemiological study of *Aspergillus fumigatus* isolates by multilocus enzyme electrophoresis. J Clin Microbiol. 34: 2559-2568.

Samson, R. A. (1994). Current systematics of the genus *Aspergillus*, p. 261-276. In K. A. Powell, A. Renwick, & J. F. Peberdy (ed.), The genus *Aspergillus* systematics. From taxonomy and genetics to industrial application. Plenum Press, New York.

Samson, R. A., E. S. Hoekstra, J. C. Frisvad & O. Filtenborg. ((1995)). Introduction to Food-Borne Fungi, 4th edition. Centraalbureau voor Schimmelcultures, Baarn.

Sokol, D. M., G. J. Demmler, & G. J. Buffone. (1992). Rapid epidemiologic analysis of cytomegalovirus by using polymerase chain reaction amplification of the L-S junction region. J Clin Microbiol. 30: 839-844.

Strand, M., T. A. Prolla, R. M. Liskay, & T. D. Petes. (1993). Destabilization of tracts of simple repetitive DNA in yeast by mutations affecting DNA mismatch repair. Nature. 365: 274-276.

Streifel, A. J., J. L. Lauer, D. Vesley, B. Juni, & F. S. Rhame. 1983. *Aspergillus fumigatus* and other thermotolerant fungi generated by hospital building demolition. Appl Environ Microbiol. 46: 375-378.

Symoens, F., M. A. Viviani, & N. Nolard. (1993). Typing by immunoblot of *Aspergillus fumigatus* from nosocomial infections. Mycoses. 36: 229-237.

Tannich, E., & G. D. Burchard. (1991). Differentiation of pathogenic fron non-pathogenic Entamoeba histolytica by restrction fragment analysis of a single gene amplified in vitro. J Clin Microbiol. 29: 250-255.

Tautz, D., & C. Schlotterer. (1994). Simple sequences. Curr Opin Genet Devel. 4: 832-837.

Van Belkum, A. (1994). DNA fingerprinting of medically important microorganisms by use of PCR. Clin Microbiol Rev. 7: 174-184.

Van Belkum, A., W. J. G. Melchers, B. E. De-Pauw, S. Scherer, W. Quint, & J. F. Meis. (1994). Genotypic characterization of sequential Candida albicans isolates from fluconazole-treated neutropenic patients. J Infect Dis. 169: 1062-1070.

Van Belkum, A., W. G. V. Quint, B. E. De Pauw, W. J. G. Melchers, & J. F. Meis. (1993). Typing of *Aspergillus* species and *Aspergillus fumigatus* isolates by interrepeat polymerase chain reaction. J Clin Microbiol. 31: 2502-2505.

Verweij, P. E., J. F. G. M. Meis, J. Sarfati, J. A. A. Hoogkamp-Korstanje, J. P. Latgé, & W. J. G. Melchers. (1996). Genotypic characterization of sequential *Aspergillus fumigatus* isolates from patients with cystic fibrosis. J Clin Microbiol. 34: 2595-2597.

Verweij, P. E., A. Van Belkum, W. J. G. Melchers, A. Voss, J. A. A. Hoogkamp-Korstanje, & J. F. G. M. Meis. (1995). Interrepeat fingerprinting of third generation cephalosporin-resistant Enterobacter cloacae isolated during an outbreak in a neonatal intensive care unit. Infect Contr Hosp Epidemiol. 16: 25-29.

Wald, A., W. Leisenring, J. A. Van Burik, & Bowden, R. A.. (1997). Epidemiology of *Aspergillus* infections in a large cohort of patients undergoing bone marrow transplantation. J Infect Dis. 175: 1459-1466.

Weissenbach, J., G. Gyapay, C. Dib, A. Virginal, J. Morissette, P. Millasseau, G. Vaysseix, & Lathrop, M. (1992). A second-generation linkage map of the human genome. Nature: 794-801.

PHENOTYPIC AND GENOTYPIC VARIABILITY WITHIN *ASPERGILLUS* SECTION *FUMIGATI*

Edit Rinyu, János Varga, Lajos Ferenczy and Zofia Kozakiewicz[1]
Department of Microbiology, Attila József University, H-6701 Szeged, Hungary and
[1]CABI Bioscience, Bakeham Lane, Englefield Green, Egham, Surrey TW20 9TY, UK

Isolates and collection strains of *Aspergillus fumigatus*, and representatives of other related species belonging to *Aspergillus* section *Fumigati* were compared for their phenotypic and genotypic features. Carbon source utilization spectra and isoenzyme patterns were useful for differentiating the species in this section. High levels of variability were detected among *N. glabra* and *A. viridinutans* strains; most of the strains of Australian origin formed distinct clusters. The mitochondrial DNA patterns, and the *Sma*I-generated repetitive DNA profiles were also characteristic for most of the species examined, except for the *A. viridinutans* and *N. glabra* strains, which displayed intraspecific variability. Amplified DNA polymorphisms and secondary metabolite profiles of these strains were also highly variable. The dendrogram produced by an unweighed pair group method from the genotypic and phenotypic data indicates a close relationship of *A. fumigatus* and *A. fischerianus* (= *N. fischeri*). Random amplified polymorphic DNA analysis was also the most useful tool for typing *A. fumigatus* isolates as compared to isoenzyme analysis, or mitochondrial DNA restriction patterns. The isoenzyme, mitochondrial DNA and repetitive DNA patterns of *A. fumigatus* var. *ellipticus* and *A. fumigatus* var. *acolumnaris* strains were the same, and their RAPD patterns were also similar to those of the other *A. fumigatus* strains. "*Aspergillus fumigatus*" strain FRR 1266 possibly represents a new asexual species of section *Fumigati*, since it exhibited unique mitochondrial DNA, repetitive DNA, amplified DNA, secondary metabolite and isoenzyme profiles.

INTRODUCTION

Section *Fumigati* is an economically important area of the genus *Aspergillus*. Teleomorphic species of this section belong to the genus *Neosartorya* (Malloch and Cain, 1972). The most important species in this section is *A. fumigatus*, which is a ubiquitous filamentous fungus in the environment, and also an important human pathogen. Several species of this section produce harmful mycotoxins such as tremorgenic toxins or gliotoxin, which displays immunosuppressive properties (Frisvad and Samson, 1990; Müllbacher and Eichner, 1984). *Neosartorya fischeri* isolates have frequently been reported as heat resistant spoilage fungi in foods (Raper and Fennell, 1965). Some species also have valuable properties for the fermentation industry; e.g. one *N. spinosa* strain has been used for the production of optically active 1,3-butanediol (US patent no. 05326705), while *N. fischeri* strains are used for the production of xylanases active at alkaline pH (Raj and Chandra, 1996). *A. fumigatus* strains are used for the production of fumagillin, which exhibits amebicidal activity (McCowen *et al.*, 1951); synthetic analogues of fumagillin were also described as suppressors of tumor growth (Ingber *et al.*, 1990). An *A. fumigatus* strain has recently also been reported to produce metabolites called the pyripyropenes, which might be useful in lowering the cholesterol levels in humans (Tomoda *et al.*, 1994).

Several studies have been carried out recently on *A. fumigatus*, and other species belonging in *Aspergillus* section *Fumigati*, mainly for taxonomic purposes (Frisvad and Samson 1990; for other references, see Croft and Varga, 1994). Our aim was to evaluate the applicability of a number of different phenotypic and genotypic approaches for species delimitation within *Aspergillus* section *Fumigati*, and to examine the intraspecific variability within some of these species.

MATERIALS AND METHODS

Strains. The strains examined are listed in Table 1. For complete lists of the strains examined see Tables 1 of Rinyu *et al.* (1995), and Varga *et al.* (1997). The identity of these strains was checked according to Raper and Fennell (1965), and Kozakiewicz (1989). Strains were maintained on malt extract agar slants. For protein and nucleic acid extractions, the strains were grown in Pontecorvo's liquid minimal medium at 30°C for 48 hours on a rotary shaker.

Isoenzyme analysis. Crude protein extracts were prepared as described previously (Kálmán *et al.* 1991). Polyacrylamide slab gel electrophoresis and detection of isoenzyme activities were carried out as detailed earlier (Rinyu *et al.*, 1995). β-arylesterase (EC 3.1.1.2), acid phosphatase (EC 3.1.3.1), NADP-dependent glutamate dehydrogenase (EC 1.4.1.4), superoxide dismutase (EC 1.15.1.1), and NADP-dependent malate dehydrogenase (EC 1.1.1.38) activities were recorded.

Carbon source utilization tests. About 80 compounds were tested as sole carbon sources in our experiments. Each compound was analyzed at concentrations of 0.2% in a minimal medium (Manczinger and Polner, 1987). Plates were inoculated with conidial suspensions (approx. 10^7 ml^{-1}) of the strains, and incubated at 30°C in the dark. Growth intensities were compared after incubation for four and seven days with a control plate not containing any carbon source.

Thin-layer chromatography (TLC) of secondary metabolites. Plates of YES solid medium (2% yeast extract, 15% sucrose, 2% agar) were inoculated with conidial suspensions of the strains, and incubated at 30°C for 10 days. TLC analyses were carried out as described previously (Varga *et al.*, 1996). The plates were developed in toluene : ethyl acetate : formic acid (5:4:1), and visualized under UV light (360 nm).

Table 1. MtDNA and rDNA types of the species of *Aspergillus* section *Fumigati* (the mtDNA and rDNA patterns are shown in Figs 2 and 3)

Strain	source	mtDNA type	rDNA type
A. brevipes (T)	NRRL 2439	8	8
A. duricaulis (T)	NRRL 4021	12	14
A. fumigatus	ATCC 32722	4	1
A. fumigatus var. *acolumnaris*	NRRL 5587	4	2
A. fumigatus (T)	NRRL 163	4	2
A. fumigatus var. *ellipticus*	NRRL 5109	4	2
"*A. fumigatus*"	FRR 1266	11	3
A. unilateralis	NRRL 577	3	6
A. viridinutans	IMI 133982	10	8
A. viridinutans	IMI 182127	10	8
A. viridinutans	IMI 280490	10	ND
A. viridinutans	IMI 306135	8	8
A. viridinutans	NRRL 576	9	8
A. viridinutans	NRRL 6106	10	4
A. viridinutans (T)	IMI 062875	9	8
N. aurata (T)	NRRL 4378	2	7
N. aurata	NRRL 4379	2	7
N. aureola (T)	NRRL 2244	7	15
N. aureola	NRRL 2391	7	15
N. fennelliae (T)	NRRL 5534	3	13
N. fennelliae (T)	NRRL 5535	3	13
N. fischeri (T)	NRRL 181	4	8
N. fischeri	NRRL A-7223	4	8
N. glabra	IMI 061450	1*	ND
N. glabra	IMI 131700	1*	ND
N. glabra	NRRL 183	1	9
N. glabra	NRRL 2163	1	9
N. glabra (T)	IMI 061447	1	9
N. hiratsukae (T)	NHL 3008	2	12
N. hiratsukae	NHL 3009	2	12
N. pseudofischeri (T)	NRRL 20748	3	8
N. pseudofischeri	NRRL 3496	3	8
N. quadricincta (T)	NRRL 2154	6	10
N. quadricincta	NRRL 2221	6	10
"*N. glabra*"	NRRL 4179	13	5
N. spathulata (T)	NHL 2947	5	8
N. spathulata (T)	NHL 2948	5	8
N. spinosa	NRRL 3435	3	11
N. spinosa (T)	NRRL 5034	3	11
N. stramenia (T)	NRRL 4652	2	7

Abbreviations: ATCC, American Type Culture Collection, Rockville, Maryland, USA; FRR, CSIRO Food Research Culture Collection, North Ryde, New South Wales, Australia; IMI, International Mycological Institute, Egham, Surrey, UK; NHL, National Institute of Hygienic Sciences, Tokyo, Japan; NRRL, Agricultural Research Service Culture Collection, Peoria, Illinois, USA. (T) type or neotype strains, * One of the bands was smaller than in the other *N. glabra* strains

Isolation and characterization of nucleic acids. Total cellular nucleic acids were isolated as described by Leach *et al.* (1986). Analysis of mitochondrial DNA (mtDNA) patterns by digesting the total DNA samples with *Hae*III was carried out as described earlier

(Varga *et al.*, 1993). Ribosomal DNA (rDNA) patterns were analysed by digesting the total DNA samples with *Sma*I, separating the fragments by electrophoresis and hybridizing the ribosomal repeat unit of *A. nidulans* (pMN1; Borsuk *et al.*, 1982) to the filters obtained after Southern blotting. Hybridization experiments were performed as described previously (Sambrook *et al.*, 1989).

Random amplified polymorphic DNA (RAPD) analyses were carried out as described previously (Williams *et al.*, 1990). Fungal DNA sequences were amplified by using the primers of the Operon random primer kit C (Operon Technologies, Inc., Alameda, California, USA). The MJ Research programmable thermal controller (Model PTC-100-60; MJ Research, Inc., Watertown, Massachusetts, USA) was programmed for 45 cycles (1 min at 92°C, 1 min at 35°C and 2 min at 72°C). The amplification products were separated by electrophoresis in 1.2% agarose gels, stained with ethidium bromide and visualized under UV light.

Statistical analysis. Statistical analysis of the data was carried out by using the SYNTAX-pc version 5.0 software package (Podani, 1993). The binomial matrix obtained from the data was used to calculate the Jaccard coefficients. The coefficient matrices were analyzed by an unweighted pair-group method using arithmetic averages (UPGMA)(Sneath and Sokal, 1973). All calculations were performed on an IBM-compatible AT computer.

RESULTS AND DISCUSSION

Carbon source utilization and isoenzyme analysis

Isoenzyme analysis and carbon source utilization patterns were found to be valuable tools for the taxonomy of *Aspergillus* section *Fumigati* (Varga *et al.*, 1997). The isoenzyme patterns of the strains were highly polymorphic. The least polymorphic patterns were observed in glutamate dehydrogenase and superoxide dismutase isoenzymes, while β-arylesterase and acid phosphatase patterns permitted differentiation between individual isolates of the same species. Similar results were obtained in our earlier studies concerning the isoenzyme patterns of *A. fumigatus* strains (Rinyu *et al.* 1995). Asexual and sexual species are scattered through the dendrogram. *A. fumigatus* strains were found to be the most closely related to *N. fischeri* strains. The heterothallic *N. fennelliae* and *N. spathulata* strains were only distantly related to other species in this section; their closest relative was found to be "*N. glabra*" NRRL 4179. "*A. fumigatus*" FRR 1266 revealed distinct carbon source utilization spectra to other *A. fumigatus* strains; utilization of galactitol by this strain, and by strains of other asexual species within this section are shown in Fig. 1.

Most species formed well-defined clusters based on their isoenzyme and carbon source utilization patterns, with the exception of *N. glabra* and *A. viridinutans* strains; isolates representing these species were highly polymorphic.

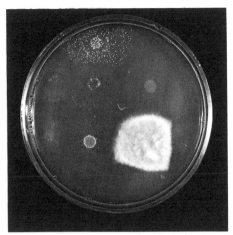

Fig. 1. Utilization of galactitol as sole carbon source by 12 strains representing the asexual species within *Aspergillus* section *Fumigati*. Only "*A. fumigatus*" FRR 1266 is able to grow on this compound. The strains tested were: *N. fischeri* NRRL 181, *A. brevipes* NRRL 2439, *A. duricaulis* IMI 217288 and NRRL 4021, *A. fumigatus* NRRL 163, NRRL 174, NRRL 5587 and NRRL 5109, "*A. fumigatus*" FRR 1266, *A. unilateralis* IMI 062876 and NRRL 577, and *A. viridinutans* IMI 133982.

Mitochondrial and nuclear DNA polymorphisms

The mtDNA patterns examined by digesting the total DNA preparations of the strains with *Hae*III were specific for most of the species (Croft and Varga, 1994). *A. fumigatus* strains gave the same mtDNA and rDNA hybridization pattern as those of the *N. fischeri* strains examined (Figs 2 and 3). This finding is in agreement with the observations of Raper and Fennell (1965) and Peterson (1992), who proposed that these taxa are closely related, based on morphological and DNA reassociation studies. "*N. glabra*" NRRL 4179, which was reported earlier to have different ascospore ornamentation and DNA reassociation kinetics from those of the other *N. glabra* strains examined (Peterson 1992), yielded different mtDNA and *Sma*I-digested repetitive DNA patterns from those of all the other *Neosartorya* strains examined. Hybridization experiments were also carried out with *Neurospora crassa* mating type genes (the A idiomorph with about 6 kb flanking sequences, or the a idiomorph flanked by about 2 kb genomic DNA on either side) to the *Eco*RI digested DNA of several teleomorphic and asexual *Aspergillus* strains. Hybridization was observed to a 1.9 kb band for both mating-type strains of *N. fennelliae* (NRRL 5534 and NRRL 5535) and "*N. glabra*" NRRL 4179, and to a 5.4 kb band of both *N. spathulata* mating-type strains (NHL 2948 and NHL 2949; Fig. 4). Hybridization was not observed to the DNA of heterothallic *Emericella heterothallica* (FGSC 251 and FGSC 252), homothallic *E. nidulans* (FGSC 513), *E. quadrilineata* (12-14 in J.H. Croft's strain collection) and *Neosartorya fischeri* (NRRL 181; lane 5), or asexual *A. ochraceus* (NRRL 405; lane 1), *A. niger* (NRRL 3122; lane 2), and *A. fumigatus* (ATCC 1022; lane 4) strains. The hybridizing DNA fragments are possibly homologous to the flanking regions of the mating type genes, since mating type idiomorphs of *Neurospora crassa* were not homologous to *N. fennelliae* DNA (Cisar *et al.* 1994). Based on these observations, "*N. glabra*" NRRL 4179 seems to be closely related to *N. fennelliae* strains. These results are

487

in agreement with those found using carbon source utilization spectra and isoenzyme analysis of these strains (Varga *et al.*, 1997). "*N. glabra*" NRRL 4179 exhibited 72% nuclear DNA relatedness to *N. fennelliae* strains as found by Peterson (1992). Some strains producing a yellow pigment (*N. aurata*, *N. stramenia*) gave the same mtDNA and rDNA patterns and their isoenzyme patterns also contained some common bands. Intraspecific variability was observed in the cases of *A. viridinutans* and *N. glabra* (Table 1; Figs 2 and 3).

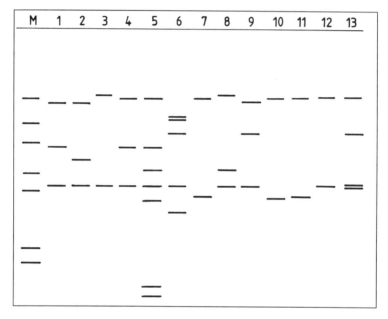

Fig. 2. *Hae*III-digested mitochondrial DNA patterns of strains belonging to *Aspergillus* section *Fumigati*. For the list of strains showing these patterns, see Table 1. M, *Hin*dIII digested lambda DNA.

Our finding that not all the *Sma*I-generated repetitive DNA bands observed on the gels showed homology to the ribosomal repeat unit of *A. nidulans* is unexpected, since similar experiments with other *Aspergillus* species did not reveal this phenomenon. The non-homologous bands observed might represent parts of the intergenic spacer region of the ribosomal repeat unit. In this case this region would seem to be highly variable in size in *Aspergillus* section *Fumigati*.

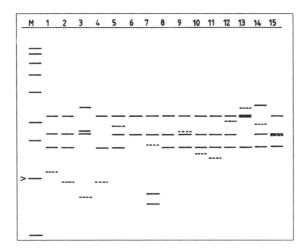

Fig. 3. *Sma*I-digested repetitive DNA patterns of strains belonging to *Aspergillus* section *Fumigati*. For the list of strains showing these patterns, see Table 1. M, 1 kb DNA ladder (the arrow indicates the 1 kb band). Bands depicted as dotted lines did not show homology to the *A. nidulans* ribosomal repeat unit.

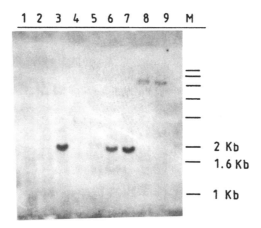

Fig. 4. Hybridization patterns obtained by using the a idiomorph of *Neurospora crassa* as probe to electrophoretically separated *Eco*RI digested DNA of different *Aspergillus* species. Homologous bands were observed in "*Neosartorya glabra*" NRRL 4179 (lane 3), *N. fennelliae* NRRL 5534 and NRRL 5535 (lanes 6 and 7), and in either *N. spathulata* mating-type strain (NHL 2948 and NHL 2949; lanes 8 and 9). M, 1 kb DNA ladder.

489

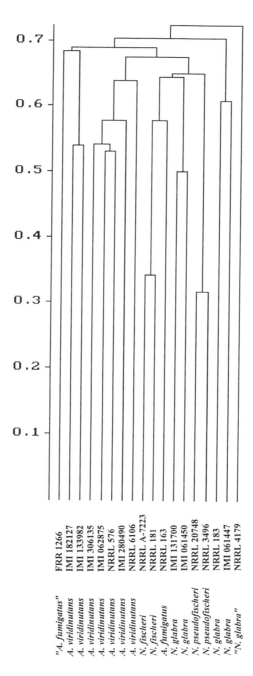

Fig. 5. UPGMA dendrogram of some strains representing species within *Aspergillus* section *Fumigati* based on both phenotypic and genotypic characters. 297 characters were taken into account. The strains are listed in Table 1. The scale represents genetic distance.

Amplified fragment length polymorphisms

In our earlier studies, RAPD analysis was found to be the most useful tool for typing *A. fumigatus* strains as compared to isoenzyme analysis, or mtDNA polymorphisms (Rinyu *et al.*, 1995). Strains of the species *A. viridinutans* and *N. glabra* were also examined by RAPD. High variability was observed among these strains (data not shown). An UPGMA clustering of the Jaccard coefficients calculated from the similarity matrix of 297 data was carried out; the dendrogram produced displays high degree of variability within these species compared to other species such as *A. fumigatus* or *N. fischeri* (Fig. 5). These results are in agreement with those of Girardin *et al.* (1995), and Samson *et al.* (1990), who detected high degrees of variability within the *N. glabra* species. In addition, the *A. viridinutans* species is also highly polymorphic compared to *A. fumigatus* as regards their phenotypic and genotypic features. Further work is in progress to examine the relationship of these strains to other species in *Aspergillus* section *Fumigati*.

CONCLUSIONS

1. Most species in *Aspergillus* section *Fumigati* are well-defined in respect of their phenotypic and genotypic features.
2. *A. fumigatus* is closely related to the teleomorphic species *N. fischeri*.
3. *N. glabra* and *A. viridinutans* species are highly polymorphic.
4. The heterothallic *Neosartorya* species are distantly related to most of the other species in section *Fumigati*; their closest relative is "*N. glabra*" NRRL 4179.
5. "*A. fumigatus*" FRR 1266 represents a new asexual species in *Aspergillus* section *Fumigati*. This *Aspergillus* strain was received from J. I. Pitt (CSIRO Food Research Laboratories, Australia). This isolate has also been morphologically re-examined by other *Aspergillus* taxonomists, and found to be a slightly unusual *A. fumigatus* isolate showing floccose colony morphology, and producing an orange mycelium on Czapek-Dox medium. Such a phenomenon, i.e. the fact that the molecular data are inconsistent with morphologically based taxonomic criteria, has been observed earlier in *Aspergillus* section *Nigri* (Kusters-van Someren *et al.*, 1991).

ACKNOWLEDGEMENTS

This work was financially supported by Hungarian Scientific Research Fund (OTKA) grants F014641, F023062 and T013044. E. Rinyu has received a grant ("Foundation for Hungarian Sciences") from the Hungarian Credit Bank. We thank R. A. Samson (CBS, Baarn, The Netherlands) and S. W. Peterson (ARS, Peoria, USA) for re-examining "*Aspergillus fumigatus*" FRR 1266. We also thank S.W. Peterson, J. I. Pitt, G. Szakács and S. Udagawa for providing us with *Aspergillus* and *Neosartorya* strains.

REFERENCES

Borsuk, P.A., Nagiec, M.M., Stepien, P.P. & Bartnik, E. (1982). Organization of the ribosomal RNA gene cluster of *Aspergillus nidulans*. Gene 17: 147-152.

Cisar, C.R., TeBeest, D.O. & Spiegel, F.W. (1994). Sequence similarity of mating type idiomorphs: a method which detects similarity among the Sordariaceae fails to detect similar sequences in other filamentous ascomycetes. Mycologia 86: 540-546.

Croft, J.H. & Varga, J. (1994). Application of RFLPs in systematics and population genetics of *Aspergilli*. In: The genus *Aspergillus*: from taxonomy and genetics to industrial applications (Eds Powell, K.A., Renwick, A. & Peberdy, J.F.) New York, Plenum Press. pp. 277-289.

Frisvad, J.C. & Samson, R.A. (1990). Chemotaxonomy and morphology of *Aspergillus fumigatus* and related taxa. In: Modern concepts in *Penicillium* and *Aspergillus* classification (Eds Samson, R. A. & Pitt, J. I.) New York, Plenum Press. pp. 201-208.

Girardin, H., Monod, M. & Latgé, J.-P. (1995). Molecular characterization of the food-borne fungus *Neosartorya fischeri* (Malloch and Cain). Appl. Environ. Microbiol. 61: 1378-1383.

Ingber, D., Fujita, T., Kishimoto, S., Sudo, K., Kanamaru, T., Brem, H. & Folkman, J. (1990). Synthetic analogues of fumagillin that inhibit angiogenesis and suppress tumour growth. Nature 348: 555-557.

Kálmán, É.T., Varga, J. & Kevei, F. (1991). Characterization of interspecific hybrids within the *Aspergillus nidulans* group by isoenzyme analysis. Can. J. Microbiol. 37: 391-396.

Kusters-van Someren, M.A., Samson, R.A. & Visser, J. (1991). The use of RFLP analysis in classification of the black Aspergilli: reinterpretation of *Aspergillus niger* aggregate. Curr. Genet. 19, 21-26.

Leach, J., Finkelstein, D.B. & Rambosek, J.A. (1986). Rapid miniprep of DNA from filamentous fungi. Fungal Genet. Newslett. 33: 32-33.

Malloch, D. & Cain, R.F. (1972). The Trichocomaceae: ascomycetes with *Aspergillus*, *Paecilomyces* and *Penicillium* imperfect states. Can. J. Bot. 50: 2613-2628.

Manczinger, L. & Polner, G. (1987). Cluster analysis of carbon source utilization patterns of *Trichoderma* isolates. System. Appl. Microbiol. 9: 214-217.

Müllbacher, A. & Eichner, R.D. (1984). Immunosuppression in vitro by a metabolite of a human pathogenic fungus. Proc. Natl. Acad. Sci. USA 81, 3835-3837.

Peterson, S.W. (1992). *Neosartorya pseudofischeri* sp. nov., and its relationship to other species in *Aspergillus* section *Fumigati*. Mycol. Res. 96: 547-554.

Podani, J. (1993). SYN-TAX-pc. Computer programs for multivariate data analysis in ecology and systematics. Version 5.0 User's Guide. Scientia Publishing, Budapest.

Raj, K.C. & Chandra, T.S. (1996). Purification and characterization of xylanase from alkali-tolerant *Aspergillus fischeri* Fxn1. FEMS Microbiol. Lett. 145: 457-461.

Raper, K.B. & Fennell, D.I. (1965). The genus *Aspergillus*. Baltimore, Williams & Wilkins.

Rinyu, E., Varga, J. & Ferenczy, L. (1995). Phenotypic and genotypic analysis of variability in *Aspergillus fumigatus*. J. Clin. Microbiol. 33: 2567-2575.

Sambrook, J., Fritsch, E.F. & Maniatis, T. (1989). Molecular cloning. A laboratory manual. Second edition. Cold Spring Harbor, New York, Cold Spring Harbor Laboratory Press.

Samson, R.A., Nielsen, P.V. & Frisvad, J.C. (1990). The genus *Neosartorya*: differentiation by scanning electron microscopy and mycotoxin profiles. In: Modern concepts in *Penicillium* and *Aspergillus* classification (Eds Samson, R. A. & Pitt, J. I.) New York, Plenum Press. pp. 455-467.

Sneath, P.H.A. & Sokal, R.R. (1973). Numerical taxonomy. San Francisco, W. H. Freeman and Co.

Tomoda, H., Hishida, H., Kim, Y.K., Obata, R., Sunazaka, T., Omura, S., Bordner, J., Guadliana, M., Dormer, P.G. & Smith, A.B. (1994). Relative and absolute stereochemistry of pyripyropene A, a potent, bioavailable inhibitor of acyl-CoA:cholesterol acyltransferase (ACAT). J. Am. Chem. Soc. 116: 12097-12098.

Varga, J., Kevei, F., Fekete, Cs., Coenen, A., Kozakiewicz, Z. & Croft J.H. (1993). Restriction fragment length polymorphisms in the mitochondrial DNAs of the *Aspergillus niger* aggregate. Mycol. Res. 97: 1207-1212.

Varga, J., Kevei, É., Rinyu, E., Téren, J. & Kozakiewicz, Z. (1996). Ochratoxin production by *Aspergillus* species. Appl. Environ. Microbiol. 62: 4461-4464.

Varga, J., Rinyu, E., Kiss, I., Botos, B. & Kozakiewicz, Z. (1997). Carbon source utilization and isoenzyme analysis as taxonomic aids among toxigenic *Neosartorya* species and their relatives. Acta Microbiol. Immunol. Hung. 44: 17-27.

Williams, J.G.K., Kubelik, A.R. Livak, K.J., Rafalski, J.A. & Tingey, S.V. (1990). DNA polymorphisms amplified by arbitrary primers are useful as genetic markers. Nucl. Acids Res. 18: 6531-6535.

492

Chapter 9

THE POTENTIAL OF
PENICILLIUM AND *ASPERGILLUS*
IN DRUG LEAD DISCOVERY

THE POTENTIAL OF *PENICILLIUM* AND *ASPERGILLUS* IN DRUG LEAD DISCOVERY

Birgitte Rømer-Rassing and Hanna Gürtler
Microbiology, Health Care Discovery, Novo Nordisk A/S, Denmark

INTRODUCTION

The Health Care Discovery Group focuses on three different sources in its drug lead discovery programme: Synthetic chemicals, combinatorial chemicals and natural products. The aim is to find compounds that can act as leads for medicinal chemistry programme. The natural product sources include micro-organisms and plants with major focus on micro-organisms.

High-quality extracts are produced and tested in a number of cellular and molecular screening assays aimed at identifying leads within our core therapy areas. For extracts showing interesting activities the active compounds are purified and the chemical structures determined. The naturally derived structures then provide starting points from which new pharmaceuticals can be designed.

The aim of Microbiology is to maximize the chemical potential of the microbial extracts. To accomplish this, our approach is a combined focus on biological and chemical diversity. The biological diversity is obtained combining the parameters ecology, geography and taxonomy. The chemical diversity is optimized using different expression conditions and different sample preparation methods (Braun *et al.*, 1993).

Why focus on *Penicillium* and *Aspergillus*?
In recent years, the screening of fungi has been focused on new or less-studied organisms, such as marine- and endophytic fungi. Part of the philosophy has been to avoid common genera like *Penicillium* and *Aspergillus* due to the risk of rediscoveries. However, with the development of new and unique screening assays, this is of less concern, as even known compounds may give rise to novel applications.

Dreyfuss and Chapela (1993) defined a so-called creativity index, meant as an easy way to point out fungal groups/genera worthwhile to investigate in the search for novel compounds. The index combines the biological diversity, represented by the estimated number of species per fungal taxa, with the chemical diversity, represented by the approximate number of known secondary metabolites produced per fungal taxa. The higher the value of the index, the more potential the fungal taxa is.

Creativity index values for selected fungal taxa are shown in Table 1. Based on present knowledge, it appears that *Penicillium* and *Aspergillus* are among the genera with the highest index of creativity in the Kingdom of Fungi.

In addition to having a high creativity index, *Penicillium* and *Aspergillus* have the advantage of being easy to handle. Due to their widespread occurrence, it is easy to get

495

potent samples; isolation can be made using simple isolation techniques, and the isolates have a high viability after preservation and storage. Furthermore, the majority of isolates belonging to *Penicillium* and *Aspergillus* has a fast growth rate and gives a good production of secondary metabolites with respect to number, diversity, and concentration.

Table 1: Creativity Indexes of Selected Fungal Taxa (M.M. Dreyfuss and I.H. Chapela, 1994).

Fungal taxa/ group	Estimated number of species	Approximate number. of known metabolites	Creativity Index
Aspergillus Eurotium, Emericella	200	525	2.6
Penicillium, Talaromyces, Eupenicillium	200	380	1.9
Trichoderma, Hypocrea	20	54	2.7
Cephalosporium-like Hyphomycetes *Acremonium, Tolypocladium, Verticillium, Monocillium, Emericellopsis*	140	116	0.8
Mucor, Rhizopus, Phycomyces	70	26	0.4
Oomycetales, Chytridiales	450	3	0.007
Yeasts	600	50	0.08
Basidiomycetes	30.000	300	0.01
Fungal species in culture	7000	4000	0.6
	1.5×10^6	?	?

Biological and chemical diversity of *Penicillium* and *Aspergillus*

Penicillium and *Aspergillus* are classified as Fungi Imperfecti, belonging to the family Trichocomaceae. Both genera cover an unusual high biological diversity, reflected by the number of species and related teleomorphic genera (see Table 2). In addition, *Penicillium* and *Aspergillus* are remarkable with their cosmopolitan presence in many kinds of ecosystems, such as the marine environment (mangrove) and soil, food- and feedstuffs (Raper and Thom, 1949; Raper and Fennell, 1965; Pitt, 1979; Ramirez, 1982; Pitt, 1985; Klich and Pitt, 1988; Samson *et al.*, 1995).

The chemical potential of *Penicillium* and *Aspergillus* is illustrated by their ability to produce compounds of many different chemical classes (Turner, 1971; Bu'Lock, 1975; Cole and Cox, 1981; Turner and Aldridge, 1983; Bennett and Ciegler, 1983; Betina, 1989; Frisvad and Filtenborg, 1989). Using the classical categorization of secondary metabolites (Turner, 1971), *Penicillium* and *Aspergillus* are seen to produce compounds of all major groups of secondary metabolites (see table no.3). Many of these have commercial relevance today; the most important being the drugs penicillin and compactin/mevinolin.

496

Table 2: Number of species in *Penicillium* and *Aspergillus*, and their teleomorphs (derived from Greuter *et al.*, 1993).

Number of Species			
Penicillium	233	Neosartorya	12
Eupenicillium	34	Chaetosartorya	3
Talaromyces	24	Fennellia	2
Hamigera	2	Hemicarpenteles	2
Aspergillus	185	Petromyces	2
Emericella	27	Sclerocleista	2
Eurotium	19	Warcupiella	1

Table 3: Fungal secondary metabolites categorized according to the precursors from which they arise (Turner, 1971; Bennett and Ciegler, 1983).

Chemical Group	Examples produced by *Penicillium* and *Aspergillus*
Secondary metabolites derived without acetate	Kojic acid
Secondary metabolites from Fatty Acids	Cyclopentanes, Brefeldin A
Polyketides	Aflatoxins, Griseofulvin, Ochratoxin A
Terpenes and Steroids	Ergosterol, Beta-Carotene
Secondary metabolites from the Citric Acid Cycle	Tenuazonic acid, Rubratoxin B
Secondary metabolites from Amino Acids	Roquefortine, Penicillic acid

Tools:

To measure the biological diversity of our fungal isolates, traditional morphological methods for identification and characterisation are used, in combination with new techniques such as RFLP and partial DNA sequencing. In addition, a special Picture Database has been developed, and macroscopic and microscopic features are measured, stored and used in subsequent image analysis.

The chemical diversity/potential of the isolates grown under various conditions is measured using an HPLC coupled to a diode array detector. The resulting chromatograms of secondary metabolites are analysed using a unique neural network programme, which includes searches against a library of microbial reference compounds and media derived compounds.

To ensure the optimal use of generated data, all observations from the biological and chemical examinations are kept in a central database, and combined with the results from the screening for various pharmaceutical activities.

Performance of *Penicillium* and *Aspergillus*: The cosmopolitan presence of *Penicillium* and *Aspergillus* makes it interesting to examine the influence of various ecological and geographical parameters. Ongoing studies have indicated a relation between such parameters and the diversity of species, including their production of secondary metabolites. These studies support the rationale of a strategy for obtaining isolates worldwide.

Using the HPLC method described above, the number, concentration (expression level) and diversity of secondary metabolites produced per fungal isolate is routinely being measured. The examinations have shown, that the relative concentration, and the average number of secondary metabolites produced under equal conditions, is considerably higher for isolates belonging to *Penicillium* and *Aspergillus* than for most of the other fungi, we work with. In addition, *Penicillium* and *Aspergillus* have the ability to produce different secondary metabolites when they are treated under different conditions, which is our definition of talented genera.

Table 4. Chemical behaviour of *Penicillium* and *Aspergillus*

No. of secondary metabolites produced per culture condition, per isolate	10-30
Total production of secondary metabolites, per isolate	30-60
Total production of secondary metabolites, per species	60-100*

* exclusive media derived components and overlap of metabolites between media/conditions.

Screening for biological activities

A high proportion of NNAS fungal extract library comes from *Penicillium*, *Aspergillus*, and their related teleomorphs. Carrying out screening of these has lead to the discovery of several new compounds with interesting activities (e.g. ZG-1494a, a novel platelet-activating factor Acetyltransferase inhibitor from *Penicillium rubrum*, and a Bis-formamidodiphenylbutadiene from *Hamigera avellanea*), as well as known compounds with new applications.

Examples of novel structures, produced by *Penicillium* and the teleomorph *Hamigera* are shown in Figure 1.

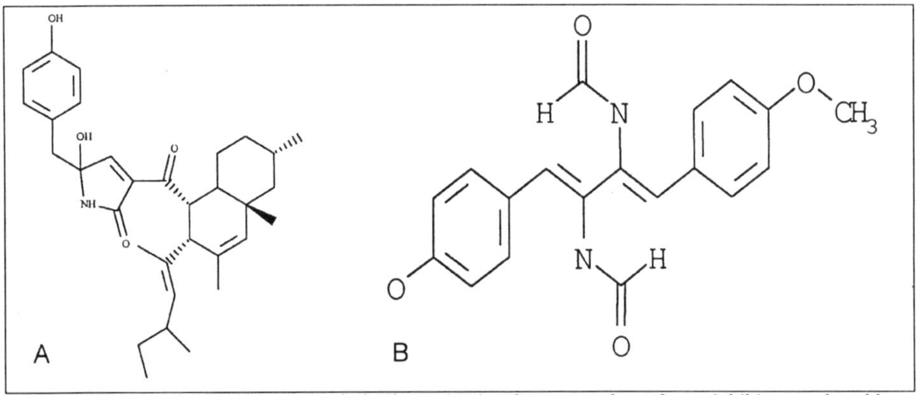

Figure 1. A. Structure of ZG-1494a, a novel platelet-activating factor acetyltransferase inhibitor produced by *Penicillium rubrum*. B. Structure of a new bis-formamidodiphenylbutadiene produced by *Hamigera avellanea*.

CONCLUDING REMARKS

As illustrated in the paragraphs above, there are many reasons to focus on *Penicillium* and *Aspergillus* in drug lead discovery. Their high creativity combined with the recent findings of new compounds and known compounds with novel activities, demonstrate that *Penicillium* and *Aspergillus* still constitute a very potential group. In addition, their cosmopolitan presence in different ecosystems and the easiness in handling, make them

498

an obvious target group. Our future strategy therefore includes a major focus on the genera *Penicillium* and *Aspergillus*.

REFERENCES

Bennett, J.W. & Ciegler, A. (Eds.) (1983) Secondary Metabolism and Differentiation in Fungi. Mycology Series, Volume 5, Marcel Dekker Inc.

Betina, V. (1989). Mycotoxins. Chemical, Biological, and Environmental Aspects Bioactive Molecules. Vol. 9, Elsevier.

Braun, D.J., Rassing, B.R., Hartjen, U. & Gürtler, H. (1994). High Quality Microbial Extracts For Screening. Poster presentation at the International Conference on Microbial Secondary Metabolism, Interlaken, Suisse.

Breinholt, J., Kjær, A., Olsen, C.E. & Rassing, B.R. (1996). A Bis-formamidodiphenylbutadiene from the fungus *Hamigera avellanea*, Acta Chemica Scandinavica, Vol. 50: 643-645

Bu¥Lock, J.D. (1975) Secondary metabolism in fungi and its relationship to growth and development. *In* The Filamentous Fungi, Vol. 1, J. E. Smith & Dr. R. Berry (Eds.). Wiley, New York.

Cole, R.J. & Cox, R.H. (1981). Handbook of Toxic Fungal Metabolites Academic Press.

Dreyfuss, M.M. & Chapela, I.H. (1994). Potential of Fungi in the Discovery of Novel, Low-Molecular Weight Pharmaceuticals. In The Discovery of Natural Products With Therapeutic Potential, V. P. Gullo. Butterworth-Heinemann, Reed Elsevier Group.

Frisvad, J.C. & Filtenborg, O. (1989). Terverticillate Penicillia: Chemotaxonomy and Mycotoxin Production. Mycologia 81: 837-861.

Greuter, W. *et al.* (Eds.) (1993). Names in Current Use in the Families Trichocomaceae, Cladoniaceae, Pinaceae, and Lemnaceae, NCU-2 Regnum Vegetabile, Vol. 128, Koeltz Scientific Books.

Klich, M.A. & Pitt, J.I. (1988). A Laboratory Guide to Common *Aspergillus* Species and their Teleomorphs. Commonwealth Scientific and Industrial Research Organization, Division of Food Processing.

Pitt, J.I. (1979). The Genus *Penicillium*. Academic Press.

Pitt, J.I. (1985). A Laboratory Guide to Common *Penicillium* Species. Commonwealth Scientific and Industrial Research Organization, Division of Food Research.

Ramirez, C. (1982). Manual and Atlas of the Penicillia. Elsevier Biomedical, Amsterdam.

Raper, K.B., & Fennell, D.I. (1965). The Genus *Aspergillus*, Baltimore, Williams and Wilkins.

Raper, K.B., & Thom, C. (1949). A Manual of the Penicillia, Baltimore, Williams and Wilkins.

Samson, R.A., Hoekstra, E.S., Frisvad, J.C. & Filtenborg, O. (1995). Introduction to food-borne fungi. Fifth edition. Centraalbureau voor Schimmelcultures, 322 pp.

Turner, W.B. & Aldridge, D.C. (1983). Fungal Metabolites II,. Academic Press, London.

Turner, W.B. (1971). Fungal Metabolites, Academic Press, London.

West, R.R., Ness, J.V., Varming, A.M., Rassing, B., Biggs, S., Gasper, S., Mckernan, P.A. & Piggot, J. (1996). ZG-1494a, a Novel Platelet-activating Factor Acetyltransferase Inhibitor from *Penicillium rubrum*, Isolation, Structure Elucidation and Biological Activity, *Journal of Antibiotics*, Vol. 49, No. 10 pp. 967-973.

INDEX

For a list of *Penicillium* and *Aspergillus* names also see pages 12-47, 51-71 and 73-79